Adaptive WCDMA

Adaptive WCDMA

Theory and Practice

Savo G. Glisic

Professor of Telecommunications
University of Oulu, Finland

WILEY

Other Wiley Editorial Offices

John Wiley & Sons Inc., 111 River Street, Hoboken, NJ 07030, USA

Jossey-Bass, 989 Market Street, San Francisco, CA 94103-1741, USA

Wiley-VCH Verlag GmbH, Boschstr. 12, D-69469 Weinheim, Germany

John Wiley & Sons Australia Ltd, 33 Park Road, Milton, Queensland 4064, Australia

John Wiley & Sons (Asia) Pte Ltd, 2 Clementi Loop #02-01, Jin Xing Distripark, Singapore 129809

John Wiley & Sons Canada Ltd, 22 Worcester Road, Etobicoke, Ontario, Canada M9W 1L1

Wiley also publishes its books in a variety of electronic formats. Some content that appears
in print may not be available in electronic books.

Library of Congress Cataloging-in-Publication Data

Glisic, Savo G.
  Adaptive WCDMA / Savo G. Glisic.
    p. cm.
  Includes bibliographical references and index.
  ISBN 0-470-84825-1 (alk. paper)
   1. Code division multiple access. I. Title.

TK5103.452 .G55 2002
621.3845'6 – dc21

                            2002033361

British Library Cataloguing in Publication Data

A catalogue record for this book is available from the British Library

ISBN 0-470-84825-1

Typeset in 10/12pt Times by Laserwords Private Limited, Chennai, India
Printed and bound in Great Britain by Antony Rowe Limited, Chippenham, Wiltshire
This book is printed on acid-free paper responsibly manufactured from sustainable forestry
in which at least two trees are planted for each one used for paper production.

To my family

Contents

Preface

This book builds a bridge between the theory and practice in the field of Wideband Code Division Multiple Access (WCDMA) technology. A joint effort from the research and academia communities has generated a significant amount of result in this field, providing a solid platform for the technology to be accepted as standard for physical layer of the third generation (3G) of mobile communications.

On one side, science is pushing toward more and more complex solutions. On the other hand, practice is forced to compromise between the complexity, reliability, cost, power consumption, size of the terminal, compatibility with the existing infrastructure and time to the market, and accept those solutions that offer the best combination of these parameters.

The focus of the book is on the implementation losses characterizing the system degradation due to imperfect implementation. This will give a picture of how much of the performance promised by theory should be expected in practical solutions based on a given technology that is not perfect, but has finite cost, power consumption, size and so on.

To estimate these losses, the current practice is predominantly to rely on large-scale simulations that simulate all possible situations in the environment (channel) and system operation. These simulations are consuming significant computational time and human resources and are producing results that are difficult to systematically analyze and interpret.

By emphasizing the need for system sensitivity modeling that takes into account a number of implementation imperfections, the book will inspire additional effort in combining theory and practice resulting in a common platform for the definition of the 'best solution'.

The material in the book is based on the author's experience in research and teaching courses in this area at universities and in industry. It is hoped that the selected material will help the readers to understand the main issues related to WCDMA, its potential and limitations and why specific solutions were chosen for the 3G standard. The book also provides a significant amount of material related to further developments and improvements in this field (beyond 3G), especially the segments on adaptive WCDMA and modifications for implementations in ad hoc networks.

The book can be used for undergraduate and postgraduate courses at universities as well as for training in industry. The material covers physical and higher layers in the network, especially adaptive radio resource management and access control. More precise suggestions for the course material selection is given in Chapter 1 of the book.

This book is devoted to my students from Finland, Europe, United States and Canada, Asia and Australia.

Oulu, 2002

Savo G. Glisic

1

Fundamentals

1.1 ADAPTIVE COMMUNICATIONS AND THE BOOK LAYOUT

In order to justify the content of the book and to make suggestions on how the book should be studied, we start with the generic block diagram of a digital communication system shown in Figure 1.1.

The standard building blocks, information source, source encoder, encryptor, channel encoder and data modulator are used to produce a narrowband signal, for example, binary phase shift keying (BPSK), quaternary phase shift keying (QPSK) or M-ary quadrature amplitude modulation MQAM carrying information content. The spreading of the signal spectra is obtained by real or complex multiplication of the narrowband signal by a code. After power amplification, the signal will be transmitted by one antenna or by multiple antennae (transmit diversity). After multipath propagation, multiple replica of the transmitted signal will reach the receiver. In a number of parallel processors (RAKE), the receiver will try to independently demodulate a number of signal replicas. The first step is signal despreading of the number of multipath components. To do so a channel estimator is needed to estimate the delays and amplitudes of these components in order to be optimally combined in coherent RAKE combiner. Prior to combining, cancelation of multiple access and multipath interference (MPI) may be performed in order to improve system performance. After signal combining, the remaining signal processing, including channel decoder, decryptor and source decoder, is performed. Separate block 'channel + network' characterizes the impact of fading, noise, network design and information broadcast from the network for control purposes.

On the basis of side information obtained either from the network or channel estimator, the *receiver configuration control* block from Figure 1.1 will put together the best possible receiver/transmitter parameters or even change the system configuration.

Coding The most powerful coding is obtained by using concatenated codes with interleavers that are known under the name *turbo codes*. The algorithm that iteratively decodes 'turbo' codes was first proposed by Berrou *et al.* [1]. It is also explained in detail by Hagenauer *et al.* [2]. A general iterative algorithm applicable to all forms of code concatenations

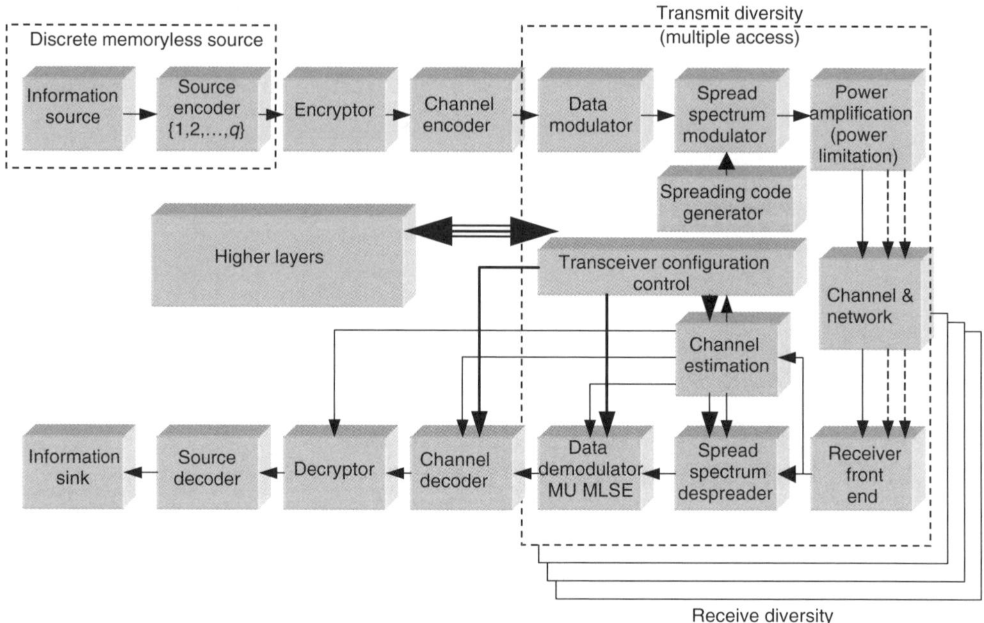

Figure 1.1 Generic block diagram of a digital communication system.

has been described by Benedetto *et al.* [3]. A number of papers have appeared on the subject of the 'turbo' iterative decoding algorithms, showing that it can be viewed as an instance of previously proposed algorithms (see, for example, Reference [4] and the extensive references therein). To avoid a huge reference list, the readers are referred to the papers and references in the *European Transactions on Telecommunications* [5], and in the *IEEE Journal on Selected Areas in Communications* [6], entirely devoted to concatenated codes and iterative decoding.

Coded modulation It has been generally accepted that modulation and coding should be combined in a single entity for improved performance. Of late, the increasing interest in mobile radio channels has led to the consideration of coded modulation for fading channels. Thus, at first blush it seemed quite natural to apply 'Ungerboeck's paradigm' of keeping coding combined with modulation even in the Rayleigh fading channel, in which the code performance depends strongly on the code minimum Hamming distance (the 'code diversity'), rather than on its minimum Euclidean distance. Several results followed this line of thought, as documented by a considerable body of work summarized and referenced in Reference [7] (see also Reference [8], Chapter 10). Under the assumption that the symbols were interleaved with a depth exceeding the coherence time of the fading process, new codes were designed for the fading channel so as to maximize their diversity.

A notable departure from Ungerboeck's paradigm was the core of Reference [9]. Schemes were designed aimed at keeping as their basic engine an off-the-shelf Viterbi decoder for the *de facto* standard, 64-state rate-1/2 convolutional code. This implied giving up the joint decoder/demodulator in favor of two separate entities.

On the basis of the latter concept, Zehavi [10] recognized that the code diversity, and hence the reliability of coded modulation over a Rayleigh fading channel, could be further improved. Zehavi's idea was to make the code diversity equal to the smallest number of distinct *bits* (rather than *channel symbols*) along any error event. This is achieved by bit-wise interleaving at the encoder output, and by using an appropriate soft-decision bit metric as an input to the Viterbi decoder. Further results along this line were recently reported in References [11–13] (for different approaches to the problem of designing coded modulation schemes for the fading channels, see References [14,15]).

Of particular interest is paper [16] based on Zehavi's findings, and in particular on his rather surprising *a priori* result that on some channels there is a downside to combining demodulation and decoding. The paper presents the theory underlying bit-interleaved coded modulation (BICM) comprehensively, and provides a general information-theoretical framework for this concept.

It also provides results for a large range of the signal constellation QPSK-256 QAM.

Adaptive coded modulation After the signal despreading point in Figure 1.1, we assume a flat-fading channel with additive white Gaussian noise (AWGN) $n(t)$ and a stationary and ergodic channel gain $\sqrt{[g(t)]}$. Let $\overline{S}$ denote the average transmit signal power, $N_0/2$ denotes the noise density of $n(t)$, B denotes the received signal bandwidth, and $\overline{g}$ denotes the average channel gain. With appropriate scaling of $\overline{S}$, we can assume that $\overline{g} = 1$. For a constant transmit power $\overline{S}$, the instantaneous received signal-to-noise ratio (SNR) is $\gamma(t) = \overline{S}g(t)/(N_0 B)$ and the average received SNR is $\overline{\gamma} = \overline{S}/(N_0 B)$. We denote the fading distribution of γ by $p(\gamma)$. If the transmit power $S(t)$ is adapted relative to $g(t)$ or, equivalently, to $\gamma(t)$, then the SNR at time t is given by

$$SNR(t) = \frac{\gamma(t)S[\gamma(t)]}{\overline{S}} = \frac{g(t)S[g(t)]}{N_0 B}$$

In accordance with Reference [17], adaptive coded modulation does not require interleaving, since error bursts are eliminated by adjusting the power, size and duration of the transmitted signal constellation, relative to the channel fading. In general, we would rather like to include the interleaver in the block 'channel encoder' in Figure 1.1. For fast fading, in which adaptation is less effective, the interleaving should help. For slow fading, in which adaptation is more effective, the interleaver cannot do much but neither does it do any damage.

However, adaptive modulation does require accurate channel estimates at the receiver, which are fed back to the transmitter with minimal latency. The effects of estimation error and feedback path delay on adaptive modulation were analyzed in Reference [18], in which it was found that an estimation error less than 1 dB and a feedback path delay less than $0.001/f_D$ results in minimal performance degradation, for $f_D = v/\lambda$ the Doppler frequency of the fading channel. The effect of estimation error and feedback path delay for adaptive coded modulation is similar, yielding the same set of requirements for minimal performance degradation. These requirements are easily met on slowly varying channels.

Another practical consideration in adaptive coded modulation scheme is how quickly the transmitter must change its constellation size. Since the constellation size is adapted to an estimate of the channel fade level, several symbol times may be required to obtain a good estimate. In addition, hardware and pulse-shaping considerations generally dictate that the constellation size must remain constant over tens to hundreds of symbols. It was shown in Reference [18] that this requirement translates mathematically to the requirement that $\overline{\tau}_j \gg T\ \forall j$, where T is the symbol for time and $\overline{\tau}_j$ is the average time when the adaptive modulation scheme continuously uses the constellation M_j. Since each constellation M_j is associated with a range of fading values called the fading region $R_j, \overline{\tau}_j$ is the average time that the fading stays within the region R_j. The value of $\overline{\tau}_j$ is inversely proportional to the channel Doppler and also depends on the number and characteristics of the different fade regions. It was shown in Reference [18] that in Rayleigh fading with an average SNR of 20 dB and a channel Doppler of 100 Hz, $\overline{\tau}_j$ ranges between 0.7 and 3.9 ms, and thus for a symbol rate of 100 ksymbols s^{-1}, the signal constellation remains constant over tens to hundreds of symbols. Similar results hold at other SNR values.

In a narrowband system, the flat-fading assumption in this model implies that the signal bandwidth B is much less than the channel coherence bandwidth $B_c = 1/T_M$, where T_M is the root-mean-square (rms) delay spread of the channel. For Nyquist pulses $B = 1/T$, so flat fading occurs when $T \gg T_M$. Combining $T \gg T_M$ and $\overline{\tau}_j \gg T$, we see that $\overline{\tau}_j \gg T \gg T_M$ must be satisfied to have both flat fading and the signal constellation constant over a large number of symbols. In general, wireless channels have rms delay spreads less than 30 μs in outdoor urban areas and less than around 1 μs in indoor environments [19]. Taking the minimum $\overline{\tau}_j = 0.7$ ms, we see that on the basis of the previous relation, rates on the order of tens of ksymbols per second in outdoor channels and hundreds of ksymbols per second in indoor channels are practical for this adaptive scheme.

For WCDMA, these conditions will be extensively discussed throughout the book, especially later on in this chapter and then in much more detail in Chapter 8.

Coset codes with adaptive modulation Reference [17] shows how the separability of code and modulation design inherent in coset codes can be used to combine coset codes with adaptive modulation. A binary encoder E, from Figure 1.1, operates on k uncoded data bits to produce $k + r$ coded bits, and then the coset (subset) selector uses these coded bits to choose one of the 2^{k+r} cosets from a partition of the signal constellation. In nonadaptive modulation dealt with in Reference [20], the *modulation* segment uses $n - k$ additional uncoded bits to choose one of the 2^{n-k} signal points in the selected coset, which is then transmitted via the modulator. These steps essentially decouple the channel coding from the modulation. Specifically, the fundamental coding gain is a function of the minimum squared distance between signal point sequences, which is determined by the encoder (E) properties and the subset partitioning, independent of the modulation. This minimum distance is given by $d_{\min} = \min\{d_s, d_c\}$, where d_s is the minimum distance between coset sequences and d_c is the minimum distance between coset points. For square MQAM signal constellations, both d_s and d_c are proportional to d_0, the minimum distance between constellation points before partitioning. The number of nearest neighbor code words also impacts the effective coding gain.

In a fading channel, the instantaneous SNR varies with time, which will cause the distance $d_0(t)$ in the received signal constellation, and, therefore, the corresponding distances $d_c(t)$ and $d_s(t)$, to vary. The basic premise for using adaptive modulation with coset codes is to keep these distances constant by varying the size $M(\gamma)$, transmit power $S(\gamma)$, and/or symbol time $T(\gamma)$ of the transmitted signal constellation relative to γ, subject to an average transmit power constraint $\overline{S}$ on $S(\gamma)$. By maintaining $\min\{d_c(t), d_s(t)\} = d_{\min}$ constant, the adaptive coded modulation exhibits the same coding gain as a coded modulation designed for an AWGN channel with minimum code word distance $d_{\min}$.

The *modulation* segment on Figure 1.1 would work as follows. The channel is assumed to be slowly fading so that $\gamma(t)$ is relatively constant over many symbol periods. During a given symbol period $T(\gamma)$, the size of each coset is limited to $2^{n(\gamma)-k}$, where $n(\gamma)$ and $T(\gamma)$ are functions of the channel SNR γ. A signal point in the selected coset is chosen using $n(\gamma) - k$ uncoded data bits. The selected point in the selected coset is one of $M(\gamma) = 2^{n(\gamma)+r}$ points in the transmit signal constellation [e.g. MQAM, M-ary phase-shift keying (MPSK)]. By using appropriate functions for $M(\gamma)$, $S(\gamma)$ and $T(\gamma)$, we can maintain a fixed distance between points in the received signal constellation $M(\gamma)$ corresponding to the desired minimum distance $d_{\min}$. The variation of $M(\gamma)$ relative to γ causes the information rate to vary, so the uncoded bits used for signal point selection must be buffered until needed. Since r redundant bits are used for the channel coding, $\log_2 M(\gamma) - r$ bits are sent over the symbol period $T(\gamma)$ for a received SNR of γ. The average rate of the adaptive scheme is thus given by

$$R = \int_{\gamma_0}^{\infty} \frac{1}{T(\gamma)} [\log_2 M(\gamma) - r] p(\gamma) \, \mathrm{d}\gamma$$

where $\gamma_0 \geq 0$ is a cutoff fade depth below which transmission is suspended ($M(\gamma) = 0$). This cutoff value is a parameter of the adaptive modulation scheme. Since γ is known to both the transmitter and the receiver, the modulation, encoding, and decoding processes are suspended while $\gamma < \gamma_0$.

At the receiver, the adaptive modulation is first demodulated, which yields a sequence of received constellation points. Then the points within each coset that are closest to these received points are determined. From these points, the maximum-likelihood coset sequence is calculated and the uncoded bits from the channel coding segment are determined from this sequence in the same manner as for nonadaptive coded modulation in AWGN. The uncoded bits from the *modulation* segment are then determined by finding the points in the maximum-likelihood coset sequence that are closest to the received constellation points and by applying standard demodulation to these points.

The adaptive modulation described above consists of any mapping from γ to a constellation size $M(\gamma)$, power $S(\gamma)$, and symbol time $T(\gamma)$ for which $d_{\min}(t)$ remains constant. Proposed techniques for adaptive modulation maintain this constant distance through adaptive variation of the transmitted power level [21], symbol time [22], constellation size [23,24], or any combination of these parameters [18,25,26]. The *modulation* segment of Figure 1.1 can use any of these adaptive modulation methods.

Adaptive coding scheme Efficient error control on time-varying channels can be performed, independent of modulation, by implementing an adaptive control system in which the optimum code is selected according to the actual channel conditions.

There are a number of burst error-correcting codes that could be used in these adaptive schemes. Three major classes of burst error-correcting codes are binary Fire block codes, binary Iwadare–Massey convolutional codes [27], and nonbinary Reed–Solomon block codes. In practical communication systems, these are decoded by hard-decision decoding methods. Performance evaluation based on experimental data from satellite mobile communication channels [28] shows that the convolutional codes with the soft-decision decoding Viterbi algorithm are superior to all the above burst error-correcting codes of the respective rates.

Superior error probability performance and availability of a wide range of code rates without changing the basic coded structure motivate the use of punctured convolutional codes [29–32] with the soft-decision Viterbi decoding algorithm in the proposed adaptive scheme. To obtain the full benefit of the Viterbi algorithm on bursty channels, ideal interleaving is assumed.

An adaptive coding scheme using incremental redundancy in a hybrid automatic-repeat-request (ARQ) error control system is reported in Reference [33]. The channel model used is binary symmetric channel (BSC) with time variable bit error probability. The system state is chosen according to the channel bit error rate (BER). The error correction is performed by shortened cyclic codes with variable degrees of shortening. When the channel BER increases, the system generates additional party bits for error correction.

An Forward Error Correction (FEC) adaptive scheme for matching the code to the prevailing channel conditions was reported in Reference [34]. The method is based on convolutional codes with Viterbi decoding and consists of combining noisy packets to obtain a packet with a code rate low enough (less than 1/2) to achieve the specified error rate. Other schemes that use a form of adaptive decoding are reported in References [35–40]. Hybrid ARQ schemes based on convolutional codes with sequential decoding on a memoryless channel were reported in References [41,42] while a Type-II hybrid ARQ scheme formed by concatenation of convolutional codes with block codes was evaluated on a channel represented by two states [43].

In order to implement the adaptive coding scheme, it is necessary again to use a return channel. The channel state estimator (CSE) determines the current channel state, on the basis of the number of erroneous blocks. Once the channel state has been estimated, a decision is made by the *reconfiguration block* whether to change the code, and the corresponding messages are sent to the encoder and locally to the decoder.

In FEC schemes, only error correction is performed, while in hybrid ARQ schemes retransmission of erroneous blocks is requested whenever the decoded data is labeled as unreliable.

The adaptive error protection is obtained by changing the code rates. For practical purposes, it is desirable to modify the code rates without changing the basic structure of the encoder and decoder. Punctured convolutional codes are ideally suited for this application. They allow almost continuous change of the code rates while decoding is done by the same decoder.

The encoded digits at the output of the encoder are periodically deleted according to the deleting map, specified for each code. Changing the number of deleted digits varies the code rate. At the receiver end, the Viterbi decoder operates on the trellis of the parent code and uses the same deleting map as in the encoder in computing path metrics [30].

The Viterbi algorithm based on this metric is a maximum-likelihood algorithm on channels with Gaussian noise since on these channels the most probable errors occur between signals that are closest together in terms of squared Euclidean distance. However, this metric is not optimal for non-Gaussian channels. The Viterbi algorithm allows use of channel state information for fading channels [44].

However, a disadvantage of punctured convolutional codes compared to other convolutional codes with the same rate and memory order is that error paths are typically long. This requires quite long decision depths of the Viterbi decoder.

A scheme with ARQ rate-compatible convolutional codes was reported in Reference [32]. In this scheme, rate-compatible codes are applied. The rate compatibility constraint increases the system throughput since in transition from higher to lower rate codes, only incremental redundancy digits are retransmitted. The error detection is performed by a cyclic redundancy check, which introduces additional redundancy.

Adaptive coding, modulation and power control While adaptive modulation (with coded or uncoded signal) and adaptive coding described earlier are conceptually well understood and elaborated, joint adaptation of coding and modulation still remains a challenge, especially from the practical point of view. The third element of the adaptation will be power control. For details on power control algorithms and extensive literature overview, the reader is referred to Chapter 6 of the book and to Reference [45]. Capacity of the cellular network with power control, including impact of power control imperfections on the system's performance, is discussed in Chapters 8 and 9.

Adaptive frequency and space domain interference cancelation Narrowband interference generated by intentional jamming (military applications) or by belonging to other systems [such as the time division multiple access (TDMA) network] may be suppressed either in frequency or space domain. Adaptive interference suppression in frequency domain is discussed in Chapter 7 with focus on possible overlay of WCDMA macro and TDMA micro cellular networks. For space domain interference suppression and capacity improvements based on adaptive antenna arrays, the reader is referred to References [46–49].

Adaptive packet length Adaptive coding combined with ARQ described earlier would require reconfiguration of layer 2 (different format for each retransmission). An additional step to be considered is to use a variable packet length including the information segment so that possibilities for additional improvements are obtained. These algorithms are discussed in Chapter 12.

Adaptive spreading factor Depending on the level of interference, an adaptive selection of the interference suppression capabilities, measured by the system processing gain, can be adopted to continuously provide the best trade-off between the BER and information rate. For the fixed bandwidth available, this is equivalent to bit rate adaptation.

Adaptation in time, space and frequency domain The concept of adaptive modulation and coding can be extended to frequency and space domain, resulting in adaptive multicarrier modulation with space diversity. For space-time coding, the reader is referred to References [50–52].

RAKE reconfiguration Coming back to Figure 1.1, the additional element of system adaptation and reconfigurability is the RAKE receiver itself. In time-varying multipath fading, the receiver will be constantly searching for the stronger components in the received signal than those being combined. Any time when such a component is found, the reassignment of the RAKE finger to the new one would take place. RAKE finger acquisition and reacquisition, and tracking in delay and space domain are discussed in Chapters 3 and 4 of the book.

Intertechnology adaptation If intertechnology roaming is assumed, and the receiver is supposed to be used in cellular and ad hoc networks, the reconfiguration in the signal format and consequently in transmitter and receiver structure would take place. A whole additional family of Code Division Multiple Access (CDMA) signal formats for application in ad hoc networks is discussed in Chapter 15. The extension of these formats to ultrawideband (UWB) technology is straightforward. The only difference is that instead of bipolar sequence, a unipolar (on–off) sequence should be used for signal spreading. For UWB technology, the reader is referred to References [53–57]. This concept can be extended to include reconfiguration of CDMA into TDMA type of receiver or reconfiguration of CDMA receiver for different standards such as the WCDMA and the cdma2000. Practical solutions are based on software radio [58].

Minimum complexity (energy consumption) adaptation In order to save energy, an adaptive receiver would be continuously trying to minimize the complexity of the receiver. For example, coding or multiuser detectors would be used only in the case in which the channel [including fading and multiple access interference (MAI)] is not good enough. So that required quality of service (QoS) cannot be provided without these components. As an example, multiuser detectors, described in Chapters 13 and 14 can be only occasionally used in the receiver. This would also require corresponding reconfiguration of the receiver. Practical solutions for such options are discussed in Chapter 17 for use in Universal Mobile Telecommunication System (UMTS) standard.

Adaptive access control Adaptation on the medium access control (MAC) layer would include access control. The access control mechanism is supposed to keep the number of simultaneously transmitting users in the network below or up to the system capacity. In WCDMA networks, this capacity varies in time as a result of the time-varying channel and the number of users in the surrounding cells. An adaptive system would continuously monitor these conditions and update the capacity threshold for access control. Adaptive algorithms based on fuzzy logic and Kalman filters are discussed in detail in Chapters 10, 11, and 12.

Adaptive routing Adaptation on the network layer would include adaptive routing in wireless network. The best available segments of the multihop rout are chosen in order to minimize retransmissions and guarantee QoS [59–74].

Adaptive source coding If adaptive routing and techniques in the physical link level control and MAC layer cannot provide the required QoS, the grade of service (GoS) can be reduced, for example, by reducing the source bit rate. Variable bit rate source encoder would be constantly adapting to the conditions in the network.

Adaptive/reconfigurable network architecture The latest concepts of telecommunications networks suggest even the evolution of network flexibility in the domain of network architecture. The communications network infrastructure would consist of a network of powerful computers and an operator would be able to rent a part of the network and establish its own network architecture depending on the market at the time. It would be able to change it in time as the market changes so that network architecture would be reconfigurable from the point of view of the operator. These issues are considered in the field of active and reprogrammable networks. To keep the list of references short, the reader is referred to Reference [75]. In ad hoc networks, the network reconfigurability adapts to the mobility and activity of the nodes [67,69,72,73].

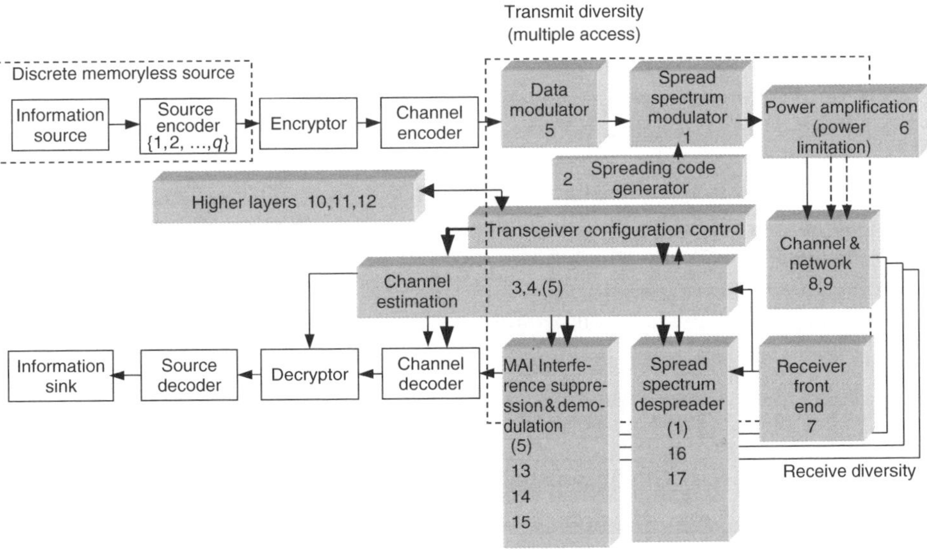

Figure 1.2　Generic block diagram of a digital communication system and book layout.

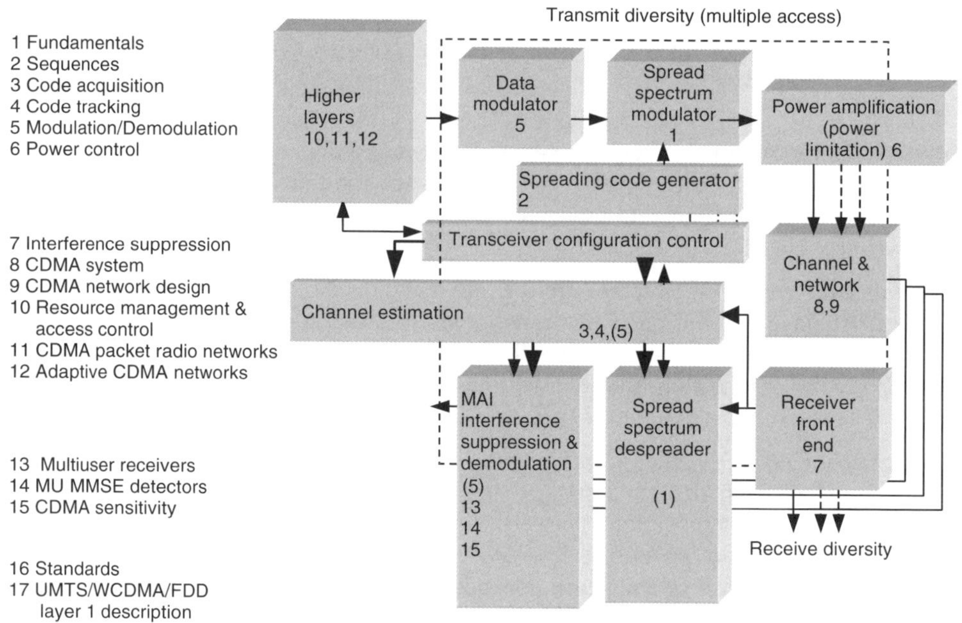

Figure 1.3   Book layout.

In this book, we cover the subsets of the problems listed above. Figure 1.2 relates to the chapters of the book and the system block diagram. Nonshaded blocks are considered as elements of the traditional communication system and are not covered in this book. For adaptive coding and modulation, the reader is referred to Reference [76]. The chapters from the book content are allocated to the respective blocks of the system, except those chapters that cover standards that cannot be allocated to specific blocks. On the left-hand side of Figure 1.3, the list of content is partitioned into four segments r – receiver, n – network, ar – advanced receiver and s – standard. This should help the reader to easily identify the specific chapters of the book. The general suggestions for the course material selections are: r – university undergraduate course on physical layer, $r + ar$ – university postgraduate course on physical layer, n – part of university undergraduate/postgraduate course on networks, $r + ar + s$ – industry course on physical layer, $n + s$ – part of industry course on networks.

1.2 SPREAD SPECTRUM FUNDAMENTALS

1.2.1 Direct sequence (DS) spread spectrum

The narrowband signal in this case is a phase-shift keying (PSK) signal of the form

$$S_n = b(t, T_m) \cos \omega t \tag{1.1}$$

where $1/T_m$ is the bit rate and $b = \pm 1$ is the information. The baseband equivalent of equation (1.1) is

$$S_n^b = b(t, T_m) \tag{1.1a}$$

Spreading operation, presented symbolically by operator $\varepsilon(\ )$, is obtained if we multiply the narrowband signal by a pseudonoise (PN) sequence (code) $c(t, T_c) = \pm 1$. The bits of the sequence are called chips and the chip rate $1/T_c \gg 1/T_m$. The wideband signal can be represented as

$$S_w = \varepsilon(S_n) = cS_n = c(t, T_c)\, b(t, T_m) \cos \omega t \tag{1.2}$$

The baseband equivalent of equation (1.2) is

$$S_w^b = c(t, T_c)b(t, T_m) \tag{1.2a}$$

Despreading, represented by operator $D(\ )$, is performed if we use $\varepsilon(\ )$ once again and band-pass filtering, with the bandwidth proportional to $2/T_m$, represented by operator $BPF(\ )$ resulting in

$$D(S_w) = BPF(\varepsilon(S_w)) = BPF(cc\, b \cos \omega t) = BPF(c^2 b \cos \omega t) = b \cos \omega t \tag{1.3}$$

The baseband equivalent of equation (1.3) is

$$D(S_w^b) = LPF(\varepsilon(S_w^b)) = LPF(c(t, T_c)c(t, T_c)b(t, T_m))$$
$$= LPF(b(t, T_m)) = b(t, T_m) \tag{1.3a}$$

where $LPF(\ )$ stands for low pass filtering. This approximates the operation of correlating the input signal with the locally generated replica of the code $\mathrm{Cor}(c, S_w)$. Nonsynchronized despreading would result in

$$D_\tau(\ );\ \mathrm{Cor}(c_\tau, S_w) = BPF(\varepsilon_\tau(S_w)) = BPF(c_\tau c\, b \cos \omega t) = \rho(\tau) b \cos \omega t \tag{1.4}$$

The baseband equivalent of equation (1.4) is

$$D_\tau(\ );\ \mathrm{Cor}(c_\tau, S_w^b) = \int_0^{T_m} c_\tau S_w^b\, \mathrm{d}t = b(t, T_m) \int_0^{T_m} c_\tau c\, \mathrm{d}t = b\rho(\tau) \tag{1.4a}$$

This operation would extract the useful signal b as long as $\tau \cong 0$, otherwise the signal will be suppressed because, as we will show in Chapter 2, $\rho(\tau) \cong 0$ for $\tau \geq T_c$. Separation of multipath components in a RAKE receiver is based on this effect. In other words, if the received signal consists of two delayed replicas of the form

$$r = S_w^b(t) + S_w^b(t - \tau)$$

the despreading process defined by equation (1.4a) would result in

$$D_\tau(\ );\ \mathrm{Cor}(c, r) = \int_0^{T_m} cr\ \mathrm{d}t = b(t, T_m) \int_0^{T_m} c(c + c_\tau)\ \mathrm{d}t = b\rho(0) + b\rho(\tau)$$

Now, if $\rho(\tau) \cong 0$ for $\tau \geq T_c$, all multipath components reaching the receiver with a delay larger then the chip interval will be suppressed. If the signal transmitted by user y is despread in receiver x, the result is

$$D_{xy}(\ );\ \mathrm{BPF}(\varepsilon_{xy}(S_w)) = \mathrm{BPF}(c_x\, c_y\, b_y \cos \omega t) = \rho_{xy}(t)\, b_y\ \cos \omega t \qquad (1.5)$$

So, in order to suppress the signals belonging to other users (multiple access interference – MAI), the cross-correlation functions should be low. In other words, if the received signal consists of the useful signal plus the interfering signal from the other user

$$r = S_{wx}^b(t) + S_{wy}^b(t) = b_x c_x + b_y c_y$$

the despreading process at the receiver of user x would produce

$$D_{xy}(\ );\ \mathrm{Cor}(c_x, r) = \int_0^{T_m} c_x r\ \mathrm{d}t = b_x \int_0^{T_m} c_x c_x\ \mathrm{d}t + b_y \int_0^{T_m} c_x c_y\ \mathrm{d}t$$
$$= b_x \rho_x(0) + b_y \rho_{xy}(0)$$

When the system is properly synchronized $\rho_x(0) \cong 1$, and if $\rho_{xy}(0) \cong 0$, the second component representing MAI will be suppressed. In addition, the size of the set of codes should be large in order to be able to allocate different codes to the large number of different users. A block diagram of the BPSK DS spread-spectrum transmitter is shown in Figure 1.4 and the receiver in Figure 1.5.

If QPSK signal is used as a narrowband signal, the general form of the transmitter will be as shown in Figure 1.6 and the receiver will be as shown in Figure 1.7.

$$S_w(t) = b_1(t)c_1(t) \cos \omega_0 t + b_2(t)c_2(t) \sin \omega_0 t \qquad (1.6)$$

For MQAM modulation, b_i would have $\log_2 M$ different values.

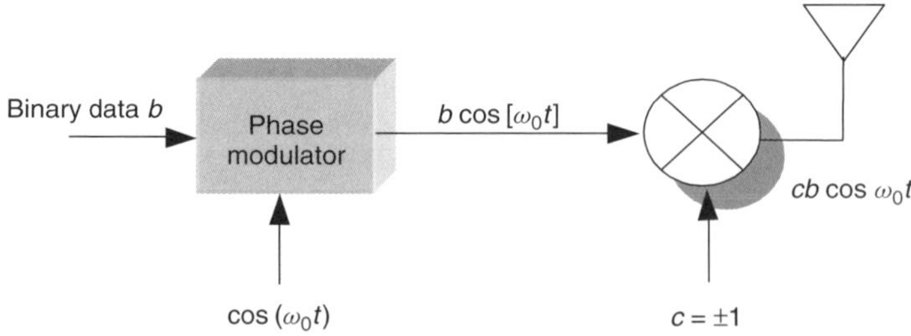

Figure 1.4 BPSK DS spread-spectrum transmitter.

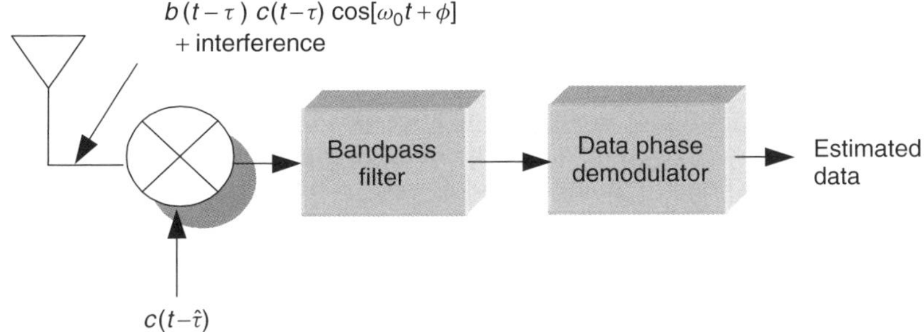

Figure 1.5 BPSK DS spread-spectrum receiver.

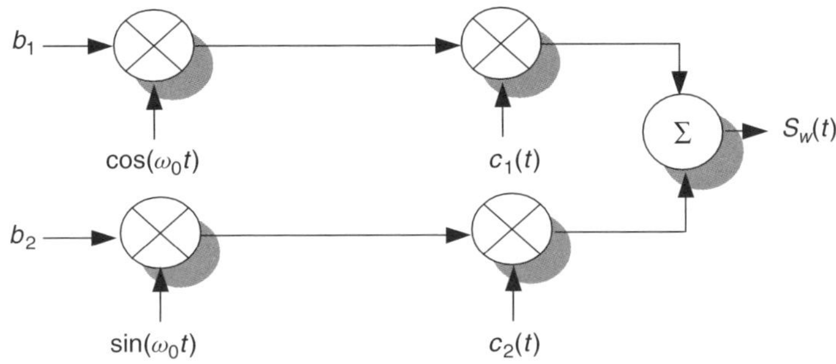

Figure 1.6 Transmitter for QPSK-DS system.

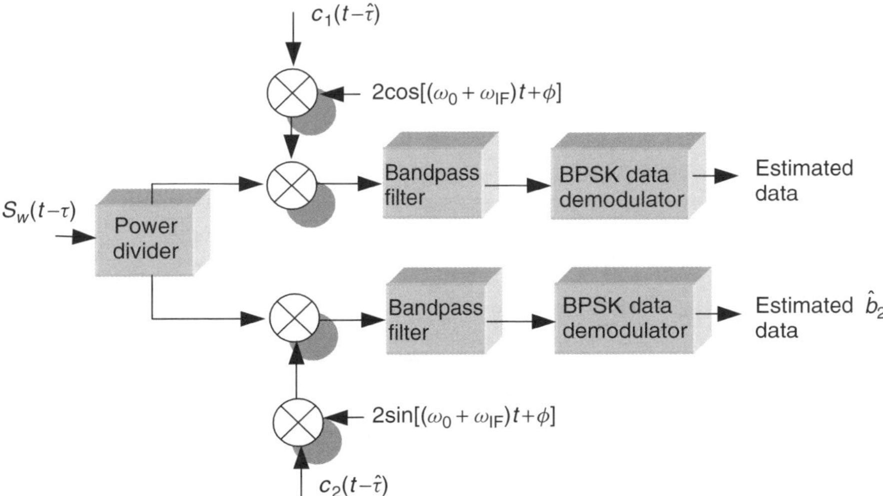

Figure 1.7 Receiver for QPSK-DS system.

If the kth transmitter sends the signal of the form given by equation (1.7) after propagation through the multipath channel, the overall received signal will have the form given by equation (1.8) where index 'lk' stands for path l of user k. As an example, the despreading process for user '$k = 1$' synchronized on path $l = 1$, will produce signal y_{11} given by equation (1.9). The first component of equation (1.9) represents a useful signal and the rest of it (double sum term) represents the MAI plus MPI. In a RAKE receiver, user $k = 1$ would separately process L signals producing $y_{l1}, l = 1, \ldots, L$. After despreading, it would have to synchronize frequency $\omega + \omega_{dlk}$ and phase θ_{lk} and after coherent demodulation get $\beta_{l1}b_1$ components to be combined in the combiner prior to final decision. The interfering terms are proportional to $\rho_{1,k}(\Delta \tau_{11,lk})$. For this reason, the codes should be designed to minimize the cross-correlation function between different users, and the autocorrelation function for $\Delta \tau \geq T_c$ to minimize the interference between the paths of the same user.

In order to improve the demodulation condition, it may use interference cancelation to remove the second term of equation (1.9) in each branch (finger) of the RAKE receiver. This problem will be discussed in Chapter 13 on multiuser detection. The block diagram of the receiver based on this concept is shown in Figures 1.8 and 1.9.

$$s_t(t) = b_k c_k \cos \omega t \tag{1.7}$$

$$r(t) = \sum_l \sum_k \beta_{lk} b_k(t - \tau_{lk}) c(t - \tau_{lk}) \cos[(\omega + \omega_{dlk})t + \theta_{lk}] \tag{1.8}$$

$$y_{11} = \beta_{11} b_1(t - \tau_{11}) \cos[(\omega_{IF} + \omega_{d11})t + \theta_{11}] +$$
$$\sum_{\substack{l \\ l,k \neq 1,1}} \sum_k \beta_{lk} b_k(t - \tau_{lk}) \rho_{1,k}(\Delta \tau_{11,lk}) \cos[(\omega_{IF} + \omega_{dlk})t + \theta_{lk}] \tag{1.9}$$

Using complex-envelope representation, shown in Figure 1.10 one can, in general, more precisely represent the oversimplified baseband equations (1.1a to 4a). The transmitted

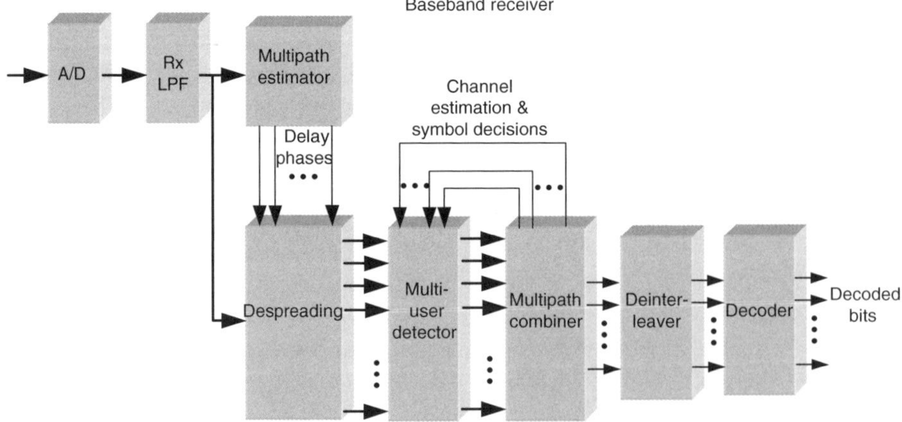

Figure 1.8   Generic receiver block diagram with optional interference cancelation stage.

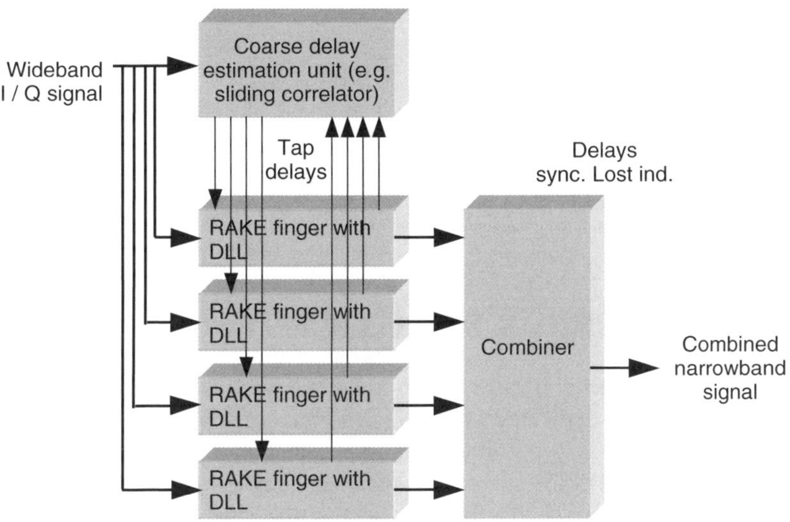

Figure 1.9 Traditional RAKE with delay lock loop (DLL) in each finger.

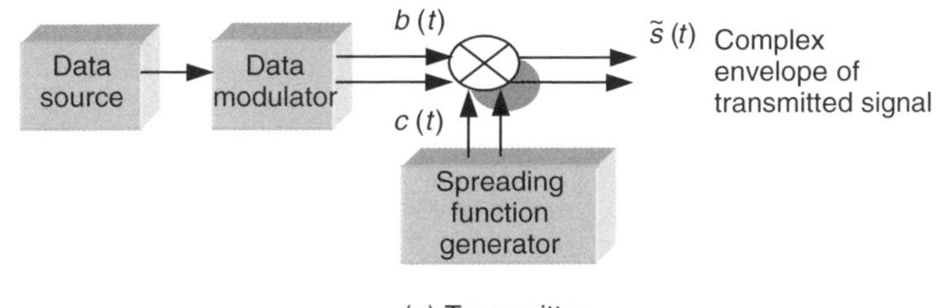

(a) Transmitter

Complex-envelope representation

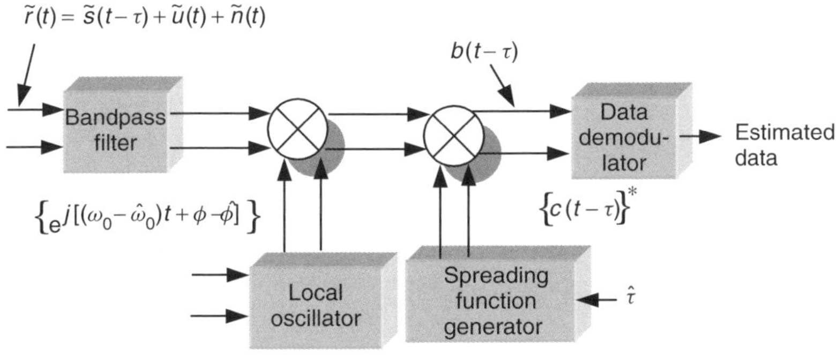

$$\tilde{r}(t) = \tilde{s}(t - \tau) + \tilde{u}(t) + \tilde{n}(t)$$

$$\left\{ e^{j[(\omega_0 - \hat{\omega}_0)t + \phi - \hat{\phi}]} \right\}$$

$$\left\{ c(t - \tau) \right\}^*$$

$$\hat{\tau}$$

(b) Receiver

Figure 1.10 Generic complex envelope model of spread spectrum modem.

signal is represented by equation (1.10). The despread complex signal is represented by equation (1.11).

$$s(t) = \mathrm{Re}\lfloor \tilde{s}(t)\mathrm{e}^{j\omega_0 t} \rfloor \tag{1.10}$$

$$\begin{aligned} b(t - \tau) = {}& b(t - \tau)c(t - \tau)c^*(t - \hat{\tau})\exp\{-j[(\omega_0 - \hat{\omega}_0)t + \varphi - \hat{\varphi}]\} \\ &+ \tilde{u}(t)c^*(t - \tau)\exp\{-j[(\omega_0 - \hat{\omega}_0)t + \varphi - \hat{\varphi}]\} \\ &+ \tilde{n}(t)c^*(t - \tau)\exp\{-j[(\omega_0 - \hat{\omega}_0)t + \varphi - \hat{\varphi}]\} \end{aligned} \tag{1.11}$$

1.3 THEORY VERSUS PRACTICE

This section provides an initial illustration on how the previous concept is implemented for multiplexing/spreading of dedicated physical data channel (DPDCH) and dedicated physical control channel (DPCCH) in universal mobile telecommunication system (UMTS). A detailed discussion of the UMTS standard is given in Chapter 17 and References [77–86]. Figure 1.11 shows the uplink DPDCH/DPCCH multiplexing and spreading for the most common case of only one DPDCH. A combination of code and IQ (In phase + Quadrature) multiplex is used, where the DPDCH and DPCCH are spread by different channelization orthogonal variable spreading factor (OVSF) codes (c_D, c_C) and mapped to an I and Q branch, respectively. The complex I + jQ signal is then scrambled by a short code C_{scramb}. A short scrambling code is used in order to simplify the future implementation of advanced receiver structures, for example, multiuser detectors. As an option, long-code scrambling may be used, in the case when the base station (BS) employs ordinary RAKE reception.

1.3.1 Multicode transmission

Additional DPDCHs can be mapped to either the I or the Q branch as illustrated in Figure 1.12. Each DPDCH should be allocated to the I or Q branch in such a way that the overall envelope variations are minimized. Any IQ imbalance is avoided with the

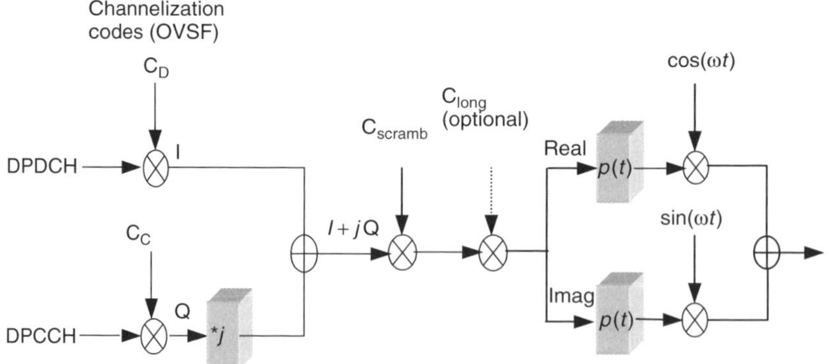

Figure 1.11 Uplink spreading and scrambling for the normal case of one DPDCH per connection.

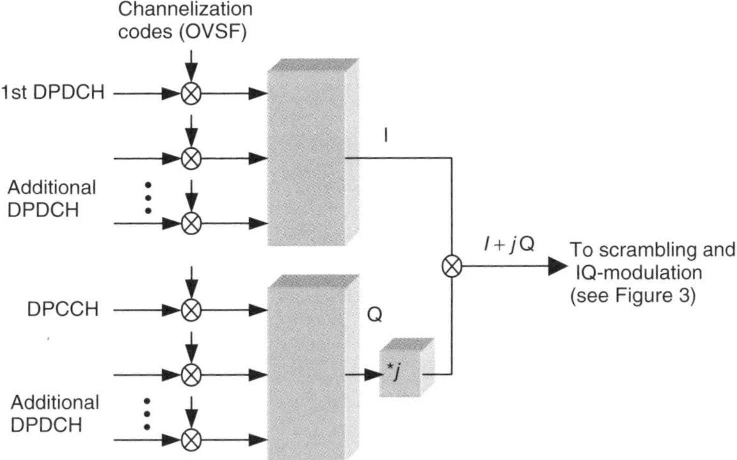

Figure 1.12 Multiplexing of multiple DPDCH on one connection (multicode transmission).

complex scrambling operation that makes the amplifier constellation similar to that with I and Q branches of equal power.

1.3.2 The downlink multiplexing and spreading

The processing is similar to that of the uplink, except that all downlink (DL) connections of a BS share a common set of short OVSF channelization codes and are jointly scrambled by a short BS unique scrambling code as shown in Figure 1.13. The BS unique scrambling code is allocated from the set of orthogonal Gold codes of length 256 chips.

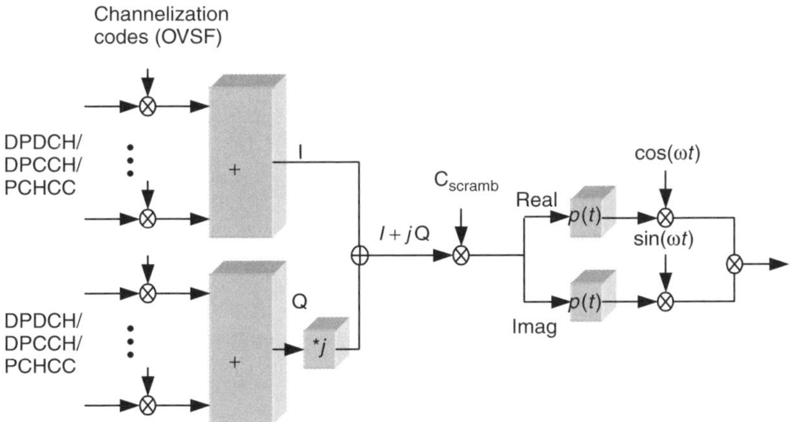

Figure 1.13 Downlink channel multiplexing and spreading.

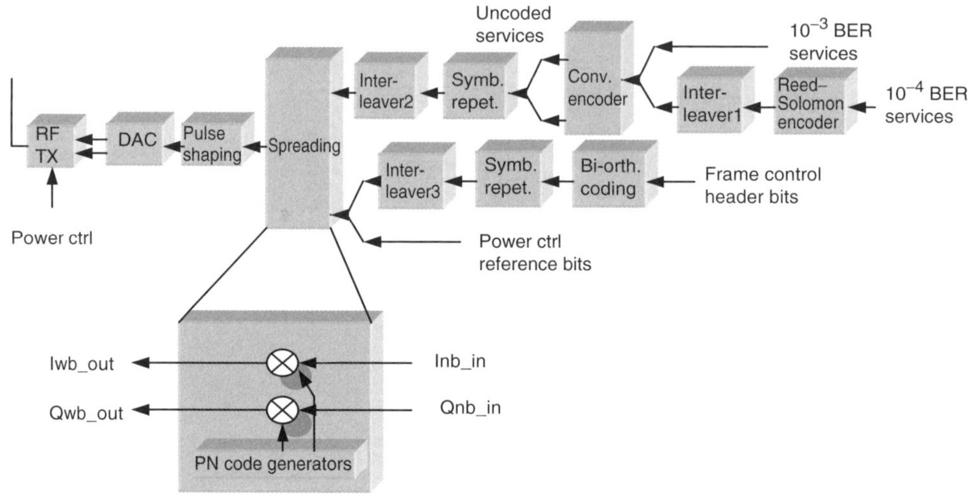

Figure 1.14 Mobile transmitter section (index wb-wideband, nb-narrowband).

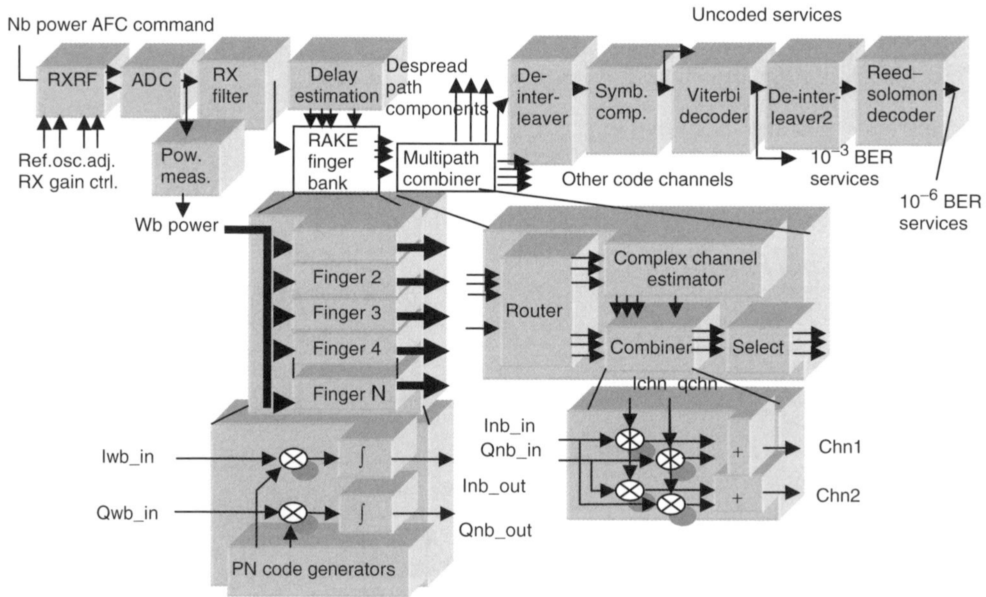

Figure 1.15 Mobile terminal receiver baseband section.

Finally, on the basis of the previous discussion, a block diagram of the mobile transmitter and receiver is shown in Figures 1.14 and 1.15, respectively. The building blocks will be discussed in detail throughout the book.

REFERENCES

1. Berrou, C. and Glavieux, A. (1996) Near optimum error-correcting coding and decoding: turbo codes. *IEEE Trans. Commun.*, **COM-44**, 1261–1271.
2. Hagenauer, J., Offer, E. and Papke, L. (1996) Iterative decoding of binary block and convolutional codes. *IEEE Trans. Inform. Theory*, **IT-42**, 429–445.
3. Benedetto, S., Divsalar, D., Montorsi, G. and Pollara, F. (1998c) Soft-input soft-output modules for the construction and distributed iterative decoding of code networks. *Eur. Trans. Telecommun.*, **9**, 155–172.
4. McEliece, R. J., MacKay, D. J. C. and Cheng, J. F. (1998) Turbo decoding as an instance of Pearl's 'Belief Propagation' algorithm. *IEEE J. Select. Areas Commun.*, **16**, 140–152.
5. Biglieri, E. and Hagenauer, J. (eds) (1995) *Eur. Trans. Telecommun.*, **6**, the whole issue.
6. Benedetto, S., Divsalar, D. and Hagenauer, J. (eds) (1998d) Concatenated coding techniques and iterative decoding: sailing toward channel capacity. *IEEE J. Select. Areas Commun.*, **16**(2), the whole issue.
7. Jamali, S. H. and Le-Ngoc, T. (1994) *Coded-Modulation Techniques for Fading Channels*. New York: Kluwer.
8. Biglieri, E., Divsalar, D., McLane, P. J. and Simon, M. K. (1991) *Introduction to Trellis-Coded Modulation with Applications*. New York: MacMillan Publishing.
9. Viterbi, A. J., Wolf, J. K., Zehavi, E. and Padovani, R. A. (1989) Pragmatic approach to trellis-coded modulation. *IEEE Commun. Mag.*, **27**, 11–19.
10. Zehavi, E. (1992) 8-PSK trellis codes for a Rayleigh channel. *IEEE Trans. Commun.*, **40**, 873–884.
11. Aoyama, A., Yamazato, T., Katayama, M. and Ogawa, A. (1994) Performance of 16-QAM with increased diversity on Rayleigh fading channels. *Proc. International Symposium on Information Theory and Its Applications*, Sydney, Australia, November 20–24, 1994, pp. 1133–1137.
12. Hansson, U. and Aulin, T. (1996) Channel symbol expansion diversity – improved coded modulation for the Rayleigh fading channel. *Presented at the International Conference on Communications, ICC '96*, Dallas, TX, June 23–27, 1996.
13. Al-Semari, S. A. and Fuja, T. (1996) Bit interleaved I-Q TCM. *ISITA '96*, Victoria, B.C., September 17–20, 1996.
14. Ventura-Traveset, J., Caire, G., Biglieri, E. and Taricco, G. (1997) Impact of diversity reception on fading channels with coded modulation. Part I: coherent detection. *IEEE Trans. Commun.*, **45**, 563–572.
15. Boutros, J., Viterbo, E., Rastello, C. and Belfiore, J.-C. (1996) Good lattice constellations for both Rayleigh fading and Gaussian channels. *IEEE Trans. Inform. Theory*, **42**, 502–518.
16. Caire, G. *et al.* (1998) Bit interleaved coded modulation. *IEEE Trans. Inform. Theory*, **44**(3), 927–945.
17. Goldsmith, A. *et al.* (1998) Adaptive coded modulation for fading channels. *IEEE Trans. Commun.*, **46**(5), 595–602.
18. Goldsmith, A. J. and Chua, S.-G. (1997) Variable-rate variable-power MQAM for fading channels. *IEEE Trans. Commun.*, **45**, 1218–1230.
19. Rappaport, T. S. (1996) *Wireless Communication Principles and Practice*. Englewood Cliffs, NJ: Prentice-Hall.
20. Forney Jr, G. D., Gallager, R. G., Lang, G. R., Longstaff, F. M. and Quereshi, S. U. (1984) Efficient modulation for band-limited channels. *IEEE J. Select. Areas Commun.*, **SAC-2**, 632–647.

21. Hayes, J. F. (1968) Adaptive feedback communications. *IEEE Trans. Commun.*, **COM-16**, 29–34.
22. Cavers, J. K. (1972) Variable-rate transmission for Rayleigh fading channels. *IEEE Trans. Commun.*, **COM-20**, 15–22.
23. Webb, W. T. and Steele, R. (1995) Variable rate QAM for mobile radio. *IEEE Trans. Commun.*, **43**, 2223–2230.
24. Kamio, Y., Sampei, S., Sasaoka, H. and Morinaga, N. (1995) Performance of modulation-level-controlled adaptive-modulation under limited transmission delay time for land mobile communications. *Proc. IEEE VTC '95*, July 1995, pp. 221–225.
25. Alamouti, S. M. and Kallel, S. (1994) Adaptive trellis-coded multiple-phased-shift keying for Rayleigh fading channels. *IEEE Trans. Commun.*, **42**, 2305–2314.
26. Matsuoka, H., Sampei, S., Morinaga, N. and Kamio, Y. (1996) Symbol rate and modulation level controlled adaptive modulation/TDMA/TDD for personal communication systems. *Proc. IEEE VTC '95*, April 1996, pp. 487–491.
27. Lin, S. and Costello, D. (1982) *Error Control Coding: Fundamentals and Applications*. Englewood Cliffs, NJ: Prentice Hall.
28. Gordon, N., Vucetic, B., Musicki, D. and Du, J. Joint error control and speech coding for 4.8 kbps digital voice transmission over satellite mobile channels. Tech. Rep., Sydney University, Sydney, Australia.
29. Cain, J. B., Clark, G. C. and Geist, J. M. (1979) Punctured convolutional codes of rate $(n - 1)/n$ and simplified maximum likelihood decoding. *IEEE Trans. Inform. Theory*, **IT-25**, 97–100.
30. Yasuda, Y., Hirata, Y., Nakamura, K. and Otani, S. (1983) Development of variable-rate Viterbi decoder and its performance characteristics. *Proc. Sixth International Conference on Digital Satellite Communications*, Phoenix, AZ, September 1983, pp. XII-24–XII-31.
31. Yasuda, Y., Kashiki, K. and Hirata, Y. (1984) High rate punctured convolutional codes for soft decision Viterbi decoding. *IEEE Trans. Commun.*, **COM-32**, 315–319.
32. Hagenauer, J. (1988) Rate-compatible punctured convolutional codes (RCPC codes) and their applications. *IEEE Trans. Commun.*, **36**, 389–400.
33. Wu, K., Lin, S. and Miller, M. (1982) A hybrid ARQ scheme using multiple shortened cyclic codes. *Proc. GLOBECOM*, Miami, FL, pp. C8.61–C8.65.
34. Chase, D. (1985) Code combining – a maximum likelihood decoding approach for combining an arbitrary number of noisy packets. *IEEE Trans. Commun.*, **COM-33**, 385–393.
35. Sovetov, B. and Stah, V. (1982) *Design of Adaptive Transmission Systems*. Leningrad: Energoizdal; in Russian.
36. Sullivan, D. (1971) A generalization of Gallager's adaptive error control scheme. *IEEE Trans. Inform. Theory*, **IT-17**, 727–735.
37. Mandelbaum, D. (1974) An adaptive-feedback coding scheme using incremental redundancy. *IEEE Trans. Inform. Theory*, **IT-20**, 388–389.
38. Vucetic, B., Drajic, D. and Perisic, D. (1988) An algorithm for adaptive error control system synthesis. *ISIT 1985*, Brighton, UK, pp. 85–94; also in Proc. IEE, Part F Feb.
39. Mandelbaum, D. M. (1975) On forward error correction with adaptive decoding. *IEEE Trans. Inform. Theory*, **IT-21**, 230–233.
40. Kallel, S. and Haccoun, D. (1988) Sequential decoding with ARQ code combining: a robust hybrid FEC/ARQ system. *IEEE Trans. Commun.*, **26**, 773–780.
41. Drukarev, A. and Costello Jr, D. J. (1983) Hybrid ARQ control using sequential decoding. *IEEE Trans. Inform. Theory*, **IT-29**, 521–535.
42. Drukarev, A. and Costello Jr, D. J. (1982) A comparison of block and convolutional codes in ARQ error control schemes. *IEEE Trans. Commun.*, **COM-30**, 2449–2455.
43. Lugand, L. and Costello Jr, D. J. (1982) A comparison of three hybrid ARQ schemes on a non-stationary channel. *Proc. GLOBECOM*, Miami, FL, pp. C8.4.1–C8.4.5.
44. Hagenauer, J. and Lutz, E. (1987) Forward error correction coding for fading compensation in mobile satellite channels. *IEEE J. Select. Areas Commun.*, **SAC-5**, 215–225.
45. Glisic, S. and Leppanen, P. (eds) (1997) *Wireless Communications; TDMA Versus CDMA*. London: Kluwer.

46. Saunders, S. (1999) *Antennas and Propagation for Wireless Communication Systems.* New York: John Wiley & Sons.
47. Winters, J. *et al.* (1994) The impact of antenna diversity on the capacity of wireless communication systems. *IEEE Trans. Commun.*, **42**(2–4), 1740–1750.
48. Marzetta, T. *et al.* (1999) Capacity of a mobile multiple-antenna communication link in Rayleigh flat fading. *IEEE Trans. Inform. Theory*, **45**(1), 139–157.
49. Foschini, G. *et al.* (1998) On the limit of wireless communication in a fading environment when using multiple antennas. *Wireless Personal Commun.*, **6**(3), 311–335.
50. Tarokh, V. *et al.* (1998) Space-time codes for high data rate wireless communication: performance criterion and code construction. *IEEE Trans. Inform. Theory*, **44**(2), 744–765.
51. Tarokh, V. *et al.* (1999) Space-time block codes from orthogonal design. *IEEE Trans. Inform. Theory*, **45**(5), 1456–1467.
52. *EURASIP J. Appl. Signal Process.*, Special issue on space-time coding and its applications-part I, **2002**(3), 2002.
53. Win, M. and Scholtz, R. (2000) Ultra-wide bandwidth time-hopping spread-spectrum impulse radio for wireless multiple access communications. *IEEE Trans. Commun.*, **48**(4), 679–689.
54. Win, M. and Scholtz, R. (1998) Impulse radio: how it works. *IEEE Commun. Lett.*, **2**(2), 36–38.
55. FCC (2002) *New Public Safety Applications and Broadband Internet Access Among Users Envisioned by FCC Authorization of Ultra Wideband Technology.* FCC first report and order, February 14, 2002, ET Docket No. 98–103, John Reed, jreed@fcc.gov. http://www.fcc.gov/Bureaus/Engineering_Technology/News-Releases/2002/nret0203.html.
56. Ramirez-Mireles, F. (2001) On the performance of ultra-wide-band signals in Gaussian noise and dense multipath. *IEEE Trans. Veh. Technol.*, **50**(1), 244–249.
57. Taylor, J. (ed.) (1995) *An Introduction to Ultra Wideband Radar Technology.* Boca Raton, FL: CRC Press.
58. *IEEE J. Select. Areas Commun.*, Special issue on "Software Radios", (4), 1999.
59. Pursley, M., Russell, H. and Wysocarski, J. (2000) Energy-efficient transmission and routing protocols for wireless multiple-hop networks and spread-spectrum radios. *EUROCOMM 2000, Information Systems for Enhanced Public Safety and Security, IEEE/AFCEA*, pp. 1–5.
60. McDonald, A. and Znati, T. (2000) A dual-hybrid adaptive routing strategy for wireless ad hoc networks. *IEEE Wireless Communications and Networking Conference, WCNC 2000*, Vol. 3, pp. 1125–1130.
61. Pursley, M., Russell, H. and Wysocarski, J. (2000) Energy-efficient routing in frequency-hop radio networks with partial-band interference. *IEEE Wireless Communications and Networking Conference, WCNC 2000*, Vol. 1, pp. 79–83.
62. Tien, T. C. and Upadhyaya, S. (2000) A local/global strategy based on signal strength for message routing in wireless mobile ad hoc networks 2000. *Proc. Academia/Industry Working Conference on Research Challenges*, pp. 227–232.
63. Tschudin, C., Lundgren, H. and Gulbrandsen, H. (2000) Active routing for ad hoc networks. *IEEE Commun. Mag.*, **38**(4), 122–127.
64. Garcia-Luna-Aceves, J. and Spohn, M. (1999) Efficient routing in packet-radio networks using link-state information. *IEEE Wireless Communications and Networking Conference*, Vol. 3, pp. 1308–1312.
65. Hettich, A. *et al.* (1999) Routing protocols for wireless ad hoc ATM networks. *2nd International Conference on ATM, ICATM '99*, pp. 49–58.
66. Ramanujan, R. *et al.* (1998) Source-initiated adaptive routing algorithm (SARA) for autonomous wireless local area networks. *Annual Conference on Local Computer Networks, LCN '98*, 23rd Proceedings, pp. 109–118.
67. Haas, Z. and Pearlman, M. (1998) The performance of a new routing protocol for the reconfigurable wireless networks. *IEEE International Conference on Communications, ICC '98 Conference Record*, Vol. 1, pp. 156–160.
68. Naghshineh, M. and Willebeek-LeMair, M. (1997) End to end QoS provisioning multimedia wireless/mobile networks using an adaptive framework. *IEEE Commun. Mag.*, **35**(11), 72–81.

69. Lin, C. *et al.* (1997) Adaptive clustering for mobile wireless networks. *IEEE J. Select. Areas Commun.*, **15**(7), 1265–1275.

70. Park, V. and Corson, M. (1997) A highly adaptive distributed routing algorithm for mobile wireless networks. *INFOCOM '97*, Proc. Vol. 3, pp. 1405–1413.

71. Gupta, P. and Kumar, P. (1997) A system and traffic dependent adaptive routing algorithm for ad hoc networks. *Proc. 36th IEEE Conference on Decision and Control*, Proc. Vol. 3, pp. 2375–2380.

72. Johnson, D. and Maltz, D. (1996) Truly seamless wireless and mobile host networking protocols for adaptive wireless and mobile networking. *IEEE Personal Commun.*, **3**(1), 34–42.

73. Roytblat, I. *et al.* (1996) Network connectivity buildup by adaptive learning. *19th Convention of Electrical and Electronics Engineers in Israel*, pp. 9–12.

74. Hortos, W. (1994) Application of neural networks to the adaptive routing control and traffic estimation of survivable wireless communication networks. *Southcon/94 Conference Record*, pp. 85–91.

75. *IEEE J. Select. Areas Commun.*, Special issue on "Active and programmable networks", **15**(3), 2001.

76. Hanzo, L. *et al.* (2002) *Adaptive Transceivers Communications*. New York: John Wiley & Sons.

77. 3GPP TS 25.308: UTRA High Speed Downlink Packet Access (HSDPA); overall description.

78. Glisic, S. and Leppanen, P. (eds) (1995) *Code Division Multiple Access Communications*. London: Kluwer.

79. Glisic, S. and Vucetic, B. (1997) *Spread Spectrum CDMA for Wireless Communications*. London: Artech House.

80. 3GPP TS 25.201: Physical layer – general description.

81. Holma, H. and Toskala, A. (2000) *WCDMA for UMTS*. New York: John Wiley & Sons.

82. Viterbi, A. J. (1995) *Principle of Spread Spectrum Communication*. Reading, MA: Addison-Wesley.

83. Prasad, R. (1996) *CDMA for Wireless Personal Communications*. London: Artech House.

84. 3GPP TS 25.101: UE Radio transmission and reception (FDD).

85. 3GPP TS 25.211: Physical channels and mapping of transport channels onto physical channels (FDD).

86. 3GPP TS 25.104: UTRA (BS) FDD; Radio transmission and reception.

2

Pseudorandom sequences

2.1 PROPERTIES OF BINARY SHIFT REGISTER SEQUENCES

Let us define a polynomial

$$h(x) = h_0 x^n + h_1 x^{n-1} + \cdots + h_{n-1} x + h_n \tag{2.1}$$

in a discrete field with two elements $h_i \in (0, 1)$ and $h_0 = h_n = 1$.

An example of a polynomial could be $x^4 + x + 1$ or $x^5 + x^2 + 1$. The coefficients h_i of the polynomial can be represented by binary vectors 10011 and 100101, or in octal notation 23 and 45 (every group of three bits is represented by a number between 0 and 7).

A binary sequence u is said to be a *sequence generated by $h(x)$* if for all integers j

$$h_0 u_j \oplus h_1 u_{j-1} \oplus h_2 u_{j-2} \oplus \cdots \oplus h_n u_{j-n} = 0$$

$$\oplus = \text{addition modulo 2} \tag{2.2}$$

If we formally change the variables,

$$\begin{aligned} j &\to j + n \\ h_0 &= 1 \end{aligned} \tag{2.3}$$

then equation (2.2) becomes

$$u_{j+n} = h_n u_j \oplus h_{n-1} u_{j+1} \oplus \cdots h_1 u_{j+n-1} \tag{2.4}$$

In this notation, u_j is the jth bit (called chip) of the sequence u. The sequence u can be generated by an n-stage binary linear feedback shift register, which has a feedback tap connected to the ith cell if $h_i = 1, 0 < i \leq n$.

Example 1

For $n = 5$, equation (2.4) becomes

$$u_{j+5} = h_5 u_j \oplus h_4 u_{j+1} \oplus h_3 u_{j+2} \oplus h_2 u_{j+3} \oplus h_1 u_{j+4} \tag{2.5}$$

For the polynomial $x^5 + x^2 + 1$, the octal representation (45), of the coefficients h_i, are

h_0	h_1	h_2	h_3	h_4	h_5
1	0	0	1	0	1

and the block diagram of the circuit is shown in Figure 2.1.

Example 2

For the polynomial $x^5 + x^4 + x^3 + x^2 + 1$, the coefficients h_i are given as

h_0	h_1	h_2	h_3	h_4	h_5	
1	1	1	1	0	1	(75)

and by using equation (2.4) one can get the generator shown in Figure 2.2.

Some of the properties of these sequences and definitions are listed below. Details can be found in the standard literature listed at the end of the chapter, especially in References [1–12]. If u and v are generated by $h(x)$, then so is $u \oplus v$, where $u \oplus v$ denotes the sequence whose ith element is $u_i \oplus v_i$. All zero state of the shift register is not allowed because for this initial state, equation (2.5) would continue to generate zero chips. For this reason, the period of u is at most $2^n - 1$, where n is the number of cells in the

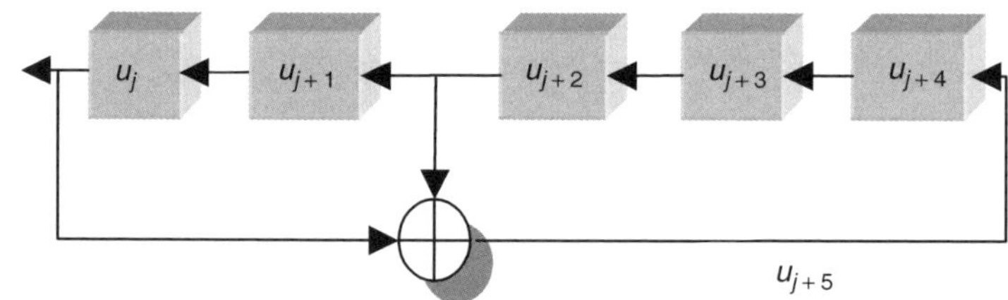

Figure 2.1 Sequence generator for the polynomial (45).

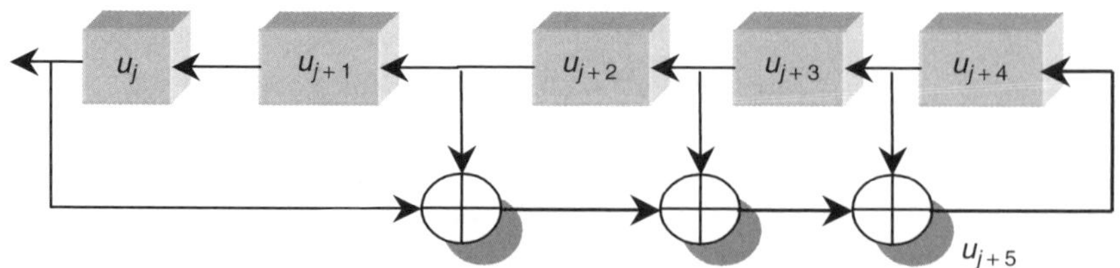

Figure 2.2 Sequence generator for the polynomial (75).

shift register, or equivalently, the degree of $h(x)$. If u denotes an arbitrary $\{0, 1\}$ – valued sequence, then $x(u)$ denotes the corresponding $\{+1, -1\}$ – valued sequence, where the ith element of $x(u)$ is just $x(u_i)$.

$$x(u_i) = (-1)_i^u \tag{2.6}$$

If T^i is a delay operator (delay for i chip periods), then we have

$$T^i(x(u)) = x(T^i u) \text{ and}$$

$$\sum x(u) = x(u_0) + x(u_1) + \cdots + x(u_{N-1})$$

$$= N^+ - N^- = (N - N^-) - N^- = N - 2N^-$$

$$= N - 2wt(u) \tag{2.7}$$

where $wt(u)$ denotes the Hamming Weight of unipolar sequence u, that is, the number of ones in u, n is the sequence period and N^+ and N^- are the number of positive and negative chips in bipolar sequence $x(u)$.

The cross-correlation function between two bipolar sequences can be represented as

$$\theta_{u,v}(l) \equiv \theta_{x(u),x(v)}(l) = \sum_{i=0}^{N-1} x(u_i)x(v_{i+l})$$

$$= \sum_{i=0}^{N-1} (-1)^{u_i}(-1)^{v_{i+l}}$$

$$= \sum_{i=0}^{N-1} (-1)^{u_i \oplus v_{i+l}}$$

$$= \sum_{i=0}^{N-1} x(u_i \oplus v_{i+l}) \tag{2.8}$$

By using equation (2.7), we have

$$\theta_{u,v}(l) = N - 2wt(u \oplus T^l v) \tag{2.9}$$

The periodic autocorrelation function $\theta_u(\,\cdot\,)$ is just $\theta_{u,u}(\,\cdot\,)$ and we have

$$\theta_u(l) = N - 2wt(u \oplus T^l u)$$

$$= N^+ - N^-$$

$$= (N - N^-) - N^-$$

$$= N - 2N^- \tag{2.10}$$

2.2 PROPERTIES OF BINARY MAXIMAL-LENGTH SEQUENCE

As it was mentioned earlier, all zero state of the shift register is not allowed because, on the basis of equation (2.4), the generator could not get out of this state. Bear in mind that the number of possible states of shift register is 2^n. The period of a sequence u generated by the polynomial $h(x)$ cannot exceed $2^n - 1$ where n is the degree of $h(x)$. If u has this maximal period $N = 2^n - 1$, it is called a maximal-length sequence or m-sequence. To get such a sequence, $h(x)$ should be a primitive binary polynomial of degree n.

Property I The period of u is $N = 2^n - 1$.

Property II There are exactly N nonzero sequences generated by $h(x)$, and they are just the N different phases of u, $Tu, T^2u, \ldots, T^{N-1}u$.

Property III Given distinct integers i and j, $0 \le i, j < N$, there is a unique integer k, distinct from both i and j, such that $0 \le k < N$ and

$$T^i u \oplus T^j u = T^k u. \tag{2.11}$$

Property IV $wt(u) = 2^{n-1} = 1/2(N + 1)$.

Property V From (2.9)

$$\theta_u(l) = \begin{cases} N, & \text{if } l \equiv 0 \mod N \\ -1, & \text{if } l \neq 0 \mod N \end{cases} \tag{2.12}$$

$\tilde{u}$ is called a characteristic m-sequence, or the characteristic phase of the m-sequence u if $\tilde{u}_i = \tilde{u}_{2i}$ for all $i \in Z$.

Property VI Let q denote a positive integer, and consider the sequence v formed by taking every qth bit of u (i.e. $v_i = u_{qi}$ for all $i \in Z$). The sequence v is said to be a *decimation* by q of u, and will be denoted by $u[q]$.

Property VII Assume that $u[q]$ is not identically zero. Then, $u[q]$ has period $N/\gcd(N, q)$, and is generated by the polynomial whose roots are the qth powers of the roots of $h(x)$ where $\gcd(N, q)$ is the greatest common divisor of the integers N and q. The tables of primitive polynomials are available in any book on coding theory. From Reference [13] we take an example of the polynomial of degree 6.

DEGREE 6

1	103F	3	127B	5	147H	7	111A
9	015	11	155E	21	007		

The letters E, F and H mean (among other things) that the polynomials 103, 147 and 155 are primitive, while the letters A and B indicate nonprimitive polynomials. Suppose that the m-sequence u is generated by the polynomial 103. Then, $u[3]$ is generated by the 127, $u[5]$ is generated by 147, $u[7]$ is generated by the 111, and so on.

$u[3]$ has period $63/\gcd(63, 3) = 21$, and thus is not an m-sequence; while $u[5]$ has period 63 and is an m-sequence. The corresponding polynomials 127 and 147 are clearly indicated as nonprimitive and primitive, respectively. $v = u[q]$ has period N if and only if $\gcd(N, q) = 1$. In this case, the decimation is called a *proper* decimation, and the sequence v is an m-sequence of period N generated by the primitive binary polynomial $\hat{h}(x)$. If, instead of u, we decimate $T^i u$ by q, we will get some phase $T^j v$ of v; that is, regardless of which of the m-sequences generated by $h(x)$ we choose to decimate, the result will be an m-sequence generated by $\hat{h}(x)$. In particular, decimating $\tilde{u}$, the characteristic phase of u, gives $\tilde{v}$, the characteristic phase of v.

Property VIII Suppose $\gcd(N, q) = 1$. If $v = u[q]$, then for all $j \geq 0$,

$$\tilde{u}[2^j q] = \tilde{u}[2^j q \bmod N] = \tilde{v}$$

and

$$u[2^j q] = u[2^j q \bmod N] = T^i v$$

for some i which depends on j.

Property VIII is also valid for $j < 0$ provided $2^j q$ is an integer. Hence, proper decimation by odd integers q gives all the m-sequence of period N. However, the following decimation by an even integer is of interest. Let $v = u[N - 1]$. Then $v_i = u_{(N-1)I} = u_{-i}$, that is, v is just a reciprocal of u.

The reciprocal m-sequence v is generated by the reciprocal polynomial of $h(x)$, that is,

$$\hat{h}(x) = x^n h(x^{-1}) = h_n x^n + h_{n-1} x^{n-1} + \cdots + h_0 \tag{2.13}$$

From Property VIII we see that a different phase of v is produced if we decimate u by $1/2(N - 1) = 2^{n-1} - 1$ instead of $(N - 1)$. Other proper decimations lead to other m-sequences. The summarized results of different decimations are shown in Figures 2.3 and 2.4 [3].

From Figure 2.3 one can see that decimation of u defined by polynomial 45 by factor $q = 3$ gives $v = u[3]$ defined by polynomial 75. All decimations by factor 3 are obtained by moving clockwise along the solid line. Decimation by factor 5 is indicated by moving clockwise along the dashed line. Moving counterclockwise along the solid lines gives decimation by factor 11 and moving counterclockwise along the dashed line gives decimation by factor 7. The same notation is valid for Figure 2.4.

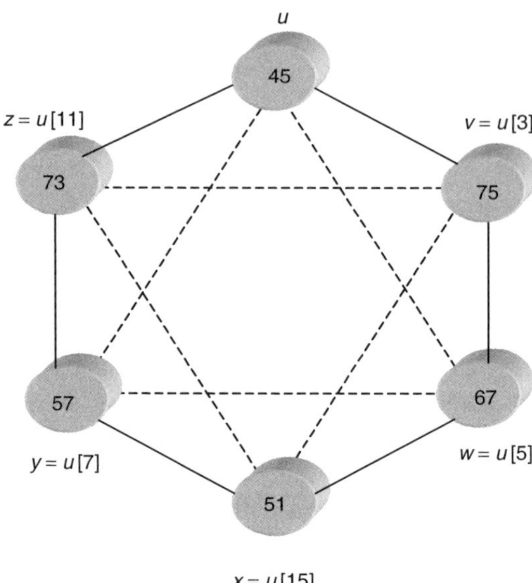

Figure 2.3 Decimation relations for m-sequences of period 31. When traversed clockwise, solid lines and dotted lines correspond to decimations by 3 and 5, respectively. Reproduced from Sarwate, S. V. and Pursley, M. B. (1980) Crosscorrelation properties of pseudorandom and related sequences. *Proc. IEEE.* Vol. 68, May 1980, pp. 593–619, by permission of IEEE.

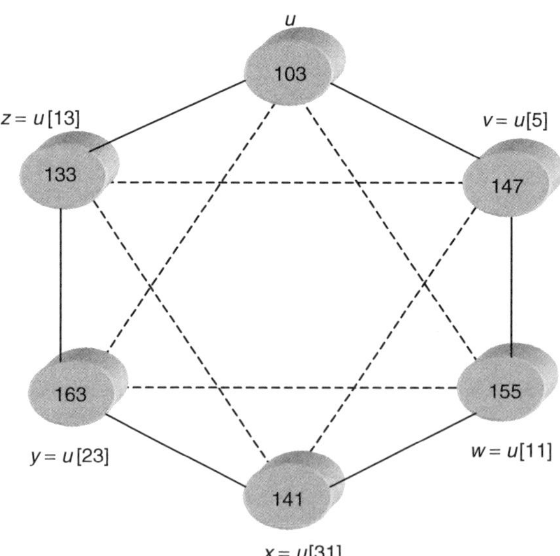

Figure 2.4 Decimation relations for m-sequences of period 63. When traversed clockwise, solid lines and dotted lines correspond to decimations by 5 and 11, respectively. Reproduced from Sarwate, S. V. and Pursley, M. B. (1980) Crosscorrelation properties of pseudorandom and related sequences. *Proc. IEEE.* Vol. 68, May 1980, pp. 593–619, by permission of IEEE.

2.2.1 Cross-correlation functions for maximal-length sequences

Cross-correlation spectra

Frequently, we do not need to know more than the set of cross-correlation values together with the number of integers l ($0 \le l < N$) for which $\theta_{u,v}(l) = c$ for each c in this set.

Theorem 1 Let u and v denote m-sequences of period $2^n - 1$. If $v = u[q]$, where either $q = 2^k + 1$ or $q = 2^{2k} - 2^k + 1$, and if $e = \gcd(n, k)$ is such that n/e is odd, then the spectrum of $\theta_{u,v}$ is three-valued [13–18] as

$$-1 + 2^{(n+e)/2} \text{ occurs } 2^{n-e-1} + 2^{(n-e-2)/2} \text{ times}$$

$$-1 \text{ occurs } 2^n - 2^{n-e} - 1 \text{ times}$$

$$-1 - 2^{(n+e)/2} \text{ occurs } 2^{n-e-1} - 2^{(n-e-2)/2} \text{ times} \qquad (2.14)$$

The same spectrum is obtained if instead of $v = u[q]$, we let $u = v[q]$. Notice that if e is large, $\theta_{u,v}(l)$ takes on large values but only very few times, while if e is small, $\theta_{u,v}(l)$ takes on smaller values more frequently. In most instances, small values of e are desirable. If we wish to have $e = 1$, then clearly n must be odd in order that n/e be odd. When n is odd, we can take $k = 1$ or $k = 2$ (and possibly other values of k as well), and obtain that $\theta(u, u[3])$, $\theta(u, u[5])$ and $\theta(u, u[13])$ all have the three-valued spectrum given in Theorem 1 (with $e = 1$).

Suppose next that $n \equiv 2 \bmod 4$. Then, n/e is odd if e is even and a divisor of n. Letting $k = 2$, we obtain that $\theta(u, u[5])$ and $\theta(u, u[13])$ both have the three-valued spectrum given in Theorem 1 (with $e = 2$).

Let us define $t(n)$ as

$$t(n) = 1 + 2^{[(n+2)/2]} \qquad (2.15)$$

where $[\alpha]$ denotes the integer part of the real number α. Then if $n \ne 0 \bmod 4$, there exist pairs of m-sequences with three-valued cross-correlation functions, where the three values are -1, $-t(n)$, and $t(n) - 2$. A cross-correlation function taking on these values is called a *preferred three-valued cross-correlation function* and the corresponding pair of m-sequences (polynomials) is called a *preferred pair of m-sequences* (polynomials).

Theorem 2 Let u and v denote m-sequences of period $2^n - 1$ where n is a multiple of 4. If $v = u[-1 + 2^{(n+2)/2}] = u[t(n) - 2]$, then $\theta_{u,v}$ has a four-valued spectrum represented as

$$-1 + 2^{(n+2)/2} \text{ occurs } (2^{n-1} - 2^{(n-2)/2})/3 \text{ times}$$

$$-1 + 2^{n/2} \text{ occurs } 2^{n/2} \text{ times}$$

$$-1 \text{ occurs } 2^{n-1} - 2^{(n-2)/2} - 1 \text{ times}$$

$$-1 - 2^{n/2} \text{ occurs } (2^n - 2^{n/2})/3 \text{ times} \qquad (2.16)$$

2.3 SETS OF BINARY SEQUENCES WITH SMALL CROSS-CORRELATION MAXIMAL CONNECTED SETS OF m-SEQUENCES

The preferred pair of m-sequences is a pair of m-sequences of period $N = 2^n - 1$, which has the preferred three-valued cross-correlation function. The values taken on by the preferred three-valued cross-correlation functions are -1, $-t(n)$, and $t(n) - 2$, where $t(n)$ is given by equation (2.15). The pair of primitive polynomials that generate a preferred pair of m-sequences is called a preferred pair of polynomials. A connected set of m-sequences is a collection of m-sequences that has the property that each pair in the collection is a preferred pair. The largest possible connected set is called the *maximal connected set* and the size of such a set is denoted by M_n. Some examples are given in Table 2.1.

Graphical representation of maximal connected sets is given in Figures 2.5 to 2.7 [3]. There are 18 maximal connected sets, and each m-sequence belongs to 6 of them.

2.4 GOLD SEQUENCES

A set of Gold sequences of period $N = 2^{n-1}$, consists of $N + 2$ sequences for which $\theta_c = \theta_a = t(n)$. A set of Gold sequences can be constructed from appropriately selected m-sequences as described below. Suppose $f(x) = h(x)\hat{h}(x)$ where $h(x)$ and $\hat{h}(x)$ have no factors in common. The set of all sequences generated by $f(x)$ is of the form $a \oplus b$

Table 2.1 Set sizes and cross-correlation bounds for the sets of all m-sequences and for maximal connected sets [3]. Reproduced from Sarwate, S. V. and Pursley, M. B. (1980) Crosscorrelation properties of pseudorandom and related sequences. *Proc. IEEE.* Vol. 68, May 1980, pp. 593–619, by permission of IEEE

n	$N = 2^n - 1$	Number of m-sequences	θ_c for set of all m-sequences	M_n	$t(n)$
3	7	2	5	2	5
4	15	2	9	0	9
5	31	6	11	3	9
6	63	6	23	2	17
7	127	18	41	6	17
8	255	16	95	0	33
9	511	48	113	2	33
10	1 023	60	383	3	65
11	2 047	176	287	4	65
12	4 095	144	1407	0	129
13	8 191	630	$\geq$703	4	129
14	16 383	756	$\geq$5631	3	257
15	32 767	1800	$\geq$2047	2	257
16	65 535	2048	$\geq$4095	0	513

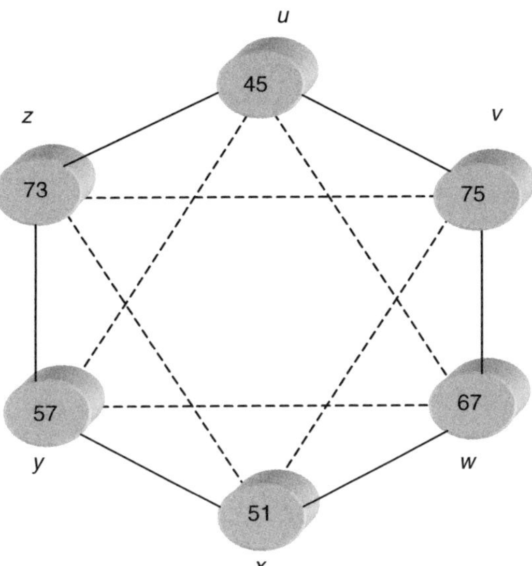

Figure 2.5 Preferred pairs of *m*-sequences of period 31. The vertices of every triangle form a maximal connected set. Reproduced from Sarwate, S. V. and Pursley, M. B. (1980) Crosscorrelation properties of pseudorandom and related sequences. *Proc. IEEE.* Vol. 68, May 1980, pp. 593–619, by permission of IEEE.

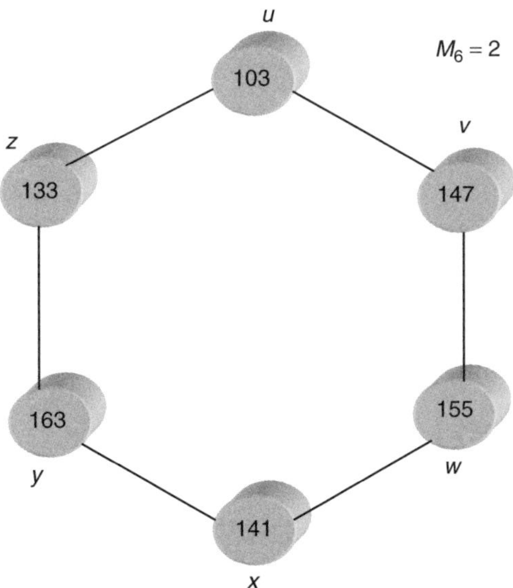

Figure 2.6 Preferred pairs of *m*-sequences of period 63. Every pair of adjacent vertices is a maximal connected set. Reproduced from Sarwate, S. V. and Pursley, M. B. (1980) Crosscorrelation properties of pseudorandom and related sequences. *Proc. IEEE.* Vol. 68, May 1980, pp. 593–619, by permission of IEEE.

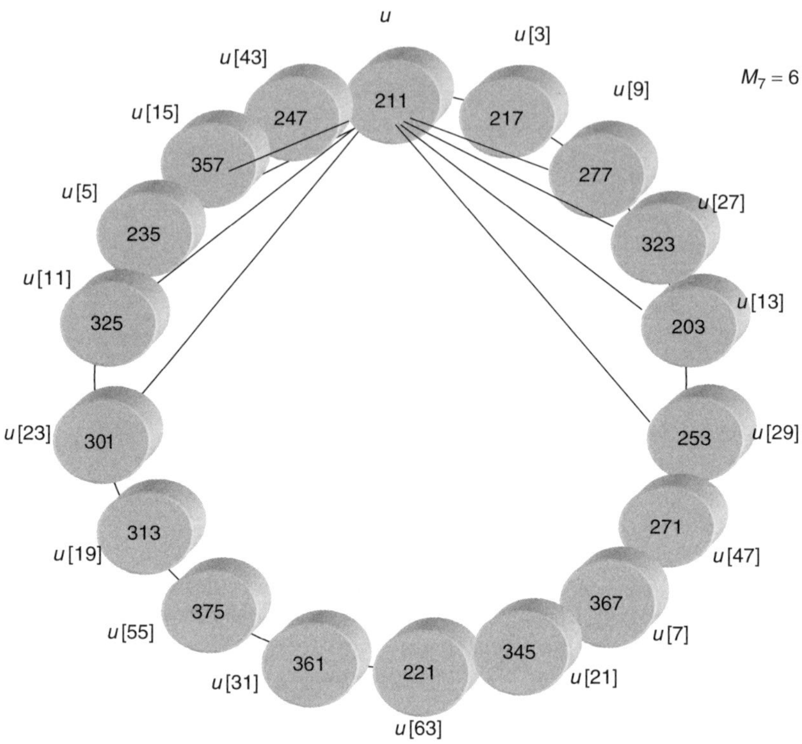

Figure 2.7 Preferred decimations for m-sequences of period 127. Every set of six consecutive vertices is a maximal connected set. Reproduced from Sarwate, S. V. and Pursley, M. B. (1980) Crosscorrelation properties of pseudorandom and related sequences. *Proc. IEEE.* Vol. 68, May 1980, pp. 593–619, by permission of IEEE.

where a is some sequence generated by $h(x)$, b is some sequence generated by $\hat{h}(x)$, and we do not make the usual restriction that a and b are nonzero sequences. We represent such a set by

$$G(u, v) \overset{\Delta}{=} \left\{ u, v, u \oplus v, u \oplus Tv, u \oplus T^2 v, \ldots, u \oplus T^{N-1} v \right\}. \tag{2.17}$$

$G(u, v)$ contains $N + 2 = 2^n + 1$ sequences of period N.

Theorem 3 Let $\{u, v\}$ denote a preferred pair of m-sequences of period $N = 2^n - 1$ generated by the primitive binary polynomials $h(x)$ and $\hat{h}(x)$, respectively. Then set $G(u, v)$ is called a set of Gold sequences. For $y, z \in G(u, v)$, $\theta_{y,z}(l) \in \{-1, -t(n), t(n) - 2\}$ for all integers l, and $\theta_y(l) \in \{-1, -t(n), t(n) - 2\}$ for all $l \neq 0 \mod N$. Every sequence in $G(u, v)$ can be generated by the polynomial $f(x) = h(x)\hat{h}(x)$.

Note that the nonmaximal-length sequences belonging to $G(u, v)$ also can be generated by adding together (term by term, modulo 2) the outputs of the shift registers corresponding to $h(x)$ and $\hat{h}(x)$. The maximal-length sequences belonging to $G(u, v)$ are, of course, the outputs of the individual shift registers.

Compare the parameter $\theta_{\max} = \max\{\theta_a, \theta_c\}$ for a set of Gold sequences to a bound due to Sidelnikov, which states that for any set of N or more *binary* sequences of period N

$$\theta_{\max} > (2N - 2)^{1/2} \tag{2.18}$$

For Gold sequences, they form an *optimal* set with respect to the bounds when n is odd. When n is even, Gold sequences are not optimal in this case.

2.5 GOLDLIKE AND DUAL-BCH SEQUENCES

Let n be even and let q be an integer such that $\gcd(q, 2^n - 1) = 3$. Let u denote an m-sequence of period $N = 2^n - 1$ generated by $h(x)$, and let $v^{(k)}, k = 0, 1, 2$, denote the result of decimating $T^k u$ by q.

Property VII of m-sequences implies that the $v^{(k)}$ are sequences of period $N' = N/3$, which are generated by the polynomial $\hat{h}(x)$ whose roots are qth powers of the roots of $h(x)$.

Goldlike sequences are defined as

$$\begin{aligned} H_q(u) = \{u, u \oplus v^{(0)}, u \oplus T v^{(0)}, \ldots, u \oplus T^{N'-1} v^{(0)}, \\ u \oplus v^{(1)}, u \oplus T v^{(1)}, \ldots, u \oplus T^{N'-1} v^{(1)}, \\ u \oplus v^{(2)}, u \oplus T v^{(2)}, \ldots, u \oplus T^{N'-1} v^{(2)}\} \end{aligned} \tag{2.19}$$

Note that $H_q(u)$ contains $N + 1 = 2^n$ sequences of period N.

For $n \equiv 0 \bmod 4$, $\gcd[t(n), 2^n - 1] = 3$ vectors $v^{(k)}$ are taken to be of length N rather than $N/3$. Consequently, it can be shown that for the set $H_{t(n)}(u)$, $\theta_{\max} = t(n)$. We call $H_{t(n)}(u)$ a set of Goldlike sequences. The correlation functions for the sequences belonging to $H_{t(n)}(u)$ take on values in the set $\{-1, -t(n), t(n) - 2, -s(n), s(n) - 2\}$ where $s(n)$ is defined (for even n only) by

$$s(n) = 1 + 2^{n/2} = \frac{1}{2}[t(n) + 1] \tag{2.20}$$

2.6 KASAMI SEQUENCES

Let n be even and let u denote an m-sequence of period $N = 2^n - 1$ generated by $h(x)$. Consider the sequence $w = u[s(n)] = u[2^{n/2} + 1]$. It follows from Property VII that w is a sequence of period $2^{n/2} + 1$, which is generated by the polynomial $h'(x)$ whose roots are the $s(n)$th powers of the roots of $h(x)$ Furthermore, since $h'(x)$ can be shown to be a polynomial of degree $n/2$, w is an m-sequence of period $2^{n/2} - 1$. Consider the sequences generated by $h(x)h'(x)$ of degree $3n/2$. Any such sequence must be of one of the forms $T^i u, T^j w, T^i u \oplus T^j w, 0 \leq i < 2^n - 1, 0 \leq j < 2^{n/2} - 1$. Thus, any sequence y of period $2^n - 1$ generated by $h(x)h'(x)$ is some phase of some sequence in the set $K_s(u)$ defined by

$$K_s(u) \overset{\Delta}{=} \{u, u \oplus w, u \oplus Tw, \ldots, u \oplus T^{2^{n/2}-2}w\} \tag{2.21}$$

This set of sequences is called the *small set of Kasami sequences* with

$$\theta = \{-1, -s(n), s(n) - 2\}$$
$$\theta_{\max} = s(n) = 1 + 2^{n/2} \tag{2.22}$$

$\theta_{\max}$ for the set $K_s(u)$ is approximately one half of the value of $\theta_{\max}$ achieved by the sets of sequences discussed previously. $K_s(u)$ contains only $2^{n/2} = (N + 1)1/2$ sequences, while the sets discussed previously contain $N + 1$ or $N + 2$ sequences.

Theorem 4 Let n be even and let $h(x)$ denote a primitive binary polynomial of degree n that generates the m-sequence u. Let $w = u[s(n)]$ denote an m-sequence of period $2^{n/2} - 1$ generated by the primitive polynomial $h'(x)$ of degree $n/2$, and let $\hat{h}(x)$ denote the polynomial of degree n that generates $u[t(n)]$. Then, the set of sequences of period N generated by $h(x)\hat{h}(x)h'(x)$, called the *large set of Kasami sequences* and denoted by $K_L(u)$ is defined as follows:

1. if $n \equiv 2 \bmod 4$, then

$$K_L(u) = G(u, v) \bigcup \left[\bigcup_{i=0}^{2^{n/2}-2} \{T^i w \oplus G(u, v)\} \right] \tag{2.23}$$

where $v = u[t(n)]$, and $G(u, v)$ is defined in equation (2.17).
2. if $n \equiv 0 \bmod 4$, then

$$K_L(u) = H_{t(n)}(u) \bigcup \left[\bigcup_{i=0}^{2^{n/2}-2} \{T^i w \oplus H_{t(n)}(u)\} \right]$$

$$\bigcup \{v^{(j)} \oplus T^k w : 0 \leq j \leq 2, 0 \leq k < (2^{n/2} - 1)/3\} \tag{2.24}$$

where $v^{(j)}$ is the result of decimating $T^j u$ by $t(n)$ and $H_{t(n)}(u)$ is defined earlier by equation (2.19).

In either case, the correlation functions for $K_L(u)$ take on values in the set $\{-1, -t(n), t(n) - 2, -s(n), s(n) - 2\}$ and $\theta_{\max} = t(n)$. If $n \equiv 2 \bmod 4$, $K_L(u)$ contains $2^{n/2}(2^n + 1)$ sequences, while if $n \equiv 0 \bmod 4$, $K_L(u)$ contains $2^{n/2}(2^n + 1) - 1$ sequences. The large set of Kasami sequences contains both the small set of Kasami sequences and a set of Gold

Table 2.2 Polynomials generating various classes of sequences of periods 31, 63, 65, 127, and [3]. Reproduced from Sarwate, S. V. and Pursley, M. B. (1980) Crosscorrelation properties of pseudorandom and related sequences. *Proc. IEEE.* Vol. 68, May 1980, pp. 593–619, by permission of IEEE

N	Polynomial	Construction	No.	Values taken on by the correlation functions
31	3 551	G	33	7–1–9
	2 373	G	33	11 7 3–1–5–9
63	14 551	G	65	15–1–17
	14 343	G	65	15 11 7 3–1–5–9–13
	12 471	H_3	64	15 7–1–9–17
	1 527	K_s	8	7–1–9
	133 605	K_L	520	15 7–1–9–17
65	10 761		63	15 11 7 3–1–5–9–13
127	41 567	G	129	15–1–17
255	231 441	G	257	31 15–1–17
	264 455	G	257	31,…, 15 11 7 3–1–5–9–13–17,…, −29
	326 161	H_{33}	256	31 15–1–17–33
	267 543	H_3	256	31 15–1–17–33
	11 367	K_s	16	15–1–17
	6 031 603	K_L	4111	31 15–1–17–33

(or Goldlike) sequences as subsets. More interestingly, the correlation bound $\theta_{\max} = t(n)$ is the *same* as that for the latter subsets. The previous discussion is summarized in Table 2.2 for some examples of codes.

2.7 JPL SEQUENCES

These sequences are constructed by combining sequence $S_1(t, T_c)$ of length L_1 and $S_2(t, T_c)$ of length L_2 with L_1, L_2 prime, as $S = S(t, T_c) = S_1(t, T_c) \oplus S_2(t, T_c)$ of length $L = L_1 \times L_2$. If the composite sequence is delayed for L_1 chips,

$$S(t - L_1 T_c, T_c) = S_1(t - L_1 T_c, T_c) \oplus S_2(t - L_1 T_c, T_c)$$

$$= S_1(t, T_c) \oplus S_2(t - L_1 T_c, T_c) \tag{2.25}$$

and summed up with its original version

$$S(t, T_c) \oplus S(t - L_1 T_c, T_c) = S_1(t, T_c) \oplus S_2(t, T_c) \oplus S_1(t - L_1 T_c, T_c) \oplus S_2(t - L_1 T_c, T_c)$$

$$= S_1(t, T_c) \oplus S_1(t, T_c) \oplus S_2(t - L_1 T_c, T_c) \oplus S_2(t, T_c)$$

$$= S_2(t - L_3 T_c, T_c) \tag{2.26}$$

The result is only a component sequence S_2. In a similar way, by delaying the composite sequence for L_2 chips a component sequence S_1 will be obtained. This can be used to synchronize sequence S of length $L_1 \times L_2$ by synchronizing separately component sequences S_1 and S_2 of length L_1 and L_2, which can be done much faster. The acquisition time is proportional to $T_{\text{acq}}(S) \sim \max[T_{\text{acq}}(S_1), T_{\text{acq}}(S_2)] \sim \max[L_1, L_2]$.

2.8 KRONCKER SEQUENCES

In this case, the component sequences $S_1(t, T_{c1})$ of length L_1 and chip intervals T_{c1} and $S_2(t, T_{c2})$ with $L_2, T_{c2} = L_1 T_{c1}$ are combined as

$$S(t, T_{c1}, T_{c2}) = S_1(t, T_{c1}) \oplus S_2(t, T_{c2}) \tag{2.27}$$

The composite sequence S synchronization is now performed in cascade, first S_1 with much faster chip rate and then S_2. Correlation of S by S_1 gives

$$F_2(S_1 \cdot S) = \rho_1 S_2 \tag{2.28}$$

and after that this result is correlated with sequence S_2. The acquisition time is proportional to $T_{\text{acq}}(S) \sim T_{\text{acq}}(S_1) + T_{\text{acq}}(S_2) \sim L_1 + L_2$.

2.9 WALSH FUNCTIONS

A Walsh function of order n can be defined recursively as follows:

$$W(n) = \begin{vmatrix} W(n/2), W(n/2) \\ W(n/2), W'(n/2) \end{vmatrix} \tag{2.29}$$

W' denotes the logical complement of W, and $W(1) = |0|$. Thus,

$$W(2) = \begin{vmatrix} 0, 0 \\ 0, 1 \end{vmatrix} \quad \text{and} \quad W(4) = \begin{vmatrix} 0, 0, 0, 0 \\ 0, 1, 0, 1 \\ 0, 0, 1, 1 \\ 0, 1, 1, 0 \end{vmatrix} \tag{2.30}$$

$W(8)$ is as follows:

$$W(8) = \begin{vmatrix} 0, 0, 0, 0, 0, 0, 0, 0 \\ 0, 1, 0, 1, 0, 1, 0, 1 \\ 0, 0, 1, 1, 0, 0, 1, 1 \\ 0, 1, 1, 0, 0, 1, 1, 0 \\ 0, 0, 0, 0, 1, 1, 1, 1 \\ 0, 1, 0, 1, 1, 0, 1, 0 \\ 0, 0, 1, 1, 1, 1, 0, 0 \\ 0, 1, 1, 0, 1, 0, 0, 1 \end{vmatrix} \tag{2.31}$$

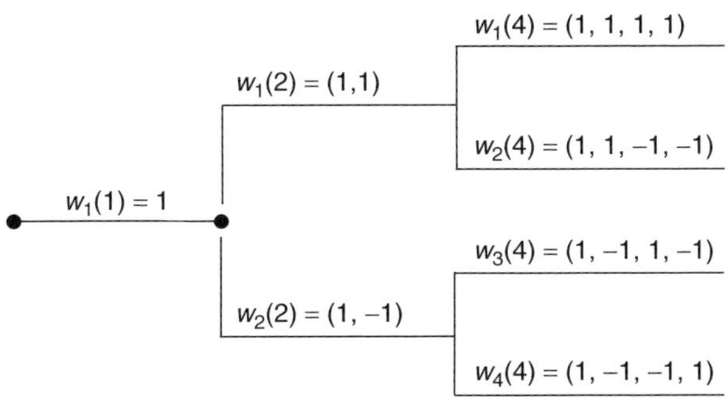

Figure 2.8 Flow graph generating OVSF codes of length 4.

One can see that any two rows from the matrix

$$w_k(n) = \{w_{k,j}(n)\},\ j = 1, \ldots, n$$
$$w_m(n) = \{w_{m,j}(n)\}$$

represent the sequences whose bipolar versions have cross-correlation equal to zero (orthogonal codes). This is valid for as long as the codes are aligned as in the matrix.

A modification of the previous construction rule is shown in Figure 2.8 producing orthogonal variable spreading factor (OVSF) sequences. At each node of the graph a code $w_k(n/2)$ of length $n/2$ produces two new codes of length n by a rule

$$w_k(n/2) \to w_{2k-1}(n) = \{w_k(n/2),\ w_k(n/2)\}$$
$$\to w_{2k}(n) = \{w_k(n/2),\ -w_k(n/2)\}$$

2.10 OPTIMUM PN SEQUENCES

If we represent the information bitstream as

$$\{b_n\} = \ldots, b_{-1}, b_0, b_1, b_2, \ldots;\ b_k = \pm 1 \tag{2.32}$$

and the sequence as a vector of chips

$$y = (y_0, y_1, \ldots, y_{N-1})\ y_k = \pm 1 \tag{2.33}$$

then the product of these two streams would create

$$\hat{y}_i = \ldots; b_{-1}y; b_0 y; b_1 y; \ldots \tag{2.34}$$

In other words, $\hat{y}$ is the Direct Sequence Spread Spectrum (DSSS) baseband signal that has as its ith element $\hat{y}_i = b_n y_k$ for all i such that $i = nN + k$ for k in the range $0 \le k \le N - 1$.

A synchronous correlation receiver forms the inner product

$$\langle \hat{y}_n, y \rangle = b_n \langle \hat{y}, y \rangle = b_n \theta_y(0) \tag{2.35}$$

If the other signal is $\hat{x}$, which is formed from the data sequence $\{b'_n\}$ and the signature sequence x (generated by a binary vector $x = (x_0, x_1, \ldots, x_{N-1})$ in exactly the same manner as $\hat{y}$ was formed from $\{b_n\}$ and y, then we have for the overall received signal

$$\hat{y} + T^{-l}\hat{x} \quad \text{where}$$

$$\hat{x} = \ldots; b'_{-1}x; b'_0 x; b'_1 x; \ldots \tag{2.36}$$

The output of a correlation receiver, which is in synchronism with y, is given by

$$z_n = \langle \hat{y}_n, y \rangle + \left[b'_{n-1} \sum_{i=0}^{l-1} x_{N-l+i} y_i + b'_n \sum_{i=l}^{N-1} x_{i-l} y_i \right] \tag{2.37}$$

Having in mind the following relations

$$\sum_{i=0}^{l-1} x_{N-l+i} y_i = \sum_{i=0}^{N-1+m} x_{i-m} y_i$$

$$\sum_{i=l}^{N-1} x_{i-l} y_i = \sum_{j=0}^{N-1-l} x_j y_{j+l} \tag{2.38}$$

and the definition of a periodic cross-correlation function $C_{x,y}$

$$C_{x,y}(l) = \begin{cases} \displaystyle\sum_{j=0}^{N-1-l} x_j y^*_{j+l}, & 0 \le l \le N - 1 \\ \displaystyle\sum_{j=0}^{N-1+l} x_{j-l} y^*_j, & 1 - N \le l < 0 \\ 0, & |l| \ge N \end{cases} \tag{2.39}$$

Equation (2.37) becomes

$$z_n = b_n \theta_y(0) + [b'_{n-1} C_{x,y}(l - N) + b'_n C_{x,y}(l)] \tag{2.40}$$

The optimum sequences should minimize the interfering term for all values of l. Further details may be found in References [19,20].

2.11 THEORY AND PRACTICE OF PN CODES

In this chapter, a heuristic approach was used to define certain classes of codes and to discuss their basic properties. The theory in this field uses mathematical tools based on discrete algebra (Galois field) for precise treatment of these problems. Only those codes that are the basis for the construction of the codes used in the existing standards are covered in this chapter. Long code in IS-95 is an m-sequence generated by a polynomial of degree $n = 42$. I and Q spreading codes are m-codes of degree 15. Channelization codes on the downlink are Walsh codes. The same codes are used on the uplink for 64-level orthogonal modulation in each mobile.

In Universal Mobile Telecommunication System (UMTS), Gold codes are used for scrambling. Different channels of the same user on the uplink are separated by using OVSF codes. A large set of Kasami codes is used in the primary and the secondary synchro channel. A number of specific issues related to the properties of the sequences are covered in References [21–46].

2.12 PN MATCHED FILTER

An important component in processing DSSS signal is the PN (pseudonoise) matched filter shown in Figure 2.9. It consists of an analog shift register with M delay elements (taps). The output of each tap is multiplied by a different chip of a PN sequence and the result is summed up to produce the output S_0 For each clock pulse (chip rate), the signal from each delay element is shifted to the right. If the input signal is baseband DSSS signal S_i presented by equation (2.41), then the output of the filter is given by S_0 in equation (2.42). In these equations, $b(t, T_m)$ is the bitstream with bit rate $1/T_m$, c is the code with chip rate $1/T_c$, and $\theta(pT_c)$ is the autocorrelation function of the code in the case when there is an offset of pT_c between the input code and the locally set coefficients of the filter. Once per sequence period MT_c, the output of the circuit will be high $Mb\theta(0) = Mb$, otherwise the output of the circuit will be low $Mb\theta(p)$. In the case of a multipath channel, each signal replica will produce a different pulse when it coincides with the filter coefficients.

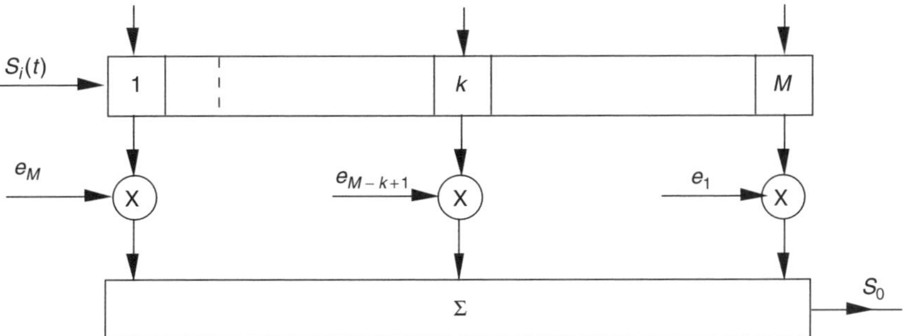

Figure 2.9 PN matched filter at the baseband.

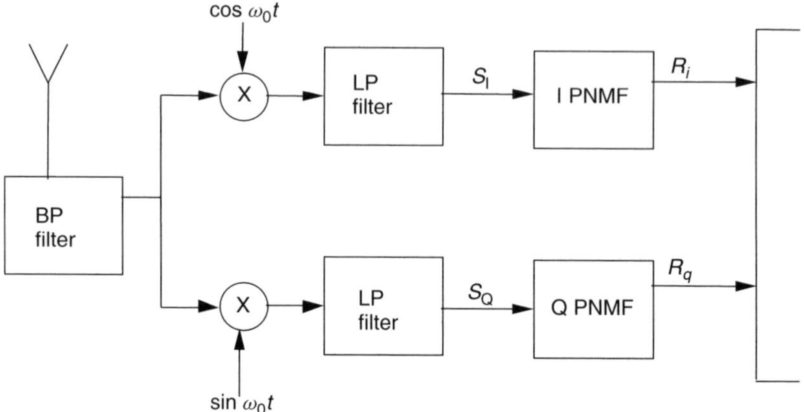

Figure 2.10 PN matched filter.

In the case of the DSSS signal presented in equation (2.43), the output of the two matched filters (I and Q) are given by equation (2.44).

$$S_i(t) = b(t, T_m) \cdot c(t, T_c) \tag{2.41}$$

$$S_0 = b(t, T_m) \sum_{k=1}^{M} c_k c_{k+p}$$

$$= M b(t, T_m) \theta(p T_c)$$

$$S_i(t) = b(t, T_m) \cdot c(t, T_e) \cos \omega t \tag{2.42}$$

$$R_i(p T_c) = M b(t, T_m) \theta(p T_c) \cos \phi(t) \tag{2.43}$$

$$R_q(p T_c) = M b(t, T_m) \theta(p T_c) \sin \phi(t) \tag{2.44}$$

The filter block diagram is given in Figure 2.10.

SYMBOLS

h – polynomial coefficients
u, v – sequences
$u[\ ]$ – decimation of u
$\theta(\ )$ – correlation function
N – code length
n – order of polynomial number of delay elements in code generator
$G(u, v)$ – Gold code
H_q – Goldlike code
gcd – greatest common divider
K_S – Kasami small set of codes

K_L – Kasami large set of codes
$W(n)$ – Walsh function
T^{-n} – delay operator (n chips)

REFERENCES

1. Ziemer, R. and Peterson, R. (1985) *Digital Communications and Spread Spectrum Systems.* New York: MacMillan Publishing.
2. Holmes, J. K. (1982) *Coherent Spread Spectrum Systems.* New York: Wiley-Interscience.
3. Sarwate, S. V. and Pursley, M. B. (1980) Crosscorrelation properties of pseudorandom and related sequences. *Proc. IEEE.* Vol. 68, May 1980, pp. 593–619.
4. Mezger, K. and Bouwens, R. J. (1972) An Ordered Table of Primitive Polynomials over GF(2) of Degrees 2 Through 19 for Use with Linear Maximal Sequence Generators. TM107, Cooley Electronics Laboratory, University of Michigan, Ann Arbor, July [AD 746876].
5. Schilling, D. L., Batson, B. H. and Pickholz, R. (1980) Spread spectrum communications. Short Course Notes, *National Telecommunication Conference.*
6. Golomb, S. W. (1967) *Shift Register Sequences.* San Francisco: Holden-Day.
7. Lindholm, J. H. (1968) An analysis of the pseudo-randomness properties of subsequences of long m-sequences. *IEEE Trans. Inform. Theory,* **14**(4), 569–576.
8. Massey, J. L. (1969) Shift-register synthesis and BCH decoding. *IEEE Trans. Inform. Theory,* **15**(1), 122–127.
9. Stiffer, J. J. (1968) Rapid acquisition sequences. *IEEE Trans. Inf. Theory.*
10. Groth, E. J. (1971) Generation of binary sequences with controllable complexity. *IEEE Trans. Inform. Theory,* **17**(3), 288–296.
11. Golomb, S. (1982) *Shift Register Sequences.* Laguna Hills CA: Aegean Park Press.
12. Glisic, S. and Vucetic, B. (1997) *CDMA for Wireless Communication.* Boston, MA: Kluwer Academic Publishers.
13. Rockwell International Corporation. (1976) *Study of Multistate PN Sequences and Their Application to Communication Systems.* Rep. (AD A025137), 1976.
14. Gold, R. (1966) Characteristic linear sequences and their coset functions. *SIAM J. Appl. Math.,* **14**, 980–985.
15. Gold, R. *Study of Correlation Properties of Binary Sequences.* Tech. Rep. AFAL-TR-66-234, AF Avionics Laboratory, Wright-Patterson AFB, OH, 1966 (AD 488858).
16. Gold, R. Optimal binary sequences for spread spectrum multiplexing. *IEEE Trans. Inform. Theory,* **IT-13**, 1967, 619–621.
17. Gold, R. *Study of Correlation Properties of Binary Sequences.* Tech. Rep. AFAL-TR-67-311, AF Avionics Laboratory, Wright-Patterson AFB, OH, 1967 (AD 826367).
18. Gold, R. Maximal recursive sequences with 3-valued recursive crosscorrelation functions. *IEEE Trans. Inform. Theory,* **IT-14**, 1968, 154–156.
19. Pursley, M. B. and Roefs, H. F. A. (1979) Numerical evaluation of correlation parameters for optimal phases of binary shift-register sequences. *IEEE Trans. Commun.,* **COM-27**, 1597–1604.
20. Pursley, M. B. and Sarwate, D. V. (1976) Bounds on aperiodic crosscorrelation for binary sequences. *Electron. Lett.,* **12**, 304–305.
21. Simon, M., Omura, J., Scholtz, R. and Levitt, B. (1985) *Spread-Spectrum Communications.* Vol. 1. New York: Computer Science Press.
22. Glisic, S. G. (1983) Power density spectrum of the product of two time displaced versions of a maximum length binary pseudonoise signal. *IEEE Trans. Commun.,* **COM-31**(2), 281–286.
23. Welch, L. R. (1974) Lower bounds on the maximum correlation of signals. *IEEE Trans. Inform. Theory,* **1T-20**, 397–399.
24. Pursley, M. B. (1977) Performance evaluation for phase-coded spread-spectrum multiple-access communication-part I: system analysis. *IEEE Trans. Commun.,* **COM-25**, 795–799.

25. Glisic, S. G. *et al.* (1987) Efficiency of digital communication system. *IEEE Trans. Commun.*, **COM-35**(6), 679–684.
26. Gordon, B., Mills, W. H. and Welch, L. R. (1962) Some new difference sets. *Can. J. Math.*, **14**, 614–625.
27. Schoulz, R. and Welch, L. (1984) GMW sequences. *IEEE Trans. Inform. Theory*, **IT-30**, 548–553.
28. Sarwate, D. V. and Pursley, M. B. (1977) Evaluation of correlations parameters for periodic sequences. *IEEE Trans. Inform.Theory*, **IT-23**, 508–513.
29. Sarwate, D. V. and Pursley, M. B. (1977) Performance evaluation for phase-coded spread-spectrum multiple access communication-part II: code sequence analysis. *IEEE Trans. Commun.*, **COM-25**, 800–803.
30. Roefs, H. F. A. (1977) *Binary Sequences for Spread-Spectrum Multiple-Access Communication*. Ph. D. Dissertation, Department of Electronic Engineering, University of Illinois, Urbana, (aslo Coordinated Science Lab. Rep. R-785).
31. Roefs, H. F. A. and Pursley, M. B. (1977) Correlation parameters of random binary sequences. *Electron. Lett.*, **13**, 488–489.
32. Roefs, H. F. A., Sarwate, D. V. and Pursley, M. B. (1977) Periodic correlation functions for sums of pairs of m-sequences, *Proc. Univ.*, Baltimore, MD, pp. 487–492.
33. Roefs, H. F. A., Sarwate, D. V. and Pursley, M. B. (1978) Crosscorrelation properties of sequences with applications to spread-spectrum multiple-access communication, *Proc. AFOSR Workshop in Communication Theory and Applications*, Provincetown, MA, pp. 88–91.
34. Sarwate, D. V. (1979) Bounds on crosscorrelation and autocorrelation of sequences. *IEEE Trans. Inform. Theory*, **IT-25**, 720–724.
35. Sarwate, D. V. and Pursley, M. B. (1976) Applications of coding theory to spread-spectrum multiple-access satellite communications, *Proc. IEEE Canadian Communications and Power Conference*, pp. 72–75.
36. Sarwate, D. V. and Pursley, M. B. (1977) New correlation identities for periodic sequences. *Electron. Lett.*, **13**(2), 48–49.
37. Sarwate, D. V. and Pursley, M. B. (1978) Hopping patterns for frequency hopped multiple-access communication. *IEEE International Conference Communications, Conference Record*, pp. 741–743.
38. Scholtz, R. A. and Welch, L. R. (1978) Group characters: sequences with good correlation properties. *IEEE Trans. Inform. Theory*, **IT-24**, 537–545.
39. Gill, W. J. and Spilker, J. J. (1963) An interesting decomposition property for the self-products of random and pseudorandom binary sequences. *IEEE Trans. Commun. Syst.*
40. Antweiler, M. and Bömer, L. (1992) Complex sequence over GF(p^m) with a two-level auto-correlation function and large linear span. *IEEE Trans. Inform. Theory*, **38**, 120–130.
41. Brynielsson, L. (1985) On the linear complexity of combined shift registers, in *Advances in Cryptology-eurocrypt '85*, Lecture Notes in Computer Science. Vol. 219. Berlin: Springer-Verlag, pp. 156–166.
42. Chan, A. H. and Games, R. On the linear span of binary sequences from finite geometries, q odd, in *Advances in Cryptology-eurocrypt '86*, Lecture Notes in Computer Science. Vol. 263. Berlin: Springer-Verlag, pp. 405–417.
43. Games, R. (1986) The geometry of m-sequences: three-valued crosscorrelations and quadrics in finite projective geometry. *SIAM J. Algebraic Discrete Methods*, **7**, 43–52.
44. MacWilliams, F. J. and Sloane, N. J. A. (1976) Pseudo-random sequences and arrays. *Proc. IEEE.*
45. Simon, M. K. *et al.* (1985) *A Unified Approach to Spread Spectrum Communications*. Rockville MD: Computer Science Press.
46. Peterson, W. W. and Weldon Jr, E. J. (1972) *Error-Correcting Codes*. 2nd edn. Cambridge MA: MIT Press.

3

Code acquisition

3.1 OPTIMUM SOLUTION

In this case, the theory starts with a simple problem where, for a received signal $r(t) = s(t, \boldsymbol{\theta}) + n(t)$, we have to estimate a generalized time invariant vector of parameters $\boldsymbol{\theta}$ (frequency, phase, delay, data, ...) of a signal $s(t, \boldsymbol{\theta})$ in the presence of Gaussian noise $n(t)$. The best that we can do is to find an estimate $\hat{\boldsymbol{\theta}}$ of the parameter $\boldsymbol{\theta}$ for which the aposterior probability $p(\hat{\boldsymbol{\theta}}/r)$ is maximum; hence the name maximum aposterior probability (MAP) estimate. In other words, the chosen estimate based on the received signal r is correct for the highest probability. Practical implementation requires us to locally generate a number of trial values $\tilde{\boldsymbol{\theta}}$, to evaluate $p(\tilde{\boldsymbol{\theta}}/r)$ for each such value and then to choose $\tilde{\boldsymbol{\theta}} = \hat{\boldsymbol{\theta}}$ for which $p(\tilde{\boldsymbol{\theta}}/r)$ is maximum. In this chapter, we focus only on code acquisition and parameter $\boldsymbol{\theta}$ will include only code delay $\boldsymbol{\theta} = \{\tau\}$ and become a scalar. Analytically, this can be expressed as

$$MAP \Rightarrow \hat{\theta} = \arg\,\max p(\tilde{\theta}/r) \tag{3.1}$$

Very often, in practice, evaluation of $p(\tilde{\theta}/r)$ in closed form is not possible. By using the Bayesian rule for the joint probability distribution function

$$p(r, \tilde{\theta}) = p(r)p(\tilde{\theta}/r) = p(\tilde{\theta})p(r/\tilde{\theta}) \tag{3.2}$$

and assuming a uniform prior distribution of θ, maximizing $p(\tilde{\theta}/r)$ becomes equivalent to maximizing $p(r/\tilde{\theta})$, a function that can be determined more easily. This algorithm is known as *maximum likelihood* (ML) estimation and can be defined analytically as

$$ML \Rightarrow \hat{\theta} = \arg\,\max p(r/\tilde{\theta}) \tag{3.3}$$

It is straightforward to show that in the case of Gaussian noise, the ML principle necessitates the search for that value of θ that would maximize the likelihood function defined as

$$\lambda(\tilde{\theta}) = \int r(t)s(t, \tilde{\theta})\,dt - \int s^2(t, \tilde{\theta})\,dt \tag{3.4}$$

where $s(t, \tilde{\theta})$ is the locally generated replica of the signal with a trial value $\tilde{\theta}$. For the given signal power, the second term in the previous equation is a constant so that the maximization is equivalent to the maximization of the first term only. This can be expressed as

$$\lambda(\tilde{\theta}) = \int r(t)s(t, \tilde{\theta})\, dt \tag{3.5}$$

Instead of searching for the maximum of $\lambda(\tilde{\theta})$ in a so-called open loop configuration, an equivalent procedure would be to find the zero of the first derivative of $\lambda(\tilde{\theta})$

$$MLT \Rightarrow \hat{\theta} = \arg\ \mathrm{zero}\frac{\partial\lambda(\tilde{\theta})}{\partial\tilde{\theta}} = \arg\ \mathrm{zero} \int r(t)\frac{\partial s(t, \tilde{\theta})}{\partial\tilde{\theta}}\, dt \tag{3.6}$$

This structure is known as the maximum likelihood tracker (MLT). In practice, the signal derivative is often approximated by the signal difference

$$\frac{\partial s(t, \tilde{\theta})}{\partial\tilde{\theta}} = \frac{1}{2\Delta\theta}\{s(t, \tilde{\theta} + \Delta\theta) - s(t, \tilde{\theta} - \Delta\theta)\} \tag{3.7}$$

where $s(t, \tilde{\theta} + \Delta\theta)$ and $s(t, \tilde{\theta} - \Delta\theta)$ are so called early and late versions of the local signal with respect to the generalized parameter θ to be estimated. This results in the so-called early–late tracker

$$ELT \Rightarrow \hat{\theta} = \arg\ \mathrm{zero}\{E(t, \tilde{\theta}) - L(t, \tilde{\theta})\} \tag{3.8}$$

where

$$\begin{aligned}
E(t, \tilde{\theta}) &= \frac{1}{2\Delta\theta} \int r(t)s(t, \tilde{\theta} + \Delta\theta)dt \\
L(t, \tilde{\theta}) &= \frac{1}{2\Delta\theta} \int r(t)s(t, \tilde{\theta} - \Delta\theta)dt
\end{aligned} \tag{3.9}$$

In the case of code synchronization, $\theta = \tau$ and the ML synchronizing receiver implied by equation (3.5) should, in principle, create all possible time-offset versions of the known code waveform, correlate all of them with the received data and choose the $\tilde{\tau}$ corresponding to the largest correlation as its estimate, $\hat{\tau}_{ML}$. Owing to the continuous range of values of τ, this is not possible in practice and some type of range quantization is necessary. The resulting candidate values are called *cells*, and the initial parameter estimation problem is translated into a multiple-hypothesis problem: to locate the cell most likely to contain the unknown offset, given this piece of data. This is exactly the *coarse code synchronization* or *code acquisition* problem, the result of which is to resolve the code phase (or the 'epoch') ambiguity within the size of the cell. Since this remaining error is typically larger than desired, further operations are required in order to reduce it to acceptable levels. This remaining part of the synchronization task, namely, that of *fine synchronization* or *code tracking*, is performed by one of the available code-tracking loops, which we discuss in the next chapter.

Once the nature and size of these cells have been determined, the next question is how to go about performing the search most successfully. Clearly, the strategy will depend on a variety of factors such as criteria of performance, degree of complexity and computational power available (directly related to cost), prior available information about the location of the correct cell and so on. A brute-force approach would try to create a bank of parallel correlation branches, each matched to a possible quantized value of the timing offset; it would then process the received waveform through all of them simultaneously, pick the largest and declare a candidate solution. Unless the uncertainty region (number of cells) is small, corresponding to either a small code period or a small initial uncertainty, such a solution (which we may call the *totally parallel solution*) becomes obviously unwieldy in complexity very quickly. We note, however, that small uncertainty regions may be encountered in a nested design, whereby a multitude of different-period codes are combined for precisely the purpose of aiding acquisition. Furthermore, neural network structures are currently being explored for this purpose, where the neural network is trained for all possible such values. Such a scheme would emulate the spirit (if not the exact statistical processing) of the above solutions.

3.2 PRACTICAL SOLUTIONS

In practice, most of the time total parallelism is out of the question when the number of cells is very large (although it appears doable for smaller uncertainty regions) and simpler solutions are necessary. One of the most familiar of such approaches is the simple technique of serial search, where the search starts from a specific cell and serially examines the remaining cells in some direction and in a prespecified order until the correct cell is found. Hence, serial search techniques do not account for any additional information gathered during the past search time, which could conceivably be used to alter the direction of search toward cells that show increased posterior likelihood of being the correct ones. A serial search starts from a cell that could be chosen totally arbitrarily (no prior information), or by some prior knowledge about a likely cell, and proceeds in a simple and easily implementable predirected manner. When the uncertainty space (collection of all possible cells) is two-dimensional (delay and frequency offset) and searching all possible cells serially appears to be very time consuming, a speedup may be achieved by employing a bank of filters, each matched to a possible Doppler offset. The same idea can be applied to the one-dimensional case (no frequency uncertainty), where now a bank of correlators may be employed, each starting from a different point of the uncertainty region. This effectively amounts to dividing the search in many parallel subsearches and therefore reducing the total search time by a proportional amount.

One should be aware that although it holds true that only one cell contains the exact delay and Doppler offsets of the incoming code, the set of desirable cells acceptable to the receiver includes a number of cells adjacent to the exact one. Indeed, the receiver will terminate acquisition and initiate tracking, the first time a cell is reached (and correctly identified), which is close enough to true synchronization so that the tracking loop can pull in and perform the remaining synchronization operation successfully. All these desirable cells are collectively called *hypothesis H_1*, and the remaining nondesirable 'out-of-sync' cells comprise *hypothesis H_0*. As an example, consider the case in which the receiver examines the code delay uncertainty in steps of half a chip time ($\delta t = T_c/2$) and there is no frequency uncertainty. Then, all four cells located in the interval $(-T_c, T_c)$ around the true delay of the incoming code are included in hypothesis H_1, since some amount of code correlation exists for each one of these cells, an amount that can initiate the code-tracking loop.

The above definition of cells and hypotheses implies that each test does not pertain to a single value of the unknown parameter τ, but rather to a range of values. It is straightforward to show that, under mild conditions and approximations pertaining to the pseudorandom nature of the code, this reformulated hypothesis testing results in a statistic (correlation) and threshold setting that do not depend on the given (tested) value of the unknown parameter (a uniformly most powerful test). This is because the threshold value is set by the desirable probability of false alarm per cell (see below), which is independent of τ under H_0.

To recapitulate, the two-dimensional time/frequency code offset uncertainty within the noisy received waveform is quantized into a number of cells, which are typically searched in a serial fashion by a correlation receiver, although parallel multiple branches are also possible. Motivated by an ML argument, the receiver creates a cross-correlation between the incoming waveform and the local code at a specific offset, whose output is used to decide whether the currently examined cell is a desirable (H_1) one. The process continues until one such cell is correctly identified. At that point, acquisition is terminated and tracking is initiated.

3.3 CODE ACQUISITION ANALYSIS

The serial code acquisition can be represented by using the signal flow graph theory. Each cell is represented by a node of a graph and transitions between the nodes depend on the outcome of the decision in a given cell. Branches connecting the nodes characterize these transitions. To motivate the operation in a transform domain, let us consider the simple model of a process represented by the graph in Figure 3.1 and evaluate the probability $p_{ac}(t)$ that the process will move from a to c in exactly t seconds.

To do this, we will introduce an additional variable τ to designate the time needed for the process to move from a to b, characterized by the probability $p_{ab}(\tau)$. The parameter

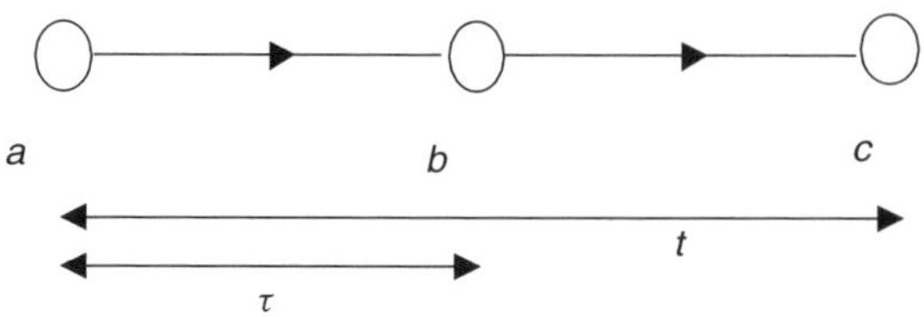

Figure 3.1 Signal flow graph for a 3-state process.

$p_{ac}(t, \tau)$ represents the joint probability that the process moves from a to c in t seconds and takes τ seconds to move from a to b. This probability can be represented as

$$p_{ac}(t, \tau) = p_{ab}(\tau)p_{bc}(t - \tau) \tag{3.10}$$

resulting in

$$p_{ac}(t) = \int p_{ac}(t, \tau)\, d\tau = \int p_{ab}(\tau)p_{bc}(t - \tau)\, d\tau$$
$$= p_{ac}(t)^* p_{bc}(t) \tag{3.11}$$

In other words, the overall probability $p_{ac}(t)$ is a convolution of the two intermode transition probabilities p_{ab} and p_{bc}. It is clear that for the graph with a large number of nodes we will have to deal with multiple convolutions giving rise to computational complexity. In this case, people being involved in electrical engineering prefer to move to a transform domain, either Laplace (s) domain for continuous variables or into z-domain for desecrate variables. This leads to using z-transform for the decision process flow graph representation and multiple convolutions will be now replaced with multiple products making the calculus much simpler. If $p_{ij}(n)$ is the probability for the process to move from node i to node j in exactly n steps, then its z-transform

$$P_{i,j}(z) = \sum_{n=0}^{\infty} z^n p_{ij}(n) \tag{3.12}$$

is called the *probability generating function*. For the analysis to follow, we will need a few relations derived from this definition. First of all, the first and the second derivative of this function can be represented as

$$\frac{\partial}{\partial z} P_{ij}(z) = \sum_{n=0}^{\infty} n p_{ij}(n) z^{n-1} \tag{3.13}$$

$$\frac{\partial^2}{\partial z^2} P_{ij}(z) = \sum_{n=0}^{\infty} n(n - 1) p_{ij}(n) z^{n-2} \tag{3.14}$$

By definition, the average number of steps to move from node i to node j is

$$\overline{n} = \sum_{n=0}^{\infty} n p_{ij}(n) = \left. \frac{\partial}{\partial z} P_{ij}(z) \right|_{z=1} \tag{3.15}$$

and the average time to do it can be represented as

$$\overline{t}_{ij} = T_{ij} = \overline{n}T = \left(\left. \frac{\partial}{\partial z} P_{ij}(z) \right|_{z=1} \right) \cdot T \tag{3.16}$$

where T is the *cell observation time* that is, the time needed to create the decision variable that will be referred to as dwell time. For the variance, we start with the definition

$$\sigma_T^2 = (\overline{n^2} - \overline{n}^2)T^2 \tag{3.17}$$

The second derivative of the generating function can be represented as

$$\frac{\partial^2}{\partial z^2} P_{ij}(z)\bigg|_{z=1} = \sum_{n=0}^{\infty} n^2 p_{ij}(n) - \sum_{n=0}^{\infty} n p_{ij}(n) = \overline{n^2} - \overline{n} \tag{3.18}$$

By using equations (3.15) and (3.18) in equation (3.17), the variance of time t_{ij} can be expressed in the following form:

$$\sigma_T^2 = \left[\frac{\partial^2 P_{ij}(z)}{\partial z^2} + \frac{\partial P_{ij}(z)}{\partial z} - \left(\frac{\partial P_{ij}(z)}{\partial z}\right)^2\right]_{z=1} T^2 \tag{3.19}$$

In what follows, we will use these few relations to analyze serial search code acquisition. In order to get an initial insight into this method, we will assume that there are q cells to be searched. Parameter q may be equal to the length of the pseudonoise (PN) code to be searched or some multiple of it. For example, if the update size is one-half chip, q will be twice the code length to be searched. Further assume that if a 'hit' (output is above threshold) is detected by the threshold detector, the system goes into a verification mode that may include both, an extended duration dwell time and an entry into a code loop tracking mode. In any event, we model the 'penalty' of obtaining a false alarm as $K\tau_d$ second and the dwell time itself as τ_d second. If a true hit is observed, the system has acquired the signal, and the search is completed. Assume that the false alarm probability P_{FA} and the probability of detection P_D are given. We will also assume that only one cell represents the synchro position. Let each cell be numbered from left to right so that the kth cell has *a priori* probability of having the signal present, given that it was not present in cells 1 through $k-1$, of

$$p_k = \frac{1}{q+1-k} \tag{3.20}$$

The generating function flow diagram is given in Figure 3.2 using the rule that at each node the sum of the probability emanating from the node equals unity. The unit time represents τ_d seconds and $K\tau_d$ seconds are represented in z-transform by z^K. Consider node 1. The *a priori* probability of having the signal present is $P_1 = 1/q$, and the probability of it not being present in the cell is $1 - P_1$. Suppose the signal was not present. Then we advance to the next node (node $1a$); since it corresponds to a probabilistic decision and not a unit time delay, no z multiplies the branch going to it. At node $1a$ a false alarm may occur, with probability $P_{FA} = \alpha$. This would require one unit of time to decide (τ_d s) and then K units of time ($K\tau_d$ s) are needed in verification mode to determine that there was a false alarm. False alarms will not occur with probability $(1 - \alpha)$. This would take one dwell time to decide and is represented by $(1 - \alpha)z$ branch going to node 2.

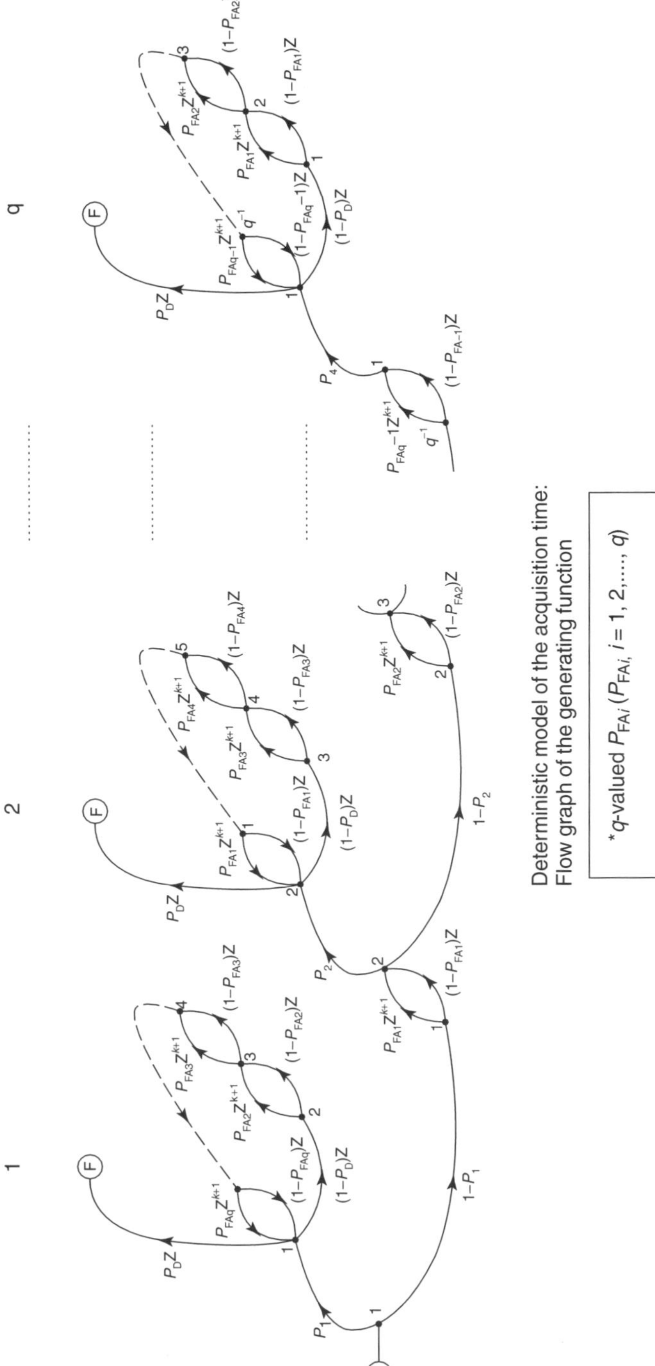

Figure 3.2 Code acquisition decision process flow graph.

Now consider the situation at node 1 when the signal is present. If a hit occurs (that is, the signal is detected), then acquisition, as we have defined it, occurs and the process is terminated in node F denoting 'finish'. If there was no hit at node 1 (the integrator output was below the threshold), which occurs with probability $1 - P_D$, one unit of time would be consumed for such a decision. This is represented by the branch $(1 - P_D)z$ leading to node 2. At node 2, in the upper left part of the diagram, either a false alarm occurs with probability α and delay $(K + 1)$, or a false alarm does not occur with a delay of 1 unit. The remaining portion of the generating function flow graph is a repetition of the portion just discussed with the appropriate node changes.

At this stage we will assume that only Gaussian noise is present so that P_{FA} and P_D are the same for each cell.

By using standard signal flow graph reduction techniques [1], one can show that the overall transfer function between nodes S (start) and F (finish) can be represented as

$$U(z) = \frac{(1 - \beta)}{1 - \beta z H^{q-1}} \frac{1}{q} \left[\sum_{l=0}^{q-1} H^l(z) \right] \tag{3.21}$$

where

$$H(z) = \alpha z^{K+1} + (1 - \alpha)z \quad \text{and} \quad \beta = 1 - P_D \tag{3.22}$$

By using equation (3.16), the mean acquisition time is given (after some algebra [1]) by

$$\overline{T} = \frac{2 + (2 - P_D)(q - 1)(1 + KP_{FA})}{2P_D} \tau_d \tag{3.23}$$

with τ_d being included in the formula to translate from our unit timescale. For the usual case, when $q \gg 1$, the mean acquisition time $\overline{T}$ is given by

$$\overline{T} = \frac{(2 - P_D)(1 + KP_{FA})}{2P_D}(q\tau_D) \tag{3.24}$$

The variance of the acquisition time is given by equation (3.19). It can be shown that the expression for σ^2 is

$$\sigma^2 = \tau_d^2 \left\{ (1 + KP_{FA})^2 q^2 \left(\frac{1}{12} - \frac{1}{P_D} + \frac{1}{P_D^2} \right) \right.$$

$$+ 6q[K(K + 1)P_{FA}(2P_D - P_D^2) \tag{3.25}$$

$$\left. + (1 + P_{FA}K)(4 - 2P_D - P_D^2)] + \frac{1 - P_D}{P_D^2} \right\}$$

In addition, when $K(1 + KP_{FA}) \ll q$, then

$$\sigma^2 = \tau_D^2(1 + KP_{FA})^2 q^2 \left(\frac{1}{12} - \frac{1}{P_D} + \frac{1}{P_D^2} \right) \tag{3.26}$$

As a partial check on the variance result, let $P_{FA} \to 0$ and $P_D \to 1$. Then we have

$$\sigma^2 = \frac{(q\tau_D)^2}{12} \qquad (3.27)$$

which is the variance of a uniformly distributed random variable, as one would expect for the limiting case. The above results provide a useful *theoretical* estimate of acquisition time for an idealized PN-type system. In *practice*, two basic modifications should be made to make the estimates reflect actual hardware or software systems. First, Doppler effects should be taken into account. The result of code Doppler is to smear the relative code phase during the acquisition dwell time, which increases or reduces the probability of detection depending on the code phase and the algebraic sign of the code Doppler rate. The Doppler also affects the effective code sweep rate, which in the extreme case can reduce it to zero to cause the search time to increase greatly. This topic will be discussed later. The second refinement to the model concerns the handover process between acquisition and tracking. Typically after a 'hit' the code-tracking loop is turned on to attempt to pull the code into tight lock. Further, often in low signal-to-noise ratio (SNR) systems in which both acquisition (pull-in) bandwidth and tracking bandwidth are used, multiple code loop bandwidths will be employed in order to soften the transition between acquisition and tracking modes. Consequently, the probability of going from the acquisition mode to the final code loop bandwidth in the tracking mode occurs with some probability less than 1. The estimation of this probability is at best a very difficult problem (although, some approximate results have been developed). At high SNRs, this probability quickly approaches 1, so it is not a problem. At low SNRs, the above formula for acquisition time should replace P_D with P_D'

$$P_D' = P_D P_{HO} \qquad (3.28)$$

with P_{HO} being the probability of handover. In the S-band shuttle system, at TRW it was found that at threshold ($C/N_0 = 51\,\text{dB Hz}$) P_{HO} varied from 0.06 to 0.5 depending upon the code Doppler. Without code Doppler P_{HO} was 0.25, which, if not taken into account in the acquisition time equation, would predict the mean acquisition time to be about four times too fast.

3.4 CODE ACQUISITION IN CDMA NETWORK

The previous Section 3.3 is limited to the case of spread-spectrum signal in Gaussian channel. In that case, the probability of false alarm in all nonsynchro cells is the same. In a communication radio network, the interfering signal is the sum of Gaussian noise and overall multiple access interference (MAI). In each cell, i, MAI has a different value so that $P_{FAi} \neq P_{FAj}$ for each $i \neq j$. In such a case, under the assumption of a static channel, the serial acquisition process can be modeled again by the graph from Figure 3.2 with P_{FA} being different for each cell. We will first deal with a simpler problem in which the probability of signal detection P_D does not depend on MAI. Besides being simpler, this model is still valid for an important class of these systems called quasi-synchronous Code Division

Multiple Access (CDMA) networks. In these networks, all users are synchronized within the range between zero delay and the position of the first significant cross-correlation peak. Examples of such systems are described for both satellite and land mobile CDMA communication systems.

The average acquisition time is obtained by using the same steps as in the previous section. The details are presented in Reference [2]. The result, after a cumbersome manipulation of very long equations can be expressed as

$$\overline{T}_{acq} = [2 + (q-1)(1 + k\overline{P}_{FA})(2 - \alpha P_D)]\frac{\tau_d}{2P_D} \tag{3.29}$$

where

$$\alpha = \frac{1 + k\rho}{1 + k\overline{P}_{FA}} \tag{3.30}$$

with

$$\rho = \frac{2}{q(q-1)}\left(\sum_{i=1}^{q}(i-1)P_{FAi}\right) \tag{3.31}$$

and

$$\overline{P}_{FA} = \frac{1}{q}\sum_{i=1}^{q}P_{FAi} \tag{3.32}$$

By inspection, we can see from equation (3.29) that the minimum average acquisition time is obtained for large values of parameter α. Besides $\overline{P}_{FA}$, this parameter also depends on the position of the cells with high P_{FAi} within the code delay uncertainty region. The set of P_{FAi}, representing the probability distribution function of P_{FA}, will be called MAI pattern or MAI profile. From equation (3.31), one can see that for a large α, the products iP_{FAi} should be large. This means larger P_{FAi} for larger i. That means that hopefully, synchronization will be acquired before we get to the region with high P_{FA} or in the case of multiple sweep of the uncertainty region, we will have smaller numbers of sweeps of the region.

In an asynchronous network, MAI takes on different values in all cells including the synchro cell so that, in general, P_D is different. In such a case, the average acquisition time becomes [2]

$$\overline{T}_{acq} = \frac{\tau_d}{2\tilde{P}_D}[2 + (1 + k\overline{P}_{FA})(q-1)(2 - \alpha\tilde{P}_D) + 2k(\overline{P}_{FA} - \overline{P}_R\tilde{P}_D)] \tag{3.33}$$

where

$$\overline{P}_{FA} = \frac{1}{q}\sum_{i=1}^{q}P_{FAi}, \quad \overline{P}_R = \frac{1}{q}\sum_{i=1}^{q}\frac{P_{FAi}}{P_{Di}}, \quad \tilde{P}_D = \left[\frac{1}{q}\sum_{i=1}^{q}\frac{1}{P_{Di}}\right]^{-1}$$

$$\alpha = \frac{1 + k\rho}{1 + k\overline{P}_{FA}} \quad \text{and} \quad \rho = \frac{2}{q(q-1)}\left(\sum_{i=1}^{q}(i-1)P_{FAi}\right) \tag{3.34}$$

Table 3.1 Mean acquisition time for different distributions of P_{FA} and P_D. Reproduced from Katz, M. and Glisic, S. (2000) Modelling of code acquisition process in CDMA networks-asynchronous networks. *IEEE J. Select. Areas Commun.*, **18**(1), 73–86, by permission of IEEE[2]

Distribution of P_{FA} and P_D	Mean acquisition time $\overline{T}_{acq}$
Case#1 $P_{FAi} = P_{FA}, \forall i \,(\text{fixed})$ $P_{Di} = P_D, \forall i \,(\text{fixed})$	$\dfrac{\tau_d}{2P_D}[2 + (1 + kP_{FA})(q - 1)(2 - P_D)]$
Case#2 $P_{FAi} = \{P_{FA1}, P_{FA2}, \ldots, P_{FAq}\}(q - \text{valued})$ $P_{Di} = P_D, \forall i \,(\text{fixed})$	$\dfrac{\tau_d}{2P_D}[2 + (1 + k\overline{P}_{FA})(q - 1)(2 - \alpha P_D)]$
Case#3 $P_{FAi} = \{P_{FA1}, P_{FA2}, \ldots, P_{FAq}\}(q - \text{valued})$ $P_{Di} = \{P_{D1}, P_{D2}, \ldots, P_{Dq}\}(q - \text{valued})$	$\dfrac{\tau_d}{2\tilde{P}_D}[2 + (1 + k\overline{P}_{FA})(q - 1)(2 - \alpha \tilde{P}_D)$ $\qquad\qquad + 2k(\overline{P}_{FA} - \overline{P}_R \tilde{P}_D)]$

It is interesting to compare the expression for mean acquisition time with previous results. Table 3.1 summarizes the results obtained for Case#1, constant P_{FA} and P_D, Case#2, q-valued P_{FA} and a constant P_D in quasi-synchronous networks and Case#3, q-valued P_{FA} and q-valued P_D in asynchronous networks.

The form of the three expressions provides an easy insight into the major differences in average acquisition times for the three cases. In the expression for case#2, when compared with case#1, P_{FA} should be replaced by $\overline{P}_{FA}$ and P_D in the numerator should be modified by a factor α given by equation (3.30). The first factor takes into account the average P_{FA} and the second modification takes into account the position of the initial search cell with respect to the distribution of P_{FAi}. In the expression for case#3, when compared with case#2, P_D should be replaced by $\tilde{P}_D$ in addition to a new term Δ that should be added to the numerator. This term can be expressed as $\Delta = 2k(\overline{P}_{FA} - \overline{P}_R \tilde{P}_D)$.

A first observation is that a sufficient condition for Δ to be zero is that P_{FA} or P_D or both of them have a constant distribution, that is, at least one of the following conditions is met: $P_{FAi} = P_{FA}, i = 1, 2, \ldots, q$ or $P_{Di} = P_D, i = 1, 2, \ldots, q$. The proof for it is straightforward from the definitions of $\overline{P}_{FA}$, $\overline{P}_R$, $\tilde{P}_D$ and Δ. Since $\overline{P}_{FA} \le \overline{P}_R$ and $\tilde{P}_D \le 1$, the sign for Δ cannot be determined without knowing the particular distributions of P_{FA} and P_D.

From the definition of $\tilde{P}_D$, one can see that $\tilde{P}_D \to \overline{P}_D$ as long as $P_{Di} \approx 1, i = 1, 2, \ldots, q$. However, it is enough that at least one P_D is small to cause a considerable reduction of the final value of $\tilde{P}_D$. The variation of $\tilde{P}_D$ also depends on the number of cells q.

Results for the normalized average acquisition time (T_{acq}/T_i) are presented in Figure 3.3. T_{acq1} is obtained by using the exact results (Case #3 in Table 3.1), T_{acq2} is the approximation where the standard expression for T_{acq} is used (Case #1 in Table 3.1) with

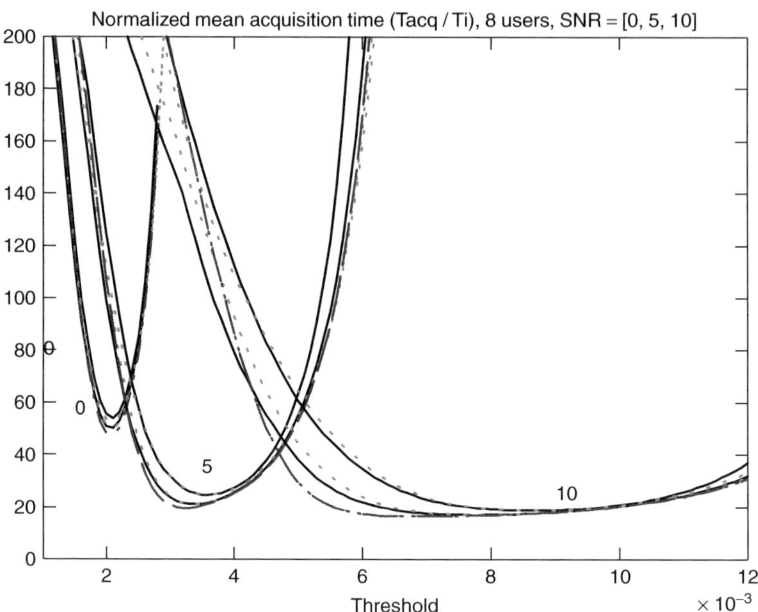

Figure 3.3   Upper and lower bounds of the mean acquisition time for 30 realizations of a random phase shift vector, K = 20, Solid line: Tacq1, Dotted line: Tacq2, Dashdot line: Tacq3 [2]. Reproduced from Katz, M. and Glisic, S. (2000) Modelling of code acquisition process in CDMA networks-asynchronous networks. *IEEE J. Select. Areas Commun.*, **18**(1), 73–86, by permission of IEEE.

$P_{FA} \Rightarrow \overline{P}_{FA}$ and $P_D \Rightarrow \overline{P}_D$ and T_{acq3} is the approximation where MAI is approximated by Gaussian noise.

3.5 MODELING OF THE SERIAL CODE ACQUISITION PROCESS FOR RAKE RECEIVERS IN CDMA WIRELESS NETWORKS WITH MULTIPATH AND TRANSMITTER DIVERSITY

The serial acquisition process of a RAKE receiver consists of two main steps. The first step, called initial acquisition, is defined as the process required to acquire the first path, corresponding to any of the available signal paths. The subsequent process required to acquire the remaining paths is referred to as postinitial acquisition.

The code delay uncertainty region will be divided into a number of cells in such a way that the delay between two adjacent cells is equal to a chip interval. The channel multipath profile will be characterized by a vector **D** (delays) as

$$\mathbf{D} = (d_1, d_2, \ldots, d_S) \tag{3.35}$$

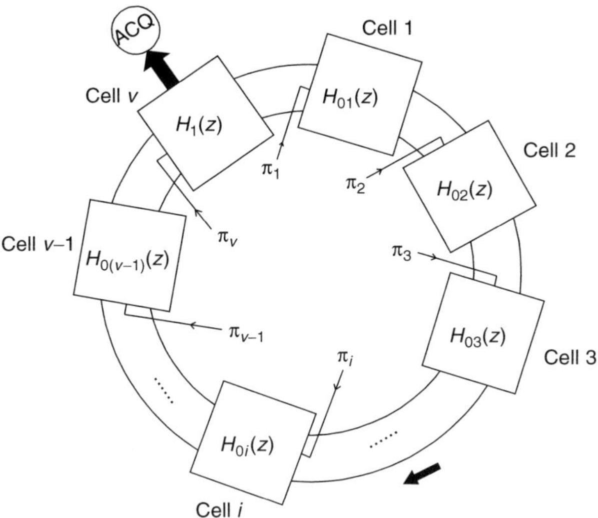

Figure 3.4 Overall decision process flow graph.

where d_l is the probability of having a multipath signal component l chip intervals after the first signal component (front end of the signal) has been received.

In order to simplify the notation, we will assume that there are $v - 1$ nonsynchro cells so that all together, with S potential synchro cells (multipath spread), the total number of cells is $v + S - 1$. The overall decision process flow diagram is shown in Figure 3.4, where nonsynchro cells are represented by $v - 1$ nodes with corresponding transfer functions $H_{0i}(z)$, $i = 1, 2, \ldots, v - 1$. Owing to MAI, $H_{0i}(z)$ is different for each cell of code delay uncertainty region.

If MAI is approximated as Gaussian noise, then $H_{0i}(z) = H_0(z)$. The vth cell represents S substates, which are potential synchro states, and its overall transfer function is $H_1(z)$. Figure 3.5 depicts the decision process flow graphs for the synchro cell v, including the first and last nonsynchro cells.

The theory for this case is available in Reference [3] and here we discuss some practical results. First of all, let us assume that the number of cells is much larger than the multipath spread, that is, $v \gg S$. In this case, the average acquisition time can be approximated by

$$\overline{T}_{acq} = \frac{2 + (2 - P_D)(v - 1)(1 + KP_{FA})}{2P_D} \tau_d \tag{3.36}$$

where

$$P_D = 1 - (1 - P_d)^L \tag{3.37}$$

and $(1 - P_d)$ represents the probability of missing one of the L available signal paths. Here we have assumed that the initial acquisition time is much longer than the postinitial acquisition time. If L_0 fingers are available, then each finger can search only v/L_0 cells, so reducing further this acquisition time by a factor $1/L_0$.

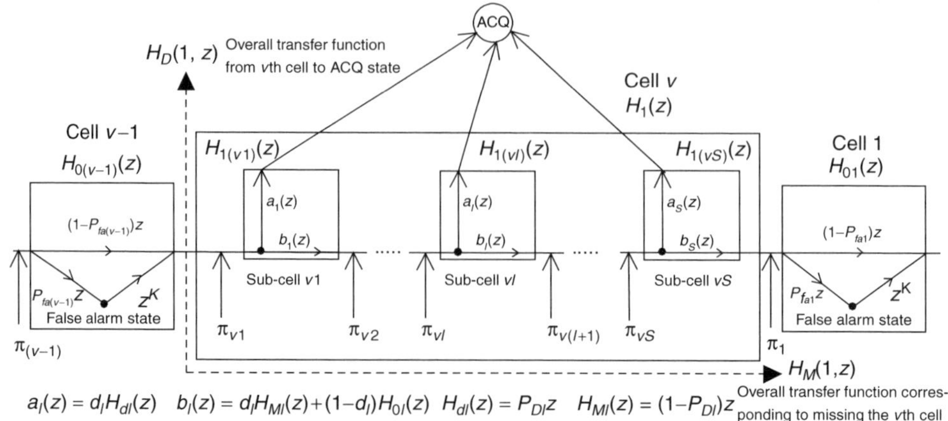

Figure 3.5 Decision process flow graph for the synchro cell (vth cell) and nonsynchro cells (e.g. first and (v − 1)th cells shown) [3]. Reproduced from Glisic, S. and Katz, M. (2001) Modeling of code acquisition process for RAKE receiver in wideband CDMA wireless networks with multipath and transmitter diversity. *IEEE J. Select. Areas Commun.*, **19**(1), 21−32, by permission of IEEE.

For macro diversity, the model is still valid with $v = 1$ so that all cells are included within S cells of the model from Figure 3.5. Within these S cells, there will be in general LM synchro cells, where M is the number of transmitters. If we assume $L = 1$ (no multipath), for L_0 available RAKE fingers, the initial search will start by partitioning the uncertainty region into L_0 segments. When one finger is synchronized, the uncertainty region will be partitioned again into equal segments among the remaining fingers. Under these conditions the average acquisition time will be approximated by

$$\overline{T}_{\text{acq}} \cong \sum_{i=0}^{L_0-1} \frac{2 + (2 - P_D(i))S(1 + KP_{\text{FA}})}{2P_D(i)(M - i)}\tau_d \tag{3.38}$$

where

$$P_D(i) = 1 - (1 - P_d)^{L-i} \tag{3.39}$$

Note that if transmit diversity is exploited (i.e. a given transmitter uses M diversity antennas), then LM synchro cells would be available at the receiving end, where the synchronization takes place. In the case of frequency nonselective channels, transmitting delayed versions of the same code from different antennas would generate an artificial multipath profile with uncorrelated components. A larger number of independent signal paths will tend to speed up the acquisition process.

In *practice*, in all existing standards on CDMA a special synchronization channel (SCH) is used for code acquisition. In wideband cdma2000, wideband IS-665 and IS-95, a pilot channel is used for these purposes. This is an unmodulated signal spread by relatively short code, which is transmitted continuously. This model is applicable directly to the systems mentioned above. For European Telecommunications Standards Institute (ETSI) Universal Mobile Telecommunication System (UMTS), a discontinuous transmission in

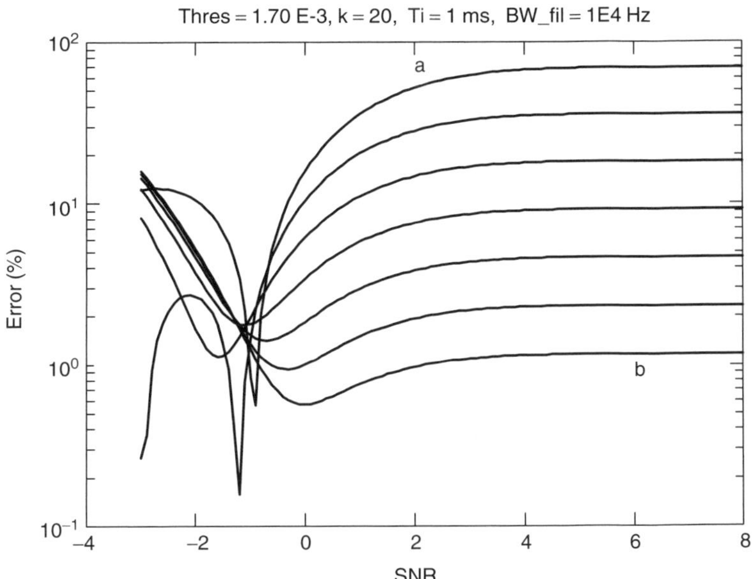

Figure 3.6 Relative error (module) between exact and approximate expressions for proposed example as a function of the SNR with the number of cells being a parameter, (v − 1) = [4; 8; 16; 32; 64; 128; 256]; (a) (v − 1) = 4 and (b) (v − 1) = 256; Thr = 1,70E-3. [3]. Reproduced from Glisic, S. and Katz, M. (2001) Modeling of code acquisition process for RAKE receiver in wideband CDMA wireless networks with multipath and transmitter diversity. *IEEE J. Select. Areas Commun.*, **19**(1), 21–32, by permission of IEEE.

the synchro channel (both primary and secondary) is used and signal detection, based on code-matched filters, is expected to be used. For these applications, the models will be discussed later in this chapter. Figure 3.6 presents the relative error in percentages defined as $\varepsilon(\%) = (\overline{T}_{acq} - T_{acq})/\overline{T}_{acq} \times 100$, where $\overline{T}_{acq}$ is the exact result [3] and T_{acq} the approximation equation (3.36).

3.6 TWO-DIMENSIONAL CODE ACQUISITION IN SPATIALLY AND TEMPORARILY WHITE NOISE

Code acquisition discussed so far dealt with the search through a discrete number of possible relative delay positions between codes, with each position being referred to as a *delay cell*. In this section, the problem is further extended to consider also the direction (or angle) of arrival of the received signal. A single-antenna receiver can resolve the signal in the delay domain while, with an antenna array, separation in the angular domain is also possible. An *angular cell* can be regarded as a 360°/m angle covered by a directional beam of antenna array. Assuming that the uncertainty region has q delay cells and m angular

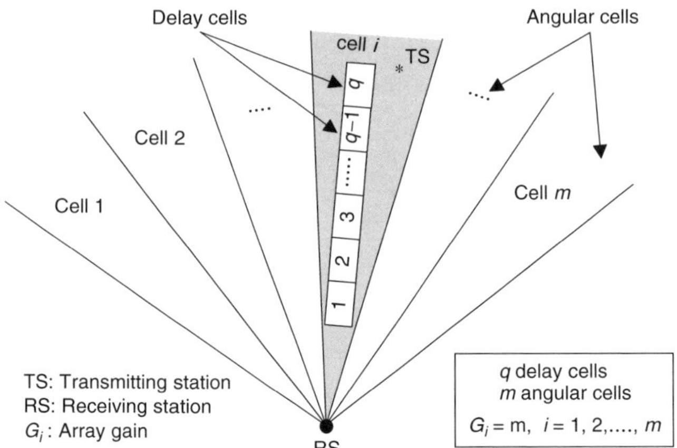

Figure 3.7 Principle of two-dimensional code acquisition.

cells, the total number of cells to be searched is $Q = qm$. This spatial and temporal partitioning of the uncertainty region is illustrated in Figure 3.7.

We assume that *a priori* probability of the synchro cell is uniformly distributed in the Q cell arrangement defining the uncertainty region. There is a direct correspondence between the spatial distribution of interference and the interference observed in the angular cells. For instance, a uniform spatial distribution of interferers will be mapped into angular cells with equal amounts of associated interfering power.

The angular division can be carried out by well-established and relatively simple beamforming techniques. Given an antenna array with m elements, an analog beamformer (e.g. Butler matrix) can be used to generate a set of m spatially orthogonal beams in fixed angular directions. A similar result can be achieved by a digital beamformer with a set of appropriate complex weighting vectors corresponding to preferred steering directions. Note that to achieve m nonoverlapping beams covering the entire spatial uncertainty region being served, a corresponding number of antenna elements is required. In order to simplify the problem formulation, it will be assumed that within each angular cell the array gain G_i corresponds to the maximum array gain, $G_i = m, i = 1, 2, \ldots, m$.

As an initial step, it is also assumed that the angular spread of the signal impinging on the antenna array is smaller than the beamwidth generated by the array. It can then be assumed that the impact of the received signal is seen only from one angular cell, occupying only one delay cell (e.g. single-path channel). The extension dealing with multipath channels is available in Reference [4]. The discussion is limited to the situation in which interference is both temporarily and spatially white. The interference power, uniformly distributed within the angular uncertainty region, is denoted by σ_I^2. The interference power in the ith sector is $\sigma_{Ii}^2, i = 1, 2, \ldots, m$, where $\sigma_{Ii}^2 = \sigma_I^2/m, i = 1, 2, \ldots, m$. The signal-to-interference ratio (SIR) for a single antenna ($\mathrm{SIR_{1D}}$) and the antenna array ($\mathrm{SIR_{2D}}$) are

$$\mathrm{SIR_{1D}} = \frac{A^2}{\sigma_I^2} = \frac{S}{\sigma_I^2} \quad \text{and} \quad \mathrm{SIR_{2D}} = \frac{(mA)^2}{(m\sigma_{Ii})^2} = m\mathrm{SIR_{1D}} \tag{3.40}$$

where the indexes 1D and 2D correspond to one-dimensional and two-dimensional search domains, A is the signal amplitude received by one antenna element and S is the corresponding signal power. For simplicity, from now on we consider SIR as SNR, and interference as noise.

3.6.1 Performance in a single-path channel

The available Q cells are serially searched in the angular and delay domains. A single synchro cell is associated with the single-path channel signal ($L = 1$). Basically, the cells could be searched by following either a fix angle/sweep delay (FASD) or a fix delay/sweep angle (FDSA) procedure. In the former approach, the search is carried out by serially searching (sweeping) through the q delay cells of a given angular cell. This (time-domain) procedure is repeated on each consecutive angular cell until the synchro cell is detected. In the latter case, a given delay cell is searched first through the m angular cells and the process is similarly repeated in the consecutive delay cells. Figure 3.8 illustrates the principles of FASD and FDSA search strategies.

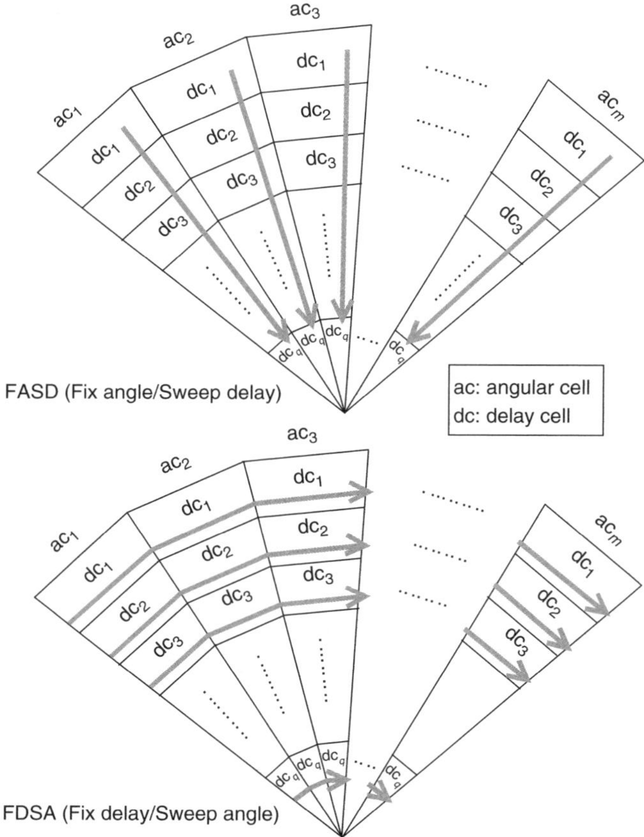

Figure 3.8 Principle of FASD and FDSA search strategies.

In the assumed scenario, since the interferers are uniformly distributed within the uncertainty region and *a priori* distribution of the synchro cell is evenly distributed, both search approaches are statistically equivalent. Thus, the mean acquisition time for these two search approaches are the same. Since the level of interference remains fixed in each angular cell (spatially white noise) and in each temporal cell (temporarily white noise), then P_{FA} and P_D are constant through the acquisition process. The conventional expression for mean acquisition time equation (3.23) can be used again with q replaced by $Q = mq$.

Owing to the directivity of the antenna array, the SNR will be improved by factor m leading to reduced values for P_{FA} and increased values for P_D, which will tend to reduce T_{MA}. On the other hand, the extended uncertainty region ($Q = mq$) implies a longer T_{MA}, as can be seen from equation (3.23). In order to get an insight into the resulting effect, some analytical results for two-dimensional acquisition performance are presented below. The conventional (one-dimensional) search through q cells ($m = 1$) is used as a reference. The effect of the number of angular cells m on T_{MA} will mainly be studied with the number of delay q cells being constant. Two threshold settings will be used for the evaluation, that is, an optimum threshold (i.e. yielding minimum T_{MA} obtained by minimizing equation (3.23) with respect to the threshold) and a threshold based on the constant false alarm rate (CFAR) principle. The details are available in Reference [4].

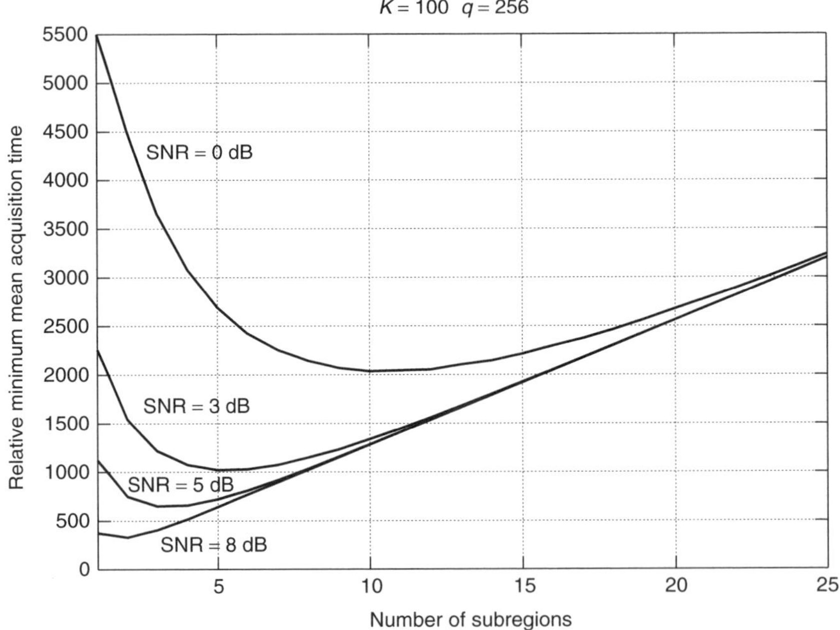

Figure 3.9   Relative mean acquisition time with optimum threshold in two-dimensional search [4]. Reproduced from Katz, M., Iinatti, J. and Glisic, S. (2000) Two-dimensional code acquisition in fixed multipath channels. *Proc. Vehicular Technology Conference*, Boston, MA, September 2000, pp. 2317–2324, by permission of IEEE.

The relative minimum mean acquisition time found for the optimal threshold is shown in Figure 3.9. The results are normalized with respect to τ_d. Since the optimum threshold depends on the SNR, which is a function of m, the optimum threshold was computed for each value of m. As it can be seen, for a given SNR there is an optimum number m for which an absolute minimum T_{MA} is obtained, for example, $m = 3$ for SNR $= 5$ dB. This is the result of the trade-off between increasing the uncertainty region length (mq) and increasing the effective signal-to-noise-ratio with larger values of m and *vice versa*.

For the performance shown in Figure 3.9, the SNR must be estimated by the receiver. This operation might be difficult, time consuming or impossible with a wideband (e.g. non-synchronized) signal. An alternative is a CFAR in which the threshold is set on the basis of the assumption that P_{FA} is constant [5,6]. Figure 3.10 presents the results for SNR $= 0$ dB when P_{FA} takes on the values 10^{-2}, 10^{-3}, and 10^{-4}. CFAR results (dashed-lines) are compared to the case of the optimum threshold (solid-line). Note that the lower bound of T_{MA} corresponds to acquisition with the optimum threshold setting. However, for an appropriate selection of P_{FA}, it is possible to obtain almost the same minimum as that obtained

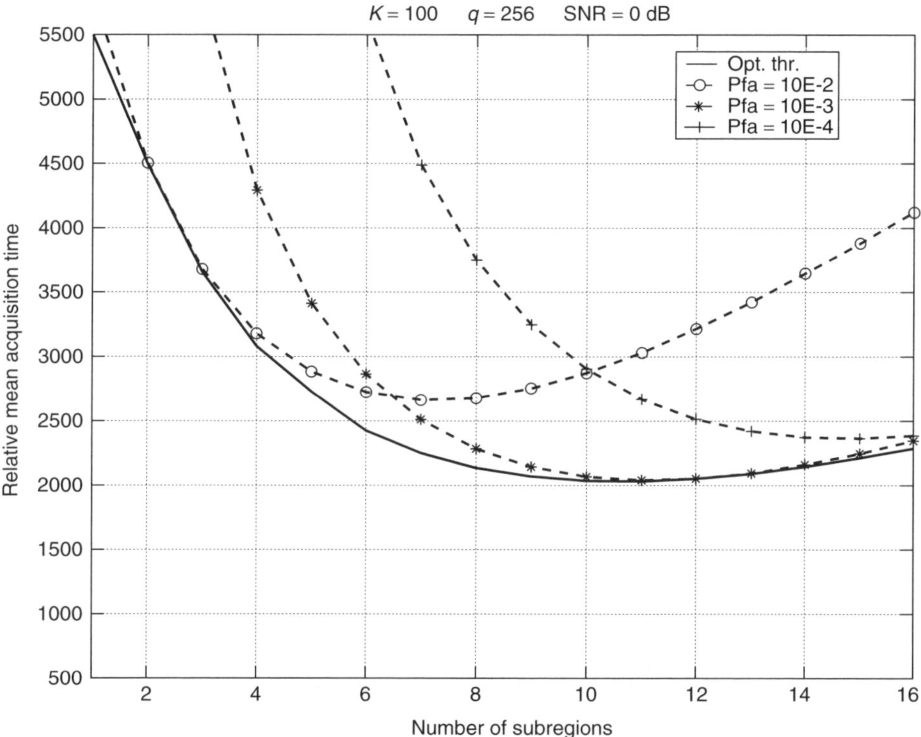

Figure 3.10 Code acquisition with optimal and CFAR threshold setting for SNR $= 0$ dB [4]. Reproduced from Katz, M., Iinatti, J. and Glisic, S. (2000) Two-dimensional code acquisition in fixed multipath channels. *Proc. Vehicular Technology Conference*, Boston, MA, September 2000, pp. 2317–2324, by permission of IEEE.

for the threshold T_{Hopt}. Performance with $P_{\text{FA}} = 10^{-3}$ is close to that with the optimum threshold. Details for performance in a multipath channel are available in Reference [4].

3.7 TWO-DIMENSIONAL CODE ACQUISITION IN ENVIRONMENTS WITH SPATIALLY NONUNIFORM DISTRIBUTION OF INTERFERENCE

In this chapter, a model for studying two-dimensional code acquisition in environments characterized by nonuniform spatial distribution of interference is presented. In order to model the nonuniform nature of spatial interference, the overall (total) angular domain is divided into a number $n \geq m$ of small angular sub-regions, each one with an associated noise power $\sigma_i^2, i = 1, 2, \ldots, n$. Figure 3.11 illustrates an arrangement of the q delay cells (dc) and m angular cells (ac) in the particular case in which $n/m = 3$ (e.g. generic jth angular cell contains three sub-regions). Note that there are $Q - 1$ nonsynchro cells and only one synchro cell. Interference will be modeled as temporarily white, that is, uniformly distributed in all delay cells of a given angular cell. It is also assumed that the autocorrelation function of the spreading code is ideal (represented by $\delta(\tau)$). The noise

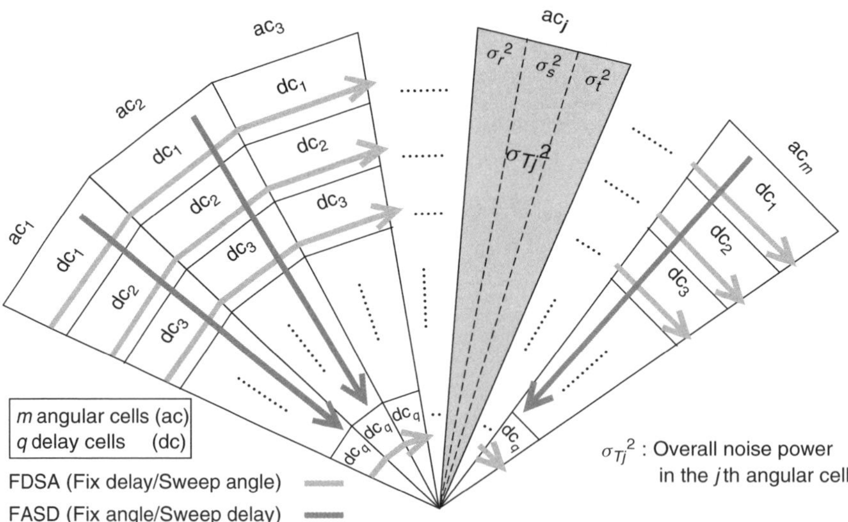

Figure 3.11 Model of cells, interference distributions and search strategies in two-dimensional code acquisition [7]. Reproduced from Katz, M., Iinatti, J. and Glisic, S. (2000) Performance of two-dimensional code acquisition in radio environments with spatially coloured interference. *Proc. 3rd International Symposium on Wireless Personal Multimedia Communications*, Bangkok, Thailand, November 2000, pp. 512–517, by permission of IEEE.

power in the jth angular cell, $j = 1, 2, \ldots, m$, denoted by σ_{Tj}^2, is

$$\sigma_{Tj}^2 = \sum_{i=(j-1)k+1}^{jk} \sigma_i^2 \tag{3.41}$$

where $k = n/m$ is an integer number. The total noise power in the uncertainty region can be computed as the sum of contributions of the n sub-regions (or equivalently the m angular cells), that is

$$\sigma_T^2 = \sum_{i=1}^{n} \sigma_i^2 = \sum_{j=1}^{m} \sigma_{Tj}^2 \tag{3.42}$$

Considering the same amount of total noise power σ_T^2 and assuming spatially white interference leads to $\sigma_i^2 = \sigma_T^2/n, i = 1, 2, \ldots, n$ and $\sigma_{Tj}^2 = \sigma_T^2/m, j = 1, 2, \ldots, m$, with $\sigma_{Tj}^2 = (n/m)\sigma_i^2$. Equations (3.41) and (3.42) for discrete distributions of interference can be generalized for a continuous function $\sigma^2(\theta)$ (i.e. $n \to \infty$) and arbitrary limits θ_a and θ_b defining an angular cell or region, as

$$\sigma_{ab}^2 = \int_{\theta_a}^{\theta_b} \sigma^2(\theta) \, d\theta \tag{3.43}$$

FASD and FDSA strategies defined in Figure 3.8, are now redefined in Figure 3.11 with additional details. Under the assumption that interference is spatially and temporarily white, every cell $C_{i,j}, i = 1, 2, \ldots, q; j = 1, 2, \ldots, m$, will have the same associated pair of false alarm probability P_{FA} and signal detection probability P_D corresponding to nonsynchro and synchro cells, respectively. It follows that the patterns of probabilities encountered during the FASD and FDSA searches are the same and hence these strategies are statistically equivalent (e.g. they exhibit the same mean acquisition time). For uniform *a priori* location distribution of the synchro cell and a nonuniform distribution of interference in the angular domain, the patterns of probabilities encountered through the search will depend on the employed strategy and consequently, different performance figures could be expected.

The mean acquisition time (T_{ma}) for a conventional time-domain serial search with different interference levels in each (delay) cell (i.e. P_{FA} and P_D change from cell to cell) is given by equation (3.33). Since the two-dimensional angle-delay search can be seen as a one-dimensional search with extended numbers of cells, the expression (3.33) can be extended and adapted to describe the two-dimensional acquisition process. In practice, the particular spatial (angular) interference profile dictates the actual shape of P_{FAj} and P_{Dj}.

3.7.1 Effect of spatial interference pattern on acquisition performance

For illustration purposes, let us look at two scenarios, that is, interference coming from (1) one angular region (pattern I) and (2) from two angular regions (pattern II), as depicted in Figures 3.12(a) and (b), respectively. The total angular uncertainty is assumed to cover

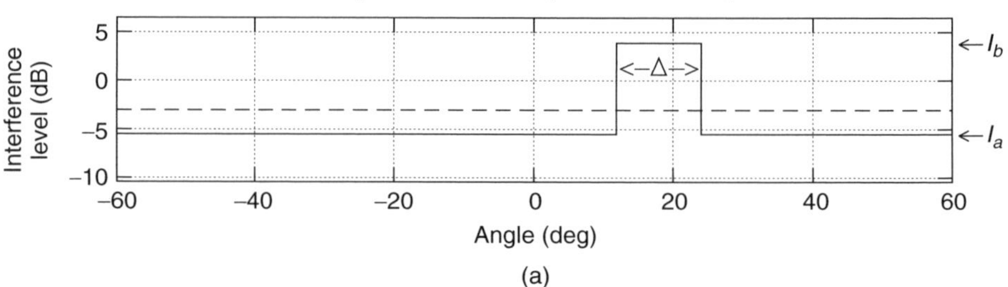

Spatial distribution of interference (solid line: pattern I, dashed line: equivalent uniform spatial distribution)

(a)

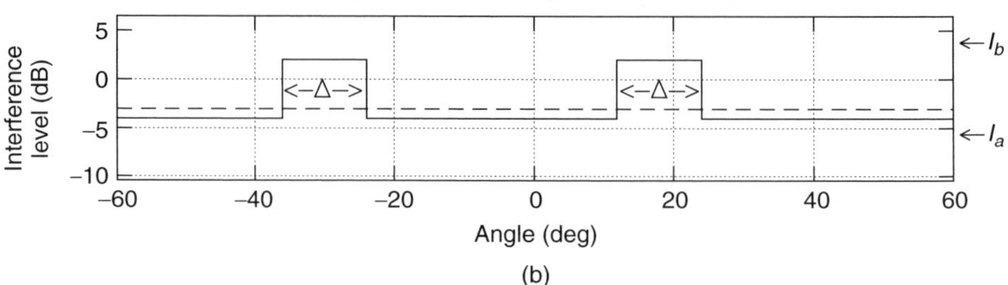

Spatial distribution of interference (solid line: pattern II, dashed line: equivalent uniform spatial distribution)

(b)

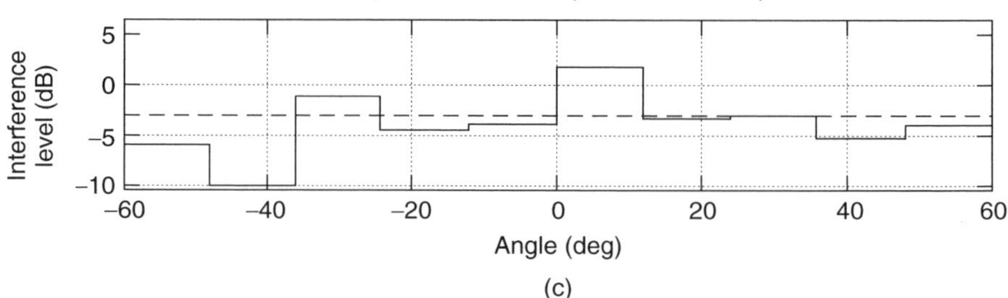

Spatial distribution of interference (solid line: pattern III, dashed line: equivalent uniform spatial distribution)

(c)

Figure 3.12 Definition of the spatially nonuniform patterns I (a), II (b) and III (c).

Table 3.2 Profile definitions used in the example

	Profile I $\Delta = 12°$, centered at $\theta_1 = +18°$	Profile II $\Delta = 12°$, centered at $\theta_2 = -30°$ and $\theta_1 = +18°$
$P_{TH}/P_{TH} = 1$	$R = 9$	$R = 4$
$P_{TH}/P_{TH} = 2$	$R = 18$	$R = 8$

$120°$ ($\pm 60°$ with $0°$ being the direction normal to the array). Pattern I is defined as a single peak of uniform power centered at $\theta_1 = +18°$ and with a width $\Delta = 12°$. It represents a strong interfering signal appearing in a particular direction affecting a particular angular cell. The ratio between interference power in a particular peak cell (I_b) and interference power in a particular low-level cell is denoted by R (see Figure 3.12). Pattern II is defined as a double peak centered at $\theta_1 = +18°$ and $\theta_1 = -30°$, also with $\Delta = 12°$. Table 3.2 summarizes the parameters defining patterns I and II where P_{TH} denotes the total power in peak interference regions and P_{TL} is the total power in angular regions with lower (uniform) interference.

Two different threshold-setting approaches are used, that is, optimum threshold leading to minimum mean acquisition times and constant false alarm rate (CFAR) setting. In all these cases, a single threshold, common to all angular cells, is used. However, since threshold level depends on SNR, a new threshold is computed for each particular selection of the number of used angular cells m. In this case, optimum thresholds are independently determined for colored and white distributions. In the CFAR case, the same threshold computed for $P_{FA} = 10^{-3}$ is used for both distributions.

Throughout this example, the number of delay cells is set to $q = 256$, the penalty factor $K = 100$, and the number of angular cells used to acquire is $m = 1, 2 \ldots, 10$. FASD search strategy is used first. A comparison between FASD and FDSA strategies will be considered next.

3.7.2 Effect of interference peak power (pattern I)

Before elaborating the results, few words should be devoted to explain how the interference pattern is seen when a different number of angular cells is used. Figure 3.13 shows an example for profile I. One can see that the interference distribution seen by the receiver depends on the relation between the interference profile itself and the employed angular partition m. Dark bars correspond to profile I while light ones show the effect of the equivalent white interference distribution, for example, interference seen by a single-antenna receiver. Figure 3.14 shows the mean acquisition time as a function of the number of cells m used in the search procedure. Other parameters are shown in the figure. One can see that when the spatial distribution of interference is nonuniform, there is a considerable degradation in performance, compared to the case of an equivalent spatially white distribution. As the peak power interference increases (i.e. larger R), the performance worsens. When $m = 1$, the performance in both interference scenarios is the same, due to the fact that the angular domain cannot be resolved with a single antenna. It is clear that searching through an angular cell with a high level of interference has an adverse effect on performance because of the fact that the process is likely to jump into time-consuming false alarm states. Similar conclusions would be obtained for pattern II of interference.

3.7.3 Approaches for improving performance of two-dimensional code acquisition

So far it has been seen that performance of two-dimensional code acquisition is considerably degraded by the presence of spatially nonuniform interference. In the worst case, and especially for very high peaks of interference, two-dimensional code acquisitions is

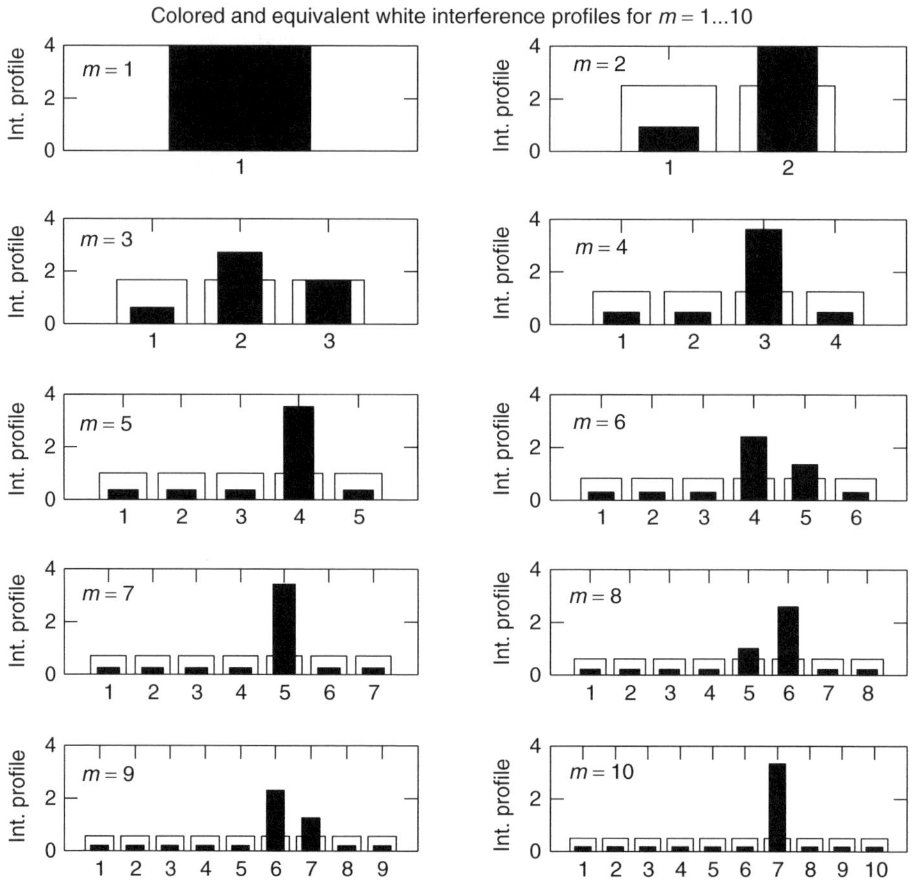

Figure 3.13 Conceptual example of the distribution of spatial interference as a function of the number of angular cells m (FASD search). Pattern I (dark bars), equivalent white distribution (light bars) [7]. Reproduced fromKatz, M., Iinatti, J. and Glisic, S. (2000) Performance of two-dimensional code acquisition in radio environments with spatially coloured interference. *Proc. 3rd International Symposium on Wireless Personal Multimedia Communications*, Bangkok, Thailand, November 2000, pp. 512–517, by permission of IEEE.

not attractive any longer since search in only the delay domain could result in shorter acquisition times. In this section, some options to improve this loss of performance is discussed. These approaches exploit the fact that the spatial distribution of interference is known *a priori* by the receiver. A uniform distribution of the synchro cell in the uncertainty region is assumed. Even though FASD and FDSA search strategies involve serial testing of the cells, the difference appears in the order that the cells are investigated. For the interference profile shown in Figure 3.15(a), (b) and (c) represent the patterns of interference seen in the search process by the following two strategies:

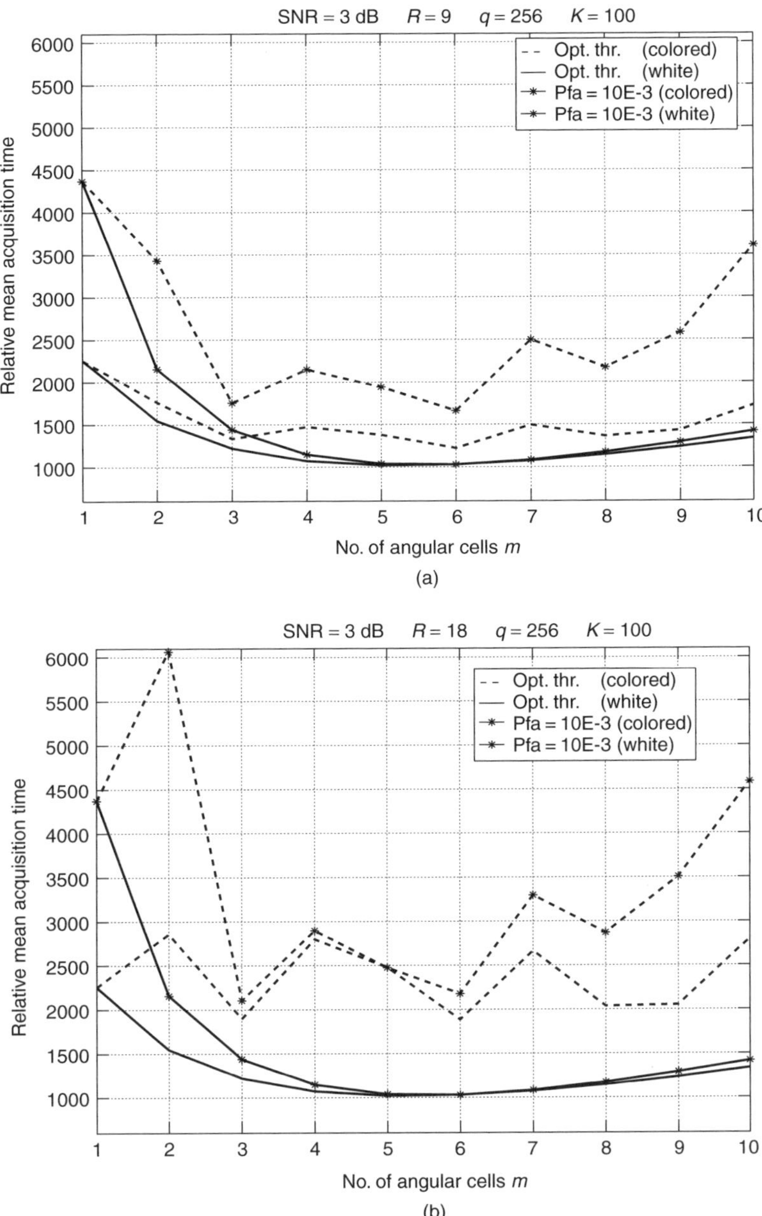

Figure 3.14 Acquisition performance (optimum and CFAR threshold setting) as a function of the peak power (pattern I), FASD search, for (a) R = 9 and (b) R = 18 [7]. Reproduced from Katz, M., Iinatti, J. and Glisic, S. (2000) Performance of two-dimensional code acquisition in radio environments with spatially coloured interference. *Proc. 3rd International Symposium on Wireless Personal Multimedia Communications*, Bangkok, Thailand, November 2000, pp. 512–517, by permission of IEEE.

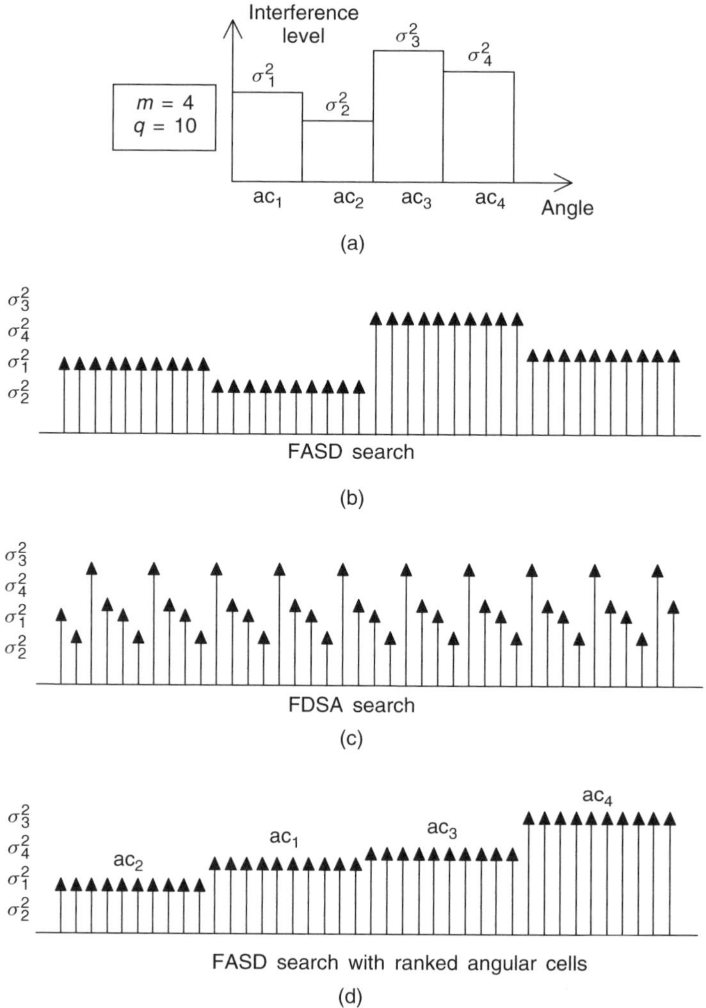

Figure 3.15 Distribution of interference according to the search strategy used.

Under the assumption that the user can be in any angular cell with the same probability, it is better to start the search process in those cells with lower interference levels.
This would speed up the process if the synchro cell turned out to be there. If not, the
search should continue in those cells with higher interference levels. Of course, since
in an FASD type of search the angular cells can be easily ordered, this strategy lends
itself to being used in such a fashion that the angular cells are ranked according to their
interference power. The two-dimensional search initiates in the angular cell with minimum interference and proceeds with the remaining angular cells in increasing order of
interference power.

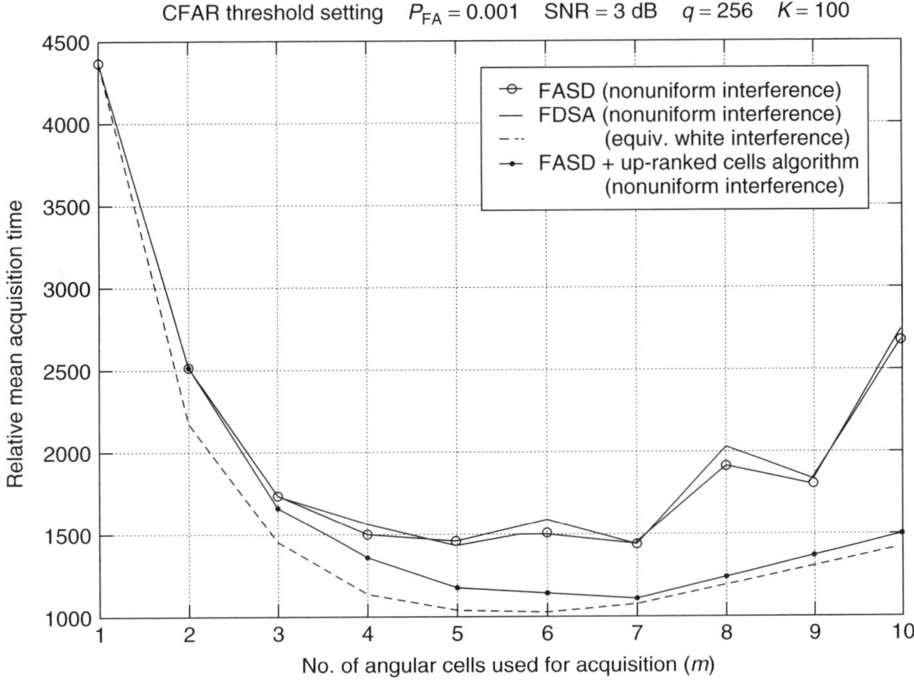

Figure 3.16   Performance of two-dimensional code acquisition with different search strategies (arbitrary nonuniform distribution, pattern III).

Figure 3.15(d) illustrates the observed interference pattern when this algorithm is applied. Analytical results based on equation (3.33) for the search using this algorithm are presented in Figures 3.16 and 3.17 for a nonuniform spatial distribution (pattern III in Figure 3.12c) and a single peak distribution (pattern I in Figure 3.12a), respectively.

As expected, the lower bound (best performance in this case) is obtained for a white interference distribution of equivalent power. As can be observed, there is a clear deterioration of acquisition performance with FASD or FDSA strategy. By using the algorithm with interference ranking, the performance is significantly improved.

It can be seen that the performance tends to approach the performance of the equivalent uniformly distributed interference case.

Similar improvements can be achieved by using adaptive integration time or threshold setting keeping CFAR. The mean acquisition time for the adaptive integration time approach is obtained after some cumbersome algebraic manipulation of the generating function, resulting in Reference [7]

$$T_{\mathrm{ma}} = \frac{1}{m P_{\mathrm{D}}}[X_1 + X_2 + X_3]\tau_d \tag{3.44}$$

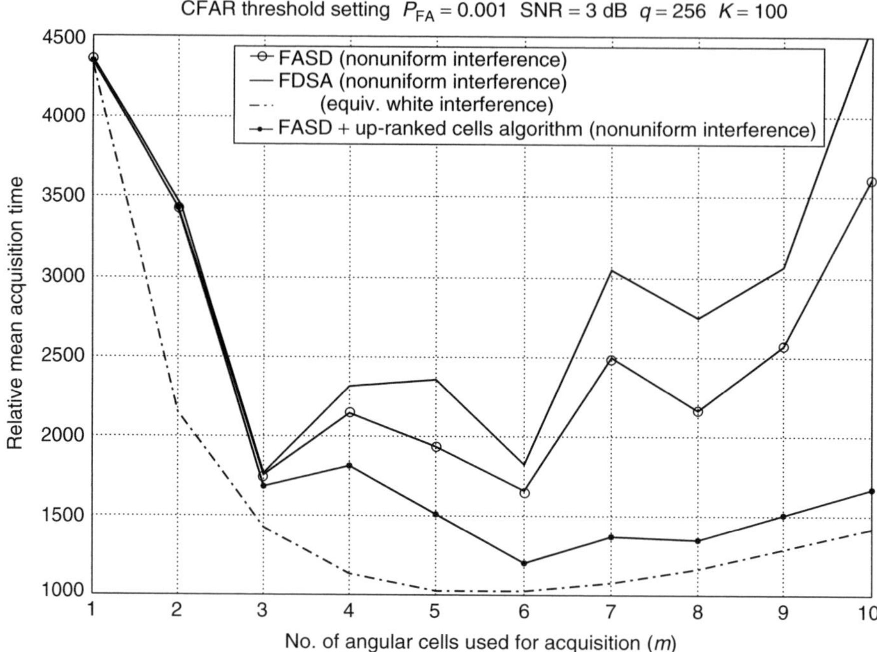

Figure 3.17 Performance of two-dimensional code acquisition with different search strategies (impulse-like spatial distribution, pattern I).

with

$$X_1 = \left[\frac{1}{2}(q + 1)(m\bar{a} - a_m) + (m - 1)\left[(2q - 1)KP_{\text{FA}} + (q - 1)a_m\right] \right.$$

$$\left. + q \sum_{k=2}^{m-1} (k - 1)a_k + \frac{1}{2}(m - 1)(m - 2)qKP_{\text{FA}} \right] \tag{3.45}$$

$$X_2 = \frac{1}{2}(q - 1)(a_m + KP_{\text{FA}}) + P_{\text{D}}a_m m \tag{3.46}$$

$$X_3 = \frac{1}{2}(1 - P_{\text{D}})\left[2qm^2\bar{a} + m(mq - 1)KP_{\text{FA}} + m(q - 1)(\bar{a} - 2a_m) - 2q \sum_{k=1}^{m-1} ka_k \right] \tag{3.47}$$

and

$$\bar{a} \stackrel{\Delta}{=} \frac{1}{m} \sum_{j=1}^{m} a_m \tag{3.48}$$

Vector τ is defined as $\tau = [a_1, a_2, \ldots, a_k, \ldots, a_m]\tau_d$, where τ_d represents a reference (or unit) dwell time and the coefficient a_k, $k = 1, 2, \ldots, m$ is a constant real number modeling

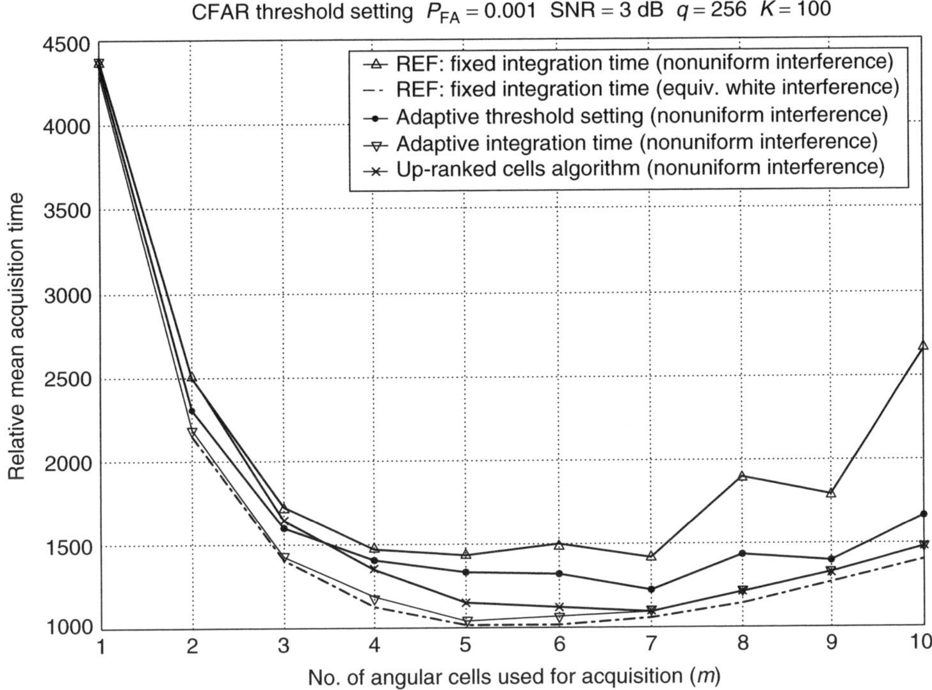

Figure 3.18 Performance comparison of two-dimensional code acquisition (FASD search) with adaptive integration time and adaptive threshold setting (arbitrary nonuniform distribution, pattern III).

the fact that each angular cell has a particular dwell time associated with it. Figures 3.18 and 3.19 present performance of the two schemes in operating scenarios defined by patterns I and III, respectively.

Constant false alarm rate (CFAR) threshold setting with $P_{FA} = 10^{-3}$ and SNR = 3 dB were used in both cases. The performance of two-dimensional code acquisition with fixed integration time, in the original environment (nonuniform interference distribution, FASD search) and in an equivalent uniform distribution of the same power is used as reference. The improvements obtained by adaptive schemes are evident.

3.8 CELL SEARCH IN W-CDMA

In this section, we discuss specific solutions for cell search in the UMTS system. The cell search itself is divided into five acquisition stages: slot synchronization, frame synchronization and scrambling code group identification, scrambling code identification, frequency acquisition and cell identification. One also should be aware that the crystal oscillators have inaccuracies in the range of 3 to 13 ppm, giving rise to a frequency error in the range of 6 to 26 kHz, when operated at 2 GHz.

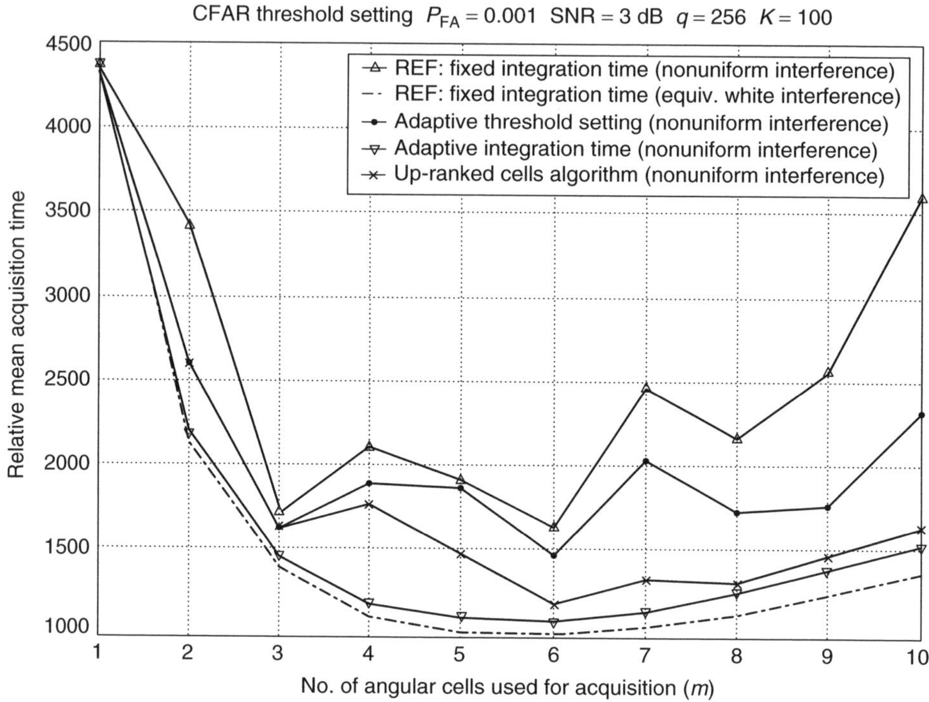

Figure 3.19 Performance comparison of two-dimensional code acquisition (FASD search) with adaptive integration time and adaptive threshold setting (impulse-like spatial distribution, pattern I).

3.8.1 Synchronization channels and cell search procedure

The UMTS standard will be discussed in detail in Chapter 17. For the purpose of this section, we need some basic details.

In W-CDMA, a cell is identified mainly by its downlink scrambling code as shown in Figure 1.13 of Chapter 1. There are 512 primary downlink scrambling codes reused throughout a system.

These 512 codes are based on length $2^{18} - 1$ Gold sequences truncated to one frame interval, which is 38 400 chips for the chip rate 3.84 Mchips/s.

To reduce the complexity of searching through the 512 downlink primary scrambling codes, the concept of code grouping and the use of code group indicator codes (GIC) were introduced in References [8,9]. The scrambling code is identified by first identifying its code group to significantly reduce the degree of code uncertainty. The complexity of cell search is further reduced by combining code group identification and frame boundary synchronization into one stage [10]. With this scheme, the time uncertainty is completely resolved when the code group identity is obtained. As a result, the complexity of identifying the scrambling code in the identified code group is significantly reduced. Schemes with further complexity reduction by increasing the number of code groups were proposed in

Reference [11]. According to Reference [12], the 512 downlink primary scrambling codes are divided into 64 groups, each of 8 codes.

To facilitate cell search, three channels are used, namely the primary synchronization channel (P-SCH), the secondary synchronization channel (S-SCH), and the common pilot channel (CPICH) [13]. The P-SCH together with the S-SCH is also referred to as the SCH. Figure 3.20 illustrates the slot and frame formats of these channels. Each frame of 38 400 chips (or 10 ms) is divided into 15 slots, each of 2560 chips (or 0.67 ms). Observe that both P-SCH and S-SCH have a 10% duty factor. The CPICH, which is used to carry the downlink common pilot symbols, is scrambled by the primary downlink scrambling code of the cell. Within each CPICH time slot, there are 10 pilot symbols, each spread by 256 chips. All symbols are quadrature phase shift keying (QPSK)-modulated, and the modulation values of the pilot symbols are known once the mobile system (MS) knows the frame boundary. The spreading sequence of CPICH is taken from the set of orthogonal variable spreading factor (OVSF) codes described in Chapter 2, maintaining mutual orthogonality between CPICH and the other downlink channels also spread by OVSF codes.

Unlike CPICH, neither the P-SCH nor the S-SCH is scrambled by the primary downlink scrambling code. Instead of the OVSF codes, other sequences of length 256 chips are used. The P-SCH sequence is transmitted once in the same position in every slot, and can thus be used for detecting the slot boundary. Furthermore, all cells use the same P-SCH sequence. As a result, only one P-SCH matched filter is needed to detect the slot boundaries of downlink signals. These filters are discussed in Section 2.12 of Chapter 2. For the purpose of this section, the code matched filter should be considered as an analog shift register with N delay elements. Outputs of these delay elements are multiplied by the filter coefficients (equal to code chips) and summed up. At the moment when the content of the delay elements coincide with the filter coefficients, the high peak of correlation function is created. This happens once per code period if the code is shifted through the shift register with the code rate.

To reduce the complexity of the P-SCH matched filter, the P-SCH sequence is derived from the Kronecker product of two sequences of length 16. The Kronecker product is

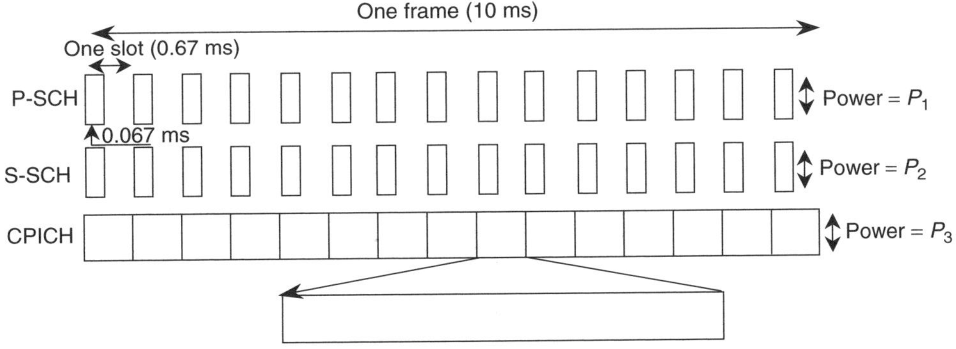

Figure 3.20   Frame/slot structures for CPICH, P-SCH, and S-SCH [14].

discussed in Chapter 1. With this property, the P-SCH matched filter can be implemented as two concatenated matched filters, each matched to one of the two constituent length 16 sequences, achieving a complexity reduction by approximately a factor of 8. S-SCH is used to identify the frame boundary and scrambling code group identity. Unlike the P-SCH sequence, the S-SCH sequences vary from slot to slot. There are 16 S-SCH sequences, mapped correspondingly to 16 S-SCH symbols, labeled from 1 to 16. A frame (15 slots) of 15 such S-SCH symbols forms a code word taken from a codebook of 64 code words. The same code word is repeated in every frame in a cell. These 64 code words correspond to the 64 code groups used throughout the system; thus a code group can be detected by identifying the code word transmitted in every S-SCH frame. Furthermore, the 64 code words are all chosen to have distinct code phase shifts, and any phase shift of a code word is different from all phase shifts of all other code words. With these properties, the frame boundary can be detected by identifying the correct starting phase of the S-SCH symbol sequence.

To maximize the minimum symbol distance of the codebook, between different cyclic shifts of the same code word or between any cyclic shifts of different code words, the use of a comma-free Reed–Solomon (RS) code was proposed [15]. For 15 slots per frame, a (15, 3) RS code over GF(16) is used. The RS code has a minimum distance of 13. To minimize cross-channel interference, the 16 S-SCH sequences and the P-SCH are mutually orthogonal [12]. Given the above SCH and CPICH, code and time synchronization can be achieved by the following stages [14]:

1. Slot boundary detection based on P-SCH (using a P-SCH matched filter).
2. Frame boundary detection and scrambling code group identification based on S-SCH (using correlators, correlating against 16 S-SCH sequences, and an RS decoder).
3. Scrambling code detection based on CPICH (using correlators, correlating against all scrambling codes in the identified code group). In the initial search, the ultimate goal is to decode the cell identity of the acquired signal. To achieve this, two extra stages are needed.
4. Frequency acquisition based on CPICH (to reduce initial frequency error so that the MS can decode the broadcast information).
5. Detecting cell identity (by reading the broadcast information).

As mentioned at the beginning of the section, the frequency acquisition step is necessary because of the large frequency error after MS powered on. Without correcting the frequency error, the cell identity, transmitted in the Broadcast Channel, cannot be decoded reliably. For the target cell search, there is no ambiguity in the mapping from downlink scrambling codes and the cell identity of a neighboring cell. Thus, identifying (and synchronizing to) the downlink scrambling code is sufficient to identify any given cell of interest.

Pipelined processing for stages 1, 2, and 3 is illustrated in Figure 3.21.

Stage 4-frequency acquisition is only activated when time and code synchronization is achieved. To minimize the delay in the pipe (no idle time in the pipe), the synchronization times used in stages 1, 2, and 3 are the same (N_t slots).

Stage 1 always generates a list of slot boundary candidates at the end of each cycle. On the basis of each of the slot boundaries detected in stage 1, stage 2 finds S-SCH and

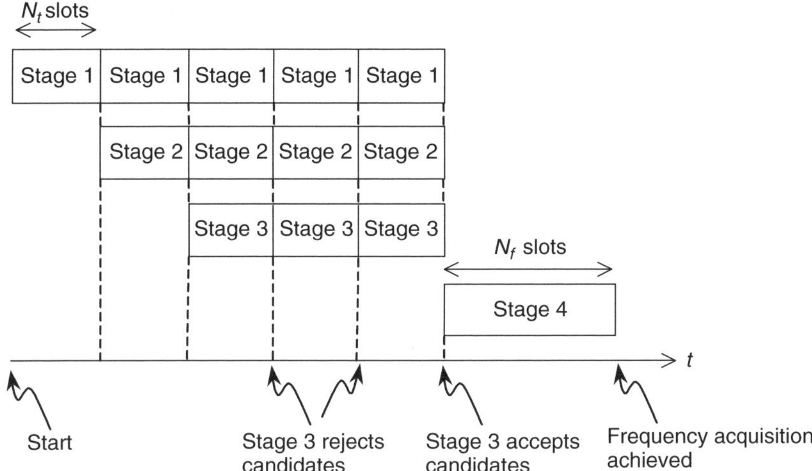

Figure 3.21 Cell search procedures [14].

performs S-SCH correlations and RS decoding. At the end of each detection cycle, stage 2 always gives a list of candidates of frame boundary code group pair to stage 3 for identification of the scrambling code. In contrast to stages 1 and 2, stage 3 only activates stage 4 when a candidate is detected with high confidence. The acquisition time can be defined as the time interval between the time when the pipelined process started and the time when stage 3 terminates the process.

A number of papers dealing with different aspects of code acquisition is added to the list of References [16–42].

REFERENCES

1. Holmes, J. K. and Chen, C. C. (1977) Acquisition time performance of PN spread-spectrum systems. *Proc. IEEE Commun.*, **COM-25**(4), 778–784.
2. Katz, M. and Glisic, S. (2000) Modelling of code acquisition process in CDMA networks-asynchronous networks. *IEEE J. Select. Areas Commun.*, **18**(1), 73–86.
3. Glisic, S. and Katz, M. (2001) Modeling of code acquisition process for RAKE receiver in wideband CDMA wireless networks with multipath and transmitter diversity. *IEEE J. Select. Areas Commun.*, **19**(1), 21–32.
4. Katz, M., Iinatti, J. and Glisic, S. (2000) Two-dimensional code acquisition in fixed multipath channels. *Proc. Vehicular Technology Conference*, Boston, MA, September 2000, pp. 2317–2324.
5. Glisic, G. S. (1988) Automatic decision threshold level control (ADTLC) in direct-sequence spread-spectrum systems based on matching filtering. *IEEE Trans. Commun.*, **36**(4), 519–527.
6. Glisic, G. S. (1991) Automatic decision threshold level control in direct-sequence spread-spectrum systems. *IEEE Trans. Commun.*, **39**(2), 187–192.
7. Katz, M., Iinatti, J. and Glisic, S. (2000) Performance of two-dimensional code acquisition in radio environments with spatially coloured interference. *Proc. 3rd International Symposium on Wireless Personal Multimedia Communications*, Bangkok, Thailand, November 2000, pp. 512–517.

8. Higuchi, K. *et al.* (1997) Fast Cell *Search Algorithm Using Long Code Masking in DS-CDMA Asynchronous Cellular Systems*. Tech. Rep. IEICE, pp. 57–62.

9. Higuchi, K. *et al.* (1997) Fast cell search algorithm in DS-CDMA mobile using long spreading codes. *Proc. IEEE 1997 Vehicular Technological Conference*, Phoenix, AZ, May, pp. 1430–1434.

10. Nyström, J. *et al.* (1998) Comparison of cell search methods for asynchronous wideband CDMA cellular system. *Proc. IEEE 1998 International Conference Universal Personal Communications*, Florence, Italy, October 1998.

11. Östberg, C. *et al.* (1998) Performance and complexity of techniques for achieving fast sector indentification in an asynchronous CDMA system. *Proc. 1st International Symposium on Wireless Personal Multimedia Communication*, Japan, November 1998, pp. 87–92.

12. 3GPP, Spreading and modulation (FDD). 3GPP Tech. Spec., TS 25.213, V3.0.0, October 1999.

13. 3GPP, Physical channels and mapping of transport channels onto physical channels (FDD). 3GPP Tech. Spec., TS 25.211, V3.0.0, October 1999.

14. 3GPP, FDD Physical layer procedures. 3GPP Tech. Spec., TS 25.214, V3.0.0, October 1999.

15. Sriram, S. and Hosur, S. (1999) Fast acquisition method for DS-CDMA systems employing asynchronous base stations. *Proc. IEEE ICC*. **3**, 1928–1932.

16. Shin, Oh-Soon and Lee, K. B. (2001) Utilization of multipaths for spread-spectrum code acquisition in frequency-selective Rayleigh fading channels. *IEEE Trans. Commun.*, **49**(4), 734–743.

17. Katz, M., Iinatti, J. and Glisic, S. (2000) Impact of spatially colored interference on two dimensional code acquisition performance. *Proc. Finnish Wireless Communications Workshop 2000 (FWCW '00)*, Oulu, Finland, May 2000, pp. 90–96.

18. Katz, M., Iinatti, J. and Glisic, S. (2000) Performance of two-dimensional code acquisition in radio environments with spatially coloured interference. *Proc. 3rd International Symposium on Wireless Personal Multimedia Communications*, Bangkok, Thailand, November 2000, pp. 512–517.

19. Comparetto, G. M. (1987) A general analysis for a dual threshold sequential detection PN acquisition receiver. *IEEE Trans. Commun.*, **35**(9), 956–960.

20. DiCarlo, D. M. and Weber, C. L. (1983) Multiple dwell serial search: performance and application to direct sequence code acquisition. *IEEE Trans. Commun.*, **31**(5), 650–659.

21. Madyastha, R. and Aazhang, B. (1995) Synchronization and detection of spread spectrum signals in multipath channels using antenna arrays. *Proc. MILCOM '95 Conference*, Vol. 3, November 1995, pp. 1170–1174.

22. Dlugos, D. and Scholtz, R. (1989) Acquisition of spread spectrum by an adaptive array. *IEEE Trans. Acoustics, Speech Signal Process.*, **37**(8), 1253–1270.

23. Hopkins, P. M. (1977) A unified analysis of pseudonoise synchronization by envelope correlation. *IEEE Trans. Commun.*, **25**, 770–778.

24. Jovanovic, V. M. (1988) Analysis of strategies for serial search spread-spectrum code acquisition-direct approach. *IEEE Trans. Commun.*, **36**, 1208–1220.

25. Jovanovic, V. M. (1992) On the distribution function of the spread-spectrum code acquisition time. *IEEE J. Select. Areas Commun.*, **10**(4), 760–769.

26. Pan, S. M., Dodds, D. E. and Kumar, S. (1990) Acquisition time distribution for spread-spectrum receiver. *IEEE J. Select. Areas Commun.*, **8**(5), 800–808.

27. Polydoros, A. and Glisic, S. (1995) Code synchronization: a review of principles and techniques, in Glisic, S. and Leppanen, P. (eds) Code Division Multiple Access Communications. Norwell, MA: Kluwer Academic Publishers, pp. 225–266.

28. Polydoros, A. and Simon, M. (1984) Generalized serial search code acquisition: the equivalent circular state diagram approach. *IEEE Trans. Commun.*, **32**(12), 1260–1268.

29. Polydoros, A. and Weber, C. L. (1984) A unified approach to serial search spread-spectrum code acquisition-part I: general theory. *IEEE Trans. Commun.*, **32**(5), 542–549.

30. Polydoros, A. and Weber, C. L. (1984) A unified approach to serial search spread-spectrum code acquisition-part II: a matched-filter receiver. *IEEE Trans. Commun.*, **32**(5), 550–560.

31. Su, Y. T. (1988) Rapid code acquisition algorithm employing PN matched filters. *IEEE Trans. Commun.*, **36**(6), 724–733.

32. Thompson, M. *et al.* (1993) Non-coherent PN code acquisition in direct sequence spread spectrum systems using a neural network. *Milcom '93, Conference Record*, Vol. 1, pp. 30–34.

33. Affes, S. and Mermelstein, P. A. (1998) New receiver structure for asynchronous CDMA: STAR-the spatio-temporal array receiver. *IEEE J. Select. Areas Commun.*, **16**(8), 1411–1422.

34. Ramos, J., Zoltowski, M. and Liu, H. (2000) Low-complexity space-time processor for DS-CDMA communications. *IEEE Trans. Signal Process.*, **48**(1), 39–52.

35. Wang, B. and Kwon, H. M. (2000) PN code acquisition with adaptive antenna array and adaptive threshold for DS-CDMA wireless communications. *Proc. IEEE GLOBECOM*, San Francisco, CA, pp. 152–156.

36. Wang, B. and Kwon, H. M. (2000) PN code acquisition using smart antenna for DS-CDMA wireless communications. *Proc. IEEE MILCOM Conference*, Los Angeles, CA, October 2000, pp. 821–825.

37. Katz, M., Iinatti, J. and Glisic, S. (2000) Two dimensional code acquisition using antenna arrays. *Proc. International Symposium on Spread Spectrum Techniques and Applications (ISSSTA 2000)*, New York, NJ, September 6–8, 2000, pp. 613–617.

38. Godara, L. (1997) Applications of antenna arrays to mobile communications. Part I: performance improvement, feasibility, and system considerations. *Proc. IEEE*, **35**, 1031–1060.

39. Polydoros, A. (1982) *On the Synchronization Aspects of Direct-Sequence Spread Spectrum Systems*. Ph. D. Dissertation, University of Southern California, Los Angeles, CA, p. 240.

40. Iinatti, J. (2000) Performance of DS code acquisition in static and fading multipath channels. *IEE Proc. Commun.*, **147**(6), 355–360.

41. Higuchi, K., Sawahashi, M. and Adachi, F. (1998) Fast cell search algorithm in inter-cell asynchronous DS-CDMA mobile radio. *IEICE Trans. Commun.*, **E81**(7), pp. 1527–1534.

42. Kim, B. and Lee, B. (2000) Distributed sample acquisition-based fast cell search in inter-cell asynchronous DS/CDMA systems. *IEEE JSAC*, **18**(8), 1455–1469.

4

Code tracking

4.1 CODE-TRACKING LOOPS

Theoretically switching from code acquisition to code tracking in this chapter means switching from open loop maximization of likelihood function (equation 3.3) to the closed-loop tracker defined by equation (3.8) of Chapter 3. A variety of practical implementation options are shown in the sequel. The baseband implementation of equation (3.8) of Chapter 3 is shown in Figure 4.1. The input signal is correlated with two locally generated, mutually delayed, replicas of the pseudonoise (PN) code. After filtering, the useful component of the control signal $e(t)$ will be proportional to

$$D_D(\delta) = R_c(\delta - \Delta/2) - R_c(\delta + \Delta/2) \tag{4.1}$$

where $R_c(\delta)$ is the auto correlation of the sequence. For the analysis of the tracking error variance, results from the standard phase lock loop theory can be used directly [1].

In Code Division Multiple Access (CDMA) system, the input signal in Delay lock loop (DLL) will be a complete Direct Sequence Spread Spectrum (DSSS) signal. In order to get rid of information, a noncoherent structure shown in Figure 4.2(a) may be used with the simplest form of the input signal

$$r(t) = s(t) + n(t) \tag{4.2}$$

and

$$s(t) = Ab(t)c(t)\cos \omega_0 t \tag{4.3}$$

It can be shown that the direct current (DC) component of $\varepsilon(t, \delta)$ is $A^2 D_\Delta(\delta)/2$ where

$$D_\Delta(\delta) \stackrel{\Delta}{=} R_c^2 \left[\left(\delta - \frac{\Delta}{2}\right) T_c\right] - R_c^2 \left[\left(\delta + \frac{\Delta}{2}\right) T_c\right] \tag{4.4}$$

The tracking error variance can be expressed as [1]

$$\tau_\delta^2 \frac{1}{2\rho_L} \left(1 + \frac{2}{\rho_{if}}\right) \tag{4.5}$$

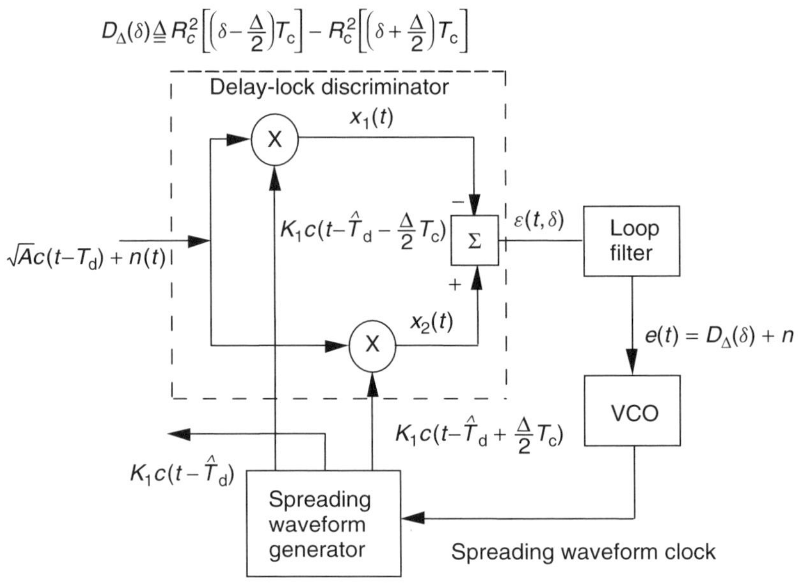

Figure 4.1 Conceptual block diagram: baseband delay-lock tracking loop.

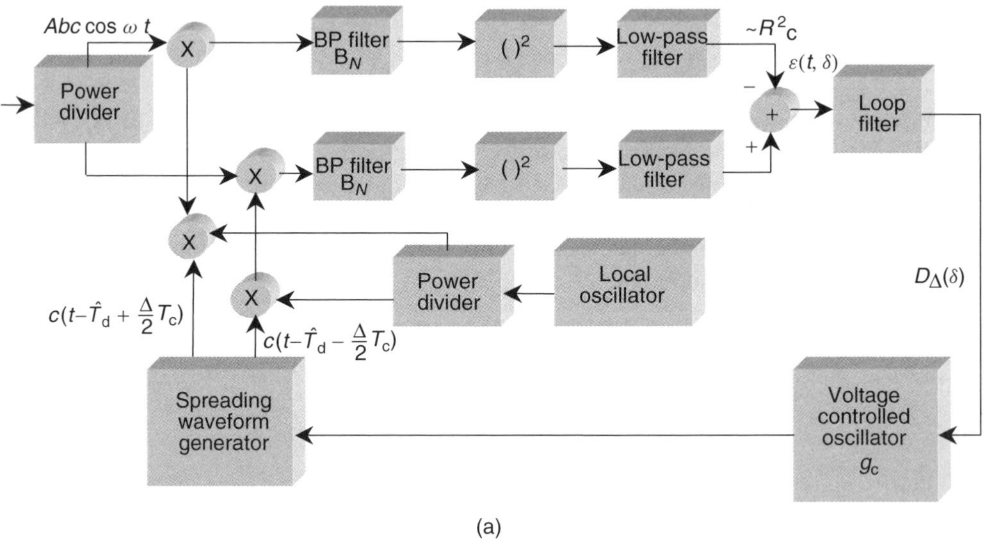

(a)

Figure 4.2 (a) Full-time early–late noncoherent code-tracking loop, (b) Noncoherent tracking loop with interference cancellation (IC) DLL/IC [2]. Reproduced from Sheen, W. and Tai, C. (1998) A noncoherent tracking loop with diversity and multipath interference cancellation for direct-sequence spread-spectrum systems. *IEEE Trans. Commun.*, **46**(11), 1516–1524, by permission of IEEE. (c) Comparisons of DLL and DLL/IC tracking loops [2].

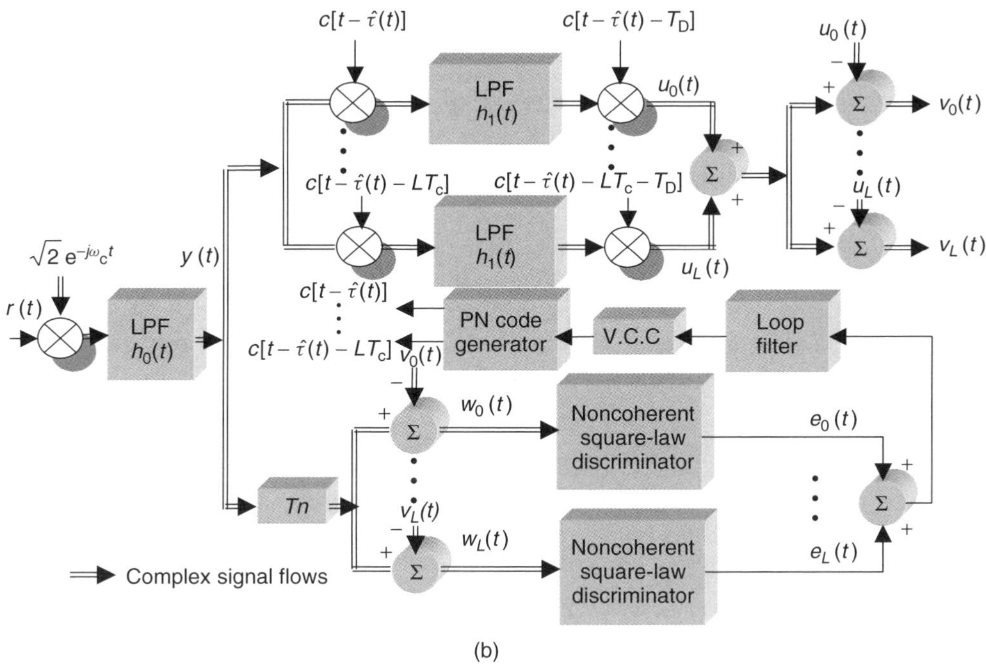

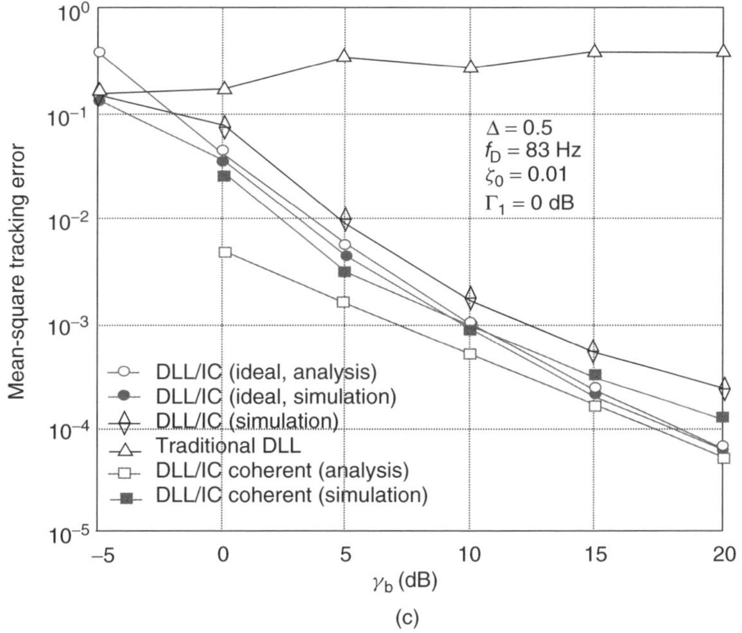

Figure 4.2 (*Continued*).

where

$$\rho_{\mathrm{L}} = \frac{2A^2}{N_0 B_{\mathrm{L}}}$$

$$\rho_{\mathrm{if}} = \frac{A^2}{N_0 B_N} \tag{4.6}$$

Parameter ρ_{L} is the loop signal-to-noise power ratio and ρ_{if} is the signal-to-noise power ratio at the output of the intermediate frequency (IF) (band pass) filter. The first term in equation (4.5) represents σ_δ^2 for a coherent loop. The second term is degradation due to the noncoherent structure. Other modifications of the code-tracking loops like τ-dither loop or double-dither loop can be seen in Reference [1].

4.1.1 Effects of multipath fading on delay-locked loops

In this section, the effects of a specular multipath fading channel on the performance of a DLL are discussed. For this type of environment, the two-path channel model becomes

$$h(\tau) = \sqrt{2P}\{\delta(\tau - \tau_1)\,\mathrm{e}^{j\theta 1} + g_2\,\mathrm{e}^{j\theta_2}\delta(\tau - \tau_1 - \tau_{\mathrm{d}})\} \tag{4.7}$$

where θ_1 is a constant phase shift, and g_2 and θ_2 are Rayleigh- and uniform-distributed random variables, respectively. When $\tau_{\mathrm{d}} = 0$, the channel becomes the familiar frequency nonselective Rician-fading model.

In order to present some quantitative results, the following important system parameters are needed: the power ratio of the main path to the second path $R \triangleq 1/E[g_2^2]$, the bit signal-to-noise ratio (SNR) (SNR in data bandwidth) $\gamma_{\mathrm{d}} \triangleq P T_{\mathrm{b}}/N_0$, the loop SNR $\gamma_{L_0} \triangleq P/N_0 B_{\mathrm{L}}|\Delta = 1$ and the ratio $\varsigma_0 = \gamma_{L_0}/\gamma_{\mathrm{d}}$ where T_{b} is the duration of an information bit, and B_{L} is the closed-loop bandwidth for the case when $g_2 = 0$. That is, $B_{\mathrm{L}} = \int_{-\infty}^{\infty} |H(f)|^2 \,\mathrm{d}f$ where $H(s)$ is the closed loop transfer function. By using the standard phase lock loop theory [3], the tracking error variance for this case has been evaluated and the results are shown in Figure 4.3. Effects of multipath fading on the normalized mean time to lose lock (MTLL) and tracking error versus early–late discriminator offsets $\Delta/2$ are shown in Figures 4.4 and 4.5, respectively.

Figures 4.4 and 4.5 demonstrate performance degradation of DLL due to the presence of multipath components. In order to improve the system performance in such an environment, some research results are reported in which multipath IC is used.

The receiver block diagram is shown in Figure 4.2(b).

For the input signal received through $L + 1$ equidistantly modeled paths, the upper half of the block diagram is used to regenerate multipath interference (MPI) for each path. In the first step, input signal $r(t)$ is correlated with $L + 1$ delayed replica of the local code to separate $L + 1$ narrowband signal components. After processing delay T_{D}, the wideband components $u_0(t), \ldots, u_L(t)$ are regenerated separately and summed up again. At this point $r(t - T_{\mathrm{D}})$ is created together with all individual components $u_l(t)$ available separately. Now in $L + 1$ branches, signal $r(t - T_{\mathrm{D}}) - u_l(t) = v_l(t)$, representing the

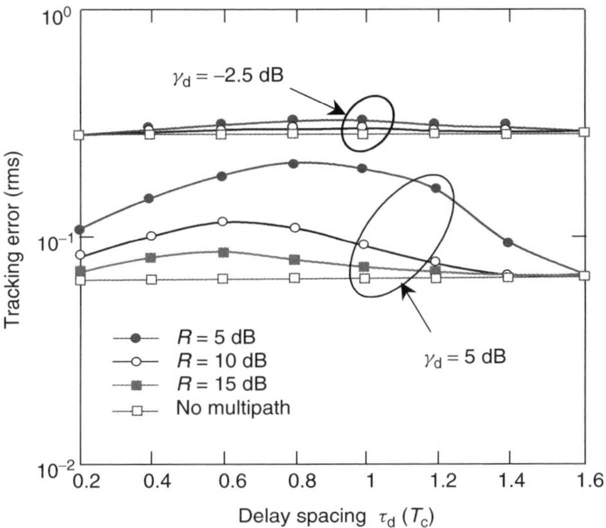

Figure 4.3 Effects of multipath fading on the tracking error performance with various delay spacings ($\Delta = 0.5$, $\zeta_0 = 100$) [3]. Reproduced from Sheen, J. W. and Stüber, G. (1994) Effects of multipath fading on delay locked loops for spread spectrum systems. *IEEE Trans. Commun.*, **42**(2/3/4), 1947–1956, by permission of IEEE.

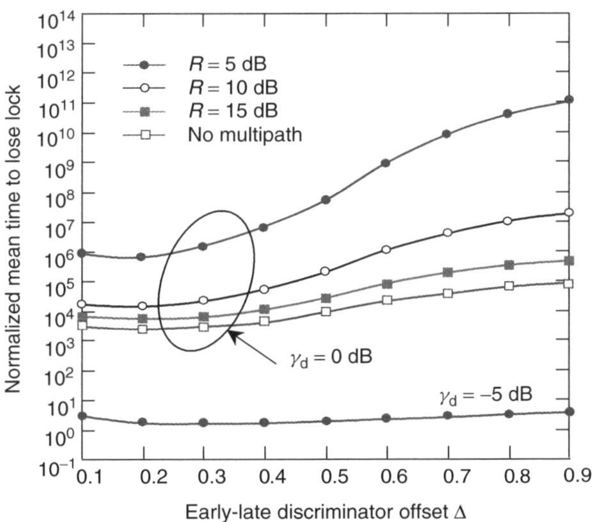

Figure 4.4 Effects of multipath fading on the MTLL performance with various early–late discriminator offsets ($\tau_d = 0.5$, $\zeta_0 = 100$) [3]. Reproduced from Sheen, J. W. and Stüber, G. (1994) Effects of multipath fading on delay locked loops for spread spectrum systems. *IEEE Trans. Commun.*, **42**(2/3/4), 1947–1956, by permission of IEEE.

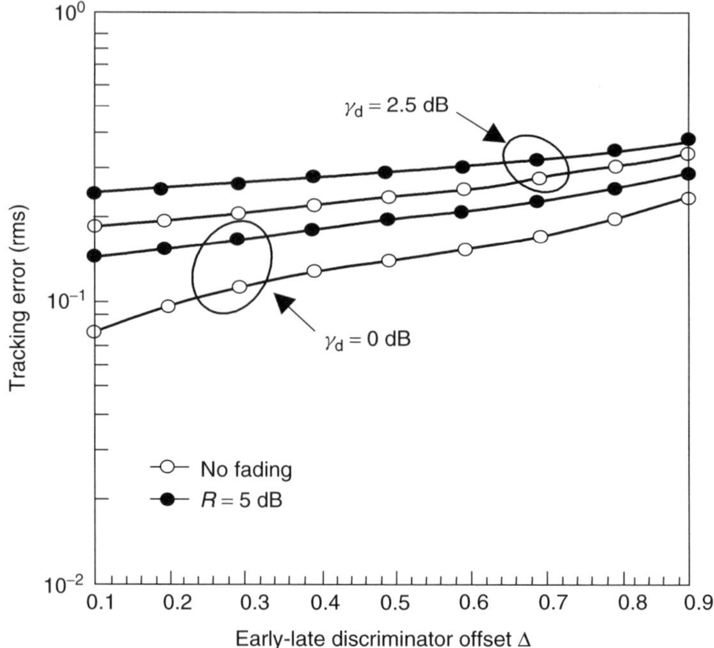

Figure 4.5 Effects of multipath fading on the tracking error performance with various delay spacings ($\Delta = 0.5, \zeta_0 = 100$) [3]. Reproduced from Sheen, J. W. and Stüber, G. (1994) Effects of multipath fading on delay locked loops for spread spectrum systems. *IEEE Trans. Commun.*, **42**(2/3/4), 1947–1956, by permission of IEEE.

interference for path l is regenerated. In the lower path of the receiver block diagram, these signals are used to generate the clean signal per path $r(t - T_D) - v_l(t) = w_0(t)$, which is used in the lth DLL to create a control signal $e_l(t)$ for the voltage controlled clock (VCC). All the individual control signals summed up represent the overall control signal for VCC. For the simulation environment defined in Table 4.1, the tracking error performance for standard DLL and DLL with IC DLL/IC is presented in Figure 4.2c. One can see that while the performance of the standard DLL is very poor, the performance of the DIL/IC loop for the appropriate signal-to-noise ratio is good. The problem of multipath IC will be visited again later in the context of multiuser detection in which in addition to the multipath the multiple access interference (MAI) will be also present at the front end of the receiver.

4.1.2 Identification of channel coefficients

After code synchronization (acquisition and tracking), signal despreading can be performed. If the processing gain is large, $T_b/T_c \geq 1$, after despreading, the received low-pass equivalent discrete time signal is

$$y_k = x_k c_k + n_k \tag{4.8}$$

Table 4.1 Simulation parameters

- PN Code: m-sequence with the generating polynomial $1 + x + x^6$
- Data modulation: binary phase shift keying (BPSK) with $R_b = 10\,\text{kb}\,\text{s}^{-1}$
- Chip rate: $630\,\text{Kb}\,\text{s}^{-1}$
- Sampling rate: 8 samples per chip period
- Total simulation time: $70\,000\ T_b$
- $\Delta = 1/2$
- Fading channel: Jake's model (independent paths with Rayleigh fading) with maximum Doppler shift of $83\,\text{Hz}$
- Low-pass filters: Elliptic filters (eighth order) with 3-dB bandwidths of $3\ R_b$ and R_b for $h_1(t)$ and $h_2(t)$, respectively
- In the simulations, the tracking range is limited to $[-T_c, T_c]$, that is, whenever $|\varepsilon| > 1$. A reacquisition process will be initiated. The initial tracking error is assumed to be $\varepsilon = 0.2$. Only the first-order loop $[F(s) = 1]$ with $\Delta = 1/2$ is to be considered for simplicity
- $\Gamma_k = E\lfloor|g_k|^2\rfloor/E\lfloor|g_0|^2\rfloor$, $\zeta_0 = B_{L_0}/R_b$ is the normalized average loop bandwidth
- Received average SNR. $\gamma_b = PT_bE\lfloor|g_0|^2\rfloor\left(1 + \sum_{k=1}^{L} \Gamma_k\right)/N_0$.

---

where c_k is the channel coefficient. At this stage we will assume that the residual fading is frequency nonselective because all multipath components are resolved in the despreading process in each finger of the RAKE receiver. If x_k is a known training symbol and if the SNR is high, a good estimate of c_k can be easily computed from equation (4.8) as

$$c_k \approx y_k/x_k \overset{\Delta}{=} \tilde{c}_k \qquad (4.9)$$

where y_k is the received signal. However, most of the received symbols are not training symbols. In these cases, the available information for estimating c_k can be based upon prediction from the past detected data bearing symbols $\overline{x}_i (i < k)$. This scheme will be referred to as decision feedback adaptive linear predictor (DFALP).

Using a standard linear prediction approach we formulate the predicted fading channel coefficient at time k as

$$\hat{c}_k = \sum_{i=1}^{N} b_i^* \tilde{c}_{k-i} \overset{\Delta}{=} \boldsymbol{b}(k)^{\text{H}}\tilde{\boldsymbol{c}}(k) \qquad (4.10)$$

where

$$\tilde{\boldsymbol{c}}(k) = (\tilde{c}_{k-1}, \tilde{c}_{k-2}, \ldots, \tilde{c}_{k-N})^{\text{T}} \qquad (4.11)$$

is a vector of past corrected channel coefficient estimates and

$$\boldsymbol{b}(k) = (b_1, b_2, \ldots, b_N)^{\text{T}} \qquad (4.12)$$

are the filter (linear predictor) coefficients at time k. The superscript T stands for transpose and H stands for Hermitian transpose. The constant N is the order of the linear predictor. The block diagram of the receiver is shown in Figure 4.6.

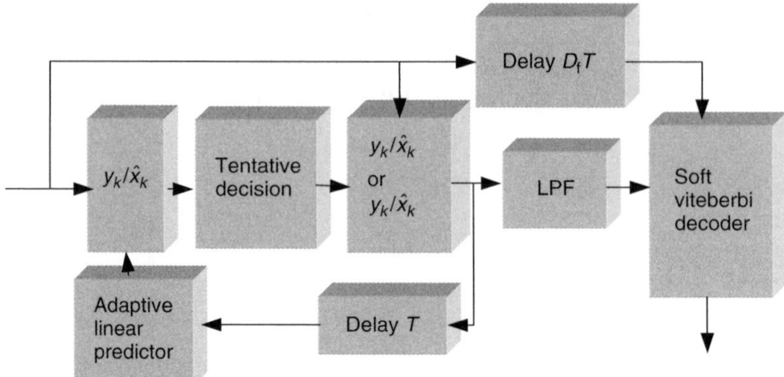

Figure 4.6 The DFALP algorithm for tracking phase and amplitude of frequency nonselective fading channels [4]. Reproduced from Liu, Y. and Blostein, S. (1995) Identification of frequency nonselective fading channels using decision feedback and adaptive linear prediction. *IEEE Trans. Commun.*, **43**(2), 1484–1492, by permission of IEEE.

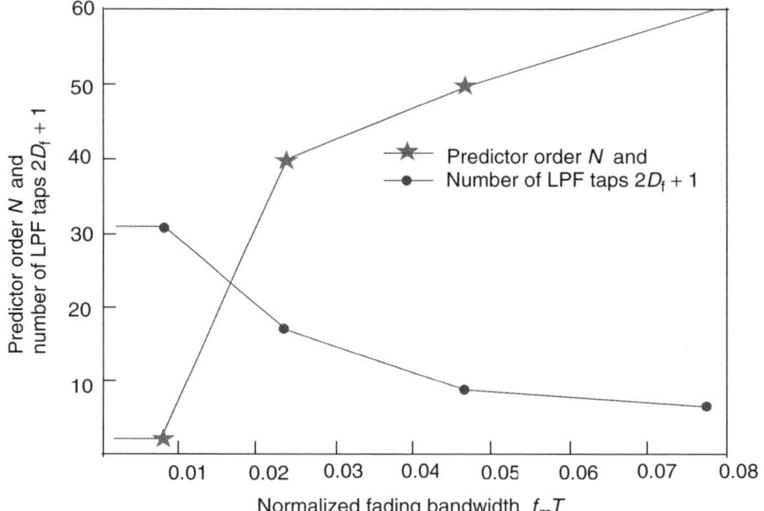

Figure 4.7 Recommended linear predictor order N and the number of LPF taps for the DFALP algorithm [4]. Reproduced from Liu, Y. and Blostein, S. (1995) Identification of frequency nonselective fading channels using decision feedback and adaptive linear prediction. *IEEE Trans. Commun.*, **43**(2), 1484–1492, by permission of IEEE.

The updating process for the filter coefficients is defined as

$$\boldsymbol{b}(k+1) = \boldsymbol{b}(k) + \mu(\tilde{c}_k - \hat{c}_k)^* \tilde{\boldsymbol{c}}(k) \tag{4.13}$$

Simulation results for predictor order N and the number of taps $2D_f + 1$ of the low-pass filter for the minimum bit error rate (BER) are shown in Figure 4.7.

4.2 CODE TRACKING IN FADING CHANNELS

The previously presented material on code tracking was based on the assumption that except for the additive white Gaussian noise the channel itself does not introduce any additional signal degradation or that only a flat frequency nonselective fading per path was present. For some applications like land and satellite mobile communications, we have to take into account the presence of severe fading due to channel dynamics. In this section we will present one possible approach to code tracking in such an environment.

4.2.1 Channel model

A channel with multipath propagation can be represented by a time-varying tapped-delay line, with impulse repose given by

$$h(\tau, t) = \sum_{l=0}^{N_\beta - 1} \beta_l(t)\delta(\tau - lT_s) \tag{4.14}$$

where T_s is the Nyquist sampling interval for the transmitted signal, N_β is the number of received signal replicas through different propagation paths and $\beta_l(t)$ represents the complex-valued time-varying channel coefficients. So, for the transmitted signal $s(t)$ the received signal $r(k)$ sampled at $t = kT_s$, will consist of N_β mutually delayed replicas that can be represented as

$$r(k) = \sum_{l=0}^{N_\beta - 1} \beta_\ell(k)s[(k - l)T_s + \tau(k)] + n(k) \tag{4.15}$$

In this equation, $n(k)$ are samples of the noise with

$$E\{n(k - i)n^*(k - j)\} = \sigma_n^2 \delta_{i,j} \tag{4.16}$$

In the RAKE receiver, each signal component is despread separately and then combined into a new decision variable for final decision. For the combining that provides maximum signal-to-noise ratio, signal components are weighted with factors β_l. So the synchronization for the RAKE receiver should provide a good estimate of delay τ and all channel intensity coefficients $\beta_l l = 0, 1, \ldots, N_\beta - 1$. The operation of the RAKE receiver will be elaborated later and within this section we will concentrate on the joint channel (β_l) and code delay (τ) estimation using the extended Kalman filter (EKF) [5,6].

For these purposes, the channel coefficients and delay are assumed to obey the following dynamic model equations.

$$\beta_\ell(k + 1) = \alpha_\ell \beta_\ell(k) + w_l(k); l = 0, 1, \ldots, N_\beta - 1$$

$$\tau(k + 1) = \zeta\tau(k) + w_\tau(k) \tag{4.17}$$

where $w_l(k)$ and $w_\tau(k)$ are mutually independent circular white Gaussian processes with variances σ^2_{wl} and σ^2_τ, respectively. In statistics, these processes are called autoregressive (AR) processes of order k, where k shows how many previous samples with indices $(k, k-1, k-2, \ldots, k-K+1)$ are included in modeling a sample with index $k+1$. In equation (4.17), the first-order AR model is used. The more the disturbances in signal are expected due to Doppler, the higher the variance of w_l and the lower α_l should be used. Variance of w_τ will not only depend on Doppler but also on the oscillator stability. A comprehensive discussion of AR modeling of wideband indoor radio propagation can be found in Reference [7].

4.2.2 Joint estimation of PN code delay and multipath using the EKF

From the available signal samples $r(k)$ given by equation (4.15) we are supposed to find the minimum variance estimates of β_l and τ These will be denoted by

$$\hat{\beta}_l(k|k) = E\{\beta_l(k)|r(k)\}$$

$$\hat{\tau}(k|k) = E\{\tau(k)|r(k)\} \tag{4.18}$$

where $r(k)$ is a vector of signal samples

$$r(k) = \{r(k), r(k-1), \ldots, r(0)\} \tag{4.19}$$

From equation (4.15) one can see that $r(k)$ is linear in the channel coefficients $\beta_l(k)$, but it is nonlinear in the delay variable $\tau(k)$. A practical approximation to the minimum variance estimator in this case is the EKF. This filter utilizes a first-order Taylor's series expansion of the observation sequence about the predicted value of the state vector, and will approach the true minimum variance estimate only if the linearization error is small.

The basic theory of extended Kalman filtering is available in textbooks [6]. Having in mind that in the delay-tracking problem, the state model is linear, while the measurement model is nonlinear, we have

$$x(k+1) = Fx(k) + Gw(k)$$

$$z(k) = H(x(k)) + n(k) \tag{4.20}$$

In this equation, $x(k)$ represents the $N_\beta + 1$ dimensional state vector and $z(k)$ is the scalar measurement $r(k)$ In terms of the previous notation we have

$$x(k) = [\tau(k), \beta_0(k), \beta_1(k), \ldots, \beta_{N_\beta-1}(k)]^{\mathrm{T}}$$

$$F = \begin{bmatrix} \zeta & 0 & & \cdots & 0 \\ 0 & \alpha_0 & & \cdots & 0 \\ 0 & 0 & \alpha_1 & \cdots & 0 \\ \vdots & \vdots & \vdots & \vdots & \vdots \\ 0 & 0 & \cdots & & \alpha_{N_\beta-1} \end{bmatrix} \tag{4.21}$$

$$\boldsymbol{w}(k) = [w_\tau(k), w_0(k), w_1(k), \ldots, w_{N-1}(k)]^{\mathrm{T}}$$

$$H[\boldsymbol{x}(k)] = \sum_{l=0}^{N-1} \beta_l(k)s[(k - l)T_{\mathrm{s}} + \tau(k)]$$

$$z(k) = r(k) = H[\boldsymbol{x}(k)] + n(k)$$

$$\boldsymbol{G} = \boldsymbol{I}$$

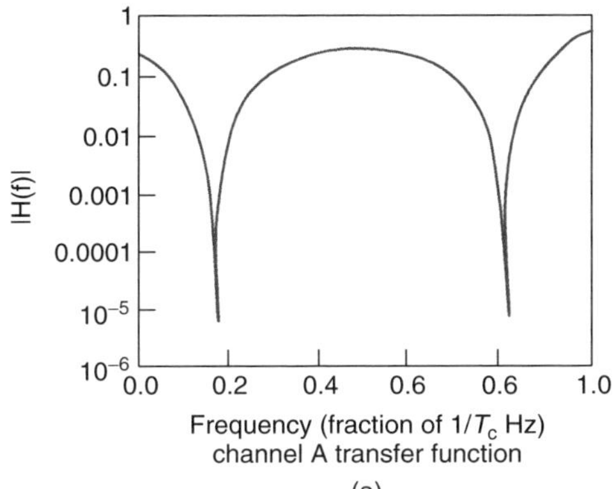

Frequency (fraction of $1/T_{\mathrm{c}}$ Hz)
channel A transfer function

(a)

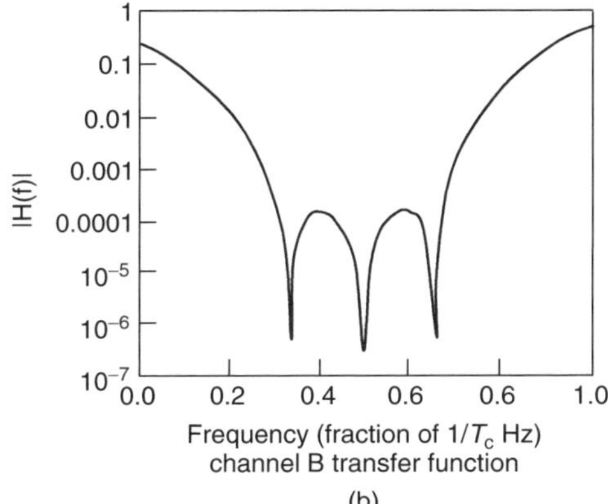

Frequency (fraction of $1/T_{\mathrm{c}}$ Hz)
channel B transfer function

(b)

Channel characteristics

Channel	N_{f} – number of paths	Channel zeros
A	3	$\omega = 1/T_{\mathrm{c}}, 5/T_{\mathrm{c}}$
B	4	$\omega = 2/T_{\mathrm{c}}, 3/T_{\mathrm{c}}, 4/T_{\mathrm{c}}$

Figure 4.8 Simulation examples – PN code in multipath [5]. Reproduced from Iltis, R. (1994) An EKF-based joint estimator for interference, multipath, and code delay in a DS spread-spectrum receiver. *IEEE Trans. Commun.*, **42**, 1288–1299, by permission of IEEE.

By using general results of the EKF theory [6], we have

$$\hat{x}(k|k) = \hat{x}(k|k-1)K(k)z(k) - H[\hat{x}(k|k-1)]$$

$$K(k) = P(k|k-1)H'(k)[H'(k)^H P(k|k-1)H'(k) + \sigma_n^2]^{-1}$$

$$P(k|k) = [I - K(k)H'(k)^H]P(k|k-1) \tag{4.22}$$

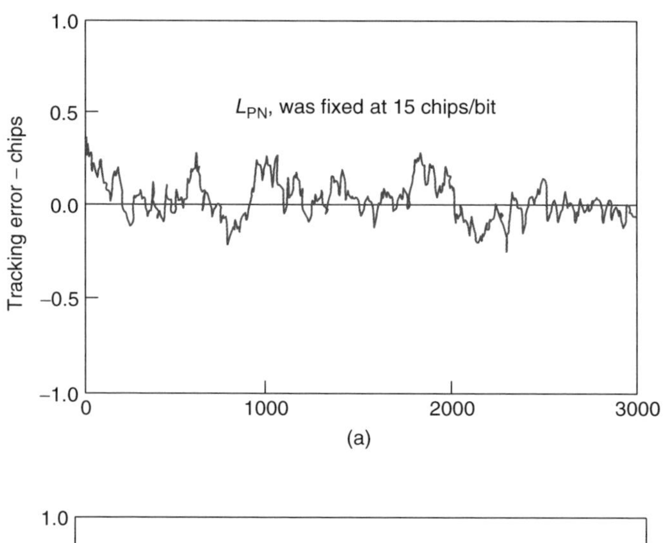

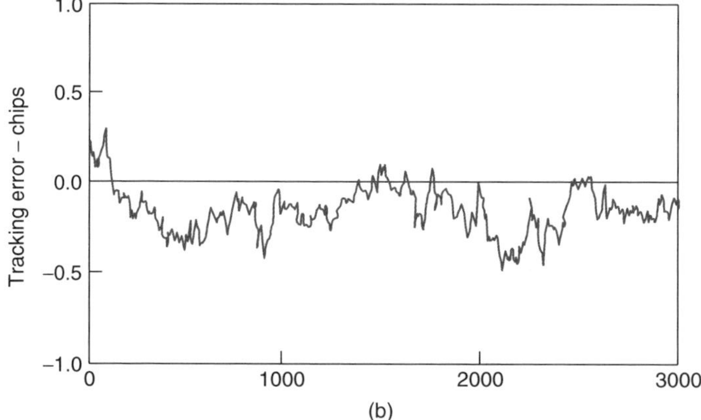

Figure 4.9 (a) Iteration number tracking error trajectory for $E_b/N_0 = 10\,\mathrm{dB}$ – Channel A, (b) Iteration number tracking error trajectory for $E_b/N_0 = 10\,\mathrm{dB}$ – N_f incorrectly assumed to be 1 – Channel A, (c) Iteration number tracking error trajectory for $E_b/N_0 = 10\,\mathrm{dB}$ – Channel B, (d) Iteration number tracking error trajectory for $E_b/N_0 = 10\,\mathrm{dB}$ – N_f incorrectly assumed to be 1 – Channel B, (e) Iteration number tracking error trajectory for $E_b/N_0 = 0\,\mathrm{dB}$ – Channel A and (f) Iteration number tracking error trajectory for $E_b/N_0 = 5\,\mathrm{dB}$ – Channel A [5]. Reproduced from Iltis, R. (1994) An EKF-based joint estimator for interference, multipath, and code delay in a DS spread-spectrum receiver. *IEEE Trans. Commun.*, **42**, 1288–1299, by permission of IEEE.

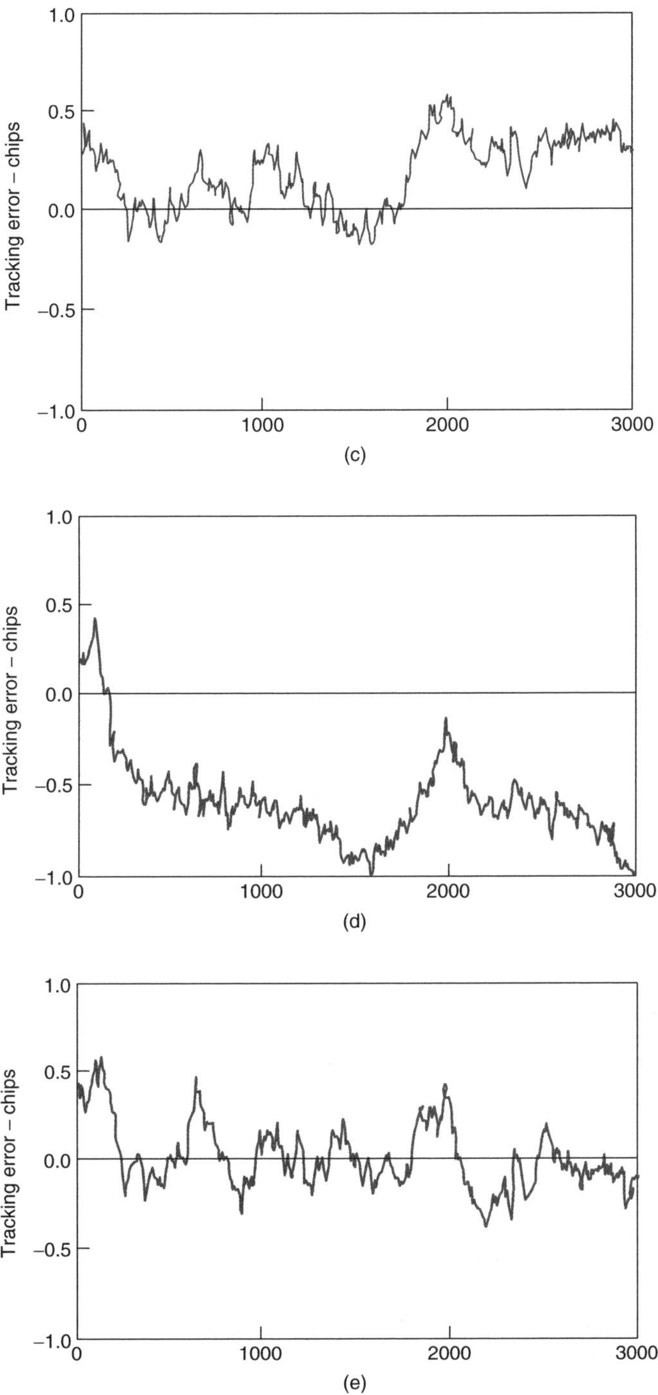

Figure 4.9 (*Continued*).

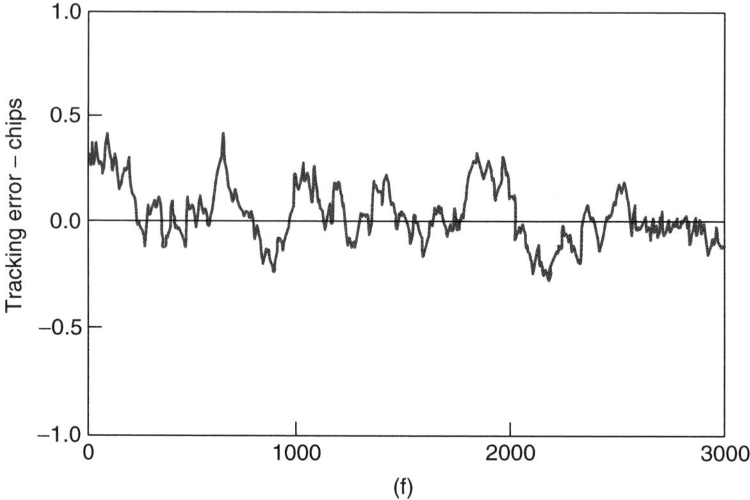

Figure 4.9 (*Continued*).

The matrix $\boldsymbol{H}'(k)$ represents the time-varying gradient of the observation scalar with respect to the one-step prediction vector.

$$\boldsymbol{H}'(k) = \left[\frac{\partial}{\partial x_1} H[\hat{\boldsymbol{x}}(k|k-1)], \frac{\partial}{\partial x_2} H[\hat{\boldsymbol{x}}(k|k-1)], \ldots, \frac{\partial}{\partial x_{N_\beta+1}} H[\hat{\boldsymbol{x}}(k|k-1)] \right]^{\mathrm{H}} \quad (4.23)$$

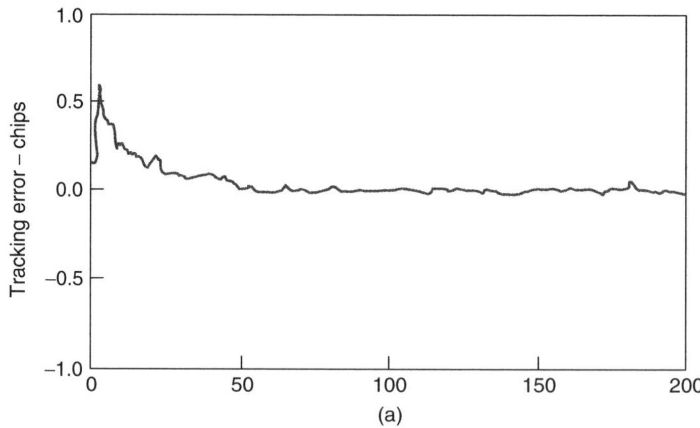

Figure 4.10 (a) Iteration number tracking error trajectory for $E_b/N_0 = 40\,\mathrm{dB}$ – Channel A, (b) Iteration number mean-square coefficient error for $E_b/N_0 = 10\,\mathrm{dB}$ – Channel A and (c) Iteration number mean-square coefficient error for $E_b/N_0 = 40\,\mathrm{dB}$ – Channel A [5]. Reproduced from Iltis, R. (1994) An EKF-based joint estimator for interference, multipath, and code delay in a DS spread-spectrum receiver. *IEEE Trans. Commun.*, **42**, 1288–1299, by permission of IEEE.

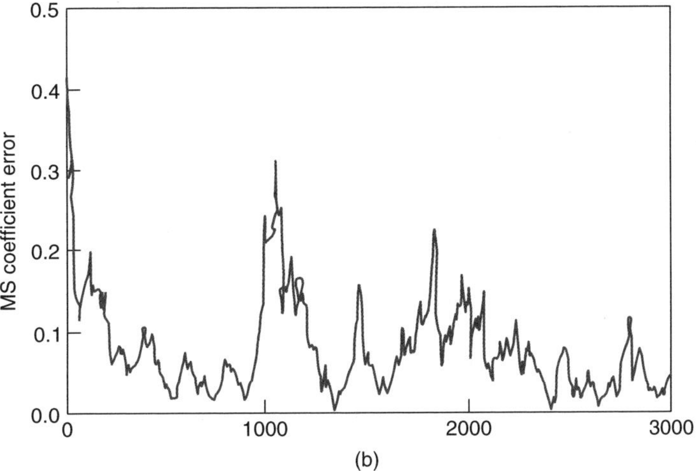

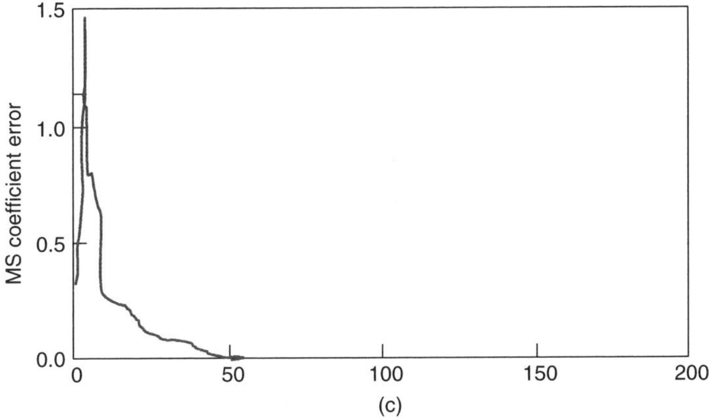

Figure 4.10 (*Continued*).

The one-step predictions of the state vector (state up-data) and error covariance matrix are given as [6]

$$\widehat{x}(k+1|k) = F\widehat{x}(k|k)$$

$$P(k+1|k) = FP(k|k)F^{H} + GQG^{T}$$

where

$$Q = \text{diag}\,[\sigma_{\tau}^2, \sigma_{w0}^2, \ldots, \sigma_{w2}^2, \ldots, \sigma_{wN_\beta - 1}^2] \tag{4.24}$$

For the two examples of the channel transfer function shown in Figure 4.8, simulation results of the tracking error are shown in Figures 4.9 and 4.10.

4.3 SIGNAL SUBSPACE-BASED CHANNEL ESTIMATION FOR CDMA SYSTEMS

In this section we present a multiuser channel estimation problem through a signal subspace-based approach [8]. For these purposes, the received signal for K users will be presented as

$$r(t) = \sum_{k=1}^{K} r_k(t) + \eta_t \quad -\infty < t < \infty \tag{4.25}$$

If the channel impulse response for user k is $h_k(t, \tau)$, we have

$$r_k(t) = h_k(t, \tau)^* s_k(t)$$
$$= \int_{-\infty}^{\infty} h_k(t, \alpha) s_k(\alpha) \, d\alpha \tag{4.26}$$

If phase-shift keying (PSK) is used to modulate the data, then the baseband complex envelope representation of the kth user's transmitted signal is given by

$$s_k(t) = \sqrt{2P_k} \, e^{j\phi_k} \sum_i e^{j(2\pi/M)m_k^{(i)}} a_k(t - iT) \tag{4.27}$$

where P_k is the transmitted power, ϕ_k is the carrier phase relative to the local oscillator at the receiver, M is the size of the symbol alphabet, $m_k^{(i)} \in \{0, 1, \ldots, M-1\}$ is the transmitted symbol, $a_k(t)$ is the spreading waveform and T is the symbol duration. The spreading waveform is given by

$$a_k(t) = \sum_{n=0}^{N-1} \prod_{T_c}(t - nT_c) \, a_k^{(n)} \tag{4.28}$$

where $\Pi_{T_c}(t)$ is a rectangular pulse, T_c is the chip duration ($T_c = T/N$) and $\{a_k^{(n)}\}$ for $n = 0, 1, \ldots, N-1$ is a signature sequence (possibly complex valued since the signature alphabet need not be binary). The chip-matched filter can be implemented as an integrate-and-dump circuit, and the discrete time signal is given by

$$r[n] = \frac{1}{T_c} \int_{nT_c}^{(n+1)T_c} r(t) \, dt \tag{4.29}$$

Thus, the received signal can be converted into a sequence of wide sense stationary (WSS) random vectors by buffering $r[n]$ into blocks of length N

$$y_i = [r(iN)r(1+iN) \cdots r(N-1+iN)]^T \in C^N \tag{4.30}$$

where the nth element of the ith observation vector is given by $y_{i,n} = r(n + iN)$. Although each observation vector corresponds to one symbol interval, this buffering was done without regard to the actual symbol intervals of the users. Since the system is asynchronous, each observation vector will contain at least the end of the previous symbol (left) and the beginning of the current symbol (right) for each user. The factors due to the power, phase and transmitted symbols of the kth user may be collected into a single complex constant $c_k^{(i)}$, for example, some constant times $\sqrt{2P_k}\,e^{j[\phi_k+(2\pi/M)m_k^{(i)}]}$ and equation (4.30) becomes

$$y_i = \sum_{k=1}^{K} [c_k^{(i-1)}\mathbf{u}_k^r + c_k^{(i)}\mathbf{u}_k^l] + \eta_i = \mathbf{A}c_i + \eta_i \tag{4.31}$$

where $\eta_i = [\eta_{i,0}, \ldots, \eta_{i,N-1}]^{\mathrm{T}} \in \mathbf{C}^N$ is a Gaussian random vector. Its elements are zero mean with variance $\sigma^2 = N_0/2T_c$ and are mutually independent.

Vectors $\mathbf{u}_k^r$ and $\mathbf{u}_k^l$ are the right side of the kth user's code vector followed by zeros, and zeros followed by the left side of the kth user's code vector, respectively. In addition, we have defined $c_i = [c_1^{(i-1)}c_1^{(i)}\cdots c_K^{(i-1)}c_K^{(i)}]^{\mathrm{T}} \in \mathbf{C}^{2K}$ and the signal matrix $\mathbf{A} = \lfloor \mathbf{u}_1^r\mathbf{u}_1^l \ldots \mathbf{u}_K^r\mathbf{u}_K^l \rfloor \in \mathbf{C}^{N \times 2K}$. We will start with the assumption that each user's signal goes through a single propagation path with an associated attenuation factor and propagation delay. We assume that these parameters vary slowly with time, so that for sufficiently short intervals the channel is approximately a linear time-invariant (LTI) system. The baseband channel impulse response can then be represented by a Dirac delta function as $h_k(t, \tau) = h_k(t) = \alpha_k\delta(t - \tau_k), \forall\tau$ where α_k is a complex-valued attenuation weight and τ_k is the propagation delay. Since there is just a single path, we assume that α_k is incorporated into $c_k^{(i)}$ and concentrate solely on the delay.

Let us define $v \in \{0, \ldots, N - 1\}$ and $\gamma \in [0, 1)$ such that $(\tau_k/T_c) \bmod N = v + \gamma$. If $\gamma = 0$, the received signal is precisely aligned with the chip matched filter and only one chip will contribute to each sample, the signal vectors become

$$\mathbf{u}_k^r = \mathbf{a}_k^r(v)$$
$$\equiv [a_k^{(N-v)}\cdots a_k^{(N-1)}0\cdots 0]^{\mathrm{T}}$$
$$\mathbf{u}_k^l = \mathbf{a}_k^l(v)$$
$$\equiv [0\cdots 0a_k^{(0)}\cdots a_k^{(N-v-1)}]^{\mathrm{T}} \tag{4.32}$$

Since the chip-matched filter is just an integrator, the samples for a nonzero γ can be represented as

$$\mathbf{u}_k^r = (1 - \gamma)\mathbf{a}_k^r(v) + \gamma\mathbf{a}_k^r(v + 1)$$
$$\mathbf{u}_k^l = (1 - \gamma)\mathbf{a}_k^l(v) + \gamma\mathbf{a}_k^l(v + 1) \tag{4.33}$$

For the more general case of a multipath transmission channel with L distinct propagation paths, the impulse response becomes a series of delta functions

$$h_k(t, \tau) = h_k(t)$$
$$= \sum_{p=1}^{L} \alpha_{k,p}\delta(t - \tau_{k,p}) \tag{4.34}$$

The signal vectors can be represented as

$$\mathbf{u}_k^r = \sum_{p=1}^{L} \alpha_{k,p}[(1 - \gamma_{k,p})\mathbf{a}_k^r(v_{k,p}) + \gamma_{k,p}\mathbf{a}_k^r(v_{k,p} + 1)]$$

$$\mathbf{u}_k^l = \sum_{p=1}^{L} \alpha_{k,p}[(1 - \gamma_{k,p})\mathbf{a}_k^l(v_{k,p}) + \gamma_{k,p}\mathbf{a}_k^l(v_{k,p} + 1)] \tag{4.35}$$

If we introduce the following notation

$$\mathbf{U}_k^r = \lfloor \mathbf{a}_k^r(0) \cdots \mathbf{a}_k^r(N - 1) \rfloor \in \mathbf{C}^{N \times N}$$

$$\mathbf{U}_k^l = [\mathbf{a}_k^l(0) \cdots \mathbf{a}_k^l(N - 1)] \in \mathbf{C}^{N \times N} \tag{4.36}$$

where the $\mathbf{a}_k$'s are as defined in equation (4.32), then the signal vectors may be expressed as a linear combination of the columns of these matrices

$$\mathbf{u}_k^r = \mathbf{U}_k^r \mathbf{h}_k$$

$$\mathbf{u}_k^l = \mathbf{U}_k^l \mathbf{h}_k \tag{4.37}$$

where $\mathbf{h}_k$ is the composite impulse response of the channel and the receiver front end, evaluated modulo, the symbol period. Thus, the nth element of the impulse response is given by

$$h_{k,n} = \sum_{j=0}^{\infty} \frac{1}{T_c} \int_{jT+nT_c}^{jT+(n+1)T_c} h_k(t)^* \prod_{T_c}(t)\, \mathrm{d}t \tag{4.38}$$

For delay spread $T_m < T/2$, at most two terms in the summation will be nonzero.

4.3.1 Estimating the signal subspace

The correlation matrix of the observation vectors is given by

$$\mathbf{R} = E[\mathbf{y}_i \mathbf{y}_i^\dagger]$$

$$= \mathbf{ACA}^\dagger + \sigma^2 \mathbf{I} \tag{4.39}$$

where $\mathbf{C} = E[\mathbf{c}_i \mathbf{c}_i^\dagger] \in \mathbf{C}^{2K \times 2K}$ is diagonal. The correlation matrix can also be expressed in terms of its eigenvector decomposition

$$\mathbf{R} = \mathbf{VDV}^\dagger \tag{4.40}$$

where the columns of $\mathbf{V} \in \mathbf{C}^{N \times N}$ are the eigenvectors of R, and D is a diagonal matrix of the corresponding eigenvalues (λ_n). Details of eigenvector decomposition are given in the appendix. Furthermore,

$$\lambda_n = \begin{cases} d_n + \sigma^2, & \text{if } n \leq 2K \\ \sigma^2, & \text{otherwise} \end{cases} \tag{4.41}$$

where d_n is the variance of the signal vectors along the nth eigenvector and we assume that $2K < N$. Since the $2K$ largest eigenvalues of $\mathbf{R}$ correspond to the signal subspace, $\mathbf{V}$ can be partitioned as $\mathbf{V} = [\mathbf{V}_S \mathbf{V}_N]$, where the columns of $\mathbf{V}_S = [\mathbf{v}_{S,1}, \ldots, \mathbf{v}_{S,2K}] \in C^{N \times 2K}$ form a basis for the signal subspace S_Y and $\mathbf{V}_N = [\mathbf{v}_{N,1}, \ldots, \mathbf{v}_{N,N-2K}] \in C^{N \times N-2K}$ spans the noise subspace N_Y. Readers less familiar with eigenvalues decomposition are referred to the appendix. Since we would like to track slowly varying parameters, we form a moving average or a Bartlett estimate of the correlation matrix based on the J most recent observations

$$\hat{\mathbf{R}}_i = \frac{1}{J} \sum_{j=i-J+1}^{i} \mathbf{y}_j \mathbf{y}_j^{\dagger} \tag{4.42}$$

It is well known [9] that the maximum-likelihood (ML) estimate of the eigenvalues and associated eigenvectors of $\mathbf{R}$ is just the eigenvector decomposition of $\hat{\mathbf{R}}_i$. Thus, we perform an eigenvalue decomposition of $\hat{\mathbf{R}}_i$ and select the eigenvectors corresponding to the $2K$ largest eigenvalues as a basis for $\hat{S}_Y$.

4.3.2 Channel estimation

Consider the projection of a given user's signal vectors into the estimated noise subspace

$$\left. \begin{array}{c} \mathbf{e}_k^r = (\mathbf{u}_k^{r\dagger} \hat{\mathbf{V}}_N)^{\mathrm{T}} \\ \mathbf{e}_k^l = (\mathbf{u}_k^{l\dagger} \hat{\mathbf{V}}_N)^{\mathrm{T}} \end{array} \right\} \in C^{N-2K} \tag{4.43}$$

If $\mathbf{u}_k^r$ and $\mathbf{u}_k^l$ both lie in the signal subspace, then their sum $\mathbf{u}_k = \mathbf{u}_k^r + \mathbf{u}_k^l$ must also be contained in $\mathbf{V}_S$. The projection of $\mathbf{u}_k$ into the estimated noise subspace

$$\tilde{\mathbf{e}}_k = (\mathbf{u}_k^{\dagger} \hat{\mathbf{V}}_N)^{\mathrm{T}} \tag{4.44}$$

is a Gaussian random vector [6] and thus has probability density function

$$p_{\tilde{\mathbf{e}}}(\tilde{\mathbf{e}}_k) = \frac{1}{\det[\pi \mathbf{K}]} \exp\{-\tilde{\mathbf{e}}_k^{\dagger} \mathbf{K}^{-1} \tilde{\mathbf{e}}_k\} \tag{4.45}$$

The covariance matrix $\mathbf{K}$ is a scalar multiple of the identity given by

$$\mathbf{K} = \frac{1}{J} \mathbf{u}_k^{\dagger} \mathbf{Q} \mathbf{u}_k \mathbf{I} \tag{4.46}$$

and

$$\mathbf{Q} = \sigma^2 \left[\sum_{k=1}^{2K} \frac{\lambda_k}{(\sigma^2 - \lambda_k)^2} \mathbf{v}_{S,k} \mathbf{v}_{S,k}^{\dagger} \right] \tag{4.47}$$

Therefore, within an additive constant, the log-likelihood function of $\tilde{\mathbf{e}}_k$ is

$$\Lambda(\tilde{\mathbf{e}}_k) = -(N - 2K)\ln(\mathbf{u}_k^\dagger \mathbf{Q}\mathbf{u}_k) - J\frac{\tilde{\mathbf{e}}_k^\dagger \tilde{\mathbf{e}}_k}{\mathbf{u}_k^\dagger \mathbf{Q}\mathbf{u}_k}$$

$$= -(N - 2K)\ln(\mathbf{u}_k^\dagger \mathbf{Q}\mathbf{u}_k)$$

$$- J\frac{\mathbf{u}_k^\dagger \mathbf{V}_N \mathbf{V}_N^\dagger \mathbf{u}_k}{\mathbf{u}_k^\dagger \mathbf{Q}\mathbf{u}_k} \tag{4.48}$$

The exact $\mathbf{V}_N$ and $\mathbf{Q}$ are unknown, but we may replace them with their estimates. The best estimates will minimize $\tilde{\mathbf{e}}_k$, which will result in the maximum of the likelihood function.

Unfortunately, maximizing this likelihood function is prohibitively complex for a general multipath channel, so we will consider only a single propagation path. In this case, the vector $\mathbf{u}_k$ is a function of only one unknown parameter: the delay τ_k. To form the timing estimate, we must solve

$$\hat{\tau}_k = \arg\max_{\tau_k \in [0, T)} \Lambda(\mathbf{u}_k) \tag{4.49}$$

Ideally, we would like to differentiate the log-likelihood function with respect to τ. However, the desired user's delay lies within an uncertainty region, $\tau_k \in [0, T]$, and $\mathbf{u}_k(\tau)$ is only piecewise continuous on this interval. To deal with these problems, we divide the uncertainty region into N cells of width T_c and consider a single cell, $c_\nu \equiv [\nu T_c, (\nu + 1)T_c)$. We again define $\nu \in \{0, \ldots, N - 1\}$ and $\gamma \in [0, 1)$ such that $(\tau/T_c) \bmod N = \nu + \gamma$, and for $\tau \in c_\nu$ the desired user's signal vector becomes

$$\mathbf{u}_k\{\tau\} = (1 - \gamma)\mathbf{u}_k(\nu) + \gamma \mathbf{u}_k(\nu + 1) \tag{4.50}$$

and

$$\frac{d}{d\tau}\mathbf{u}_k(\tau) = \mathbf{u}_k(\nu + 1) - \mathbf{u}_k(\nu)$$

$$= \text{a constant} \tag{4.51}$$

Thus, within a given cell, we can differentiate the log-likelihood function and solve for the maximum in closed form. We then choose whichever of the N-solutions that yields the largest value for equation (4.48). Details can be found in Reference [8].

Under certain conditions, it may be possible to simplify this algorithm. Note that maximizing the log-likelihood function (4.48) is equivalent to maximizing

$$\Lambda(\tilde{\mathbf{e}}_k) = -\frac{N - 2K}{J}\ln(\mathbf{u}_k^\dagger \mathbf{Q}\mathbf{u}_k)$$

$$- \frac{\mathbf{u}_k^\dagger \mathbf{V}_N \mathbf{V}_N^\dagger \mathbf{u}_k}{\mathbf{u}_k^\dagger \mathbf{Q}\mathbf{u}_k} \tag{4.52}$$

As $J \to \infty$, the leading term goes to zero; thus, for large observation windows, we can use the following approximation:

$$\Lambda(\tilde{\mathbf{e}}_k) \approx -\frac{\mathbf{u}_k^\dagger \mathbf{V}_N \mathbf{V}_N^\dagger \mathbf{u}_k}{\mathbf{u}_k^\dagger \mathbf{Q} \mathbf{u}_k} \tag{4.53}$$

This yields a much simpler expression for the stationary points [8].

The MUSIC (multiple signal classification) algorithm is equivalent to equation (4.53) when one only maximizes the numerator and ignores the denominator, that is, one assumes $\mathbf{u}_k^\dagger \mathbf{Q} \mathbf{u}_k$ is equal to one in equation (4.52) or (4.53). This yields an even simpler approximation for the log-likelihood function

$$\Lambda(\tilde{\mathbf{e}}_k) \approx -\mathbf{u}_k^\dagger \mathbf{V}_N \mathbf{V}_N^\dagger \mathbf{u}_k \tag{4.54}$$

which further simplifies the solution for the stationary points [8].

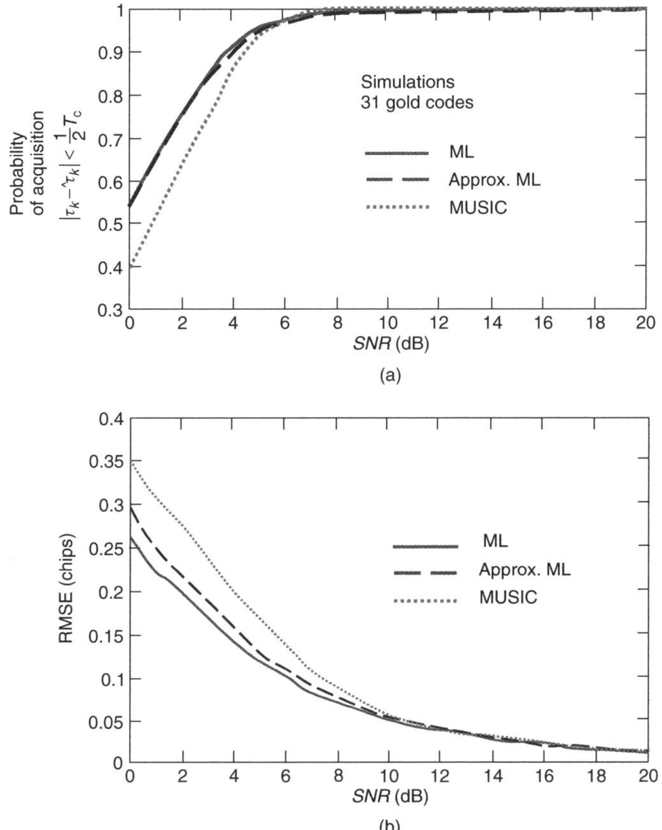

Figure 4.11 (a) Probability of acquisition for the maximum-likelihood (ML) estimator, the approximate ML, and the MUSIC algorithm [$K = 5$, $N = 31$, $J = 200$, $MAI = 20\,$dB] and (b) Root mean-squared error (RMSE) of the delay estimate in chips for the ML estimator, the approximate ML, and the MUSIC algorithm [$K = 5$, $N = 31$, $J = 200$, $MAI = 20\,$dB]. Reproduced from Bensley, J. S. and Aazhang, B. (1996) Subspace-based channel estimation for code division multiple access communications & systems. *IEEE Trans. Commun.*, **44**(8), 1009–1020, by permission of IEEE.

For illustration purposes, the simulation results for five users with length 31 Gold codes are presented in Figures 4.11 to 4.13.

A single desired user was acquired and tracked in the presence of strong MAI. The power ratio between each of the four interfering users and the desired user is designated the MAI level.

We first compare the true log-likelihood estimate (equation 4.48) with the large observation window approximation (equation 4.53) and the MUSIC algorithm (equation 4.54). This is done for a window size of 200 symbols and with a varying SNR. Figure 4.11(a)

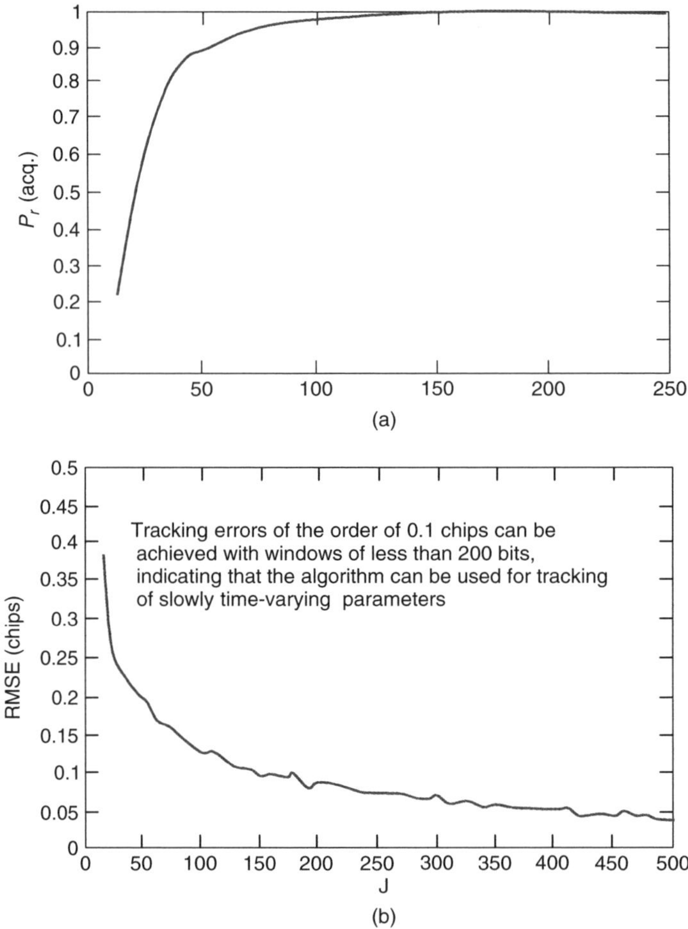

Figure 4.12   (a) Probability of acquisition and (b) root mean-squared error (RMSE) of timing estimate in chips of the subspace-based maximum-likelihood estimator for varying window size [$N = 31$, $SNR = 8\,\mathrm{dB}$, $K = 5$, $MAI = 20\,\mathrm{dB}$] [8]. Reproduced from Bensley, J. S. and Aazhang, B. (1996) Subspace-based channel estimation for code division multiple access communications & systems. *IEEE Trans. Commun.*, **44**(8), 1009–1020, by permission of IEEE.

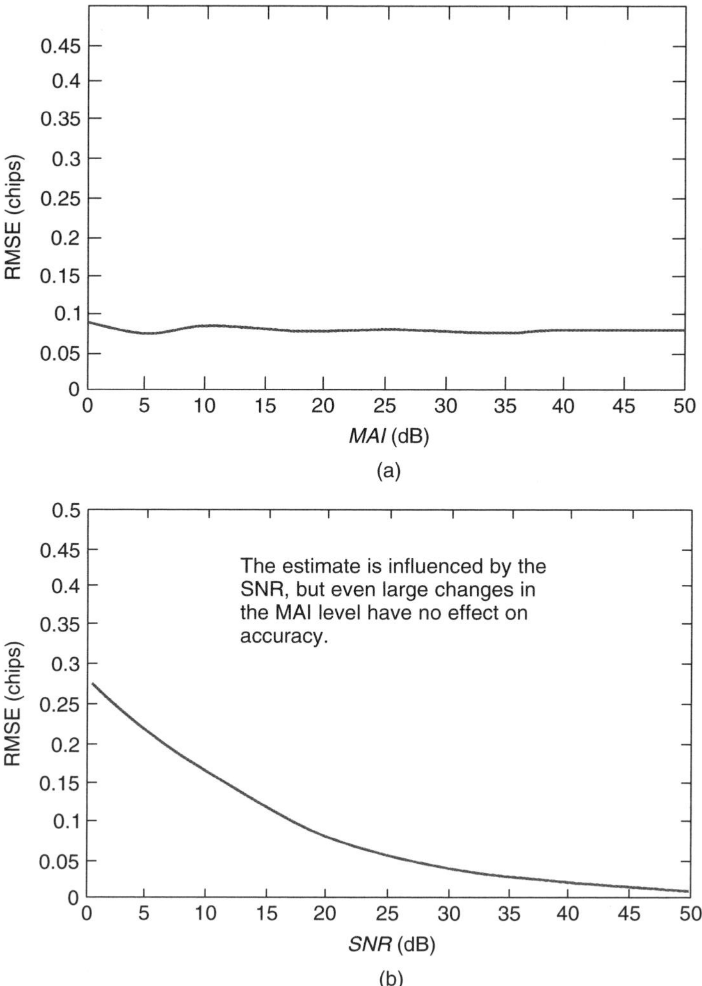

Figure 4.13 RMSE of the subspace-based maximum-likelihood estimator for varying (a) MAI level [$K = 5$, $N = 31$, $J = 200$, $SNR = 8\,\text{dB}$] and (b) SNR values [$K = 5$, $N = 31$, $J = 200$, $MAI = 20\,\text{dB}$].

shows the probability of acquisition for each method, in which acquisition id is defined as $|\tau_k - \hat{\tau}_k| < \frac{1}{2}T_c$. Using the approximate log-likelihood function results in almost no drop in performance. Furthermore, when the SNR is poor, both probabilistic approaches considerably outperform the MUSIC algorithm. In Figure 4.11(b), we compare the RMSE of the delay estimate once acquisition has occurred, that is, after processing enough symbols to reach within half of one chip. The approximate log-likelihood function experiences a slight increase in error at low SNR, but again both probabilistic methods do better than MUSIC.

The same parameters as a function of the window size are shown in Figure 4.12. One can say that for $J > 100$ the performance curve settles down to steady-state values. The RMSE versus MAI and SNR are shown in Figure 4.13. One can see that for an extremely wide range of near-far effect, the performance is good.

4.4 TURBO PROCESSOR AIDED RAKE RECEIVER SYNCHRONIZATION FOR UMTS W-CDMA

In this section we discuss how much of the theory presented in this chapter so far, and under what conditions, can be implemented in a practical system.

4.4.1 Signal model

In W-CDMA (UMTS FDD Mode), the data transmission is organized in frames of 10 ms, each divided into 15 slots. Details are elaborated in Chapter 17. The slot structure in the uplink given in Figure 4.14(a) consists of both data bits (dedicated physical data channel – DPDCH) and control information (dedicated physical control channel – DPCCH): The number of data bits per slot N_{d} depends on the data rate of the link. The number of control bits per slot is fixed at 10. It consists of pilot symbols for channel estimation, transmit power control (TPC) bits and transport frame indicator (TFI) bits. For the numerical analysis, the number of pilot bits per slot $N_{\mathrm{p}} = 6$ is used, if not stated otherwise. Figure 4.14(b) shows the spreading and modulation. The data bits $d(i)$ of the DPDCH are spread with an orthogonal variable spreading factor (OVSF) sequence $s_{\mathrm{d}}(k)$ (channelization sequence for data) with chip rate of 3.84 Mcps. The control bits $d_{\mathrm{c}}(i)$ of the DPCCH are spread by code $s_{\mathrm{c}}(k)$ (channelization code for control channel, orthogonal to $s_{\mathrm{d}}(k)$). The power of the control channel is adjusted by a gain factor to establish a variable control-to-data channel power ratio (CDR) of $\beta = p_{\mathrm{c}}/p_{\mathrm{d}}$. Note that the effective available power in the data channel is reduced by factor $p_{\mathrm{d}}/(p_{\mathrm{c}} + p_{\mathrm{d}}) = 1/(1 + \beta) = 1/V$. The chip streams of data and control channel are I/Q multiplexed and scrambled with the complex scrambling sequence $s_{\mathrm{scr}}(k)$ Thus, the transmit signal can be represented as

$$t(k) = s_{\mathrm{scr}}(k)\{s_{\mathrm{d}}(k)d(\lfloor k/N \rfloor) + j\sqrt{\beta}s_{\mathrm{c}}(k)d_{\mathrm{c}}(\lfloor k/N \rfloor)\} \tag{4.55}$$

The channel impulse response is defined as

$$h(t, \tau) = \sum_{l} C^{(l)}(t)\delta(t - \tau_l) \tag{4.56}$$

For illustration purposes, we will use the channel model specified by Rec.ITU-R M.1225 and summarized in Table 4.2. For the channel coefficient correlation function, we use Jack's model

$$\rho_{\mathrm{c}}(\tau) = J_0(\omega_D \tau) \tag{4.57}$$

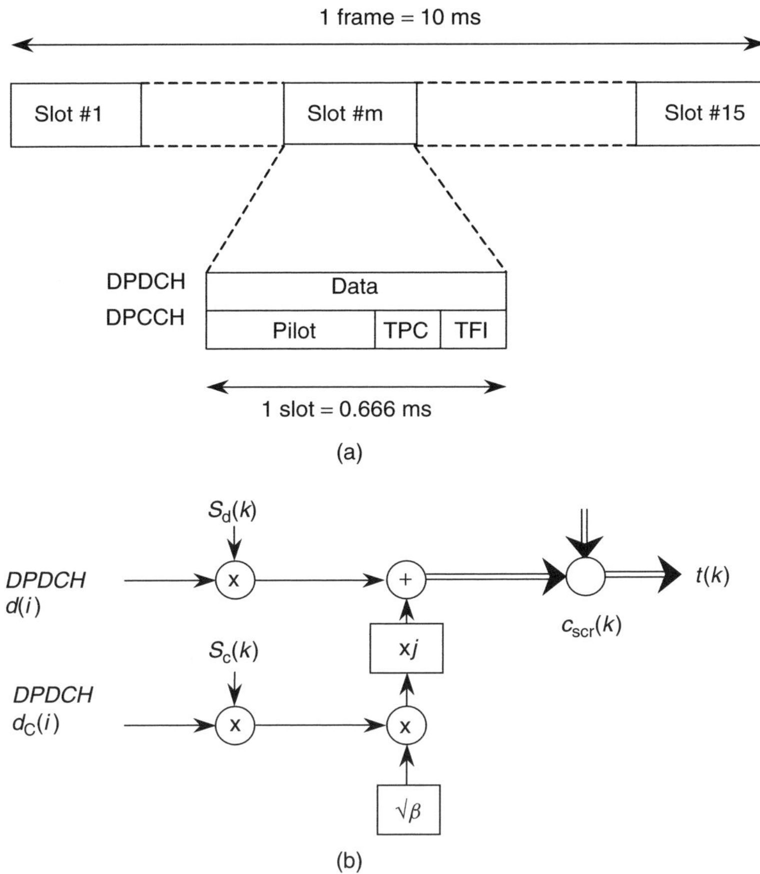

Figure 4.14 The system model: (a) W-CDMA uplink frame structure and (b) W-CDMA uplink spreading and modulation.

Table 4.2 Vehicular test environment, high antenna, tapped-delay line parameters

Tap L	Channel A relative delay (ns)	average power $\overline{|C^{(l)}|^2}/|C^{(0)}|^2$ (dB)	Channel B relative delay (ns)	average power $\overline{|C^{(l)}|^2}/|C^{(0)}|^2$ (dB)	Doppler spectrum
1	0	0.0	0	−2.5	classic
2	310	−1.0	300	0	classic
3	710	−9.0	8900	−12.8	classic
4	1090	−10.0	12 900	−10.0	classic
5	1730	−15.0	17 100	−25.2	classic
6	2510	−20.0	20 000	−16.0	classic

In simulations, this model is approximated by filtering Gaussian noise using the second-order Buterworth filter, a method used very often in practice. For the Kalman filter analysis, we also use the first-order AR model.

4.4.2 The receiver model

The general block diagram of the RAKE receiver is shown in Figure 4.15. A direct-sequence spread-spectrum (DSSS) signal after propagation through a multipath channel

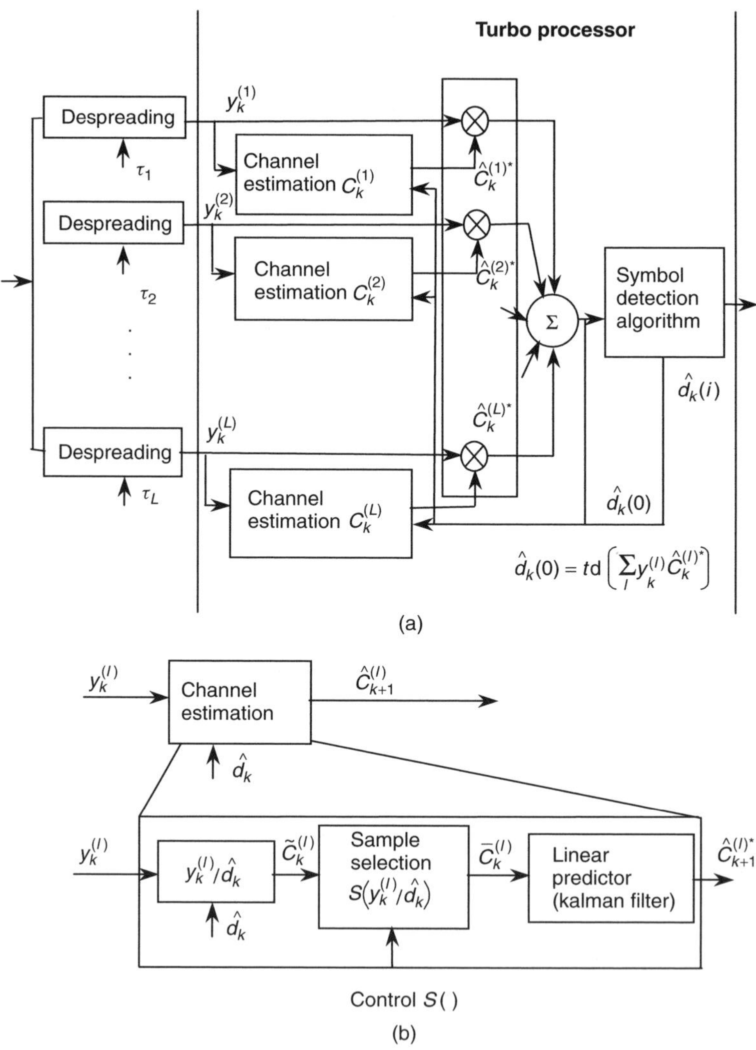

Figure 4.15 Receiver block diagram: (a) generic block diagram of Rake receiver and (b) The decision feedback adaptive linear predictor (ALP) algorithm for tracking a complex multipath coefficient in one RAKE finger.

will be despread in L RAKE fingers. Because of the relatively high-processing gain, the output of the lth despreading circuit will have the form

$$y_k^{(l)} = C_k^{(l)} \cdot d_k + n_k^{(l)} \tag{4.58}$$

Channel estimation in the presence of high-level MAI will be discussed in Chapter 14.

This operation is performed separately in both the data and the control channels. We assume perfect code synchronization per finger. In equation (4.58) k is the sampling index, $l = 1, \ldots, L$ is the path index, $C_k^{(l)}$ is the complex channel coefficient; $d_k = d(k)$ for data channel and $d_k = d_c(k)$ for control channel. Parameter $n_k^{(l)}$ is the overall noise in the lth RAKE finger including residual MPI, MAI, inter-channel interference (ICI) and thermal noise. All together, this component will be approximated as Gaussian noise with zero mean and variance $\sigma^{(l)2}$. Signal-to-noise ratio in each finger will be designated as

$$\wp^{(l)} = \overline{|C_k^{(l)}|^2}/\sigma^{(l)2} \tag{4.59}$$

As it was explained above, in the DPCCH channel the pilot symbols are used to facilitate the channel estimate. A sequence of N_p bits is periodically inserted into the control channel stream and used as a preamble for channel estimation. In the remaining interval, the tentative decisions $\hat{d}_k(0)$ or the ith decision iteration $\hat{d}_k(i)$ can be used to remove the modulation from the signal components $y_k^{(l)}$.

In order to further improve the signal-to-noise ratio in the channel estimator, both channels (DPCCH and DPDCH) can be used for channel estimation. In the DPCCH, the pilot symbol will be multiplexed with the data symbols, whereas the DPDCH will contain data only. This will be referred to as the joint channel estimation. The initial tentative decision, in each channel, will be in general obtained as

$$\hat{d}_k(0) = t\,d \left(\sum_l y_k^{(l)} \hat{C}_k^{(l)*} \right)$$

$$= \min_{d^{(m)}} \left| \sum_l y_k^{(l)} \hat{C}_k^{(l)*} - \sum_l d_k^{(m)} |\hat{C}_k^{(l)}|^2 \right| \tag{4.60}$$

where $d^{(m)} \in D, m = 1, 2, \ldots, M$ are all possible symbols from the signal constellation set. Our interest will be focused on BPSK modulation where $d^{(m)} = \pm 1$ (for Universal Mobile Telecommunication System (UMTS) model) and quadrature phase shift keying (QPSK) where $d^{(m)} = \pm 1 \pm j$ (for more general analysis). For these two examples, equation (4.60) can be approximated as

$$\hat{d}_k(0) = \mathrm{sgn} \left(\sum_l y_k^{(l)} \hat{C}_k^{(l)*} \right) \tag{4.61a}$$

for BPSK and

$$\hat{d}_k(0) = \mathrm{sgn} \left[\mathrm{Re} \left(\sum_l y_k^{(l)} \hat{C}_k^{(l)*} \right) \right] + j\,\mathrm{sgn} \left[\mathrm{Im} \left(\sum_l y_k^{(l)} \hat{C}_k^{(l)*} \right) \right] \tag{4.61b}$$

for QPSK signals. If the joint channel estimation is used, then $\hat{C}_k^{(l)}$ represents the sum of estimates obtained in data and control channels.

$$\hat{C}_k^{(l)} = \left[\hat{C}_{k,d}^{(l)} + \alpha(\beta, T_d)\hat{C}_{k,c}^{(l)}\right]/2 \tag{4.62}$$

Combining parameter $\alpha(\beta, T_d)$ depends on the signal power ratio β in the two channels and the bit rate in the data channel. Obtaining the subsequent decision iterations $\hat{d}_k(i)$ and their use in the turbo processor will depend on the demodulation/decoding algorithm. Every iteration is supposed to reduce BER P_e, which makes the channel estimate better and that in turn further improves the P_e in the next iteration. The maximum number of iterations for a given k will be I.

4.4.3 Channel estimation algorithm and turbo processor

In each iteration of the turbo processor, the algorithm consists of the following steps:

1. The decisions $\hat{d}_k = \hat{d}_k(i)$, $i = 0, 1, \ldots, I$ are used to generate channel samples $\tilde{C}_k^{(l)}$ as

$$\tilde{C}_k^{(l)} = y_k^{(l)}/\hat{d}_k \tag{4.63}$$

If $\hat{d}_k$ is correct ($\hat{d}_k = d_k$), and if the noise is negligible, from equations (4.58) and (4.63) we have $\tilde{C}_k^{(l)} \cong C_k^{(l)}$.

2. If d_k belongs to the preamble, the above conclusion is correct but during the data transmission $\hat{d}_k$ might occasionally be incorrect. Because of the effect of incorrect decisions we expect larger gain from joint estimation for lower data rates and higher signal-to-noise ratios. In this solution we invest an additional effort to recognize when this happens and use a preselection of the channel samples prior to further processing. In general, this function is designated as $S(\ )$ so that we have

$$\overline{C}_k^{(l)} = S(\tilde{C}_k^{(l)}) \tag{4.64}$$

The preselection algorithms are discussed later.

3. The preselected samples $\overline{C}_k^{(l)}$ are used in the predictor to generate $\hat{C}_{k+1}^{(l)}$. Two types of predictor are analyzed, linear predictor (moving average and Wiener filter) and Kalman filter estimator. In the case of Wiener filter, a possibility to reduce the length of the filter by additional LPF is also considered. In this case only, the prediction $\hat{C}_{k+1}^{(l)}$ is additionally filtered out and used to generate a channel sample for the next sampling instant. This is then used in the $(k + 1)$th sampling moment to generate data estimate $\hat{d}_{k+1}(0)$ in accordance with equation (4.60) and the process is repeated again by using equations (4.63) and (4.64), representing Step 1 and Step 2, respectively. Once again, for joint estimation $\hat{C}_k^{(l)} = [\hat{C}_{k,d}^{(l)} + \alpha(\beta, T_d)\hat{C}_{k,c}^{(l)}]/2$. These steps are presented in Figure 4.15(b). Depending on the demodulation/decoding algorithm, the turbo processor may operate on a symbol-by-symbol basis including Steps 1 and 2 or on a block-by-block basis including Steps 1, 2 and 3.

4.4.4 Channel sample preselection & modification

Here we define a number of possible sample selection modification functions $S(\ )$. Different choices are offered for a trade-off between the complexity and performance. For these algorithms, we define the selection decision variables.

$$\Delta d_k = \min_{d^{(m)}} \text{ value of } \left| \sum_l y_k^{(l)} \hat{C}_k^{(l)*} - \sum_l d_k^{(m)} |\hat{C}_k^{(l)}|^2 \right| \tag{4.65}$$

$$\Delta C_k^{(l)} = |\tilde{C}_{k+1}^{(l)} - \tilde{C}_k^{(l)}| \tag{4.66}$$

$$\sum \Delta C_k^{(l)} = |\tilde{C}_{k+1}^{(l)} - \tilde{C}_k^{(l)}| - |\tilde{C}_{k+1}^{(l)} + \tilde{C}_k^{(l)}| \tag{4.67}$$

The sample selection/modification algorithms can use any of these decision variables. We will designate this variable as $\upsilon_k^{(l)}$. On the basis of these definitions, the following sample preselection/modification functions are defined for each path l from the subset of paths L_s.

Hard decision

$$\#1 \quad \overline{C}_k^{(l)} = S_1(\tilde{C}_k^{(l)}) = \begin{cases} \tilde{C}_k^{(l)}; & \text{if} \quad \upsilon_k^{(l)} < th; \forall l \in L_s \\ 0; & \text{otherwise} \end{cases} \tag{4.68}$$

When a large disturbance is detected, the corresponding channel sample is removed from the input to the estimator.

Interpolation

$$\#2 \quad \overline{C}_k^{(l)} = S_2(\tilde{C}_k^{(l)}) = \begin{cases} \tilde{C}_k^{(l)}; & \text{if} \quad \upsilon_k^{(l)} < th; \forall l \in L_s \\ \tilde{C}_{k-1}^{(l)}; & \text{otherwise} \end{cases} \tag{4.69}$$

When a large disturbance is detected, the corresponding channel sample is replaced by the previous sample.

Substitution

$$\#3 \quad \overline{C}_k^{(l)} = S_3(\tilde{C}_k^{(l)}) = \begin{cases} \tilde{C}_k^{(l)}; & \text{if} \quad \upsilon_k^{(l)} < th; \forall l \in L_s \\ \hat{C}_k^{(l)}; & \text{otherwise} \end{cases} \tag{4.70}$$

When a large disturbance is detected, the corresponding channel sample is replaced by the estimation of the sample generated in the previous sampling interval.

Alternation (for BPSK modulation)

$$\#4 \quad \overline{C}_k^{(l)} = S_4(\tilde{C}_k^{(l)}) = \begin{cases} \tilde{C}_k^{(l)}; & \text{if} \quad \upsilon_k^{(l)} < th; \forall l \in L_s \\ -\tilde{C}_k^{(l)}; & \text{otherwise} \end{cases} \tag{4.71}$$

In the case when a large disturbance is detected, the corresponding channel sample is replaced by the inverted sample. The method can be used only for BPSK modulation. In order to improve signal-to-noise ratio at the estimator, we can use a joint estimator defined by

$$\overline{C}_k^{(l)} = \left[\overline{C}_{k,\mathrm{d}}^{(l)} + \alpha(\beta, T_\mathrm{d}) \overline{C}_{k,\mathrm{c}}^{(l)} \right] / 2 \tag{4.72}$$

Each of the preselection algorithms defined above can be used for each component in equation (4.72). In the previous definition, th is a threshold parameter to be optimized. L_s is the subset of channel multipath indexes. If equation (4.67) is used, then $th = 0$. If $v_k^{(l)}$ is defined by equation (4.65), and $L_\mathrm{s} = \{1\}$, then we have an individual decision for each channel. If $v_k^{(l)}$ is defined by equation (4.65), and L_s is any subset of size $|L_\mathrm{s}| > 1$, then we have a collective decision for all channel. In summary, for the Algorithm #1, the selection function $S_1(\ )$ simply does or does not forward the sample for further processing, hence the name *hard decision*. If $v_k^{(l)} > th$, it is an indication that the symbol estimation was probably incorrect and that the sample should not be taken for further processing. If this happened, Algorithm #2 would use the previous sample (interpolation). Algorithm #3 would replace such a sample with its prediction (substitution). Algorithm #4 would go even further and change the sign of the sample (alternate) before forwarding it for further processing. The last option is possible for BPSK modulation.

4.4.5 Channel sample prediction

For channel sample prediction, we will analyze three options:

1. Moving average filter (MAF): This algorithm is simple, but for higher Doppler it will start to average out the signal itself. To avoid this the filter length (number of taps) should be reduced, which would degrade the averaging of noise.

2. Linear prediction algorithm: Optimum solution (Wiener filter) is supposed to improve the performance but has two limitations. It is derived for a stationary signal and has an increased complexity (matrix inversion). Imperfect data removal will make the resulting signal at the output of the preselector nonstationary. The iterative approximation would result in an least mean square (LMS) algorithm. As we will see later, incorrect data removal would seriously disturb the operation of such a filter.

3. Kalman filter: This algorithm takes into account the channel dynamics and is an optimal solution [6] in the minimum mean square error (MMSE) sense. Incorrect data removal will degrade the performance of this solution too, but much less than that in the case of the LMS algorithm.

Channel prediction by MAF

The filter is producing

$$\hat{C}_k = \frac{1}{K} \sum_{i=1}^{K} (C_{k-i} + n_{k-i}) \tag{4.73}$$

Channel prediction by Wiener filter

For a transversal filter with coefficients

$$\mathbf{W}_k = (w_k, w_{k-1}, w_{k-2}, \ldots, w_{k-L}) \tag{4.74}$$

and channel sample vector

$$\mathbf{C}_k = (C_k, C_{k-1}, C_{k-2}, \ldots, C_{k-L}) \tag{4.75}$$

the steady-state tracking error variance (MMSE) is given as

$$\sigma_\varepsilon^2 = E\{|C_k|^2\} - \mathbf{P}^{\mathrm{T}}\mathbf{W}_0 = \sigma_c^2 - \mathbf{P}^{\mathrm{T}}\mathbf{W}_0 \tag{4.76}$$

where $\mathbf{W}_0$ is the optimum solution for the prediction coefficients obtained from

$$-2\mathbf{P} + 2\mathbf{R}\mathbf{W}_0 = 0 \tag{4.77}$$

and the vector $\mathbf{P}$ and the matrix $\mathbf{R}$ are defined as

$$\mathbf{P}^{\mathrm{T}} = E[C_k C_k]; \quad \mathbf{R} = E[\mathbf{C}_k \mathbf{C}_k^{\mathrm{T}}] = [\rho(k - m)]; \quad k, m = 1, \ldots, L \tag{4.78}$$

Channel prediction by Kalman filter

Here we use the channel model represented by the first-order AR process

$$C_k = \rho C_{k-1} + n_{ck} \tag{4.79}$$

where n_{ck} is the modeling error (zero mean Gaussian variable with variance $\sigma_c^2(1 - \rho^2)$). For the Kalman filter, the estimation error is the solution to the Ricatti equation [8], which in this case can be expressed as

$$\sigma_\varepsilon^2 = \frac{\sigma^2[\rho^2\sigma_\varepsilon^2 + (1 - \rho^2)\sigma_c^2]}{\rho^2\sigma_\varepsilon^2 + (1 - \rho^2)\sigma_c^2 + \sigma^2} \tag{4.80}$$

For further elaboration of the impact of channel modeling on prediction, see Reference [10].

4.4.6 BER analysis

For the BER, we use the standard results for the diversity of order L [11]. Expressions for SNR_f per finger should be further modified by modifying the equivalent noise to include interference between different paths and different users.

As a further elaboration of these equations, for BPSK modulation in Rayleigh-fading channel with Lth-order diversity and flat multipath intensity profile, the bit error probability can be presented in the form [11]

$$BER = \left[\frac{1}{2}(1-\mu)\right]^L \sum_{k=0}^{L-1} \binom{L-1+k}{k} \left[\frac{1}{2}(1+\mu)\right]^k \tag{4.81}$$

where

$$\mu = \sqrt{\frac{SNR_f}{1 + SNR_f}} \tag{4.82}$$

The SNR per bit is $L \cdot SNR_f/k$, where $k = \log_2 M$, M is the signal constellation size ($M = 2$ for BPSK and $M = 4$ for QPSK). For nonflat multipath intensity profiles,

$$BER = \frac{1}{2} \sum_{k=1}^{L} \pi_k \left[1 - \sqrt{\frac{\overline{\gamma}_k(1-\rho_r)}{2 + \overline{\gamma}_k(1-\rho_r)}}\right] \tag{4.83}$$

where $\rho_r = -1$ for BPSK and $\rho_r = 0$ for QPSK signals, and

$$\pi_k = \prod_{\substack{i=1 \\ i \neq k}}^{L} \frac{\overline{\gamma}_k}{\overline{\gamma}_k - \overline{\gamma}_i} \tag{4.84}$$

$$\gamma_b = \frac{E_b}{N_0} \sum_{k=1}^{L} \alpha_k^2 = \sum_{k=1}^{L} \gamma_k$$

Each γ_k is chi-square-distributed with two degrees of freedom

$$p(\gamma_k) = \frac{1}{\overline{\gamma}_k} e^{-\gamma_k/\overline{\gamma}_k} \tag{4.85}$$

where $\overline{\gamma}_k$ is the average SNR for the kth path, defined as

$$\overline{\gamma}_k = \frac{E_b}{N_0} E(\alpha_k^2) \tag{4.86}$$

4.4.7 Impact of power control

Reverse link power control will change the received signal statistics and the optimal solution (Wiener filter or Kalman predictor) would require an additional step in order to find out the new correlation function or to eliminate the impact of power control on the signal.

In Figure 4.16, we present a set of BER curves versus the receiver speed for the input SNR per path, $SNR = -1\,\text{dB}$ and $SNR = 5\,\text{dB}$ and the AR channel model defined by equations (4.79) and (4.57). All performance becomes worse with the increase of the receiver speed. Kalman performs the best. For larger lengths of the smoother ($N_s = 30$), the performance is better for small speeds but when the speed is increased the performance becomes significantly worse than that for the short smoother ($N_s = 7$).

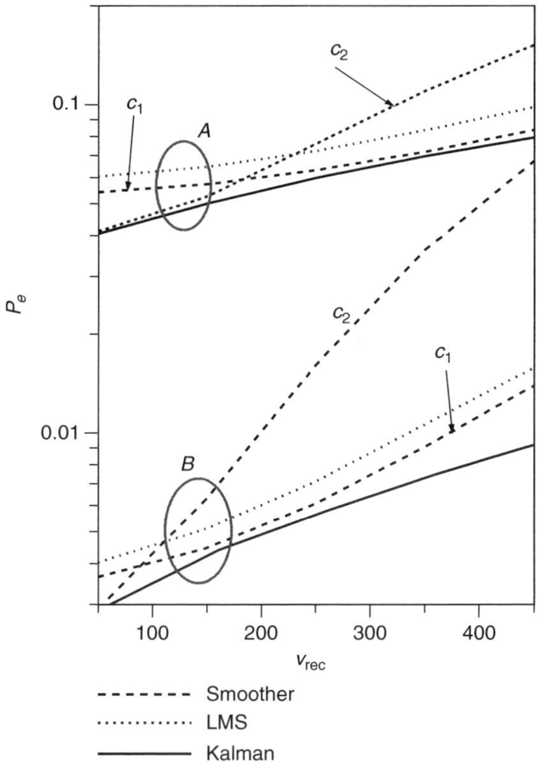

Figure 4.16 Error probability as a function of the receiver speed.

$A - SNR = -1\,\mathrm{dB}$ (average input SNR per path)
$B - SNR = 5\,\mathrm{dB}$

$c_1 - N_s = 7$
$c_2 - N_s = 30$

Sample preselection/modification function: *Substitution*
fading model: *AR model with equal power per path (equations 4.79 and 4.57)*

Decision variable $\sum \Delta C_k^{(l)} = |\tilde{C}_{k+1}^{(l)} - \tilde{C}_k^{(l)}| - |\tilde{C}_{k+1}^{(l)} + \tilde{C}_k^{(l)}|$

$L = 3$ – The number of RAKE fingers
$l = 3$ – The number of fading paths
$N_{\mathrm{LMS}} = 5$

In Figure 4.17, the previous phenomenon is examined in more detail. BER versus the smoother length N_s is presented for different speeds of the receiver. As expected, the optimum length of the smoother is smaller for larger speed. In addition, it was demonstrated that the joint estimator, using the signal samples from both the data and the control channels, performs better.

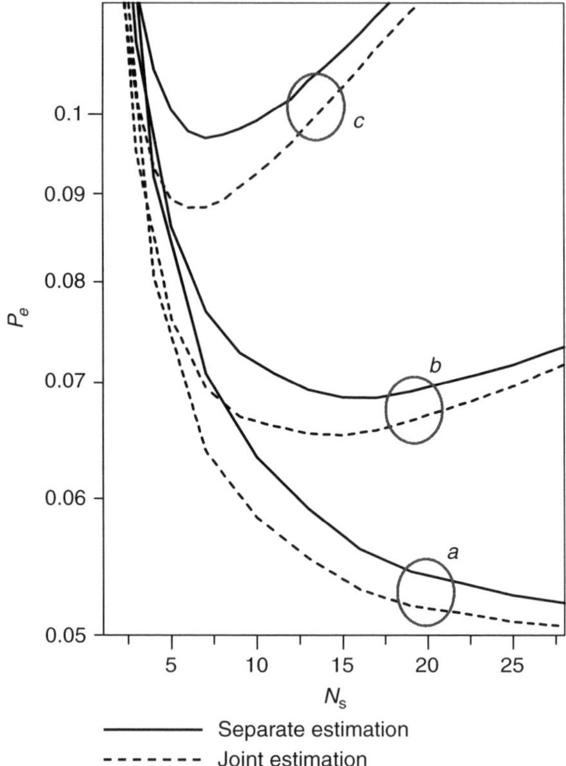

Figure 4.17 Error probability as a function of smoother length.

$a - v_r = 75\,\text{km}\,\text{h}^{-1}$
$b - v_r = 200\,\text{km}\,\text{h}^{-1}$
$c - v_r = 400\,\text{km}\,\text{h}^{-1}$

$SNR = -1\,\text{dB}$

Sample preselection/modification function: *Substitution* with $th = 0$
Decision variable $\sum \Delta C_k^{(l)} = |\tilde{C}_{k+1}^{(l)} - \tilde{C}_k^{(l)}| - |\tilde{C}_{k+1}^{(l)} + \tilde{C}_k^{(l)}|$
Fading model: *AR model with power allocation per path from Table 4.2*

$L = 3$ – The number of RAKE fingers
$l = 3$ – The number of fading paths

Figure 4.18 illustrates the BER performance versus the receiver speed for the case with and without turbo processing. It is assumed that the decoding process reduces the BER rate to zero so that only asymptotic results are presented. This makes the analysis independent of the type of decoding process used in the system. The curves are compared with the theoretical minimum for a known channel.

A number of specific solutions related to channel estimation are given in References [12–28].

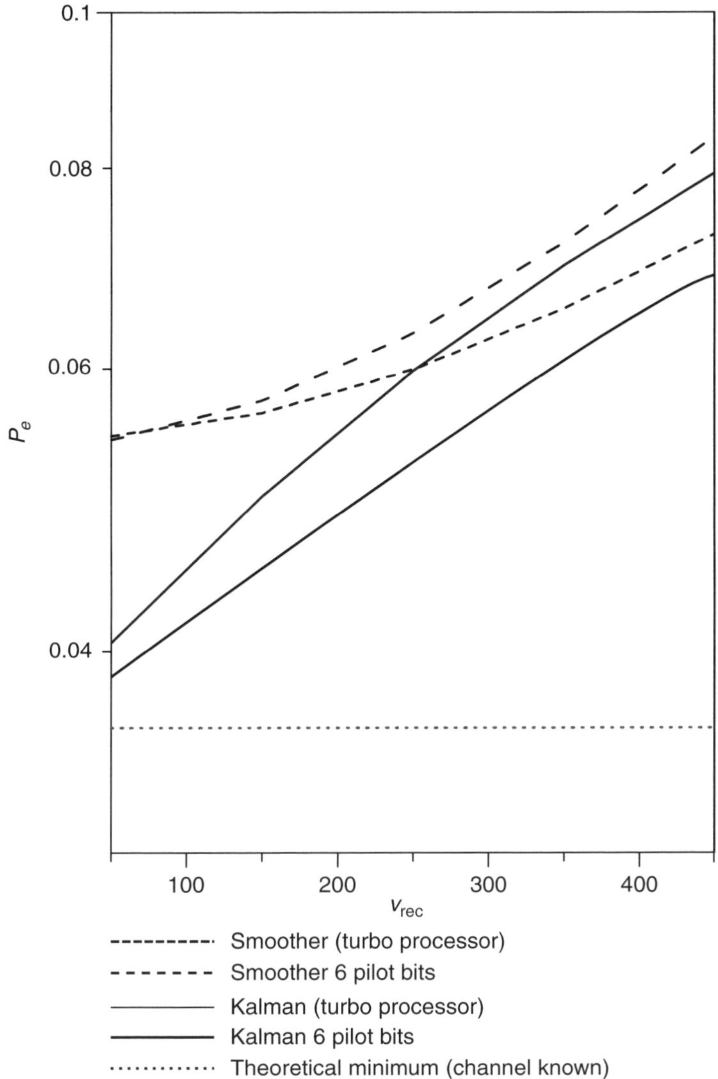

Figure 4.18 Error probability as a function of receiver speed.

$SNR = -1\,\mathrm{dB}$

Sample preselection/modification function: *Substitution*
Fading model: *AR model with equal power per path*
Decision variable $\sum \Delta C_k^{(l)} = |\tilde{C}_{k+1}^{(l)} - \tilde{C}_k^{(l)}| - |\tilde{C}_{k+1}^{(l)} + \tilde{C}_k^{(l)}|$

Joint estimation

$L = 3$ – The number of RAKE fingers
$l = 3$ – The number of fading paths
$N_s = 7$

APPENDIX: LINEAR AND MATRIX ALGEBRA

Definitions

Consider an $m \times n$ matrix $\mathbf{R}$ with elements r_{ij}, $i = 1, 2, \ldots, m$; $j = 1, 2, \ldots, n$. A shorthand notation for describing $\mathbf{R}$ is

$$[\mathbf{R}]_{ij} = r_{ij}$$

The transpose of $\mathbf{R}$, which is denoted by $\mathbf{R}^{\mathrm{T}}$, is defined as the $n \times m$ matrix with elements $\mathbf{r}_{ji}$ or

$$[\mathbf{R}^{\mathrm{T}}]_{ij} = r_{ji}$$

A square matrix is one for which $m = n$. A square matrix is symmetric if $\mathbf{R}^{\mathrm{T}} = \mathbf{R}$. The rank of a matrix is the number of linearly independent rows or columns, whichever is less. The inverse of a square $n \times n$ matrix is the square $n \times n$ matrix $\mathbf{R}^{-1}$ for which

$$\mathbf{R}^{-1}\mathbf{R} = \mathbf{R}\mathbf{R}^{-1} = \mathbf{I}$$

where $\mathbf{I}$ is the $n \times n$ identity matrix. The inverse will exist if and only if the rank of $\mathbf{R}$ is n. If the inverse does not exist, then $\mathbf{R}$ is singular. The determinant of a square $n \times n$ matrix is denoted by $\det(\mathbf{R})$. It is computed as

$$\det(\mathbf{R}) = \sum_{j=1}^{n} r_{ij} C_{ij}$$

where

$$C_{ij} = (-1)^{i+j} M_{ij}$$

M_{ij} is the determinant of the submatrix of $\mathbf{R}$ obtained by deleting the ith row and jth column and is termed the minor of $\mathbf{r}_{ij}$. C_{ij} is the cofactor of $\mathbf{r}_{ij}$. Note that any choice of i for $i = 1, 2, \ldots, n$ will yield the same value for $\det(\mathbf{R})$. A quadratic form Q is defined as

$$Q = \sum_{i=1}^{n} \sum_{j=1}^{n} r_{ij} x_i x_j$$

In defining the quadratic form it is assumed that $r_{ji} = r_{ij}$. This entails no loss in generality since any quadratic function may be expressed in this manner. Q may also be expressed as

$$Q = \mathbf{x}^{\mathrm{T}}\mathbf{R}\mathbf{x}$$

where $\mathbf{x} = [x_1 x_2 \ldots x_n]^{\mathrm{T}}$ and $\mathbf{R}$ is a square $n \times n$ matrix with $r_{ji} = \mathrm{r}_{ij}$ or $\mathbf{R}$ is a symmetric matrix.

A square $n \times n$ matrix $\mathbf{R}$ is positive-semidefinite if $\mathbf{R}$ is symmetric and

$$\mathbf{x}^{\mathrm{T}}\mathbf{R}\mathbf{x} \geq 0$$

for all $\mathbf{x} \neq \mathbf{0}$. If the quadratic form is strictly positive, then $\mathbf{R}$ is positive-definite. When referring to a matrix as positive-definite or positive-semidefinite, it is always assumed that the matrix is symmetric. The trace of a square $n \times n$ matrix is the sum of its diagonal elements or

$$\mathrm{tr}(\mathbf{R}) = \sum_{i=1}^{n} r_{ii}$$

A partitioned $m \times n$ matrix $\mathbf{R}$ is one that is expressed in terms of its submatrices. An example is the 2×2 partitioning

$$\mathbf{R} = \begin{bmatrix} \mathbf{R}_{11} & \mathbf{R}_{12} \\ \mathbf{R}_{21} & \mathbf{R}_{22} \end{bmatrix}$$

Each 'element' $\mathbf{R}_{ij}$ is a submatrix of $\mathbf{R}$. The dimensions of the partitions are given as

$$\begin{bmatrix} k \times l & k \times (n-l) \\ (m-k) \times l & (m-k) \times (n-l) \end{bmatrix}$$

Special matrices

A diagonal matrix is a square $n \times n$ matrix with $r_{ij} = 0$ for $i \neq j$ or all elements off the principal diagonal are zero. A diagonal matrix appears as

$$\mathbf{R} = \begin{bmatrix} r_{11} & 0 & \cdots & 0 \\ 0 & r_{22} & \cdots & 0 \\ \vdots & \vdots & \ddots & \vdots \\ 0 & 0 & \cdots & r_{nn} \end{bmatrix}$$

A diagonal matrix will sometimes be denoted by diag $(r_{11}, r_{22}, \ldots, r_{nn})$. The inverse of a diagonal matrix is found by simply inverting each element on the principal diagonal. A generalization of the diagonal matrix is the square $n \times n$ block diagonal matrix

$$\boldsymbol{R} = \begin{bmatrix} \boldsymbol{R}_{11} & \mathbf{0} & \ldots\ldots\ldots\ldots & \mathbf{0} \\ \mathbf{0} & \boldsymbol{R}_{22} & \ldots\ldots\ldots\ldots & \mathbf{0} \\ \cdot & & & \\ \mathbf{0} & \mathbf{0} & \ldots\ldots\ldots\ldots\ldots & \boldsymbol{R}_{kk} \end{bmatrix}$$

in which all submatrices $\boldsymbol{R}_{ii}$ are square and the other submatrices are identically zero. The dimensions of the submatrices need not be identical. For instance, if $k = 2$, $\boldsymbol{R}_{11}$ might have dimension 2×2, while $\boldsymbol{R}_{22}$ might be a scalar. If all $\boldsymbol{R}_{ii}$ are nonsingular, then the inverse is easily found as

$$\boldsymbol{R}^{-1} = \begin{bmatrix} \boldsymbol{R}_{11}^{-1} & \mathbf{0} & \ldots\ldots\ldots & \mathbf{0} \\ \mathbf{0} & \boldsymbol{R}_{22}^{-1} & \ldots\ldots\ldots & \mathbf{0} \\ \cdot & & & \\ \mathbf{0} & \mathbf{0} & \ldots\ldots\ldots\ldots & \boldsymbol{R}_{kk}^{-1} \end{bmatrix}$$

Also, the determinant is

$$\det(\mathbf{R}) = \prod_{i=1}^{n} \det(\mathbf{R}_{ii})$$

A square $n \times n$ matrix is orthogonal if

$$\mathbf{R}^{-1} = \mathbf{R}^{\mathrm{T}}$$

For a matrix to be orthogonal, the columns (and rows) must be orthonormal or if

$$\mathbf{R} = [\mathbf{r}_1 \quad \mathbf{r}_2 \quad \cdots \quad \mathbf{r}_n]$$

where $\mathbf{r}_i$ denotes the ith column, the conditions

$$\mathbf{r}_i^T \mathbf{r}_j = \begin{array}{ll} 0 & \text{for } i \neq j \\ 1 & \text{for } i = j \end{array}$$

must be satisfied.

An idempotent matrix is a square $n \times n$ matrix that satisfies

$$\mathbf{R}^2 = \mathbf{R}$$

This condition implies that $\mathbf{R}^l = \mathbf{R}$ for $l \geq 1$. An example is the projection matrix

$$\mathbf{R} = \mathbf{H}(\mathbf{H}^{\mathrm{T}}\mathbf{H})^{-1}\mathbf{H}^{\mathrm{T}}$$

where $\mathbf{H}$ is an $m \times n$ full rank matrix with $m > n$.

A square $n \times n$ Toeplitz matrix is defined as

$$[\boldsymbol{R}]_{ij} = r_{i-j}$$

or

$$\boldsymbol{R} = \begin{bmatrix} r_0 & r_{-1} & r_{-2} & \cdots & r_{-(n-1)} \\ r_1 & r_0 & r_{-1} & \cdots & r_{-(n-2)} \\ \vdots & \vdots & \vdots & \vdots & \vdots \\ r_{n-1} & r_{n-2} & r_{n-3} & \cdots & r_0 \end{bmatrix}$$

Each element along a northwest–southeast diagonal is the same. If in addition, $r_{-k} = r_k$, then $\mathbf{R}$ is symmetric to Toeplitz.

Matrix manipulation and formulas

Some useful formulas for the algebraic manipulation of matrices are summarized in this section. For $n \times n$ matrices $\mathbf{R}$ and $\mathbf{P}$, the following relationships are useful.

$$(\mathbf{RP})^{\mathrm{T}} = \mathbf{P}^{\mathrm{T}}\mathbf{R}^{\mathrm{T}}$$

$$(\mathbf{R}^{\mathrm{T}})^{-1} = (\mathbf{R}^{-1})^{\mathrm{T}}$$

$$(\mathbf{RP})'^{-1} = \mathbf{P}^{-1}\mathbf{R}^{-1}$$

$$\det(\mathbf{R}^{\mathrm{T}}) = \det(\mathbf{R})$$

$$\det(c\mathbf{R}) = c^{n}\det(\mathbf{R}) \quad (c \text{ a scalar})$$

$$\det(\mathbf{RP}) = \det(\mathbf{R})\det(\mathbf{P})$$

$$\det(\mathbf{R}^{-1}) = \frac{1}{\det(\mathbf{R})}$$

$$\mathrm{tr}(\mathbf{RP}) = \mathrm{tr}(\mathbf{PR})$$

$$\mathrm{tr}(\mathbf{R}^{\mathrm{T}}\mathbf{P}) = \sum_{i=1}^{n}\sum_{j=1}^{n}[\mathbf{R}]_{ij}[\mathbf{P}]_{ij}$$

For vectors $\mathbf{x}$ and $\mathbf{y}$, we have

$$\mathbf{y}^{\mathrm{T}}\mathbf{x} = \mathrm{tr}(\mathbf{xy}^{\mathrm{T}})$$

It is frequently necessary to determine the inverse of a matrix analytically. To do so one can make use of the following formula:

The inverse of a square $n \times n$ matrix is

$$\mathbf{R}^{-1} = \frac{\mathbf{C}^{\mathrm{T}}}{\det(\mathbf{R})}$$

where $\mathbf{C}$ is the square $n \times n$ matrix of cofactors $\mathbf{R}$. The cofactor matrix is defined by

$$[\mathbf{C}]_{ij} = (-1)^{i+j}M_{ij}$$

where M_{ij} is the minor of r_{ij} obtained by deleting the ith row and jth column of $\mathbf{R}$.

Another formula that is quite useful is the matrix inversion lemma

$$(\mathbf{R} + \mathbf{PCD})^{-1} = \mathbf{R}^{-1} - \mathbf{R}^{-1}\mathbf{P}(\mathbf{DR}^{-1}\mathbf{P} + \mathbf{C}^{-1})^{-1}\mathbf{DR}^{-1}$$

where it is assumed that $\mathbf{R}$ is $n \times n$, $\mathbf{P}$ is $n \times m$, $\mathbf{C}$ is $m \times m$, and $\mathbf{D}$ is $m \times n$ and that the indicated inverses exist. A special case known as Woodbury's identity results for $\mathbf{P}$ and $n \times 1$ column vector $\mathbf{u}$, $\mathbf{C}$ a scalar of unity and $\mathbf{D}$ a $1 \times n$ row vector $\mathbf{u}^{\mathrm{T}}$. Then

$$(\mathbf{R} + \mathbf{uu}^{\mathrm{T}})^{-1} = \mathbf{R}^{-1} - \frac{\mathbf{R}^{-1}\mathbf{uu}^{\mathrm{T}}\mathbf{R}^{-1}}{1 + \mathbf{u}^{\mathrm{T}}\mathbf{R}^{-1}\mathbf{u}}$$

Partitioned matrices may be manipulated according to the usual rules of matrix algebra by considering each submatrix as an element. For multiplication of partitioned matrices, the submatrices that are multiplied together must be conformable. As an illustration, for 2×2 partitioned matrices

$$RP = \begin{bmatrix} R_{11} & R_{12} \\ R_{21} & R_{22} \end{bmatrix} \begin{bmatrix} P_{11} & P_{12} \\ P_{21} & P_{22} \end{bmatrix}$$

$$= \begin{bmatrix} R_{11}P_{11} + R_{12}P_{21} & R_{11}P_{12} + R_{12}P_{22} \\ R_{21}P_{11} + R_{22}P_{21} & R_{21}P_{12} + R_{22}P_{22} \end{bmatrix}$$

The transposition of a partitioned matrix is formed by transposing the submatrices of the matrix and applying T to each submatrix. For a 2×2 partitioned matrix,

$$\begin{bmatrix} R_{11} & R_{12} \\ R_{21} & R_{22} \end{bmatrix}^{\mathrm{T}} = \begin{bmatrix} P_{11}^{\mathrm{T}} & P_{21}^{\mathrm{T}} \\ P_{12}^{\mathrm{T}} & P_{22}^{\mathrm{T}} \end{bmatrix}$$

The extension of these properties to arbitrary partitioning is straightforward. Determination of the inverses and determinants of partitioned matrices is facilitated by employing the following formulas. Let $\mathbf{R}$ be a square $n \times n$ matrix partitioned as

$$\mathbf{R} = \begin{bmatrix} \mathbf{R}_{11} & \mathbf{R}_{12} \\ \mathbf{R}_{21} & \mathbf{R}_{22} \end{bmatrix} = \begin{bmatrix} k \times k & k \times (n-k) \\ (n-k) \times k & (n-k) \times (n-k) \end{bmatrix}$$

Then,

$$\mathbf{R}^{-1} = \begin{bmatrix} (\mathbf{R}_{11} - \mathbf{R}_{12}\mathbf{R}_{22}^{-1}\mathbf{R}_{21})^{-1} & -(\mathbf{R}_{11} - \mathbf{R}_{12}\mathbf{R}_{22}^{-1}\mathbf{R}_{21})^{-1}\mathbf{R}_{12}\mathbf{R}_{22}^{-1} \\ -(\mathbf{R}_{22} - \mathbf{R}_{21}\mathbf{R}_{11}^{-1}\mathbf{R}_{12})^{-1}\mathbf{R}_{21}\mathbf{R}_{11}^{-1} & (\mathbf{R}_{22} - \mathbf{R}_{21}\mathbf{R}_{11}^{-1}\mathbf{R}_{12})^{-1} \end{bmatrix}$$

$$\det(\mathbf{R}) = \det(\mathbf{R}_{22}) \det(\mathbf{R}_{11} - \mathbf{R}_{12}\mathbf{R}_{22}^{-1}\mathbf{R}_{21})$$

$$= \det(\mathbf{R}_{11}) \det(\mathbf{R}_{22} - \mathbf{R}_{21}\mathbf{R}_{11}^{-1}\mathbf{R}_{12})$$

where the inverses of $\mathbf{R}_{11}$ and $\mathbf{R}_{22}$ are assumed to exist.

Theorems

Some important theorems are summarized in this section.

1. A square $n \times n$ matrix $\mathbf{R}$ is invertible (nonsingular) if and only if its columns (or rows) are linearly independent or, equivalently, if its determinant is nonzero. In such a case, $\mathbf{R}$ is full rank. Otherwise, it is singular.
2. A square $n \times n$ matrix $\mathbf{R}$ is positive-definite if and only if
   (a) it can be written as

$$\mathbf{R} = \mathbf{C}\mathbf{C}^{\mathrm{T}}$$

   where $\mathbf{C}$ is also $n \times n$ and is full rank and hence invertible, or

(b) the principal minors are all positive. (The ith principal minor is the determinant of the submatrix formed by deleting all rows and columns with an index greater than i.) If $\mathbf{R}$ can be written as in the previous equation, but $\mathbf{C}$ is not full rank or the principal minors are only nonnegative, then $\mathbf{R}$ is positive-semidefinite.

3. If $\mathbf{R}$ is positive-definite, then the inverse exists and may be found from the previous equation as

$$\mathbf{R}^{-1} = (\mathbf{C}^{-1})^{\mathrm{T}}(\mathbf{C}^{-1})$$

4. Let $\mathbf{R}$ be positive-definite. If $\mathbf{P}$ is an $m \times n$ matrix of full rank with $m \leq n$, then $\mathbf{PRP}^{\mathrm{T}}$ is also positive-definite.
5. If $\mathbf{R}$ is positive-definite (positive-semidefinite), then
   (a) the diagonal elements are positive (nonnegative),
   (b) the determinant of $\mathbf{R}$, which is a principal minor, is positive (nonnegative).

Eigendecomposition of matrices

An eigenvector of a square $n \times n$ matrix $\mathbf{R}$ is an $n \times 1$ vector $\mathbf{v}$ satisfying

$$\mathbf{R}\mathbf{v} = \lambda\mathbf{v}$$

for some scalar λ, which may be complex. λ is the eigenvalue of $\mathbf{R}$ corresponding to the eigenvector $\mathbf{v}$. It is assumed that the eigenvector is normalized to have unit length or $\mathbf{v}^{\mathrm{T}}\mathbf{v} = 1$. If $\mathbf{R}$ is symmetric, then one can always find n linearly independent eigenvectors, although they will not, in general be unique. An example is the identity matrix for which any vector is an eigenvector with eigenvalue 1. If $\mathbf{R}$ is symmetric, then the eigenvectors corresponding to distinct eigenvalues are orthonormal or $\mathbf{v}_i^{\mathrm{T}}\mathbf{v}_j = \delta_{ij}$ and the eigenvalues are real. If, furthermore, the matrix is positive-definite (positive-semidefinite), then the eigenvalues are positive (nonnegative). For a positive-semidefinite matrix, the rank is equal to the number of nonzero eigenvalues.

The defining previous relation can also be written as

$$\mathbf{R}[\mathbf{v}_1\, \mathbf{v}_2 \ldots \mathbf{v}_n] = [\lambda_1\mathbf{v}_1\, \lambda_2\mathbf{v}_2 \ldots \lambda_n\mathbf{v}_n]$$

or

$$\mathbf{R}\mathbf{V} = \mathbf{V}\Lambda$$

where

$$\mathbf{V} = [\mathbf{v}_1\mathbf{v}_2 \ldots \mathbf{v}_n]$$

$$\Lambda = \mathrm{diag}(\lambda_1, \lambda_2, \ldots, \lambda_n)$$

If $\mathbf{R}$ is symmetric so that the eigenvectors corresponding to distinct eigenvalues are orthonormal and the remaining eigenvectors are chosen to yield an orthonormal eigenvector set, then $\mathbf{V}$ is an orthonormal matrix. As such, its inverse is $\mathbf{V}^{\mathrm{T}}$, so that the previous equation becomes

$$\mathbf{R} = \mathbf{V}\Lambda\mathbf{V}^{\mathrm{T}}$$
$$= \sum_{i=1}^{n} \lambda_i \mathbf{v}_i \mathbf{v}_i^{\mathrm{T}}$$

Also, the inverse is easily determined as

$$\mathbf{R}^{-1} = \mathbf{V}^{\mathrm{T}-1}\Lambda^{-1}\mathbf{V}^{-1}$$
$$= \mathbf{V}\Lambda^{-1}\mathbf{V}^{\mathrm{T}}$$
$$= \sum_{i=1}^{n} \frac{1}{\lambda_i} \mathbf{v}_i \mathbf{v}_i^{\mathrm{T}}$$

A final useful relationship follows as

$$\det(\mathbf{R}) = \det(\mathbf{V})\det(\Lambda)\det(\mathbf{V}^{-1})$$
$$= \det(\Lambda)$$
$$= \prod_{i=1}^{n} \lambda_i$$

REFERENCES

1. Glisic, S. and Vucetic, B. (1997) *Spread Spectrum CDMA Systems for Wireless Communications*. Artech House, London.
2. Sheen, W. and Tai, C. (1998) A noncoherent tracking loop with diversity and multipath interference cancellation for direct-sequence spread-spectrum systems. *IEEE Trans. Commun.*, **46**(11), 1516–1524.
3. Sheen, J. W. and Stüber, G. (1994) Effects of multipath fading on delay locked loops for spread spectrum systems. *IEEE Trans. Commun.*, **42**(2/3/4), 1947–1956.
4. Liu, Y. and Blostein, S. (1995) Identification of frequency nonselective fading channels using decision feedback and adaptive linear prediction. *IEEE Trans. Commun.*, **43**(2), 1484–1492.
5. Iltis, R. (1994) An EKF-based joint estimator for interference, multipath, and code delay in a DS spread-spectrum receiver. *IEEE Trans. Commun.*, **42**, 1288–1299.
6. Kay, S. (1993) *Fundamentals of Statistical Signal Processing-Estimation Theory*. New York: Prentice Hall.
7. Howard, S. and Pahlavan, K. (1992) Autoregressive modeling of wide band indoor radio propagation. *IEEE Trans. Commun.*, **40**(9), 1540–1552.
8. Bensley, J. S. and Aazhang, B. (1996) Subspace-based channel estimation for code division multiple access communications & systems. *IEEE Trans. Commun.*, **44**(8), 1009–1020.
9. Muirhead, R. (1982) *Aspects of Multivariate Statistical Theory*. New York: John Wiley & Sons.
10. Baltersee, J. *et al.* (2000) Performance analysis of phasor estimation algorithms for a FDD/ UMTS RAKE receiver. *ISSSTA 2000*, September 6–8, 2000, pp. 476–480.
11. Proakis, J. (1998) *Digital Communications*. New York: McGraw-Hill.
12. Xie, Z., Rushforth, C., Short, R. and Moon, T. (1993) Joint signal detection and parameter estimation in multiuser communications. *IEEE Trans. Commun.*, **41**, 1208–1216.

13. Aazhang, B., Paris, B. and Orsak, G. (1992) Neural networks for multiuser detection in code-division multiple-access communications. *IEEE Trans. Commun.*, **40**, 1212–1222.
14. Iltis, R. A. and Mailaender, L. (1994) An adaptive multiuser detector with joint amplitude and delay estimation. *IEEE J. Select. Areas Commun.*, **12**(5), 774–785.
15. Iltis, R. (1990) Joint estimation of PN code delay and multipath using the extended Kalman filter. *IEEE Trans. Commun.*, **38**, 1677–1685.
16. Iltis, R. and Fuxjaeger, A. (1991) A digital DS spread-spectrum receiver with joint channel and Doppler shift estimation. *IEEE Trans. Commun.*, **39**, 1255–1267.
17. Juntti, M. and Glisic, S. (1997) Advanced CDMA for wireless communications, in Glisic, S. G. and Leppänen, P. A. (eds) *Wireless Communications: TDMA Versus CDMA*. Dordrecht, The Netherlands: Kluwer, pp. 447–490.
18. Mämmelä, A. and Kaasila, V. P. (1997) Smoothing and interpolation in a pilot symbol assisted diversity system. *Int. J. Wireless Inform. Networks*, **4**(3), 205–214.
19. Haeb, R. and Meyr, H. (1989) Systematic approach to carrier recovery and detection of digitally phase modulated signals on fading channels. *IEEE Trans. Commun.*, **37**(7), 748–754.
20. Clark, A. P. and Harun, R. (1986) Assessment of Kalman-filter channel estimators for an HF radio link. *IEE Proc.*, **133**, 513–521.
21. Aghamohammadi, A. and Meyr, H. (1991) A new method for phase synchronization and automatic gain control of linearly-modulated signals on frequency-flat fading channels. *IEEE Trans. Commun.*, **39**, 25–29.
22. Lo, N. W. K., Falconer, D. D. and Sheikh, A. U. H. (1991) Adaptive equalization and diversely combining for mobile radio using interpolated channel estimates. *IEEE Trans. Veh. Technol.*, **40**(3), 636–645.
23. Fechtel, S. and Meyr, H. (1994) Optimal parametric feedforward estimation of frequency selective fading radio channels. *IEEE Trans. Commun.*, **42**(2/3/4), 1639–1650.
24. Cavers, J. K. (1991) An analysis of pilot symbol assisted modulation for Rayleigh fading channels. *IEEE Trans. Veh. Technol.*, **40**(4), 686–693.
25. Gooch, R. P. and Harp, J. C. (1988) Blind channel identification using the constant modulus adaptive algorithm. *Proc. 1988 IEEE International Conference of Communication*, Philadelphia, PA, June 12–15, 1988, pp. 75–79.
26. Hatzinakos, D. and Nikias, C. L. (1989) Estimation of multipath channel response in frequency selective channels. *IEEE J. Select. Areas Commun.*, **SAC-7**, 12–19.
27. Shalvi, O. and Weinstein, E. (1990) New criteria for blind deconvolation of nonminimum phase systems (channels). *IEEE Trans. Inform. Theory*, **IT-36**, 312–321.
28. Tugnait, J. (1994) Blind estimation of digital communication channel impulse response. *IEEE Trans. Commun.* **COM-42**, 1606–1616.

5

Modulation and demodulation

5.1 MAXIMUM LIKELIHOOD ESTIMATION

We start again with the ML principle defined in Section 3.1 of Chapter 3. After the signal despreading, vector of parameters $\boldsymbol{\theta}$ to be estimated includes timing of the received symbols τ_0, phase of the received carrier θ_0, frequency offset of the received signal ν_0, amplitude of the signal A_0 and data symbols a_n

$$\boldsymbol{\theta}(\tau_0, \theta_0, \nu_0, A_0, a_n) \tag{5.1}$$

After despreading, the narrowband signal can be represented as

$$r(t) = s(t, \boldsymbol{\theta}) + w(t) \tag{5.2}$$

The likelihood becomes

$$L(\tilde{\boldsymbol{\theta}}) = C_1 \exp\left[-\frac{C_2}{N_0} \int_{T_0} |(r(t) - s(t, \tilde{\boldsymbol{\theta}})|^2 \mathrm{d}t\right] \tag{5.3}$$

In the sequel, we will use a linear-modulated complex-signal format given by

$$s(t, \tilde{\boldsymbol{\theta}}) = A_0 \exp(j\tilde{\theta}_0) \sum \lfloor \tilde{a}_0 h(t - nT - \tilde{\tau}_0) + j\tilde{b}_n h(t - nT - \varepsilon T - \tilde{\tau}_0)\rfloor \tag{5.4}$$

where $h(\ )$ is the pulse shape and for $\varepsilon = 0$ or $1/2$ we have quadrature phase shift keying (QPSK) or offset QPSK (OQPSK) signals, respectively. The likelihood function defined by equation (3.5) now becomes

$$\lambda(\boldsymbol{\theta}) = R(T_0, \tilde{\boldsymbol{\theta}}) = \mathrm{Re}\left[\int_0^{T_0} r(t)s^*(t, \tilde{\boldsymbol{\theta}})\, \mathrm{d}t\right] \tag{5.5}$$

If we define the filters matched to the pulse shape in I and Q channel as

$$p(n, \tilde{\tau}) = \int_{-\infty}^{\infty} r(t)h(t - nT - \tilde{\tau})\, \mathrm{d}t$$

$$q(n, \tilde{\tau}, \varepsilon) = \int_{-\infty}^{\infty} r(t)h(t - nT - \varepsilon T - \tilde{\tau})\, \mathrm{d}t \tag{5.6}$$

then equation (5.5) becomes

$$R(N, \tilde{\boldsymbol{\theta}}) = \mathrm{Re}\left[\exp(-j\tilde{\theta}) \sum_{n=1}^{N} \tilde{a}_n p(n, \tilde{\tau})\right] + \mathrm{Im}\left[\exp(-j\tilde{\theta}) \sum_{n=1}^{N} \tilde{b}_n q(n, \tilde{\tau}, \varepsilon)\right] \quad (5.7)$$

In the special, important case of nonstaggered signals ($\varepsilon = 0$), we find $q = p$. If we define $c_n = a_n + jb_n$, the correlation integral becomes

$$R(N, \tilde{\boldsymbol{\theta}}) = \mathrm{Re}\left[\exp(-j\tilde{\theta}) \sum_{n=1}^{N} \tilde{c}_n^* p(n, \tilde{\tau})\right] \quad (5.8)$$

5.1.1 Phase and frequency correction: phase rotations and NCOs

For a given phase error $\theta(n)$, the complex signal sample (sampling index n) $z_{\mathrm{in}}(n)$ is corrected by multiplying the sample by a complex correlation factor $\exp(j\theta(n))$ as follows:

$$z_{\mathrm{in}}(n) = x_{\mathrm{in}}(n) + j y_{\mathrm{in}}(n)$$
$$z_0(n) = z_{\mathrm{in}}(n) \times \exp(j\theta(n)) \quad (5.9)$$

By using $\exp(j\theta) = \cos\theta + j\sin\theta$, we get

$$z_0 = x_{\mathrm{in}} \cos\theta - y_{\mathrm{in}} \sin\theta + j(x_{\mathrm{in}} \sin\theta + y_{\mathrm{in}} \sin\theta) \quad (5.10)$$

The operation is known as phase rotation and the block diagram for the realization of equation (5.10) is shown in Figure 5.1.

Frequency corrections (translations) can be performed by the same circuitry but now the phase correction will change in time. For the frequency error v, the correction becomes

$$z_0 = z_{\mathrm{in}} \exp(j2\pi nvT_{\mathrm{s}}) \quad (5.11)$$

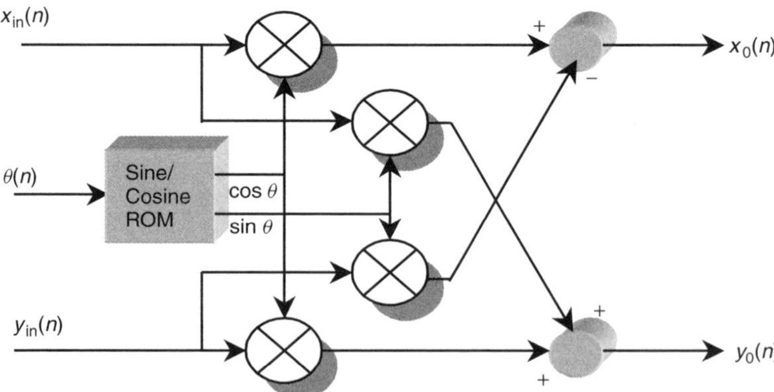

Figure 5.1 Phase rotation.

A simultaneous phase rotation and frequency translation is performed as

$$\exp[j\,(2\pi n v T_{\mathrm{s}} + \theta)] \tag{5.12}$$

In the next section we will focus on the problem of detecting phase and frequency error. The circuit from Figure 5.1 will be used for error corrections, given an error value of phase or frequency.

5.2 FREQUENCY-ERROR DETECTION

We start again with the likelihood function in the following form:

$$L(\tilde{\boldsymbol{\theta}}) = \exp\left[\frac{2C_2}{N_0} \int_{T_0} \mathrm{Re}[r(t)s^*(t,\tilde{\boldsymbol{\theta}})]\,dt\right] \tag{5.13}$$

To emphasize the existence of frequency error, the signal defined by equation (5.4) is rewritten as

$$s^*(t,\tilde{\boldsymbol{\theta}}) = \sum_{n=-\infty}^{\infty} [\tilde{a}_n h(t - nT - \tilde{\tau}) - j\tilde{b}_n h(t - nT - \varepsilon T - \tilde{\tau})]\exp[-j(\tilde{\theta} + 2\pi\tilde{v}t)] \tag{5.14}$$

In this case equation (5.13) becomes

$$L(\tilde{\boldsymbol{\theta}}) \cong \prod_{m=0}^{N-1} \exp\left\{\frac{2C_2}{N_0}\tilde{a}_m\,\mathrm{Re}\left[\int_{-\infty}^{\infty} r(t)h(t - mT - \tilde{\tau})e^{-j(\tilde{\theta}+2\pi\tilde{v}t)}\,dt\right]\right\}$$

$$\times \exp\left\{\frac{2C_2}{N_0}(-\tilde{b}_m)\,\mathrm{Im}\left[\int_{-\infty}^{\infty} r(t)h(t - mT - \varepsilon T - \tilde{\tau})e^{-j(\tilde{\theta}+2\pi\tilde{v}t)}\,dt\right]\right\} \tag{5.15}$$

$$L(\tilde{\boldsymbol{\theta}}) \cong \prod_{m=0}^{N-1} \exp\left\{\frac{2C_2}{N_0}\tilde{a}_m\,\mathrm{Re}[p(m)]\right\} \times \exp\left\{\frac{2C_2}{N_0}(-\tilde{b}_m)\,\mathrm{Im}[q(m)]\right\} \tag{5.16}$$

Joint maximization of equation (5.16) with respect to all the parameters would be rather complex for practical implementation. To remove data from equation (5.16), we use averaging of the function. For M-ary modulation this can be represented as

$$L_{a,b} = \prod_{m=0}^{N-1}\left\{\sum_{i=1}^{m}\frac{1}{M}\exp\left[\frac{2C_2}{N_0}a_i\,\mathrm{Re}[p(m)]\right]\right\}$$

$$\times \left(\sum_{i=1}^{m}\frac{1}{M}\exp\left\{-\frac{2C_2}{N_0}b_i\,\mathrm{Im}[q(m)]\right\}\right) \tag{5.17}$$

In the simple case of binary modulation we have

$$L_{a,b} = \prod_{m=0}^{N-1} \cosh\left\{\frac{2C_2}{N_0} \mathrm{Re}[p(m)]\right\} \cdot \cosh\left\{\frac{2C_2}{N_0} \mathrm{Im}[q(m)]\right\} \qquad (5.18)$$

For nonoffset QPSK modulation, $q(m) = p(m)$ and equation (5.18) becomes

$$L_{a,b} = \prod_{m=0}^{N-1} \cosh\left\{\frac{2C_2}{N_0} \mathrm{Re}[p(m)]\right\} \cdot \cosh\left\{\frac{2C_2}{N_0} \mathrm{Im}[p(m)]\right\} \qquad (5.19)$$

By taking the logarithm of equation (5.19), we have

$$\Lambda_{a,b} \underline{\Delta} \ln[L_{a,b}] = \sum_{m=0}^{N-1} \left(\prod_{m=0}^{N-1} \ln \cosh\left\{\frac{2C_2}{N_0} \mathrm{Re}[p(m)]\right\} + \ln \cosh\left\{\frac{2C_2}{N_0} \mathrm{Im}[p(m)]\right\}\right) \qquad (5.20)$$

The following approximations are used at this point:

$$\ln \cosh(x) \cong \frac{x^2}{2}, \; |x| \ll 1$$
$$\cong |x|, \; |x| \gg 1 \qquad (5.21)$$

For the small value of the argument we have

$$\Lambda_{a,b} \cong C_3 \sum_{m=0}^{N-1} (\{\mathrm{Re}[p(m)]\}^2 + \{\mathrm{Im}[p(m)]\}^2)$$
$$= C_3 \sum_{m=0}^{N-1} |p(m)|^2 \qquad (5.22)$$

Equation (5.22) can be maximized by changing $\tilde{v}$ in $p(m)$ in the open loop search. By taking the derivative of equation (5.20), we get the tracker for the QPSK signal.

$$\frac{\partial \Lambda_{a,b}}{\partial \tilde{v}} = \sum_{m=0}^{N-1} \frac{2C_2}{N_0} \mathrm{Re}[p_v(m)] \tanh\left\{\frac{2C_2}{N_0} \mathrm{Re}[p(m)]\right\}$$
$$+ \sum_{m=0}^{N-1} \frac{2C_2}{N_0} \mathrm{Im}[p_v(m)] \tanh\left\{\frac{2C_2}{N_0} \mathrm{Im}[p(m)]\right\} \qquad (5.23)$$

where $p_v(m) \underline{\Delta} \partial p(m)/\partial \tilde{v}$. A sample of frequency-error detector control signal is

$$u_v(n) = \mathrm{Re}[p_v(n)] \tanh\left\{\frac{2C_2}{N_0} \mathrm{Re}[p(n)]\right\} + \mathrm{Im}[p_v(n)] \tanh\left\{\frac{2C_2}{N_0} \mathrm{Im}[p(n)]\right\} \qquad (5.24)$$

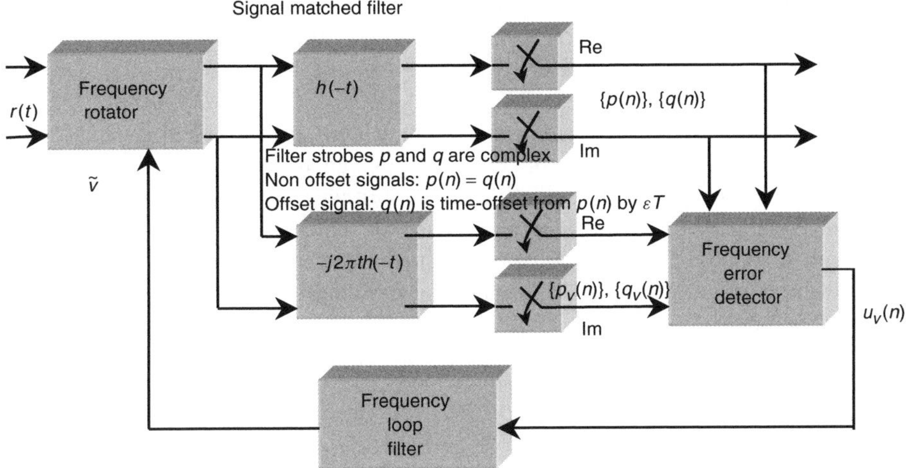

Figure 5.2 ML-derived frequency detector.

and its time average $U_v = E_n[u_v(n)]$ is called the detector characteristic or S-curve. The block diagram realizing equation (5.24) is shown in Figure 5.2.

5.2.1 QPSK tracking algorithm: practical version

Further simplification is obtained if we use

$$\tanh(x) \cong x, |x| \ll 1$$
$$\cong \operatorname{sgn}(x), |x| \gg 1 \tag{5.25}$$

which for nonoffset QPSK results in

$$u_v(n) = \operatorname{Re}[p(n)]\operatorname{Re}[p_v(n)] + \operatorname{Im}[p(n)]\operatorname{Im}[p_v(n)]$$
$$= \operatorname{Re}[p(n)p_v^*(n)] \tag{5.26}$$

5.2.2 Time-domain example: rectangular pulse

After considerable labor, the S-curve defined as $u_v(n) = E_n[U_v(n)]$ for a unit-amplitude rectangular pulse in time domain and random data is found to be [1]

$$U_v(\Delta v, \Delta \tau) = C_c A_0 \left\{ \Delta\tau \frac{\sin \pi \Delta v \Delta\tau}{\pi \Delta v^2} \left(\cos \pi \Delta v \Delta\tau - \frac{\sin \pi \Delta v \Delta\tau}{\pi \Delta v \Delta\tau} \right) \right.$$
$$+ (T - \Delta\tau) \frac{\sin \pi \Delta v(T - \Delta\tau)}{\pi \Delta v^2}$$
$$\left. \times \left[\cos \pi \Delta v(T - \Delta\tau) - \frac{\sin \pi \Delta v(T - \Delta\tau)}{\pi \Delta v(T - \Delta\tau)} \right] \right\} \tag{5.27}$$

For random data the S-curve is shown in Figure 5.3.

If the data pattern is assumed to rotate by 90° from one symbol to the next, that is, $c_{m+1} = jc_m$, the S-curve is as given in Figure 5.4.

For binary phase shift keying (BPSK) dotting signal where $c_{m+1} = -c_m$, the S-curve is as given in Figure 5.5.

One should notice that for random data, S-curve demonstrates regular shape with the slope decreasing with the timing error. Even for nonsynchronized systems with $\Delta\tau/T = 0.5$, the system would operate. For rotating data the impact of timing is larger. For dotting signal, if the timing error becomes large, the S-curve not only has a reduced slope, which is equivalent to reducing the signal-to-noise ratio in the loop, but also changes the sign resulting in the devastating effects of generating the control signal that would cause loss of synchronization.

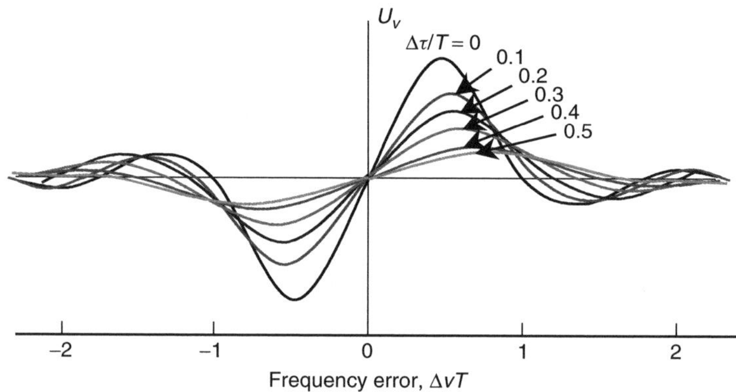

Figure 5.3 Frequency detector S-curves for rectangular pulses (nonoffset signal).

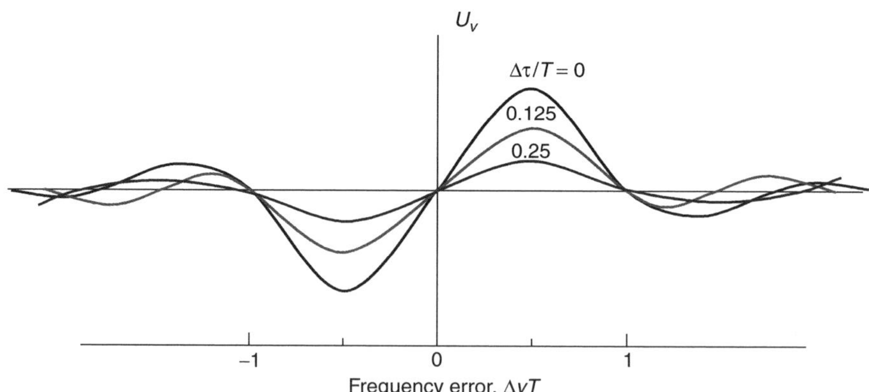

Figure 5.4 S-curve for rotating pattern: square pulses.

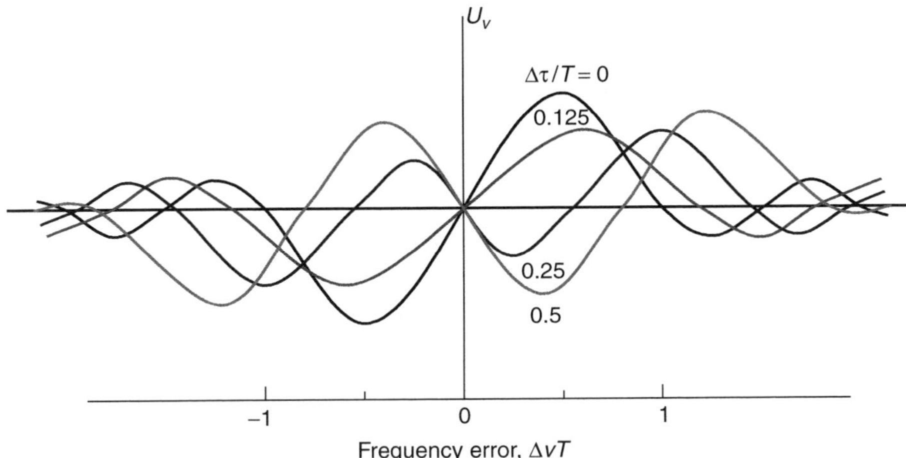

Figure 5.5 S-curve for BPSK dotting pattern; square pulses.

5.3 CARRIER PHASE MEASUREMENT: NONOFFSET SIGNALS

In this case possible solutions will depend very much on a number of parameters. Regarding the signal format, there will be differences for single amplitude [M-ary phase-shift keying (MPSK)] versus multiamplitude [M-ary quadrature amplitude modulation (MQAM)] or offset versus nonoffset signal. Different representations such as rectangular versus polar representation of phase error or parallel versus serial representation of signal (offset only) will result in different solutions. Additional knowledge such as clock timing (clock-aided) or data or decisions [data-aided (DA) or decision-directed (DD)] will also be of great importance. Configurations such as feedforward (FF) versus feedback (FB) will also offer different advantages and drawbacks.

5.3.1 Data-aided (DA) operation

In this case a preamble c_n and timing τ_0 are available and equation (5.8) becomes

$$R(N, \tilde{\theta}) = \mathrm{Re}\left[\exp(-j\tilde{\theta}) \sum_{n=1}^{N} c_n^* p(n) \right] \tag{5.28}$$

At the maximum point, $\partial R/\partial \theta$ vanishes and we have

$$\mathrm{Im}\left[\exp(-j\hat{\theta}) \sum_n c_n^* p(n) \right] = 0 \Leftrightarrow \sum_n \mathrm{Im}[\exp(-j\hat{\theta}) c_n^* p(n)] = 0 \tag{5.29}$$

and we define again the sample of the S-curve as

$$u_\theta(n) = \mathrm{Im}\left[\exp(-j\tilde{\theta})\sum_n c_n^* p(n)\right] \tag{5.30}$$

Under the ideal conditions the matched filter output pulse becomes

$$p(n) = A_0 c_n \exp(j\theta_0) + v(n) \tag{5.31}$$

where $v(n)$ is a sample of noise (assumed zero-mean Gaussian) and A_0 is signal amplitude. Averaging gives the S-curve as

$$U_0 = E\lfloor\mathrm{Im}\{\exp(-j\tilde{\theta})c_n^*[A_0 c_n\exp(j\theta_0) + v(n)]\}\rfloor$$
$$= A_0 E|c_n|^2\,\mathrm{Im}\{\exp[j(\theta_0 - \tilde{\theta})]\} = A_0 E|c_n|^2\sin\Delta\theta \tag{5.32}$$

where $\Delta\theta = \theta_0 - \tilde{\theta}$.

5.3.2 Decision-directed (DD) operation

If preamble is not available, detected data can be used instead, resulting in a DD solution. Implementation of equation (5.30) for such a case is shown in Figure 5.6.

Specific DD algorithm

If we represent the output of the phase rotator $p(n)\mathrm{e}^{-j\tilde{\theta}}$ as the complex sequence $\{x(n), y(n)\}$, the output of the decision algorithm as $\hat{c}_n = \hat{a}_n + j\hat{b}_n = \mathrm{sgn}(x_n) + j\,\mathrm{sgn}(y_n)$

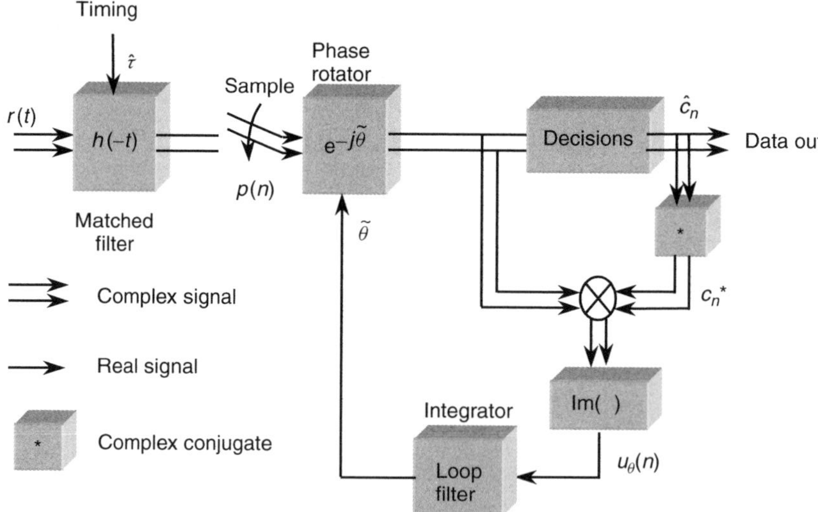

Figure 5.6   Decision-directed carrier tracking.

where $\mathrm{sgn}(v) = +1$ (-1) if v is greater than 0 $(v < 0)$, then equation (5.30) is given by

$$u_n(n) = \mathrm{Im}\lfloor(x_n + jy_n)(\hat{a}_n + j\hat{b}_n)\rfloor = y_n\mathrm{sgn}(x_n) - x_n\mathrm{sgn}(y_n) \tag{5.33}$$

This is known as four-phase hard-limiting Costas detector that is so widely used in QPSK systems.

Rectangular representation

If we use the following steps:

- $\exp(-j\tilde{\theta}) \rightarrow$ a rectangular representation in equation (5.28)

$$\lambda(\tilde{\theta}) = \mathrm{Re}\left[(\cos\tilde{\theta} - j\sin\tilde{\theta})\sum_{n=1}^{N}\hat{c}_n^* p(n)\right] \tag{5.34}$$

differentiate with respect to $\tilde{\theta}$
- bring all the expressions into the summation sign
- take the real part of the derivative
- the ML estimate $\hat{\theta}$ occurs for the value of $\tilde{\theta}$ at which the derivative goes to zero

$$\sum_{n=1}^{N}\sin\hat{\theta}\,\mathrm{Re}[\hat{c}_n^* p(n)] - \sum_{n=1}^{N}\cos\hat{\theta}\,\mathrm{Im}[\hat{c}_n^* p(n)] = 0$$

Solving for the angle gives

$$\hat{\theta}(n) = \arctan\left\{\frac{\displaystyle\sum_{i=n-M}^{n-1}\mathrm{Im}[\hat{c}_i^* p(i)]}{\displaystyle\sum_{i=n-M}^{n-1}\mathrm{Re}[\hat{c}_i^* p(i)]}\right\} \tag{5.35}$$

Implementation of equation (5.35) is shown in Figure 5.7.

5.3.3 Nondecision-aided measurements

Why might one choose to avoid DD measurements? First of all there are some circumstances, such as acquisition intervals or low signal-to-noise ratios, for which data decisions are of poor quality and should not be used. One can show that in BPSK the equivalent signal-to-noise ratio in such systems will be reduced by factor $(1-2P_e)^2$ where P_e is the bit error rate. Omitting the decision operation might reduce equipment complexity. (Not likely to be a good reason in a digital implementation. Indeed, we find digital DD methods are often simpler than non-DD methods.) In this case the likelihood function will be averaged out with respect to data.

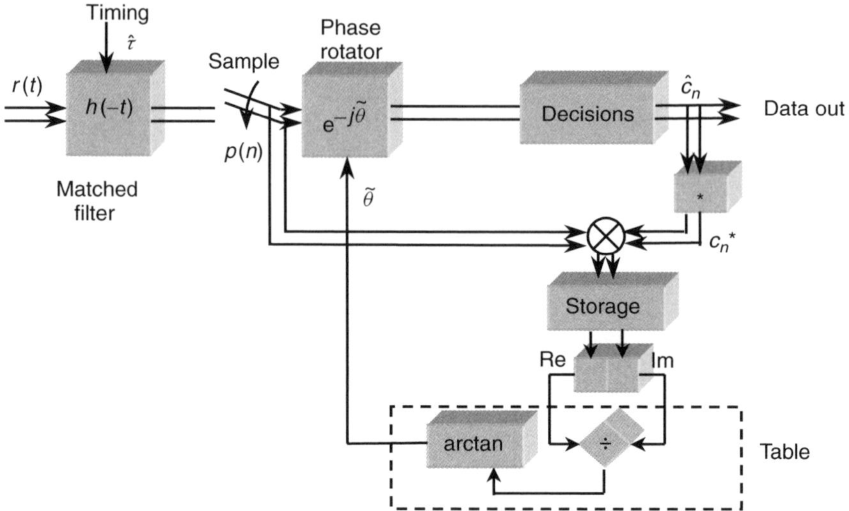

Figure 5.7 DD arctan phase recovery.

The starting point is equation (5.18) in the form

$$L(\tilde{\theta}) = C_3 \exp\left\{\frac{2C_2}{N_0} \mathrm{Re}\left[e^{-j\tilde{\theta}} \sum_{n=1}^{N} c^*(n)p(n)\right]\right\} \Leftrightarrow$$

$$L(\tilde{\theta}) = C_3 \prod_{n=1}^{N} \exp\left\{\frac{2C_2}{N_0} \mathrm{Re}\left[e^{-j\tilde{\theta}} \sum_{n=1}^{N} c^*(n)p(n)\right]\right\} \tag{5.36}$$

Averaging with respect to data results in

$$L_c(\tilde{\theta}) = E_c[L(\tilde{\theta})] \quad L_c(\tilde{\theta}) = C_3 \prod_{n=1}^{N} \xi(n) \tag{5.37}$$

For BPSK, $c(n) = a(n) + j0$, where $a(n) = \pm 1$ and equation (5.37) gives

$$\xi(n) = \cosh\left\{\frac{2C_2}{N_0} \mathrm{Re}[e^{-j\tilde{\theta}}p(n)]\right\} \tag{5.38}$$

The log-averaged likelihood function is

$$\Lambda_c(\tilde{\theta}) = \ln C_3 + \sum_{n=1}^{N} \ln \cosh\left\{\frac{2C_2}{N_0} \mathrm{Re}[e^{-j\tilde{\theta}}p(n)]\right\} \tag{5.39}$$

and its derivative can be represented as

$$\frac{\partial \Lambda_c}{\partial \tilde{\theta}} = \frac{2C_2}{N_0} \sum_N \mathrm{Im}[(e^{-j\tilde{\theta}} p(n)] \tanh\left\{\frac{2C_2}{N_0} \mathrm{Re}[e^{-j\tilde{\theta}} p(n)]\right\} \qquad (5.40)$$

The sample of the S-curve becomes

$$u(n) = \frac{2C_2}{N_0} \mathrm{Im}[e^{-j\tilde{\theta}} p(n)] \tanh\left\{\frac{2C_2}{N_0} \mathrm{Re}[e^{-j\tilde{\theta}} p(n)]\right\}$$

$$= \frac{2C_2}{N_0} y(n) \tanh\left[\frac{2C_2}{N_0} x(n)\right] \qquad (5.41)$$

and its implementation is shown in Figure 5.8.

For a FF operation we start with

$$e^{-j\hat{\theta}} = \cos\hat{\theta} - j \sin\hat{\theta}$$

$$p(n) = p_1(n) + j p_2(n) \qquad (5.42)$$

$$\tanh x \cong x \text{ for small } x$$

and equation (5.41) gives

$$\frac{\partial u}{\partial \theta} \to C_4 \sum_N \left[p_1 p_2 \cos 2\hat{\theta} - \frac{1}{2}(p_1^2 - p_2^2) \sin 2\hat{\theta}\right] = 0 \qquad (5.43)$$

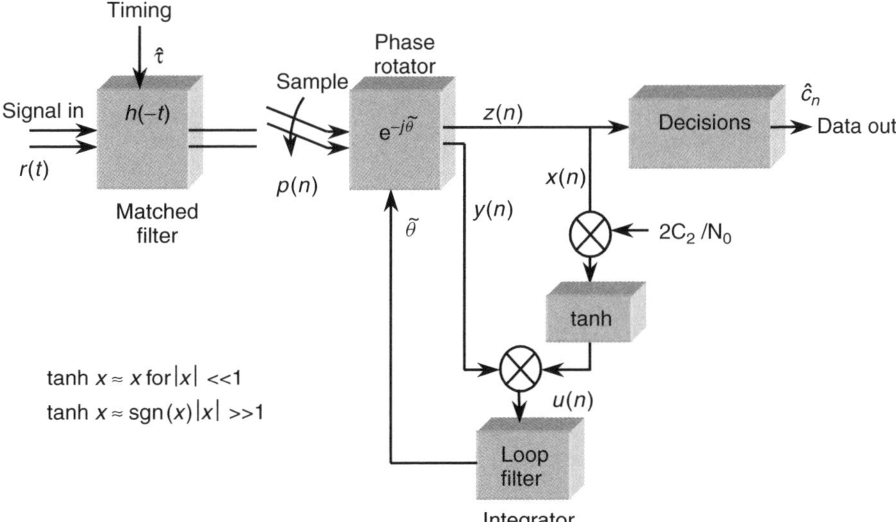

Figure 5.8 NDA BPSK tracker [maximum likelihood estimation (MLE) carrier estimator].

Solving the last line for $\hat{\theta}$ gives the desired result:

$$\hat{\theta} = \frac{1}{2}\arctan\frac{2\sum_{N}p_1 p_2}{\sum_{N}p_1^2 - p_2^2} \tag{5.44}$$

Additional manipulation gives

$$p^2(n) = p_1^2(n) - p_2^2(n) + j2p_1(n)p_2(n) \tag{5.45}$$

$$\hat{\theta} = \frac{1}{2}\arctan\frac{\sum_{N}\text{Im}[p^2(n)]}{\sum_{N}\text{Re}[p^2(n)]} \tag{5.46}$$

Implementation of equation (5.44) is given in Figure 5.9.

5.3.4 QPSK tracker

By following the same reasoning that led to the BPSK tracker, the QPSK phase-error detector becomes

$$u(n) = \frac{C_2\sqrt{2}}{N_0}y(n)\tanh\left[\frac{C_2\sqrt{2}}{N_0}x(n)\right] - \frac{C_2\sqrt{2}}{N_0}x(n)\tanh\left[\frac{C_2\sqrt{2}}{N_0}y(n)\right]$$

$$\cong C_4[y(n)\text{sgn}x(n) - x(n)\text{sgn}y(n)] \tag{5.47}$$

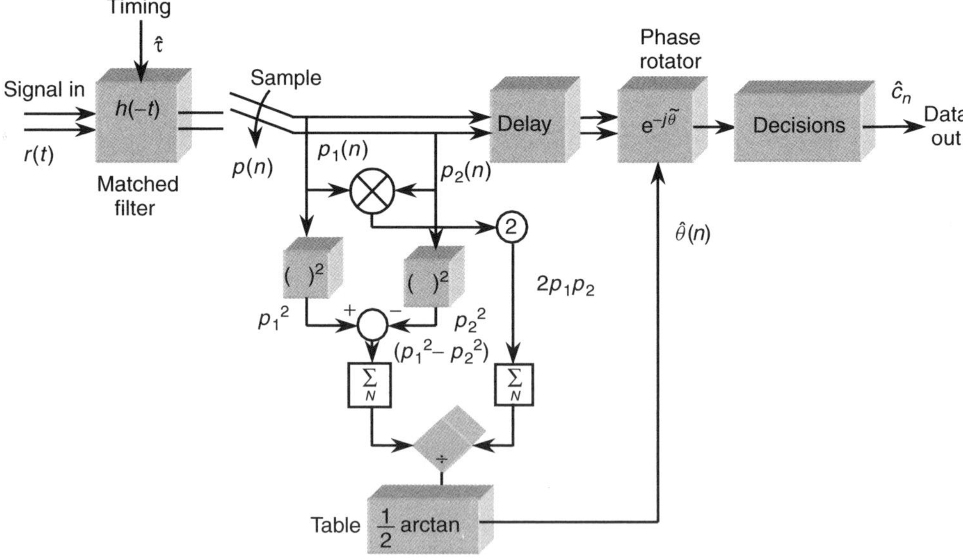

Figure 5.9 Feedforward NDA ML estimator for BPSK phase.

This result is equivalent to the difference between the outputs of two BPSK Costas detectors. If we use the approximation

$$\tanh x \cong x - \frac{x^3}{3} \tag{5.48}$$

$$u(n) \cong C_5[x(n)y^3(n) - y(n)x^3(n)] \tag{5.49}$$

where x and y are the rotated samples of the matched filter

$$p(n)\mathrm{e}^{-j\tilde{\theta}} = x(n) + jy(n)$$

Further manipulation results in

$$u(n) \cong [x(n)y(n)]\lfloor y^2(n) - x^2(n)\rfloor \tag{5.50}$$

5.3.5 QPSK feedforward algorithm

Following the methods introduced for BPSK, we set the derivative of the likelihood function to zero and solve for $\hat{\theta}$. Using the cubic approximation for $\tanh(x)$ leads to the solution

$$\hat{\theta} = \frac{1}{4} \arctan \frac{\sum_N 4p_1 p_2(p_1^2 - p_2^2)}{\sum_N (p_1^2 - p_2^2)^2 - 4p_1^2 p_2^2} \tag{5.51}$$

Equation (5.51) with some manipulations similar to those represented by equations (5.45) and (5.46) can be represented as

$$\hat{\theta} = \frac{1}{4} \arctan \frac{\sum_N \mathrm{Im}[p^4(n)]}{\sum_N \mathrm{Re}[p^4(n)]} \tag{5.52}$$

This algorithm is a digital equivalent of the X4 multiplier that is found often in analog carrier synchronizers.

5.3.6 NDA extension to MPSK

The result from the previous section is extended to the MPSK signal, by Viterbi and Viterbi (V&V) [2] who suggested the following transformation of the output of the matched filter:

$$\gamma(n) = F[\rho(n)] \exp[jM\psi(n)] \tag{5.53}$$

where ρ and ψ are the amplitude and the phase of the polar representation of $p(n)$. As the generalization of equation (5.52) carrier phase is estimated as

$$\hat{\theta} = \frac{1}{M} \arctan \frac{\sum_N \mathrm{Im}[\gamma(n)]}{\sum_N \mathrm{Re}[\gamma(n)]} \tag{5.54}$$

The implementation of equation (5.54) is shown in Figure 5.10.

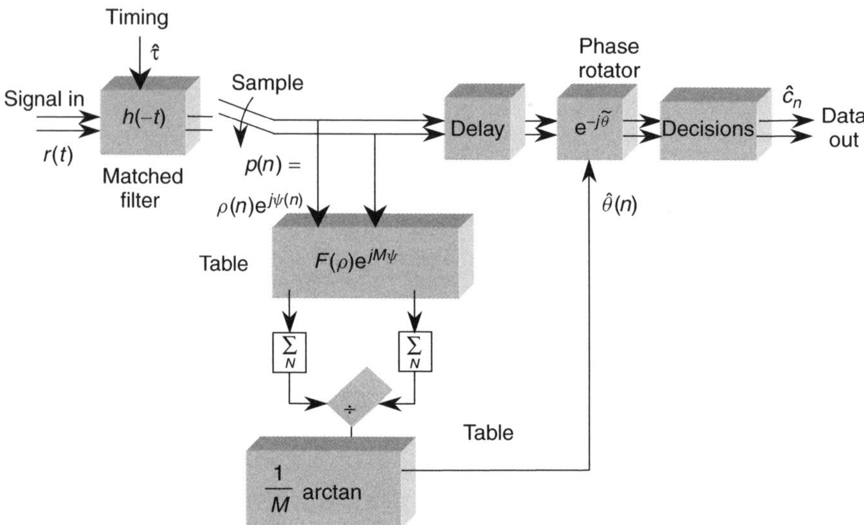

Figure 5.10   Viterbi feedforward NDA estimator for MPSK phase.

5.4 PERFORMANCE OF THE FREQUENCY AND PHASE SYNCHRONIZERS

A uniform representation of the input signal is used by modifying equation (5.14) as follows:

$$S(t, \tilde{\boldsymbol{\theta}}) = \left[\sum_m a_m h_I(t - mT - \tau) + j \sum_m b_m h_Q(t - mT - \tau) \right] e^{j\theta} + n(t) \qquad (5.55)$$

where the pulse shape $h(t)$ is defined as

$$\text{BPSK}: \quad h_I(t) = h(t), h_Q(t) = 0$$

$$S(t, \tilde{\boldsymbol{\theta}}) = \sum_m a_m h(t - mT - \tau)e^{j\theta} + n(t) \qquad (5.56)$$

$$\text{QPSK}: \quad h_I(t) = h_Q(t) = h(t)$$

$$S(t, \tilde{\boldsymbol{\theta}}) = \sum_m (a_m + jb_m)h(t - mT - \tau)e^{j\theta} + n(t) \qquad (5.57)$$

$$\text{OQPSK}: \quad h_I(t) = h(t), h_Q(t) = h(t - T/2)$$

$$S(t, \tilde{\boldsymbol{\theta}}) = \left[\sum_m a_m h(t - mT - \tau) + j \sum_m b_m h(t - mT - T/2 - \tau) \right]$$

$$\times e^{j\theta} + n(t) \qquad (5.58)$$

The output of the pulse-matched filter can be represented as

$$g(t) = \int_{-\infty}^{+\infty} h(t+u)h(u)\,\mathrm{d}u = h(t) * h(-t) \tag{5.59}$$

For

$$g(t) = \frac{\sin(\pi t/T)}{\pi t/T} \frac{\cos(\alpha \pi t/T)}{1-(2\alpha t/T)^2} \tag{5.60}$$

we have

$$G(f) = \begin{cases} T & |f| \le \dfrac{1-\alpha}{2T} \\[2ex] \dfrac{T}{2}\left\{1 - \sin\left[\dfrac{\pi T}{\alpha}\left(|f| - \dfrac{1}{2T}\right)\right]\right\} & \dfrac{1-\alpha}{2T} < |f| \le \dfrac{1+\alpha}{2T} \\[2ex] 0 & \dfrac{1+\alpha}{2T} < |f| \end{cases} \tag{5.61}$$

$$H(t) = \sqrt{G(t)}$$

Pulses $p(m)$ and $g(m)$ used in the previous sections can be represented as

$$p(m : \tilde{\tau}) = [S(t, \tilde{\boldsymbol{\theta}}) * h_I(-t)]_{t=mT+\tilde{\tau}}$$

$$q(m : \tilde{\tau}) = [S(t, \tilde{\boldsymbol{\theta}}) * h_Q(-t)]_{t=mT+\tilde{\tau}}$$

For the three modulation formats considered in this section, we have

- BPSK : $q(m : \tilde{\tau}) = 0$
- QPSK : $q(m : \tilde{\tau}) = p(m : \tilde{\tau})$
- OQPSK : $q(m : \tilde{\tau}) = p(m : \tilde{\tau} + T/2) = p(m + 1/2 : \tilde{\tau})$

The spectra of $h(t)$ is presented in Figure 5.11.

One should be aware that the tracking error variance for the linearized tracking system is proportional to loop noise to signal power ratio $\sigma_\theta^2 \propto B_L N(fT)/\text{slope}^2 = B_L N(fT)/S$. Noise power is proportional to noise density and loop bandwidth and signal power to the square of the slope of the equivalent S-curve. We will also use notation $\text{slope}^2 = S$. These parameters are shown in Figures 5.12 to 5.23.

The noise power spectral density of the decision-directed maximum likelihood (DDML) detector for BPSK and QPSK signals is shown in Figure 5.12 with E_s/N_0 being a parameter. The same results for OQPSK signal is shown in Figure 5.13.

The slope of the phase error discriminator S-curve for different modulations is shown in Figure 5.14.

The normalized noise power spectral density for the DDML scheme is shown in Figures 5.15 and 5.16. To get the tracking error variance results, from Figure 5.15, should be multiplied by the loop bandwidth.

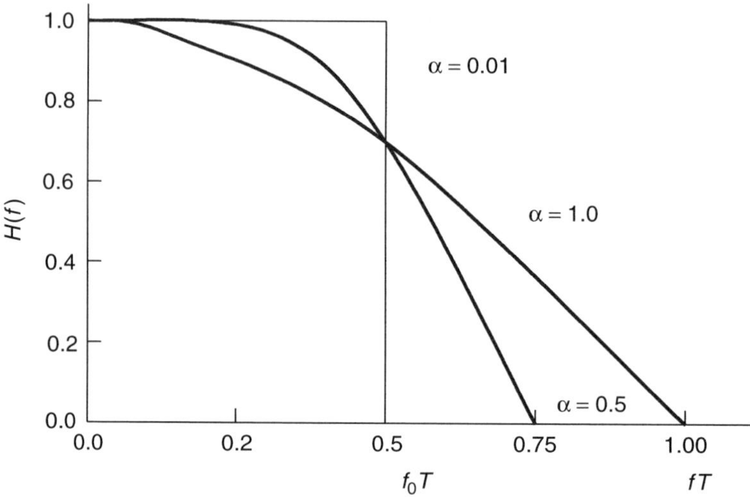

Figure 5.11 Transfer function of root-raised cosine filter with roll-off factor (α) as the parameter.

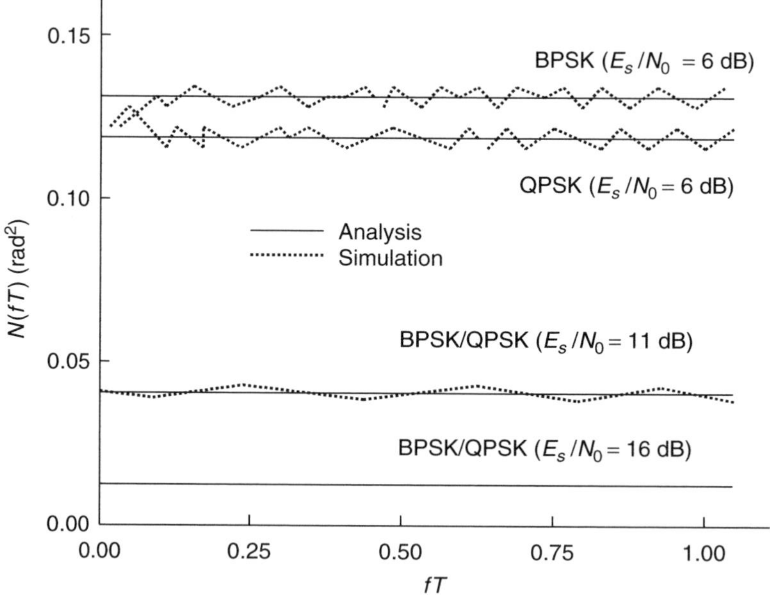

Figure 5.12 Noise power spectral density of DDML for BPSK and QPSK.

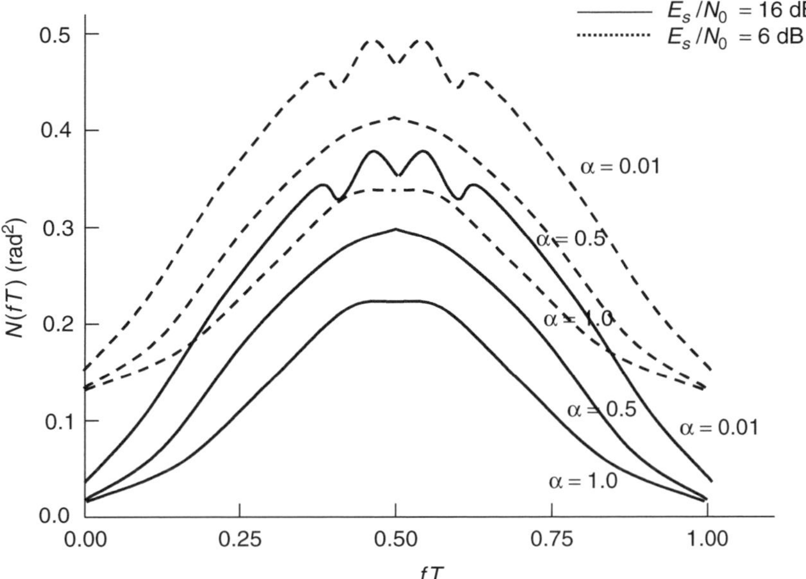

Figure 5.13   Noise power spectral density of DDML for OQPSK.

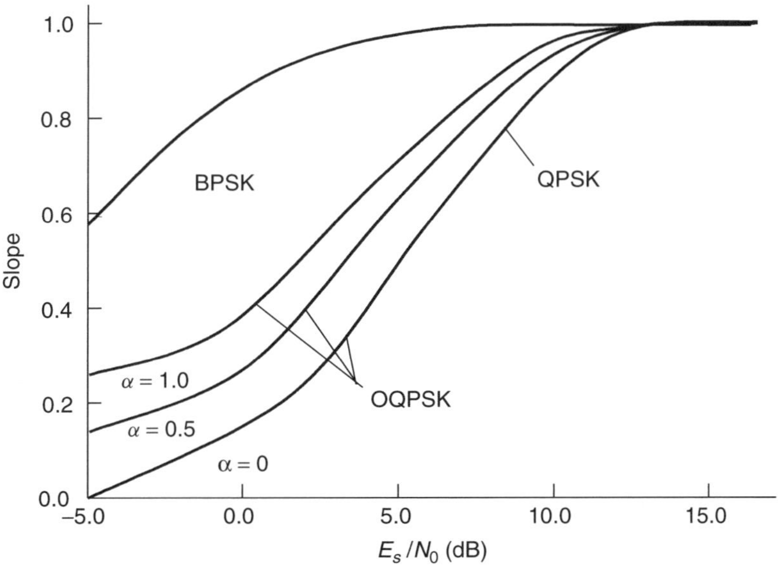

Figure 5.14   Decision-directed maximum likelihood feedback (DDMLFB) phase detector slopes.

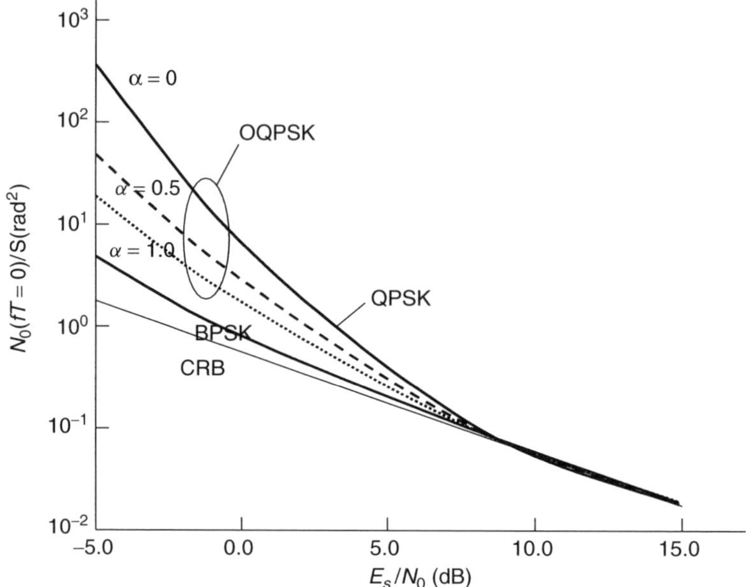

Figure 5.15   Normalized noise power spectral density for DDML, logarithmic scale.

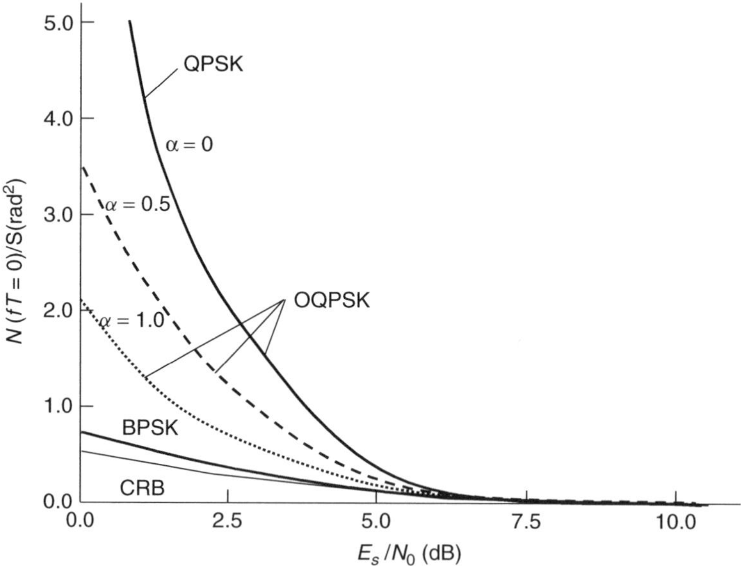

Figure 5.16   Normalized power spectral density $N_0(fT = 0)/\mathrm{Slope}^2$ for DDML, linear scale.

                                                   141

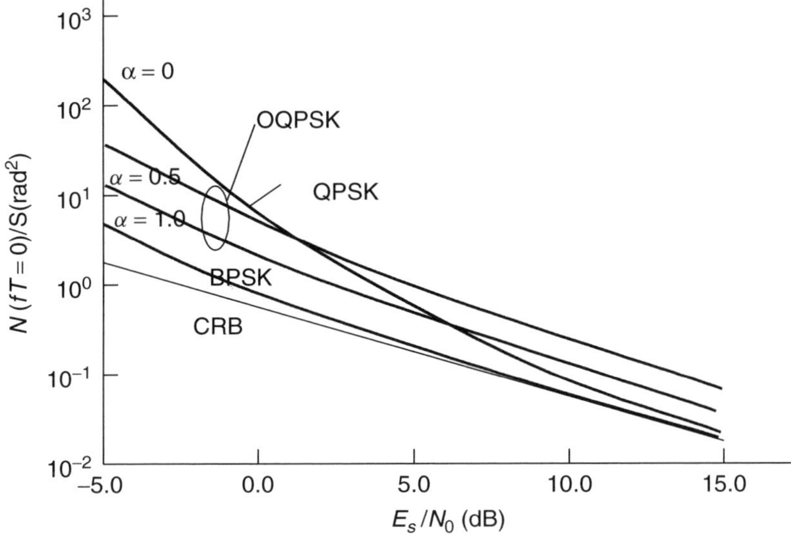

Figure 5.17 Normalized noise power spectral density for the NDAML, logarithmic scale.

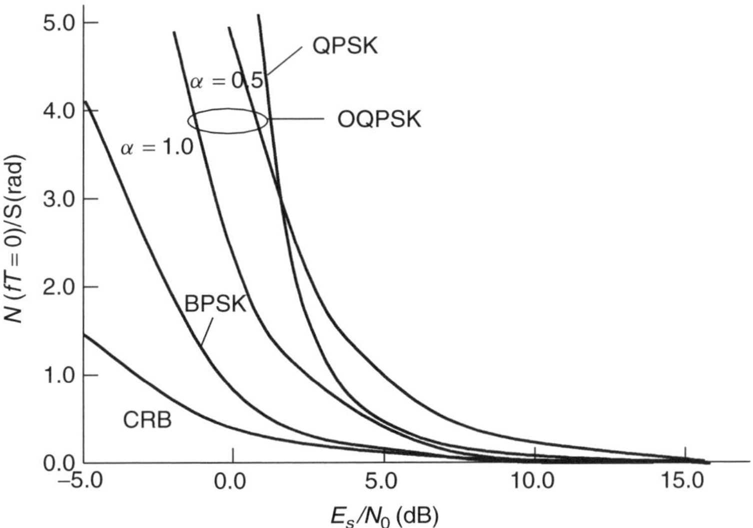

Figure 5.18 Normalized noise power spectral density for the NDAML, linear scale.

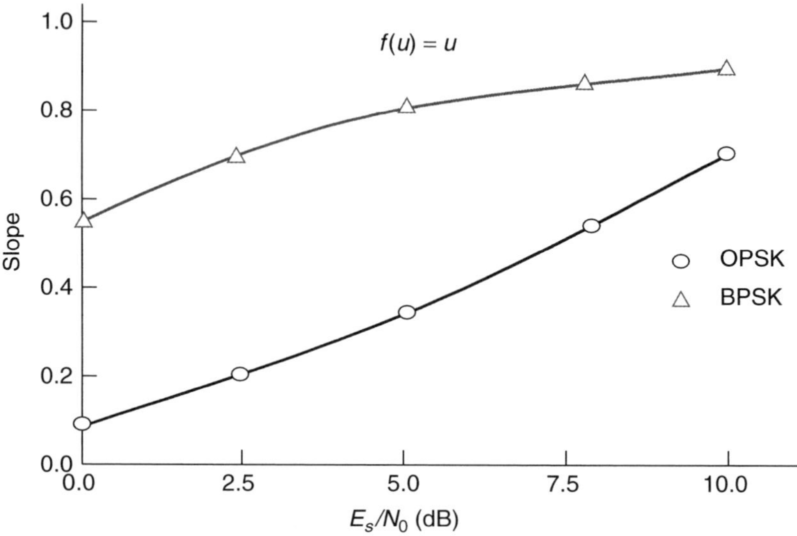

Figure 5.19 V&VFB phase detector slopes for BPSK and QPSK, nonlinearity $f(u) = u$.

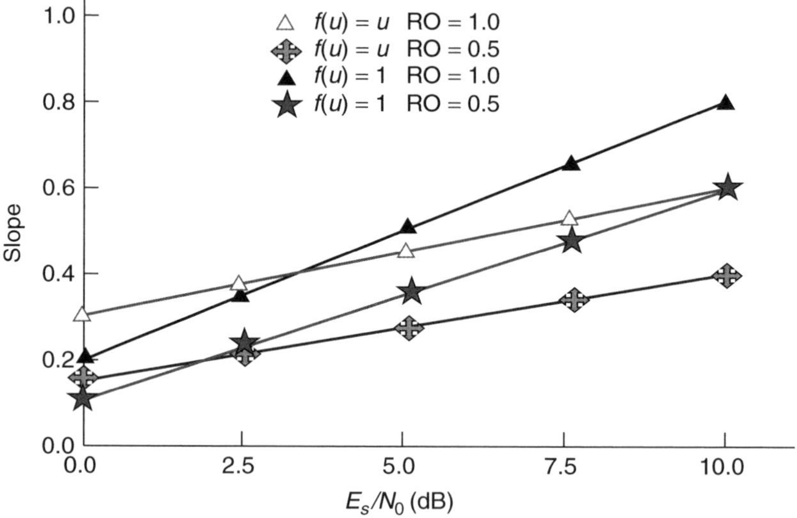

Figure 5.20 V&VFB phase detector slopes for OQPSK, nonlinearity $f(u) = u$ and $f(u) = 1$.

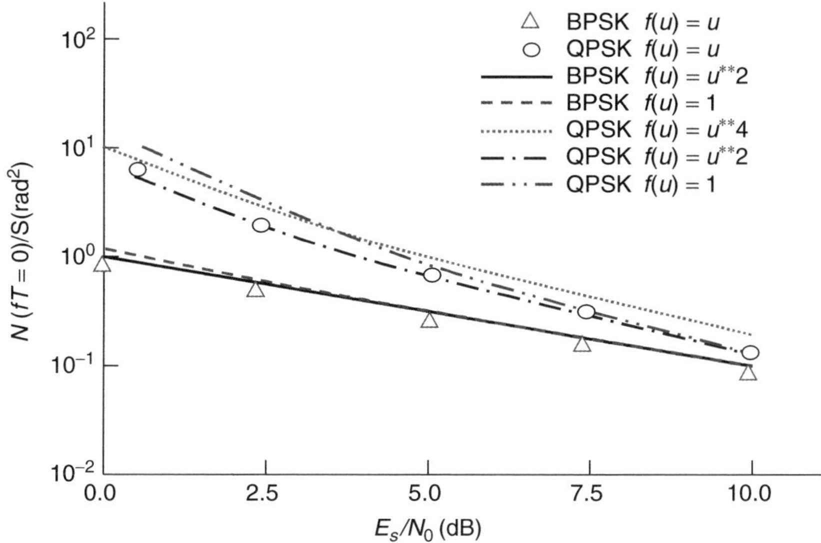

Figure 5.21 Normalized noise power spectral density of V&V for BPSK and QPSK, logarithmic scale.

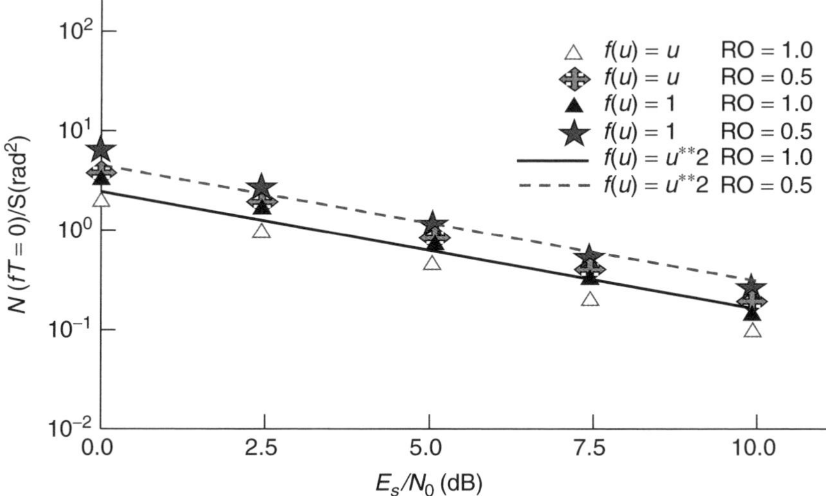

Figure 5.22 Normalized noise power spectral density of V&V for OQPSK, logarithmic scale.

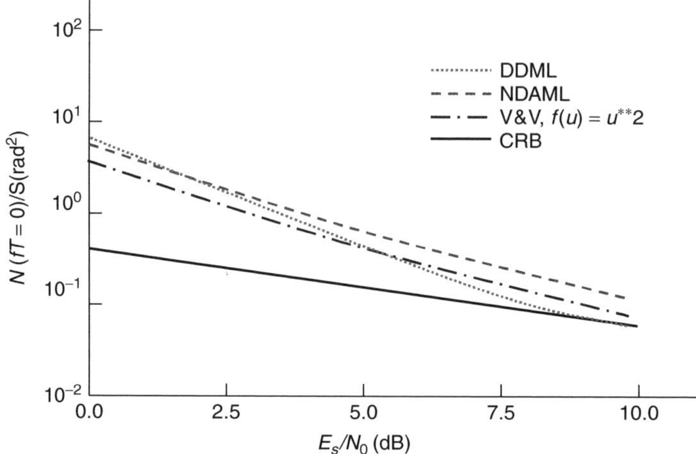

Figure 5.23   Normalized tracking error variance (σ_θ^2/B_L) DDML, NDAML and V&V with $f(u) = u^2$ for QPSK, logarithmic scale.

Table 5.1 Tracking error variance at moderate E_s/N_0
$$\mathrm{Var} \cong (2B_L T)A(N_o/2E_s) = A \cdot \mathrm{CRB}$$
carrier synchronization

Degradation [dB] from CRB		
DDML, NDAMLB/Q, V&VB/Q (same as CRB)		
$\alpha = 0$	A = 1	0
$\alpha = 0.5$	A = 1	0
$\alpha = 1$	A = 1	0
DNAMLO, DNASQFFTO		
$\alpha = 0$	A = ∞	∞
$\alpha = 0.5$	A = 4	6
$\alpha = 1$	A = 2	3
DNAMLTB (continuous-time operation)		
$\alpha = 0$	A = 1	0
$\alpha = 0.5$	A = 1.061	0.26
$\alpha = 1$	A = 1.111	0.45

The results for NDAML logarithm are shown in Figures 5.17 and 5.18. The same set of results for V&V algorithm is shown in Figures 5.19 to 5.22.

Comparison of different algorithms is shown in Figure 5.23 and Table 5.1. In the table the results are compared to Cramer-Rao Bound (CRB), which is the best achievable result.

A number of specific solutions and results related to carrier estimation are given in References [3–32].

SYMBOLS

$s(t, \boldsymbol{\theta})$ – signal
$\boldsymbol{\theta}$ – vector of unknown parameters
$R(,)$ – likelihood function
$h(t)$, $g(t)$, $G(f)$ – pulse shape
$p(,)$ – output of pulse-matched filter (I channel)
$q(,)$ – output of pulse-matched filter (Q channel)
a_n – data (I channel)
b_n – data (Q channel)
z_n – complex signal envelope
x, y – real, imaginary part of z
T_s – sampling period
V – frequency error
θ – phase error
$p_v = \partial p / \partial v$
c – complex data $a + jb$
DD – decision-directed
DA – data-aided
$L(\ )$ – loglikelihood function
NDA – nondata aided
$S(fT)$ – power spectra
ML – maximum likelihood
V&V – Viterbi and Viterbi
FB – feedback
FF – feedforward

REFERENCES

1. Gardner, F. M. (1990) *Frequency Detectors for Digital Demodulators via Maximum-Likelihood Derivation*. Final report (part II) to ESA contract No. 8022/88/NL/DG.
2. Viterbi, A. J. and Viterbi, A. M. (1983) Nonlinear estimation of PSK-modulated carrier phase with application to burst digital transmission. *IEEE Trans. Inform. Theory*, **IT-29**, 543–551.
3. de Buda, R. (1972) Coherent demodulation of frequency-shift keying with low deviation ratio. *IEEE Trans. Commun.*, **COM-20**, 429–435.
4. Falconer, D. D. and Salz, J. (1977) Optimal reception of digital data over the Gaussian channel with unknown delay and phase jitter. *IEEE Trans. Inform. Theory*, **IT-23**, 117–126.
5. Franks, L. E. and Bubrouski, J. B. (1974) Statistical properties of timing jitter in a PAM timing recovery scheme. *IEEE Trans. Commun.*, **COM-22**, 913–920.
6. Franks, L. E. (1980) Carrier and bit synchronization in data communication – a tutorial review. *IEEE Trans. Commun.*, **COM-28**, 1107–1121.
7. Gardner, F. M. (1986) A BPSK/QPSK timing-error detector for sampled receivers. *IEEE Trans. Commun.*, **COM-34**, 423–429.
8. Gardner, F. M. (1988) *Demodulator Reference Recovery Techniques Suited for Digital Implementation*. Final report to ESA contract No. 6847/86/NL/DG.
9. Gardner, F. M. (1990) *Timing Adjustment via Interpolation in Digital Demodulators*. Final report (part I) to ESA contract No. 8022/88/NL/DG.
10. Ascheid, G. and Meyr, H. (1982) Cycle slips in phase-locked loops: a tutorial survey. *IEEE Trans. Commun.*, **COM-30**, 2228–2241.

11. Kobayashi, H. (1971) Simultaneous adaptive estimation and decision algorithm for carrier modulated data transmission systems. *IEEE Trans. Commun.*, **C0M-19**, 268–280.
12. Lindsey, W. C. and Meyr, H. (1977) Complete statistical description of the phase-error process generated by correlative tracking systems. *IEEE Trans. Inform. Theory*, **IT-23**, 194–202.
13. Max, J. (1960) Quantization for minimum distortion. *IRE Trans. Inform. Theory*, **IT-6**, 7–12.
14. Mengali, U. (1977) Joint phase and timing acquisition in data transmission. *IEEE Trans. Commun.*, **COM-25**, 1174–1185.
15. Meyr, H. (1975) Nonlinear analysis of correlative tracking systems using renewal process theory. *IEEE Trans. Commun.*, **COM-23**, 192–203.
16. Meyers, M. H. and Franks, L. E. (1980) Joint carrier phase and symbol timing recovery for PAM systems. *IEEE Trans. Commun.*, **COM-28**, 1121–1129.
17. Meyr, H. and Popken, L. (1980) Phase acquisition statistics for phase-locked loops. *IEEE Trans. Commun.*, **COM-28**, 1365–1372.
18. Meyr, H. and Ascheid, G. (1990) *Synchronization in Digital Communications, Volume I: Phase, Frequency Locked Loops and Amplitude Control*. New York: John Wiley & Sons.
19. Moeneclaey, M. (1983) A comparison of two types of symbol synchronizers for which self-noise is absent. *IEEE Trans. Commun.*, **COM-31**, 329–334.
20. Moeneclaey, M. (1984) Two maximum-likelihood synchronizers with superior tracking performance. *IEEE Trans. Commun.*, **COM-32**, 1178–1185.
21. Moeneclaey, M. (1985) The influence of phase-dependent loop noise on the cycle slipping of symbol synchronizers. *IEEE Trans. Commun.*, **COM-33**, 1234–1239.
22. Moeneclaey, M., Stamak, S. and Meyr, H. (1988) Cycle slips in synchronizers subject to smooth narrowband loop noise. *IEEE Trans. Commun.*, **COM-36**, 867–874.
23. Moeneclaey, M. and Mengali, U. (1990) Sufficient conditions on trellis-coded modulation for code-independent synchronizer performance. *IEEE Trans. Commun.*, **COM-38**, 595–601.
24. Mueller, K. H. and Muller, W. (1976) Timing recovery in digital synchronous data receivers. *IEEE Trans. Commun.*, **COM-24**, 516–530.
25. Oerder, M. and Meyr, H. (1988) Digital filter and square timing recovery. *IEEE Trans. Commun.*, **COM-36**, 605–612.
26. Ryan, C. R., Hambley, A. R. and Voght, D. E. (1980) 760 Mbit/s serial MSK microwave modem. *IEEE Trans. Commun.*, **COAZ-28**, 771–777.
27. Ryter, D. and Meyr, H. (1978) Theory of phase tracking systems of arbitrary order: statistics of cycle slips and probability distribution of the state vector. *IEEE Trans. Inform. Theory*, **IT-24**, 1–7.
28. Simon, M. K. and Lindsey, W. C. (1977) Optimum performance of suppressed carrier receivers with Costas loop tracking. *IEEE Trans. Commun.*, **C0M-25**, 215–227.
29. Simon, M. K. (1978) Tracking performance of Costas loops with hard-limited in-phase channel. *IEEE Trans. Commun.*, **COM-26**, 420–432.
30. Simon, M. K. (1978) Optimum receiver structures for phase-multiplexed receivers. *IEEE Trans Commun.*, **COM-26**, 865–872.
31. Simon, M. K. (1979) On the optimality of the MAP estimation loop for carrier phase tracking BPSK and QPSK signals. *IEEE Trans. Commun.*, **C011Z-27**, 158–165.
32. Ascheid, G., Oerder, M., Stahl, J. and Meyr, H. (1989) An all digital receiver architecture for bandwidth efficient transmission at high data rates. *IEEE Trans. Commun.*, **COM-37**, 804–813.

6

Power control

6.1 ALGORITHMS

In Chapter 8, we will show that the Code Division Multiple Access (CDMA) network capacity depends significantly on the so-called *near–far* effect. From the very beginning, theory and practice of CDMA were aware of this fact. All practical systems use Power control (PC) to reduce this effect. PC is more efficient in the system optimized for speech, such as IS-95. In a multimedia network such as Universal Mobile Telecommunication System (UMTS) in which different signals levels are used for different data rates, additional solutions like multiuser detectors are used.

In IS-95, every mobile station attempts to adjust its transmission power so that signals received at a base station are at the same, minimum level at which good quality communication can still be provided. Both the closed and open loop methods are used. The closed-loop includes two different loops, that is, a relatively fast inner and a slow outer loop. In addition to the data signals, every base station transmits a so-called pilot signal, which is an unmodulated signal [1] used at the mobile stations for PC, synchronization and demodulation as a power level, phase, frequency and time reference. In the open loop method, a mobile station measures the average received total power and adjusts its transmission power to be inversely proportional to the received power. In the initial phase of the call, the average received pilot signal power is measured. The open loop algorithm is presented in Reference [2]. The mobile station transmission power is a certain constant divided by the received total power. The constant value used depends on several base station parameters, such as antenna gain, the number of active users, transmission power, required signal-to-interference ratio (SIR) and interference caused by other base stations. The base station informs the mobile stations before transmission about the value of that constant.

Open loop PC can be nonlinear [3]. The purpose of nonlinearity is to allow fast response (maximum control speed of $10\,\text{dB}\,\text{ms}^{-1}$) for negative corrections, but slow response (maximum control speed of $1\,\text{dB}\,\text{ms}^{-1}$) for positive corrections. When attenuation is suddenly decreased, the mobile station quickly decreases the transmission power in order not to cause additional interference to other users. The extra interference would diminish the system capacity. Since the separation of the reverse and forward-link frequency bands far exceeds the coherence bandwidth, Rayleigh fades in different links correlate poorly with each other. Since the open loop method cannot estimate reverse-link fading, open loop PC cannot be accurate. Its inaccuracy is as much as 10 dB.

In order to compensate for reverse-link fading, a closed-loop method is required. In the closed-loop method, a base station measures (measurement time 1.25 ms) the average received power [1] or the SIR and compares it to a threshold. As a result of the comparison, the base station sends a power-control command to the mobile station, the size of which is nominally 0.5 to 1.0 dB, by puncturing one data bit every 1.25 ms. The bit rate in the feedback is then 800 bps. The closed loop employs delta modulation (DM), that is, after a control delay of about 1.25 ms, the power-control command adjusts the previous transmission power of the mobile station up or down by a fixed step. PC commands are thus extracted and integrated at the mobile station. The part of the closed-loop method discussed above is called an inner loop and will be discussed in detail in the next section. In an outer loop, a base station measures the frame-error rate (FER) of each mobile station, according to which it adjusts the threshold so that the FER is maintained in the required region (e.g. smaller than 1%). The outer loop algorithm is presented in Reference [4]. The outer loop acts more slowly than the inner loop since its updates are once per every 20 ms frame. The outer loop algorithm discussed above is a fixed-step variable threshold algorithm, which uses fixed-size steps in adjusting the target threshold. The improved variable-step variable threshold method is proposed in Reference [5]. Final PC is completed when closed-loop control commands are added to open loop PC.

The dynamic range of the received power can be reduced, and thus facilitate the task of PC, by using a diversity receiver. In Reference [6], functioning of PC is analyzed when a mobile station is in a soft handoff region. In soft handoff, the mobile station is connected simultaneously to several base stations, and it can use lower transmission power. The mobile station transmission power is increased only when all the base stations request it. Otherwise, the transmission power is reduced. The performance of the CDMA system can also be improved by interleaving and channel coding [7,8]. PC and interleaving are complementary methods since with low velocities interleaving is not efficient but PC performs accurately. With high velocities, it is difficult for PC to compensate for the channel effects while, on the other hand, interleaving operates more effectively. In delay insensitive data traffic, in addition to channel coding, an automatic repeat request (ARQ) protocol can be used to achieve a very low bit error rate (BER) value [9]. In Reference [10], a CDMA system with soft PC is proposed, in which the processing gain and code rate are controlled according to the variation of the channel. Since the proposed adaptive processing gain and code rate technique equivalently control the received signal-to-noise ratio (SNR) per bit to the constant value, the conventional PC, which adjusts the received carrier-to-interference ratio (CIR) to be constant, is no longer needed. In Reference [11], a convolutionally coded hybrid DS/SFH (direct sequence/slow frequency hopping) CDMA system using PC is presented. It is shown using simulations that much less accurate PC is required when the DS/SFH CDMA, instead of the pure DS/CDMA system, is employed. The reason for this is that the hybrid system is less susceptible to the near–far problem than the DS/CDMA system. The hybrid system, with selection diversity and without PC, is even better suited to solve the near–far problem than a

DS/CDMA system with accurate PC and an even higher order of diversity [12]. The near–far self-resistant CDMA network concept is discussed in Chapter 15 of this book.

Field tests have been carried out for IS-95 DS/CDMA system in varying environments [7]. The performance of PC in particular has been examined. It appeared that mobile stations in the CDMA system used, on the average, 20 to 30 dB lower transmission power than mobile stations in the analog American mobile phone system (AMPS). The inaccuracy of PC was observed to approximate a lognormal distribution with a standard deviation of about 2.5 dB when normal mobile station velocities and small enough FER values (smaller than 1%) are used [13,14].

The details of power-control implementation, IS-95 will be discussed in Chapter 17 and can be seen in Reference [15]. In Reference [16], the influence of average PC, voice activity detection and micro- and macrodiversity to cellular DS/CDMA systems were studied. The performance of PC of the cellular CDMA system when the channel model includes propagation loss and Rayleigh fading is discussed in Reference [17]. The mobile station transmission power was proportional to the fourth power of the distance. The capacity of the microcellular CDMA system was evaluated using simulations in Reference [18] when IS-95 type, fixed-step adjustment, closed-loop PC – FSAPC (only inner loop, i.e. no FER measurement), was used. The channel model included long-term attenuation and Rayleigh fading. Furthermore, in Reference [19] simulation results for single-cell and multicell DS/CDMA systems employing FSAPC were combined with coding bounds to obtain quasi-analytic estimates of the reverse-link capacity, over both frequency-nonselective and frequency-selective fading channels.

Ariyavisitakul and Chang simulated the performance of closed-loop PC (only inner loop) in both fixed (FSAPC) and variable-step (VSAPC) cases over a Rayleigh fading multipath channel [20]. The variable-step was implemented by removing a hard quantizer in the step-generation process. The bit rate of PC commands was assumed to be at least 10 times the Doppler frequency in order for PC to function effectively (see also Reference [21]). In the single user case, they realized that the performances of the FSAPC and VSAPC were approximately equal when a diversity order of two was used. The same conclusion with the performance comparison between FSAPC and VSAPC was also drawn in Reference [22], especially when the number of tap coefficients in the RAKE receiver was greater than two. In Reference [22], bit rates of FSAPC and VSAPC were equal. That is, in the variable-step scheme, the logic pattern of many successive stored command bits was taken into consideration when adjusting the mobile station's transmission power. FSAPC was not very sensitive to control command errors occurring in the feedback channel [20,22]. In the case of no diversity, the performance of VSAPC was noticed to be superior to that of FSAPC according to Reference [19].

The effect of feedback delay on FSAPC was simulated in Reference [23]. The influence of the delay was diminished by estimating the received power by a linear predictor based on the recursive least-squares (RLS) algorithm. The performance with high ($>50\,\mathrm{km\,h^{-1}}$) mobile station velocities, using estimation based on the RLS algorithm, was better than with conventional PC with power measurement by straight averaging. In cellular systems, the interference power received at the base station was noticed to be larger in the cases of FSAPC and ideal PC (tracks fading accurately) than with ideal average PC [20]. This is due to the effects of power command errors and/or the interference peaking caused by the perfect tracking of deep fades. The use of fast PC is, however, reasonable since interleaving is inefficient if the average PC employed is slow.

Performances of FSAPC and adaptive fuzzy proportional-plus-integral (PI) PC were simulated and compared in Reference [24]. Parameter P in fuzzy PI control extends the bandwidth improving response to changes, and it also prevents the system from becoming unstable. Term I attempts to force the steady-state error to zero. Fixed-step adjustment control is a slight modification of the integral (I) control. Fuzzy PI PC was observed to achieve a shorter rise time, smaller overshoot and smaller rms tracking error. Chang and Wang modified the rule base to also take into account a control delay [25]. The drawback of fuzzy PC is that the channel behavior has to be estimated in advance when constructing the rule base. In neural network-based PC, the channel behavior can be learned adaptively on line during the control process; these algorithms will be discussed later in this chapter.

The optimal PC in the multimedia CDMA system, in which many kinds of information (e.g. voice, image and data) are transferred simultaneously, is analyzed in Reference [26]. Data rate and required communication quality, and thus the PC of each media, depend on transmitted information. A method is proposed by which increasing (decreasing) the transmission power of media with high (low) transmission rates or small (large) processing gains attempts to improve the BER. Data service is bursty in nature. This makes its PC more difficult than the PC of voice calls since channel conditions change between consecutive packets and are difficult to predict. Fortunately, the capacity is more sensitive to the power-control errors of voice service than those of data service. Zhuang has derived an upper bound for the BER for the packetized multimedia CDMA system using optimal PC, diversity and convolutional coding with ARQ protocol for delay insensitive traffic [9]. Using a fixed-rate channel coder and PC in a CDMA system can be seen as one solution for performing unequal error protection (UEP) for different traffic types [27].

6.2 CLOSED-LOOP POWER CONTROL IN DS-CDMA CELLULAR SYSTEM: PROBLEM DEFINITION

Closed-loop PC is a topic covered to a great extent by the control theory. For this reason, in this book we will limit ourselves to the problem definition and literature survey, rather than going into details of the control theory itself, which is available in numerous textbooks. The general block diagram of the closed-loop PC used for this application is shown in Figure 6.1.

Let us start from the point in the loop marked by P_n^t, representing the mobile unit transmit power at the sampling instant with index n. In the loglinear model, presented in Figure 6.1, the received power R_n will be equal to the sum of channel losses A_n and P_n^t.

The base station will be estimating R_n in order to find out what kind of correction is needed. This estimation will be incorrect and the estimation error power is N_n. All together B such samples will be averaged out in order to remove the impact of noise on the overall process. After that, the result is compared with 'the desired received power' P_n^* and a sample of error signal is created.

Different ways of generating reference level P_n^* will be discussed later. This error is transmitted on the downlink and after propagation delay of D samples the error signal

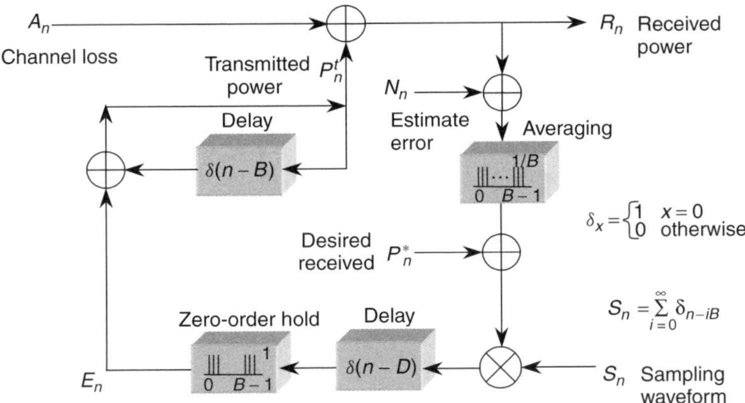

Figure 6.1 Loglinear power-control model [19]. Reproduced from Chockalingam, A., Dietrich, P., Milstein, L. B. and Rao, R. R. (1998) Performance of closed loop power control in DS-CDMA cellular systems. *IEEE Trans. Veh. Technol.*, **47**(3), 774–789, by permission of IEEE.

E_{n-D} will be added to P_n^t to generate a new power level at the mobile transmitter.

$$P_{n+1}^t = P_n^t + E_{n-D} \tag{6.1}$$

One can see that delay B, due to signal processing, is known to both mobile and base station and will be compensated in the signal processing. Delay D due to propagation will not be compensated, which will cause performance degradation depending on the Doppler rate. A simplified model from Figure 6.1 is presented in Figure 6.2.

Some of the results of the analysis of the loop behavior are shown in Figures 6.3 to 6.6. First of all, the received signal power covariance function will be changed dramatically

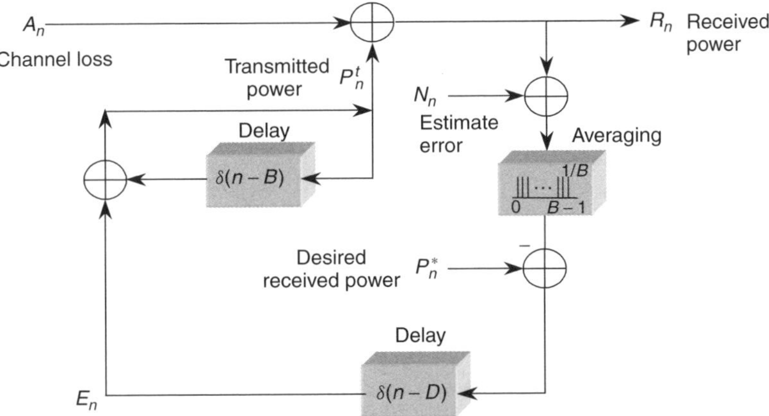

Figure 6.2 Simplified loglinear power-control model [19] Reproduced from Chockalingam, A., Dietrich, P., Milstein, L. B. and Rao, R. R. (1998) Performance of closed loop power control in DS-CDMA cellular systems. *IEEE Trans. Veh. Technol.*, **47**(3), 774–789, by permission of IEEE.

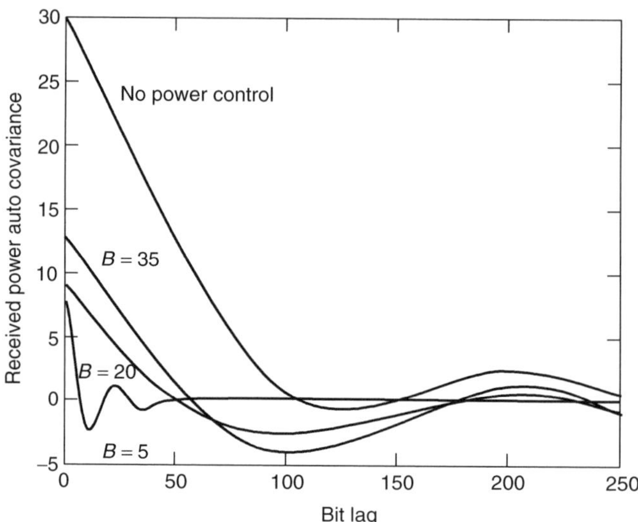

Figure 6.3 Effect of averaging interval B on the received power autocovariance for $f_d = 25\,\text{Hz}$ and $D = 5$ [19]. Reproduced from Chockalingam, A., Dietrich, P., Milstein, L. B. and Rao, R. R. (1998) Performance of closed loop power control in DS-CDMA cellular systems. *IEEE Trans. Veh. Technol.*, **47**(3), 774–789, by permission of IEEE.

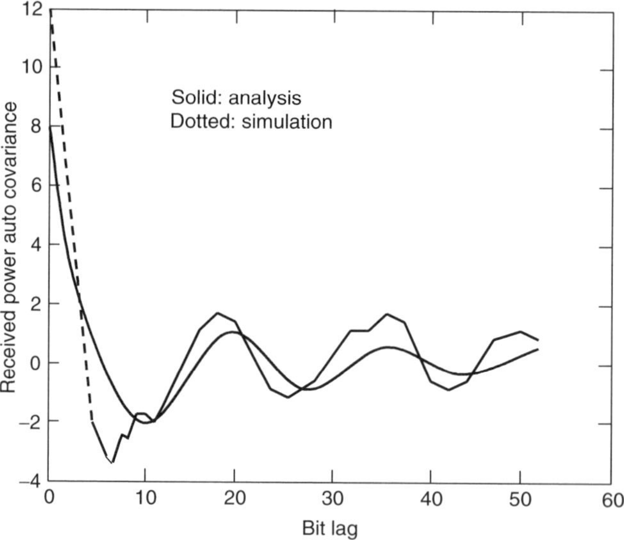

Figure 6.4 Comparison of received power autocovariance functions as predicted by analysis and simulation for $f_d = 25\,\text{Hz}$, $B = 20$, $E_b/N_0 = 10\,\text{dB}$, $D = 5$ and $T_b = 1/8000\,\text{s}$ [19]. Reproduced from Chockalingam, A., Dietrich, P., Milstein, L. B. and Rao, R. R. (1998) Performance of closed loop power control in DS-CDMA cellular systems. *IEEE Trans. Veh. Technol.*, **47**(3), 774–789, by permission of IEEE.

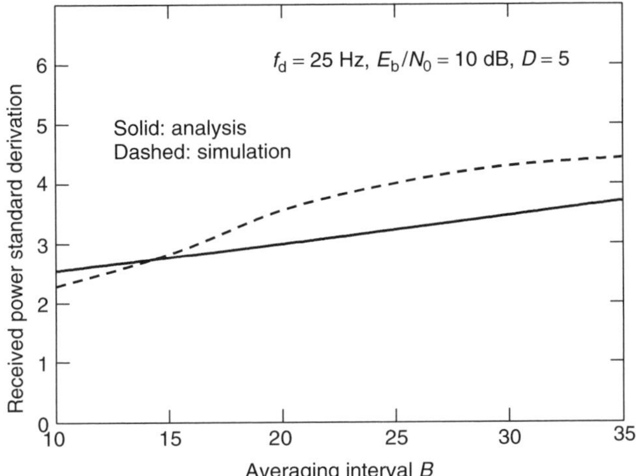

Figure 6.5 Comparison of received power standard derivation as predicted by analysis and simulation for $f_d = 25$ Hz, $B = 20$, $E_b/N_0 = 10$ dB, $D = 5$ and $T_b = 1/8000$ s [19]. Reproduced from Chockalingam, A., Dietrich, P., Milstein, L. B. and Rao, R. R. (1998) Performance of closed loop power control in DS-CDMA cellular systems. *IEEE Trans. Veh. Technol.*, **47**(3), 774–789, by permission of IEEE.

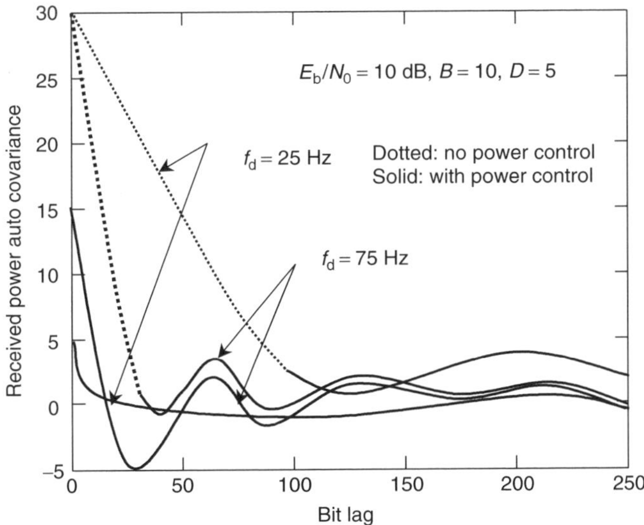

Figure 6.6 The effect on the received power autocovariance function as a result of increasing Doppler frequency, f_d (Hz). $B = 10$, $D = 5$, $E_b/N_0 = 10$ dB and $T_b = 1/8000$ s [19]. Reproduced from Chockalingam, A., Dietrich, P., Milstein, L. B. and Rao, R. R. (1998) Performance of closed loop power control in DS-CDMA cellular systems. *IEEE Trans. Veh. Technol.*, **47**(3), 774–789, by permission of IEEE.

for different B and f_d. *No power control* curve corresponds to the Jack's channel model. This will bring a new problem to channel estimation algorithms that require knowledge of the channel correlation coefficients like Wiener or Kalman estimator. The received signal power standard deviation is shown in Figure 6.5 and these results can be used later as a rough indication of the power-control error. From Figure 6.6 one can see that for larger Dopplers the difference in received signal power statistics between the controlled and uncontrolled signal is reduced. In order to analyze some additional issues, a system with the following set of parameters is assumed:

1. The simulated system has an information rate of 8 kbps, such that a B value of 20 corresponds to a 400-Hz update rate, 10 corresponds to 800 Hz, 5 corresponds to 1.6 kHz and so on.
2. D value of 20 corresponds to a loop delay of 2.5 ms, 10 corresponds to 1.25 ms, and so on. The P^* value is set to provide the desired E_b/N_0.
3. One should be aware that the inverse algorithm implementations need additional bandwidth on the return channel to carry the power-control step size, in addition to the power up/down command.

BER for such a system is presented in Figure 6.7. The set of parameters is shown in the figure itself. One can see that inverse control, which assumes that a precise analogue value of error E_n is transmitted, is the best. One should be aware that this would require additional bandwidth to transmit such information.

Figure 6.8 demonstrates how, for a fixed Doppler, the BER reduces with increasing the PC updating rate. The impact of vehicular speed is shown in Figure 6.9. The larger the

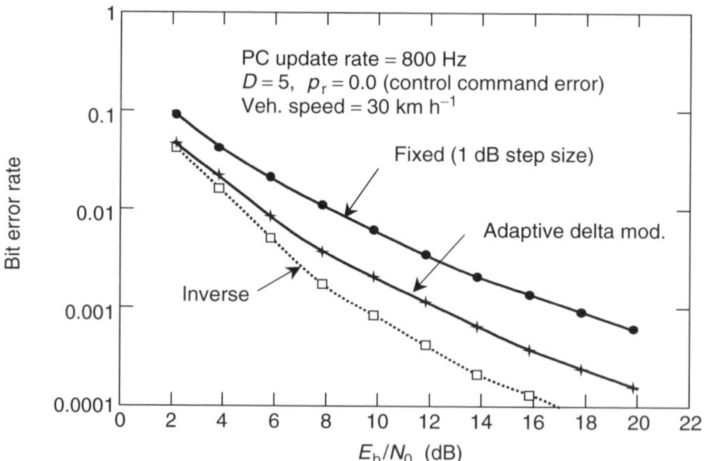

Figure 6.7 Comparison of the BER performance of fixed-step size, adaptive delta modulation, and reverse algorithm, flat Rayleigh fading, $P^* = E_b/N_0$. Update rate = 800 Hz [19]. Reproduced from Chockalingam, A., Dietrich, P., Milstein, L. B. and Rao, R. R. (1998) Performance of closed loop power control in DS-CDMA cellular systems. *IEEE Trans. Veh. Technol.*, **47**(3), 774–789, by permission of IEEE.

speed, the less effective the PC and larger the bit error rate. Bit error rate will be larger if delay D is larger as shown in Figure 6.10 because the correction term becomes less and less relevant. The impact of the correction command error p_r is shown in Figure 6.11. One can see that even the error of the order of 10% can be tolerated.

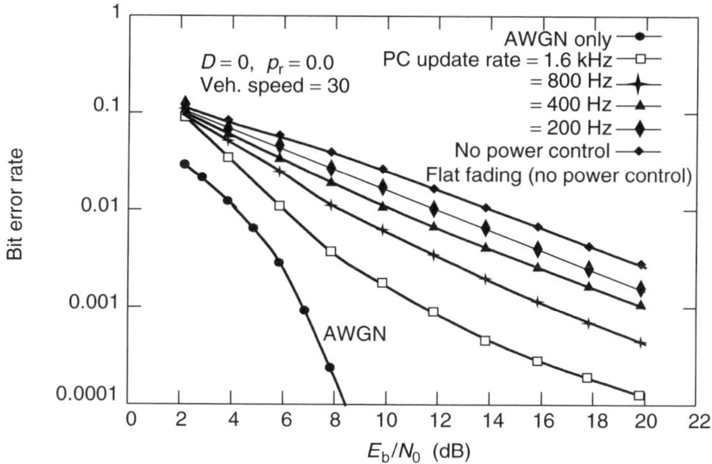

Figure 6.8   Bit error rate versus E_b/N_0 as a function of power-control update rate, flat Rayleigh fading, $P^* = E_b/N_0$, $\Delta = 1\,$dB [19]. Reproduced from Chockalingam, A., Dietrich, P., Milstein, L. B. and Rao, R. R. (1998) Performance of closed loop power control in DS-CDMA cellular systems. *IEEE Trans. Veh. Technol.*, **47**(3), 774–789, by permission of IEEE.

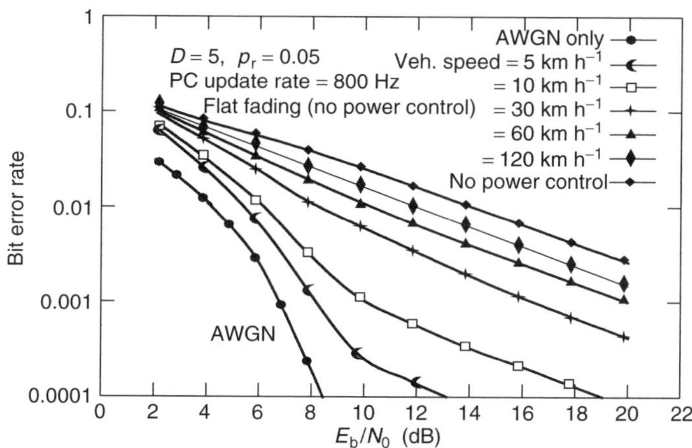

Figure 6.9   Bit error rate versus E_b/N_0 as a function of vehicle speed, flat Rayleigh fading, $P^* = E_b/N_0$, $\Delta = 1\,$dB, update rate $= 800\,$Hz [19]. Reproduced from Chockalingam, A., Dietrich, P., Milstein, L. B. and Rao, R. R. (1998) Performance of closed loop power control in DS-CDMA cellular systems. *IEEE Trans. Veh. Technol.*, **47**(3), 774–789, by permission of IEEE.

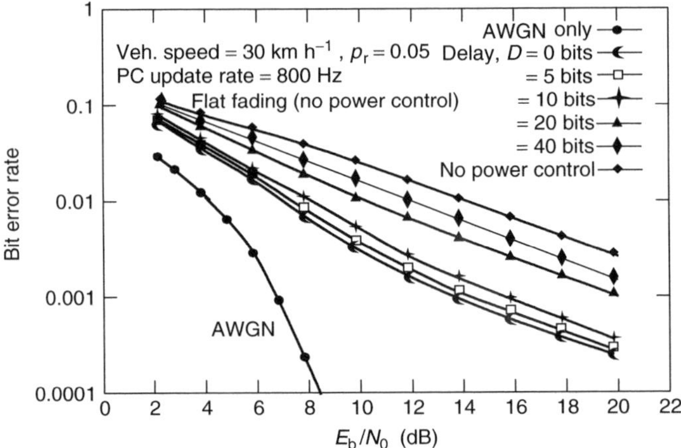

Figure 6.10 Bit error rate versus E_b/N_0 as a function of return channel delay, flat Rayleigh fading, $P^* = E_b/N_0$, $\Delta = 1\,\mathrm{dB}$, update rate $= 800\,\mathrm{Hz}$ [19]. Reproduced from Chockalingam, A., Dietrich, P., Milstein, L. B. and Rao, R. R. (1998) Performance of closed loop power control in DS-CDMA cellular systems. *IEEE Trans. Veh. Technol.*, **47**(3), 774–789, by permission of IEEE.

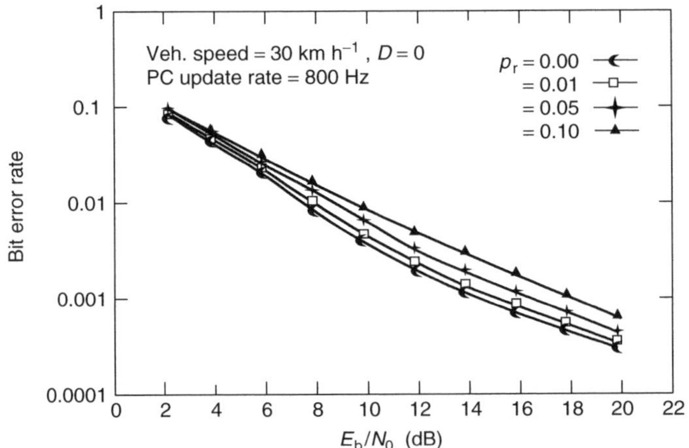

Figure 6.11 Bit error rate versus E_b/N_0 as a function of return channel error rate (p_r), flat Rayleigh fading, $P^* = E_b/N_0$, $\Delta = 1\,\mathrm{dB}$, update rate $= 800\,\mathrm{Hz}$ [19]. Reproduced from Chockalingam, A., Dietrich, P., Milstein, L. B. and Rao, R. R. (1998) Performance of closed loop power control in DS-CDMA cellular systems. *IEEE Trans. Veh. Technol.*, **47**(3), 774–789, by permission of IEEE.

6.3 REFERENCE POWER LEVEL

Since the measurement of the average received power in practice is very difficult, power-control based on SIR (the effect of noise is assumed to be negligible) is preferable [20].

In addition, SIR, not the received power, determines the bit error probability of the user. Utilizing SIR, both the near–far problem and the control of multiple-access interference (MAI) is addressed [28]. Methods for estimating SIR are proposed, for example, in References [22,29–31]. A power-control algorithm was proposed in Reference [32] in which a BER value, instead of SIR, was estimated as a quality measure. PC schemes in which transmitters adapt their power to meet at the receiver some signal quality target, instead of received power target, are called quality-based PC. If the variations of the interference level are not fast compared to the signal changes, the performance of power-control methods based on average power or SIR measurement are quite similar. This is also the case when the number of simultaneous users in a system is small. In that case, all the users reduce their transmission powers, and thermal noise dominates over MAI. It is usually assumed that when there are a large number of simultaneous users, PC of a single user does not affect the total interference power much. That is, with a large number of simultaneous users, the performance of power-control methods based on average power or SIR measurement should also be similar. The simulations in Reference [20], however, showed remarkable changes in the interference levels, even though there were several tens of simultaneous active mobile stations in a base station service area.

FSAPC based on SIR measurement was studied via analysis and simulations in Reference [33]. The closed-loop method used is otherwise similar to that in Reference [20] except that the SIR is measured instead of the average power. It is difficult to analyze power-control on the basis of the SIR measurement since PC of each user affects the PC of all the other users. The change in transmission power of any user has an effect on other users' received interference levels, and thus on the SIR values, according to which transmission powers are adjusted. PC based on SIR was observed to be stable in these simulations. A better system performance was obtained for PC based on SIR than that based on the average power. This is because of the interference adaptation capability of SIR-based PC. The performance, however, was quite dependent on where each user's target threshold was set. Furthermore, in the cellular CDMA system using SIR-based PC, the SIR values of many users were noticed to decrease significantly when the number of users exceeded the capacity limit. This is opposite to the CDMA system employing PC based on the average power where soft degradation in the capacity takes place. In Reference [34], VSAPC was studied in such a way that the knowledge of both the received power and the SIR was exploited. Simulations showed that the performance of this PC was better than with PC based on SIR only.

Su and Shieh [35] compared the performances of PC on the basis of DM, modified adaptive delta modulation (ADM) and differential pulse code modulation (DPCM). The performances of ADM and DPCM control, which use variable-step sizes, were better than that of DM control. DPCM control, however, requires more than one command bit, and ADM control needs an intelligent step size controller. VSAPC with PCM realization was studied in Reference [36]. Either the average power or the SIR was measured. Also, the effect of the loop delay on the performance was investigated. The performance of PCM-based PC appeared to be better than with FSAPC. On the other hand, PCM-based PC is more prone to PC command errors, which occur in the feedback channel. Furthermore, the performance of SIR-based PC was better than the performance of average power-based PC, but it was not as stable as the power-based PC. A system with the SIR-based power-control mechanism is inherently unstable because, in general, most mobile stations must adjust their transmission powers toward their maximum limitations. Simple upper bounds of stability for the SIR target threshold were derived in Reference [30]. Setting the desired target threshold too high or too low in the SIR-based scheme will significantly degrade the system performance [36] (this was also noted in Reference [33]). The optimal target threshold depends on many factors, such as the number of users, loop delay, control mode (dynamic range of adjustment power) and minimum step size. In Reference [37], a nonlinear control system approach was invoked in order to study the stability and convergence properties of FSAPC and VSAPC when coupling between different users was taken into account. This will be discussed later in this chapter in more detail. Su and Geraniotis proposed a closed-loop power-control algorithm, which uses an optimal minimum mean square error (MMSE) quantizer at the receiver and a loop filter at the transmitter [38]. The loop filter is included in order to smooth the distorted feedback information and exploit its memory. In conventional FSAPC, the loop filter at a transmitter contains only one tap.

In the early work, Aien focused on satellite communication systems, and laid the foundation for PC based on SIR by introducing the term SIR balancing for the power-control strategy, with which all the users aim to get the same (balanced) SIR [39]. The proposed algorithm was based on solving the eigenvalue problem. This algorithm is actually optimal in a sense that there exist no other power vectors yielding a higher SIR for all receivers [28,40].

These results were extended and applied to spread spectrum cellular radio systems in References [41–43]. Zander analyzed transmitter PC for cellular systems in References [40,44]. The analysis is especially applicable to time division multiple access (TDMA) and frequency division multiple access (FDMA) systems since PC was employed in order to control interference from each mobile station to mobile stations located in other cells that used the same radio channel (cochannel). The target was to maximize the smallest CIR in the cochannel cells. The assumptions made are not very realistic for CDMA systems. Zander assumes that orthogonal channels are used, thus neglecting the effect of the near–far problem. In optimal PC (in interference limited systems), the probability that the CIR of a randomly chosen mobile station is smaller than the threshold, that is, outage probability, is minimized [44]. The optimal algorithm is very complex since a central controller has to know the attenuation values of every user in the cellular system at every time instant. Furthermore, the central controller simultaneously adjusts the transmission powers of all the users. In optimal (brute force) PC, it is first determined whether the maximum achievable CIR of all mobile stations exceeds the target threshold for the (normalized) link gain matrix, for which the maximum CIR can be calculated as an eigenvalue problem. If the target is achievable, optimal transmission powers can be obtained as an eigenvector of the largest real eigenvalue of the link gain matrix. In the opposite case, the algorithm tries to fulfill the CIR requirement by removing (in practice, by dropping a call) one mobile station. If this does not help, every combination of two mobile stations, then three and so on, is tried, until the requirement is satisfied. In the suboptimal stepwise removal algorithm (SRA), one mobile station at a time is removed until the CIR values exceed the threshold. Note that straightforward CIR balancing, without mobile station removals, may be disastrous since all links may drop below the target threshold.

Wu extended Zander's analysis to be applicable in CDMA systems, and he presented an optimal power-control algorithm for cellular CDMA systems [45]. That is, the performance upper bounds for all types of power-control algorithms for cellular CDMA systems, assuming the SIR threshold is given, were evaluated. In practice, each link has its individual varying SIR threshold at any moment. Thus, the optimal power-control algorithm is not really optimal for the practical mobile radio environment. Furthermore, the concept of soft capacity is not inherent in the optimal PC. These two phenomena were also stated in Reference [33] for SIR-based power-control schemes. Wu also presented a suboptimal sequential algorithm, the performance of which was demonstrated by simulations to be better than with Zander's SRA algorithm.

In distributed PC, only the knowledge of the CIR of each mobile station is required. In References [40,46,47], suboptimal distributed power-control algorithms for narrowband systems are presented. The algorithm proposed in Reference [47] converges much faster than the algorithms in References [40,46], which are special cases of the first algorithm. Also, the performance of the distributed algorithm proposed in Reference [21] is better than that with the algorithms in References [40,46]. The last algorithm is its special case. The distributed algorithms described in References [40,46] are efficient in CDMA systems also, when not considering SIR estimation errors. In these algorithms, it was assumed that the transmission power is sufficiently high in order to allow thermal noise to be neglected. These algorithms are actually not fully distributed, but a normalization procedure in transmission powers based on global information is required. A fully distributed algorithm, where the inclusion of thermal noise in the definition of interference avoids the use of the normalization procedure, was introduced in Reference [48]. Instead of using a constant target threshold, it is beneficial to tune its value according to the mobile station transmission power so that the target SIR is decreased when the mobile station increases its transmission power [49,50]. Then, the probability of the target not being reached, though the mobile station's transmission power is at a maximum level, is minimized. It was shown in Reference [28] that the algorithms in References [21] and [49] can yield an unstable system when subject to a small time delay. Ulukus and Yates proposed stochastic PC in which matched filter outputs, instead of exact knowledge of SIR, are required [51].

In previous analyses for PC, users were assumed to firmly belong to a certain base station's service area. Algorithms for combined base station selection and PC are proposed in References [50,52]. The total reverse-link transmission power is minimized subject to maintaining an individual target CIR for each mobile station. This minimization occurs over the set of power vectors and base station assignments. In Reference [53], it is shown that the capacity can be increased significantly over that presented in References [50,52] by applying joint PC, base station assignment and beamforming. Finally, Hanly [52] extends his previous approach by removing the cellular structure and allowing each mobile station to be jointly decoded by all the receivers in the network.

6.4 FEEDBACK CONTROL LOOP ANALYSIS

Feedback control loop theory is well established and widely used. For this reason we do not go much into the details, but rather refer the reader to the numerous literatures available in this field.

Even the definitions of feedback methods vary in the literature (see, e.g. References [54–56]). We categorize the methods according to Reference [55]. Feedback communication systems are divided into sequential and nonsequential systems. In the sequential system, the decision times are not fixed *a priori* since a receiver updates the likelihood ratio, compares it to a set threshold, and makes the final decision only when the threshold is exceeded. If, in the sequential system, a receiver feeds back only the decision time, we have a synch feedback. Nonsequential systems use fixed-length transmission blocks, and the decision times are fixed. Note that the feedback link in this section is also typically delayless if not otherwise stated [see equation (6.1)].

Turin [57] compared the performance of sequential and nonsequential systems when uncertainty feedback or information feedback was employed. That is, a receiver continuously sends information to a transmitter on the basis of what has been received. Feedback information is analog, for example, *a posteriori* probabilities of the transmitted data symbol or (in PC) the channel state values. In a decision feedback method, fed back information is digital, and it can consist of tentative decisions, such as which symbol is the most likely symbol at a given time. Thus, tentative decisions are sent to the transmitter before the final decision. According to Reference [54], in a decision feedback or post-decision feedback method, the receiver does not send information to the transmitter until the final decision has been made. Digital information can also be a decision whether to adjust the transmission power up or down, as in FSAPC. Information feedback usually needs larger bandwidth than decision feedback, but the potential performance improvement is also bigger when compared to the system with no feedback. Several feedback methods can be used simultaneously, and the system performance can be improved further at the expense of increased system complexity. In fading channels, systems often use error detection channel coding. When a receiver detects an error (probably due to deep fading in the channel), the transmitter is informed by using feedback to repeat the transmission. This method is called an ARQ feedback method.

Schalkwijk and Kailath proposed in 1966 a coding scheme with feedback based on a stochastic approximation procedure in the case of an additive white Gaussian noise (AWGN) channel and no bandwidth constraint [58]. The use of feedback simplifies coding and decoding significantly. Schalkwijk extended the analysis to band-limited signals also [59], where he showed that his feedback scheme is apparently the first deterministic coding procedure (with or without feedback) to achieve the Shannon capacity. The capacity is the same with and without feedback, as stated previously in Reference [60]. The error probabilities achieved, however, are considerably different. Schalkwijk proposed the optimal feedback method over an AWGN channel, and he showed that some proposed feedback methods presented in the literature are actually special cases of his method, for example, the schemes presented in Reference [58]. In this iterative center-of-gravity scheme, a signal, that is, the center of gravity of the signal structure, is subtracted optimally from the transmitted signal. At the receiver the same signal is added to the noisy received signal. The transmission power is thus decreased considerably without affecting the error probability of the system. Note that if we have noisy feedback, the channel capacity cannot be achieved while having a finite SNR in the feedback link. In that case, a transmitter should use a weighted sum of feedback information to average out feedback noise to a certain extent. The performances of many proposed suboptimal feedback methods are, of course, poorer than the performance of the center-of-gravity scheme, but less bandwidth is required in the feedback link in systems employing them [61]. Butman [62] discussed a rather general formulation of linear feedback communication systems, in which the additive noise could be also colored. When reverse-link noise is colored, the channel capacity can be increased by using noiseless feedback. In particular, feedback may increase the capacity of a Gaussian channel by at most a factor of two [63]. Practical constraints, such as maximum power limitation, were shown to significantly reduce, in the idealized conditions, calculated feedback communication systems' performance presented in the literature. Other practical constraints are, for example, noise in feedback, a delay and bandwidth constraint.

The center-of-gravity scheme is no longer optimal in a fading channel [64]. At a receiver, we cannot compensate for the effect of the signal subtracted at a transmitter, since the channel state is not known exactly. Hayes derived the optimal transmission powers (energies) as a function of known channel state values for the coherent antipodal and noncoherent orthogonal system over a Rayleigh fading multipath channel [65]. In optimal PC, the average error probability of the system was minimized when the average transmission power was fixed. Only the sum of the squared attenuation values needed to be fed back to the transmitter for the purpose of PC. It appeared that the influence of optimal PC to the system performance was significant with small average error probabilities, or with large average SNRs.

Cavers analyzed the optimal variation of the data rate with the assumption of known channel state values [66]. In variable-rate transmission, the transmission power is constant, but the data rate is adjusted such that the average error probability is minimized when the average data rate is fixed. With noiseless and delayless feedback and unconstrained maximum data rate, the average probability of error for binary signaling and incoherent detection appeared to be the same as that for a nonfading channel. Cavers found that the transmitted energy per bit can, however, increase infinitely even though the average energy per bit is finite. Cavers also discussed the effects of bandwidth limitation, feedback delay, length of data rate change period and a finite number of transmission rates on the system performance. When the ratio of maximum rate and average rate was assumed to be two, the performance loss was shown to be 0.9 dB compared to the unconstrained bandwidth case. Srinivasan showed that the performance with the constraint on the bandwidth can be improved by controlling, instead of the transmission rate only, either the rate or the transmission power, depending on whether the channel gain is above or below a certain threshold [67]. That is, whenever the data rate saturates at the upper bounds of the rate, the transmission power is varied according to an optimized control rule. The constraint of the number of data rates did not significantly affect the system performance [66]. In contrast, the delays (feedback delay and nonzero rate change period) had a significant effect on the performance.

Hentinen analyzed both the optimal control of power and the data rate when the channel state values were assumed to be known [68]. He showed that Cavers' result, in which the average probability of error for binary signaling and incoherent detection is the same as that for a nonfading channel, is valid for a wide class of modulation schemes. Furthermore, the performance of orthogonal signals is even better over a Rayleigh fading channel with rate control than over the equivalent nonfading channel. Rate control appeared to be superior to PC. In optimal PC, the ratio of maximum power and average power, likewise in optimal control of data rate the ratio of maximum rate and average rate was shown to be large. When the ratio of maximum rate and average rate was assumed to be two, the performance loss was less than 1 dB compared to the case when the maximum rate was not constrained. This corresponds to the result obtained in Reference [66]. The performance decreased significantly, however, when the ratio was reduced to below two. Hentinen also considered suboptimal control of the data rate, and he noticed that by controlling both the power and rate simultaneously, the system performance could be improved further compared to the case of varying only the rate. Hentinen showed that with simultaneous control of power and rate there is no optimal control rule for finite power and data rate.

When we vary the data rate, a large buffer is required in practice at both the transmitter and the receiver. In all the above cases, when the data rate has been varied, it has been assumed that the buffer size is infinite. A finite buffer size impairs the system performance. Buffer control methods have been proposed in References [67–69], in which it is shown that in order to achieve a certain performance, the size of the buffer can be decreased by taking the queue length in the buffer into account.

In all the above methods, the channel state values are assumed to be known. In order to estimate the channel state by one-shot maximum aposterior probability (MAP) (or MMSE) estimator, in addition to an antipodal data symbol, Srinivasan used a constant-power, known pilot symbol in a time-multiplexed form in each frame [70]. Thus, each frame included only one data symbol in addition to a pilot symbol. The channel state estimates were used to optimally adjust the transmission power (energy) or to suboptimally vary the data rate, respectively. The transmission power was evaluated numerically as a function of the channel state by minimizing the average error probability of the pilot symbol system when the average transmitted data symbol energy was the same as the energy of the pilot symbol. The performance of the pilot symbol system employing feedback PC was compared to the cases when PC is not used and when optimal PC with known channel state values [65] is employed. Again, the (suboptimal) control of data rate with an infinite buffer size was superior to PC. For simple implementation, a binary nonsequential decision feedback system was proposed in Reference [71] in which the receiver communicates an initial message estimate to the transmitter generated over a part of the signaling interval. The transmitter transmits, over the rest of the interval, either no more energy or a signal with increased energy, depending on whether the initial message estimate was correct. The transmitter is thus not required to be adaptive to channel conditions. Also, channel coding can benefit from the fed back channel state values. For example, the code rate can be changed adaptively as a function of the channel state [72]. A system was proposed in Reference [73], in which information is transmitted simultaneously via several independent channels, in each of which the code rate used depends on the instantaneous channel state.

6.5 NONLINEAR POWER CONTROL

The nonlinear up/down power-control algorithm can be represented by rewriting equation (6.1) as

$$P(n-1) = P(n) + \mathrm{d}\Psi[P^* + I + P(n) - A(n)] \qquad (6.2)$$

where d is the adaptation step, $A(n)$ the channel losses and the nonlinear term Ψ is defined as

$$\Psi(x) = \begin{cases} 1 & \text{if } x \geq 0 \\ -1 & \text{otherwise} \end{cases} \qquad (6.3)$$

Block diagram for equation (6.2) is shown in Figure 6.12.

This model is analyzed in Reference [37]. For shadow fading, empirical studies have shown that $a(n)$ follows a lognormal distribution. This implies $A(n)$ is Gaussian. A simple and realistic model of $A(n)$ is a Gaussian process with the correlation given as

$$R_A(n) = \sigma_A^2 \xi^{(vT/D)|n|} \qquad (6.4)$$

where ξ is the correlation between two signal samples separated by a spatial distance of D, T the sampling period and v the speed of the mobile, which gives the distance covered by the mobile in a sample interval.

Different channels are characterized by different values of ξ, D and v. Some experimental values for different environments can be found in the experimental studies of References [74,75]. Also note that in equation (6.2), we can combine A with I and P^* by defining

$$B(n) = -A(n-1) + I + P^* \qquad (6.5)$$

Note that $B(n)$ is still a Gaussian process with the same covariance as $A(n)$. The state equation is simplified as

$$P(n+1) = P(n) + \mathrm{d}\Psi[B(n+1) - P(n)] \qquad (6.6)$$

Some results of the analysis presented in Reference [37] are shown in Figures 6.13 to 6.15.

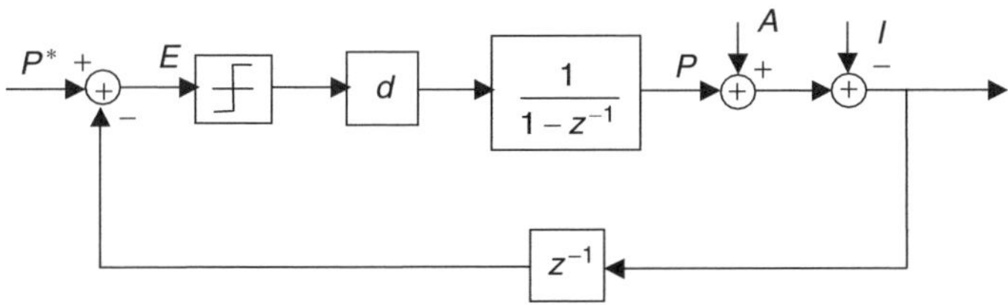

Figure 6.12 Block diagram for the up/down power-control algorithm.

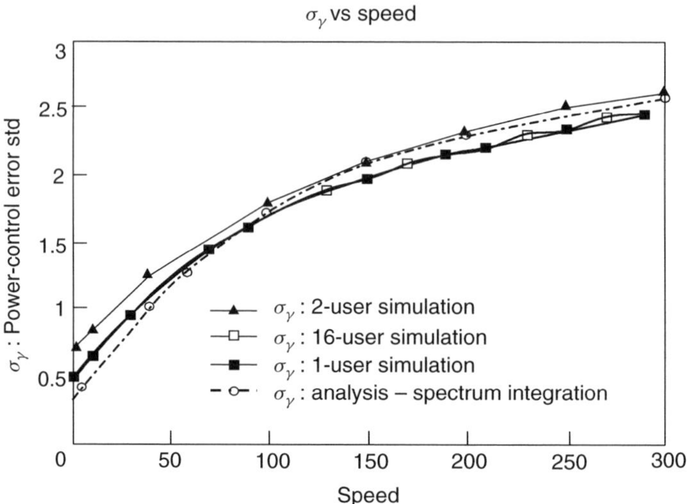

Figure 6.13 Power-control error standard deviation σ_γ versus v ($d = 0.5\,\text{dB}$, $D = 1\,\text{m}$, $\xi = 0.1$, $T = 1.25\,\text{ms}$, $\sigma_A = 3\,\text{dB}$) [37]. Reproduced from Song, L., Mandayam, N. B. and Gajic, Z. (1999) Analysis of an up/down power control algorithm for the CDMA reverse link: a nonlinear control system approach. *Proc. Conference on Information Sciences and Systems*, Baltimore, MD, pp. 119–124, by permission of IEEE.

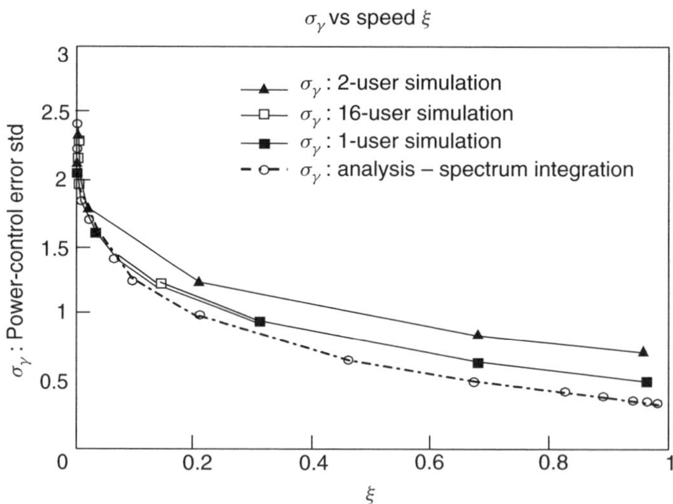

Figure 6.14 Power-control error standard deviation σ_γ versus ξ ($d = 0.5\,\text{dB}$, $D = 1\,\text{m}$, $v = 60\,\text{km}\,\text{h}^{-1}$, $T = 1.25\,\text{ms}$, $\sigma_A = 3\,\text{dB}$) [37]. Reproduced from Song, L., Mandayam, N. B. and Gajic, Z. (1999) Analysis of an up/down power control algorithm for the CDMA reverse link: a nonlinear control system approach. *Proc. Conference on Information Sciences and Systems*, Baltimore, MD, pp. 119–124, by permission of IEEE.

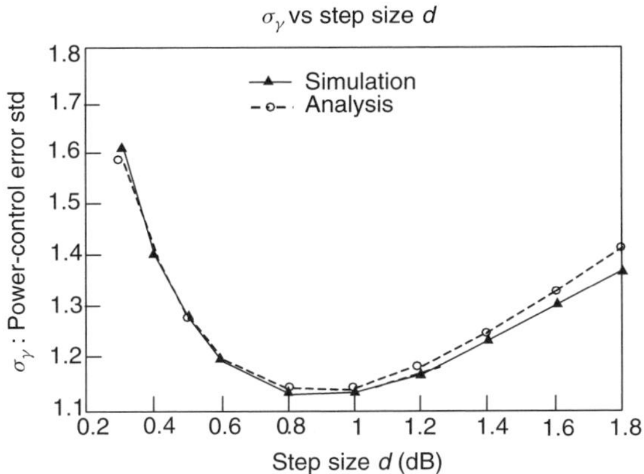

Figure 6.15 Power-control error standard deviation σ_γ versus step size d ($v = 60\,\mathrm{km\,h^{-1}}$, $D = 1\,\mathrm{m}$, $\xi = 0.1$, $T = 1.25\,\mathrm{ms}$, $\sigma_A = 3\,\mathrm{dB}$) [37]. Reproduced from Song, L., Mandayam, N. B. and Gajic, Z. (1999) Analysis of an up/down power control algorithm for the CDMA reverse link: a nonlinear control system approach. *Proc. Conference on Information Sciences and Systems*, Baltimore, MD, pp. 119–124, by permission of IEEE.

6.6 FUZZY LOGIC POWER CONTROL

In this section, we present one more example of nonlinear power-control loop. For a perfect (noiseless) measurement of the received power at time $t - \tau$ seconds, and the power adjustment command sent to the mobile's power actuator directly without being corrupted by any forward-link channel noise, the ratio of the signal standard deviation of controlled (σ_c) and uncontrolled system (σ_{uc}) is [76]

$$\sigma_c/\sigma_{uc} \geq \eta = \sqrt{1 - R^2(\tau)} \tag{6.7}$$

The minimum reduction factor is equal to 0.25 when $\tau = 1$ ms and the maximum Doppler frequency is 40 Hz (e.g. around 900 MHz, 30 mph). The value of η becomes large by increasing the time delay τ since $R(\tau)$ becomes smaller. For example, at $\tau = 4$ ms, the minimum reduction factor η is 0.89. In the extreme case, η approaches unity when $R(\tau)$ becomes zero by letting τ be infinity. In this section we will use a modified model as represented in Figure 6.16. For the purpose of the analysis, the equivalent scheme is shown in Figure 6.17.

A conventional PI control algorithm is given by

$$\Delta p(t) = k_p e(t) + k_I \int e(t)\,\mathrm{d}t \tag{6.8}$$

where $\Delta p(t)$ is a control action at time instant t and $e(t)$ equals the set point minus the process output (power error) k_p and k_I are scaling gain factors. In digital implementation, its incremental form is written as

$$p_{k+1} = p_k + \Delta p_{k+1}$$
$$\Delta p_{k+1} = k_p \Delta e_k + k_I e_k \tag{6.9}$$

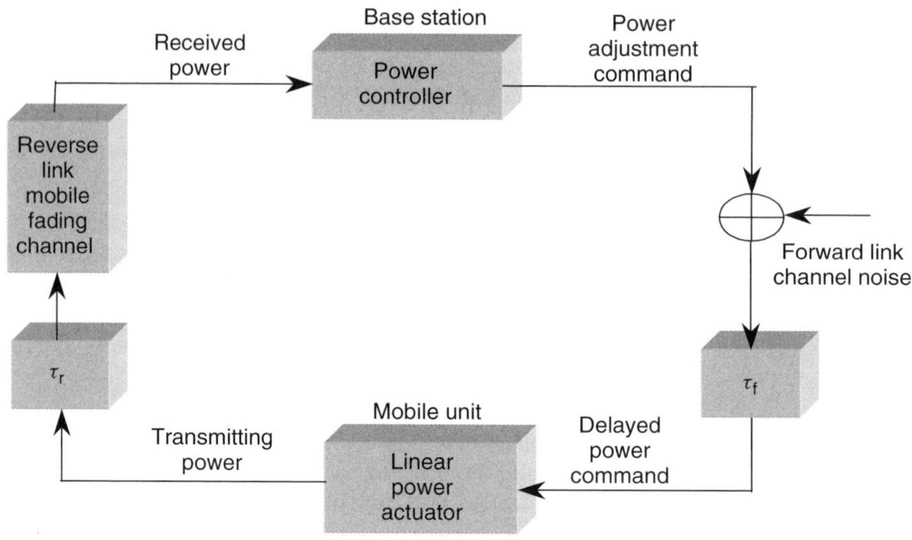

Figure 6.16 Overall schematics of a closed loop power-control system with reverse- and forward-link delays τ_r, and τ_f and a mobile power actuator.

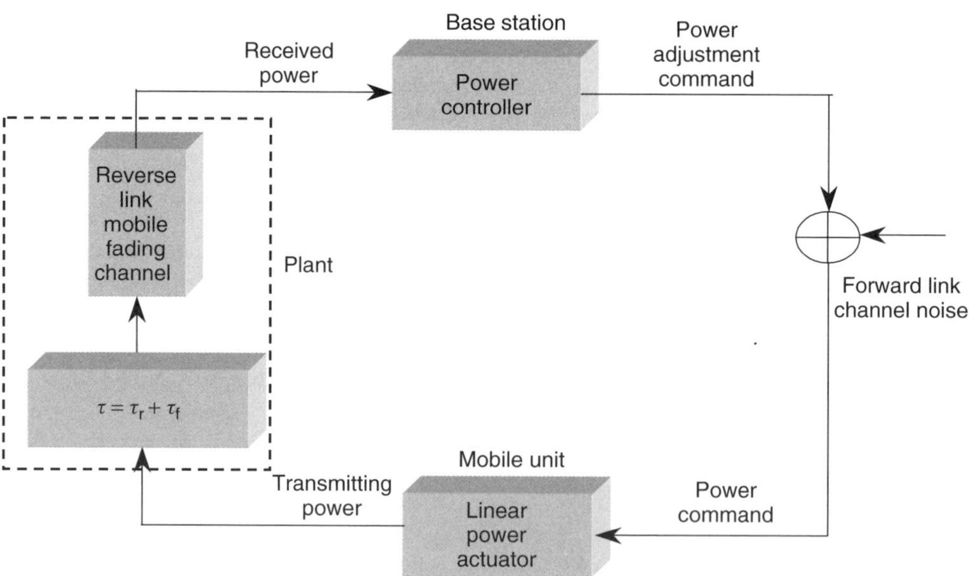

Figure 6.17 Equivalent closed loop power-control system with a standard control scheme and a new reverse-link delay $\tau = \tau_r + \tau_f$.

where Δe_k equals the current error minus the last error and p_{k+1} and Δp_{k+1} are, respectively, the control and incremental control actions for the next time interval. A practical fuzzy PI control is defined as

$$p_{k+1} = p_k + \Delta p_{k+1}$$

$$\Delta p_{k+1} = F\{k_I e_k, k_p \Delta e_k\} \tag{6.10}$$

$F\{\cdot, \cdot\}$ denotes the fuzzy function that acts on the rules of the form R_i: *if* $(k_I e)$ is A_i and $(k_p \Delta e)$ is B_i *then* Δp is C_i where (A_i, B_i, C_i) are linguistic terms. For these definitions, the scheme shown in Figure 6.17 becomes more detailed as shown in Figure 6.18.

The derivation presented in the sequel is very much based on Reference [25]. In this field we use the following terminology.

The two input variables, e and Δe, and the output control variable, Δp, where e, Δe and Δp are the received power error, power error change and transmitting control power increment, respectively. The range of values (ROV) e, Δe and Δp are assumed to be $E = \{e| -18\,\text{dB} \le e \le 18\,\text{dB}\}$, $\Delta E = \{\Delta e| -12\,\text{dB} \le \Delta e \le 12\,\text{dB}\}$, and $\Delta P = \{\Delta p| -6\,\text{dB} \le \Delta p \le 6\,\text{dB}\}$, respectively. In the standard fuzzy logic terminology, ROV is called the *universe of discourse*. Associated term sets, $T(E)$, $T(\Delta E)$ and $T(\Delta P)$ are identical and given by {LP (large positive), MP (medium positive), SP (small positive), ZE (zero), SN (small negative), MN (medium negative), LN (large negative)}. There are 343 possible combinations of the terms generating a maximum possible 343 rules of the form indicated earlier. The membership functions relating the discrete values within ROV and associated term set are shown in Figure 6.19.

For the modeling of the control algorithm, we start with a possible outlook of the received signal power shown in Figure 6.20.

The envelope within region I can be modeled as a portion of the step response of a second-order system. The envelope belonging to region II is also characterized by a portion of the step response of another second-order system with large overshoot. As a conclusion, we assume that any fading process can be modeled as a piecewise second-order system. A combination of the primitive curves generated by second-order systems with different local performance indexes can approximate the envelope of any fading process.

Let us now represent a segment of the curve from Figure 6.20 as shown in Figure 6.21. The overall response is divided into four areas:

$$A_1 : e > 0 \text{ and } \Delta e < 0 \qquad A_2 : e < 0 \text{ and } \Delta e < 0$$

$$A_3 : e < 0 \text{ and } \Delta e > 0 \qquad A_4 : e > 0 \text{ and } \Delta e > 0 \tag{6.11}$$

For the control rule we will use the error, which is the difference between the set value and the response, the slope of the response at crosspoints called crossover index c, and the maximum value m of the error.

For different areas of the curve, a set of values for crossover index c, and parameter m are defined in Figures 6.22 and 6.23, respectively.

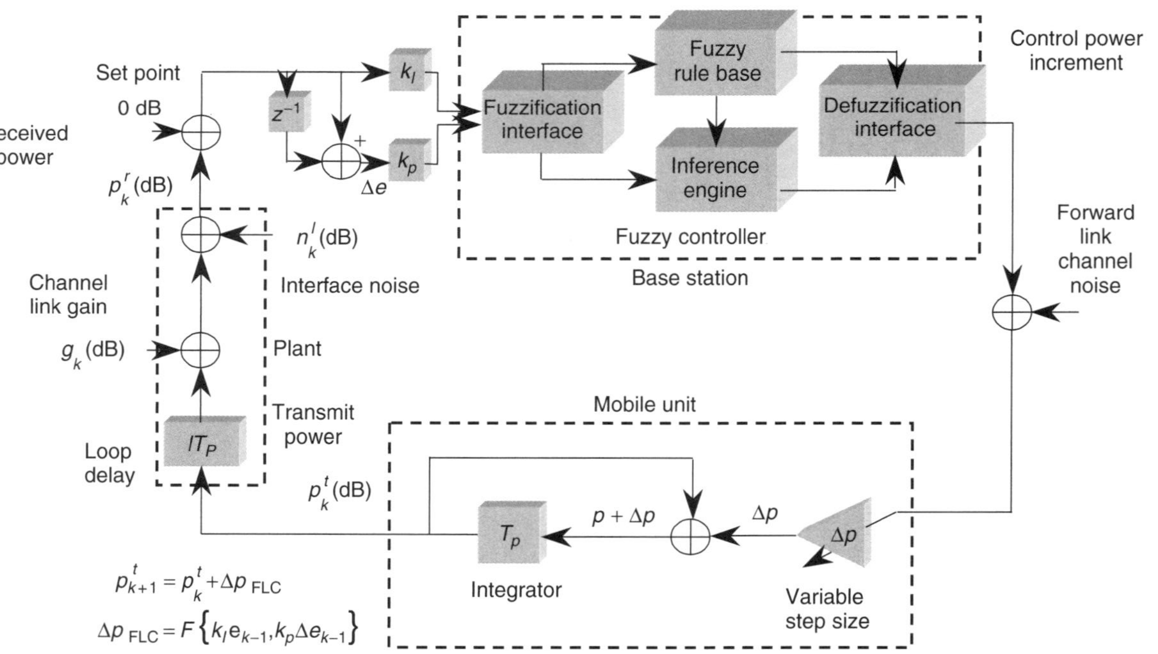

Figure 6.18 Adaptive fuzzy power-control system for CDMA mobile radio channels, where z^{-1} denotes the delay operator [25]. Reproduced from Chang, P. R. and Wang, B. C. (1996b) Adaptive fuzzy proportional integral power control for a CDMA system with time delay. *IEEE J. Select. Areas Commun.*, **14**(9), 1818–1829, by permission of IEEE.

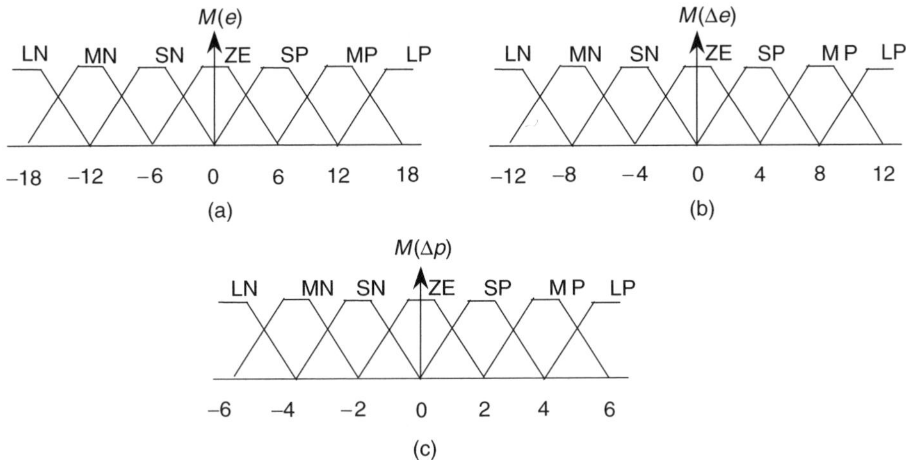

Figure 6.19 Membership function: (a) error e (dB), (b) error change Δe (dB) and (c) power increment Δp (dB).

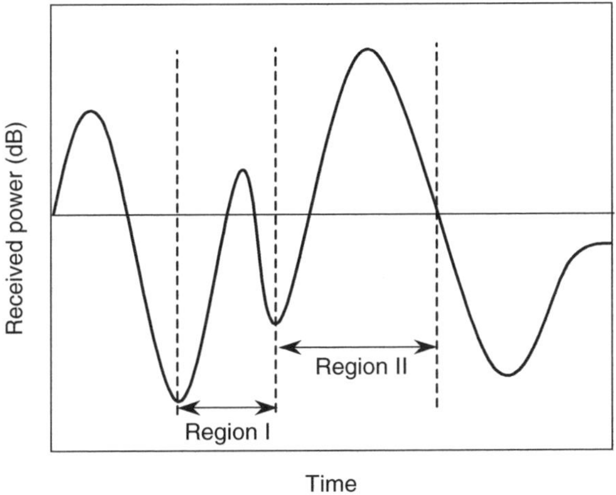

Figure 6.20 A typical fading power signal.

The crossover index c_i^j for identifying the slope behavior of the response across the set point ($e \approx 0$) is defined as

$$c_1^{\,1} : (e > 0 \rightarrow e < 0) \text{ and } \Delta e \lll 0$$

$$c_2^{\,1} : (e < 0 \rightarrow e > 0) \text{ and } \Delta e \ggg 0$$

$$c_1^{\,2} : (e > 0 \rightarrow e < 0) \text{ and } \Delta e \ll 0$$

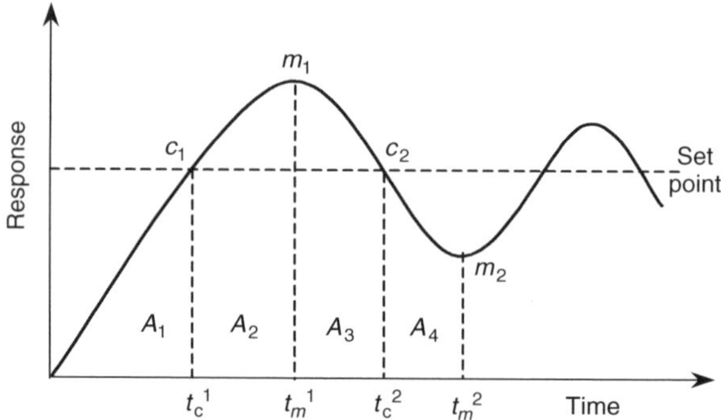

Figure 6.21 General behavior of second-order system response: response areas [25]. Reproduced from Chang, P. R. and Wang, B. C. (1996b) Adaptive fuzzy proportional integral power control for a CDMA system with time delay. *IEEE J. Select. Areas Commun.*, **14**(9), 1818–1829, by permission of IEEE.

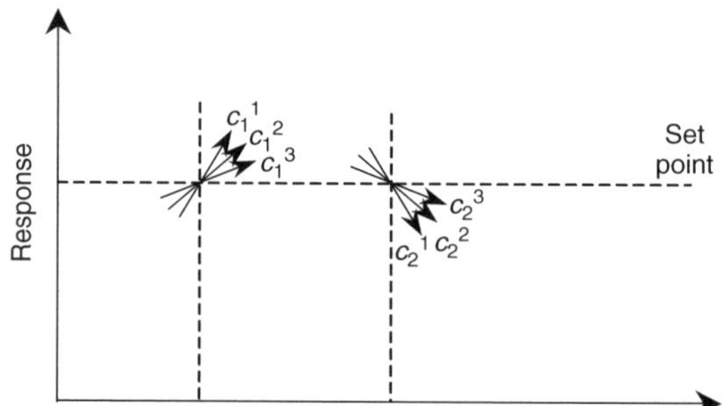

Figure 6.22 Crossover points with six different index values [25]. Reproduced from Chang, P. R. and Wang, B. C. (1996b) Adaptive fuzzy proportional integral power control for a CDMA system with time delay. *IEEE J. Select. Areas Commun.*, **14**(9), 1818–1829, by permission of IEEE.

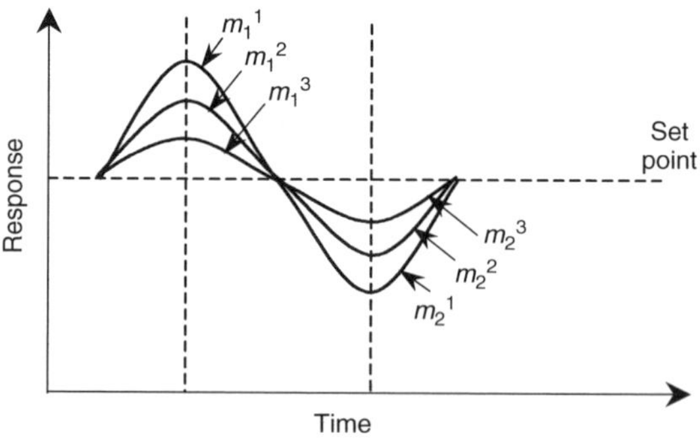

Figure 6.23 Maximum–minimum points with six different index values [25]. Reproduced from Chang, P. R. and Wang, B. C. (1996b) Adaptive fuzzy proportional integral power control for a CDMA system with time delay. *IEEE J. Select. Areas Commun.*, **14**(9), 1818–1829, by permission of IEEE.

$$c_2{}^2 : (e < 0 \rightarrow e > 0) \text{ and } \Delta e \gg 0$$

$$c_1{}^3 : (e > 0 \rightarrow e < 0) \text{ and } \Delta e < 0$$

$$c_1{}^1 : (e < 0 \rightarrow e > 0) \text{ and } \Delta e > 0 \tag{6.12}$$

The minimum–maximum index for representing the amount of overshoot and undershoot is defined as

$$m_1{}^1 : \Delta e \approx 0 \text{ and } e \lll 0$$

$$m_2{}^1 : \Delta e \approx 0 \text{ and } e \ggg 0$$

$$m_1{}^2 : \Delta e \approx 0 \text{ and } e \ll 0$$

$$m_2{}^2 : \Delta e \approx 0 \text{ and } e \gg 0$$

$$m_1{}^3 : \Delta e \approx 0 \text{ and } e < 0$$

$$m_2{}^3 : \Delta e \approx 0 \text{ and } e > 0 \tag{6.13}$$

The mapping of the time domain response in phase plane (error state) space defined by e, m and c is shown in Figure 6.24.

In area A_1, the control rules should shorten the rise time when e is large and prevent the overshoot in A_2 when e is close to zero. A positive large control increment is required to drive the closed-loop response toward the set point generating an improvement in the rise time when e is large, and the control increment is zero or negative in order to prevent the overshoot when the response approaches the set point. In area A_2, the control rules should decrease the overshoot around the peak above the set point. The control increment must be negative. Control rules for area A_3 and A_4 are dual to those listed above. The control increment for A_4 is positive in order to prevent the overshoot around the peak below the set point. A negative large control increment is required for A_3 when $|e|$ is far away from zero. The above observations are summarized in Table 6.1 as a control rule.

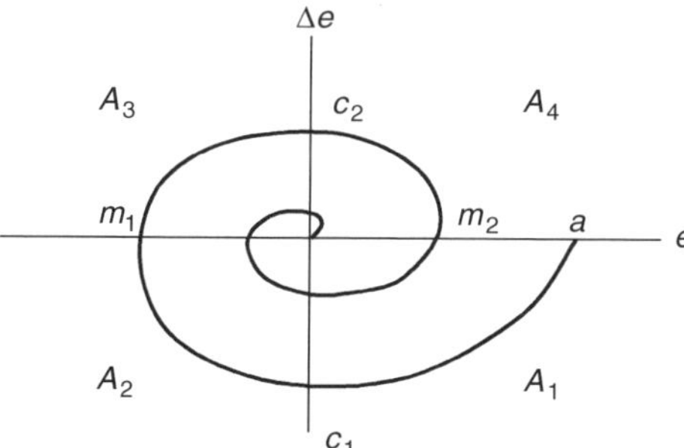

Figure 6.24 The mapping of the time domain response in phase plane (error state) space [25]. Reproduced from Chang, P. R. and Wang, B. C. (1996b) Adaptive fuzzy proportional integral power control for a CDMA system with time delay. *IEEE J. Select. Areas Commun.*, **14**(9), 1818–1829, by permission of IEEE.

Table 6.1 Rule base frame for phase plane method

		LN	MN	SN	ZE	SP	MP	LP
					e			
	LP				$c_2{}^1$			
	MP		A_3		$c_2{}^2$		A_4	
	SP				$c_2{}^3$			
Δe	ZE	$m_1{}^1$	$m_1{}^2$	$m_1{}^3$	ZE	$m_2{}^3$	$m_2{}^2$	$m_2{}^1$
	SN				$c_1{}^3$			
	MN		A_2		$c_1{}^2$		A_1	
	LN				$c_1{}^1$			

R_i : IF e is ZE and Δe is ZE THEN Δp is ZE

Table 6.2 Fuzzy PI control rule table for dealing with the fading process [25]

		LN	MN	SN	ZE	SP	MP	LP
					e			
	LP	ZE	SP	MP	MP	MP	MP	LP
	MP	SN	ZE	SP	MP	MP	MP	LP
	SP	MN	SN	ZE	SP	MP	MP	LP
Δe	ZE	LN	MN	SN	ZE	SP	MP	LP
	SN	LN	LN	MN	SN	ZE	SP	MP
	MN	LN	LN	LN	MN	SN	ZE	SP
	LN	LN	LN	LN	LN	MN	SN	ZE

The general rule from Table 6.1 is elaborated in Table 6.2.

In a real system with propagation delay, there will be an offset between the real signal power and the one seen by the system as shown in Figure 6.25.

The modified control rule is now given in Table 6.3. Let us go back to the parameters specified in Figure 6.19. e, $-18\,$dB $\sim$ to $18\,$dB is mapped into 13 integer quantization levels, -6 to 6. A membership matrix table is a discretization of membership function and can be defined by assigning grade of membership values to each quantization level. It includes the error, error change and control power increment variables. Each table consists of seven terms, including LP, MP, SP, ZE, SN, MN and LN, and each set consists of 13 quantization levels, labeled as -6, -5, $\ldots$, 6. All error, error change and control power increment variables are quantized to these 13 levels. The discrete ROV, membership matrix table, and the control rules of Table 6.2 are combined to form a decision table for the fuzzy controller. The decision table is shown in Table 6.4.

The performance of the original decision table can be significantly improved by introducing a new decision table to perform fine control. The fine decision table would be active with finer quantization levels when $(e, \Delta e)$ falls within a predetermined nested

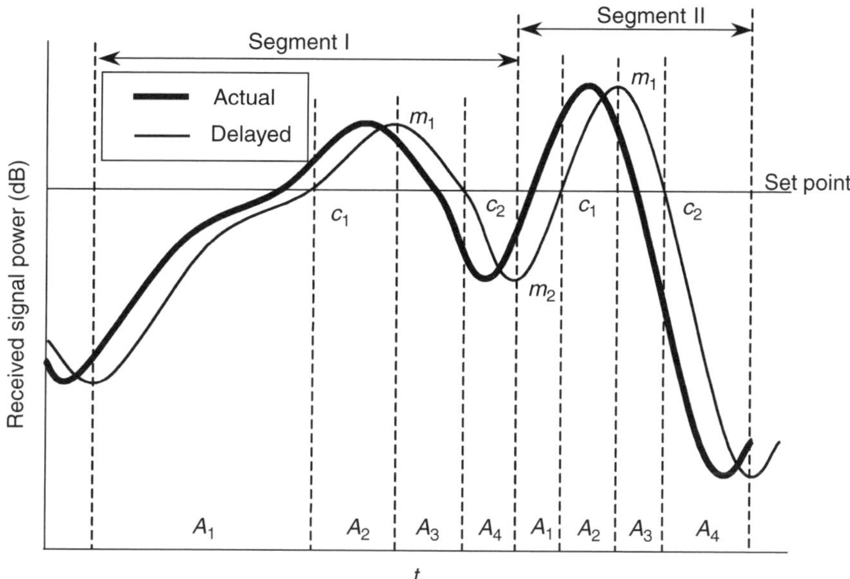

Figure 6.25 The effects of deadtime for a fading process [25]. Reproduced from Chang, P. R. and Wang, B. C. (1996b) Adaptive fuzzy proportional integral power control for a CDMA system with time delay. *IEEE J. Select. Areas Commun.*, **14**(9), 1818–1829, by permission of IEEE.

Table 6.3 The modified control rule

				e			
	LN	MN	SN	ZE	SP	MP	LP
LP	SP	SP	MP	MP	MP	MP	LP
MP	ZE	SP	SP	MP	MP	MP	LP
SP	SN	ZE	SP	SP	MP	MP	LP
Δe ZE	LN	MN	SN	ZE	SP	MP	LP
SN	LN	LN	LN	MN	SN	SN	ZE
MN	LN	LN	LN	MN	SN	SN	ZE
LN	LN	LN	LN	LN	MN	SN	SN

region. When $(e, \Delta e)$ is outside the nested region, Table 6.4 carries out the coarse control. In this application, the nested region is chosen as $-3\,\text{dB}$ to $3\,\text{dB}$ for e and $-6\,\text{dB}$ to $6\,\text{dB}$ for Δe. The limit of Δp is set between $-3.6\,\text{dB}$ and $3.6\,\text{dB}$. Corresponding term sets are {SP, ZE, SN} for e, {MP, SP, ZE, SN, MN} for Δe, and {MP, SP, ZE, SN, MN} for Δp. The associated decision table is shown in Table 6.5. In summary, the coarse table is used to achieve the fast response. The fine table produces a minimum steady-state error

Table 6.4 Decision look-up table for coarse control

e

		−6	−5	−4	−3	−2	−1	0	1	2	3	4	5	6
	−6	−6	−6	−6	−6	−6	−6	−6	−5	−4	−3	−2	−2	−2
	−5	−6	−6	−6	−6	−6	−5	−5	−4	−3	−3	−2	−1	−1
	−4	−6	−6	−6	−6	−6	−5	−4	−3	−2	−2	−2	−1	0
	−3	−6	−6	−6	−5	−5	−4	−3	−3	−2	−1	−1	0	1
	−2	−6	−6	−6	−5	−4	−3	−2	−2	−2	−1	0	1	2
	−1	−6	−5	−5	−4	−3	−2	−1	0	0	1	2	3	4
Δe	0	−6	−5	−4	−3	−2	−1	0	1	2	3	4	5	6
	1	−4	−3	−2	−1	0	0	1	2	3	4	4	5	6
	2	−2	−1	0	1	2	2	2	3	4	4	4	5	6
	3	−1	0	1	1	2	3	3	3	4	4	4	5	6
	4	0	1	2	2	2	3	4	4	4	4	4	5	6
	5	1	1	2	3	3	3	4	4	4	4	4	5	6
	6	2	2	2	3	4	4	4	4	4	4	4	5	6

Table 6.5 Decision look-up table for fine control

e

		−6	−5	−4	−3	−2	−1	0	1	2	3	4	5	6
	−6	−6	−6	−6	−6	−5	−5	−5	−5	−5	−5	−5	−5	−5
	−5	−6	−5	−5	−5	−5	−4	−4	−4	−4	−4	−4	−4	−4
	−4	−5	−5	−4	−4	−4	−4	−3	−3	−3	−3	−3	−3	−3
	−3	−4	−4	−4	−3	−3	−3	−3	−2	−2	−1	−1	−1	−1
	−2	−3	−3	−3	−3	−2	−2	−2	−1	−1	−1	0	0	0
	−1	−3	−2	−2	−2	−2	−1	−1	0	0	0	0	0	1
Δe	0	−2	−1	−1	−1	−1	0	0	0	1	1	1	1	2
	1	−1	0	0	0	0	0	1	1	2	2	2	2	3
	2	0	0	0	1	1	1	2	2	2	3	3	3	3
	3	1	1	1	1	2	2	3	3	3	3	4	4	4
	4	3	3	3	3	3	3	3	4	4	4	4	5	5
	5	4	4	4	4	4	4	4	4	4	4	4	5	5
	6	5	5	5	5	5	5	5	5	5	5	5	5	5

with a magnitude comparable to the width of the nested region.

To demonstrate the performance of the system, we assume the following scenario [25]:

1. The service area consists of 19 hexagonal shaped cells, that is, the desired cell is surrounded by two tiers of interfering cells.

2. All cells contain the same number of active mobile units, and the positions of the active mobile units within each cell are uniformly distributed with a density of K users per base station.
3. Interference reduction techniques such as cell sectorization and voice activity detection are not considered. It is believed that the improvement from these effects can be introduced through multiplicative factors.
4. Each user scans signals from the closest base stations and decides to communicate with the base station that has the largest local-mean signal power. This local-mean signal power was determined from path loss proportional to the fourth power of the propagation distance and simulated lognormal shadow fading with standard deviation of 8 dB.

The following system parameters are assumed:

- The spreading bandwidth is 1.25 MHz and the user data rate is 8 kb s^{-1}, which give a processing gain of approximately 22 dB.
- The required energy per bit to interference spectral density ratio, E_b/I_0 is selected as 7 dB (reverse link).
- The required receiver front end SIR threshold, SIR$_{th}$ is found to be -15 dB (reverse link).
- The sampling time period is set at $T_p = 1.25$ ms.
- For long-term fading, the path propagation loss exponent, α is assumed to be 4 and the standard deviation for shadowing is set at 8 dB.
- For short-term fading, $f_D T_p$ is uniformly distributed between 0.01 and 0.05, where f_D denotes the Doppler rate.
- The density order, m of Nakagami distribution is assumed to be either 2 or 4.
- For simplicity, it is assumed that the power adjustment command from the base station is not corrupted by the forward-link channel noise.
- There are two sets of control gains for fuzzy PI power-control system.
- One set of control gains, for example, $\{k_{p1}, k_{I1}\}$ is used for a coarse control, to speed up transient response.
- When the error falls within the preset limit, the second set of gains $\{k_{p2}, k_{I2}\}$ is used for fine control, which can smooth the response around the set point.
- In the experiments, k_{p1}, k_{I1}, k_{p2}, k_{I2} are chosen as 1/2, 1/3, 1 and 2, respectively.

The result for a given set of parameters is shown in Figure 6.26. One can see that Fuzzy PI control would reach the set point faster and, in general, maintain the set level with less error.

The tracking error root mean square (RMS) can be seen more accurately from Tables 6.6 and 6.7 for a given set of parameters. In general, one can see that a fuzzy controller would operate better.

The better tracking performance will result in better capacity, which is shown in Figure 6.27. Capacity results will be discussed later in much more detail.

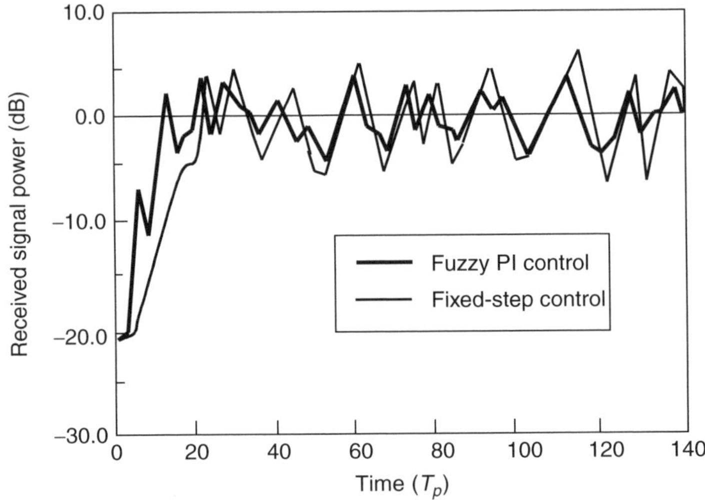

Figure 6.26 Comparison of the waveforms of the received signals achieved by the fuzzy PI control and 1 dB fixed-step control when $\tau = 2T_p$, $m = 4$, $f_D T_p = 0.05$, and the desired mobile unit is initially placed at position that causes a 20-dB path loss [25]. Reproduced from Chang, P. R. and Wang, B. C. (1996b) Adaptive fuzzy proportional integral power control for a CDMA system with time delay. *IEEE J. Select. Areas Commun.*, **14**(9), 1818–1829, by permission of IEEE.

Table 6.6 Comparison of RMS tracking error achieved by fuzzy PI control and fixed-step control when $m = 2$ or 4, $f_D T_p = 0.05$ and $\tau = T_p$ or $2T_p$ or $3T_p$ [25]

	$m = 2$			$m = 4$		
τ (time delay)	T_p	$2T_p$	$3T_p$	T_p	$2T_p$	$3T_p$
Fuzzy PI	3.84	4.53	5.52	3.38	3.95	4.78
Fixed step	6.05	6.76	7.44	5.45	6.08	6.96

Table 6.7 Comparison of RMS tracking error achieved by fuzzy PI control and fixed-step control when $m = 2$, $\tau = 2T_p$ and $f_D T_p = 0.025$, 0.0375 or 0.05 [25]

	$m = 2$	$m = 4$		
	Fuzzy PI	Fixed step	Fuzzy PI	Fixed step
$f_D T_p = 0.025$	3.8	6.16	3.52	5.81
$f_D T_p = 0.0375$	4.15	6.61	3.70	6.02
$f_D T_p = 0.05$	4.53	6.76	3.95	6.06

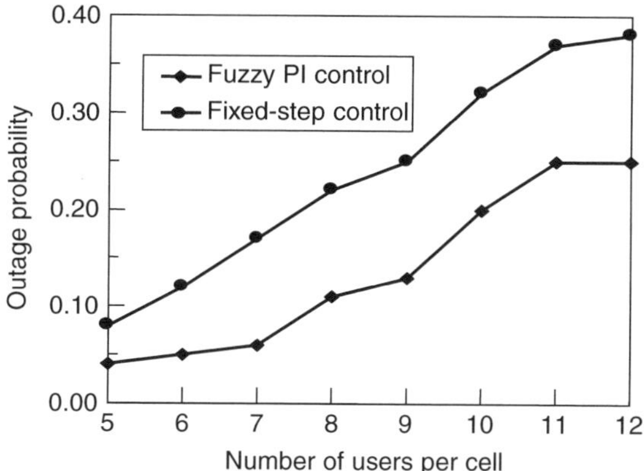

Figure 6.27 Comparison of outage probabilities against the number of users per cell achieved by fuzzy PI control and 1 dB fixed-step control when $m = 2$, $\tau = 2T_\mathrm{p}$, and $SIR_{th} = -15\,\mathrm{dB}$ [25]. Reproduced from Chang, P. R. and Wang, B. C. (1996b) Adaptive fuzzy proportional integral power control for a CDMA system with time delay. *IEEE J. Select. Areas Commun.*, **14**(9), 1818–1829, by permission of IEEE.

6.7 IMPERFECT POWER CONTROL IN CDMA SYSTEMS

The inaccuracy of PC is caused by a large change rate of the channel (e.g. high Doppler frequency), control delays, nonideal channel estimation, power-control command errors occurring in the feedback link and restricted dynamic range of the transmitter. Power-control errors increase the error probability exponentially, whereas the effect of processing gain and the number of users on the error probability is linear. Power-control errors increase the error probability in diversity systems also, since the error is the same in all diversity branches.

To analyze this issue, we represent the received signal as follows:

$$s(t) = \sum_{k=1}^{KM} A\lambda_k y(k) d_k(t - \tau_k) c_k(t - \tau_k) \mathrm{e}^{j\phi_k} = \sum_k s_k(t) \tag{6.14}$$

$$y(k) = \begin{cases} 1, & k \leq K \\ \left(\dfrac{r_{mk}}{r_{0k}}\right)^2 10^{(\xi_{0k}-\xi_{mk})/20}, & k > K \end{cases} \tag{6.15}$$

Factor $r_{mk}^2 10^{\xi_{mk}/20}$ is due to the PC of the kth user in the mth cell to compensate for propagation loss to its own base station, and the factor $(1/r_{0k}^2)10^{-\xi_{0k}/20}$ represents the distance loss and shadowing suffered from the same signal traveling to the base station of interest.

A detailed derivation of equation (6.15) will be given in Chapter 8. K is the number of users and M is the number of interfering cells. Parameter λ_k characterizes the power-control error ($\lambda_k = 1$ no error). For the multipath channel represented by impulse response,

$$h(\tau, t) = \sum_{l=1}^{L} \alpha_l(t) e^{j\psi_l(t)} \delta(\tau - lT_c) \tag{6.16}$$

the received signal is

$$r'(t) = \sum_{l=1}^{L} \sum_{k=1}^{KM} \alpha_{lk} e^{j\psi_{lk}} s_k(\tau - lT_c) + n_w(t) \tag{6.17}$$

The optimum receiver, maximum ratio combining coherent RAKE, is shown in Figure 6.28. For the evaluation of bit error probability we define [77]

$$\gamma_L \equiv \sum_{l=1}^{L} \frac{\alpha_l^2}{c} = \sum_{l=1}^{L} \gamma_l \tag{6.18}$$

$c \equiv 2c' + \eta_0/E_b$ normalized interference plus noise density implicitly defined by equation (6.19). γ_l is the instantaneous signal to interference + noise ratio (SINR) for each resolvable path. The $\{\gamma_l\}$ are independent exponential random variables. When they are identically distributed, that is, when every path has the same average SINR γ_c defined as

$$\overline{\gamma}_c = E\{\alpha^2\} \Bigg/ \left((2/N)E\{\alpha^2\} \left\{ (K - 1) + \sum_{k=K+1}^{KM} E\left[y(k)^2\right] \right\} + \eta_0/E_b \right) \tag{6.19}$$

With $E\{\alpha_l^2\} = E\{\alpha^2\}$, for all l, the channel has a constant multipath intensity profile (MIP); otherwise, the channel has a variable MIP.

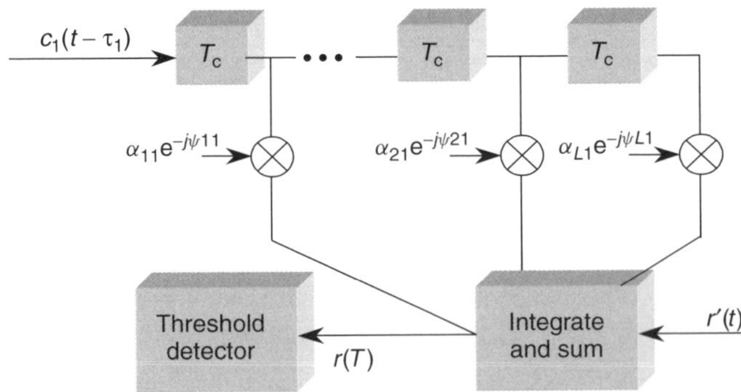

Figure 6.28   The optimum receiver. Reproduced from Kong, N. and Milstein, L. B. (1995) Performance of multicell CDMA with power control error. *Proc. IEEE Military Communications Conference*, San Diego, CA, pp. 513–517, by permission of IEEE.

When the channel has a variable MIP, the density function of γ_L, $f_v(\gamma_L)$ can be found through the Fourier transform of its characteristic function, which is a product of the characteristic function of each independent γ_l. In particular,

$$f_v(\gamma_L) = \sum_{l=1}^{L} \frac{C_l}{\overline{\gamma}_l} e^{-(\gamma_L/\overline{\gamma}_l)}$$

where

$$C_l = \prod_{\substack{i=1 \\ i \neq l}}^{L} \frac{\overline{\gamma}_l}{\overline{\gamma}_l - \overline{\gamma}_i} \tag{6.20}$$

By using the standard expression for BER and its averaging with the above probability density function (pdf) for γ_L, we get for the average bit error rate

$$P(e) = \frac{1}{2} - \frac{1}{2} \sum_{l=1}^{L} C_l E_\lambda \left\{ \frac{1}{\sqrt{1 + \dfrac{1}{\lambda^2 \overline{\gamma}_l}}} \right\} \tag{6.21}$$

Constant CIR $\quad P(e) = \binom{2L-1}{L} \left(\dfrac{1}{4\overline{\gamma}_c} \right)^L e^{(1/2)Lb^2\sigma_e^2} \tag{6.22}$

where σ_e^2 is power-control mean square error $E_1[\lambda - 1]^2 = \sigma_e^2$ and $b = \ln 10/10$

$$\text{Nonconstant CIR \& high SINR } P(e) = \binom{2L-1}{L} \left(\prod_{l=1}^{L} \frac{1}{4\overline{\gamma}_l} \right) e^{(1/2)Lb^2\sigma_e^2} \tag{6.23}$$

For frequency-nonselective fading, the result is valid when $L = 1$

$$P(e/\lambda) = \frac{1}{2} - \frac{1}{2} \frac{1}{\sqrt{1 + \dfrac{1}{\lambda^2 \overline{\gamma}_l}}} \tag{6.24}$$

For small standard deviations of the power control error (PCE), and for large SINR

$$P(e) \approx \frac{e^{(1/2)b^2\sigma_e^2}}{4\overline{\gamma}} \tag{6.25}$$

An example of bit error probability evaluation is shown in Figure 6.29 for a given set of parameters and exponential MIP.

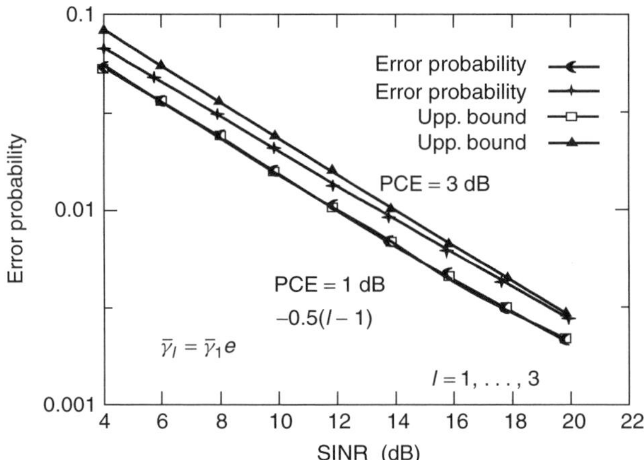

Figure 6.29 Error probability compared to upper bound as a function of SINR for different PCE [77]. Reproduced from Kong, N. and Milstein, L. B. (1995) Performance of multicell CDMA with power control error. *Proc. IEEE Military Communications Conference*, San Diego, CA, pp. 513–517, by permission of IEEE.

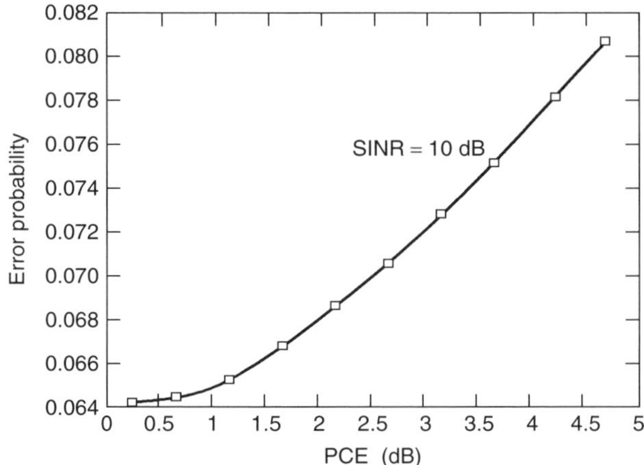

Figure 6.30 Error probability as a function of PCE [77]. Reproduced from Kong, N. and Milstein, L. B. (1995) Performance of multicell CDMA with power control error. *Proc. IEEE Military Communications Conference*, San Diego, CA, pp. 513–517, by permission of IEEE.

Error probability as a function of power-control error is given in Figure 6.30.

For the system with diversity, probability of error is given in Figure 6.31.

If we want the inaccuracy of PC to be at most ± 1 dB with 90% probability when the mobile station velocity is $3\,\mathrm{km\,h^{-1}}$ (with carrier frequency $f_c = 850\,\mathrm{MHz}$, Doppler frequency becomes $f_d \approx 2.4\,\mathrm{Hz}$), the feedback delay can be at most 0.5 ms [78].

Since the CDMA system is interference limited, the system capacity is maximized by minimizing each mobile to other mobile interference. By capacity, we mean here the

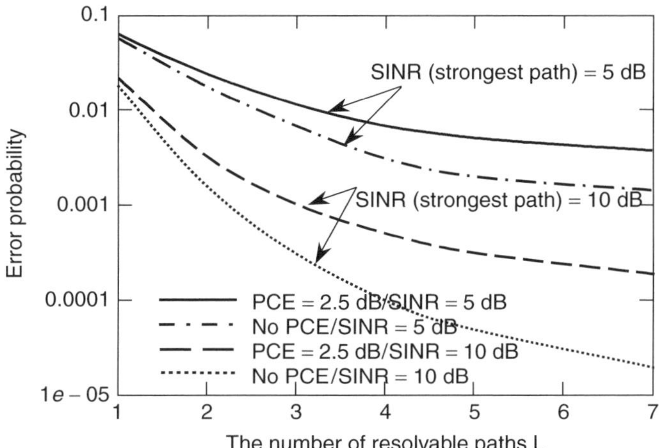

Figure 6.31 Error probability as a function of diversity with different PCE [77]. Reproduced from Kong, N. and Milstein, L. B. (1995) Performance of multicell CDMA with power control error. *Proc. IEEE Military Communications Conference*, San Diego, CA, pp. 513–517, by permission of IEEE.

maximum number of simultaneous active users per cell. Interference is minimized when mobile stations transmit with the minimum possible power by which good quality communication is achieved. Thus, PC significantly affects the capacity of the CDMA system. In this section, we introduce papers that present studies on how much imperfect PC reduces the CDMA system capacity. Wu has shown analytically that the capacity is decreased by 2.3 dB when the standard deviation of the lognormally distributed, received SIR is 1 dB [79]. Since a variety of system models and parameters are used in the following papers, numerical results are not directly comparable. We can notice from the results, however, that even a small error in PC reduces the capacity considerably. In this paper, it was assumed that the received power is a lognormally distributed random variable. The inaccuracy of PC is modeled by the logarithmic standard deviation of the received power, denoted by σ. In Reference [80], the effect of imperfect PC on the capacity was analyzed in both a single cell and the cellular CDMA system. In the case of the single cell, the influences of both voice activity detection and processing gain were also taken into account. Jensen and Prasad established a simple channel model in which the received power was inversely proportional to the fourth power of the distance [81]. The capacity was determined as the maximum number of users in a cell with which each received signal's SIR, at the base station, is at least (with 99% probability) larger than 7 dB. When the power-control error was $\sigma = 1$ dB, the capacity of the cellular system was observed to decrease about 50 to 60% compared to the capacity of the system using ideal $\sigma = 0$ dB) PC. Also, the throughput and delay with imperfect PC were investigated for data communications. According to Reference [82], the system capacity diminishes 35% with 5% probability when the standard deviation is $\sigma = 1$ dB. Correspondingly, reduction of the capacity is 50% with $\sigma = 2$ dB. The channel model included long-term attenuation.

The influence of imperfect PC was introduced in the analytical expression of the CDMA system MAI in Reference [83]. In a single cell system, the capacity was defined as the number of users whose error probabilities are at most 0.001. When the standard deviation was $\sigma = 1$ dB, the capacity decreased by approximately 15%. Similarly, when $\sigma = 1.4$ dB, the reduction was about 30%, and almost 60% when $\sigma = 2$ dB.

Kudoh investigated the effect of PC on the system capacity by establishing a simulation model [84]. The channel model included long-term attenuation. The power-control error was, again, assumed to follow lognormal distribution. When the standard deviation of the error was $\sigma = 1$ dB, the capacity (number of users whose average SIR is greater than 7 dB with 99% probability) was decreased by 31% compared to the system employing ideal PC.

Correspondingly, the reduction was 61% with $\sigma = 2$ dB and 81% with $\sigma = 3$ dB. Also, the effects of a finite dynamic range of PC, restricted base station diversity and nonuniform user distribution on the capacity were considered.

In the channel model presented in Reference [85], only propagation loss was taken into account, and the received power was inversely proportional to the fourth power of distance. In this case, when the standard deviation was $\sigma = 2$ dB, the capacity was observed to decrease by over 50%.

The effect of imperfect PC on the Erlang capacity of the cellular CDMA system was investigated in Reference [86]. The Erlang capacity was defined as an average number of users in a cell when the received total interference power at a base station is (with 99% probability) at most 10 dB greater than the background noise power. When the standard deviation was chosen, according to field tests, to be $\sigma = 2.5$ dB, the Erlang capacity decreased by 20% compared to the system using ideal PC. Note that σ now indicates the standard deviation of errors in tracking Rayleigh fading. These errors have a far smaller impact than errors in tracking long-term fading (as was the case in the papers discussed previously).

6.8 ADAPTIVE COMMUNICATIONS

Previous discussion was focused on the problem of how the system capacity in CDMA network can be maximized by using adaptive PC. This concept has been extended to the possibility of adapting other parameters of the system too, in order to maximize the system capacity. This has attracted a significant interest of information theory too.

Recently, in Reference [87], the optimal adaptive transmission scheme was derived that achieves the Shannon capacity for a fading channel. Channel state values were assumed to be known. The modulation and coding strategy that achieves this capacity is a multiplexing technique whereby the coding and modulation transmitted over the channels are optimized for instantaneous fade levels [87,88]. The resulting transmission scheme is both variable power and variable rate. The power adaptation in this scheme is a 'water-filling' strategy. In particular, when the channel is favorable, more power is allocated for transmission. Conversely, when the channel is not as good, less power is transmitted. If the channel quality drops below a certain threshold, the channel is not used for that transmission. Note that if the transmission power optimization is performed in order to obtain, not the maximum capacity, but the best error exponent, the optimal power method is different from the water-filling strategy [89]. It was realized that the capacity difference between the optimal scheme and the constant power variable-rate scheme was a small fraction of a decibel for most types of fading [88]. Note that the effect of power adaptation is pronounced in the multiuser case, in which power adaptation affects the interference on other users [90]. On the other hand, Lau [91] proved that the capacity of the constant-power variable-rate scheme over a Rayleigh fading channel is the same as with the constant-power constant-rate technique (only the error exponent can be increased, i.e. the error probability can be decreased faster with the code block length by adjusting the rate only). Goldsmith [87] had shown that this is the case for independent and identically distributed (iid.) fading. In Reference [92], a capacity of a Rayleigh fading channel with Lth-order independent antenna diversity and maximal ratio combining was evaluated. It was found that a moderate diversity order ($L = 3$) is sufficient to approach the AWGN capacity by less than 1 dB. In Reference [93], it was demonstrated that by using instantaneous (instead of average) BER constraints, one can obtain constant-power variable-rate policy that achieves rates comparable to the optimum variable-power variable-rate scheme.

Furthermore, in Reference [94], optimal constant-rate transmission schemes for a block-fading channel with a strict transmission delay constraint were studied under the assumption, again, that both the transmitter and receiver have perfect channel state information. In a block-fading channel, the block of several symbols undergoes the same channel state defined by the fading gain. A code word spans a group of certain amount of blocks, referred as a frame. The number of blocks in a frame determines the interleaving depth, and is also considered to be a measure of the overall transmission delay. It should be emphasized that Goldsmith [87,88] discussed the capacity in a delay-unconstrained (ideal interleaving) case, and the resulting transmission scheme was variable power and variable rate. The important observation in Reference [94] was that no variable-rate coding (or 'multiple-codebook' transmission) is required in order to achieve the capacity, but that a constant-rate (single-codebook) variable-power scheme is sufficient. Recall that when the performance criterion was bit error probability (and no channel coding was included), the effect of rate adjustment was superior to PC [66]. Also, the 'delay-limited' capacity (a delay-constrained case) over a Rayleigh fading channel was shown to be only 2.5 dB away from the (delay-unlimited) capacity for high rates in the case of two independent blocks in a frame and no diversity. Note that the delay-limited capacity is zero when only one block is included in a frame (no interleaving case) and there is no diversity.

Two suboptimal variable-power constant-rate schemes using channel inversion and truncated channel inversion were discussed in Reference [88]. Channel inversion adapts the transmission power to maintain a constant received SNR. This form of power adaptation greatly simplifies the coding and modulation for the fading channel since the channel with inversion appears as an AWGN channel to the encoder and decoder. However, this technique suffers a large power penalty since most of the average signal power is used to compensate for deep fades. In fact, the capacity for channel inversion in Rayleigh fading is zero. Truncated channel inversion maintains a constant received SNR unless fading falls below a given cutoff level, at which point a signal outage is declared and no signal is sent. The capacity of this truncated policy with optimized cutoff level was shown to exhibit a power loss relative to optimal variable-power variable-rate policy [87] of 1–2 dB in

Rayleigh fading. However, the corresponding outage probability can be quite high. Thus, constant-rate transmission with truncated channel inversion approximates a packet radio protocol, with bursts of high-speed data when the channel is favorable and idle times in between. Note that the capacity penalties with different suboptimal schemes compared to the optimal transmission strategy are diminished in the diversity case with an increasing diversity order [92]. In particular, the channel inversion rather than the truncated channel inversion method can be a better choice when diversity is available.

The spectral efficiency of the variable-power variable-rate M-ary quadrature amplitude modulation (MQAM) modulation scheme was also derived and compared to the fading channel capacity presented above in Reference [88]. The power-control scheme in MQAM modulation has the same form as the optimal power-control strategy, which achieves this capacity [88]. There is a constant power gap between the spectral efficiency of the MQAM modulation scheme and the channel capacity, and this gap is a simple function of the required BER. The variable-power variable-rate MQAM modulation scheme exhibits up to 20 dB of gain relative to nonadaptive transmission, in which both the transmission power and the rate are constant. Note that in Reference [93] it was demonstrated that a simultaneous optimization of rate and power adaptation under the instantaneous BER constraint actually yields a constant-power policy when there is no restriction on the available rates. To improve the spectral efficiency, the coding scheme can be superimposed on top of adaptive modulation. Achievable rates for adaptive trellis-coded MQAM have been investigated in Reference [95]. At low BERs, a simple four-state trellis code yields an asymptotic coding gain of 3 dB, and an eight-state code yields a gain of 3.6 dB. Finally, adaptive coded modulation comes within 6 dB of the Shannon capacity of the fading channel with adaptive transmission using a 128-state code. Thus, the constant gap between the spectral efficiency of adaptive modulation and Shannon capacity cannot be fully closed. This discrepancy between Shannon capacity and achievable rates arises from the lack of complexity and implementation constraints inherent to Shannon theory. The authors in Reference [95], however, believe that by using powerful turbo codes, adaptive coded modulation will come quite close to the Shannon capacity of fading channels.

In practice, a wireless channel varies over time, which results in a different channel at the time of data transmission than at the time of channel estimation. Goeckel characterized the effects of this channel variation on the adaptive signaling paradigm [96], and used this characterization to design adaptive signaling schemes that are effective for the time-varying channel. He considered uncoded MQAM and trellis-coded modulation systems with low mobility. The proposed scheme was robust. That is, neither the Doppler frequency nor the exact shape of the autocorrelation function of the channel fading process needed to be known. Either a single noiseless outdated fading estimate was available at the transmitter, or multiple estimates were employed in order to achieve spectral efficiency gains for systems operating over channels that exhibit higher rates of variation. In the case of multiple outdated fading estimates, the data rate of the robust adaptive signaling method can be greatly reduced, and the adaptive signaling, in which the autocorrelation function is known should be called for.

Although adaptive modulation techniques increase the spectral efficiency (bps Hz^{-1}) of a single channel, these techniques may also increase cochannel interference levels in a cellular system [88]. Adaptive modulation may therefore reduce the area spectral efficiency of a cellular system, defined as its average bps $Hz^{-1}\,km^{-2}$. Indeed, while channel inversion can significantly reduce the spectral efficiency of a single user relative to optimal adaptation, this type of inversion is necessary in CDMA cellular systems without multiuser detection to reduce the near–far effect, or because of a signal dynamics restriction in detection. Truncated channel inversion is most effective for channels with large power fluctuations, and for channels with large background noise, in which multiuser interference is not the dominant source of errors. In Reference [90], the Shannon capacity region of the forward-link channel (corresponding to a single isolated cell) is obtained in fading and AWGN for time-division, frequency-division and code-division multiple access. The maximum capacity region is achieved by using a variable-power variable-rate multiresolution code division with successive decoding. However, the capacity region of the different spectrum sharing techniques is the same if all users have the same transmission power and fading distribution. Spread-spectrum code division with successive interference cancellation also maximizes spectral efficiency, although bandwidth expansion will result in some rate penalty. The optimality of this multiuser method is, however, only valid for Shannon capacity bounds, in which the probability of decoding error is asymptotically small.

SYMBOLS

n – sampling index
P_n^t – transmitted power
P_n^* – reference power
E_n – power error
A_n – channel losses
R_n – received power
D – delay
B – error averaging period
f_d, f_D – Doppler frequency
E_b/N_0 – signal-to-noise ratio
PC – power control
Δ – power step
p_r – power command bit error
$\tau_{(r,f)}$ – propagation delay (reverse, forward)
$u(t)$ – control action
$e(t)$ – error signal
$F\{\ \}$ – fuzzy function
Δp – power-control step
LP, MP, SP – large, medium, small (positive)
ZE – zero
SN, MN, LN – small, medium, large (negative)
$M(\cdot)$ – membership function
m_a^b – membership element (amplitude)
c_a^b – membership element (slope)
$T(\cdot)$ – associated term set

A_k – region (set of values)

K – number of users

M – Nakagami distribution parameter

T_p – power-control updating interval

λ – power-control error

f_s – bandwidth of the transmitted real bandpass signal

$(\Delta f)_c$ – the channel coherence bandwidth

L – number of resolvable paths, and $L = [f_s/(\Delta f)_c]$

$\xi \sim N(0, \sigma_s^2)$ – shadowing in decibels

r_{mk} – the distance from the kth user in the mth cell to its own base station

r_{0k} – the distance from the kth user in the mth cell to the base station of interest

α_{lk} – the Rayleigh fading r.v. for the kth user and lth path

N – processing gain, defined as $N = T/T_c$, where T_c is the chip duration and T is the symbol duration

$r(T)$ – the RAKE receiver decision variable

$g_1(T)$ – multipath interference from the user-of-interest

$g_2(T)$ – multipath and multiuser interference from all other users

γ_l – instantaneous SINR for the lth resolvable path

$\lambda' = 1/\lambda^2$

REFERENCES

1. Salmasi, A. and Gilhousen, S. (1991) On the system design aspects of code division multiple access (CDMA) applied to digital cellular and personal communications networks. *Proc. IEEE Vehicular Technology Conference*, St. Louis, MN, pp. 57–62.
2. Soliman, S., Wheatley, C. and Padovoani, R. (1992) CDMA reverse link open loop power control. *Proc. IEEE Global Telecommunication Conference*, Orlando, FL, pp. 69–73.
3. Gilhousen, K. S., Padovani, R. and Wheatley, C. E. (1991a) *Method and Apparatus for Controlling Transmission Power in a CDMA Cellular Mobile Telephone System*. US Patent 5.056.109. App. 433.031, Qualcomm Inc.
4. Sampath, A., Kumar, P. S., and Holtzman, J. M. (1997) On setting reverse link target SIR in a CDMA system. *Proc. IEEE Vehicular Technology Conference*, Phoenix, AZ, pp. 929–933.
5. Won, S. H., Kim, W. W. and Jeong, I. M. (1997) Performance improvement of CDMA power control in variable fading environments. *Proc. SouthEastCon '97*, Blacksburg, VA, pp. 241–243.
6. Ling, F., Love, B. and Wang, M. M. (1997) Behavior and performance of power controlled IS-95 reverse-link under soft handoff. *Proc. IEEE Vehicular Technology Conference*, Phoenix, AZ, pp. 924–928.
7. Viterbi, A. J. and Padovani, R. (1992) Implications of mobile cellular CDMA. *IEEE Commun. Mag.*, **30**(12), 38–41.
8. Simpson, F. and Holtzman, J. M. (1993) Direct sequence CDMA power control, interleaving, and coding. *IEEE J. Select. Areas Commun.*, **11**(7), 1085–1095.
9. Zhuang, W. (1997) Channel coding and power control for DS/CDMA multimedia wireless communications. *Proc. IEEE Global Telecommunication Conference*, Phoenix, AZ, pp. 604–608.
10. Abeta, S., Sampei, S. and Morinaga, N. (1996) Channel activation with adaptive coding rate and processing gain control for cellular DS/CDMA systems. *Proc. IEEE Vehicular Technology Conference*, Atlanta, GA, pp. 1115–1119.
11. Yamazato, T., Shinkaji, Y., Katayama, M. and Ogawa, A. (1994) Near-far problem of hybrid DS/SFH-SSMA with multi-level power control. *Proc. IEEE International Symposium on Information Theory and its Applications*, Sydney, Australia, pp. 109–113.

12. Grujev, S., Rooimans, R. G. A. and Prasad, R. (1996) Hybrid DS/SFH CDMA system with near-far effect and imperfect power control. *Proc. IEEE International Symposium on Spread Spectrum Techniques and Applications*, Mainz, Germany, pp. 329–333.
13. Padovani, R. (1994) Reverse link performance of IS-95 based cellular systems. *IEEE Personal Commun.*, **1**(3), 28–34.
14. Viterbi, A. J., Viterbi, A. M. and Zehavi, E. (1993) Performance of power-controlled wideband terrestrial digital communication. *IEEE Trans. Commun.*, **41**(4), 559–569.
15. Pichna, R. and Wang, Q. (1996) *The Mobile Communications Handbook*. New York: CRC Press, pp. 370–380.
16. Stuber, G. L. and Kchao, C. (1992) Analysis of a multiple-cell DS/CDMA cellular mobile radio system. *IEEE J. Select. Areas Commun.*, **10**(4), 669–679.
17. Tonquz, O. K. and Wang, M. M. (1994) Cellular CDMA networks impaired by Rayleigh fading: system performance with power control. *IEEE Trans. Veh. Technol.*, **43**(3), 515–526.
18. Jalali, A. and Mermelstein, P. (1994) Effects of diversity, power control, and bandwidth on the capacity of microcellular CDMA systems. *IEEE J. Select. Areas Commun.*, **12**(5), 952–961.
19. Chockalingam, A., Dietrich, P., Milstein, L. B. and Rao, R. R. (1998) Performance of closed loop power control in DS-CDMA cellular systems. *IEEE Trans. Veh. Technol.*, **47**(3), 774–789.
20. Ariyavisitakul, S. and Chang, L. F. (1993) Signal and interference statistics of a CDMA system with feedback power control. *IEEE Trans. Commun.*, **41**(11), 1626–1634.
21. Lee, T.-H. and Lin, J.-C. (1996) A fully distributed power control algorithm for cellular mobile systems. *IEEE J. Select. Areas Commun.*, **14**(4), 692–697.
22. Lee, C.-C. and Steele, R. (1996) Closed-Loop power control in CDMA systems. *IEE Proc.-Part F*, **143**(4), 231–239.
23. Nikolai, D. and Kammeyer, K.-D. (1996) Noncoherent RAKE receiver with optimum weighted combining and improved closed-Loop power control. *Proc. IEEE International Symposium on Spread Spectrum Techniques and Applications*, Mainz, Germany, pp. 239–243.
24. Chang, P. R. and Wang, B. C. (1996a) Adaptive fuzzy power control for CDMA mobile radio systems. *IEEE Trans. Veh. Technol.*, **45**(2), 225–236.
25. Chang, P. R. and Wang, B. C. (1996b) Adaptive fuzzy proportional integral power control for a CDMA system with time delay. *IEEE J. Select. Areas Commun.*, **14**(9), 1818–1829.
26. Wu, J. and Kohno, R. (1996) A wireless multimedia CDMA system based on transmission power control. *IEEE J. Select. Areas Commun.*, **14**(4), 683–691.
27. Yun, L. C. and Messerschmitt, D. G. (1995) Variable quality of service in CDMA systems by statistical power control. *Proc. IEEE International Conference on Communications*, Seattle, WA, pp. 713–719.
28. Gunnarsson, F. (2000) *Power Control in Cellular Radio Systems: Analysis, Design and Estimation*. Doctoral Thesis, Linköpings Universitet, Linköping, Sweden, p. 245.
29. Chang, L. F. and Ariyavisitakul, S. (1991) Performance of power control method for CDMA radio communication systems. *Electron. Lett.*, **27**(11), 920–922.
30. Kim, J.-H. Huang, G. M. and Georghiades, C. N. (1999) Stability upper bounds for reverse link power control of CDMA systems. *Proc. IEEE Vehicular Technology Conference*, Houston, TX, pp. 2139–214.
31. Seo, S., Dohi, T. and Adachi, F. (1998) SIR-based transmit power control of reverse link for coherent DS-CDMA mobile radio. *IEICE Trans. Commun.*, **E81-B**(7), 1508–1516.
32. Kumar, P. S., Yates, R. D. and Holtzman, J. (1995) Power control based on bit error rate (BER) measurements. *Proc. IEEE Military Communications Conference*, San Diego, CA, pp. 617–620.
33. Ariyavisitakul, S. (1994) Signal and interference statistics of a CDMA system with feedback power control – part II. *IEEE Trans. Commun.*, **42**, 597–605.
34. Yang, Y.-J. and Chang, J.-F. (1996) A strength and SIR combined adaptive power control for CDMA mobile radio channels. *Proc. IEEE International Symposium on Spread Spectrum Techniques and Applications*, Mainz, Germany, pp. 1167–1171.
35. Su, S.-L. and Shieh, S.-S. (1995) Reverse-link power control strategies for CDMA cellular network. *Proc. IEEE International Symposium on Personal, Indoor, and Mobile Radio Communications*, Toronto, Canada, pp. 461–465.

36. Chang, C. J., Lee, J. H. and Ren, F. C. (1996) Design of power control mechanisms with PCM realization for the uplink of a DS/CDMA cellular mobile radio system. *IEEE Trans. Veh. Technol.*, **45**(3), 522–530.

37. Song, L., Mandayam, N. B. and Gajic, Z. (1999) Analysis of an up/down power control algorithm for the CDMA reverse link: a nonlinear control system approach. *Proc. Conference on Information Sciences and Systems*, Baltimore, MD, pp. 119–124.

38. Su, H.-J. and Geraniotis, E. (1999) Adaptive closed-loop power control with quantized feedback and loop filtering. *Proc. Conference on Information Sciences and Systems*, Baltimore, MD, pp. 130–135.

39. Aien, J. M. (1973) Power balancing in systems employing frequency reuse. *COMSAT Tech. Rev.*, **3**(2), 277–300.

40. Zander, J. (1992a) Distributed cochannel interference control in cellular radio systems. *IEEE Trans. Veh. Technol.*, **41**(3), 305–311.

41. Nettleton, R. W. (1980) Traffic theory and interference management for a spread spectrum cellular mobile radio system. *Proc. IEEE International Conference on Communications*, Seattle, WA.

42. Alavi, H. and Nettleton, R. W. (1982) Downstream power control for a spread spectrum cellular mobile radio system. *Proc. IEEE Global Telecommunication Conference*, Miami, FL, pp. 84–88.

43. Nettleton, R. W. and Alavi, H. (1983) Power control for a spread spectrum cellular mobile radio system. *Proc. IEEE Vehicular Technology Conference*, Toronto, Canada, pp. 242–246.

44. Zander, J. (1992b) Performance of optimum transmitter power control in cellular radio systems. *IEEE Trans. Veh. Technol.*, **41**(1), 57–62.

45. Wu, Q. (1999) Performance of optimum transmitter power control in CDMA cellular mobile systems. *IEEE Trans. Veh. Technol.*, **48**(2), 571–575.

46. Grandhi, S. A., Vijayan, R. and Goodman, D. J. (1994) Distributed power control in cellular radio systems. *IEEE Trans. Commun.*, **42**, 226–228.

47. Leung, Y.-W. (1996) Power control in cellular networks subject to measurement error. *IEEE Trans. Commun.*, **44**(7), 772–775.

48. Foschini, G. J. and Miljanic, Z. (1993) A simple distributed autonomous power control algorithm and its convergence. *IEEE Trans. Veh. Technol.*, **42**(4), 641–646.

49. Almgren, M., Andersson, H. and Wallstedt, K. (1994) Power control in a cellular system. *Proc. IEEE Vehicular Technology Conference*, Stockholm, Sweden, pp. 833–837.

50. Yates, R. D. and Huang, C.-Y. (1995) Integrated power control and base station assignment. *IEEE Trans. Veh. Technol.*, **44**(3), 638–644.

51. Ulukus, S. and Yates, R. D. (1998a) Adaptive power control and MMSE interference suppression. *Wireless Networks*, **4**(6), 489–496; Correction.

52. Hanly, S. V. (1996) Capacity and power control in spread spectrum macrodiversity radio networks. *IEEE Trans. Commun.*, **44**(2), 247–256.

53. Rashid-Farrokhi, R., Tassiulas, L. and Liu, K. J. R. (1998) Joint optimal power control and beamforming in wireless networks using antenna arrays. *IEEE Trans. Commun.*, **46**(10), 1313–1323.

54. Schalkwijk, J. P. M. (1969) Recent development in feedback communication. *Proc. IEEE*, **57**(7), 1242–1249.

55. Lucky, R. W. (1973) A survey of the communication theory literature: 1968–1973. *IEEE Trans. Inform. Theory*, **IT-19**(5), 725–739.

56. Ural, A. T. and Haddad, A. H. (1972) A binary sequential communication scheme with information feedback. *IEEE Trans. Commun.*, **COM-20**(6), 423–429.

57. Turin, G. L. (1966) Comparison of sequential and nonsequential detection systems with uncertainty feedback. *IEEE Trans. Inform. Theory*, **12**(1), 5–8.

58. Schalkwijk, J. P. M. and Kailath, T. (1966) A coding scheme for additive noise channels with feedback – part I: no bandwidth constraint. *IEEE Trans. Inform. Theory*, **IT-12**(2), 172–182.

59. Schalkwijk, J. P. M. (1966) A coding scheme for additive noise channels with feedback – part II: band-limited signals. *IEEE Trans. Inform. Theory*, **IT-12**(2), 183–189.

60. Shannon, C. E. (1956) The zero error capacity of a noisy channel. *IRE Trans. Inform. Theory*, **IT-2**, 8–19.
61. Kramer, A. J. (1969) Improving communication reliability by use of an intermittent feedback channel. *IEEE Trans. Inform. Theory*, **IT-15**(1), 52–60.
62. Butman, S. (1969) A general formulation of linear feedback communications systems with solutions. *IEEE Trans. Inform. Theory*, **IT-15**(3), 392–400.
63. Ebert, P. M. (1970) The capacity of the Gaussian channel with feedback. *Bell Syst. Tech. J.*, **49**(8), 1705–1712.
64. Glave, F. E. (1972) Communication of fading dispersive channels with feedback. *IEEE Trans. Inform. Theory*, **IT-18**(1), 142–150.
65. Hayes, J. F. (1968) Adaptive feedback communications. *IEEE Trans. Commun.*, **16**(2), 29–34.
66. Cavers, J. K. (1972) Variable-rate transmission for Rayleigh fading channels. *IEEE Trans. Commun.*, **COM-24**(1), 15–22.
67. Coutts, R. P. and Davis, B. R. (1976) Buffer requirements for intermittent data transmission over a Rayleigh fading channel. *IEEE Trans. Commun.*, **24**(10), 1122–1129.
68. Cavers, J. K. and Lee, S. K. (1976) A simple buffer control for variable-rate communication systems. *IEEE Trans. Commun.*, **COM-24**(9), 1045–1048.
69. Cavers, J. K. (1977) Buffer control for transmission of blocked data over fading channels. *IEEE Trans. Commun.*, **COM-25**(5), 496–502.
70. Srinivasan, R. (1981) Feedback communications over fading channels. *IEEE Trans. Commun.*, **29**(1), 50–57.
71. Srinivasan, R. (1975) *Feedback Communication Systems for Time-Varying Channels*. Ph. D. Thesis, University of Aston, Birmingham, UK, p. 105.
72. Alamouti, S. M. and Kallel, S. (1994) Adaptive trellis-coded multiple-phase-shift keying for Rayleigh fading channels. *IEEE Trans. Commun.*, **42**(6), 2305–2314.
73. Kousa, M. A. and Turner, L. F. (1993) Multichannel adaptive forward error-correction system. *IEE Proc. – Part I*, **140**(5), 357–364.
74. Gudmundson, M. (1991) Analysis of handover algorithm. *Proc. VTC '91*, Vol. 1, May 1991, pp. 537–541.
75. Gudmundson, M. (1991) Correlation model for shadow fading in mobile radio systems. *Electron. Lett.*, **27**, 2145–2146.
76. Holtzman, J. M. (1992) CDMA power control for wireless network, in Nanda, S. and Goodman, D. J. (eds) *Third Generation Wireless Information Network*, Boston, MA: Kluwer, pp. 299–311.
77. Kong, N. and Milstein, L. B. (1995) Performance of multicell CDMA with power control error. *Proc. IEEE Military Communications Conference*, San Diego, CA, pp. 513–517.
78. Larsson, A. and Maseng, T. (1996) A statistical analysis of the power control error in fast Rayleigh fading. *Proc. IEEE Vehicular Technology Conference*, Atlanta, GA, pp. 1140–1144.
79. Wu, Q., Wu, W.-L. and Zhou, J.-P. Distributed power control in CDMA cellular mobile systems. *IEEE Trans. Veh. Technol.*; submitted for publication.
80. Prasad, R., Jansen, M. and Kegel, A. (1992) *Cellular DS/CDMA Systems with Imperfect Power control, Part I: Reverse Link*. Tech. Rep., COST 231 TD (98) 48, Leeds, UK.
81. Jansen, M. G. and Prasad, R. (1995) Capacity, throughput, and delay analysis of a cellular DS/CDMA system with imperfect power control and imperfect sectorization. *IEEE Trans. Veh. Technol.*, **44**(1), 67–74.
82. Falciasecca, G., Gaiani, G., Missiroli, M., Murator, F., Palestini, V. and Riva, G. *Influence of Propagation Parameters and Imperfect Power Control on Cellular CDMA Capacity*. Tech. Rep., CSELT, Vol. XX, No. 6, 1992.
83. Cameron, C. and Woerner, B. (1996) Performance analysis of CDMA with imperfect power control. *IEEE Trans. Commun.*, **44**(7), 777–781.
84. Kudoh, E. (1993) On the capacity of DS/CDMA cellular mobile radios under imperfect transmitter power control. *IEICE Trans. Commun.*, **E76-B**(8), 886–893.
85. Newson, P. and Heath, M. R. (1994) The capacity of spread spectrum CDMA system for cellular mobile radio with consideration of system imperfections. *IEEE J. Select. Areas Commun.*, **12**(12), 673–684.

86. Viterbi, A. M. and Viterbi, A. J. (1993) Erlang capacity of a power controlled CDMA system. *IEEE J. Select. Areas Commun.*, **11**(6), 892–899.

87. Goldsmith, A. J. and Varaiya, P. P. (1997) Capacity of fading channels with channel side information. *IEEE Trans. Inform. Theory*, **43**(6), 1986–1992.

88. Goldsmith, A. J. and Chua, S.-G. (1997) Variable-rate variable-power MQAM for fading channels. *IEEE Trans. Commun.*, **45**(10), 1218–1230.

89. Ahmed, W. K. M. and McLane, P. J. (1999) On the error exponent for memoryless flat fading channels with channel-state-information feedback. *IEEE Commun. Lett.*, **3**(2), 49–51.

90. Goldsmith, A. J. (1997) The capacity of downlink fading channels with variable rate and power. *IEEE Trans. Veh. Technol.*, **46**(3), 569–580.

91. Lau, V. K. N. (1999) Channel capacity and error exponents of variable rate adaptive channel coding for Rayleigh fading channels. *IEEE Trans. Commun.*, **47**(9), 1345–1356.

92. Alouini, M.-S. and Goldsmith, A. (1999) Capacity of Rayleigh fading channels under different adaptive transmission and diversity-combining techniques. *IEEE Trans. Veh. Technol.*, **48**(4), 1165–1181.

93. Köse, C. and Goeckel, D. L. (1999) On power adaptation in adaptive signaling systems. *Proc. Conference on Information Sciences and Systems*, Baltimore, MD, pp. 103–108.

94. Caire, G., Taricco, G. and Biglieri, E. (1999) Optimum power control over fading channels. *IEEE Trans. Inform. Theory*, **45**(5), 1468–1489.

95. Goldsmith, A. J. and Chua, S.-G. (1998) Adaptive coded modulation for fading channels. *IEEE Trans. Commun.*, **46**(5), 595–602.

96. Goeckel, D. L. (1999) Adaptive coding for time-varying channels using outdated fading estimates. *IEEE Trans. Commun.*, **47**(6), 844–855.

7

Interference suppression and CDMA overlay

7.1 NARROWBAND INTERFERENCE SUPPRESSION

To get an initial insight into the problem, we assume that the received signal after frequency down conversion has the form

$$x(i) = b \cdot c(i) + J(i) + n(i) \tag{7.1}$$

where at sampling instant iT_c, b is data, $c(i)$ is the code, $J(i)$ is the narrowband interference, T_c is the chip interval and $n(i)$ is the Gaussian noise. The receiver structure is shown in Figure 7.1. For the two types of filters, from Figures 7.2 and 7.3 we define vectors of input samples and filter taps as follows:

$$X_{i1} \triangleq [x_i, x_{i-1}, x_{i-2}, \dots, x_{i-L}]^{\mathrm{T}}$$

$$X_{i2} \triangleq [x_{i+M}, x_{i+M-1}, \dots, x_{i+1}, x_{i-1}, \dots, x_{i-M}]^{\mathrm{T}}$$

$$W_1 \triangleq [a_1, a_2, \dots, a_L]^{\mathrm{T}}$$

$$W_2 \triangleq [a_{-M}, a_{-M+1}, \dots, a_{-1}, a_1, \dots, a_M]^{\mathrm{T}} \tag{7.2}$$

where T stands for transpose. With this notation, the filter output signal can be represented as

$$y_{if} = x_{if} - W_f^{\mathrm{T}} \cdot X_{if} \tag{7.3}$$

where $f = 1$ for one-sided filter (1SF) and $f = 2$ for two-sided filter (2SF). In the sequel index, f can be dropped for simplicity whenever this does not cause any ambiguity. If the interfering signal is stronger than the sum of Gaussian noise and useful signal, then the whole process can be interpreted as the estimation of $J(i)$ in the presence of an equivalent noise. In this case, equation (7.3) can be interpreted as the estimation error.

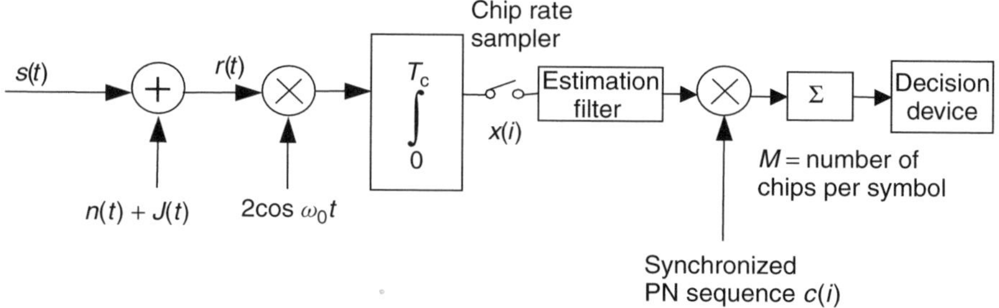

Figure 7.1   Receiver block diagram.

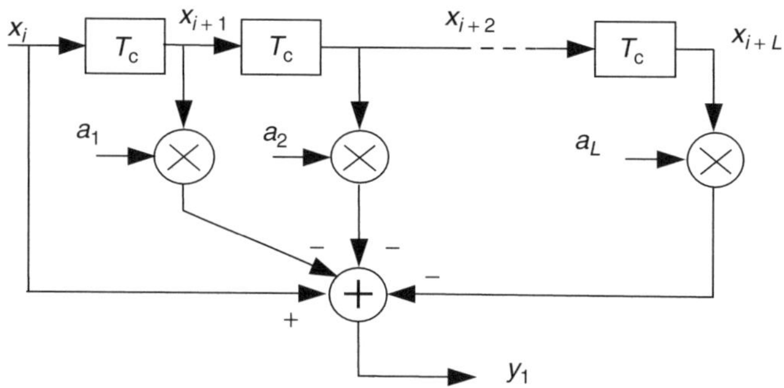

Figure 7.2   Single-sided transversal filter. Linear prediction filter.

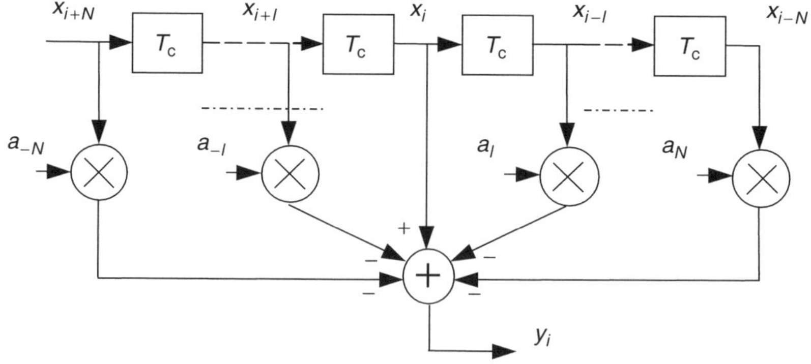

Figure 7.3   Two-sided transversal filter.

The filter coefficients will be evaluated from the condition that the Mean-Square Error (*mse*) of the estimation is minimized. So, we first evaluate

$$y_i^2 = x_i^2 - 2x_i X_i^{\mathrm{T}} W + W^{\mathrm{T}} X_i X_i^{\mathrm{T}} W \tag{7.4}$$

The mean value can be represented as

$$\xi = E[y_i^2] = E[x_i^2] - 2E[x_i X_i^{\mathrm{T}}] W + W^{\mathrm{T}} E[X_i X_i^{\mathrm{T}}] W$$

$$\stackrel{\Delta}{=} E[x_i^2] - 2P^{\mathrm{T}} W + W^{\mathrm{T}} R W \tag{7.5}$$

where

$$P^{\mathrm{T}} \stackrel{\Delta}{=} E[x_i X_i^{\mathrm{T}}]$$

$$R \stackrel{\Delta}{=} E[X_i X_i^{\mathrm{T}}] = [\rho_x(k - m)]; k, m = 1, \ldots, M \tag{7.6}$$

where $\rho_x(k - m)$ is the signal covariance function. To minimize the estimation error, the filter tap weights are obtained from

$$\frac{\partial E[y_i^2]}{\partial a_{kf}} = 0 \quad \begin{cases} k_2 = -M, \ldots, -1, 1, \ldots, M \\ k_1 = 1, \ldots, L \end{cases} \tag{7.7}$$

It is straightforward to show that equation (7.7) results in

$$-2P + 2R W_0 = 0$$

$$W_0 = R^{-1} P \tag{7.8}$$

where W_0 is the optimum tap weight vector. This equation is well known as the Wiener–Hopf equation. By taking z-transform of equation (7.3), the filter transfer function can be represented as

$$A_1(z) = 1 - \sum_{k=1}^{L} a_k z^{-k}$$

$$A_2(z) = 1 - \sum_{\substack{k=-M \\ k \neq 0}}^{M} a_k z^{-k} \tag{7.9}$$

The signal-to-noise ratio (SNR) improvement factor G is defined as the ratio of the output SNR to the input SNR.

$$G = \frac{(SNR)_{\mathrm{out}}}{(SNR)_{\mathrm{in}}} \tag{7.10}$$

7.2 GENERALIZATION OF NARROWBAND INTERFERENCE SUPPRESSION

In the previous section, it was shown that the optimum filter coefficients depend on the input signal correlation. So, if the interfering signal correlation function is specified, the closed-form solution for the SNR improvement factor can be obtained. This will be illustrated in this section by modeling the interference as a narrowband first-order autoregressive process [1,2]. At the sampling instant iT_c, after ideal frequency down conversion, the filter input signal, for these purposes, can be represented again by equation (7.1).

We assume that instead of the chip-matched filter in Figure 7.1 only a low-pass filter of bandwidth proportional to $1/T_c$ is used to limit the noise. The interfering signal $\{J(i)\}$ is assumed to be a wide sense stationary stochastic process with zero mean and covariance sequence $\{\rho_i(k)\}$. At this point, we introduce notation $\Re(a, b)$ to be a set of integers between a and b including a and b and $\Re_0(a, b)$, the same set excluding zero. The filter output signal can be represented as

$$y(i) = \sum_{l \in \Re} h(l)x(i - l) \tag{7.11}$$

where $\Re$ is $\Re(0, M)$ or 1SF and $\Re(-M, M)$ for 2SF

$$h(l) = \begin{cases} -a_l, l \neq 0 \\ a_0 = 1 \end{cases} \tag{7.12}$$

and a_l is defined by equation (7.2). By substituting equation (7.1) into equation (7.11), we have

$$y(i) = C_0(i) + J_0(i) + n_0(i) \tag{7.13}$$

Decision variable U at the input of decision device in Figure 7.1 is formed by multiplying the filter output signal by code and can be resolved in three components

$$U = \sum_{i=1}^{N} y(i)c(i)$$

$$= \sum_{i=1}^{N} C_0(i)c(i) + \sum_{i=1}^{N} J_0(i)c(i) + \sum_{i=1}^{N} n_0(i)c(i)$$

$$= U_1 + U_2 + U_3 \tag{7.14}$$

Under the assumption that signal noise and narrowband interference are mutually independent, we have for the average values

$$E[U_1] = b \cdot N, \ E[U_2] = E[U_3] = 0 \tag{7.15}$$

and bearing in mind that $b^2 = 1$, we have for the variance

$$\text{var } U_1 = N \sum_{m \in \Re_0} h^2(m)$$

$$\text{var } U_2 = N \sum_{m_1, m_2 \in \Re} h(m_1)h(m_2)\rho_i(m_2 - m_1)$$

$$\text{var } U_3 = N \sum_{m_1, m_2 \in \Re} h(m_1)h(m_2)\rho_n(m_2 - m_1) \tag{7.16}$$

where $\rho_i(\ )$ and $\rho_n(\ )$ are covariance functions of the interfering signal and the noise signal, respectively. For the covariance functions, we have

$$\text{cov}\{U_i, U_j\} = 0 \quad i \neq j \tag{7.17}$$

The signal-to-noise ratio at the filter output can be expressed as

$$(SNR)_0 \triangleq \frac{E^2[U]}{\text{var}[U]}$$

$$= \frac{N}{\displaystyle\sum_{m \in \Re_0} h^2(m) + \sum_{m_1, m_2 \in \Re} h(m_1)h(m_2)[\rho_i(m_2 - m_1) + \rho_n(m_2 - m_1)]} \tag{7.18}$$

When no suppression filter is used, $h(0) = 1$, and $h(l) = 0$ for $l \neq 0$, and we have

$$(SNR)_{n_0} = \frac{N}{\rho_i(0) + \rho_n(0)} \tag{7.19}$$

The improvement factor in the performance due to the use of the filter is then the ratio of equations (7.18 and 7.19)

$$G = \frac{\rho_i(0) + \rho_n(0)}{\displaystyle\sum_{m \in \Re_0} h^2(m) + \sum_{m_1, m_2 \in \Re}^{M \quad M} h(m_1)h(m_2)[\rho_i(m_2 - m_1) + \rho_n(m_2 - m_1)]} \tag{7.20}$$

7.2.1 Examples of the interfering signal

For the signal $x(i)$ given by equation (7.1), the covariance function $\rho(i)$ can be expressed as

$$\rho(i) = \delta_c(i) + \rho_i(i) + \rho_n(i) \tag{7.21}$$

where $\delta_c(i)$, the Kronecker delta, is the covariance sequence of the pseudonoise (PN) code. For $\rho_n(i)$ and $\rho_i(i)$, we will assume

$$\rho_n(i) = \sigma_n^2 \delta_c(i)$$
$$\rho_i(i) = \sigma_i^2 \alpha^{|i|}; 0 < \alpha < 1 \tag{7.22}$$

where σ_n^2 and σ_i^2 are the noise variance and the interference variance, respectively. The power spectral density function $\phi_i(\omega)$ is obtained by the Fourier transform of $\rho_i(i)$ as

$$\phi_i(\omega) = \frac{(1-\alpha^2)\sigma_i^2/2\pi}{|1-\alpha\exp(j\omega)|^2}; -\pi \le \omega \le \pi$$
$$= \frac{(1-\alpha^2)\sigma_i^2/2\pi}{1+\alpha^2 - 2\alpha\cos\omega} \tag{7.23}$$

and parameter α will characterize the shape of the spectra. The larger the α, the narrower the spectra, and *vice versa*. Bearing in mind equations (7.21 and 7.22), we have

$$\rho(i) = (1 + \sigma_n^2)\delta(i) + \sigma_i^2 \alpha^{|i|} \tag{7.24}$$

It is straightforward to show that the Wiener–Hopf equation (7.8) for this case becomes

$$a_i(1 + \sigma_n^2) + \sigma_i^2 \sum_{m \in \Re_0} a_m \alpha^{|i-m|} = \sigma_i^2 \alpha^{|i|} \tag{7.25}$$

Solving the filter coefficients from this system of equations is conceptually straightforward, but rather cumbersome and tedious work. Without going into any further details one can show that using equation (7.25) to evaluate coefficients a_i and then substituting equation (7.22) in equation (7.23) we have for the filter improvement factor

$$G_{\mathrm{ISF}} = \frac{\sigma_n^2 + \sigma_i^2}{\sigma_n^2 + \sigma_i^2(1-\alpha^2) \cdot \dfrac{(1-\alpha\beta) + (\alpha-\beta)\beta^{2M+1}}{(1-\alpha\beta)^2 - (\alpha-\beta)^2\beta^{2M}}}$$

$$\beta = \gamma - \sqrt{\gamma^2 - 1}$$

$$\gamma = \frac{1}{2\alpha}\left[(1+\alpha^2) + \frac{\sigma_i^2(1-\alpha^2)}{1+\sigma_n^2}\right] \tag{7.26}$$

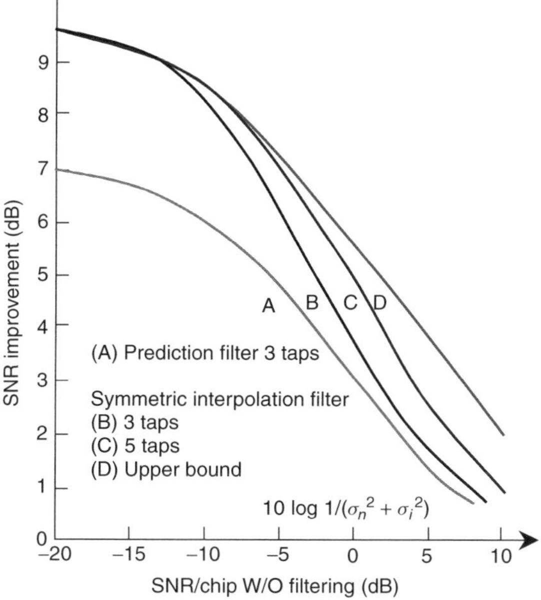

Figure 7.4 Improvement factor for a first-order autoregressive interference with $\alpha = 0.9$; $\sigma_n^2 = 0$.

for the single-sided filter (1SF) and

$$G_{2SF} = \frac{\sigma_n^2 + \sigma_i^2}{\sigma_n^2 + \sigma_i^2(1 - \alpha^2)\dfrac{(1 - \alpha\beta) + (\alpha - \beta)\beta^{2M+1}}{(1 - \alpha\beta)(1 + \alpha^2 - 2\alpha\beta) - (\alpha - \beta)(2\alpha - \beta - \alpha^2\beta)\beta^{2M}}}$$

(7.27)

for the two-sided filter (2SF), where β and γ are the same as in equation (7.26). As an illustration, Figure 7.4 presents several curves for the filter improvement factor G with the given set of the signal and filter parameters (Wiener optimum W/O). Curve D, designated as upper bound, is obtained for $M \to \infty$.

For the analysis of the mutual influence of Code Division Multiple Access (CDMA) and narrowband communications network, we will assume the interfering signal to occupy a multiple frequency band that can be represented as

$$\phi_i(\omega) = \begin{cases} \sigma_i^2/2\pi p, & \omega \in A_j, j = 1, \ldots, J \\ 0, & \omega \notin \bigcup_{j=1}^{J} A_j \end{cases}$$

(7.28)

where the intervals $A_j s$ are disjoint and their total length $\sum_{j=1}^{J} |A_j| = 2\pi p$ for some $0 < p < 1$. The jammer occupies a pth fraction of the signal band. By using the same procedure as in the previous case, numerical results are shown in Figure 7.5 for $p = 20\%$.

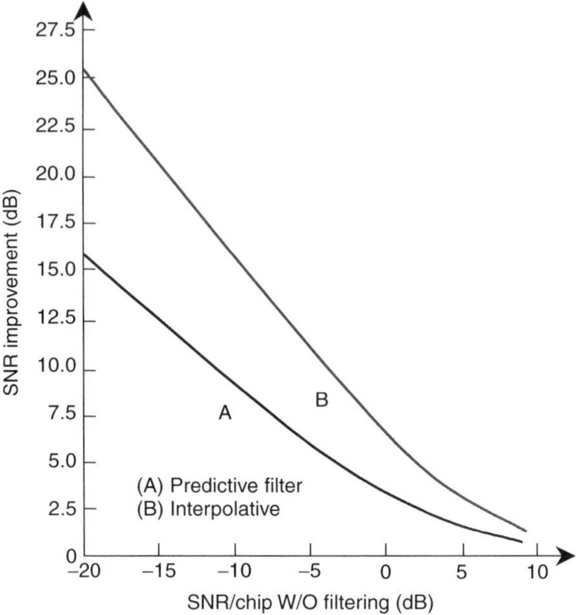

Figure 7.5 Upper bounds on improvement factor for a multiband interference with 20% bandwidth occupancy.

7.3 RECURSIVE SOLUTIONS FOR THE FILTER COEFFICIENTS

For the evaluation of the optimum filter coefficients, defined by equation (7.8) a matrix inversion is required. This is a computationally intensive operation, and for practical applications a form of recursive algorithm is preferred. An option is to solve equation (7.8) by using the recursive procedure. An example is Levinson's algorithm that can be found in textbooks on signal processing. Another option is to build up a recursive algorithm that will evaluate an improved set of filter coefficients in each step. Within this section, we will discuss the method of steepest descent and the least mean square (LMS) algorithm. The method of steepest descent uses gradients of the performance surface in seeking its minimum. For this reason, we will first extend a little bit of theory presented in this section.

7.3.1 The gradient and the Wiener solution

The gradient of the *mse* function defined by equation (7.7) will be denoted as

$$\nabla = -2P + 2RW \tag{7.29}$$

When we set the gradient to zero, we get the optimal Wiener–Hopf solution defined by equation (7.8). Putting back equations (7.8) to (7.5) gives the minimum *mse*

$$\xi_{\min} = E[x_i^2] - P^{\mathrm{T}} W_0 \tag{7.30}$$

Now, if equation (7.30) is used back in equation (7.5) we have

$$\xi = \xi_{\min} + (W - W_0)^{\mathrm{T}} R (W - W_0) \tag{7.31}$$

This can be further expressed as

$$\xi = \xi_{\min} + V^{\mathrm{T}} R V \tag{7.32}$$

where

$$V \overset{\Delta}{=} W - W_0 \tag{7.33}$$

is the difference between W and the optimal values W_0. Differentiation of equation (7.32) gives another form of the gradient

$$\nabla = 2 R V \tag{7.34}$$

If Q is the orthonormal modal matrix of symmetric and positive-definite matrix R and Λ is its diagonal matrix of eigenvalues

$$\Lambda = \mathrm{diag}[\lambda_1, \lambda_2, \ldots, \lambda_n] \tag{7.35}$$

then we can write

$$R = Q \Lambda Q^{-1} = Q \Lambda Q^{\mathrm{T}} \tag{7.36}$$

Now equation (7.32) becomes

$$\xi = \xi_{\min} + V^{\mathrm{T}} Q \Lambda Q^{-1} V \tag{7.37}$$

If we use notation

$$V' \overset{\Delta}{=} Q^{-1} V \rightarrow V = Q V' \tag{7.38}$$

equation (7.37) can be expressed as

$$\xi = \xi_{\min} + V'^{\,\mathrm{T}} \Lambda V' \tag{7.39}$$

and the primed coordinates are therefore the principal axes of the quadratic surface. In the same way, we may apply transformation (7.38) to vector W itself to get

$$W' = Q^{-1} W \rightarrow W = Q W' \tag{7.40}$$

7.3.2 The steepest descent algorithm

The method of steepest descent updates the filter coefficients in accordance with

$$W_{i+1} = W_i + \mu(-\nabla_i) \tag{7.41}$$

where μ is a convergence factor that controls the stability and the rate of adaptation and ∇_i is the gradient at the ith iteration. Using the equations (7.34–7.40) in equation (7.41) we have

$$V'_{i+1} = (I - 2\mu\Lambda)V'_i \tag{7.42}$$

which after successive iterations for V'_i becomes

$$V'_i = (I - 2\mu\Lambda)^i V'_{\text{in}} \tag{7.43}$$

where V'_{in} is the initial difference between W and W_0

$$V'_{\text{in}} = W'_{\text{in}} - W'_0 \tag{7.44}$$

From equation (7.43) one can see that for each component k of the vector V', the transients will be geometric with the geometric ratio

$$r_k = (1 - 2\mu\lambda_k) \tag{7.45}$$

For convergence, it is necessary that

$$|r_{\text{max}}| = |1 - 2\mu\lambda_{\text{max}}| < 1 \tag{7.46}$$

leading to the conditions

$$\begin{aligned} 1 - 2\mu\lambda_{\text{max}} &> 1 \\ 1 - 2\mu\lambda_{\text{max}} &< 1 \end{aligned} \tag{7.47}$$

which results into

$$1/\lambda_{\text{max}} > \mu > 0 \tag{7.48}$$

In order to determine the time constant of the transients, an exponential envelope is fitted to a geometric sequence. If the time is normalized to the iteration cycle time, constant τ_k can be determined from

$$r_k = (1 - 2\mu\lambda_k) \cong \exp\left(-\frac{1}{\tau_k}\right) = 1 - \frac{1}{\tau_k} + \frac{1}{2!\tau_k^2} - \frac{1}{3!\tau_k^3} + \cdots$$
$$\cong 1 - \frac{1}{\tau_k} \tag{7.49}$$

leading to

$$\tau_k \cong \frac{1}{2\mu\lambda_k} \tag{7.50}$$

On the basis of this, the time constant for the process can be defined as the maximum value of parameter τ_k

$$\tau = \max_k \tau_k = \frac{1}{2\mu\lambda_{\text{min}}} \tag{7.51}$$

For theoretical analysis, the steepest descent can be regarded as a feedback process in which the gradient plays the role of the vector error signal. The feedback model can be described by the following set of equations

$$W_i \Rightarrow W_{i+1}| \quad \text{delayed one iteration}$$

$$W_{i+1} = W_i + \mu(-\nabla_i + n_\nabla)$$

$$\nabla_i = 2R(W_i - W_0) = 2RV \tag{7.52}$$

Gradient noise n_∇ takes into account all imperfections in the gradient vector estimation.

7.3.3 The LMS algorithm

In practice, ∇_j is not known and has to be estimated. So, the algorithm defined by equation (7.41) becomes

$$W_{i+1} = W_i + \mu(-\hat{\nabla}_i) \tag{7.53}$$

where $-\hat{\nabla}_i$ is an estimate of the true gradient ∇_i. When the gradient estimate is obtained as the gradient of the square of a single error sample, we end up with the LMS algorithm. By taking a derivative of equation (7.4) we have

$$\hat{\nabla}_i = 2y_i y_i' = -2y_i X_i \tag{7.54}$$

and equation (7.53) becomes

$$W_{i+1} = W_i + 2y_i X_i \tag{7.55}$$

In some papers y_i is denoted as ε_i and the previous equation gets the more familiar form

$$W_{i+1} = W_i + 2\varepsilon_i X_i \tag{7.56}$$

The condition defined by equations (7.46–7.48) is necessary and sufficient for convergence of the LMS algorithm. Although theoretically correct, in practice, these equations are not of much use because the individual eigenvalues are rarely known. Since tr R is the total input power to the weights, a generally known quantity, and since tr $R > \lambda_{max}$ as R is positive definite, condition (7.46–7.48) can be replaced by

$$1/\text{tr } R > \mu > 0 \tag{7.57}$$

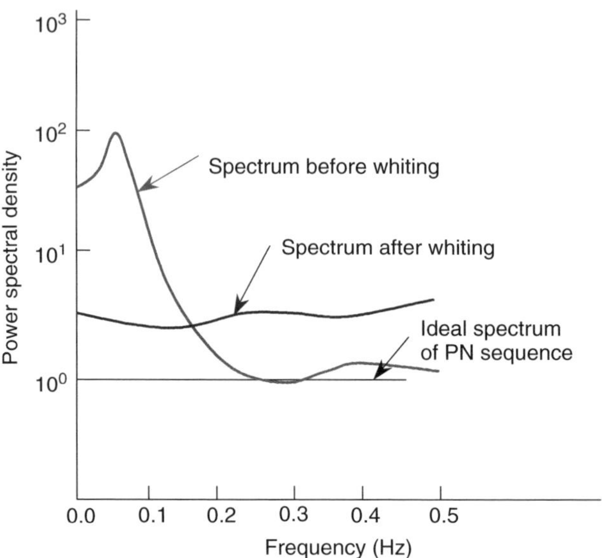

Figure 7.6 Power spectra comparison for $M = 100$ evenly spaced tones $\{f_m = m/1000, m = 1, \ldots, 100\}$, $SNR = -11\,$dB. Amplitude of the tone $c_m = 0.5$, standard deviation of the white noise $\sigma = 0.5$, order of the whitening filter $N = 4$.

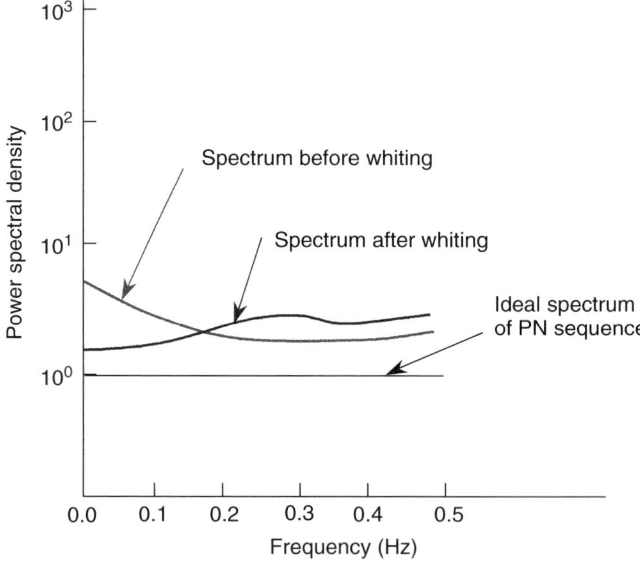

Figure 7.7 Power spectra comparison for $M = 100$ evenly spaced tones $\{f_m = m/1000, m = 1, \ldots, 100\}$, $SNR = -2\,$dB. Amplitude of the tone $c_m = 0.1$, standard deviation of the white noise $\sigma = 1.0$, order of the whitening filter $N = 9$.

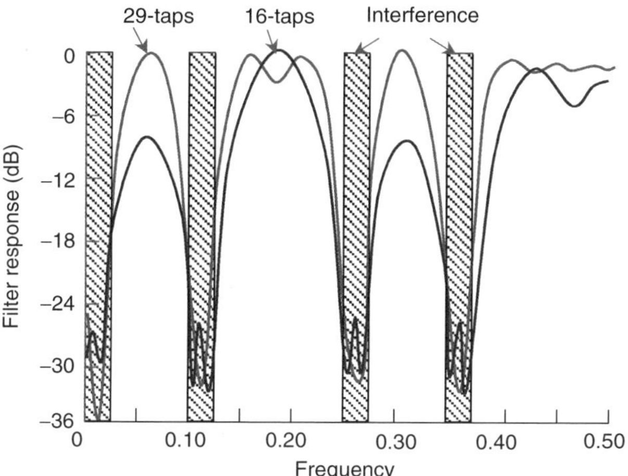

Figure 7.8 The frequency response characteristics of the filters with 16-tap and 29-tap predictors with four bands of interference and a signal-to-interference ratio (SIR)/chip of -20.

As an illustration, for the multiple band interfering signal, represented by equation (7.58) where M is the number of interferes, the results are shown in Figures 7.6 to 7.8. Because of the fact that narrowband interference is removed in this process, and that the remaining signal is a useful signal plus noise that has white spectra, the process is called whitening.

$$X_k = UP_k + \sum_{m=1}^{M} c_m \cos(2\pi f_m k \Delta t + \Phi_m) + n_k$$

$$c_m = [2P(f_m)\Delta f]^{1/2}$$

$$f_m = m\Delta f \tag{7.58}$$

A number of specific solutions related to interference suppression are given in References [1–53].

7.4 THE LEARNING CURVE AND ITS TIME CONSTANT

In the transient period, the error y_i is nonstationary as the weight vector adapts toward W_0. From equation (7.39), we have

$$\xi_i = \xi_{\min} + \mathbf{V}_i'^{\mathrm{T}}\Lambda\mathbf{V}_i' \tag{7.59}$$

The *mse* ξ_i is a function of the iteration number i, obtained by averaging over the ensemble of possible outcomes of ξ_i at iteration i. Using equation (7.39) in equation (7.59), we have

$$\xi_i = \xi_{\min} + \mathbf{V}_{\mathrm{in}}'^{\mathrm{T}}\Lambda(I - 2\mu\Lambda)^{2i}\mathbf{V}_{\mathrm{in}}'$$

$$= \xi_{\min} + \mathbf{V}_{\mathrm{in}}^{\mathrm{T}}\Lambda(I - 2\mu R)^i R(I - 2\mu R)^i V_{\mathrm{in}} \tag{7.60}$$

As long as the adaptive process is convergent, which is defined by equations (7.46–7.48), the previous relation will give

$$\lim_{i \to \infty} \xi_i = \xi_{\min} \tag{7.61}$$

The geometric decay in ξ_i going from ξ_{in} in $\xi_{\min}$ will, for the kth vector element, have a geometric ratio of r_k^2 where r_k is given by equations (7.46–7.48). The time constant will be

$$\tau_{kmse} \overset{\Delta}{=} \frac{1}{2}\tau_k = \frac{1}{4\mu\lambda_k} \tag{7.62}$$

The curve obtained by plotting *mse* against the number of iterations i is called the 'learning curve'.

7.4.1 Gradient and weight vector noise

The estimated gradient given by equation (7.54) can be represented as a sum of the true gradient ∇_i and the gradient estimation noise $\boldsymbol{n}_{\nabla i}$ as

$$\hat{\nabla} = -2y_i X_i = \nabla_i + \boldsymbol{n}_{\nabla i} \tag{7.63}$$

When $W_i = W_0$, the true gradient is zero, but the algorithm would still make an estimate in accordance with equation (7.54), which will be now equal to the estimation noise

$$\boldsymbol{n}_{\nabla i} = -2y_i X_i \tag{7.64}$$

When W_i is represented in $2M + 1$ dimensional vector space, for $W_i = W_0$, error y_i and x_i are orthogonal (uncorrelated). If they are assumed zero mean Gaussian, y_i and x_i are then statistically independent. For such a case, the covariance of $\boldsymbol{n}_{\nabla i}$ is

$$\begin{aligned}
\text{Cov}[\boldsymbol{n}_{\nabla i}] &= E[\boldsymbol{n}_{\nabla i}\boldsymbol{n}_{\nabla i}^{\text{T}}] \\
&= 4E[y_i^2 X_i X_i^{\text{T}}] \\
&= 4E[y_i^2] \cdot E[X_i X_i^{\text{T}}] \\
&= 4\xi R
\end{aligned} \tag{7.65}$$

When $W_i = W_0$, $\xi = \xi_{\min}$ and we have

$$\text{Cov}[\boldsymbol{n}_{\nabla i}] = 4\xi_{\min} R \tag{7.66}$$

In the tracking mode, $W_i \cong W_0$, and we approximate the gradient noise as stationary and uncorrelated with covariance given by equation (7.66). Projecting the gradient noise similar to equation (7.38) gives

$$\boldsymbol{n}_{\nabla i}' = Q^{-1}\boldsymbol{n}_{\nabla i} \tag{7.67}$$

and its variance becomes

$$
\begin{aligned}
\text{Cov}[\boldsymbol{n}'_{\nabla i}] &= E[\boldsymbol{n}'_{\nabla i}\boldsymbol{n}'^{\text{T}}_{\nabla i}] \\
&= E[\boldsymbol{Q}^{-1}\boldsymbol{n}_{\nabla i}\boldsymbol{n}^{\text{T}}_{\nabla i}\boldsymbol{Q}] \\
&= \boldsymbol{Q}^{-1}Cov[\boldsymbol{n}_{\nabla i}]\boldsymbol{Q} \\
&= 4\xi_{\min}\boldsymbol{Q}^{-1}\boldsymbol{R}\boldsymbol{Q} \\
&= 4\xi_{\min}\Lambda
\end{aligned}
\tag{7.68}
$$

From equations (7.66 and 7.68), components of $\boldsymbol{n}_{\nabla i}$ are correlated with each other, while those of $\boldsymbol{n}'_{\nabla i}$ are mutually uncorrelated and can therefore be handled more easily. By using a similar procedure and starting with the second line of equation (7.52), one can show

$$
\boldsymbol{V}'_{i+1} = \boldsymbol{V}'_i + \mu(-2\Lambda\boldsymbol{V}'_i + \boldsymbol{n}'_{\nabla i})
$$
$$
\text{Cov}[\boldsymbol{V}'_i] = \mu\xi_{\min}\boldsymbol{I}
\tag{7.69}
$$

The derivation is based on the fact that near the minimum point of the error surface in steady state (tracking mode), the mean of $\mathbf{V}'_i$ is zero.

7.4.2 Misadjustment due to gradient noise

The *mse*, represented by equation (7.59), has two components. For $\mathbf{W}_i \cong \mathbf{W}_0$, $\xi_i = \xi_{\min}$, but owing to random noise in the weight vector ξ_i will be increased by the factor called 'excess *mse*' designated as ξ_{ex} and given as

$$
\xi_{\text{ex}} = \mathbf{V}'^{\text{T}}_i \Lambda \mathbf{V}'_i
\tag{7.70}
$$

The average excess *mse* is

$$
\begin{aligned}
E[\xi_{\text{ex}}] &= E[\boldsymbol{V}'^{T}_i \Lambda \boldsymbol{V}'_i] \\
&= \sum_{k=1}^{n} \lambda_k E[\vartheta'^2_{ki}]
\end{aligned}
\tag{7.71}
$$

where n is the number of filter taps ($n = L$ for 1SF and $n = 2M$ for 2SF), and ϑ'_{ki} is the kth component of V'_i. In the tracking mode, $E[\mathbf{V}'_i] = 0$ and using equation (7.69) in equation (7.71) gives

$$
E[\vartheta'^2_{ki}] = \mu\xi_{\min}, \ \forall_k
\tag{7.72}
$$

By using this in equation (7.71), we have

$$
E[\xi_{\text{ex}}] = \mu\xi_{\min}\sum_{k=1}^{n} \lambda_k = \mu\xi_{\min}\text{tr}\,\boldsymbol{R}
\tag{7.73}
$$

Now we define the 'misadjustment' due to gradient noise as the ratio of the average excess *mse* to the minimum *mse*

$$M \triangleq \frac{E[\xi_{ex}]}{\xi_{min}} \tag{7.74}$$

Using equation (7.73), we have

$$M = \mu \text{tr } \boldsymbol{R} \tag{7.75}$$

This formula is derived for the tracking mode in which $\boldsymbol{W}_i \cong \boldsymbol{W}_0$. In practice, this works for as long as M is less than 0.25. One should be aware of the relation between the misadjustment and the speed of adaptation. From equations (7.73 and 7.75) we have

$$M = \mu \text{tr } \boldsymbol{R} = \mu \sum_{k=1}^{n} \lambda_k = \mu n \lambda_a \tag{7.76}$$

where λ_a is the average of the eigenvalue. From equation (7.62) we have

$$\lambda_k = \frac{1}{4\mu\tau_{kmse}} \Rightarrow \lambda_a = \frac{1}{4\mu} \left(\frac{1}{\tau_{kmse}}\right)_a \tag{7.77}$$

Using this in equation (7.76) gives

$$M = \frac{n}{4} \left(\frac{1}{\tau_{kmse}}\right)_a \tag{7.78}$$

which in a special case where all eigenvalues are equal becomes

$$M = \frac{n}{4\tau_{kmse}} \tag{7.79}$$

Since transients settle in (4–5) time constants, we can say that the misadjustment equals the number of filter taps divided by the settling time. Most of the time, in practice a 10% misadjustment would be satisfactory for many engineering applications. Operation with $M = 0.1$ can generally be achieved with an adaptive settling time equal to ten times the length of the filter.

7.4.3 Misadjustment due to nonstationary environment

In this case, the tracking error is due to both the effects of gradient noise and weight-vector lag caused by time variation of the input signal parameters. The weight-vector error $\Delta \boldsymbol{W}$ can be expressed as

$$\begin{aligned}
V_i = \Delta \boldsymbol{W}_i &= \boldsymbol{W}_i - \boldsymbol{W}_{0i} \\
&\equiv (\boldsymbol{W}_i - E[\boldsymbol{W}_i]) + (E[\boldsymbol{W}_i] - \boldsymbol{W}_{0i}) \\
&= \Delta \boldsymbol{W}_{gi} + \Delta \boldsymbol{W}_{li} \\
&= \boldsymbol{V}_{gi} + \boldsymbol{V}_{li}
\end{aligned} \tag{7.80}$$

The first component is due to gradient noise and the second component takes into account weight-vector lag due to nonstationary input signal. W_{0i} stands for the optimum (Wiener) weight vector, which is now different from iteration to iteration due to the changes of the input signal parameters. The expectations are averages over the ensemble. Weight-vector error causes an excess *mse*. The ensemble average excess *mse* at the ith iteration given by equations (7.70 and 7.71) now becomes

$$E[\xi_{ex}]_i = E[(W_i - W_{0i})^{\mathrm{T}}R(W_i - W_{0i})]$$
$$= E[V_i^{\mathrm{T}}RV_i] \tag{7.81}$$

If V_i is replaced by $V_{gi} + V_{li}$, the previous equation becomes

$$E[\xi_{ex}]_i = E[V_{gi}^{\mathrm{T}}RV_{gi}] + E[V_{li}^{\mathrm{T}}V_{li}] + 2E[V_{gi}^{\mathrm{T}}RV_{li}] \tag{7.82}$$

Since W_{0i} is constant over the ensemble, one can show by expanding the last form of the previous equation that

$$2E[V_{gi}^{\mathrm{T}}RV_{li}] = 0 \tag{7.83}$$

so that equation (7.82) becomes

$$E[\xi_{ex}]_i = E[\xi_{exg}]_i + E[\xi_{exl}]_i \tag{7.84}$$

where

$$E[\xi_{exg}]_i = E[V_{gi}^{\mathrm{T}}RV_{gi}] = E[V_{gi}'^{\mathrm{T}}\Lambda V'_{gi}]$$
$$E[\xi_{exl}]_i = E[V_{li}^{T}RV_{li}] = E[V_{li}'^{\mathrm{T}}\Lambda V'_{li}] \tag{7.85}$$

Up to now we have resolved the weight-vector error in two components. The first component is caused by the propagation of gradient noise and the second one by the response of the adaptive process to the random variations of W_{0i} caused by a nonstationary input signal. In what follows, we will show that increasing the time constant of the adaptive process diminishes the propagation of gradient noise but at the same time increases the lag error that results from the random changes in W_{0i}. Equation (7.83) shows that the propagation of gradient noise in the linear feedback system representing the adaptive process is not affected by the variability of W_{0i}. So, equation (7.73) can be used for evaluation of the first term of equation (7.85). In order to evaluate the second term of equation (7.85), we need the statistics of $V'_{li} = E[W'_i] - W'_{0i}$. By combining equations (7.34 and 7.41) we have

$$W_{i+1} = W_i + \mu(-\nabla_i)$$
$$= W_i + \mu[-2R(W_i - W_{0i})] \tag{7.86}$$

which gives

$$W_{i+1} - (I - 2\mu R)W_i = 2\mu RW_{0i} \tag{7.87}$$

After premultiplication of both sides by $\boldsymbol{Q}^{-1}$ we have an alternative form

$$\boldsymbol{W}'_{i+1} - (\boldsymbol{I} - 2\mu\Lambda)\boldsymbol{W}'_i = 2\mu\Lambda\boldsymbol{W}'_{0i} \tag{7.88}$$

Although time variant, all components of $\boldsymbol{W}_{0i}$ are assumed stationary, ergodic, independent and first-order Markov. They all have the same variances and the same autocorrelation functions. In addition to this, since $\boldsymbol{W}'_{0i} = \boldsymbol{Q}^{-1}\boldsymbol{W}_{0i}$ and $\boldsymbol{Q}^{-1}$ is orthonormal, all components of $\boldsymbol{W}'_{0i}$ are independent and have the same autocorrelation functions as the components of $\boldsymbol{W}'_{0i}$. On the basis of these assumptions, equation (7.88), having diagonal form and a driving function whose components are independent, may be treated as an array of n independent first-order linear difference equations. So, if the z-transform of $\boldsymbol{W}'_i$ is $\boldsymbol{W}'(z)$, from equation (7.88) we have

$$z\boldsymbol{W}'(z) - (\boldsymbol{I} - 2\mu\Lambda)\boldsymbol{W}'(z) = 2\mu\Lambda\boldsymbol{W}'_0(z) \tag{7.89}$$

which gives for $\boldsymbol{W}'(z)$ the following expression

$$\boldsymbol{W}'(z) = 2\mu\Lambda(z\boldsymbol{I} - \boldsymbol{I} + 2\mu\Lambda)^{-1}\boldsymbol{W}'_0(z) \tag{7.90}$$

From this relation we get the weight tracking error $(\boldsymbol{W}'_i - \boldsymbol{W}'_{0i})$ as

$$\boldsymbol{W}'(z) - \boldsymbol{W}'_0(z) = [2\mu\Lambda(z\boldsymbol{I} - \boldsymbol{I} + 2\mu\Lambda)^{-1} - \boldsymbol{I}]\boldsymbol{W}'_0(z) \tag{7.91}$$

The transfer function defined as the ratio of the weight error vector to the optimum weight vector becomes

$$\begin{aligned}
\boldsymbol{T}(z) &= \frac{\boldsymbol{W}'(z) - \boldsymbol{W}'_0(z)}{\boldsymbol{W}'_0(z)} \\
&= 2\mu\Lambda(z\boldsymbol{I} - \boldsymbol{I} + 2\mu\Lambda)^{-1} - \boldsymbol{I}
\end{aligned} \tag{7.92}$$

Since this equation is diagonal, its kth component gives

$$\begin{aligned}
T_k(z) &= 2\mu\lambda_k(z - 1 + 2\mu\lambda_k)^{-1} - 1 \\
&= \frac{z^{-1} - 1}{1 - (1 - 2\mu\lambda_k)z^{-1}}
\end{aligned} \tag{7.93}$$

$T_k(z)$ has zero at $z = 1$ and a pole at $z = 1 - 2\mu\Lambda_k = r_k$. If, as an example, we assume that each component of $\boldsymbol{W}_{0i}$ is obtained when independent stationary ergodic white noise of variance σ^2 is filtered by a one-pole filter having transfer function $1/(1 - az^{-1})$, the overall transfer function for each component of the process T_{kg} can be represented as

$$\begin{aligned}
T_{\text{kg}}(z) &= \frac{z^{-1} - 1}{(1 - az^{-1})(1 - (1 - 2\mu\lambda_k)z^{-1})} \\
&= \frac{z^{-1} - 1}{(1 - az^{-1})(1 - r_k z^{-1})} \\
&= \frac{\frac{1-a}{a-r_k}}{1 - az^{-1}} + \frac{\frac{r_k-1}{a-r_k}}{1 - r_k z^{-1}}
\end{aligned} \tag{7.94}$$

By inversion of this equation into time domain, we get the sampled impulse response of this transfer function. After that, the variance of the lag error of the kth component of the primed weight vector can be computed as the sum of the squares of the samples of the impulse response multiplied by σ^2. The sum of the squares is given as

$$\sum \mathrm{sq}(k) = \sum_{i=0}^{\infty} \left[\frac{1-a}{a-r_k} a^i + \frac{r_k-1}{a-r_k} r_k^i \right]^2$$

$$= \left(\frac{1}{a-r_k} \right)^2 \left[\left(\frac{1-a}{1+a} \right) + \left(\frac{1-r_k}{1+r_k} \right) + \frac{2(1-a)(r_k-1)}{1-ar_k} \right] \quad (7.95)$$

From equation (7.40) we have

$$\tau_k = \frac{1}{2\mu\lambda_k} = \frac{1}{1-r_k} \quad (7.96)$$

From the process-generating function

$$\tau_{W_0} = \frac{1}{1-a} \quad (7.97)$$

A region of interest would be where

$$\tau_{W_0} \gg \tau_k, \forall k \quad (7.98)$$

In other words, the value of μ is set so that the response times of the adaptive weights are short compared to the time constant of the nonstationary. Under these conditions, equation (7.95) reduces to

$$\mathrm{Cov} V'_{li} \big|_{\tau_W \gg \tau_k} = \frac{\sigma^2}{2} \mathrm{diag}(\tau_k) = \frac{\sigma^2}{4\mu} \Lambda^{-1} \quad (7.99)$$

and by using equation (7.85), we have

$$E[\xi_{\mathrm{exl}}]_i = \frac{\sigma^2}{2} \sum_{k=1}^{n} \tau_k \lambda_k = \frac{n\sigma^2}{4\mu} \quad (7.100)$$

The misadjustment due to lag is

$$M_l = \left(\frac{n\sigma^2}{4\xi_{\mathrm{min}}} \right) \frac{1}{\mu} \quad (7.101)$$

Using equations (7.84, 7.75 and 7.101), the total misadjustment is

$$M = \mu \mathrm{tr}\, R + \left(\frac{1}{\mu} \right) \frac{n\sigma^2}{4\xi_{\mathrm{min}}} \quad (7.102)$$

Solving the equation

$$\frac{\partial M}{\partial \mu} = 0 \tag{7.103}$$

gives the optimal μ as

$$\mu_0 = \sqrt{\frac{n\sigma^2}{4\xi_{\min}\mathrm{tr}\,\boldsymbol{R}}} \tag{7.104}$$

7.5 PRACTICAL APPLICATIONS: CDMA NETWORK OVERLAY

The most important application of the theory shown in this chapter is military systems. In the presence of the extremely high level of interference due to jamming, the receiver has to apply this kind of preliminary interference suppression in order to get to the operating point where processing gain would be enough to handle the residual interference. Besides these applications, some research projects are also looking into a possibility to overlay a wideband W-CDMA and the existing narrowband systems. In the sequel, we will present such a system with parameters that are mainly based on Reference [54].

7.5.1 Application scenario

In many locations, the capacity of a large macrocell is insufficient to support all the demands for service. In an analogue system [like AMPS (American mobile phone system)] with the 12.5 MHz allocated for each cellular service in an area, there are a possible 416 channels, each 30 kHz wide. With a 7-frequency reuse pattern, this becomes 59 channels per cell. In high traffic environments, the cell is usually subdivided into three 120 sectors, with 18 or 19 channels per sector, plus three control/access channels. In order to provide additional capacity for hot spots at, say, shopping malls, office plazas, transportation hubs, and so on, microcells serving limited areas are desirable. A possible scenario is shown in Figure 7.9.

7.5.2 Cell parameters

The macrocell antenna has height H_a, which is considerably higher than that of the microcell, H_c. In the example, a simple two-ray propagation model is used. This leads to propagation loss proportional to R^{-2} before and R^{-4} after a breakpoint distance given by $4h_t h_r/\lambda$, in which h_t and h_r are the transmit and receive antenna heights, respectively. The microcell radius $R_C = 4H_m H_C/\lambda$ was selected, in which H_m is the mobile antenna height. This is primarily for ease of calculation and may not be the optimum design. With

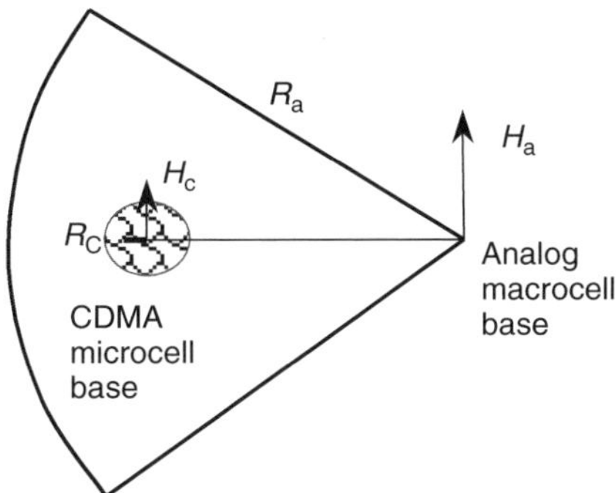

Figure 7.9 Microcell underlay geometry.

the wavelength $\lambda = 1$ ft. and $H_m = 5$ ft., we have $R_C = 20 H_C$. The effects of shadow loss are ignored for analytical simplicity.

7.5.3 System parameters

For the number of analog users $M = 0$, 8, and 16, the following parameters of the system were considered:

Microcell radius $R_C = 0.1 R_A$
Macrocell antenna height 4X microcell antenna height ($\delta = 0.25$)

CDMA bandwidth	$B_C = 10\,\text{MHz}$
Analog bandwidth	$B_A = 15\,\text{kHz}$
Spread spectrum gain	$K = 666$
Chip rate	$f_C = 8\,\text{Mcps}$
Bit rate	$f_b = 8\,\text{kbps}$
Processing gain	$G = 1000$
Channel activity factor	$\alpha = 0.75$ (with overhead)
Required mobile E_b/N_0	$\Gamma_{CM} = 4.5\,\text{dB}$
Required analog mobile carrier-to-interference ratio (CIR)	$\Gamma_{AM} = 17\,\text{dB}$ due to CDMAinterference

The required CDMA mobile E_b/N_0 of 4.5 dB assumes the use of the interleaved, rate 1/2 convolutionally encoded data with constraint length 7. Fast closed-loop power control is assumed to compensate for multipath fades, resulting in an approximately constant signal level. In that case, a 3 dB E_b/N_0 is required to provide an acceptable bit error rate (BER) of 0.001. This leaves a 1.5-dB margin for implementation loss and power control inaccuracies.

Figure 7.10 represents possible additional capacity in microcells versus the normalized distance of the microcell base station from the base station of the macrocell [54]. One can see a significant capacity achievable in this system. Figure 7.11 shows the power ratio of CDMA microcell-to-analog macrocell base stations. Figures 7.12 and 7.13 show significant improvement in the capacity if notching of the analog user is used.

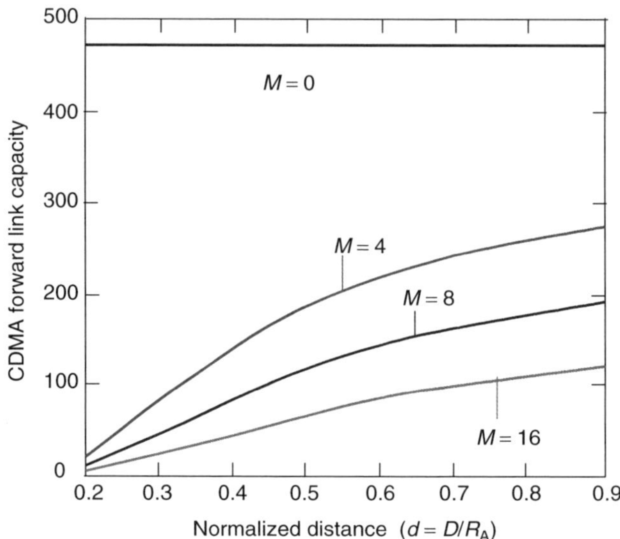

Figure 7.10 CDMA microcell forward link capacity versus normalized distance from macrocell base, for various macrocell usage.

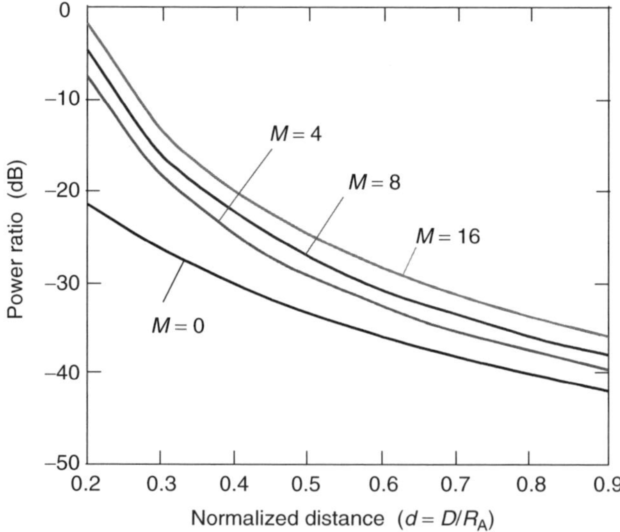

Figure 7.11 CDMA microcell-to-analog macrocell base transmit power ratio versus normalized distance from macrocell base, for various macrocell usage.

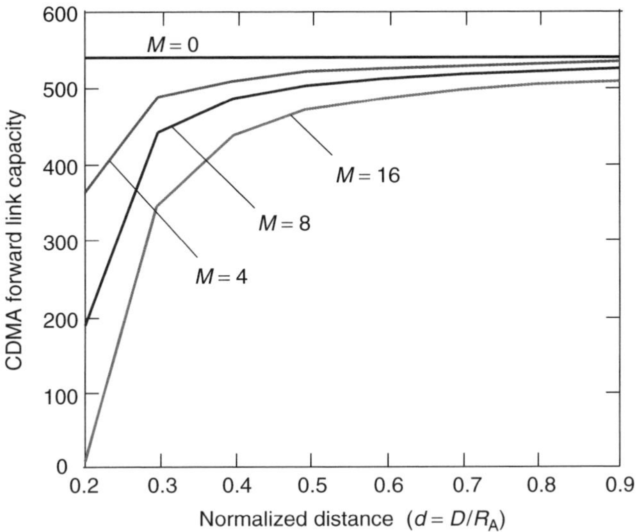

Figure 7.12 CDMA microcell forward link capacity with fixed power versus normalized distance, with 20 dB notch filter depth.

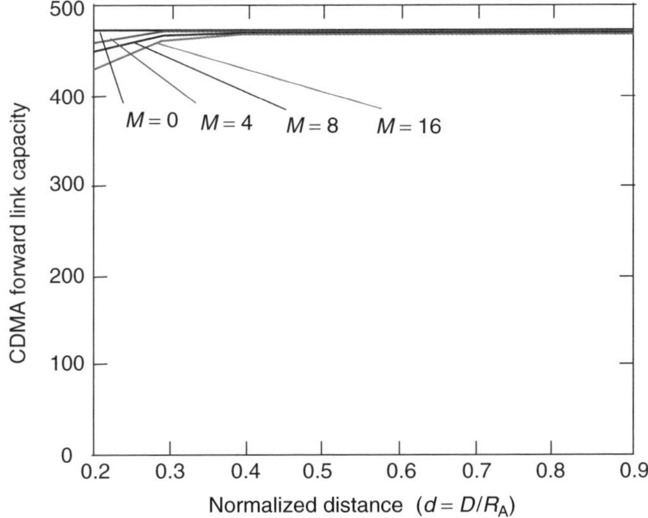

Figure 7.13 CDMA microcell forward link capacity versus normalized distance from the macrocell base, for various macrocell usage, with 30 dB notch filter depth.

REFERENCES

1. Masry, E. (1985) Closed-form analytical results for the rejection of narrowband interference in PN spread-spectrum systems – part ll: linear interpolation filters. *IEEE Trans. Commun.*, **COM-33**, 10–19.
2. Masry, E. and Milstein, L. B. (1986) Performance of DS spread-spectrum receivers employing interference-suppression filter under a worst-case jamming condition. *IEEE Trans. Commun.*, **COM-34**, 13–21.
3. Amoroso, F. (1983) Adaptive A/D converter to suppress CW interference in DSPN spread-spectrum communications. *IEEE Trans. Commun.*, **COM-31**, 1117–1123.
4. Amoroso, F. and Bricker, J. L. (1986) Performance of the adaptive A/D converter in combined CW and Gaussian interference. *IEEE Trans. Commun.*, **COM-34**, 209–213.
5. Schilling, D. L. *et al.* (1993) Broadband CDMA overlay. *Proc. IEEE 43rd VTS Conference*, Secancus, NJ, May 18–20, 1993, pp. 452–455.
6. Milstein, L. B. *et al.* (1992) On the feasibility of a CDMA overlay for personal communications networks. *IEEE ISAC*, **10**, 655–667.
7. Alexander, S. T. (1986) *Adaptive Signal Processing*. New York: Springer-Verlag.
8. Baier, P. W. and Friederichs, K. J. (1985) A nonlinear device to suppress strong interfering signals with arbitrary angle modulation in spread-spectrum receivers. *IEEE Trans. Commun.*, **COM-33**, 300–302.
9. Bouvier Jr, M. J. (1978) The rejection of large CW interferers in spread spectrum systems. *IEEE Trans. Commun.*, **COM-28**, 254–256.
10. Das, P., Milstein, L. B. and Webster, R. T. (1976) Application of SAW chirp transform filter in spread spectrum communication systems. *6th European Microwave Conference*, September, 1976, pp. 261–266.
11. Gersho, A. (1975) Charge coupled devices: the analog shift register comes of age. *IEEE Commun. Mag.*, **13**, 27–32.
12. Gevargiz, J., Rosenmann, M., Das, P. and Milstein, L. B. (1984) A comparison of weighted and nonweighted transform domain processing systems for narrowband interference excision. *IEEE Military Communications Conference*, October, 1984, pp. 32.3.1–32.3.4.
13. Gevargiz, J., Das, P. and Milstein, L. B. (1985) Implementation of a transform domain processing radiometer for DS spread spectrum signals with adaptive narrowband interference exciser. Presented at the *IEEE International Conference on Communications*, June, 1985.
14. Gervargiz, J., Das, P., Milstein, L. B., Moran, J. and Mckee, O. (1986) Implementation of DS-SS intercept receiver with an adaptive narrowband interference exerciser using transform domain processing and time weighting. *IEEE Military Communications Conference*, October, 1986, pp. 20.1.1–20.1.5.
15. Gevargiz, J., Das, P. and Milstein, L. B. (1986) Performance of a transform domain processing DS intercept receiver in the presence of finite bandwidth interference. *IEEE Global Telecommunications Conference*, December, 1986, pp. 21.5.1–21.5.5.
16. Giordano, A. A. and Hsu, F. M. (1985) *Least Square Estimation with Applications to Digital Signal Processing*. New York: Wiley-Interscience.
17. Guilford, J. and Das, P. (1985) The use of the adaptive lattice filter for narrowband jammer rejection in DS spread spectrum systems. *Proc. IEEE International Conference on Communications*, June 22–26, 1985, pp. 822–826.
18. Helstrom, C. W. (1960) *Statistical theory of Signal Detection*. New York: Pergamon Press.
19. Hsu, F. M. and Giordano, A. A. (1978) Digital whitening techniques for improving spread-spectrum communications performance in the presence of narrowband jamming and interference. *IEEE Trans. Commun.*, **COM-26**, 209–216.
20. Iltis, R. A. and Milstein, L. B. (1978) Performance analysis of narrowband interference rejection techniques in DS spread-spectrum systems. *IEEE Trans. Commun.*, **COM-26**, 209–216.

21. Iltis, R. A. and Milstein, L. B. (1985) An approximate statistical analysis of the Widrow LMS algorithm with application to narrowband interference rejection. *IEEE Trans. Commun.*, **COM-33**, 121–130.

22. Ketchum, J. W. and Proakis, J. G. (1982) Adaptive algorithms for estimating and suppressing narrowband interference in PN spread-spectrum systems. *IEEE Trans. Commun.*, **COM-30**, 913–924.

23. Ketchum, J. W. (1984) Decision feedback techniques for interference cancellation in PN spread-spectrum communication systems. *IEEE Military Communications Conference*, October, 1984, pp. 3951–3955.

24. Li, L. and Milstein, L. B. (1982) Rejection of narrowband interference in PN spread-spectrum systems using transversal filters. *IEEE Trans. Commun.*, **COM-30**, 925–928.

25. Li, L. and Milstein, L. B. (1983) Rejection of CW interference in QPSK systems using decision-feedback filters. *IEEE Trans. Commun.*, **COM-31**, 473–483.

26. Li, Z., Yuan, H. and Bi, G. (1987) Rejection of multi-tone interference in PN spread spectrum systems using adaptive filters. *IEEE International Conference on Communications*, June, 1987, pp. 2451–2455.

27. Lin, F. and Li, L. M. (1987) Rejection of finite-bandwidth interference in QPSK systems using decision-feedback filters. *IEEE International Conference on Communications*, June, 1987, pp. 2461–2465.

28. Masry, E. (1985) Closed-form analytical results for the rejection of narrowband interference in PN spread-spectrum systems – part 1: linear prediction filters. *IEEE Trans. Commun.*, **COM-32**, 888–896.

29. Milstein, L. B. and Das, P. (1977) Spread spectrum receiver using acoustic surface wave technology. *IEEE Trans. Commun.*, **COM-25**(8), 841–847.

30. Milstein, L. B. and Das, P. (1979) Surface acoustic wave devices. *IEEE Commun. Mag.*, **17**(5), 25–33.

31. Milstein, L. B. and Das, P. (1980) An analysis of a real-time transform domain filtering digital communication system, part 1: narrowband interference rejection. *IEEE Trans. Commun.*, **COM-28**, 816–824.

32. Milstein, L. B., Das, P. K. and Gevargiz, J. (1982) Processing gain advantage of transform domain filtering DS spread spectrum systems. *Military Communications Conference*, October, 1982, pp. 2121–2124.

33. Milstein, L. B. and Das, P. K. (1983) An analysis of a real-time transform domain filtering digital communication system – part ll: wideband interference rejection. *IEEE Trans. Commun.*, **COM-31**, 21–27.

34. Milstein, L. B. and Iltis, R. A. (1986) Signal processing for interference rejection in spread-spectrum communications. *IEEE ASSP Mag.*, 1–31.

35. Mostafa, A. E. S., Abdel-Kader, M. and El-Osmany, A. (1983) Improvements of anti-jam performance of spread-spectrum systems. *IEEE Trans. Commun.*, **COM-31**, 803–808.

36. Nudd, G. R. and Otto, O. W. (1975) Chirp signal processing using acoustic surface wave filters. *Ultrasonics Symposium Proceedings*, p. 350.

37. Ogawa, J., Cho, S. J., Morinaga, N. and Namekawa, T. (1981) Optimum detection of M-ary PSK signal in the presence of CW interference. *Trans. /ECE lapan*, **E64**, 800–806.

38. Otto, O. W. (1972) Real-time Fourier transform with a surface wave convolver. *Electron. Lett.*, **8**, 623.

39. Papoulisr, A. (1965) *Probability, Random Variables and Stochastic Processes*. New York: McGraw-Hill, pp. 218–220.

40. Pergal, F. I. (1987) Adaptive threshold A/D conversion techniques for interference rejection in DSPN receiver applications. *IEEE Military Communications Conference*, October, 1987, pp. 471–477.

41. Pickholtz, R. L., Schilling, D. L. and Milstein, L. B. (1982) Theory of spread-spectrum communications – a tutorial!. *IEEE Trans. Commun.*, **COM-30**, 855–884.

42. Proakis, J. G. (1983) *Digital Communications*. New York: McGraw-Hill.

43. Rosenmann, M., Gevargiz, M. J., Das, P. K. and Milstein, L. B. (1983) Probability of error measurement for an interference resistant transform domain processing receiver. *IEEE Military Communications Conference*, October, 1983, pp. 638–640.
44. Saulnier, G. I., Das, P. and Milstein, L. B. (1984) Suppression of narrowband interference in a PN spread-spectrum receiver using a CTD-based adaptive filter. *IEEE Trans. Commun.*, **COM-32**, 1227–1232.
45. Saulnier, G. I., Das, P. and Milstein, L. B. (1985) An adaptive digital suppression filter for direct sequence spread-spectrum communications. *IEEE J. Select. Areas Commun.*, **SAC-3**(5), 676–686.
46. Saulnier, G. I., Das, P. and Milstein, L. B. (1985) Suppression of narrowband interference on a direct sequence spread spectrum receiver in the absence of carrier synchronization. *IEEE Military Communications Conference*, October, pp. 13–17.
47. Saulnier, G. J., Yum, K. and Das, P. (1987) The suppression of tone jammers using adaptive lattice filtering. *IEEE International Conference on Communications*, June, 1987, pp. 2441–2445.
48. Shklarsky, D., Das, P. K. and Milstein, L. B. (1979) Adaptive narrowband interference suppression. *National Telecommunications Conference*, November, 1979, pp. 1521–1524.
49. Simon, M. K., Omura, J., Scholtz, R. A. and Levitt, B. K. (1985) *Spread Spectrum Communications*. Vols. I–III. Rockville, MD: Computer Science Press.
50. Takawira, F. and Milstein, L. B. (1986) Narrowband interference rejection in PN spread spectrum systems using decision feedback filters. *IEEE Military Communications Conference*, October, 1986, pp. 2041–2045.
51. Wang, Y.-C. and Milstein, L. B. (1988) Rejection of multiple narrowband interference in both BPSK and QPSK DS spread-spectrum systems. *IEEE Trans. Commun.*, **COM-36**, 195–204.
52. Widrow, B. *et al.* (1975) Adaptive noise canceling: Principles and applications. *Proc. IEEE*, **63**, 1692–1716.
53. Widrow, B. and Stearns, S. D. (1985) *Adaptive Signal Processing*. Englewood Cliffs, NJ: Prentice Hall.
54. Grieco, D. M. (1994) The capacity achievable with a broadband CDMA microcell underlay to an existing cellular macrosystem. *IEEE JSAC*, **12**(4), 744–750.

8

CDMA network

In this chapter, we initiate discussion on CDMA network capacity. The issue will be revisited again later in Chapter 13 to include additional parameters in a more comprehensive way.

8.1 CDMA NETWORK CAPACITY

For initial estimation of CDMA network capacity, we start with a simple example of single cell network with n users and signal parameters defined as in the list above.

If α_i is the power ratio of user i and the reference user with index 0, and N_i is the interference power density produced by user i defined as

$$\alpha_i = P_i/P_0, \quad i = 1, \ldots, n-1$$

$$N_i = P_i/R_c = P_i T_c = \alpha_i P_0 T_c \tag{8.1}$$

then the energy per bit per noise density in the presence of n users is

$$\left(\frac{E_b}{N_0}\right)_n = \frac{E_b}{N_0 + \sum_{i=1}^{n-1} N_i} \tag{8.2}$$

If $(E_b/N_0)_R$ is the *required* single-user E_b/N_0 necessary to make the n-user signal-to-noise ratio (SNR), namely, $(E_b/N_0)_n$ equal to $(E_b/N_0)_1$, then we have

$$\left(\frac{E_b}{N_0}\right)_n = \frac{(E_b/N_0)_R}{1 + G^{-1}(E_b/N_0)_R \left(\sum_{i=1}^{n-1} \alpha_i\right)}$$

$$= \left\{ (E_b/N_0)_R^{-1} + G^{-1} \left(\sum_{i=1}^{n-1} \alpha_i\right) \right\}^{-1} \tag{8.3}$$

where $G \triangleq T_b/T_c = R_c/R_b$ is the so-called *system processing gain*. At the point where $(E_b/N_0)_n = (E_b/N_0)_1$, equation (8.3) gives

$$\left(\frac{E_b}{N_0}\right)_R = \frac{(E_b/N_0)_1}{1 - G^{-1}(E_b/N_0)_1 \left(\sum_{i=1}^{n-1} \alpha_i\right)} \tag{8.4}$$

and the degradation factor DF can be represented as

$$DF = \frac{(E_b/N_0)_R}{(E_b/N_0)_1} = \frac{1}{1 - G^{-1}(E_b/N_0)_1 \left(\sum_{i=1}^{n-1} \alpha_i\right)} \tag{8.5}$$

For n equal-power users, and no coding we have $\alpha_i = 1$ for all i, and equation (8.5) becomes

$$DF = \frac{1}{1 - (n-1)G^{-1}(E_b/N_0)_1}$$

$$\left(\frac{E_b}{N_0}\right)_n = \frac{(E_b/N_0)_R}{1 - (n-1)G^{-1}(E_b/N_0)_R} \tag{8.6}$$

For large values of $(E_b/N_0)_R$,

$$\lim_{(E_b/N_0)_R \to \infty} \left(\frac{E_b}{N_0}\right)_n = \frac{G}{(n-1)}, \quad n \geq 2 \tag{8.7}$$

This is the largest value that the SNR $= (E_b/N_0)_n$ can attain. With this motivation, we define the *multiple-access capability factor* (MACF) as $G/(n-1)$ normalized by the SNR, $(E_b/N_0)_n$.

$$MACF \triangleq \frac{G}{(n-1)} \left(\frac{E_b}{N_0}\right)_n^{-1} \tag{8.8}$$

which can also be expressed as

$$MACF \triangleq \frac{G}{(n-1)} \left(\frac{E_b}{N_0}\right)_n^{-1} = \frac{G}{(n-1)} \left(\frac{E_b}{N_0}\right)_R^{-1} \tag{8.9}$$

As long as the desired SNR, namely, $(E_b/N_0)_n$, is such that the left-hand side is greater than or equal to one, we can achieve that SNR by appropriately adjusting $(E_b/N_0)_R$ in the right-hand side. If the left-hand side is less than one, however, no value of $(E_b/N_0)_R$ will give the desired value of $(E_b/N_0)_n$. An example of the system performance is shown in Figure 8.1. One can see that for $G = 100$ and 1000 the maximum number of users that can be accommodated with finite DF is 10 and 100, respectively. In other words, the

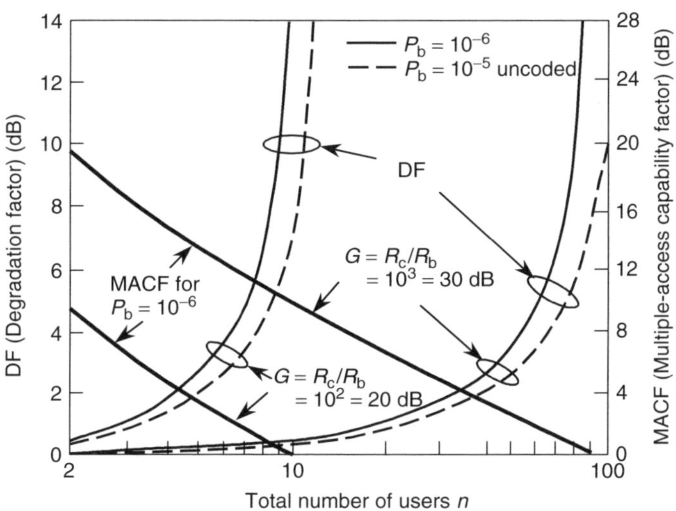

Figure 8.1 System performance for n equal-power users.

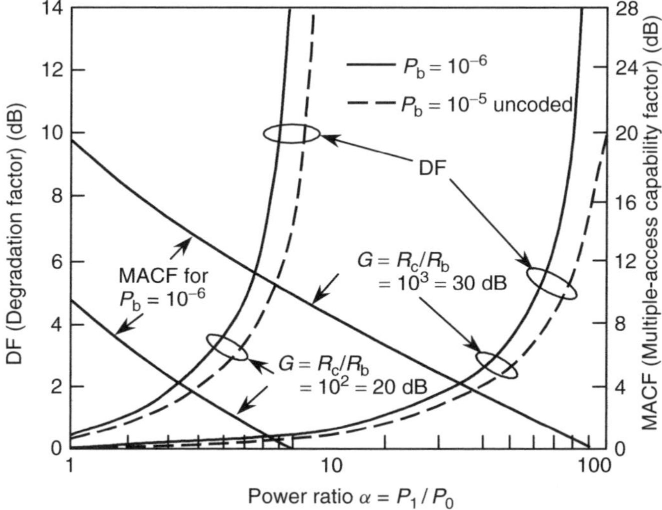

Figure 8.2 System performance for two users of unequal power.

system capacity C (maximum number of users) is about 10% of the processing gain in the system, $C \cong 0.1G$.

If we now assume $n = 2$ users of different powers, and set $\alpha = P_1/P_0$ The DF becomes

$$\mathrm{DF} = [1 - \alpha G^{-1}(E_b/N_0)_2]^{-1} \tag{8.10}$$

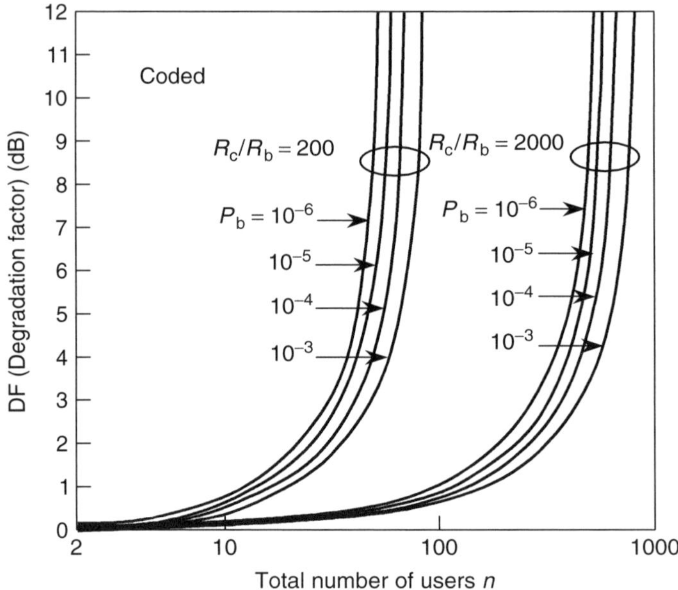

Figure 8.3 Degradation factor versus total number of users with $K = 7$, $R = 1/2$ convolutional coding and Viterbi decoding with soft decisions.

It shows that the performance is equivalent to n users for the equal-power example when we substitute $\alpha = n - 1$. In other words, having two users one of which is α times stronger is equivalent to having additional $(n - 1)$ users of the same power.

This is to be expected, particularly since we have modeled additional users as adding more broadband noise. This is the first time where we explicitly demonstrate the importance of near–far effect and the role of power control discussed in Chapter 6. These results are demonstrated in Figure 8.2. Figure 8.3 demonstrates the same results for the system with coding. In general, more coding would require less S/N ratio for the same performance, which means that more users can be brought into the system, $C \cong 0.4G$.

8.2 CELLULAR CDMA NETWORK

In this section, we extend our analysis on a whole cellular network. In such a network users communicate through a central point, the base station (BS) placed usually in the middle of an area called cell. The link between the mobile and BS is called reverse or uplink and between the BS and mobile is called forward or downlink. These two links may be separated in frequency, which is referred to as frequency division duplexing (FDD) or in time, referred to as time division duplexing (TDD). The basic block diagram of the system transmitter is shown in Figure 8.4 and the network layout, composed of a collection of cells is shown in Figure 8.5.

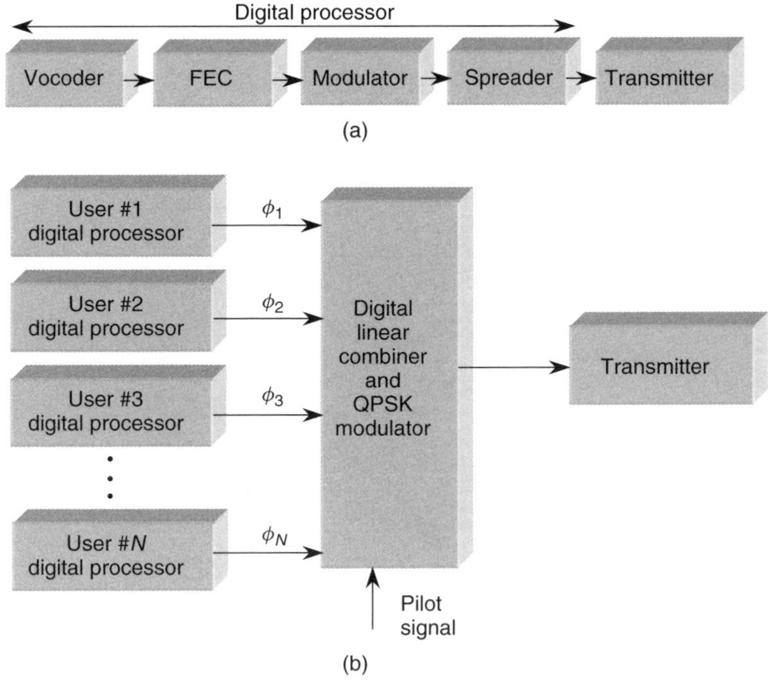

Figure 8.4 Cellular system simplified block diagram: (a) reverse link subscriber processor/transmitter, (b) forward link cell-site processor/transmitter.

For the initial discussion we assume single cell scenario and existence of:

1. Pilot signal in the forward (cell-site-to-subscriber) direction.
2. Initial power control by the mobile, based on the level of detected pilot signal. The mobile adjusts its output power inversely to the total signal power it receives.

This, plus closed loop control, described in Chapter 4, should justify the assumption that at the BS all received signals have the same power S. Under this assumption SNR, and energy per bit per noise density in the network with N users can be expressed as

$$SNR = \frac{S}{(N-1)S} = \frac{1}{N-1} \tag{8.11}$$

$$E_{\rm b}/N_0 = \frac{S/R}{(N-1)S/W} = \frac{W/R}{N-1} \cong \frac{G}{N} \tag{8.12}$$

If the presence of thermal noise is also taken into account, we have

$$E_{\rm b}/N_0 = \frac{W/R}{(N-1)+(\eta/S)} \tag{8.13}$$

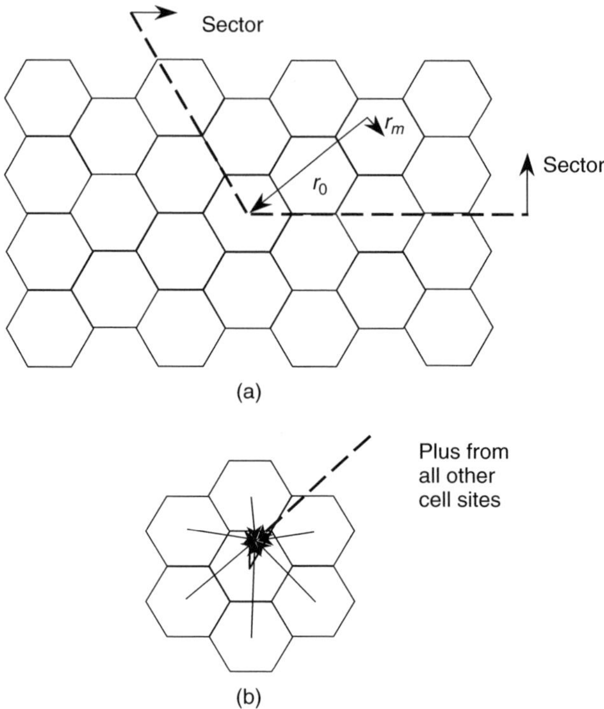

Figure 8.5   Cell geometrics: (a) reverse link geometry, (b) forward link geometry.

For a given E_b/N_0, required for a certain bit error rate (BER), the number of users is

$$N = 1 + \frac{W/R}{E_b/N_0} - \frac{\eta}{S} \cong \frac{G}{E_b/N_0} \tag{8.14}$$

where R is the bit rate, W is the bandwidth proportional to chip rate, G is the processing gain $G = W/R$ and η is Gaussian noise (thermal noise) power density. This very simple expression shows that the system capacity measured in number of users is inversely proportional to E_b/N_0 required for a certain quality of service (QoS). This explains why the equipment in a CDMA network should use everything available in the modern signal processing technology to keep this level as low as possible. Powerful coding, antenna diversity and advanced signal processing including multiuser detectors are considered for these applications. In order to extend the previous analysis on a network of cells we make the following assumptions:

For the reverse direction, noncoherent reception and dual antenna diversity are used. The required $E_b/N_0 = 7\,\text{dB}$ (constraint length 9, rate 1/3 convolution code) [1].

The forward link employs coherent demodulation by the pilot carrier. Multiple transmitted signals are synchronously combined. Its performance in a single cell system will be much superior to that of the reverse link. For a multiple-cell system, however, other cell interference will tend to equalize performance in the two directions.

Using directional antennas at the cell site both for receiving and transmitting signals is assumed. With three antennas per cell site, each having $120°$ effective beamwidths, the interference sources seen by any antenna are approximately one-third of those seen by an omnidirectional antenna. Using three sectors, the number of users per cell is $N = 3N_S$.

If voice activity is monitored and a signal is transmitted only if there is a signal at the output of the microphone, the level of interference will be in average reduced, and equation (8.13) becomes

$$\frac{\overline{E_b}}{N_0} = \frac{W/R}{(N_S - 1) \propto +(\eta/S)} \tag{8.15}$$

where 'the voice activity factor' $\propto = 3/8$.

8.2.1 Reverse link power control

Prior to any transmission, each of the subscribers monitors the total received signal power from the cell site. According to the power level it detects, it transmits at an initial level that is as much below (above) a nominal level in decibels as the received pilot power level is above (below) its nominal level. Experience has shown that this may require a dynamic range of control on the order of 80 dB. Further refinements in power level in each subscriber can be controlled by the cell site, depending on the power level it receives from the subscriber (20 dB dynamics). For these purposes a closed loop power control of the type described in Chapter 4 is used. In multiple-cell CDMA the interference level from subscribers in the other cells varies not only according to the attenuation in the path to the subscriber's cell site, but also inversely to the attenuation from the interfering user to his own cell site. This may increase, or decrease, the interference to the desired cell site through power control by that cell site.

8.2.2 Reverse link capacity for multiple-cell CDMA

The generally accepted model for propagation is as follows:

- The path loss between the subscriber and the cell site is proportional to $10^{(\xi/10)}r^{-4}$.
- r is the distance from the subscriber to the cell site.
- ξ is a Gaussian random variable with standard deviation $\sigma = 8$ and with zero mean.
- Within a single cell the propagation may vary from inverse square law, very close to the cell antenna, to as great as the inverse of 5.5 power, far from the cell in a very dense urban environment such as Manhattan.

The cell geometry is shown in Figure 8.5

In order to reach its own BS with power level S, the user with index m would have to transmit power P_m. This can be represented as

$$S = P_m \left(\frac{10^{\xi_m/10}}{r_m^4} \right) \tag{8.16}$$

This signal will at the same time represent interference at the reference site that can be represented as

$$I(r_0, m) = P_m \left(\frac{10^{\xi_0/10}}{r_0^4} \right)$$

(8.17)

By substituting P_m from equation (8.16) to equation (8.17) we have

$$\frac{I(r_0, r_m)}{S} = \left(\frac{10^{(\xi_0/10)}}{r_0^4} \right) \left(\frac{r_m^4}{10^{(\xi_m/10)}} \right)$$

$$= \left(\frac{r_m}{r_0} \right)^4 10^{(\xi_0 - \xi_m)/10} \leq 1$$

(8.18)

ξ_0 and ξ_m are independent so that the difference has zero mean and variance $2\sigma^2$.

Signal to noise ratio

E_b/N_0 given by equation (8.13) in the reverse link now becomes

$$E_b/N_0 = \frac{W/R}{\sum_{i=1}^{N_s-1} \chi_i + (I/S) + (\eta/S)}$$

(8.19)

where the first term in the nominator represents intracell interference with

$$\chi_i = \begin{cases} 1, & \text{with probability } \propto \\ 0, & \text{with probability } 1 - \propto \end{cases}$$

(8.20)

Parameter I represents other (multiple) cell user interference approximated as Gaussian random variable with $E(I/S) \leq 0.247 N_s$ and $\text{var}(I/S) \leq 0.078 N_s$ [1]. Parameters W/R and S/η, are constants.

Outage probability

If we define

$$P = \text{Pr}(BER < 10^{-3}) = \text{Pr}(E_b/N_0 \geq 5)$$

(8.21)

then the system outage probability is defined as

$$1 - P = \text{Pr}(BER > 10^{-3}) = \text{Pr}\left(\sum_{i=1}^{N_s} \chi_i + I/S > \delta \right)$$

where

$$\delta = \frac{W/R}{E_b/N_0} - \frac{\eta}{S}, \quad E_b/N_0 = 5$$

(8.22)

Since the random variable χ_i has binomial distribution and I/S is a Gaussian variable, the averaging gives

$$1 - P = \sum_{k=0}^{N_s-1} \Pr\left(I/S > \delta - k \,\Big|\, \sum x_i = k\right) \Pr\left(\sum x_i = k\right)$$

$$= \sum_{k=0}^{N_s-1} \binom{N_s-1}{k} \alpha^k (1-\alpha)^{N_s-1-k} Q\left(\frac{\delta - k - 0.274 N_s}{\sqrt{0.078 N_s}}\right) \qquad (8.23)$$

This equation is represented graphically in Figure 8.6 for the system parameters from the standard IS-95. The standard is presented in more detail in Chapter 16. If we accept outage probability of 1%, the system capacity becomes 37 for the sector that represents $37/(W/R) \cong 20\%$ of processing gain, $0.2G$. For Universal Mobile Telecommunication System (UMTS) standard, this number would be modified by two factors. From equation (8.14) the capacity in UMTS would be three times larger (G_w) owing to the three times larger chip rate. This effectively is not a gain because with three IS-95 systems in the same bandwidth, the capacity would be also increased three times.

The real improvement would come from the fact that by using three times larger chip rate, the multipath resolution would be better and the RAKE receiver (with gain G_{RAKE}) would be more effective, requiring lower E_b/N_0. These issues will be discussed later.

At this point it would be worth comparing the capacity of CDMA and time division multiple access (TDMA) system [like global system of mobile communication (GSM)]. GSM uses 200 kHz bandwidth for 8 users. In the band of 1.2 MHz (6 times 200 kHz) it would be possible to accommodate $6 \times 8 = 48 \cong 50$ users. One should be aware that the frequency reuse factor in TDMA network would be 7 as opposed to 1 in CDMA network which makes the normalized equivalent capacity of GSM in 1.2 MHz bandwidth $50/7 \cong$ 7 as opposed to 37 obtained in CDMA network.

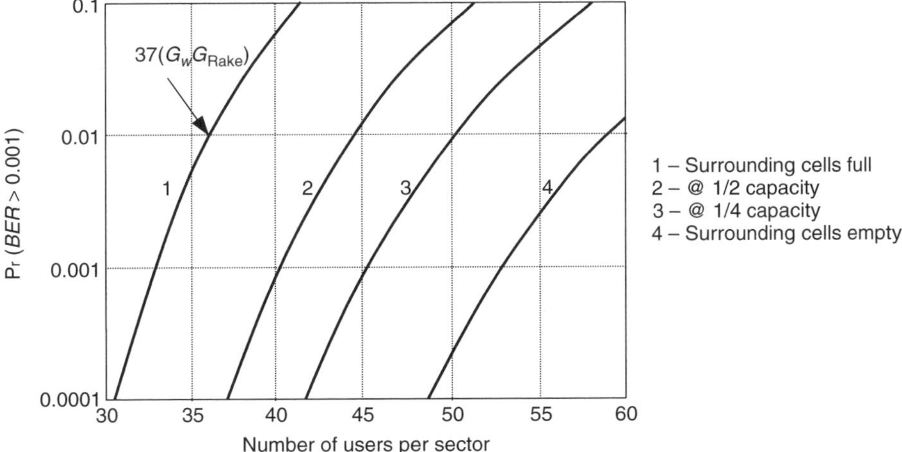

Figure 8.6   Reverse link capacity/sector ($W = 1.25$ MHz, $R = 8$ kbps, voice activity $= 3/8$).

For a fair comparison, one should be aware that GSM codec uses 13 kbit as opposed to 8 used in the previous calculus for CDMA, which reduces 37 by a factor of 8/13. The intention of this discussion is not to offer at this stage a final statement about the capacity but rather to give some initial elements relevant for this discussion. The numbers will be modified throughout the following chapters. They will be increased by a number of sophisticated algorithms for signal processing and also reduced by a number of sources of degradation, due to imperfections in the implementation of these algorithms.

8.2.3 Multiple-cell forward link capacity with power allocation

We assume that measurement by the mobile of its relative SNR, defined as the ratio of the power from its own cell-site transmitter to the total power received, is available.

Measurements can be transmitted to the selected (largest power) cell site when the mobile starts to transmit. On the basis of these two measurements, the cell site has reasonably accurate estimates of S_{T_1} and $\sum_{i=1}^{K} S_{T_i}$, where

$$S_{T_1} > S_{T_2} > \cdots > S_{T_K} > 0 \tag{8.24}$$

are the powers received by the given mobile from the cell-site sector facing it. S_{T1} is the total power transmitted from the cell site. The remainder of S_{T1} as well as the other cell-site powers are received as noise. Thus for user i, E_b/N_0 can be lower bounded by

$$\left(\frac{E_b}{N_0}\right)_i \geq \frac{\beta \emptyset_i S_{T_1}/R}{\left[\left(\sum_{j=1}^{K} S_{T_j}\right) + \eta\right]/W} \tag{8.25}$$

There is inequality because the interference includes the useful signal too. β is the fraction of the total cell-site power devoted to subscribers ($1 - \beta$ is devoted to the pilot). $\emptyset_i$ is the fraction of this devoted to subscriber i. From equation (8.25) we have

$$\emptyset_i \leq \frac{(E_b/N_0)_i}{\beta W/R}\left[1 + \left(\frac{\sum_{j=2}^{K} S_{T_j}}{S_{T_1}}\right)_i + \frac{\eta}{(S_{T_1})_i}\right] \tag{8.26}$$

where

$$\sum_{i=1}^{N_s} \emptyset_i \leq 1 \tag{8.27}$$

Outage probability

The relative received cell-site power measurements are defined as

$$f_i \triangleq \left(1 + \sum_{j=2}^{K} S_{T_j}/S_{T_1}\right)_i, \quad i = 1, \ldots, N_S \tag{8.28}$$

and from equation (8.27) we have

$$\sum_{i=1}^{N_\mathrm{s}} f_i \leq \frac{\beta W/R}{E_\mathrm{b}/N_0} - \sum_{i=1}^{N_\mathrm{s}} \frac{\eta}{S_{\mathrm{T}_1}} \triangleq \delta' \tag{8.29}$$

If we take $\beta = 0.8$ to provide 20% of the transmitted power in the sector to the pilot signal and use the required $E_\mathrm{b}/N_0 = 5\,\mathrm{dB}$ to ensure $BER \leq 10^{-3}$, then the outage probability can be represented as

$$1 - P = \Pr(BER > 10^{-3}) = \Pr\left(\sum_{i=1}^{N_\mathrm{s}} f_i > \delta'\right) \tag{8.30}$$

8.2.4 Histogram of forward power allocation

By using the propagation model [1] defined as

$$10^{(\xi_k/10)} r_k^{-4}$$
$$k = 0, 1, 2, \ldots, 18 \tag{8.31}$$

for each sample, the 19 values were ranked to determine the maximum (S_{T_1}), after which the ratio of the sum of all other 18 values to the maximum was computed to obtain $f_i - 1$. This was repeated 10 000 times per point for each of 65 equally spaced points on the triangle. From this, the histogram of $f_i - 1$ was constructed and the results are shown in Figure 8.7.

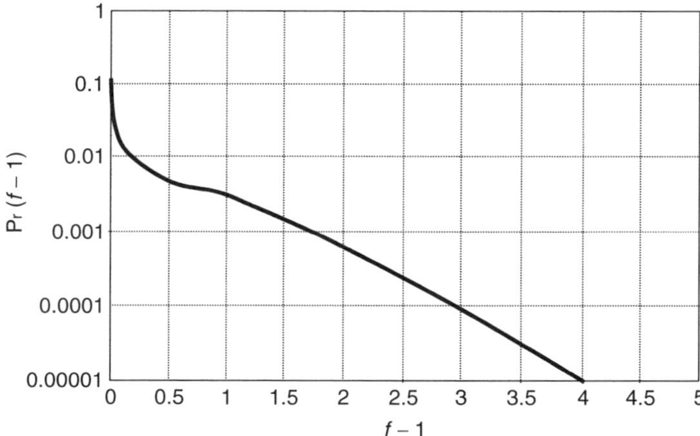

Figure 8.7 Histogram of forward power allocation [1]. Reproduced from Gilhousen, K. S., Jacobs, I. M., Padovani, R., Viterbi, A. J., Weaver, L. A. and Wheatley, C. E. (1991) On the capacity of a cellular CDMA system. *IEEE Trans. Veh. Technol.*, **40**(2), 303–312, by permission of IEEE.

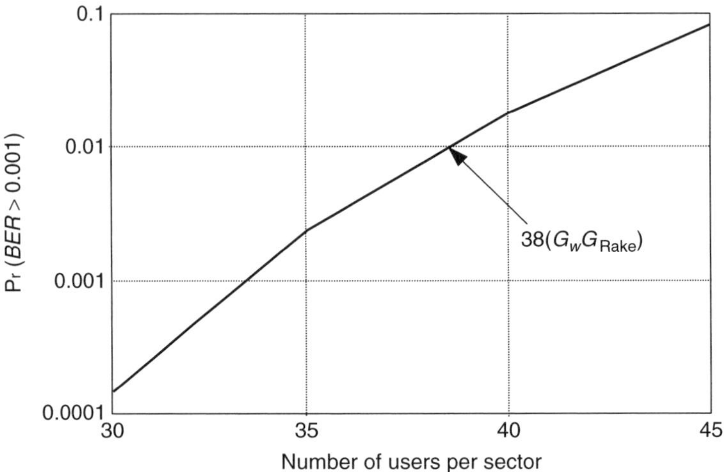

Figure 8.8   Forward link capacity/sector ($W = 1.25\,\text{MHz}$, $R = 8\,\text{kbps}$, voice activity $= 3/8$, pilot power $= 20\%$).

Outage probability

From this histogram the Chernoff upper bound is obtained as

$$1 - P \leq \min_{s>0} E \exp\left[s \sum_{i=1}^{N_s} f_i - s\delta' \right]$$

$$= \min_{s>0} \left[(1-\alpha) + \alpha \sum_{k} P_k \exp(sf_k) \right]^{N_s} e^{-s\delta'} \tag{8.32}$$

where E stands for expectation, P_k is the probability (histogram value) that f_i falls in the kth interval. The result of the minimization over s based on the histogram is shown in Figure 8.8. The results are obtained for IS-95 systems parameters. Discussion for UMTS standard is already presented in the section for reverse link capacity.

8.3 IMPACT OF IMPERFECT POWER CONTROL

We start with the cellular network shown in Figure 8.9. The signal received from cell j in mobile i can be represented as

$$I_{ij} = P_{\text{p}} \cdot r_{ij}^{-n} \cdot 10^{(\xi/10)} \tag{8.33}$$

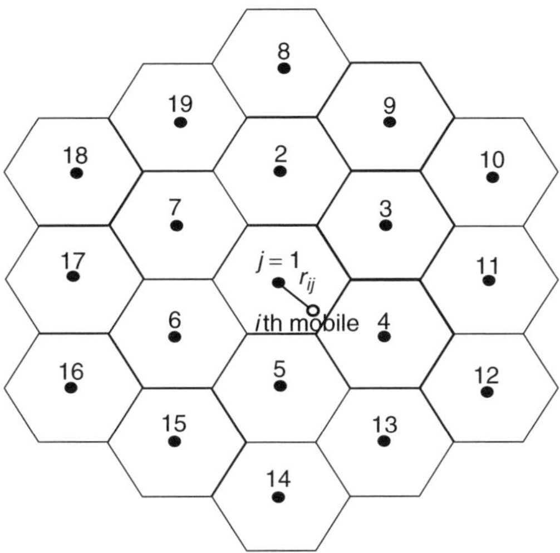

Figure 8.9 Hexagonal cell layout.

where P_p is the transmitted pilot signal power of a BS, r_{ij} is the distance between ith mobile and jth BS, n is the propagation constant, ξ is a random variable corresponding to shadowing, which is lognormally distributed with a mean of 0 dB and standard deviation of σ_s dB. The ith mobile transmitter transmits signal to a BS whose pilot signal power received by the mobile receiver satisfies

$$A_i = \max_j(I_{ij}) \qquad (8.34)$$

If in equation (8.18) power S is not perfectly controlled and the real received power can be represented as

$$S_i \rightarrow S10^{\delta_i/10} \qquad (8.35)$$

where δ_i (in decibels) denotes the control error in the transmitter power, then by using the same steps as in Section 8.2 instead of Figure 8.6 we get the results shown in Figure 8.10 [2,3].

Set of parameters used to generate Figure 8.10 is: power control error $10^{\delta_i/10}$ has a lognormal pdf with a standard deviation of σ_E, spread-spectrum bandwidth is 1.25 MHz, the information bit rate was 8 kbps, the speech activation factor α was 3/8, the required E_b/N_0 was 7 dB, the values of the propagation constant n and shadowing standard deviation σ_s used are $n = 4$ and $\sigma_s = 8$ dB. One can see significant losses due to imperfect power control.

8.3.1 Forward link TPC

We assume that all the transmitted signals (including pilot signal) arrive at the i-mobile station with power $P_M(i, j)$ of

$$P_M(i, j) = P_t \cdot r_{ij}^{-n} \cdot 10^{(\xi/10)} \qquad (8.36)$$

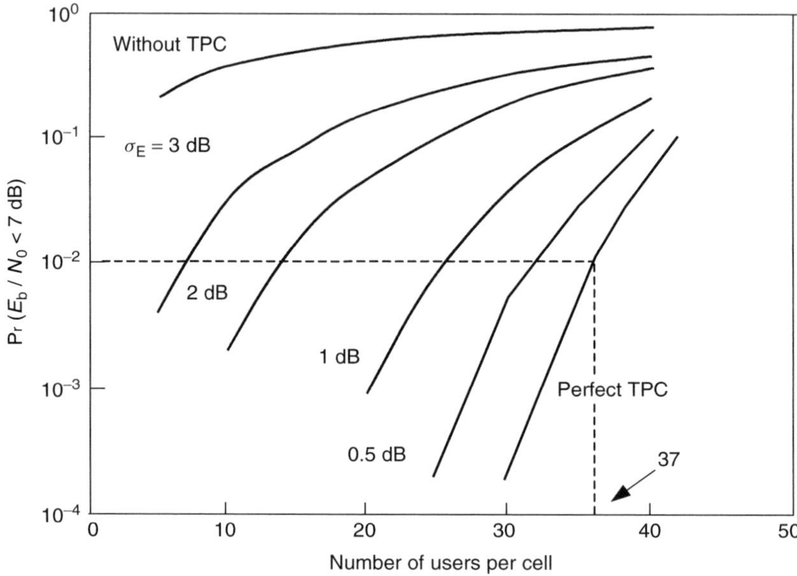

Figure 8.10   Reverse link capacity under imperfect TPC.

where P_t is the total transmitted power (including pilot signal) from BS j. Assume that before the forward link transmitter power control (TPC) is performed, $\phi_i \times 100\%$ of the jth BS transmitter power was assigned to communicate with the ith mobile station.

The interference signal power is the sum of each of the signal powers arriving at the mobile receiver except that of the desired signal. If $\beta \times 100\%$ of the total transmission power is used for signal transmission to all the mobile stations communicating with the jth BS ($1 - \beta$ is used for the pilot signal transmission), the received signal-to-interference ratio (SIR) of the ith mobile receiver can be expressed as

$$\frac{1}{SIR} \cong \frac{\sum_{j=1}^{m} P_{\mathrm{M}}(i, j)}{\beta \cdot \phi_i \cdot P_{\mathrm{M}}(i, j)} \tag{8.37}$$

If the power ratio of the ith receiver (ϕ_i) is modified

$$\phi_i' \cong \frac{\phi_i}{\sum_{i=1}^{n} \psi_i \cdot \phi_i} \tag{8.38}$$

by the forward link TPC for all mobile receivers, they receive their desired signals with the smaller SIR. Because of control error, the power ratio ϕ_i' deviates from its correct value as

$$\phi_i' = \frac{10^{(\delta_i/10)} \cdot \phi_i}{\sum\limits_{i=1}^{n} \psi_i \cdot 10^{(\delta_i/10)} \cdot \phi_i} \tag{8.39}$$

where δ_i (in decibel) denotes the control error in the transmitter power assignment. The required forward link communication quality is realized if

$$\frac{1}{SIR} \cong \frac{\sum\limits_{j=1}^{m} P_{\mathrm{M}}(i, j)}{\beta \cdot \phi_i' \cdot P_{\mathrm{M}}(i, j)} \leqq \frac{W}{R} \frac{1}{E_b/N_0} \tag{8.40}$$

The outage probability calculated with the same procedure as in Section 8.2 is now represented in Figure 8.11.

In the analysis shown in Section 8.2, parameters $E(I/S)$ and $\mathrm{var}(I/S)$ were calculated under the assumption that the users were uniformly distributed within the cell. If the distribution is modified, for example, as shown in Figure 8.12, the outage probability will be modified accordingly as shown in the same figure. The new distribution from Figure 8.12 means that the users from surrounding cells are concentrated within the belt of width a_r close to the reference cell.

In equation (8.20) parameters $E(I/S)$ and $\mathrm{var}(I/S)$ were calculated under the assumption that the propagation factor $n = 4$ and the standard deviation of shadowing $\sigma = 8\,\mathrm{dB}$. If n and σ are changed in a certain range, these parameters will change as shown in Figures 8.13 and 8.14.

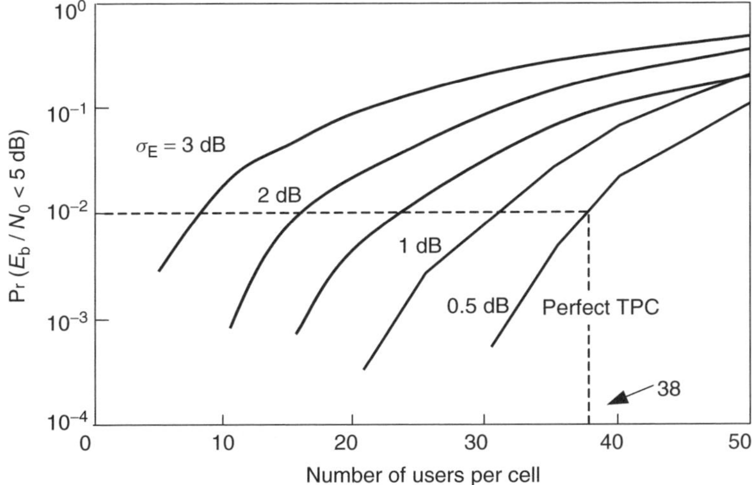

Figure 8.11 Forward link capacity under imperfect TPC. Required E_b/N_0 of 5 dB. The power ratio of $1 - \beta = 0.2$.

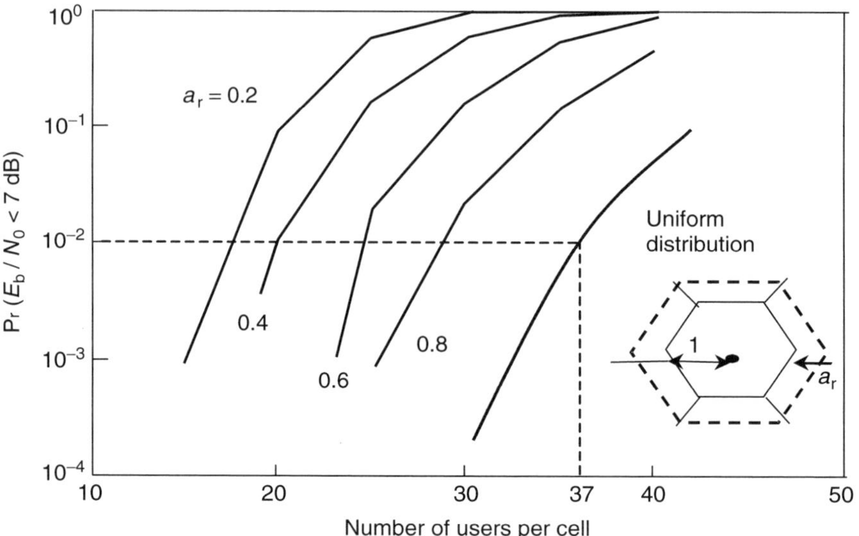

Figure 8.12 Reverse link capacity under nonuniform user distribution.

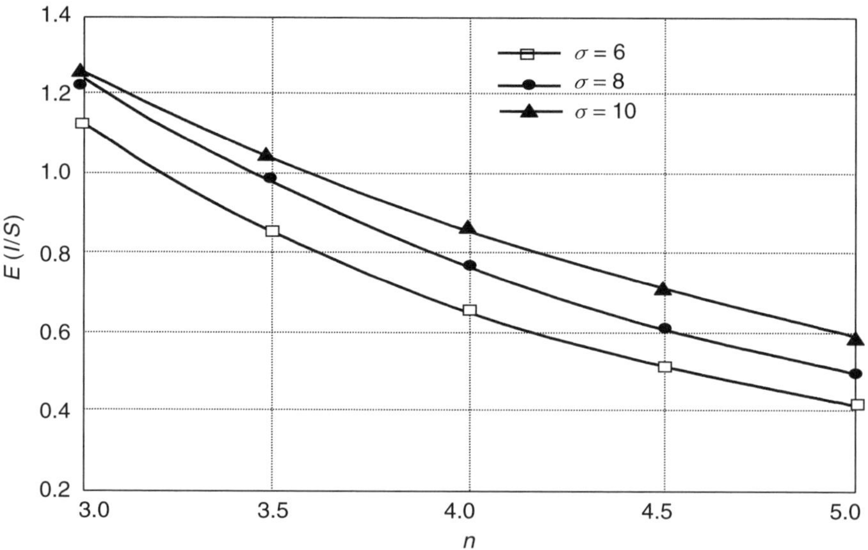

Figure 8.13 Mean value of the external interference (normalized to the number of users per cell) versus propagation factor n, with the standard deviation of the lognormal shadowing, σ, as parameter.

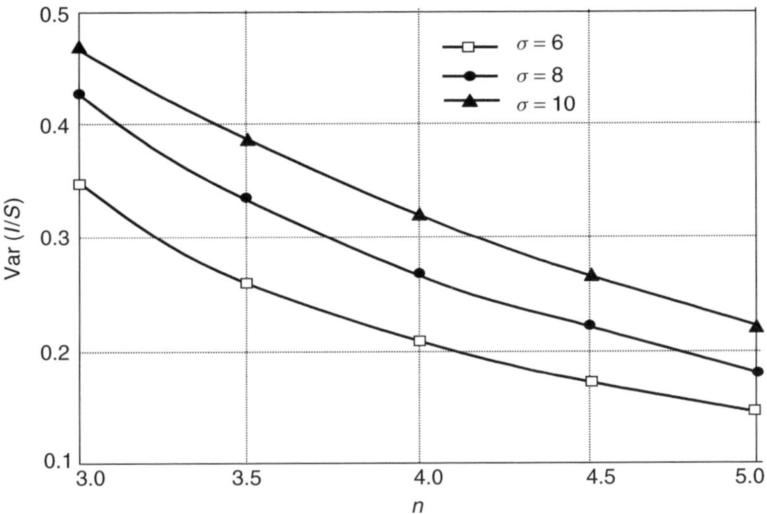

Figure 8.14 Variance of the external interference (normalized to the number of users per cell) versus propagation factor n, with the standard deviation of the lognormal shadowing, σ, as parameter.

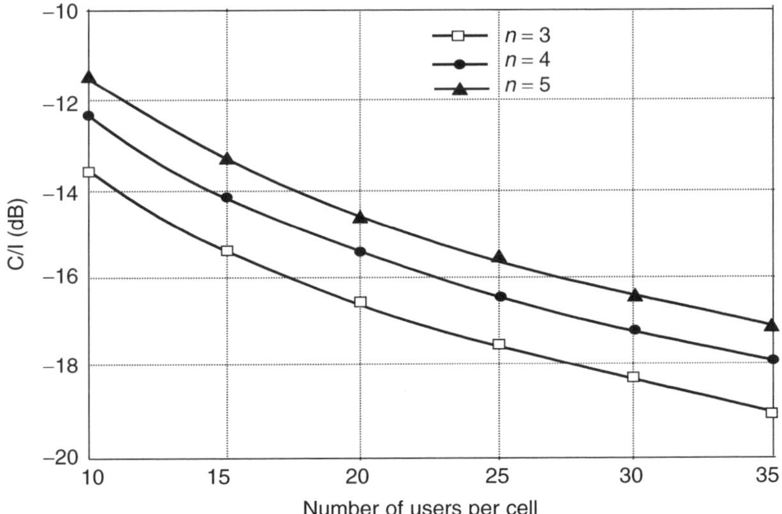

Figure 8.15 Carrier to interference ratio at the cell-site receiver versus the number of users per cell, with standard deviation of the lognormal shadowing equal to 6 dB and outage probability 10%.

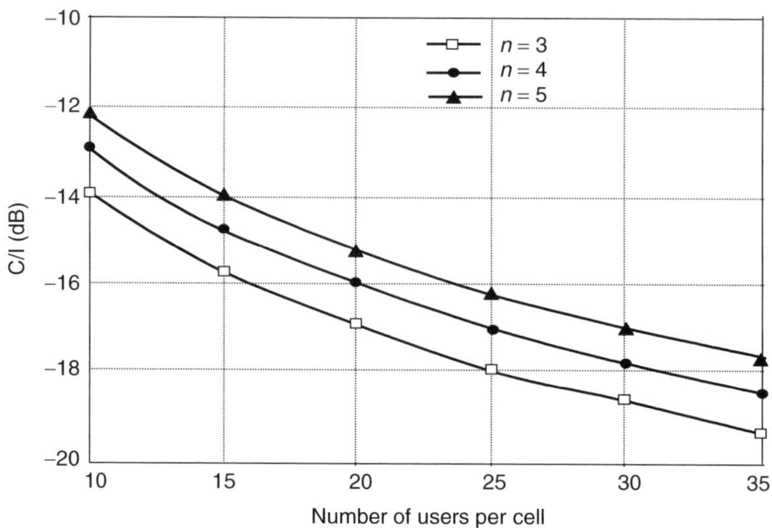

Figure 8.16   Carrier-to-interference ratio at the cell-site receiver versus the number of users per cell, with standard deviation of the lognormal shadowing equal to 10 dB and outage probability 10%.

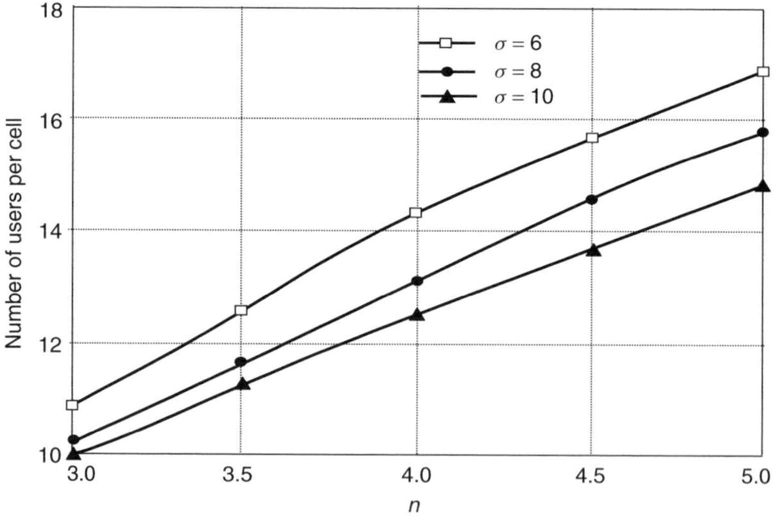

Figure 8.17   Number of users per cell versus the propagation factor, α, with the standard deviation of the lognormal shadowing, σ, as parameter and outage probability 10%. The processing gain is assumed to be equal to 128 and the required E_b/N_0 equals 7 dB.

The impact of variation in n and σ on carrier-to-interference ratio, and the number of users in a cell are shown in Figures 8.15 to 8.17.

8.4 CHANNEL MODELING IN CDMA NETWORKS

In general, fading channel can be characterized by multipath propagation and the impulse response of such a channel can be represented as

$$h(t, \tau) = \sum_{k=0}^{N(\tau)-1} c_k(t)\delta(\tau - \tau_k(t))e^{j\theta_k(t)}$$

$$\omega_{\mathrm{D}}(t) = \partial\theta(t)/\partial t \tag{8.41}$$

where $N(\tau)$ is the number of paths, $c_k(t)$ is the path intensity coefficient and τ_k and θ_k its delay and phase. Different channel coefficients can vary in time as shown in Figure 8.18.

8.4.1 Distribution of the arrival time sequence

In theory, different functions are used for the distribution of the arrival time sequence such as

- Standard Poisson model
- Modified Poisson – the $\Delta-K$ model
- Modified Poisson-nonexponential interarrivals
- The Neyman–Scott clustering model
- The Gilbert's burst model
- The pseudo-Markov model

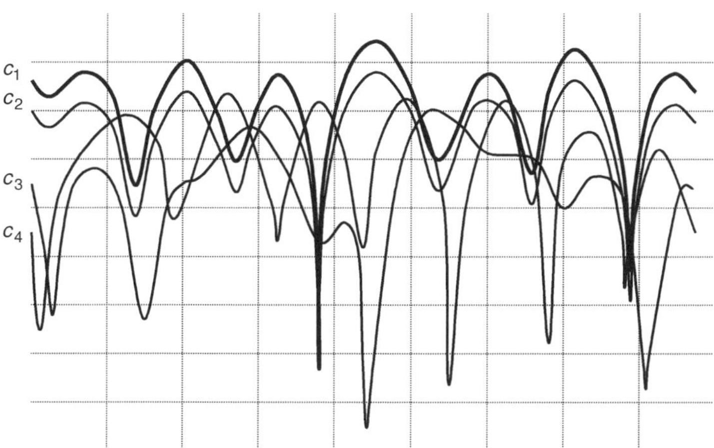

Figure 8.18 Variation of channel coefficients in time.

Table 8.1 Suburban area

| | Probability | | |
	Excess delay (μs)		
Number of paths	0–0.78	0–1.56	0–6.24
2	0.12	0.1	0.08
4	0.2	0.18	0.18
6	0.1	0.11	0.13

Table 8.2 Urban area

| | Probability | | |
	Excess delay (μs)		
Number of paths	0–0.78	0–1.56	0–6.24
4	0.05	0.02	0
6	0.17	0.05	0.02
8	0.25	0.1	0.04
10	0.02	0.12	0.06
12	0	0.11	0.08
14	0	0.08	0.08
16	0	0.03	0.08

8.4.2 Distribution of the number of paths

Probability of finding N paths (echos) in the delay window (excess delay) is given in Tables 8.1 and 8.2 for suburban and urban areas, respectively. These probabilities are presented graphically in Figures 8.19 and 8.20. The correlation between the paths is presented in Figure 8.21.

This gives you a rough picture of how many fingers of a RAKE receiver will be used and with what probability.

8.4.3 The mean excess delay and the root mean square (RMS) delay spread

The expected delays for different environments are

- 20–50 ns – small and medium size office buildings
- <100 ns – university buildings
- 30–300 ns – factory environments
- $<1\,\mu$s – rural area
- 1–5 μs – suburban area
- 10–20 μs – urban area
- $<100\,\mu$s – (rarely) mountainous/hilly regions.

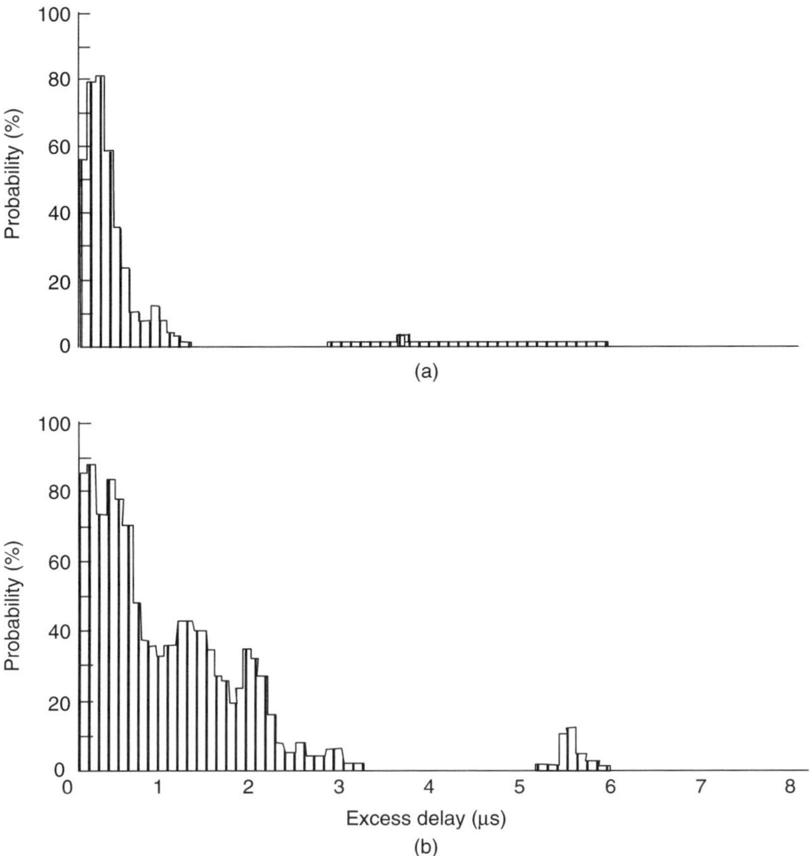

Figure 8.19 Probability of path occurrence: (a) suburban locality, (b) urban locality.

8.4.4 The path loss

For a macrocell, the path losses are modeled as

$$10^{\xi/10} r^{-n} \tag{8.42}$$

ξ is a Gaussian variable with standard deviation $\sigma = 8$ and zero mean and $n = 2$ (rural) to 5.5 (urban).

For indoor communications model r^{-n} is used with $2 < n < 12$.

r	n
1–10 m	2
10–20 m	3
20–40 m	6
>40 m	12

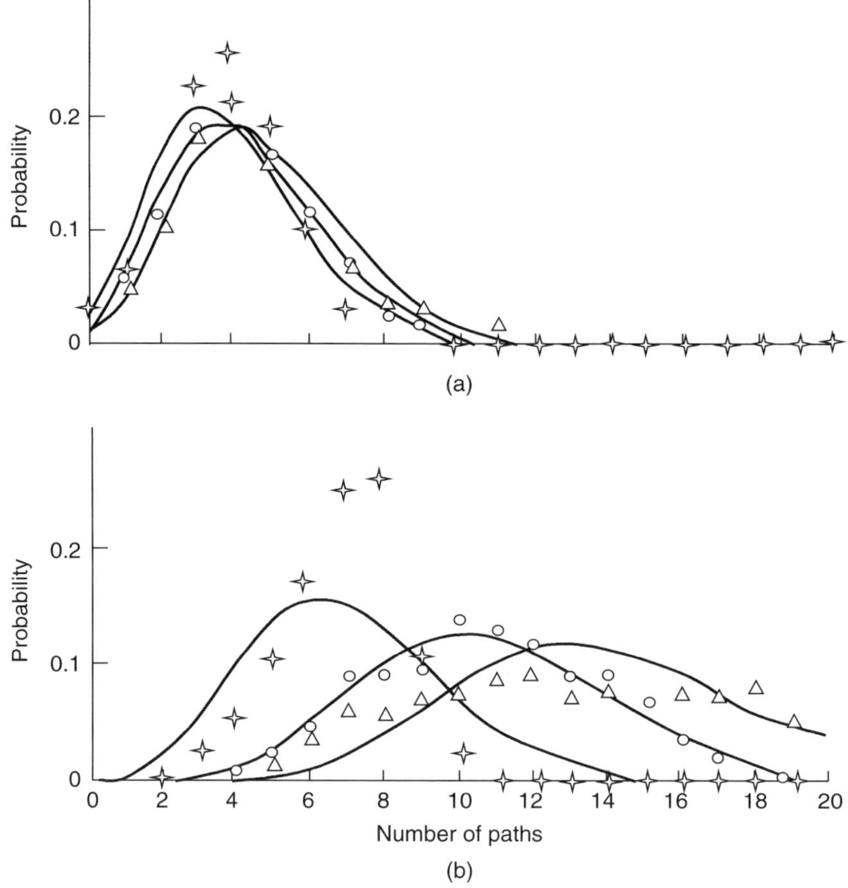

Figure 8.20 Echo path-number distributions: (a) suburban, (b) urban. Theoretical cumulative excess delay intervals are $0-0.78\,\mu s$, $0-1.56\,\mu s$, $0-6.24\,\mu s$.

8.4.5 Voice activity factor

The voice statistics are shown in Table 8.3.

On the basis of this, the voice activity factor is in the range

$$\alpha = \frac{\text{talk spurt}}{\text{pause} + \text{talk spurt}} \cong 0.4 - 0.5 \tag{8.43}$$

For the relative channel coefficient intensities we use CODIT (COde DIvision Test bed) model [4]. For macro-, micro- and picocells the results are shown in Tables 8.4 to 8.6, respectively. The results are also shown graphically in Figures 8.22 to 8.24, respectively.

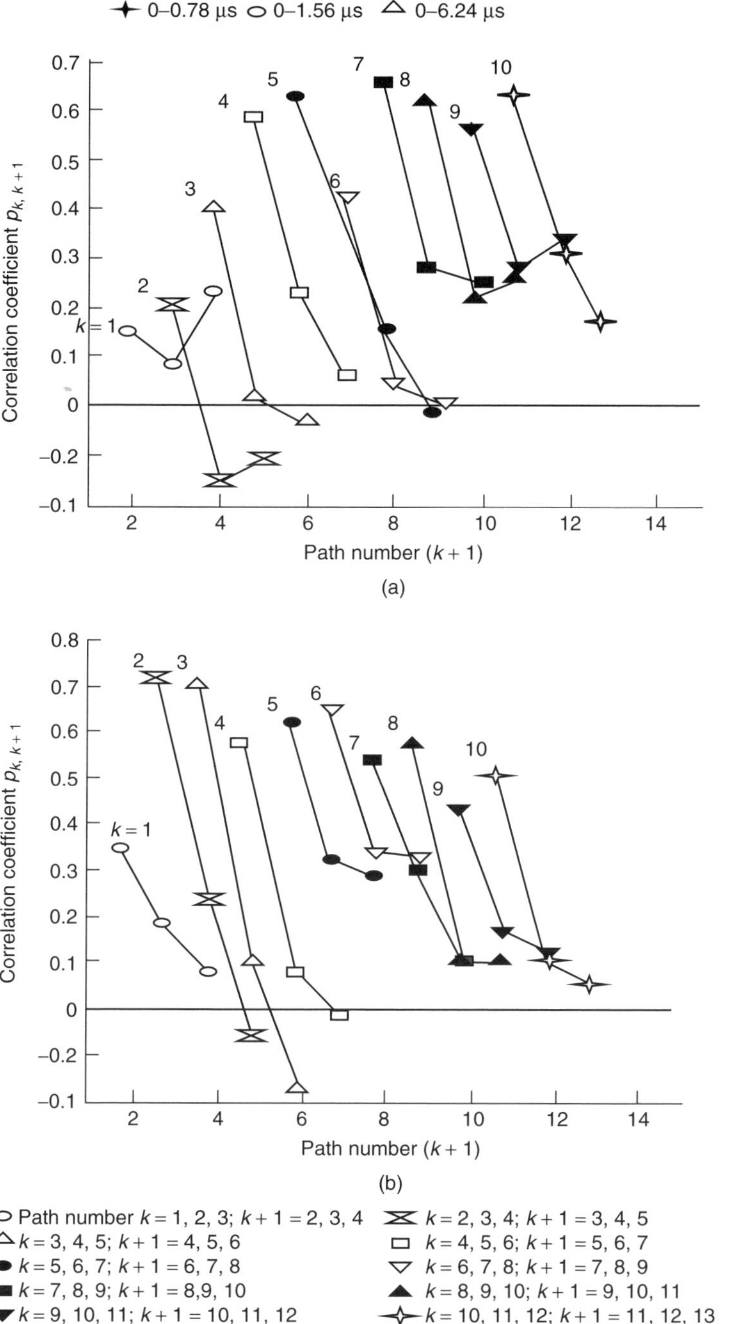

Figure 8.21 Correlation coefficients of echo strengths: (a) urban, (b) suburban.

Table 8.3 Average talk spurts and pauses based on the
study by Brady

Threshold	$-45\,$dB	$-40\,$dB	$-35\,$dB
Talk spurt (ms)	1311	1125	902
Pause (ms)	1695	1721	1664

Table 8.4 CODIT channel model realization in COST 207 format

Tap	Relative delay (ns)	Relative power (dB)	Doppler spectra
1	100	-3.2	CLASS
2	200	-5.0	CLASS
3	500	-4.5	CLASS
4	600	-3.6	CLASS
5	850	-3.9	CLASS
6	900	0.0	CLASS
7	1050	-3.0	CLASS
8	1350	-1.2	CLASS
9	1450	-5.0	CLASS
10	1500	-3.5	CLASS

Macrocellular channel

In the table CLASS refers to Jack's classical model with channel correlation function
$\rho(\tau) = J_0(w_\mathrm{D}\tau)$, where w_D is the Doppler and J_0 is the zero-order Bessel function.
These results are obtained with the signal bandwidth of 20 MHz, so that the maximum
resolution between paths is 50 ns. In UMTS, the chiprate is 3.84 Mchips and these results
will be modified by combining a number of paths into one equivalent path. This will be
discussed later in the book.

Microcellular channels

Table 8.5 CODIT microcell channel model using COST 207 format

Tap	Delay (ns)	Average power (dB)	Doppler spectrum	Ricean factor (dB)
1	0	-2.3	RICE	-7.3
2	0	0.0	RICE	-3.5
3	0	-13.6	CLASS	$-$
4	50	-3.6	RICE	-3.5
5	50	-8.1	CLASS	
6	100	-10.0	CLASS	
7	1700	-12.6	RICE	-2.2

Picocellullar channels

Table 8.6 CODIT picocell channel model using COST 207 format

Tap	Relative delay (ns)	Relative power (dB)	Doppler spectra
1	0	−3.6	CLASS
2	50	0.0	CLASS
3	100	−3.2	CLASS

8.4.6 Static path loss models

Macrocells

The path losses are characterized by equation (8.42). Values for the loss exponent n are in the range from 3.0 to 5.0 depending on the environment. Value $n = 3.6$ is used in the CODIT model. In addition, there is the shadowing effect. A Gaussian random variable, ξ (dB), is used for modeling this long-term loss, (see equation (8.42)). In CODIT project [4] the proposal is to use a mean and variance as follows:

$$\langle \xi \rangle = 0\,\text{dB}$$

$$\sigma_\xi = 6\,\text{dB} \qquad (8.44)$$

The resulting path loss in dB is computed as

$$L_{\text{macro}} = \xi + 3.6 \cdot 10 \log(r)\,(\text{dB}) \qquad (8.45)$$

Microcells

In this case, a three-slope path loss model is used as follows:

$$L_{\text{LoS1}} = L_b + 20 \cdot n_{\text{LoS1}} \cdot \log(x/R_b) \quad x \leq R_b,\ \text{LoS}$$

$$L_{\text{LoS2}} = L_b + 40 \cdot n_{\text{LoS2}} \cdot \log(x/R_b) \quad x > R_b,\ \text{LoS}$$

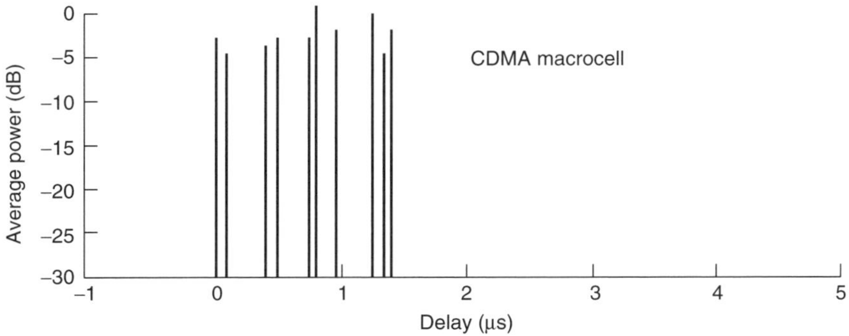

Figure 8.22 Impulse responses of the CODIT macrocell channel model.

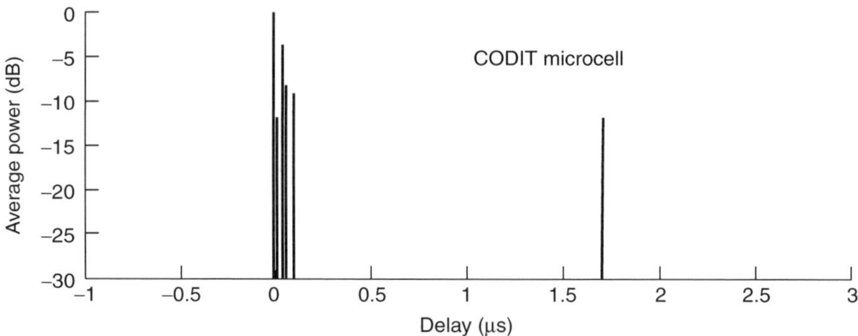

Figure 8.23   Impulse responses of the CODIT microcell channel model.

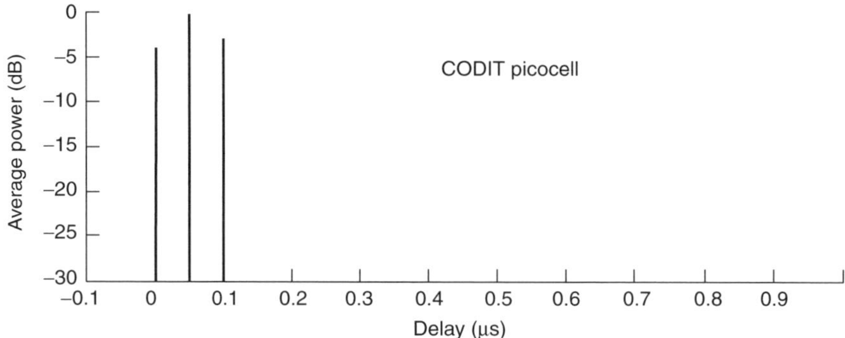

Figure 8.24   Impulse responses of the CODIT picocell channel model.

$$L_{n\text{LoS}} = L_{\text{LoS}}(x_{\text{corner}}) + L_{\text{corner}}$$

$$+ 10 \cdot n_{n\text{LoS}} \cdot \log(x/x_{\text{corner}}) \quad n\text{LoS} \tag{8.46}$$

In equation (8.46) $n_{\text{LoS1}}, n_{\text{LoS2}}$, and $n_{n\text{LoS}}$ denote each segment slope, where LoS and nLoS refer to line of sight and nonline of sight. Distance from the transmitter to the receiver is measured along the street path, x. If in a nLoS situation the distance from the transmitter to the corner is x_{corner}, the breakpoint distance is given as

$$R_{\text{b}} = \frac{4 \cdot h_{\text{b}} \cdot h_{\text{m}}}{\lambda} \tag{8.47}$$

where h_{b} and h_{m} are the heights of base and mobile stations, respectively.

$$L_{\text{b}} = \left| 20 \cdot \log \left(\frac{\lambda^2}{8 \cdot \pi \cdot h_{\text{b}} \cdot h_{\text{m}}} \right) \right|$$

$$L_{\text{corner}} = -0.1 \, w_{\text{s}} + 0.05 x_{\text{corner}} + 20$$

$$n_{n\text{LoS}} = -0.05 \, w_{\text{s}} + 0.02 x_{\text{corner}} + 4 \tag{8.48}$$

For the microcellular scenario $w_s = 30$ m. As a typical example let us assume wavelength of $\lambda = 0.15$ m. If the mobile station (MS) and BS antenna heights are defined as $h_b h_m = 11.25$ m^2, then we have

$$R_b = 300 \text{ m}$$

$$L_b = 82 \text{ dB}$$

$$L_{\text{corner}} = 17 + 0.05 x_{\text{corner}}$$

$$n_{n\text{LoS}} = 2.5 + 0.02 x_{\text{corner}}$$

$$n_{\text{LoS1}} = 1$$

$$n_{\text{LoS2}} = 2 \tag{8.49}$$

$$L_{\text{LoS1}} = 82 + 20 \cdot \log\left(\frac{x}{300}\right) \quad x \leq 300, \text{LoS}$$

$$L_{\text{LoS2}} = 82 + 40 \cdot \log\left(\frac{x}{300}\right) \quad x > 300, \text{LoS}$$

$$L_{n\text{LoS}} = L_{\text{LoS}}(x_{\text{corner}}) + 17 + 0.05 x_{\text{corner}}$$

$$+ (25 + 0.2 x_{\text{corner}}) \log\left(\frac{x}{x_{\text{corner}}}\right) n\text{LoS} \tag{8.50}$$

As in the case of macrocells, the shadowing effect is modeled by $10^{\xi/10}$, where ξ is again a Gaussian random variable with

$$\langle \xi \rangle = 0 \text{ dB} \quad \text{and} \quad \sigma_\xi = 4 \text{ dB} \tag{8.51}$$

Picocells

In this case the Motley–Keenan model is used

$$L_{\text{pico}} = L_0 + 10n \log(x) + \sum_{j=1}^{J} N_{w_j} \cdot L_{w_j} + \sum_{i=1}^{I} N_{f_i} \cdot L_{f_i} \tag{8.52}$$

L_0 denotes the loss at the reference point (at 1 m) and n is the power decay index. x represents the transmitter to receiver path length. N_{wj} and N_{fi} denote the number of walls and floors, respectively, of different kinds that are traversed by the transmitted signal. L_{wj} (dB) and L_{fi} (dB) represent their corresponding losses factors. Typical values for these parameters are

$$L_0 = 37 \text{ dB}$$

$$n = 2$$

$$L_f = 20 \text{ dB}$$

$$L_w = 3 \text{ dB} \tag{8.53}$$

This yields

$$L_{\text{pico}} = 37 + 20\log(x) + 3N_{\text{w}} + 20N_{\text{f}}\,(\text{d}B) \tag{8.54}$$

Unlike macro- and microcellular environments, no shadowing is modeled in picocells. This model is widely accepted for modeling the picocellular path loss. It yields to excessive losses predictions when a corridor is involved. This is due to a certain waveguide effect present in corridors.

8.4.7 Dynamic path loss models

Dynamic lognormal shadowing

The shadowing effects are usually modeled with a lognormal law. Dynamic behavior is controlled by a proper correlation (autoregressive) model

$$\text{sh}(n) = \sqrt{(1 - \rho^2)} \cdot a(n) + \rho \cdot \text{sh}(n - 1) \tag{8.55}$$

where $a(n)$ is the sample of a lognormal law. The correlation factor ρ is in the range between 0 and 1. Evaluation of the correlation factor is based on propagation measurements. The concept of a *decorrelation length* distance where the autocorrelation (or autocovariance) function of the long-term fading is equal to 0.5 is used. Decorrelation lengths have been evaluated for each environment (20 m in rural areas and 5 m in metropolitan areas). In the system simulation the decorrelation profile is introduced so that mobiles move with steps equal to the decorrelation length. Shadowing samples are generated at each movement with a correlation factor equal to 0.5. A linear or cosine interpolation is then used to ensure a smooth transition between consecutive path loss plus shadowing values. This simplified scheme is thought of as being a good compromise between computation time and accuracy of the correlated shadowing model.

8.4.8 Microcellular corners transition model

The static path loss model for microcells is not valid anymore for a dynamic analysis. This is due to the rough behavior of this model in the corners transitions. The transition distance x_{t} is a distance beyond which the MS quits the corner transition situation (newly introduced slope) and goes into a deep nLoS region. The signal loss at the corner is not constant any longer. In this case losses are calculated as

$$\begin{aligned}
L_{\text{corner}}(x) = \ &L_{\text{LoS}}(x_{\text{corner}}) \\
&+ \frac{L_{\text{LoS}}(x_{\text{t}}) - L_{\text{LoS}}(x_{\text{corner}})}{\log\left(\dfrac{x_{\text{t}}}{x_{\text{corner}}}\right)} \cdot \log\left(\frac{x}{x_{\text{corner}}}\right)
\end{aligned} \tag{8.56}$$

The corner transition condition

$$L_{\text{corner}}(x) \leq L_{n\text{LoS}}(x)$$
$$x \leq x_{\text{t}}$$
$$x_{\text{t}} = 2\,w_{\text{s}} \tag{8.57}$$

The combination of equations (8.46) and (8.56) results in

$$L_{\text{LoS}_1} = L_{\text{b}} + 20 \cdot n_{\text{LoS}_1} \cdot \log\left(\frac{x}{R_{\text{b}}}\right) \quad x \leq R_{\text{b}}, \text{LoS}$$

$$L_{\text{LoS}_2} = L_{\text{b}} + 20 \cdot n_{\text{LoS}_2} \cdot \log\left(\frac{x}{R_{\text{b}}}\right) \quad x > R_{\text{b}}, \text{LoS}$$

$$L_{\text{corner}}(x) = L_{\text{LoS}}(x_{\text{corner}})$$
$$+ \frac{L_{\text{LoS}}(x_{\text{t}}) - L_{\text{LoS}}(x_{\text{corner}})}{\log\left(\frac{x_{\text{t}}}{x_{\text{corner}}}\right)} \cdot \log\left(\frac{x}{x_{\text{corner}}}\right) \quad x \leq x_{\text{t}}, n\text{LoS}$$

$$L_{n\text{LoS}} = L_{\text{LoS}}(x_{\text{corner}}) + L_{\text{corner}} + 10 \cdot n_{n\text{LoS}} \cdot \log\left(\frac{x}{x_{\text{corner}}}\right) n\text{LoS} \tag{8.58}$$

Equation (8.50) now becomes

$$L_{\text{LoS}_1} = 82 + 20 \cdot \log\left(\frac{x}{300}\right) \quad x \leq 300, \text{LoS}$$

$$L_{\text{LoS}_2} = 82 + 40 \cdot \log\left(\frac{x}{300}\right) \quad x > 300, \text{LoS}$$

$$L_{\text{corner}}(x) = L_{\text{LoS}}(x_{\text{corner}})$$
$$+ \frac{L_{\text{LoS}}(x_{\text{t}}) - L_{\text{LoS}}(x_{\text{corner}})}{\log\left(\frac{x_{\text{t}}}{x_{\text{corner}}}\right)} \cdot \log\left(\frac{x}{x_{\text{corner}}}\right) \quad x \leq x_{\text{t}}, n\text{LoS}$$

$$L_{n\text{LoS}} = L_{\text{LoS}}(x_{\text{corner}}) + L_{\text{LoS}}(x_{\text{t}}) + 0.05x_{\text{corner}}$$
$$+ (25 + 0.2x_{\text{corner}}) \log\left(\frac{x}{x_{\text{corner}}}\right) n\text{LoS} \tag{8.59}$$

All results for the propagation losses discussed in this section will be used in the next chapter to analyze the network (cell) coverage, which is the main input parameter for CDMA network design and deployment.

8.4.9 Mobility

Mobility models for outdoor environments

Here we start with the following assumptions: The macrocell layout is based on the classic hexagonal grid. The microcell is based on the Manhattan grid with pedestrian and vehicular users moving along the streets. The users are completely free to move in the whole service area in the macrocell environment. Movements are restricted to the street line in the microcell layout.

Mobility model for the simulation of outdoor macrocellular environments

The movement of a mobile is modeled with snapshots of a realistic trajectory, with a short observation interval. Every new position is calculated according to the mobile speed and its old direction. The new direction is randomly generated with a small variation with respect to the present one. Mobile speed is kept fixed during the whole simulation, but can be different for different mobiles. A realistic trajectory for a slow mobile (such as a pedestrian) can comprise also sharp curves, while a fast mobile (such as a car) can go along quasi-linear paths. This behavior can be obtained considering the centrifugal acceleration of a trajectory and setting a maximum value. A relation can be found between speed and maximum allowed variation in direction.

Centrifugal acceleration

Referring to Figure 8.25, the approximate expression for speed and acceleration as a function of moving during observation time is

$$v = \frac{\Delta x}{\Delta t} \quad a_{\mathrm{c}} = \frac{v^2}{r} = \left(\frac{\Delta x^2}{\Delta t^2}\right)\frac{1}{r} \tag{8.60}$$

The variation of direction as a function of speed and observation time can be expressed as

$$\sin\frac{\Delta\theta}{2} = \frac{\Delta x}{r} \Rightarrow \Delta\theta = 2\arcsin\left(\frac{a_{\mathrm{c}}\Delta t}{2v}\right) \tag{8.61}$$

The variation of direction is generated as a Gaussian random variable with zero mean. The variance is chosen according to the maximum acceleration the mobile is supposed to have

$$\mathrm{erfc}\left(\frac{\Delta\theta}{\sigma\sqrt{2}}\right) = p[\Delta\theta > \Delta\theta] \tag{8.62}$$

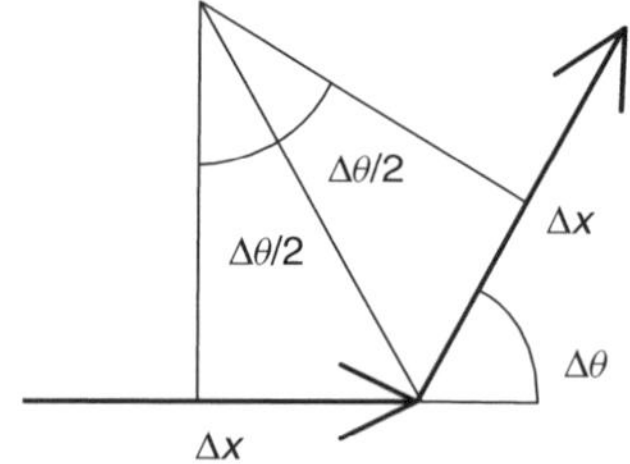

Figure 8.25 Centrifugal acceleration.

where erfc is the complementary error function:

$$\mathrm{erfc}(x) = \frac{2}{\sqrt{\pi}} \int_x^\infty e^{-t^2} dt \tag{8.63}$$

The maximum value for acceleration is $9.81\,\mathrm{m\,s^{-2}}$. For exceeding the probability by less than 1% the possible trajectory is shown in Figure 8.26. The model cannot take care of any topological information (streets, corners, etc.).

Mobility model for outdoor microcellular environment

The movement of a mobile in a microcellular environment depends on the user speed and has the main constraint of street layout – fixed directions with standard rotation of $90°$ when turning at a street corner. The mobility model can be implemented with the same formulas adopted for a macrocell, with specific constraints on the permitted values of each parameter. Whenever the user is generated, its movement is characterized by the user speed, two possible directions along the street with equal probability and three possible changes in direction $(0°, \pm90°)$ at the street corner area, again with the same probability.

Mobility model for indoor environments

The steady-state model is characterized by the following assumptions:

- The SIG5 indoor scenario [5]
- 12 rooms and one corridor, divided into 18 areas as shown in Figure 8.27.

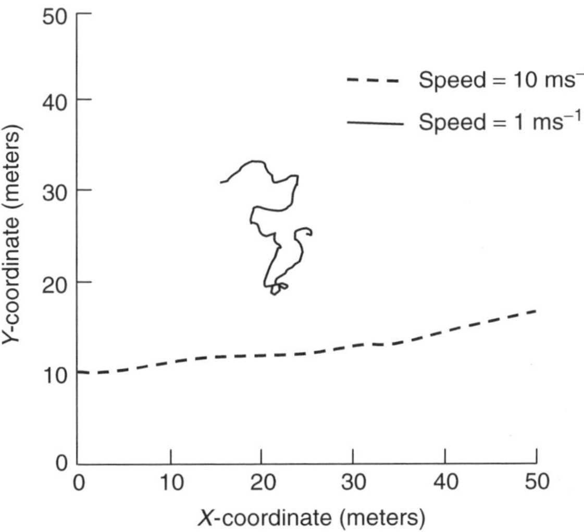

Figure 8.26 Examples of generated trajectories.

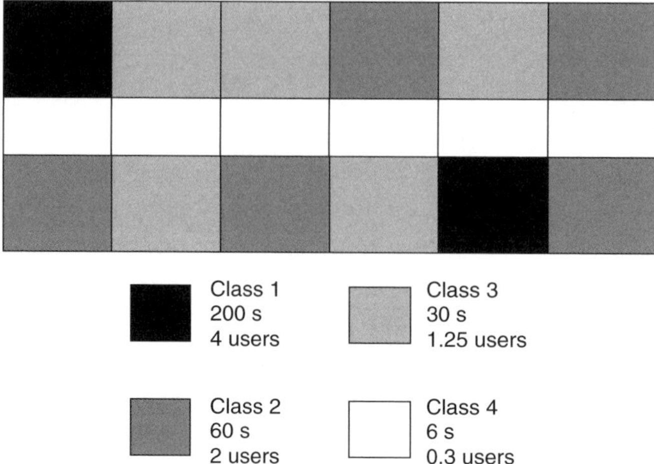

Figure 8.27 Example of a possible scenario in which the users are grouped in four classes: each class is characterized by given values for the mean crossing time and the average number of mobile terminals.

- 12 square areas, 5 m × 5 m, for the 12 rooms;
- 6 rectangle areas, 5 m × 3 m, for the corridor.

In each area users have similar behavior and distribution

- mean crossing time of an exponential distribution
- average number of mobile terminals.

The environment should be organized by grouping the areas in few classes.

- As a first step, only two classes could be considered.
- A class for the rooms, with high mean crossing time (600 s) and average number of mobile terminals (7% of total number of users).
- A class for the corridor, with lower value for those parameters (10 s and 2.7% of total number of users, respectively).
- The mean crossing time represents the average time spent by a user within an area and it is used as the mean value of an exponential probability density function.
- The transition probability matrix P_{ij} governs the movement of the users.
- The generic element p_{ij} gives the probability that a user leaving the area i is going into an adjacent area j.
- The matrix P_{ij} is generated by an algorithm resolving the following equations:

$$\sum_{\substack{j=1 \\ j \neq i}}^{M} \phi_{ij} = \phi_i, \quad \sum_{\substack{i=1 \\ i \neq j}}^{M} \phi_{ij} = \phi_j \tag{8.64}$$

$$\phi_{ij} \geq c_{ij}$$

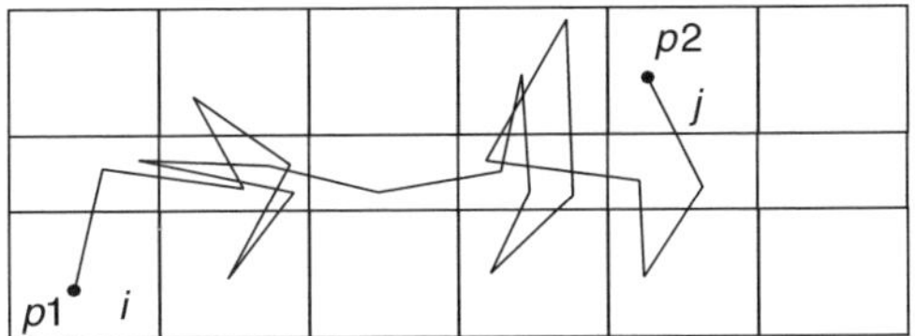

Figure 8.28   Example of an MS path.

where, for each area i

$\phi_i = N_i / T_i$

N_i = average number of mobile terminals in area i,

T_i = mean crossing time of area i

M = number of areas (18 in this case),

ϕ_{ij} = flow from area i to area j,

c_{ij} = constant, $p_{ij} = \phi_{ij}/\phi_i$.

For the generation of the matrix $\boldsymbol{P}_{ij}$, the following parameters are needed:

- total number of areas representing the environment under study (18 in this case);
- a matrix $\boldsymbol{A}_{ij}$ representing the adjacencies between the areas: this matrix also allows to indicate the position of walls (there is no adjacency between two areas separated by a wall with no door);
- average number of mobile terminals per area (depends on the class to which the area belongs) and
- average area crossing times (depends on the class to which the area belongs).

The obtained matrix $\boldsymbol{P}_{ij}$, when used in the simulation program to move the mobile terminals from area to area of environment, according to the defined crossing times T_i, produces the distribution of mobile terminals N_i. Figure 8.28 shows the possible path of a mobile terminal during the simulation time. The destination points within each area are chosen randomly.

8.5 RAKE RECEIVER

Time-varying multipath channel represented by equation (8.41) will be further specified by an explicit representation of Doppler as

$$h_{\mathrm{c}}(\tau;t) = \sum_{i=1}^{L} \beta_i(t)\delta(\tau - \tau_i)\exp j[\Delta\omega_i(\tau - \tau_i) + \theta_i(t)] \tag{8.65}$$

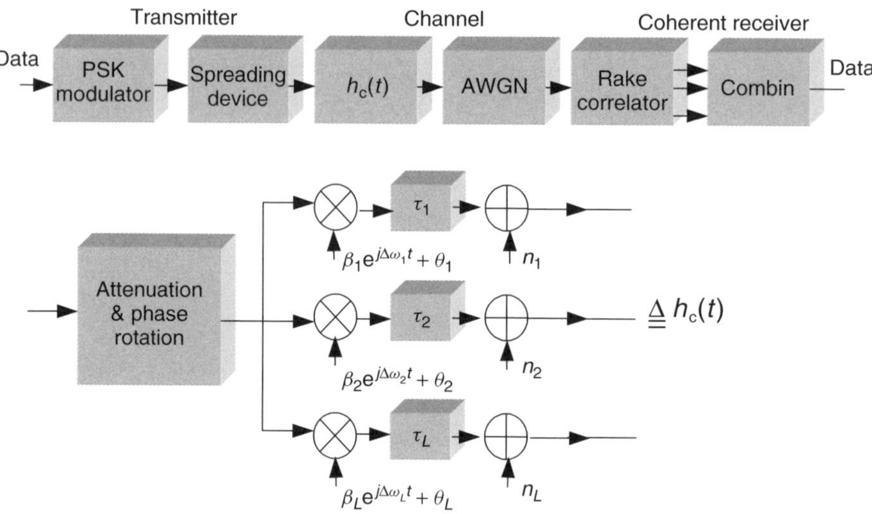

Figure 8.29 Transmission model in the baseband.

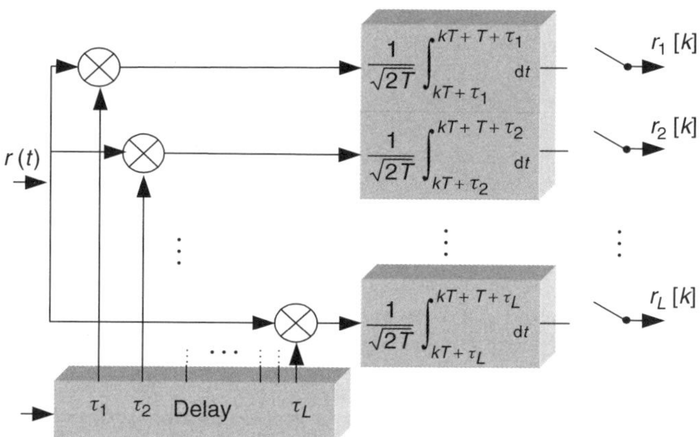

Figure 8.30 Functional block diagram of the RAKE-correlator of Figure 8.29 with
L-despreading arms.

where L is the total number of paths and $\beta_i(t)$, $\Delta\omega_i(t)$, $\theta_i(t)$, $\tau_i(t)$ are the ith path amplitude, frequency shift, phase and delay, respectively. The transmission model is shown in
Figure 8.29 with an additional elaboration of the channel impulse response $h_c(t)$.

Elaboration of the RAKE receiver from Figure 8.29 is shown in Figure 8.30. The
receiver will synchronize L replicas of the local code $c(t)$ to each incoming path with
delay τ_i and despread the signal received through the path generating the variable r_i.
The possible paths that will be picked up in this process are shown in Figure 8.31 for

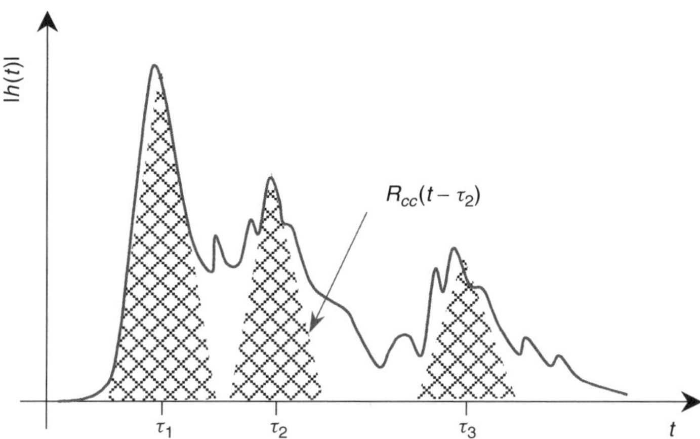

Figure 8.31 Magnitude of a typical channel impulse response $h(t)$ with a possible placing of $L = 3$ arms for data demodulation.

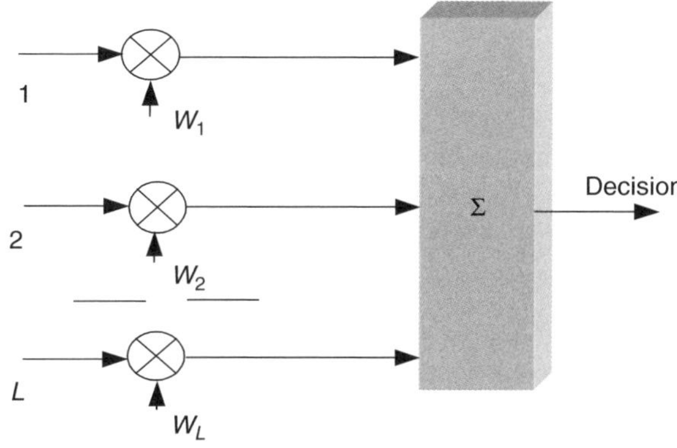

Figure 8.32 Diversity combiner.

$L = 3$. In the next step, variables r_i will be combined after being weighed with a certain coefficient W_i as shown in Figure 8.32.

Weight values that maximize E_b/N_0 [maximum ratio combiner (MRC)] are given as

$$
W_i = \frac{\beta_i \left(N_0 + I_a \sum_{j=2}^{L} \beta_j^2 \right)}{\beta_1 \left(N_0 + I_a \sum_{\substack{j=1 \\ j \neq i}}^{L} \beta_j^2 \right)}
\tag{8.66}
$$

For W_i we need to know N_0. When N_0 is difficult to estimate, a suboptimal solution is to use $W_i = \beta_i/\beta_1$. BER for Lth order diversity was already discussed in Chapter 6. equations (6.21) and (6.22) give the BER for nonconstant and constant multipath intensity profile (MIP). The gain obtained from MRC expressed as

$$G_{\text{MRC}} = \frac{\text{SNR}_{\text{com}}}{\text{SNR}} \Rightarrow N_{\text{MRC}} = G_{\text{MRC}}N \tag{8.67}$$

used in equation (8.14) is a direct indication of how much the network capacity will be increased. By using equation (8.12) one can see that for the case of L paths with correlated equal-power signal and uncorrelated noise we have
L paths, no RAKE

$$\frac{E_b}{N_0} \cong \frac{G_p}{LN}; \text{reference} \tag{8.68}$$

L paths, L combined

$$\frac{E_b}{N_0} \cong \frac{G_p}{N}; G_{\text{MRC}} = L \tag{8.69}$$

L paths, M combined

$$\frac{E_b}{N_0} \cong \frac{M}{L}\frac{G_p}{N}; G_{\text{MRC}} = M \tag{8.70}$$

M_a antennas, L paths, M combined/antenna

$$\frac{E_b}{N_0} \cong \frac{M_a M}{L}\frac{G_p}{N}; G_{\text{MRC}} = M_a M \tag{8.71}$$

If the noise is correlated, the improvement factor when the number of fingers doubled is $2/(1 + \rho)$, where ρ is the correlation coefficient between the noise samples in the two paths. Implementation of the system presented by equation (8.71) for two antennas ($M_a = 2$) and no multipath ($L, M = 1$) is shown in Figure 8.33.

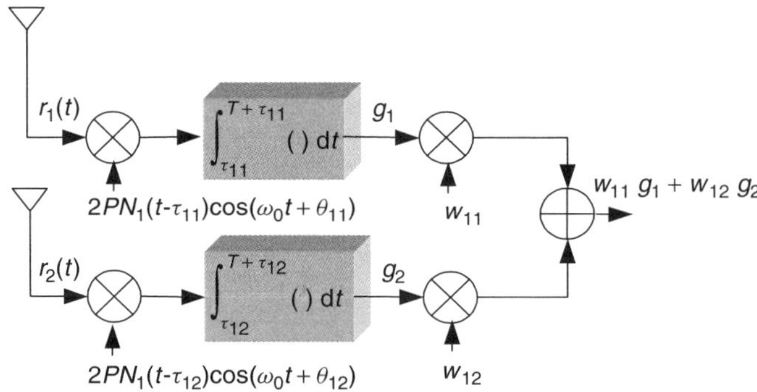

Figure 8.33 Antenna diversity receiver.

The gains suggested by equations (8.69)–(8.71) are obtained under rather artificial scenario. For a better insight into the possible gains we assume the following:

- The cell-of-interest is surrounded by 24 other cells.
- The variance of the fade caused by shadowing is assumed to be 6 dB.
- All cell shapes are squares.
- All mobile units are assumed to be uniformly distributed in the cells.
- The number of multipaths that the channel can resolve is $L = 3$.
- The channel has an exponential MIP with

$$E\{\gamma_1\} = E\{\gamma_l\}e^{-\beta(l-1)}, \quad l = 1, 2, \ldots, L \tag{8.72}$$

or

$$\alpha_l = \alpha_1 e^{-\beta(l-1)}, \quad l = 1, 2, \ldots, L \tag{8.73}$$

where β is a decay constant. We consider three types of receivers: (a) selection combiner S_1 chooses the largest path, (b) S_2 receiver selects two of the largest and combines them and MRC combines all three parts by using the MRC principle.

Figures 8.34 to 8.36 present error probability as a function of the number of users in the cell. The results are obtained by using techniques presented in Section 8.2. One can see from these figures that for constant MIP it makes sense to combine more and more paths because all these paths are equally relevant. The larger the decay factor β, the less it makes sense to use the third component. Figures 8.37 to 8.39 represent the same results versus SNR.

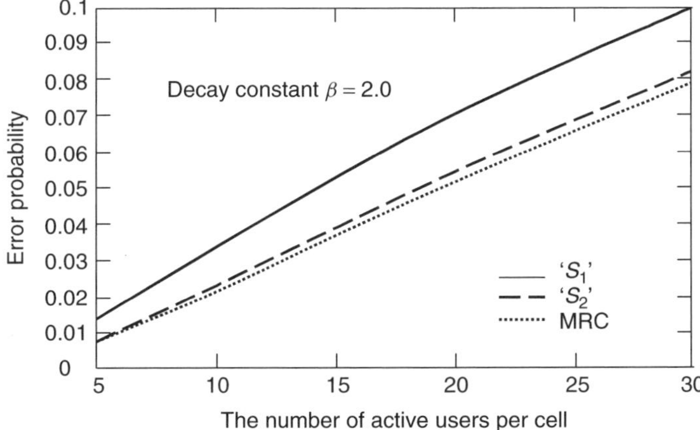

Figure 8.34 Error probability comparison of three different combining techniques as a function of the number of CDMA users for an exponential MIP.

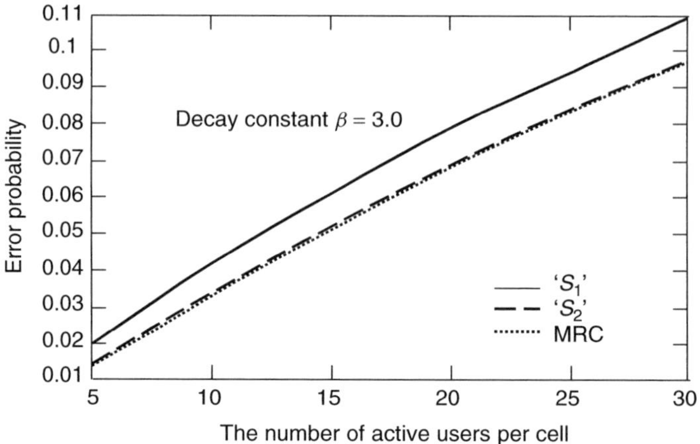

Figure 8.35 Error probability comparison of three different combining techniques as a function of the number of CDMA users for an exponential MIP with a higher decay rate.

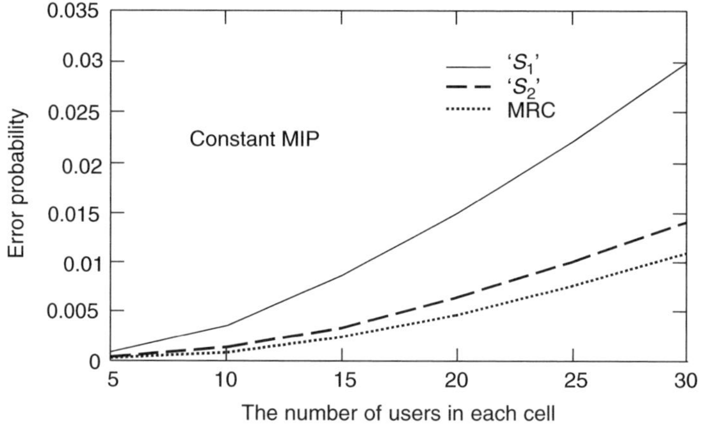

Figure 8.36 Error probability comparison of three different combining techniques as a function of the number of CDMA users for a constant MIP.

8.6 CDMA CELLULAR SYSTEM WITH ADAPTIVE INTERFERENCE CANCELLATION

If in equation (8.12) we can cancel interfering components originating from own and surrounding cells, then we can increase the number of users in these cells. The problem of multiple-access interference (MAI) cancellation will be discussed in depth in Chapters 11 and 12. Here we start with a simple solution in order to get a first insight into the

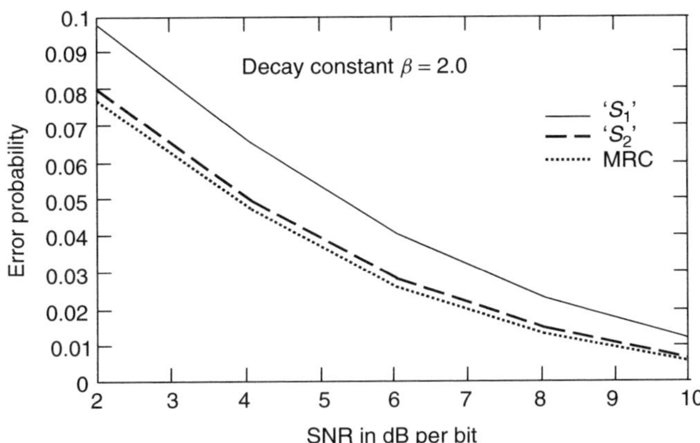

Figure 8.37 Error probability comparison of three different combining techniques as a function of SNR for an exponential MIP.

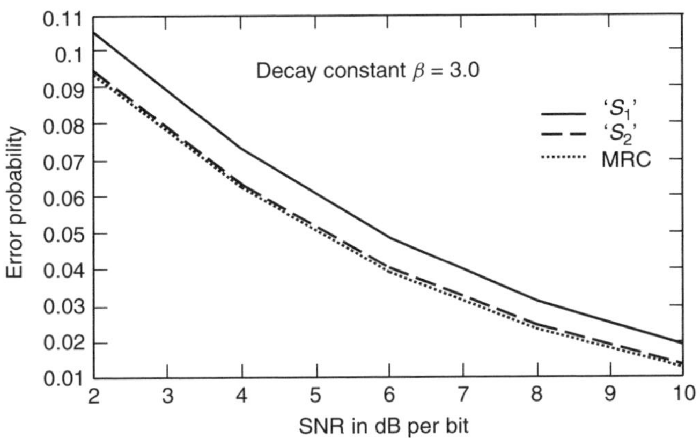

Figure 8.38 Error probability comparison of three different combining techniques as a function of the SNR for an exponential MIP with a higher decay rate.

possibility of improving the system capacity by using this approach. For this purpose we assume a very simple signal format at the input of the receiver (in the baseband)

$$r = \sum_k b_k C_k \tag{8.74}$$

where user k uses code C_k. At the receiver (assume the reference receiver with no index) r will be correlated with code C to produce

$$y(t) = \int_0^T rC \, \mathrm{d}t = b + \sum_k b_k R_k + n \tag{8.75}$$

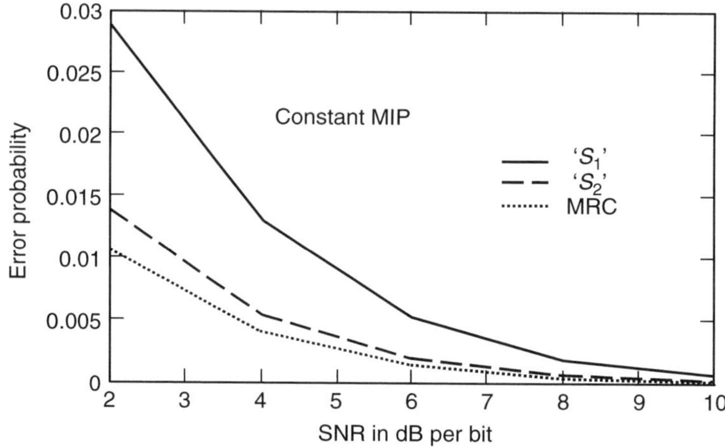

Figure 8.39 Error probability comparison of three different combining techniques as a function of the SNR for a constant MIP.

where R_k is the cross-correlation between the reference code $C = (c_1, c_2, \ldots, c_{T/T_c})$ and code C_k. The receiver makes preliminary estimation of bit

$$\hat{b} = \text{sgn } y(t) \tag{8.76}$$

and then uses $\varepsilon(t) = y(t) - \hat{b}$ in the minimum mean square error (MMSE) algorithm to change chips in C as long as the bit takes to minimize $\varepsilon^2(t)$.

For differential modulation and partial matched filter with mT/T_c taps, the receiver block diagram is shown in Figure 8.40. By minimizing ΣR_k we are actually orthogonalizing the codes, hence the name code orthogonalizing filter (COF). BER for the system with COF and with standard MF is shown in Figure 8.41. A significant improvement in performance is evident. For a system with antenna diversity as shown in Figure 8.42 and COF and the parameters listed in Table 8.7, BER curves are shown in Figure 8.43. One can see that system performance with 24 users and COF get close to the performance of the system with only one user (no MAI).

A network with 19 cells is simulated with parameters shown in Table 8.8. Performance results, outage probability versus number of users per cell are shown in Figure 8.44. For outage probability of 5%, an increase in capacity from 5 to 14 can be seen.

As discussed in Section 8.2, the system capacity given by equation (8.14) depends on the required E_b/N_0. For the system specified in Table 8.8, the results are shown in Figure 8.45 and Table 8.9.

One can see that a system using adaptive interference cancellation significantly increases the system capacity. For this reason, in Chapters 11 and 12 we will discuss this technology in much more detail.

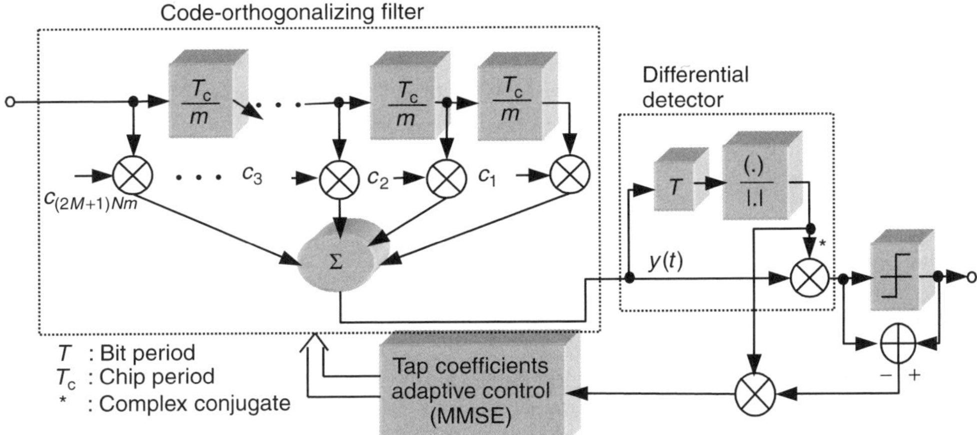

Figure 8.40 CDMA adaptive interference canceler (AIC) [6]. Reproduced from Yoshida, S. and Ushirokawa, A. Capacity evaluation of CDMA-AIC: CDMA cellular system with adaptive interference cancellation. *1995 Fourth IEEE International Conference on Universal Personal Communications Record*, pp. 148–152, by permission of IEEE.

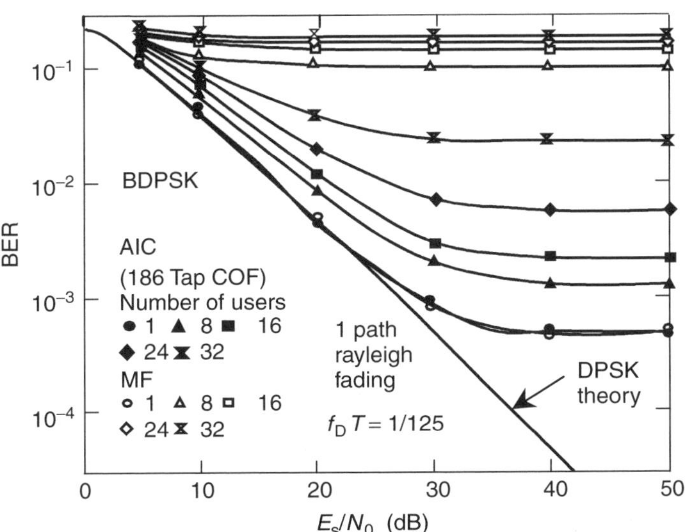

Figure 8.41 BER performance in fast-fading channel [6]. Reproduced from Yoshida, S. and Ushirokawa, A. Capacity evaluation of CDMA-AIC: CDMA cellular system with adaptive interference cancellation. *1995 Fourth IEEE International Conference on Universal Personal Communications Record*, pp. 148–152, by permission of IEEE.

Table 8.7 Receiver simulation parameters [6]. Reproduced from Yoshida, S. and Ushirokawa, A. Capacity evaluation of CDMA-AIC: CDMA cellular system with adaptive interference cancellation. *1995 Fourth IEEE International Conference on Universal Personal Communications Record*, pp. 148–152, by permission of IEEE

Modulation scheme	BPSK/QPSK (0.2 roll-off factor)
Bit rate	8 kbps (BPSK)/16 kbps (QPSK)
FEC scheme	Convolutional coding ($K = 9$, $R = 1/2$) (20 ms block interleaving)
Symbol rate	16 kbps (BPSK/QPSK)
Spreading codes	Gold sequences (31 code length)
Chip rate	496 kHz
Number of users	1, 8, 16, 24, 32
Transmission channel	1 path Rayleigh fading ($f_D = 128$ Hz) independent among users
Signal received timing	Uniformly distributed among users with 0.5 chip resolution
Received signal power	Equal among users
Sampling rate	992 kHz (double chip rate)

Note: FEC – forward error correction.

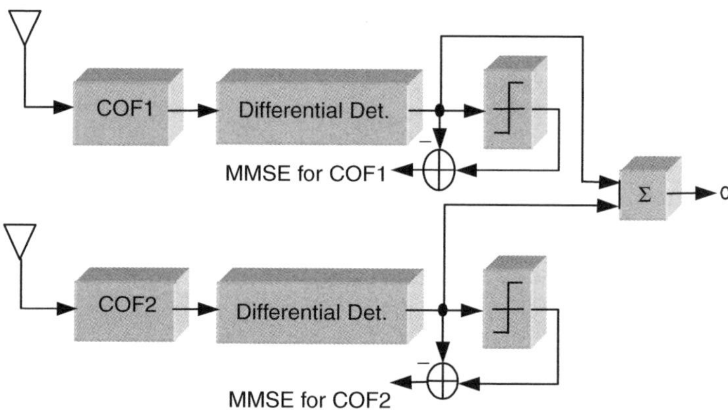

Figure 8.42 AIC with 2-branch space diversity.

8.7 DIVERSITY HANDOVER IN DS-CDMA CELLULAR SYSTEMS

When a mobile is moving from one cell to another, it should switch connection from one BS to another. This process is called handover. One should be aware that at the region where handover is about to happen, a mobile talking to its BS is using pretty high power level in order to reach the BS with the proper level. Because of that, at the same time it is producing a high level of interference in the next cell. If the mobile can talk to both BS simultaneously then it would require less power per BS and could reduce its level giving way to additional mobile to be admitted in the cell. Figure 8.46 illustrates capacity improvements (capacity/processing gain) versus required SIR. The largest capacity is obtained for an isolated cell (single cell). The lowest capacity is in the network with selection combining which means that the mobile is talking to the BS with strongest signal (select the strongest). If MRC is used to combine two or more BS signals, the capacity would increase.

In practice, not all signals will be combined but rather the signals that are less than a threshold Δ lower than the strongest one. In other words,

- A mobile station compares the levels of signals received from surrounding cell sites for cell selection.
- The BS whose differential level from the maximal level is within a certain threshold are selected as site diversity BS.

A possible scenario of received signals is shown in Figure 8.47.

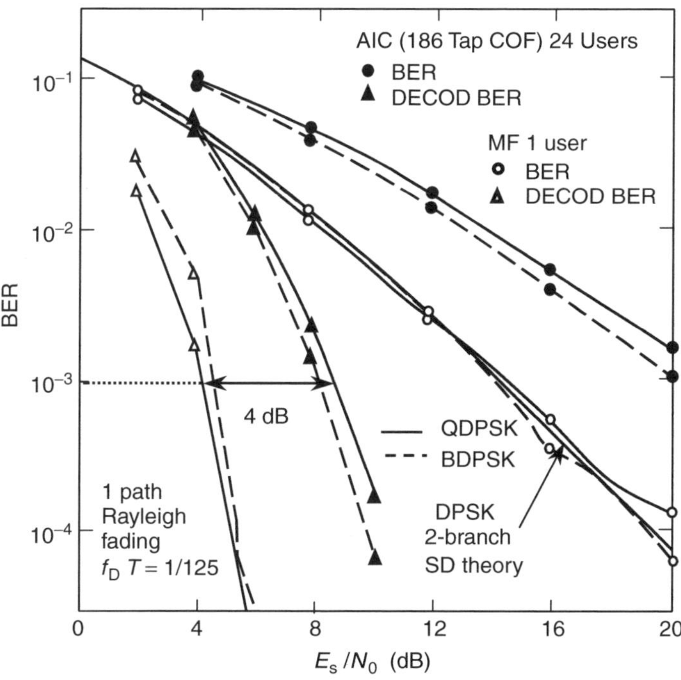

Figure 8.43 BER performance with 2-branch space diversity [6]. Reproduced from Yoshida, S. and Ushirokawa, A. Capacity evaluation of CDMA-AIC: CDMA cellular system with adaptive interference cancellation. *1995 Fourth IEEE International Conference on Universal Personal Communications Record*, pp. 148–152, by permission of IEEE.

Table 8.8 System simulation parameters [6]. Reproduced from Yoshida, S. and Ushirokawa, A. Capacity evaluation of CDMA-AIC: CDMA cellular system with adaptive interference cancellation. *1995 Fourth IEEE International Conference on Universal Personal Communications Record*, pp. 148–152, by permission of IEEE

Cell layout	19 hexagonal cells
MS location	Uniformly distributed in a cell
Spread bandwidth	0.5 MHz
Bit rate	8 kbps (BPSK)/16 kbps (QPSK)
Spreading factor	62 (BPSK)/31 (QPSK)
Path loss	$\rho = 4$
Shadowing	Lognormally distributed ($\sigma = 8$ dB)
Fading	Rayleigh distributed
Voice activity factor	$\alpha = 50\%, 100\%$
Power control error	$\sigma_e = 0$ dB, 1.5 dB

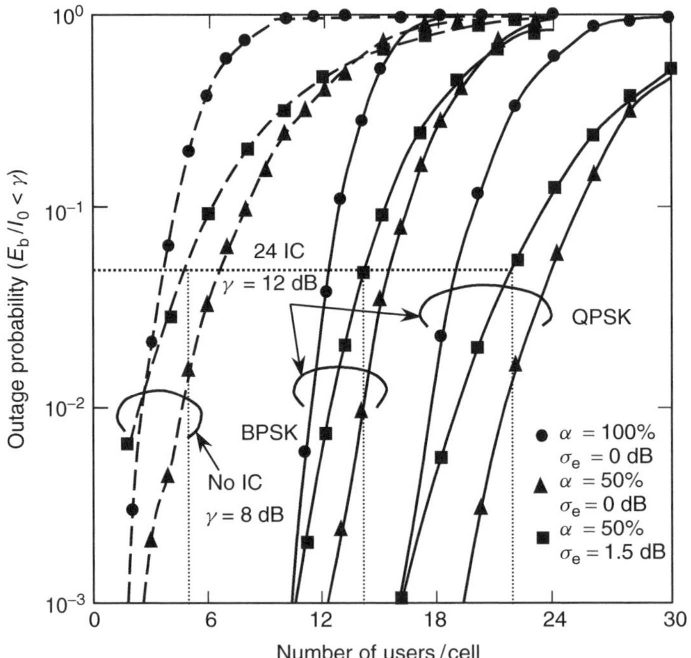

Figure 8.44 Outage probability versus number of users/cell[6]. Reproduced from Yoshida, S. and Ushirokawa, A. Capacity evaluation of CDMA-AIC: CDMA cellular system with adaptive interference cancellation. *1995 Fourth IEEE International Conference on Universal Personal Communications Record*, pp. 148–152, by permission of IEEE.

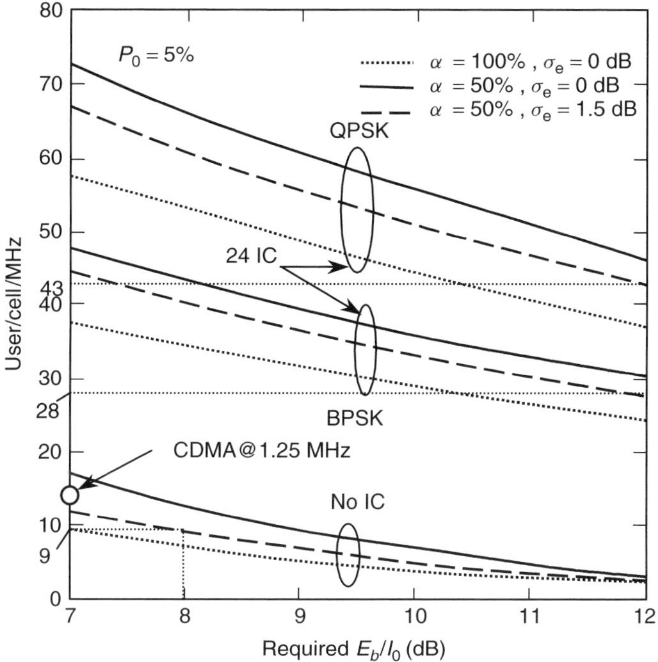

Figure 8.45 User capacity versus required Eb/I0 [6]. Reproduced from Yoshida, S. and Ushirokawa, A. Capacity evaluation of CDMA-AIC: CDMA cellular system with adaptive interference cancellation. *1995 Fourth IEEE International Conference on Universal Personal Communications Record*, pp. 148–152, by permission of IEEE.

Table 8.9 User capacity comparison (3 sectors/cell) [6]. Reproduced from Yoshida, S. and Ushirokawa, A. Capacity evaluation of CDMA-AIC: CDMA cellular system with adaptive interference cancellation. *1995 Fourth IEEE International Conference on Universal Personal Communications Record*, pp. 148–152, by permission of IEEE

Schemes	Users/cell/MHz	×AMPS	F, reuse
FDMA (AMPS)	4.8	1.0	7 cells
TDMA (D-AMPS)	14.4	3.0	7 cells
CDMA@0.5 MHz	$9 \times 2.55 = 26$	4.8	1 cell
CDMA@1.25 MHz, $\alpha = 50\%$	$14 \times 2.55 = 36$	7.4	
(CDMA@1.25 MHz, $\alpha = 40\%$)	$(18 \times 2.55 = 46)$	(9.6)	
CDMA-AIC@0.5 MHz	$43 \times 2.55 = 110$	22.8	1 cell

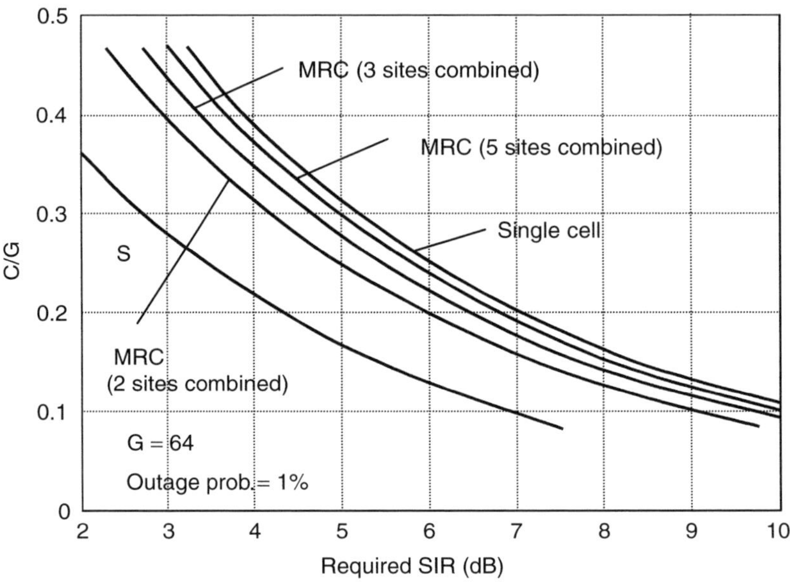

Figure 8.46   Reverse link system capacity versus required SIR [7]. Reproduced from Nakano, E., Umeda, N. and Ohno, K. Performance of diversity handover in DS-CDMA cellular systems. *1995 Fourth IEEE International Conference on Universal Personal Communications Record*, pp. 421–425, by permission of IEEE.

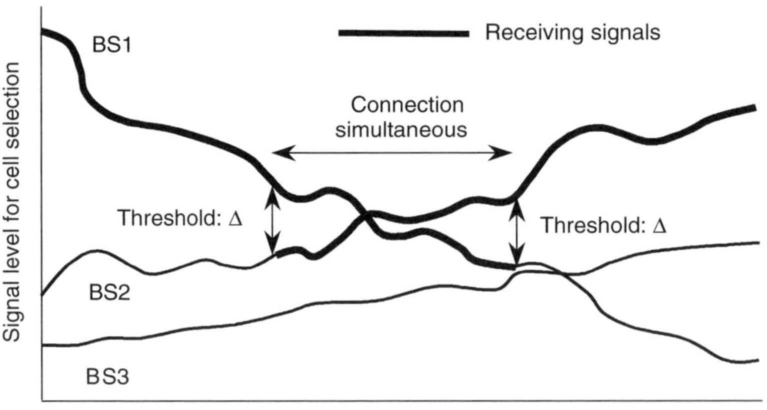

Figure 8.47   Signal levels from different base stations seen by a mobile receiver [7]. Reproduced from Nakano, E., Umeda, N. and Ohno, K. Performance of diversity handover in DS-CDMA cellular systems. *1995 Fourth IEEE International Conference on Universal Personal Communications Record*, pp. 421–425, by permission of IEEE.

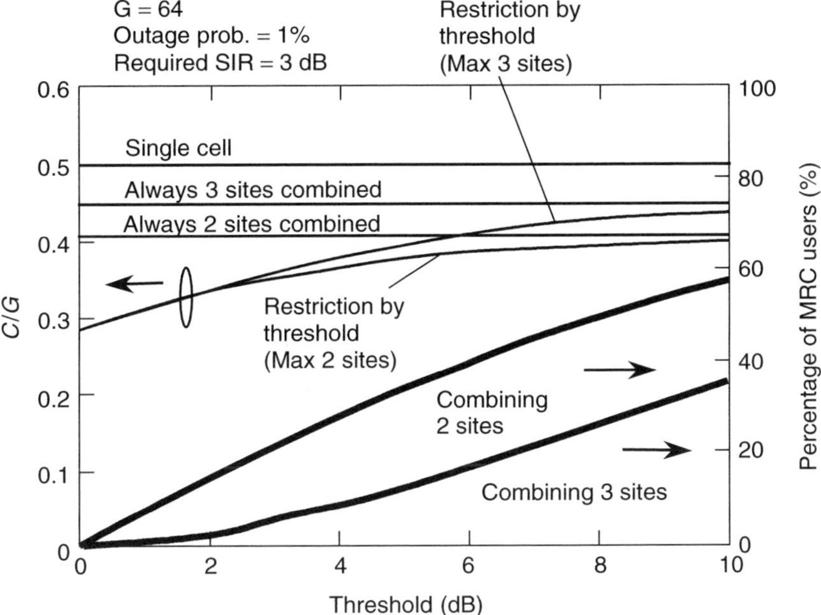

Figure 8.48 Reverse link system capacity with threshold restriction [7]. Reproduced from Nakano, E., Umeda, N. and Ohno, K. Performance of diversity handover in DS-CDMA cellular systems. *1995 Fourth IEEE International Conference on Universal Personal Communications Record*, pp. 421–425, by permission of IEEE.

Capacity versus the threshold Δ is shown in Figure 8.48. One can see that for Δ equals 10 dB there is not much difference if the signals are combined all the time as opposed to the restriction that only signals with differential level less than Δ are combined.

8.7.1 Forward link system capacity

Normalized capacity in this case is presented in Figure 8.49. Ideal SC means that only one BS, the one with minimum propagation loss, transmits the signal to the mobile station. In the ideal model, site diversity gain is small, so the total transmit power becomes larger than that when each BS transmits without diversity. This increases the interference power and reduces system capacity. In case of MRC (two sites combined), forward link system capacity is 70% of that with ideal SC. Figure 8.50 represents C/G with threshold restrictions. Practical implementation of this system would require signals for the reverse link to be combined somewhere in the network. In general this would not impose traffic problem because from the BS up, links are not critical from the point of view of capacity. Still, it seems more feasible to implement this approach to combine the signals from different sectors of the same BS.

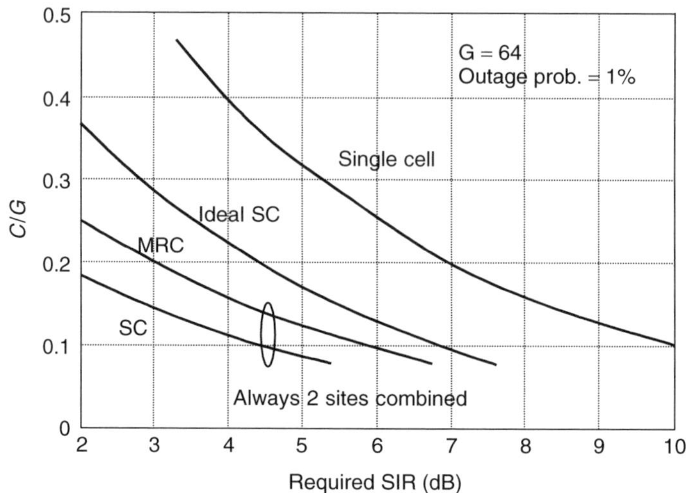

Figure 8.49   Forward link system capacity versus required SIR [7].Reproduced from Nakano, E., Umeda, N. and Ohno, K. Performance of diversity handover in DS-CDMA cellular systems. *1995 Fourth IEEE International Conference on Universal Personal Communications Record*, pp. 421–425, by permission of IEEE.

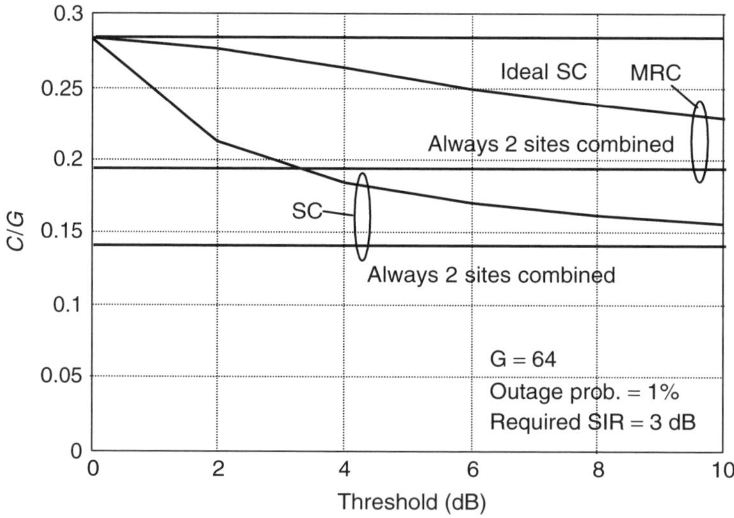

Figure 8.50   Forward link system capacity with threshold restriction [7]. Reproduced from Nakano, E., Umeda, N. and Ohno, K. Performance of diversity handover in DS-CDMA cellular systems. *1995 Fourth IEEE International Conference on Universal Personal Communications Record*, pp. 421–425, by permission of IEEE.

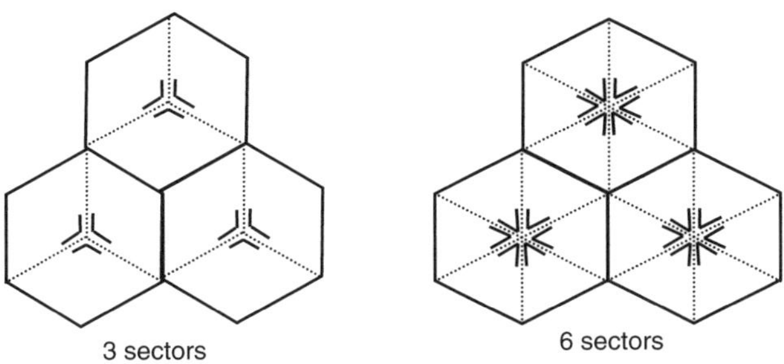

Figure 8.51 Sector structure [7]. Reproduced from Nakano, E., Umeda, N. and Ohno, K. Performance of diversity handover in DS-CDMA cellular systems. *1995 Fourth IEEE International Conference on Universal Personal Communications Record*, pp. 421–425, by permission of IEEE.

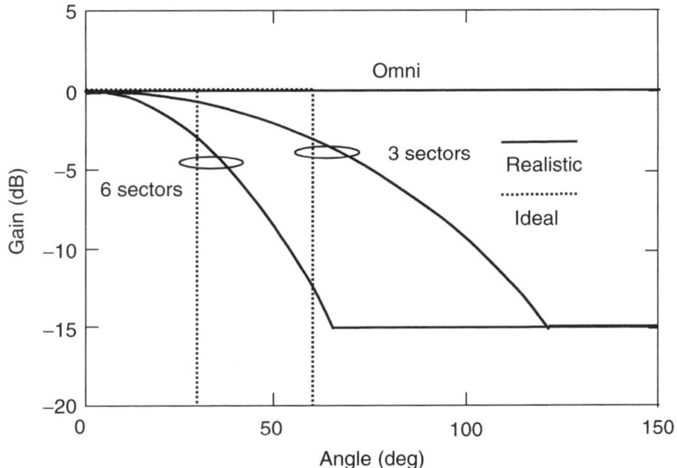

Figure 8.52 Gain of sector antennas [7]. Reproduced from Nakano, E., Umeda, N. and Ohno, K. Performance of diversity handover in DS-CDMA cellular systems. *1995 Fourth IEEE International Conference on Universal Personal Communications Record*, pp. 421–425, by permission of IEEE.

In UMTS standard, this is the specified option. The cell sectorization and antenna beams used are shown in Figure. 8.51. and 8.52, respectively.

In this scenario, we use two different strategies.

Strategy 1: SC for all combining nodes.
Strategy 2: MRC between sectors in the common cell, SC for the others.

The results are shown in Figures 8.53 to 8.55 and Table 8.10.

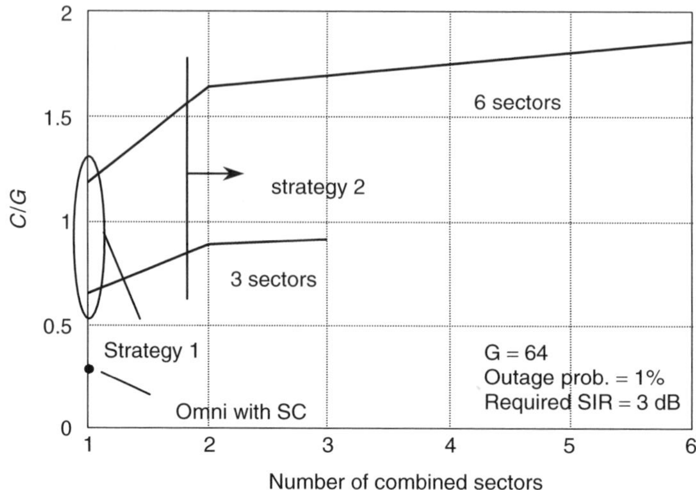

Figure 8.53 Reverse link system capacity with sectorization [7]. Reproduced from Nakano, E., Umeda, N. and Ohno, K. Performance of diversity handover in DS-CDMA cellular systems. *1995 Fourth IEEE International Conference on Universal Personal Communications Record*, pp. 421–425, by permission of IEEE.

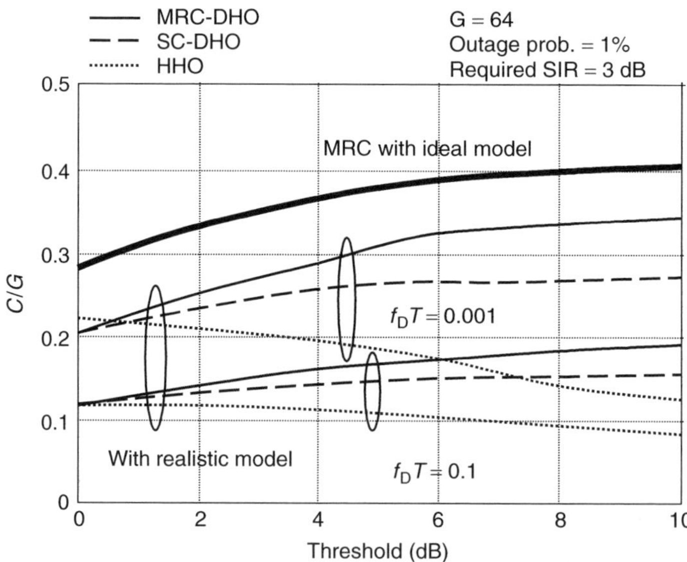

Figure 8.54 Reverse link system capacity with threshold restriction (realistic model) [7]. Reproduced from Nakano, E., Umeda, N. and Ohno, K. Performance of diversity handover in DS-CDMA cellular systems. *1995 Fourth IEEE International Conference on Universal Personal Communications Record*, pp. 421–425, by permission of IEEE.

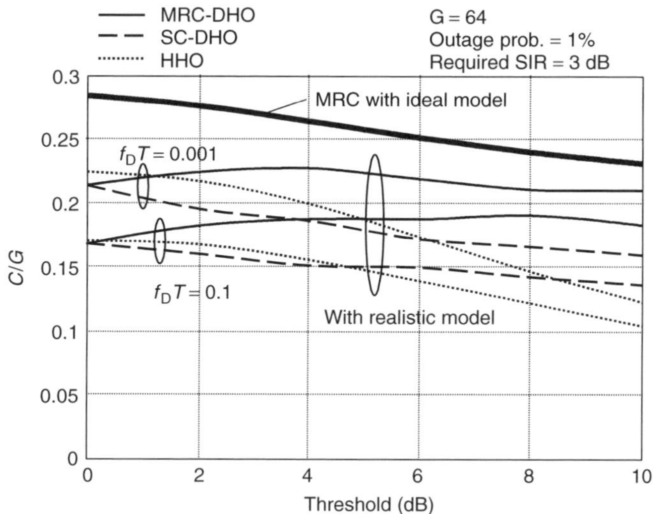

Figure 8.55 Forward link system capacity with threshold restriction (realistic model) [7]. Reproduced from Nakano, E., Umeda, N. and Ohno, K. Performance of diversity handover in DS-CDMA cellular systems. *1995 Fourth IEEE International Conference on Universal Personal Communications Record*, pp. 421–425, by permission of IEEE.

Table 8.10 Standard deviation of received SIR with threshold = 10 dB [7]

	$f_D T$	SD in reverse link (dB)	SD in forward link (dB)
HHO	0.1	3.65	2.20
	0.001	0.75	0.72
SC-DHO	0.1	3.28	1.31
	0.001	0.58	0.55
MRC-DHO	0.1	3.13	1.07
	0.001	0.55	0.53

Note: HHO – Hard hand over; DHO – Diversity hand over.

A number of additional issues related to CDMA cellular systems are discussed in References [8–20].

SYMBOLS

Section 8.1

CDMA – code division multiple access
$\alpha_i = P_i/P_0$ – power ratio of user i and referent user 0
P_i – power of user i
n – number of users

N_i – interference i power density
$(E_b/N_0)_1$ – signal-to-noise ratio with one user in the network
$(E_b/N_0)_n$ – signal-to-noise ratio with n users in the network
$(E_b/N_0)_R$ – required signal-to-noise ratio with one user in the network to guarantee the required quality of service
T_b, T_c – bit, chip interval
R_b, R_c – bit, chip rate
G – processing gain
DF – degradation factor
MACF – multiple-access capability factor

Section 8.2

ϕ_i – portion of downlink power for user i
S – signal power
N, N_s – number of users per cell, sector
R_c, R – chip, bit rate
$W = 1/R_c$
G – processing gain
η – noise density
$\propto$ – voice activity factor
r – distance
ξ – parameter of lognormal distribution (zero mean Gaussian variable)
S_{T_i} – total power transmitted by cell site
I – interference power
β – fraction of the total cell-site power devoted to subscribers, $(1-\beta$ is devoted to the pilot)
$\emptyset_i$ – fraction of β devoted to subscriber i
K – number of interfering cell sites

Section 8.3

P_p – transmitted pilot signal power of a BS
r_{ij} – distance between ith mobile and jth BS
n – propagation constant
ξ – random variable corresponding to shadowing lognormally distributed with mean of 0 dB and standard deviation of σ_s dB
δ_i – (in decibel) control error in the transmitter power
σ_E – power control error standard deviation
σ_s, σ – shadowing standard deviation
P_t – total transmitted power (including pilot signal) from BS
$\phi_i \cdot 100\%$ – of the jth BS transmitter power is assigned to communicate with the ith mobile station
a_r – radius modification factor for nonuniform user distribution
μ, v^2 – overall interference mean value and variance

Section 8.4

h – channel impulse response
N – number of paths
c_k, τ_k, θ_k – intensity, delay, phase, respectively, of path k

ω_D – Doppler
r, x – distance
ξ – shadowing variable
n, α – propagation constant
L – losses
R_b – two segment propagation break distance
h_b, h_m – base, mobile station height
L_0 – loss at the reference point (at 1 m)
N_{wj} and N_{fi} – number of walls and floors of different kinds that are traversed by the transmitted signal
L_{wj} (dB) and L_{fi} (dB) – corresponding loss factors
erfc – complementary error function
$\phi_i = N_i / T_i$
N_i – average number of mobile terminals in area i
T_i – mean crossing time of area i
M = number of areas
ϕ_{ij} – flow from area i to area j
$p_{ji} = \phi_{ij} / \phi_i$

Section 8.5

h_c – channel impulse response
L – total number of paths
$\beta_i(t)$, $\Delta\omega_i(t)$, $\theta_i(t)$, $\tau_i(t)$ – the ith path amplitude, frequency shift, phase and delay, respectively
G_p – processing gain
K – number of users
c, PN – code
W_i – maximum ratio combiner coefficients
M – number of combined paths
M_a – number of antennas
α – flat fading channel coefficient
b – data (bits)
e – code error correction capabilities
η_0 – Gaussian noise spectral density
σ^2 – variance of Rayleigh faded signal
n – propagation factor

Section 8.6

T, T_c – bit, chip interval
c – code
AIC – adaptive interference cancellation
MF – matched filter
f_D – Doppler
COF – code orthogonalization filter
SD – selection diversity
α – voice activity factor
σ_e – power control error variance
ρ – propagation coefficient
σ – shadowing variance
γ – signal to noise ratio

REFERENCES

1. Gilhousen, K. S., Jacobs, I. M., Padovani, R., Viterbi, A. J., Weaver, L. A. and Wheatley, C. E. (1991) On the capacity of a cellular CDMA system. *IEEE Trans. Veh. Technol.*, **40**(2), 303–312.
2. Kudoh, E. (NTT) (1993) On the capacity of DS/CDMA cellular mobile radios under imperfect transmitter power control. *IEICE Trans. Commun.*, **E76-B**(8), 886–893.
3. Kudoh, E. and Matsumoto, T. (NTT) (1992) Effects of power control error on the system user capacity of DS/CDMA cellular mobile radios. *IEICE Trans. Commun.*, **E75-B**(6), 524–529.
4. Jimenez, J. (ed.) Final Propagation Model. CODIT Deliverable R2020/TDE/PS/P/040/b1., June, 1994.
5. Valdivia, G. and Perez, V. Summary of reference channel models for SIG 5 common evaluations. CODIT/TDE/CA-036/1.0.
6. Yoshida, S. and Ushirokawa, A. Capacity evaluation of CDMA-AIC: CDMA cellular system with adaptive interference cancellation. *1995 Fourth IEEE International Conference on Universal Personal Communications Record*, pp. 148–152.
7. Nakano, E., Umeda, N. and Ohno, K. Performance of diversity handover in DS-CDMA cellular systems. *1995 Fourth IEEE International Conference on Universal Personal Communications Record*, pp. 421–425.
8. Viterbi, A. M. and Viterbi, A. J. (Qualcomm) (1993) Erlang capacity of a power controlled CDMA system. *IEEE J. Select. Areas Commun.*, **11**(6), 892–900.
9. Viterbi, A. J., Viterbi, A. M. and Zehavi, E. (Qualcomm) (1993) Performance of power-controlled wideband terrestrial digital communication. *IEEE Trans. Commun.*, **41**(4), 559–569.
10. Erceg, V., Ghassemzadeh, S., Taylor, M., Li, D. and Schilling, L. (1992) Urban/suburban out of sight propagation modelling. *IEEE Commun. Mag.*, **30**, 677–684.
11. Joe, W., Marquis, A., Juy, M. and Benoit, G. Analytical Microcell Path Loss Model at 2,2 GHz COST 231 TD(93)56. Source: France.
12. Espinel, I., Lozano, J. L., Ruiz-Boque, S., Casadevall, F. and Agusti, R. Propagation Measurements and Models for Microcells at 1 900 MHz COST 231 TD(93) 17. Source: Spain.
13. Lee, W. C. Y. (1991) Overview of cellular CDMA. *IEEE Trans. Veh. Technol.*, **40**, 291–302.
14. Gollreiter, R. and Mohr, W. Additional CC Propagation Models for Application in the Simulated Testbeds. ATDMA Internal Document: R2084/ESG/CC3/IN/I/0351al.
15. Gollreiter, R. Channel Models Issue 2. R2084/ESG/CC3/DS/P/0291b1. 15104/94.
16. Strasser, G. (ed.) Propagation Models Issue 1. R2084/ESG/CC3/DS/P/012/b1. 14/04/93.
17. Viterbi, A. J. and Padovani, R. (1992) Implications of mobile cellular CDMA. *IEEE Commun. Mag.*, **30**(12), 38–41.
18. Gollreiter, R. Reference channel models for SIG 5 cornrnon evaluations. Issue 1.0. R2084/ESG/CC3/R/I/031/al.
19. Prasad, R., Kegel, A. and Jansen, M. G. (1992) Effect of imperfect power control on cellular CDMA system. *Electron. Lett.*, **28**(9), 848–849.
20. Kudoh, E. and Matsumoto, T. (1992) Effect of transmitter power control imperfections on capacity in DS/CDMA cellular mobile radios. *ICC '92*, pp. 237–242.

9

CDMA network design

9.1 BASIC SYSTEM DESIGN PHILOSOPHY

In Code Division Multiple Access (CDMA) systems, the capacity increase is based on how much interference the desired signal can tolerate. Prior to despreading, the signal level of a desired signal is always below the interference level. All the users have to share the same radio channel. If one user takes more power than needed, then the others will suffer and the system capacity will be reduced.

In analog and TDMA systems, the most important key element is the carrier to interference ratio (C/I). There are two different kinds of C/I. One is the measured (C/I), which is used to indicate the voice quality in the system. The higher the measured value, the better it is. The other is called the specified (C/I)s, which is the required value for a specified performance of the cellular system. For example, the (C/I)s in the American mobile phone system (AMPS) is 18 dB. Since in analog and TDMA systems, owing to the spectral and geographical separations, the interference (I) is much lower than the received signal (C), sometimes we can utilize field strength meter to measure C to determine the coverage of each cell. The field strength meter therefore becomes a useful tool in designing the TDMA system.

In CDMA all the traffic channels are served solely by a single radio channel in every cell. In an m-voice channel cell, one of the m traffic channels is the desired channel and the remaining $m - 1$ traffic channels are the interference channels. In this case, at the receiver front end (prior to despreading) the interference is much stronger than the desired channel. C/I is hard to obtain by using the signal strength meter that will receive more interference than the desired signal. The key elements in designing a CDMA system are different from the key element in designing a TDMA system. We can design the CDMA system based on the specified E_b/I_0

$$\frac{C}{I} = \left(\frac{E_b}{I_0}\right) \cdot \left(\frac{R_b}{B}\right) \cdot \eta \Leftrightarrow \frac{E_b}{I_0} = G \cdot \frac{C}{I\eta} \tag{9.1}$$

The left-hand side of equation (9.1) is derived from the right-hand side specifying that the signal-to-noise ratio (SNR) after despreading is G times higher than the input SNR. Values of E_b/I_0 for the forward-link channels and for the reverse-link channels are different because of the different modulation schemes. In general, there will be two different requirements for C/I. One $(C/I)_F$ for the forward-link channels and the other $(C/I)_R$ for the reverse-link channels. In Chapter 8 we used $(E_b/I_0)_R = 7\,\text{dB}$ and $(E_b/I_0)_F = 5\,\text{dB}$. So, for a given E_b/I_0 the network design should make sure that the required C/I is guaranteed in each spot of the coverage area. In the first step, we start with a simple, very much approximative approach to the problem in order to get the very first initial insight into the system parameters. In the next iteration we will come up with a more detailed analysis.

9.1.1 Uniform cell-size scenario

For the forward link a worst-case scenario is used to find the relation among the transmitted powers of cell sites. The position of the mobile for this case is shown in Figure 9.1.

If we assume that the signal propagation losses can be approximated as R^{-4} (shadowing ignored at this stage) and if R is the cell parameter, then C/I at the mobile front end can be represented as

$$\frac{C}{I} = \frac{\alpha_1 R^{-4}}{I(s) + I(a) + I(i) + I(d)} \tag{9.2}$$

where

$$I(s) = I(\text{self cell}) = \alpha_1(m_1 - 1)R^{-4}$$
$$I(a) = I(2 \text{ adjacent cells}) = (\alpha_2 m_2 + \alpha_3 m_3)R^{-4}$$
$$I(i) = I(3 \text{ intermediate cells}) = \beta(2R)^{-4}$$
$$I(d) = I(6 \text{ distant cells}) = \gamma(2.633R)^{-4}$$

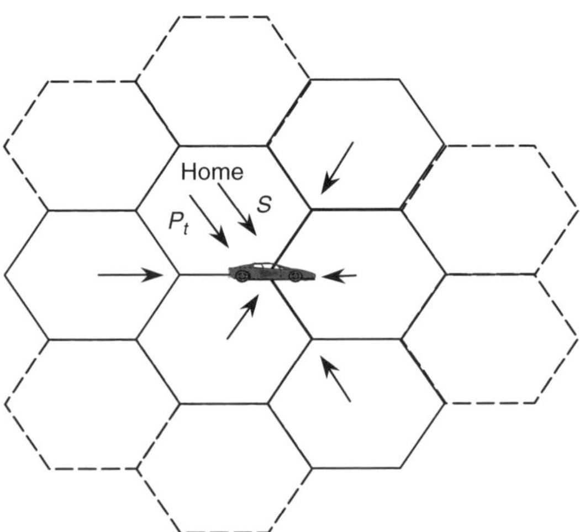

Figure 9.1   CDMA system and its interference (from a forward link).

$\alpha_i (1, 2, 3)$ is the transmitted power of each voice channel in the cell
m_i is the number of channels per cell
β and γ are transmitted powers of the combined adjacent cells at a distance $2R$ and $2.633R$, respectively.

By solving the equation we get m_1 as follows:

$$m_1 = \left(\frac{1}{C/I} + 1 \right) - \left[\frac{\alpha_2 m_2 + \alpha_3 m_3}{\alpha_1} \right] - \frac{\beta}{\alpha_1}(2)^{-4} - \frac{\gamma}{\alpha_1}(2.633)^{-4} \qquad (9.3)$$

If there is no adjacent cell interference, $\alpha_2 = \alpha_3 = \beta = \gamma = 0$ in equation (9.3) and we have

$$m_1 = \frac{1}{C/I} + 1 \qquad (9.4)$$

For $C/I = -17\,\text{dB}$, we have $m_1 = 51$.
   If there is no interference other than from the two close-in interfering cells then

$$\alpha_1 = \frac{\alpha_2 m_2 + \alpha_3 m_3}{[1/(C/I)] + 1 - m_1} \qquad (9.5)$$

If $C/I = -17\,\text{dB}$, $m_1 = 30$, $m_2 = 25$ and $m_3 = 15$, then

$$\alpha_1 = \frac{25\alpha_2 + 15\alpha_3}{51 - 30} = 1.19\alpha_2 + 0.714\alpha_3 \qquad (9.6)$$

which gives the relationship among α_1, α_2 and α_3.
   If the total transmitted power P in each cell site is $P_1 = \alpha_1 m_1$, $P_2 = \alpha_1 m_2$, $P_3 = \alpha_3 m_3$, when m_1, m_2, m_3 are given, then P_1, P_2 and P_3 are the maximum transmitted powers of three cells.

$$\left(\frac{1}{C/I} + 1 \right) \cdot \frac{P_1}{m_1} = P_1 + P_2 + P_3 \qquad (9.7)$$

Following the same derivation steps

$$\left(\frac{1}{C/I} + 1 \right) \frac{P_2}{m_2} = P_1 + P_2 + P_3$$

$$\left(\frac{1}{C/I} + 1 \right) \frac{P_3}{m_3} = P_1 + P_2 + P_3 \qquad (9.8)$$

The relationship of three maximum transmitted powers of three cells are

$$\frac{P_1}{m_1} = \frac{P_2}{m_2} = \frac{P_3}{m_3}$$

Deduced from the equation, a design criterion that will be used in general for a CDMA system of N cells can be expressed as

$$\frac{P_i}{m_i} = \frac{P_j}{m_j} = \text{constant}$$

For the reverse link the received signal from a desired mobile unit at the home cell site is C. Each signal of other m_1 channels received at the home site is also C (owing to power control). The interference power of certain mobile units, say $r \cdot m_1$, from the two adjacent cells comes from the cell boundary (see the worst case scenario in Figure 9.2). Because of the power control in each adjacent cell, the interference coming from the adjacent cell for each voice channel would roughly be C at the home cell site. So we have

$$\frac{C}{I} = \frac{C}{(m_1 - 1) \cdot C + r_{12} \cdot m_2 C + r_{13} \cdot m_3 C} = \frac{1}{m_1 - 1 + r_{12}m_2 + r_{13}m_3} \tag{9.9}$$

r_{12} and r_{13} are a portion of the total number of voice channels in adjacent cells that will interfere with the desired signal at the home cell, which is Cell 1. The worst-case scenario is when

$$m_1 + r_{12} \cdot m_2 + r_{13} \cdot m_3 \leq \frac{1}{C/I} + 1$$

$$r_{21}m_1 + m_2 + r_{23}m_3 \leq \frac{1}{C/I} + 1$$

$$r_{31}m_1 + r_{32}m_2 + m_3 \leq \frac{1}{C/I} + 1 \tag{9.10}$$

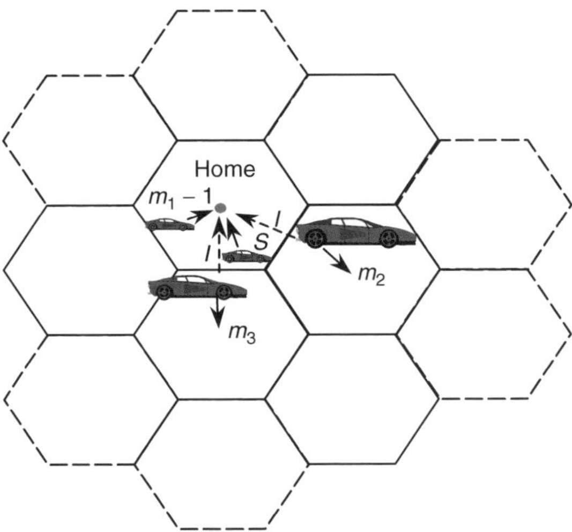

Figure 9.2 CDMA system and its interference (from a reverse-link scenario).

r depends on the size of the overlapped region in the adjacent cell and can be reasonably assumed to be 1/6 (which is 0.166) if the system is properly designed. If $C/I = -17\,\text{dB}$, which is 50^{-1} and $r_{12} = r_{13} = 0.166$, then

$$m_1 + 0.166 \cdot (m_2 + m_3) \le 51 \tag{9.11}$$

This is a relationship among the number of voice channels in each cell, m_1, m_2 and m_3. From the reverse-link scenario, we can check to see whether all the conditions expressed in the equations can be met. The unknowns in these conditions come from the demanded voice channels, m_1, m_2 and m_3. Then, on the basis of the forward-link equations, we can determine the maximum transmitted power of each cell.

9.1.2 Nonuniform cell scenario

We may first assign the number of voice channels m in each cell owing to requirements from demographical data. Then we may calculate the total transmit power on the forward-link channel in each cell from the worst-case scenario as shown in Figure 9.3.

The $(C/I)_\text{F}$ received at vehicle 1 is

$$\left(\frac{C_1}{I_1}\right)_\text{F} = \frac{\alpha_1 R_1^{-4}}{(m_1 - 1)\alpha_1 R_1^{-4} + \alpha_2 m_2 R_2^{-4} + \alpha_3 m_3 R_3^{-4} + I_{a1}} \tag{9.12}$$

I_a is the interference coming from other interfering cells besides these three cells. This component is usually very small as compared to the two terms and can be neglected.

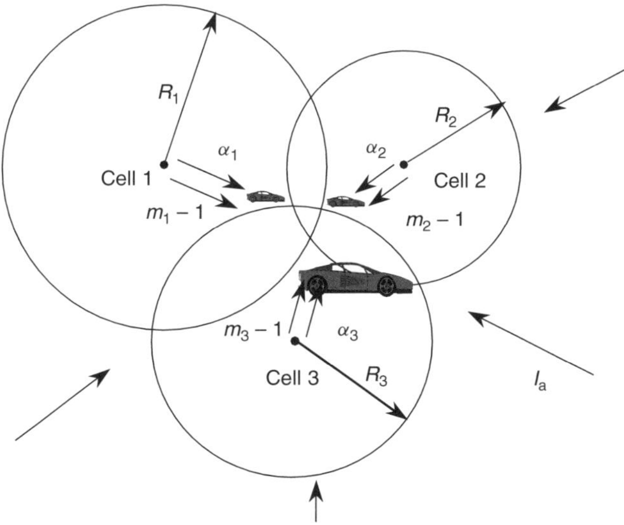

Figure 9.3 The worst-case scenario on a forward-link channel.

$(C_2/I)_F$ received at vehicles 2 and 3 can be expressed as

$$\left(\frac{C_2}{I_2}\right)_F = \frac{\alpha_2 R_2^{-4}}{(m_2 - 1)\alpha_2 R_2^{-4} + \alpha_1 m_1 R_1^{-4} + \alpha_3 m_3 R_3^{-4} + I_{a2}} \tag{9.13}$$

$$\left(\frac{C_3}{I_3}\right)_F = \frac{\alpha_3 R_3^{-4}}{(m_3 - 1)\alpha_3 R_3^{-4} + \alpha_1 m_1 R_1^{-4} + \alpha_2 m_2 R_2^{-4} + I_{a3}} \tag{9.14}$$

If

$$\left(\frac{C_1}{I_1}\right)_F = \left(\frac{C_2}{I_2}\right)_F = \left(\frac{C_3}{I_3}\right)_F = \left(\frac{C}{I}\right)_F \quad \text{and} \quad I_{a1} = I_{a2} = I_{a3} = 0 \tag{9.15}$$

gives

$$\alpha_1 m_1 + \alpha_2 m_2 \left(\frac{R_2}{R_1}\right)^{-4} + \alpha_3 m_3 \left(\frac{R_3}{R_1}\right)^{-4} = \alpha_1 \left[\frac{1}{(C/I)_F} + 1\right] = \alpha_1 \cdot G$$

$$\alpha_1 m_1 \left(\frac{R_1}{R_2}\right)^{-4} + \alpha_2 m_2 + \alpha_3 m_3 \left(\frac{R_3}{R_2}\right)^{-4} = \alpha_2 \cdot G$$

$$\alpha_1 m_1 \left(\frac{R_1}{R_3}\right)^{-4} + \alpha_2 m_2 \left(\frac{R_2}{R_3}\right)^{-4} + \alpha_3 m_3 = \alpha_3 \cdot G \tag{9.16}$$

Solving these equations gives

$$\alpha_1 R_1^{-4} = \alpha_2 R_2^{-4} = \alpha_3 R_3^{-4} \tag{9.17}$$

Assume that the minimum values of α_1, α_2 and α_3 will be α_1^0, α_2^0 and α_3^0, respectively, then we have

$$\alpha_1 \geq \alpha_1^0 = C_0 R_1^{+4}/k_1$$

$$\alpha_2 \geq \alpha_2^0 = C_0 R_2^{+4}/k_2$$

$$\alpha_3 \geq \alpha_3^0 = C_0 R_3^{+4}/k_3 \tag{9.18}$$

where C_0 is the required signal received level at the vehicle location and k_i is a constant gain related to the antenna heights at the cell sites. Now the total transmit power of each cell site will be

$$P_1 = m_1 \alpha_1$$

$$P_2 = m_2 \alpha_2$$

$$P_3 = m_3 \alpha_3 \tag{9.19}$$

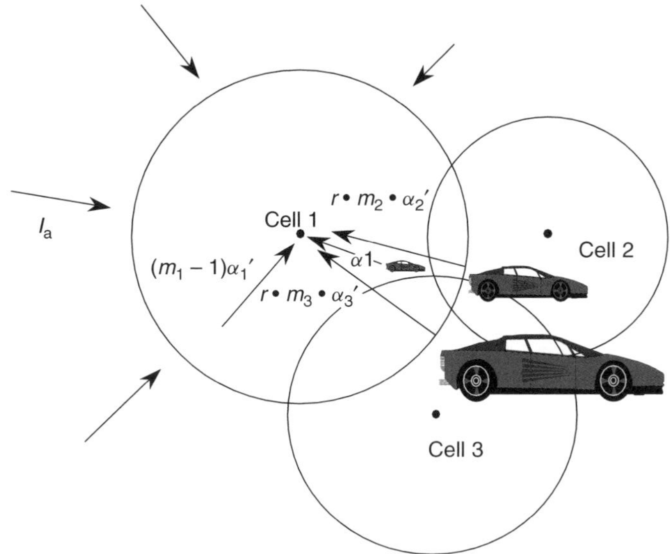

Figure 9.4 The worst-case scenario for reverse link.

The worst-case scenario for reverse link is depicted in Figure 9.4. On the basis of the power control algorithm, all the signals will be the same on reaching the cell site.

$$\left(\frac{C_1}{I_1}\right)_R \geq \frac{\alpha_1' \cdot R_1^{-4}}{(m_1 - 1)\alpha_1' R_1^{-4} + r_{12}m_2\alpha_2' R_1^{-4} + r_{13}m_3\alpha_3' R_1^{-4} + \dot{I}_{a1}} \tag{9.20}$$

where α_1', α_2' and α_3' are the power of individual channels transmitted back to their corresponding cell sites. r_{12} and r_{13} are the portion of the total number of voice channels in the adjacent cell that will interfere with the desired signal at cell 1. $\dot{I}_{a1}$ is the interference coming from other users in other cells that are not cell 2 and cell 3, which is a relatively small value and can be neglected. Similarly we have

$$\left(\frac{C_2}{I_2}\right)_R \geq \frac{\alpha_2' R_2^{-4}}{r_{21} \cdot m_1\alpha_1' R_2^{-4} + (m_2 - 1)\alpha_2' \cdot R_2^{-4} + r_{23} \cdot m_3\alpha_3' R_2^{-4}}$$

$$\left(\frac{C_3}{I_3}\right)_R \geq \frac{\alpha_3' R_3^{-4}}{r_{31}m_1\alpha_1' R_3^{-4} + r_{32}m_2\alpha_2 R_3^{-4} + (m_3 - 1)\alpha_3 \cdot R_3^{-4}} \tag{9.21}$$

where r is the percentage of total channels from the interfering cell received by the home site. Simplifying the equations gives

$$\left(\frac{I}{C}\right)_R \geq (m_1 - 1) + r_{12}m_2\frac{\alpha_2'}{\alpha_1'} + r_{13}m_3\frac{\alpha_3'}{\alpha_1'}$$

$$\left(\frac{I}{C}\right)_R \geq r_{21}m_1\frac{\alpha_1'}{\alpha_2'} + (m_2 - 1) + r_{23}m_3\frac{\alpha_3'}{\alpha_2'}$$

$$\left(\frac{I}{C}\right)_R \geq r_{31}m_1\frac{\alpha_1'}{\alpha_3'} + r_{32}m_2\frac{\alpha_2'}{\alpha_3'} + (m_3 - 1) \tag{9.22}$$

where

$$\left(\frac{C}{I}\right)_R = \left(\frac{C_1}{I_1}\right)_R = \left(\frac{C_2}{I_2}\right)_R = \left(\frac{C_3}{I_3}\right)_R \tag{9.23}$$

The minimum values of α_1', α_2' and α_3' can be defined as follows:

$$\alpha_1' \geq \alpha_1^0 = \frac{C_0 R_1^4}{k_1}$$

$$\alpha_2' \geq \alpha_2^0 = \frac{C_0 R_2^4}{k_2}$$

$$\alpha_3' \geq \alpha_3^0 = \frac{C_0 R_3^4}{k_3} \tag{9.24}$$

where R_1, R_2 and R_3 are the radii of the three cells and k is a constant gain related to the antenna heights at the cell sites. Now equation (9.23) becomes

$$\left(\frac{I}{C}\right)_R \geq (m_1 - 1) + r_{12}m_2\left(\frac{R_2}{R_1}\right)^4 + r_{13}m_3\left(\frac{R_3}{R_1}\right)^4$$

$$\left(\frac{I}{C}\right)_R \geq r_{21}m_1\left(\frac{R_1}{R_2}\right)^4 + (m_2 - 1) + r_{23}m_3\left(\frac{R_3}{R_2}\right)^4$$

$$\left(\frac{I}{C}\right)_R \geq r_{31}m_1\left(\frac{R_1}{R_3}\right)^4 + r_{32}m_2\left(\frac{R_2}{R_3}\right)^4 + m_3 - 1 \tag{9.25}$$

which gives the basic design equation for the relation between the network parameters. From this we have

$$m_1, m_2 \text{ or } m_3 < \frac{1}{(C/I)_R} + 1 \tag{9.26}$$

9.2 CDMA NETWORK PLANNING

In this section we provide more details on network planning and dimensioning. The approach is based on References [1–5]. WCDMA radio network dimensioning is the process through which the possible configurations and the amount of network equipment is estimated, on the basis of the operator's requirements related to the following:

Coverage, which includes coverage regions, area type information, propagation conditions.

Capacity, which includes spectrum available, subscriber growth forecast, traffic density information.

Quality of Service, which includes area location probability (coverage probability), blocking probability, end user throughput.

Dimensioning activities include radio link budget and coverage analysis, capacity estimation, estimations on the amount of sites and base station hardware, radio network controllers (RNCs), equipment at different interfaces and core network elements (i.e. circuit-switched domain and packet-switched domain core networks).

9.2.1 Radio link budgets and coverage efficiency

The interference margin is needed in the link budget because of the loading of the cell by other users. The load factor, which will be later related to $(E_b/N_0)_R$ defined in equation (8.2) of Chapter 8, affects the coverage. The more loading is allowed in the system, the larger is the interference margin needed in the uplink, and the smaller is the coverage area. For coverage-limited cases a smaller interference margin is suggested, while in capacity-limited cases a larger interference margin should be used. In the coverage-limited cases the cell size is limited by the maximum allowed path loss in the link budget, and the maximum air interference capacity of the base station site is not used. Typical values for the interference margin in the coverage-limited cases are 1.0 to 3.0 dB, corresponding to 20 to 50% loading. Some headroom is needed in the mobile station transmission power for maintaining adequate closed-loop fast power control. This applies especially to slow-moving pedestrian mobiles in which fast power control is able to effectively compensate the fast fading. Typical values for fast fading margin are 2.0 to 5.0 dB for slow-moving mobiles.

Handovers – soft or hard – give a gain against slow fading (lognormal fading) by reducing the required lognormal fading margin. This is because the slow fading is partly uncorrelated between the base stations, and by making handover the mobile can select a better base station. Soft handover gives an additional macro diversity gain against fast fading by reducing the required E_b/N_0 relative to a single radio link, owing to the effect of macro diversity combining, as explained in Chapter 8, Section 8.7. The total soft handover gain is assumed to be between 2.0 and 3.0 dB in the examples given below, including the gain against slow and fast fading. The following system assumptions given in Tables 9.1 and 9.2 will be used in this section [1–5].

On the basis of this assumption, the link budget for three different services is shown in Tables 9.3 to 9.5.

Table 9.1 Assumptions for the mobile station

	Speech terminal	Data terminal
Maximum transmission power	21 dBm	24 dB
Antenna gain	0 dBi	2 dBi
Body loss	3 dB	0 dB

Table 9.2 Assumption for the base station

Noise figure	5.0 dB
Antenna gain	18 dBi (three-sector base station)
E_b/N_0 requirement	Speech: 5.0 dB
	144-kbps real-time data: 1.5 dB
	384-kbps non-real-time data: 1.0 dB
Cable loss	2.0 dB

Table 9.3 Reference link budget of adaptive multirate (AMR) 12.2-kbps voice service ($120\,\mathrm{km\,h^{-1}}$, in-car users, vehicular A type channel, with soft handover)

12.2-kbps voice service ($120\,\mathrm{km\,h^{-1}}$, in-car)

Transmitter (mobile)		
Max. mobile transmission power (dBm)	21	A
Body loss (dB)	3	B
Equivalent isotropic radiated power (dBm)	18	$c = a + b$
Receiver (base station)		
Thermal noise density (dBm Hz^{-1})	-174	d
Base station receiver noise figure (dB)	5	e
Receiver noise density (dBm Hz^{-1})	-169	$f = d + e$
Receiver noise power (dBm)	$-103,2$	$g = f + 10^* \log(3840000)$
Interference margin (dB)	3	h
Receiver interference power (dBm)	$-103,2$	$i = 10^* \log(10^{**}[(g + h)/10 - 10^{**}(g/10)]$
Total effective noise + interference (dBm)	$-100,2$	$j = 10^* \log[10^{**}(g/10) + 10^{**}(i/10)]$
Processing gain (dB)	25	$k = 10^* \log(3840/12.2)$
Required E_b/N_0 (dB)	5	l
Receiver sensitivity (dBm)	$-120,2$	$m = l - k + j$
Base station antenna gain (dBi)	18	n
Cable loss in the base station (dB)	2	o
Fast fading margin (dB)	0	p
Max. path loss (dB)	154,2	$q = c - m + n - o - p$
Coverage probability (%)	95	
Lognormal fading constant (dB)	7	
Propagation model exponent	3,52	
Lognormal fading margin (dB)	7,3	r
Soft handover gain (dB), multicell	3	s
In-car loss (dB)	8	t
Allowed propagation loss for cell range (dB)	141,9	$u = q - r + s - t$

Table 9.4 Reference link budget of 144-kbps real-time data service ($3\,\mathrm{km\,h^{-1}}$, indoor user covered by outdoor base station, vehicular A type channel, with soft handover)

144-kbps voice service ($120\,\mathrm{km\,h^{-1}}$, in-car)

Transmitter (mobile)		
Max. mobile transmission power (dBm)	24	a
Mobile antenna gain [dBm]	2	b
Body loss [dB]	0	c
Equivalent isotropic radiated power (dBm)	26	$d = a + b - c$
Receiver (base station)		
Thermal noise density (dBm Hz^{-1})	−174	e
Base station receiver noise figure (dB)	5	f
Receiver noise density (dBm Hz^{-1})	−169	$g = e + f$
Receiver noise power (dBm)	−103,2	$h = g + 10^* \log(3840000)$
Interference margin (dB)	3	i
Receiver interference power (dBm)	−103,2	$j = 10^* \log(10^{**}[(h + i)/10 - 10^{**}(h/10)]$
Total effective noise + interference (dBm)	−100,2	$k = 10^* \log[10^{**}(h/10) + 10^{**}(j/10)]$
Processing gain (dB)	14,3	$l = 10^* \log(3840/144)$
Required E_b/N_0 (dB)	1,5	m
Receiver sensitivity (dBm)	−113	$n = m - l + k$
Base station antenna gain (dBi)	18	o
Cable loss in the base station (dB)	2	p
Fast fading margin (dB)	4	q
Max. path loss (dB)	154,2	$r = d - n + o - p - q$
Coverage probability (%)	80	
Lognormal fading constant (dB)	12	
Propagation model exponent	3,52	
Lognormal fading margin (dB)	4,2	s
Soft handover gain (dB), multicell	2	t
Indoor loss (dB)	15	u
Allowed propagation loss for cell range (dB)	133,8	$v = r - s + t - u$

It was assumed in Table 9.3 that mobile antenna gain is omnidirectional.

The *coverage efficiency* of WCDMA is defined by the average coverage area per site, in square kilometers per site, for a predefined reference propagation environment and supported traffic density. From the link budgets above, the cell range R can be readily calculated for a known propagation model, like those defined in Chapter 8.

The propagation model describes the average signal propagation in that environment, and it converts the maximum allowed propagation loss in decibels to the maximum cell range in kilometers.

Table 9.5 Reference link budget of non-real-time 384-kbps real-time data service ($3\,\mathrm{km\,h^{-1}}$, outdoor user, vehicular A type channel, no soft handover)

384-kbps non-real-time data, no soft handover

Transmitter (mobile)

Max. mobile transmission power (dBm)	24	a
Mobile antenna gain (dBm)	2	b
Body loss (dB)	0	c
Equivalent isotropic radiated power (dBm)	26	$d = a + b - c$

Receiver (base station)

Thermal noise density ($\mathrm{dBm\,Hz^{-1}}$)	-174	e
Base station receiver noise figure (dB)	5	f
Receiver noise density ($\mathrm{dBm\,Hz^{-1}}$)	-169	$g = e + f$
Receiver noise power (dBm)	$-103{,}2$	$h = g + 10^* \log(3840000)$
Interference margin (dB)	3	i
Receiver interference power (dBm)	$-103{,}2$	$j = 10^* \log(10^{**}((h + I)/ 10 - 10^{**}(h/10))$
Total effective noise + interference (dBm)	$-100{,}2$	$k = 10^* \log(10^{**}(h/10) + 10^{**}(j/10))$
Processing gain (dB)	10	$l = 10^* \log(3840/144)$
Required E_b/N_0 (dB)	1	m
Receiver sensitivity (dBm)	$-109{,}2$	$n = m - l + k$
Base station antenna gain (dBi)	18	o
Cable loss in the base station (dB)	2	p
Fast fading margin (dB)	4	q
Max. path loss (dB)	$147{,}2$	$r = d - n + o - p - q$
Coverage probability (%)	95	
Lognormal fading constant (dB)	7	
Propagation model exponent	$3{,}52$	
Lognormal fading margin (dB)	$7{,}3$	s
Soft handover gain (dB), multicell	0	t
Indoor loss (dB)	0	u
Allowed propagation loss for cell range (dB)	$139{,}9$	$v = r - s + t - u$

Example – with the fine tuning of (8.45) propagation model for an urban macrocell with base station antenna height of 30 m, mobile antenna height of 1.5 m and carrier frequency of 1950 MHz [1–5],

$$L = 137.4 + 35.2 \log_{10}(R) \tag{9.27}$$

L is the path loss in dB and R is the range in kilometer. In this case propagation coefficient $n = 3.52$ is used. For suburban areas an additional area correction factor of 8 dB is used.

$$L = 129.4 + 35.2 \log_{10}(R) \tag{9.28}$$

So, the cell range of 12.2-kbps speech service with 141.9-dB path loss in Table 9.3 in suburban area would be 2.3 km. The range of 144 kbps indoors with parameters from Table 9.4 would be 1.4 km. Once the cell range R is determined, the site area, which is also a function of the base station sectorization configuration, can then be derived. For a cell of hexagonal shape covered by an omnidirectional antenna, the coverage area can be approximated as 2.6 R^2.

9.2.2 Load factors and spectral efficiency

The second phase consists of estimating the amount of supported traffic per base station site. When the frequency reuse is 1, the system is typically interference-limited. For this purpose modification of equation (8.12) gives

$$\left(\frac{E_b}{N_0}\right)_j = G_j \cdot \frac{\text{Signal power of user } j}{\text{Total receiver power (excl. own signal)}} \tag{9.29}$$

where G_j is the processing gain of user j. This can be represented as

$$\left(\frac{E_b}{N_0}\right)_j = \frac{W}{\alpha_j R_j} \cdot \frac{P_j}{I_{\text{total}} - P_j} \tag{9.30}$$

In equation (9.30) the following notation is used. W is the chip rate, P_j is the receiver signal power from user j, α_j is the activity factor of user j, R_j is the bit rate of user j and I_{total} is the total receiver wideband power including thermal noise power in the base station. From equation (9.30) we have

$$P_j = \frac{1}{1 + [W/(E_b/N_0)_j \cdot R_j \cdot \alpha_j]} I_{\text{total}} \tag{9.31}$$

By defining $P_j = L_j \times I_{\text{total}}$, we obtain *the uplink load factor L_j* of one connection

$$L = \frac{1}{1 + [W/(E_b/N_0)_j \cdot R_j \cdot \alpha_j]} \tag{9.32}$$

It defines in what proportion connection j participates in the overall interfering signal. *The total receiver interference*, excluding the thermal noise P_N, can be represented as

$$I_{\text{total}} - P_N = \sum_{j=0}^{N} P_j = \sum_{j=1}^{N} L_j \cdot I_{\text{total}} \tag{9.33}$$

The noise rise is defined as

$$\text{Noise rise} = \frac{I_{\text{total}}}{P_N} = \frac{1}{1 - \sum_{j=1}^{N} L_j} = \frac{1}{1 - \eta_{\text{UL}}} \tag{9.34}$$

and the overall uplink load factor as

$$\eta_{\mathrm{UL}} = \sum_{j=1}^{N} L_j \tag{9.35}$$

When η_{UL} becomes close to 1, the corresponding noise rise approaches infinity and the system has reached its pole capacity. We leave it to the reader to elaborate the relation between the degradation factor introduced in Chapter 8, Section 8.1 and the load factor defined by equation (9.35). In the expression for load factor the interference from the other cells must be taken into account by the ratio of other-cell to own-cell interference, i:

$$i = \frac{\text{other cell interference}}{\text{own cell interference}} \tag{9.36}$$

The uplink load factor now becomes

$$\eta_{\mathrm{UL}} = (1+i) \cdot \sum_{j=1}^{N} L_j = (1+i) \cdot \sum_{j=1}^{N} \frac{1}{1 + [W/(E_b/N_0)_j \cdot R_j \cdot \alpha_j]} \tag{9.37}$$

The load equation predicts the amount of noise rise over thermal noise due to interference. From equation (9.34) the noise rise is equal to $-10\log_{10}(1 - \eta_{\mathrm{UL}})$. The interference margin in the link budget must be equal to the maximum planned noise rise. The required E_b/N_0 can be derived from link level simulations and from measurements. It includes the effect of the closed-loop power control and soft handover. The effect of soft handover is measured as the macro diversity combining gain relative to the single-link E_b/N_0 result. The other-cell to own (serving)-cell interference ratio i is a function of cell environment or cell isolation (e.g. macro/micro, urban/suburban) and antenna pattern (e.g. omni, 3-sector or 6-sector). The parameters are further explained in Table 9.6.

The load equation is commonly used to make a semianalytical prediction of the average capacity of a WCDMA cell, without going into system-level capacity simulations. This load equation can be used for the purpose of predicting cell capacity and planning noise rise in the dimensioning process. For a classical all-voice-service network, where all N users in the cell have a low bit rate R, equation (8.15) of Chapter 8 is valid and we have

$$\frac{W}{(E_b/N_0) \cdot R \cdot \alpha} \gg 1 \tag{9.38}$$

So, the uplink load equation can be approximated and simplified to

$$\eta_{\mathrm{UL}} = \frac{E_b/N_0}{W/R} \cdot N \cdot \alpha \cdot (1+i) \tag{9.39}$$

By using equation (9.39) in equation (9.34), an example for uplink noise rise is shown in Figure 9.5 for data service, assuming an E_b/N_0 requirement of 1.5 dB and $i = 0.65$. The noise rise of 3.0 dB corresponds to a 50% load factor and the noise rise of 6.0 dB

Table 9.6 Parameters used in uplink load factor calculation [1] Reproduced from Holma, H. and Toskala, A. (2000) *WCDMA for UMTS*. New York: John Wiley & Sons, by permission of IEEE

	Definitions	Recommended values
N	Number of users per cell	
	Activity factor of user j at physical layer	0.67 for speech, assumed 50% voice activity and DPCCH overhead during DTX 1.0 for data
E_b/N_0	Signal energy per bit divided by noise spectral density that is required to meet a predefined Quality of Service (e.g. bit error rate). Noise includes both thermal noise and interference	Dependent on service, bit rate, multipath fading channel, receive antenna diversity, mobile speed, etc.
W	WCDMA chip rate	3.84 Mcps
	Bit rate of user j	Dependent on service
(i)	Other-cell to own-cell interference ratio seen by the base station receiver	Macrocell with omnidirectional antennas: 55%

Note: DPCCH – Dedicated physical control channel.

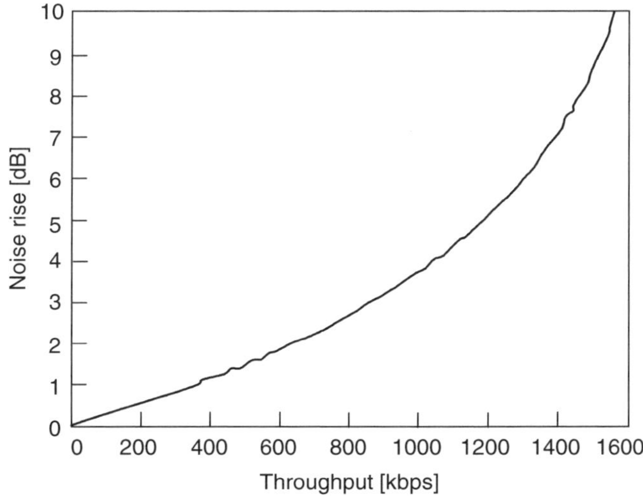

Figure 9.5 Uplink noise rise as a function of uplink data throughput [1]. Reproduced from Holma, H. and Toskala, A. (2000) *WCDMA for UMTS*. New York: John Wiley & Sons, by permission of IEEE.

to a 75% load factor. Instead of showing the number of users N, we show the total data throughput per cell of all simultaneous users. In this example, a throughput of 860 kbps can be supported with 3.0-dB noise rise and 1300 kbps with 6.0-dB noise rise.

For downlink the load factor becomes

$$\eta_{\mathrm{DL}} = \sum_{j=1}^{N} \alpha_j \cdot \frac{(E_{\mathrm{b}}/N_0)_j}{W/R_j} \cdot [(1 - \mathrm{orth}_j) + i_j] \qquad (9.40)$$

where orth_j is the orthogonality of user j and $-10 \log_{10}(1-\eta_{\mathrm{UL}})$ is the noise rise over thermal noise due to multiple access interference. The downlink load factor η_{DL} exhibits very similar behavior to the uplink load factor η_{UL}, in the sense that when approaching unity, the system reaches its pole capacity and the noise rise over thermal goes to infinity. For illustration purposes we use parameters presented in Table 9.7 [1].

Table 9.7 Parameters used in downlink load factor calculation [1]. Reproduced from Holma, H. and Toskala, A. (2000) *WCDMA for UMTS*. New York: John Wiley & Sons, by permission of IEEE

	Definitions	Recommended values
N	Number of users per cell = number of users per cell * (1+ soft handover overhead)	
	Activity factor of user j at physical layer	0.67 for speech, assumed 50% voice activity and DPCCH overhead during DTX 1.0 for data
E_{b}/N_0	Signal energy per bit divided by noise spectral density that is required to meet a predefined Quality of Service (e.g. bit error rate). Noise includes both thermal noise and interference	Dependent on service, bit rate, multipath fading channel, receive antenna diversity, mobile speed, etc.
W	WCDMA chip rate	3.84 Mcps
	Bit rate of user j	Dependent on service
	Orthogonality of channel of user j	Dependent on the multipath propagation 1: fully orthogonal 1-path channel 0: no orthogonality
	Ratio of other-cell to own-cell base station power, received by user j	Each user sees a different ij, depending on its location in the cell and lognormal shadowing
	Average orthogonal factor in the cell	ITU vehicular A channel: $\sim$60% ITU pedestrian A channel: $\sim$90%
	Average ratio of other-cell to own-cell base station power received by user. Own-cell interference is here wideband	Macrocell with omnidirectional antennas: 55%

Note: DPCCH – Dedicated physical control channel.

The effect of noise rise due to interference is added to this minimum power and the total represents the transmission power required for a user at an 'average' location in the cell.

$$P_{\mathrm{B}} = \frac{N_{\mathrm{rf}} \cdot W \cdot \overline{L} \cdot \sum_{j=1}^{N} \alpha_j \dfrac{(E_{\mathrm{b}}/N_0)_j}{W/R_j}}{1 - \eta_{\mathrm{DL}}} \tag{9.41}$$

where N_{rf} is the noise spectral density of the mobile receiver front end.

$$N_{\mathrm{rf}} = k \cdot T + NF = -174.0\,\mathrm{dBm} + NF \quad (\text{for } T = 290\,\mathrm{K}) \tag{9.42}$$

where k is the Boltzmann constant of $1.381 \times 10^{-23}\,\mathrm{J\,K^{-1}}$, T is the temperature in Kelvin and NF is the mobile station receiver noise figure with typical values of 5 to 9 dB. *The load factor* can be approximated by its average value across the cell, that is,

$$\overline{\eta_{\mathrm{DL}}} = \sum_{j=1}^{N} \alpha_j \cdot \frac{(E_{\mathrm{b}}/N_0)_j}{W/R_j} \cdot [(1 - \overline{\mathrm{orth}}) + \overline{i}] \tag{9.43}$$

In both uplink and downlink the air interface load affects the coverage but the effect is not exactly the same.

The maximum path loss, that is, coverage, as a function of the load for both uplink and downlink is shown in Figure 9.6. A three-sector site is assumed, and the throughputs are shown per sector per 5-MHz carrier. The uplink is calculated for 144-kbps data and the link budget is shown in Table 9.4. An other-cell to own-cell interference ratio $i = 0.65$ is used. In the downlink, orth $= 0.6$ and E_{b}/N_0 of 5.5 dB are assumed, giving a pole capacity of 820 kbps per cell. No transmit diversity is assumed in this E_{b}/N_0. The base station transmission power is assumed to be 10 W and additionally cable loss is taken into account. The effect of downlink common channels is included in the downlink calculations, that is, part of 10 W is allocated for downlink common channels. The uplink pole capacity in this example is 1730 kbps per cell, (out of the scope of the figure).

In the downlink the coverage (maximum path loss) depends more on the load than in the uplink. The reason is that in the downlink the maximum transmission power is the same 10 W regardless of the number of users and is shared between the downlink users, while in the uplink each additional user has its own power amplifier. Therefore, even with low load in the downlink, the coverage decreases as a function of the number of users. We note that with the above assumptions the coverage is clearly limited by the uplink for a load below 650 kbps, while the capacity is downlink-limited. This is why we continue with the discussion of coverage on the uplink and the discussion of capacity on the downlink. One should remember that in third generation networks, traffic can be asymmetric between uplink and downlink, and the load can be different in these links.

In Figure 9.6 a base station maximum power of 10 W is assumed. The question is how much could we improve the downlink coverage and capacity by using more power, such as 20 W. The difference in downlink coverage and capacity between 10 and 20 W base station output powers is shown in Figure 9.7. If we increase the downlink power by 3.0 dB, we can allow 3.0-dB higher maximum path loss regardless of the load. The

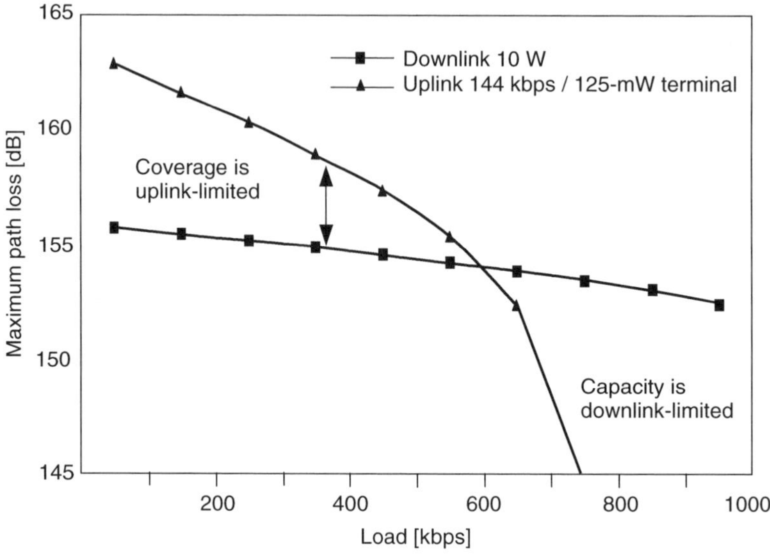

Figure 9.6 Example coverage versus capacity relation in downlink and uplink in macrocells [1]. Reproduced from Holma, H. and Toskala, A. (2000) *WCDMA for UMTS*. New York: John Wiley & Sons, by permission of IEEE.

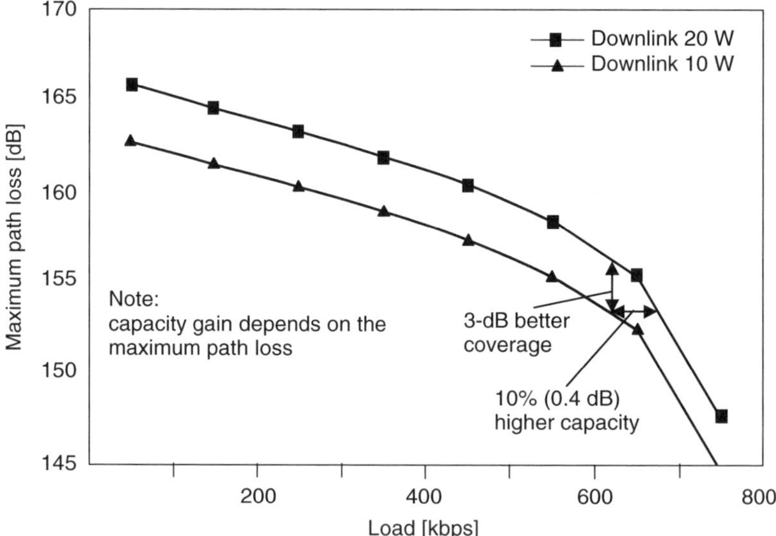

Figure 9.7 Effect on base station output to downlink capacity and coverage [1]. Reproduced from Holma, H. and Toskala, A. (2000) *WCDMA for UMTS*. New York: John Wiley & Sons, by permission of IEEE.

capacity improvement is smaller than the coverage improvement because of the load curve. If we now keep the downlink path loss fixed at 153 dB, which is the maximum uplink path loss with 3 dB interference margin, the downlink capacity can be increased by only 10% (0.4 dB) from 680 to 750 kbps. Increased downlink transmission power is an inefficient approach to increase downlink capacity, since the available power does not affect the pole capacity. Assume we had 20-W downlink transmission power available. Splitting the downlink power between two frequencies would increase downlink capacity from 750 kbps to 2×680 kbps $= 1360$ kbps, that is, by 80%. The splitting of the downlink power between two carriers is an efficient approach to increase the downlink capacity without any extra investment in power amplifiers. The power splitting approach requires that the operator's frequency allocation allows the use of two carriers in the base station.

9.3 SPECTRAL EFFICIENCY OF WCDMA

The spectral efficiency of WCDMA can be defined either by the number of simultaneous calls of some defined bit rates as in Chapter 8 or more appropriately in third-generation systems, by the aggregated physical layer throughput supported in each cell per 5-MHz carrier, measured in kbps per cell per carrier. Spectral efficiency is a function of radio environment, user mobility and location, services and quality of service and propagation conditions. The variation can be quite large (e.g. 50–100%). Therefore, most system simulations that attempt to offer some indication of the average spectral efficiency of WCDMA reflect only the results for some predefined cell conditions and user behavior.

9.3.1 Soft capacity

Erlang capacity

The traffic density can be measured in Erlang:

$$\text{Traffic density [Erlang]} = \frac{\text{Call arrival rate [calls per second]}}{\text{Call departure rate [calls per second]}} \tag{9.44}$$

If the capacity is hard-blocked, that is, limited by the amount of time slots in TDMA, the Erlang capacity can be obtained from the Erlang B model. If the maximum capacity is limited by the amount of interference in the air interface, it is referred to as a soft capacity, since there is no single fixed value for the maximum capacity. For a soft capacity–limited system, the Erlang capacity cannot be calculated from the Erlang B formula, since it would give too pessimistic results. The total channel pool is larger than just the average number of channels per cell, since the adjacent cells share part of the same interference, and therefore more traffic can be served with the same blocking probability. The lesser the interference from the neighboring cells, the more will be the channels available in the middle cell. With a low number of channels per cell, that is, for high bit rate real-time data users, the average loading must be quite low to guarantee low blocking probability. Since the average loading is low, there is typically extra capacity available in the neighboring cells. This capacity can be borrowed from adjacent cells; therefore the interference sharing gives soft capacity. Soft capacity is important for high bit rate real-time data users, for example, for video connections. It can also be obtained in global systems of mobile communications (GSM) if the air interface capacity is limited by the amount of interference instead of the number of time slots; this assumes low frequency reuse factors in GSM with fractional loading.

A detailed presentation of soft capacity in packetized CDMA network will be presented in the next chapter. Here we provide an initial presentation based mainly on References [1–5].

In the soft capacity calculations, presented below, it is assumed that the number of subscribers is the same in all cells but the connections start and end independently. In addition, the call arrival interval follows a Poisson distribution. This approach can be used in dimensioning when calculating Erlang capacities. There is an additional soft capacity in WCDMA if also the number of users in the neighboring cells is smaller. WCDMA soft capacity is defined as the increase of Erlang capacity with soft blocking compared to that with hard blocking with the same maximum number of channels per cell on average with both soft and hard blocking:

$$\text{Soft capacity} = \frac{\text{Erlang capacity with soft blocking}}{\text{Erlang capacity with hard blocking}} - 1 \tag{9.45}$$

The wideband power-based admission control strategy gives soft blocking and soft capacity. Uplink soft capacity can be approximated on the basis of the total interference at the base station. This total interference includes both own-cell and other-cell interference. Therefore, the total channel pool can be obtained by multiplying the number of channels per cell in the equally loaded case by $1 + i$, which gives the single isolated cell capacity, since

$$
\begin{aligned}
i + 1 &= \frac{\text{other-cell interference}}{\text{own-cell interference}} + 1 \\
&= \frac{\text{other-cell interference} + \text{own-cell interference}}{\text{own-cell interference}} \\
&= \frac{\text{isolated-cell capacity}}{\text{multicell capacity}}
\end{aligned}
\tag{9.46}
$$

The basic Erlang B formula is then applied to this larger channel pool (= interference pool). The Erlang capacity obtained is then shared equally between the cells. These steps should be summarized as follows:

1. Calculate the number of channels per cell, N, in the equally loaded case and on the basis of the uplink load factor solve equation (9.37).
2. Multiply that number of channels by $1 + i$ to obtain the total channel pool in the soft blocking case.
3. Calculate the maximum offered traffic from the Erlang B formula.
4. Divide the Erlang capacity by $1 + i$.

For illustration purposes parameters from Table 9.8 were used [1].

The results are shown in Table 9.9. One can see that capacity gain increase for the services with higher bit rates and can reach as much as 28% for 144 kbps. The trunking efficiency shown in Table 9.9 is defined as the hard-blocked capacity divided by the number of channels. The lower the trunking efficiency, the lower is the average loading, the more capacity can be borrowed from the neighboring cells, and the more soft capacity is available. Figure 9.8 illustrates results from Table 9.9. More detailed discussion on soft capacity will be given in the next chapter.

Table 9.8 Assumptions in soft capacity calculations

Bit rates	Speech: 12.2 kbps
	Real-time data: 16–144 kbps
Voice activity	Speech 67%
	Data 100%
E_b/N_0	Speech: 4 dB
	Data 16–32 kbps: 3 dB
	Data 64 kbps: 2 dB
	Data 144 kbps: 1.5 dB
(i)	0.55
Noise rise	3 dB (= 50% load factor)
Blocking probability	2%

Table 9.9 Soft capacity calculations in the uplink

Bit rate (kbps)	Channels per cell	Hard-blocked capacity (Erlang)	Trunking efficiency (%)	Soft-blocked capacity	Soft capacity
12,2	60,5	50.8	84	50.8	5
16	39,0	30.1	77	32.3	7
32	19,7	12.9	65	14.4	12
64	12,5	7.0	59	8.2	17
144	6,4	2.5	39	3.2	28

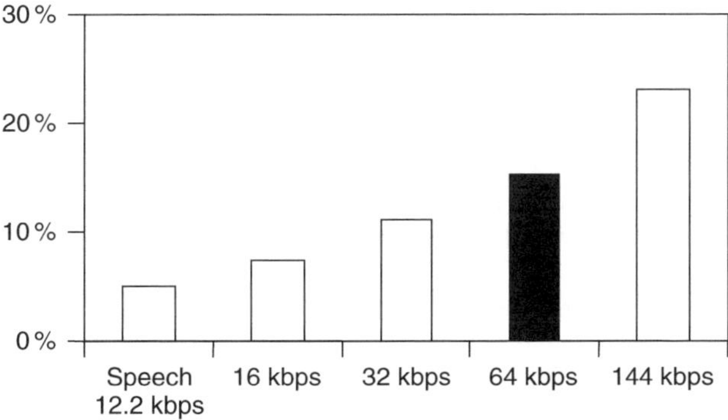

Figure 9.8 Soft capacity as a function of bit rate for real-time connections.

A number of additional issues related to cell capacity and coverage are presented in References [6–22].

SYMBOLS

C – signal power (carrier)

I – interference power

m – number of voice channels

G – processing gain

R_b – bit rate

B – bandwidth

η – voice activity factor

F, R (index) – forward, reverse

E_b/I_0 – energy per bit per overall interference density

α – power per user

R – cell radius

P – base station overall transmit power

r_{ik} – portion of users from cell k interfering in cell i

REFERENCES

1. Holma, H. and Toskala, A. (2000) *WCDMA for UMTS*. New York: John Wiley & Sons.
2. Wacker A. *et al.* (1999) Static simulator for studying WCDMA radio network planning issues. *Proceedings of VTC '99*, Houston, TX, pp. 2436–2440.
3. Pace, A. and Valentini, L. (2000) System level performance evaluation of UTRA-FDD (UMTS terrestrial radio access-frequency division duplex). *The 11th IEEE International Symposium on Personal, Indoor and Mobile Radio Communications, PIMRC 2000*, Vol. 1, pp. 343–347.
4. Ketchum, J., Wallace, M. and Walton, R. (1996) CDMA network deployment of 8 kbps and 13 kbps voice services. *5th IEEE International Conference on Universal Personal Communications*, Vol. 1, pp. 179–183.
5. Xia, H. and Siu, F. (1996) System design aspects of CDMA personal communications services. *IEEE Vehicular Technology Conference*, Vol. 3, pp. 1647–1651.
6. Law, A. M. and McComas, M. G. (1994) Simulation software for communications networks: the state of the art. *IEEE Commun. Mag.*, **32**, 44–50.
7. Glisic, S. and Leppanen, P. (eds) (1995) *Code Division Multiple Access Communications*. Kluwer, Dordrecht, The Netherlands.
8. Spilling, A. G., Nix, A. R., Beach, M. A. and Harrold, T. J. (2000) Self-organisation in future mobile communications. *Electron. Commun. Eng. J.*, **12**(3), 133–147.
9. Steele, R. (1990) Deploying personal communication networks. *IEEE Commun. Mag.*, **28**, 12–15.
10. Fernandes, J. and Garcia, J. (2000) Cellular coverage for efficient transmission performance in MBS. *Vehicular Technology Conference 2000, IEEE VTC Fall*, Vol. 5, pp. 2225–2232.
11. El-Jabu, B. and Steele, R. (1999) Aerial platforms: a promising means of 3G communications. *Vehicular Technology Conference*, Vol. 3, pp. 2104–2108.
12. Ganesh, R. (1999) Impact of adding sites on PN offset planning in CDMA networks. *IEEE International Conference on Personal Wireless Communication*, pp. 446–450.
13. Hristov, H., Feick, R. and Grote, W. (2001) Improving indoor signal coverage by use of through-wall passive repeaters. *Antennas and Propagation Society, 2001 IEEE International Symposium*, Vol. 2, pp. 158–161.

14. Faruque, S. (1998) Science, engineering and art of cellular network deployment. *The Ninth IEEE International Symposium on Personal, Indoor and Mobile Radio Communications*, Vol. 1, pp. 313–317.
15. Fagen, D., Aksu, A. and Giordano, A. (1997) A case study of CDMA and PCS-1900 using the GRANET/sup TM/radio planning tool. *IEEE International Conference on Personal Wireless Communications*, pp. 505–509.
16. Rodrigues, R. C., Mateus, G. R. and Loureiro, A. A. F. (2000) On the design and capacity planning of a wireless local area network. *Proc. IEEE NOMS*, September, 2000, pp. 335–348.
17. Hortos, W. (1996) Analysis of the deployment of transportable base stations in personal communication services networks with expanded user location features. *Southcon/96, Conference Record*, pp. 104–111.
18. Tan, M., Lee, J., Xu, H., Introne, J. and Matheus, C. (2000) Wireless usage analysis for capacity planning and beyond: a data warehouse approach. *Proc. IEEE NOMS*, September, 2000, pp. 905–917.
19. Steele, R., Whitehead, J. and Wong, W. (1995) System aspects of cellular radio. *IEEE Commun. Mag.*, **33**(1), 80–87.
20. Liang, J.-W. and Paulraj, A. J. (1995) On optimizing base station antenna array topology for coverage extension in cellular radio networks. *IEEE 45th Vehicular Technology Conference*, Vol. 2, pp. 866–870.
21. Goldburg, M. and Roy, R. (1994) The impacts of SDMA on PCS system design. *Third Annual International Conference on Universal Personal Communications*, pp. 242–246.
22. Sarnecki, J., Vinodrai, C., Javed, A., O'Kelly, P. and Dick, K. (1993) Microcell design principles. *IEEE Commun. Mag.*, **31**(4), 76–82.

10

Resource management and access control

10.1 POWER CONTROL AND RESOURCE MANAGEMENT FOR A MULTIMEDIA CDMA WIRELESS SYSTEM

10.1.1 System model and analysis

In this section we assume N different classes of users in the system characterized by the following set of parameters [1]

Transmitted power vector	$P = [P_1, P_2, \ldots, P_N]$
Vector of rates	$R = [R_1, R_2, \ldots, R_N]$
Vector of required E_b/N_0s	$\Gamma = [\gamma_1, \gamma_2, \ldots, \gamma_N]$
Power limits	$p = [p_1, p_2, \ldots, p_N]$
Rate limits	$r = [r_1, r_2, \ldots, r_N]$
Channel gains vector $\mathbf{h}$	

Energy per bit per noise density E_b/N_0 of each user can be represented as

$$\left(\frac{E_b}{N_0} \right)_i = \frac{W}{R_i} \frac{h_i P_i}{\sum_{j \neq i} h_j P_j + \eta_0 W} \tag{10.1}$$

The resource management has to provide quality of service (QoS) for each user that can be represented as

$$\frac{W}{R_i} \frac{h_i P_i}{\sum_{j \neq i} h_j P_j + \eta_0 W} \geq \gamma_i \quad i = 1, \ldots, N \tag{10.2}$$

Power and rate constraints can be defined as

$$0 < P_i \leq p_i \quad R_i \geq r_i \quad i = 1, \ldots, N \tag{10.3}$$

As optimization criteria, we can have

1. Minimum total transmitted power. For this criterion we also find the maximum number of users of each class that can be simultaneously supported while meeting their constraints.
2. Maximum sum of the rates (overall throughput).

10.1.2 Minimizing total transmitted power

For a single cell system, let P be the transmitted power vector. The problem we are looking at is defined by

$$\text{Minimize}_{P,R} \quad \sum_{i=1}^{N} P_i \tag{10.4}$$

subject to constraints P, given by equation (10.3). An easily proven observation about the solution is

1. at the optimal solution all QoS constraints are met with equality,
2. the optimal power vector is one that achieves all rate constraints with equality.

The optimum rate vector is $R^* = [r_1, r_2, \ldots, r_N]$. If we write equation (10.2) for each user with equality, we have

$$\frac{W}{r_i} \frac{h_i P_i^*}{\displaystyle\sum_{j \neq i} h_j P_j^* + \eta_0 W} = \gamma_i \quad \forall i = 1, \ldots, N \tag{10.5}$$

By using $r_i' = r_i \gamma_i$, this yields the matrix equation

$$\mathbf{A}P^* = \eta_0 W \mathbf{1} \tag{10.6}$$

where

$$\mathbf{A} = \begin{pmatrix} \dfrac{W h_1}{r_1'} & -h_2 & \cdots & -h_N \\[2ex] -h_1 & \dfrac{W h_2}{r_2'} & \cdots & -h_N \\[2ex] \vdots & \vdots & & \vdots \\[2ex] -h_1 & -h_2 & \cdots & \dfrac{W h_N}{r_N'} \end{pmatrix} \tag{10.7}$$

$[P_1^*, P_2^*, \ldots, P_N^*]^\mathrm{T}$ is the optimal power vector and $\mathbf{1} = [1, 1, \ldots, 1]^\mathrm{T}$ is an all-ones vector. By elementary row operations (subtraction of each row from the next), this reduces to the following equation in P_1^*

$$\left(\frac{W}{r_1'} + 1\right) h_1 P_1^* \left[1 - \sum_{j=1}^{N} \frac{1}{\dfrac{W}{r_j'} + 1}\right] = \eta_0 W \tag{10.8}$$

Positivity of P^* implies the following condition:

$$\sum_{j=1}^{N} \frac{1}{\dfrac{W}{r_j'} + 1} < 1 \tag{10.9}$$

If this condition is satisfied for a set of rates and E_b/N_0 requirements, the powers can be obtained. By solving for the powers and imposing power constraints, equation (10.8) gives

$$\sum_{j=1}^{N} \frac{1}{\dfrac{W}{r_j'} + 1} \leq 1 - \frac{\eta_0 W}{\min_i \left[p_i h_i \left(\dfrac{W}{r_i'} + 1\right)\right]} \qquad i = 1, \ldots, N \tag{10.10a}$$

This equation now determines feasibility of a set of rates, QoS requirements and power constraints. By solving for the powers and imposing power constraints, equation (10.8) gives

$$\sum_{j=1}^{N} \frac{1}{\dfrac{W}{r_j'} + 1} \leq 1 - \frac{\eta_0 W}{\min_i \left[p_i h_i \left(\dfrac{W}{r_i'} + 1\right)\right]} \qquad i = 1, \ldots, N \tag{10.10b}$$

This equation now determines feasibility of a set of rates, QoS requirements and power constraints.

10.1.3 Capacity (number of users) of a cell in the multimedia case

Consider K classes of users. For any class i, γ_i the QoS requirement, r_i the rate required and p_i the upper bound on the power are all fixed. N_i represents the number of simultaneous users of class i. The channel gains for the users of class i are given as

$$\mathbf{h_i} = [h_i^1, h_i^2, \ldots, h_i^{N_i}]^\mathrm{T} \tag{10.11}$$

Equation (10.10b) in this case becomes

$$\sum_{j=1}^{K} \frac{N_j}{\frac{W}{r'_j} + 1} \leq 1 - \frac{\eta_0\, W}{\min_i \left\{ p_i \left[\min_j (h_i^j) \left(\frac{W}{r'_i} + 1 \right) \right] \right\}} \tag{10.12}$$

10.1.4 Maximizing sum of rates

The system tries to give each user the best throughput possible within the specified constraints. For a received power vector $\mathbf{Q}$, this is defined as

$$\text{Maximize}_{Q,R} \sum_{i=1}^{N} R_i \tag{10.13}$$

Subject to

$$\frac{W}{R_i} \frac{Q_i}{\sum_{j \neq i} Q_j + \eta_0\, W} \geq \gamma_i \quad i = 1, \ldots, N$$

$$0 < Q_i \leq q_i \quad R_i \geq r_i \quad i = 1, \ldots, N \tag{10.14}$$

From equations (10.13 and 10.14), the problem can be written as

$$\text{Maximize}_Q \sum_{i=1}^{N} \frac{W}{\gamma_i} \frac{Q_i}{\sum_{j \neq i} Q_j + \eta_0\, W} \tag{10.15}$$

subject to

$$0 < Q_i \leq q_i \quad \frac{W}{\gamma_i} \frac{Q_i}{\sum_{j \neq i} Q_j + \eta_0\, W} \geq r_i \quad i = 1, \ldots, N \tag{10.16}$$

The objective function is nonlinear, whereas the constraints are linear functions of the variables. Efficient methods analogous to linear programming exist for solving such problems. Furthermore, the objective function is convex in each of the variables. This restricts the search to the surfaces of the polyhedron defined by the constraints. One technique that can be used is the gradient projection method, which is well elaborated in textbooks [2]. A solution to the rate maximization problem exists if and only if a solution to the minimum total power problem exists. The equation is first checked. If it is satisfied, the corresponding power vector is chosen as the initial iterate to the gradient projection method. The similar rate constraint is the first included in the active set. It was found that the method converges quickly to the solution [3]. However, with initial guesses containing the maximum rate constraint in the active set, the algorithm converged to a local minimum. With other initial guesses, this problem was avoided.

10.1.5 Example of capacity evaluation for minimum power problem

Consider a system with two classes of service, voice and data. The parameters of the system are [1]

Item	Symbol	Value
Bandwidth	W	1.25 MHz
Voice rate	R_v	8 kbps
Data rate	R_d	4, 8, 20 kbps
(E_b/N_0) voice	γ_v	5 (7 dB)
(E_b/N_0) data	γ_d	12, 10, 5
Max. power voice	p_v	0.5 W
Max. power data	p_d	0.3, 0.5, 0.6 W
Min. channel gain voice	h_v	0.25
Min. channel gain data	h_d	0.25
AWGN spectral density	η_0	10^{-6}

With these parameters and equations (10.4 to 10.12), the maximum number of data users of each class is found for a given number of active voice users in the network. These results are shown in Figures 10.1 and 10.2. These results can be used for data access control in the system.

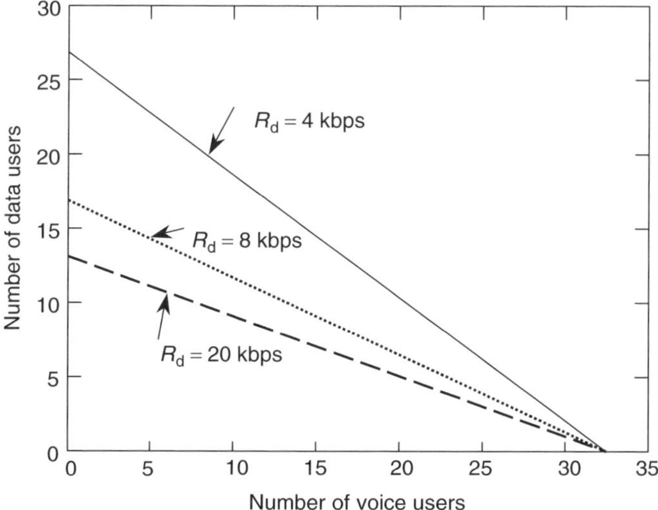

Figure 10.1   Capacity curves for unconstrained power case. Parameters $R_v = 8$ kbps, $\gamma_v = 5$. For data, three cases $\gamma_d = 12$, $R_d = 4$ kbps; $\gamma_d = 10$, $R_d = 8$ kbps and $\gamma_d = 5$, $R_d = 20$ kbps [1]. Reproduced from Sampath, A., Kumar, P. S. and Holtzman, J. M. (1995) Power control and resource management for a multimedia CDMA wireless system. *Proc. PIMRC*, Vol. 1, pp. 21–25, by permission of IEEE.

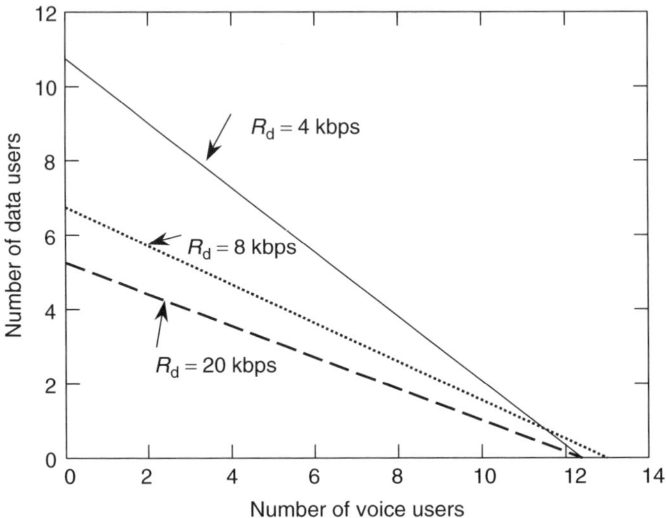

Figure 10.2 Capacity curves for constrained power case. Parameters $R_v = 8$ kbps, $p_v = 0.5$ W and $\gamma_v = 5$ for voice. For data, three cases $\gamma_d = 12$, $R_d = 4$ kbps; $p_d = 0.3$ W; $\gamma_d = 10$, $R_d = 8$ kbps, $p_d = 0.5$ W and $\gamma_d = 5$, $R_d = 20$ kbps, $p_d = 0.6$ W [1]. Reproduced from Sampath, A., Kumar, P. S. and Holtzman, J. M. (1995) Power control and resource management for a multimedia CDMA wireless system. *Proc. PIMRC*, Vol. 1, pp. 21–25, by permission of IEEE.

10.2 ACCESS CONTROL OF IN DATA INTEGRATED VOICE/DATA CDMA SYSTEMS

The problem from the previous section is now elaborated in more detail in the case when only voice and one class of data users are used. Equation (10.9) still holds

$$\sum_{j=1}^{N} \frac{1}{\dfrac{W}{r_j \gamma_j} + 1} < 1 \tag{10.17}$$

If we use the following notation for the bit rates for voice (R_V) and data (R_D), respectively, when active, then inequality from the above equation can be written as [3]

$$S = \frac{v}{a_V} + \frac{d}{a_D} < 1 \tag{10.18}$$

where v is the number of *active* voice users and d is the number of data users who transmit and

$$a_V = (W/R_V \gamma_V + 1)$$
$$a_D = (W/R_D \gamma_D + 1) \tag{10.19}$$

S is called the load. If we discretize the timescale into slots, then equation (10.19) becomes

$$S(n) = \frac{V(n)}{a_V} + \frac{D(n)}{a_D} < 1 \qquad (10.20)$$

$V(n)$ is the number of active voice users in the nth slot. Through access control, the number $D(n)$ can be dynamically controlled. More data users can be allowed to transmit when the voice load $V(n)$ is low and less when the voice load is heavy. This is the motivation behind access control, and for the same target outage probability, more data calls can be admitted into the system than if no access control scheme was used. The penalty lies in introducing delay for the packets of data since they may have to wait to be transmitted.

In practice, power control is not perfect, and also, power control loops are designed to adjust the power of users on an individual basis, on the basis of current conditions for that user. Dynamic range limitations at the base station (BS) receiver require that the total received power be limited. Of interest is to maintain the total received power Z to within around 10 dB of the background noise power. In a probabilistic access control scheme, permission probability for data is varied on the basis of either measuring S or Z. If S (or Z) is less than the limit, the permission probability is increased, and it is reduced if otherwise.

The performance measure presented in the sequel is very much based on Reference [3]. As defined, the probability of outage (P_{out}) is the fraction of time that the outage condition is violated. Since no retransmission for voice is possible, it is designed to keep this probability low, nominally around 1%. For data users, outage probability is important too since it affects throughput. When the outage condition is violated, data packets are errored but can be retransmitted subsequently. In addition, mean access delay for data (D_A) and goodput for data (G) are also considered. For these purposes the voice activity is modeled with the two-state process shown in Figure 10.3. The system model in the presence of K_v voice users is shown in Figure 10.4.

$$P(i/j) \text{ is } P\{V(n+1) = i/V(n) = j\}$$

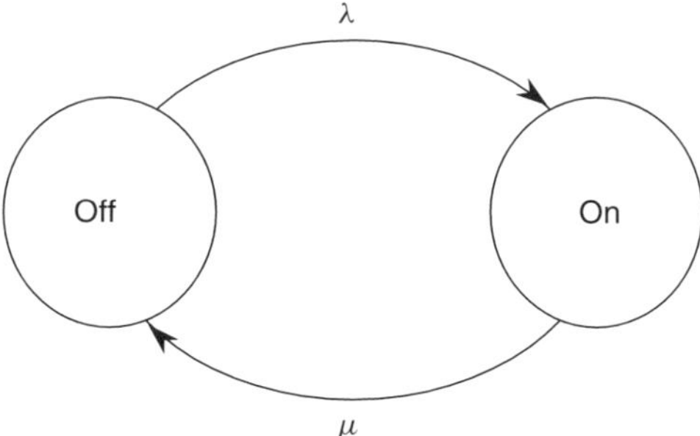

Figure 10.3 Two-state model for voice activity.

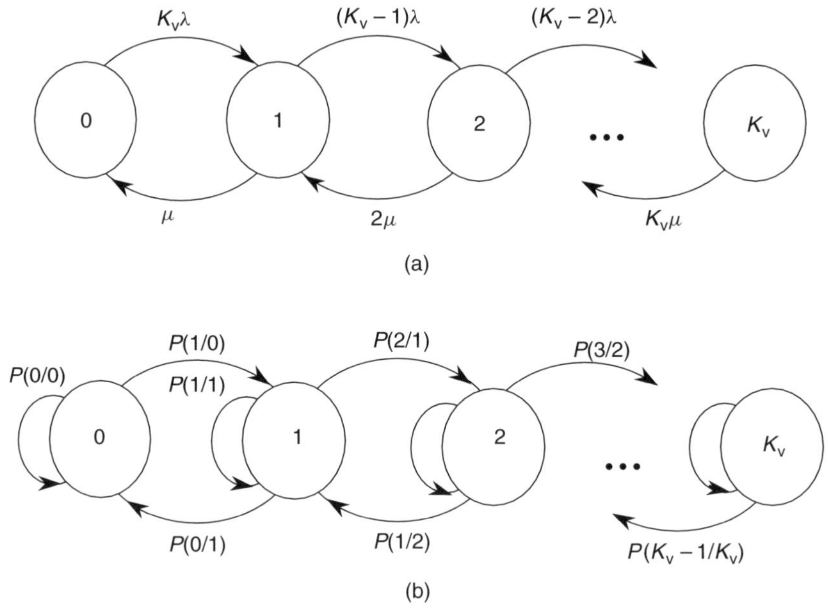

Figure 10.4   (a) Continuous-time Markov chain that represents the cumulative voice process. (b) Approximate discrete-time Markov chain for the cumulative voice process.

The time epochs correspond to positive integer multiples of the slot duration d. It is assumed that d is small compared to $1/\lambda$ and $1/\mu$, so that the probabilities of two or more events in a slot are negligible. Under these assumptions, we can use the Markov model for the process. Time spent in state k before making a transition to state $(k + 1)$ is exponentially distributed with mean $1/\tilde{\lambda}_k = 1/\lambda(K_V - k)$. The transition time to state $(k - 1)$ is exponentially distributed with mean $1/\tilde{\mu}_k = 1/(\mu k)$. The probability of remaining in state k is one minus the sum of two previous probabilities. From the theory we have

$$P\{V(n + 1) = k | V(n) = k\} = \exp(-\lambda_k d) \cdot \exp(-\mu_k d) = \exp(-(\tilde{\lambda}_k + \tilde{\mu}_k)d) \quad (10.21)$$

This is the probability that there will now be new arrival or new departure. The probability that there will be exactly one arrival and one departure is neglected. So, we have

$$P[V(n + 1) = k + 1 | V(n) = k] = \frac{\tilde{\lambda}_k}{\tilde{\lambda}_k + \tilde{\mu}_k} \{1 - \exp[-(\tilde{\lambda}_k + \tilde{\mu}_k)d]\}$$

$$P[V(n + 1) = k - 1 | V(n) = k] = \frac{\tilde{\mu}_k}{\tilde{\lambda}_k + \tilde{\mu}_k} \{1 - \exp[-(\tilde{\lambda}_k + \tilde{\mu}_k)d]\} \quad (10.22)$$

The stationary probability of state k can be obtained from the state equations for the model from Figure 10.4 and the solution is

$$\Pi_V(k) = \frac{\binom{K_V}{k}(\lambda/\mu)^k}{\sum_{j=0}^{K_V}\binom{K_V}{j}(\lambda/\mu)^j} \tag{10.23}$$

A simple data model with a fixed number of admitted data calls K_D and each offering exactly one packet per slot is used. If a packet is blocked by the access control scheme or errored on the channel, it remains in the buffer to be transmitted. No new packets are generated until that packet is delivered. Transmitted packets are at rate R_D bits s^{-1}. This model would be adequate for services such as file transfer, e-mail and store-and-forward facsimile. The result can be extended to other data models. A Poisson model is believed to represent short message service (SMS) very well. Although the analysis gets complicated in this case, the qualitative results from the simple data model still hold. Interactive data service can be modeled as a queue of packets at each source with an arrival process into the queue. All results from the fixed data model directly apply in this case, with an additional stability condition to ensure that none of the queue lengths become unbounded.

10.2.1 Access control under perfect power control

Assumptions

There is a slotted system for the reverse link. No processing or feedback delay of the permission probability is considered. All voice users share a common target signal-to-interference ratio (SIR), as do the data users. Whenever power control is feasible, transmit power assignment that gives each user its desired SIR is made. No limits on total received power at the BS or transmit power limits at the mobile station are considered. Other cell interference is incorporated as background noise. With no received power limits at the BS or the transmit power limits at the mobile station, other cell interference only leads to a scaling of the received powers and does not affect the feasibility condition for power control.

No access control

From equation (10.20) the load in the nth slot is

$$S(n) = \frac{V(n)}{a_V} + \frac{K_D}{a_D} \tag{10.24}$$

Probability of outage is given by

$$P_{\text{out}} = \lim_{n\to\infty} P\{S(n) \geq 1\} \tag{10.25}$$

Hence, from equations (10.23 and 10.25) we have

$$P_{\text{out}} = \sum_{j=[a_{\text{V}}(1-K_{\text{D}}/a_{\text{D}})]}^{K_{\text{D}}} \frac{\binom{K_{\text{V}}}{j}(\lambda/\mu)^j}{\sum_{l=0}^{K_{\text{V}}}\binom{K_{\text{V}}}{l}(\lambda/\mu)^l} \tag{10.26}$$

where $[x]$ is the smallest integer greater than or equal to x. Since no access control is used, the mean access delay for data D_{A} is zero.

Average throughput

A data packet that is transmitted in a slot in which the outage condition is violated is errored and must be retransmitted. A packet transmitted in a slot in which the outage condition is met is received error-free. Let $D_{\text{S}}(n)$ be the random variable that represents the number of successful data packets (over all data users) in the nth slot. Then

$$D_{\text{S}}(n) = \begin{bmatrix} K_{\text{D}}, & \text{if } S(n) < 1 \\ 0, & \text{if } S(n) \geq 1 \end{bmatrix} \tag{10.27}$$

The expected value of $D_{\text{S}}(n)$ represents the goodput for data

$$G = \lim_{n\to\infty} E[D_{\text{S}}(n)] = K_{\text{D}} \lim_{n\to\infty} P[S(n) < 1] = K_{\text{D}}(1 - P_{\text{out}}) \tag{10.28}$$

The goodput per user is simply

$$\frac{G}{K_{\text{D}}} = (1 - P_{\text{out}}) \tag{10.29}$$

Access control based on prediction

This control is based on the following steps:

1. Measure $V(n)$, the number of active voice users in the nth slot.
2. Predict the number of active voice users in the $(n + 1)$th slot

$$\widehat{V}(n + 1) = \begin{cases} V(n + 1) & \text{for perfect prediction} \\ E[V(n + 1)|V(n), V(n - 1), V(n - 2), \ldots,] \\ \text{for MMSE prediction} \end{cases} \tag{10.30}$$

3. Compute permission probability for data such that

$$P\left[\frac{\widehat{V}(n + 1)}{a_{\text{V}}} + \frac{D(n + 1)}{a_{\text{D}}} < 1\right] = \delta \tag{10.31}$$

where $D(n + 1)$ is the number of data users who transmit in the $(n + 1)$th slot and $(1 - \delta)$ is the outage probability requirement. The minimum mean square error (MMSE) predicted value for the number of active voice users can be represented as [3]

$$\widehat{V}_{\text{MMSE}}(n + 1) = E[V(n + 1)|V(n), V(n - 1), V(n - 2), \ldots,]$$

$$= E[V(n + 1)|V(n)]$$

$$= \sum_{j=-1}^{1} [V(n) + j]P[V(n + 1) = V(n) + j|V(n)] \qquad (10.32)$$

By using equations (10.20 to 10.22) we have

$$\widehat{V}_{\text{MMSE}}(n + 1) = V(n) + \frac{\tilde{\lambda}_{\text{V}}(n) - \tilde{\mu}_{\text{V}}(n)}{\tilde{\lambda}_{\text{V}}(n) + \tilde{\mu}_{\text{V}}(n)}\{1 - \exp[-(\tilde{\lambda} + \tilde{\mu})d]\}$$

where $\qquad \tilde{\lambda}_{\text{V}}(n) = \lambda[K_{\text{V}} - V(n)] \quad$ and $\quad \tilde{\mu}_{\text{V}}(n) = \mu V(n) \qquad (10.33)$

Permission probability for the $(n + 1)$th slot P is implicitly defined by

$$P\widehat{V}(n + 1) = \max_{0<p\leq1} \arg\left\{ P\left[\frac{\widehat{V}(n + 1)}{a_{\text{V}}} + \frac{D(n + 1)}{a_{\text{D}}} < 1 \right] = \delta \right\}$$

$$= \max_{0<p\leq1} \arg\left[\sum_{j=0}^{\{a_{\text{D}}[1-(\widehat{V}(n+1)/a_{\text{V}})]\}} \binom{K_{\text{D}}}{j} p^{j}(1 - p)^{K_{\text{D}}-j} = \delta \right] \qquad (10.34)$$

In other words, the data users should transmit with maximum probability p for which the above condition is satisfied.

If the upper limit is $\lfloor a_{\text{D}}(1 - [\widehat{V}(n + 1)/a_{\text{V}}])\rfloor \geq K_{\text{D}}$, then $P_{\widehat{V}(n+1)} = 1$. If $\lfloor a_{\text{D}}(1 - [\widehat{V}(n + 1)/a_{\text{V}}])\rfloor < 0$, then $P_{\widehat{V}(n+1)} = 0$. The actual performance of the control depends on $V(n + 1)$.

A simple access control scheme

The access control methods based on prediction are useful as benchmarks and upper bounds on performance. However, they are not very useful in practice. In the sequel we present a simple, real-time access control scheme originally by Viterbi [4].

If in the nth slot the persistence state parameter is $j(n)$, each data user, independent of other users, transmits with probability $\pi^{j(n)} = p_{\text{t}}(n)$ and refrains from transmitting probability $1 - \pi^{j(n)}$. The parameter $\pi(0 < \pi < 1)$ is fixed and known to the data users.

The persistence parameter is broadcast to all users by the BS. The persistence parameter in the $(n + 1)$th slot is assigned as follows:

$$j(n + 1) = \begin{cases} j(n) + K, & \text{if } S(n) \geq \Omega \\ j(n) - 1, & \text{if } S(n) < \Omega \end{cases} \tag{10.35}$$

where $\Omega \leq 1$ is the threshold used to trigger access control.

10.2.2 Access control under imperfect power control

The relevant outage condition can be written in terms of the received SIR as

$$Z = \frac{1}{G_V} \sum_{i=1}^{K_V} v_i \varepsilon_i^V + \frac{1}{G_D} \sum_{i=1}^{K_D} \zeta_i \varepsilon_i^D \leq (1 - \eta) \tag{10.36}$$

$v(\cdot)$ and $\zeta(\cdot)$ are $\{0, 1\}$ activity indicators for voice and data, G_V and G_D are the processing gains, ε^V and ε^D are the SIRs for voice and data, respectively, assumed to be independent and lognormal. $1/\eta$ is the maximum tolerable received power to background noise density. In a multicell system, the measured received power at the BS will include interference from other cells. Access control will respond to changes in other cell interference, as it should. Access control schemes limit the probability of outage by reducing the permission probability for data when the load Z is 'high' and increasing the permission probability when the load is 'low'. Persistence parameter access control is now triggered by $Z(n)$ rather than $S(n)$. For illustration purposes the following parameters are used [3].

Parameter	Symbol	Value
Bandwidth	W	1.25 MHz
Voice and data rates	R_V, R_D	9.6 kbps
Perf. PC-SIR voice, data	γ_V, γ_D	7 dB
Imp. PC-mean SIR voice, data	m_V, m_D	7 dB
Std. Dav. SIR voice, data	σ_V, σ_D	2.5 dB
Permission parameter	π	0.95
Persistence step	K	Variable
Slot duration	d	0.02 s
Voice mean ON time	$1/\mu$	1.0 s
Voice mean OFF time	$1/\lambda$	1.5 s
Outage threshold for imperfect PC	$(1 - \eta)$	0.9

The system performance is shown in Figures 10.5 to 10.7. One can see from Figure 10.5 that the system throughput (goodput) can be increased by a proper choice of K.

Probability of outage will be reduced by a larger K as shown in Figure 10.6 but the larger K will also increase the access delay as shown in Figure 10.7.

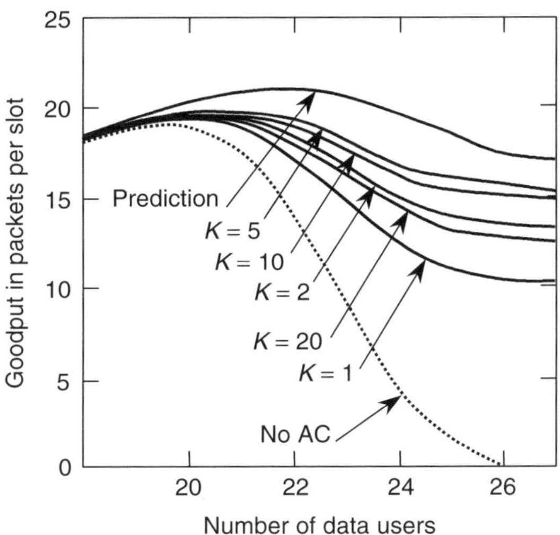

Figure 10.5 Goodput for data users for different control schemes under perfect power control. Parameters are $K_V = 10$, $\pi = 0.95$ and $\Omega = 1.0$.

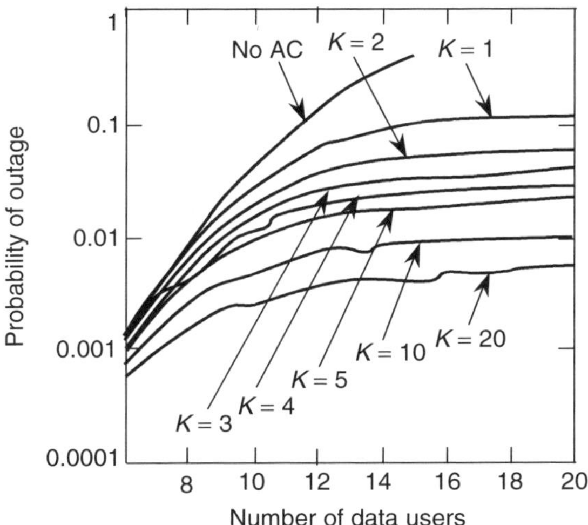

Figure 10.6 Outage versus data load for $K_V = 10$ and $K = 1 - 5$, and 20 under imperfect power control. The outage threshold $\omega = 1 - P_{out}$ was 0.9, and the access control threshold Ω was 0.7.

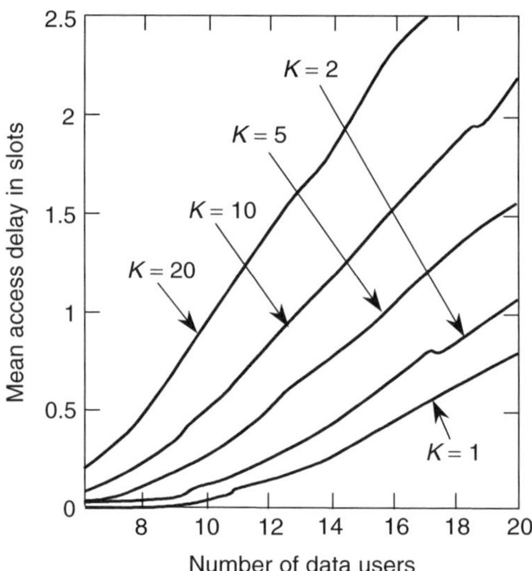

Figure 10.7 Mean access delay versus data load for $K_V = 10$ and $K = 1, 2, 5, 10$ and 20. The outage threshold $\omega = 1 - P_{\text{out}}$ was 0.9 and the access control threshold Ω was 0.7.

10.3 DELTA MODULATION–BASED PREDICTION FOR ACCESS CONTROL IN INTEGRATED VOICE/DATA CDMA SYSTEMS

The prediction scheme used in the previous section for estimating residual capacity had only theoretical value for setting up the upper limit on system performance. For practical application a modified delta modulation (MDM) for estimating the residual capacity can be used. Two different access protocols, MDM with scheduled access (MDM-S) and MDM with random access (MDM-R), will be discussed. Results are compared to those of the persistence state–based access control, shown in the previous section and it is shown that both MDM-S and MDM-R perform better. The approach is very much based on Reference [5]. By choosing δ to be a very small value (e.g. $\delta = 0.001$), the condition (10.18 or 10.24) is guaranteed to be true if

$$S(n) = \frac{v(n)}{a_v} + \frac{d(n)}{a_d} \le 1 - \delta \tag{10.37}$$

which gives

$$d(n) \le a_d(1 - \delta) - \frac{a_d}{a_v} v(n) \tag{10.38}$$

$d(n)$ is the ideal residual capacity of the system, (remaining capacity) after the voice contribution is subtracted. Violation of equation (10.38) constitutes an outage.

Since $v(n)$ is a random variable, only an estimate $\overline{d(n)}$ as a function of $v(n-1)$ can be derived. The access control protocol attempts to schedule exactly $\lfloor \overline{d(n)} \rfloor$ users at each time slot. To maintain QoS, the fraction of the time the outage condition is violated should be very small, typically 1%. The outage may be caused by (1) imperfections in estimating the residual capacity; (2) imperfections in scheduling the desired number of data users and (3) imperfections in power control.

Several modifications for DM are necessary in order to guarantee $\overline{d(n)} \leq d(n)$. First of all a guard margin equal to the step of modulation (Δ) is used, such that the function to be

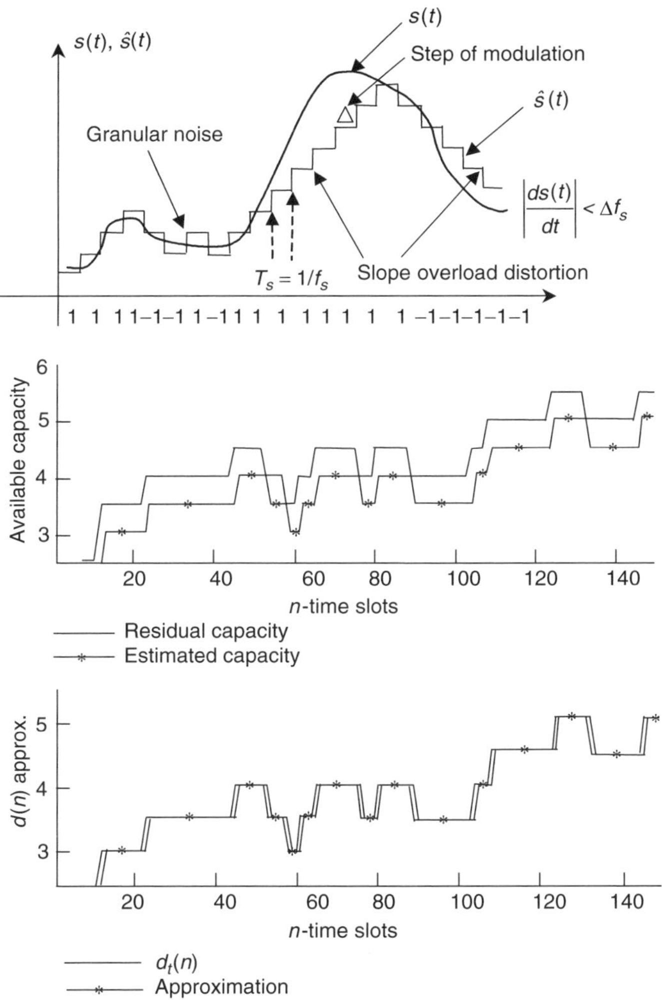

Figure 10.8 Delta modulation: staircase approximation of an analog signal and algorithm for imperfect power control (simulation).

approximated becomes $d_t(n) = d(n) - \Delta$, where $d(n)$ is computed as the maximum value for which equation (10.38) holds. DM and its approximation are illustrated in Figure 10.8. Then the steps of the DM algorithm are modified as follows: (1) For the first time slot, no access procedure is used: at the end of the first time slot, the number of active voice users is measured $[v(1)]$ and the value $d(1)$ is computed by using equation (10.38) and definition of $d_t(n)$. So, $\overline{d(2)}$ is initialized as $\overline{d(2)} = d(1) - \Delta$. So, $\overline{d(2)}$ users are allowed to transmit in the second time slot. (2) At the end of each time slot n, $n > 1$ the following steps are taken. The number of active voice users in the current time slot $v(n)$ is measured and $d_t(n) = d(n) - \Delta$ is computed.

$$\text{If } \overline{d(n)} > d(n) - \Delta, d(\overline{n+1}) = \overline{d(n)} - \Delta$$

$$\text{If } \overline{d(n)} < d(n) - \Delta, d(\overline{n+1}) = \overline{d(n)} + \Delta$$

$$\text{If } \overline{d(n)} = d(n) - \Delta, d(\overline{n+1}) = \overline{d(n)} \tag{10.39}$$

An access procedure is used to allow $\lfloor d(\overline{n+1}) \rfloor$ users to transmit in the next time slot.

10.3.1 MDM-S protocol

For MDM-S, the BS will maintain a round fair list with all data users admitted in the system and will signal users to transmit or not using a 1-bit access flag: one if the user is allowed to transmit in the next time slot, zero if not. A drawback for this procedure is that the access bit cannot be broadcast; a possible solution will be for the users to listen to a dedicated fraction of a time slot to extract the proper access bit. The algorithm guarantees $\overline{d(n)} \leq d(n)$; thus the access control protocol will never schedule more users than the available residual capacity $d(n)$. In the case of perfect power control, the access procedure causes no outage, that is, MDM-S gives zero outage probability.

10.3.2 MDM-R protocol

At each time slot n an access probability $p(n)$ to all data users will be

$$p(n) = \frac{\left\lfloor \overline{d(n)} \right\rfloor}{K_d} \tag{10.40}$$

The performance can be fine-tuned by introducing a probability tolerance (t_p) parameter.

$$p(n) = \frac{\left\lfloor \overline{d(n)} \right\rfloor - t_p}{K_d} \tag{10.41}$$

A positive value of t_p has the effect of decreasing the access probability, which results in a smaller probability of outage. For an imposed outage value, decreasing $p(n)$ gives larger capacity for the system. The penalty is an increase in the data access delay. For illustration purposes the following set of simulation parameters is used [5]. $W = 1.23\,\text{MHz}$, $G_v = 128$, $G_d = 64$, $R_v = 9.6\,\text{kbps}$, $R_d = 19.2\,\text{kbps}$, $1/\lambda = 1.0\,\text{s}$, $1/\mu = 1.5\,\text{s}$, $d = 0.02\,\text{s}$, $\gamma_v = \gamma_d = 7\,\text{dB}$ and $K_v = 10$. The results are compared with the protocol defined by equation (10.35) called *persistent state algorithm* (PSA) and shown in Figures 10.9 to 10.12. From Figure 10.9 one can see that the outage probability for PSA is slightly better than for MDM protocols. On the other hand, delay and goodput characteristics are better for MDM protocol as shown in Figures 10.10 and 10.11. The gain in goodput is shown in Figure 10.12. As much as 40% better goodput can be achieved by using MDM protocols.

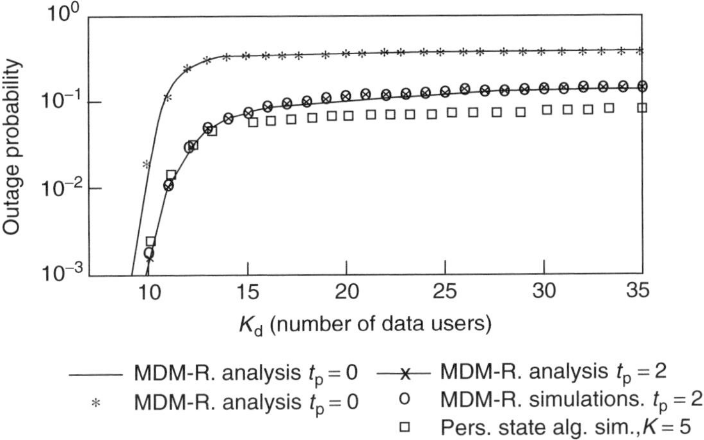

Figure 10.9 Outage probability versus the number of data users.

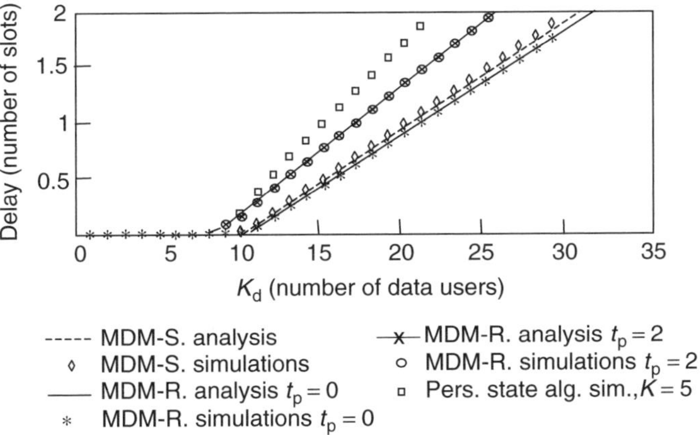

Figure 10.10 Delay versus number of data users.

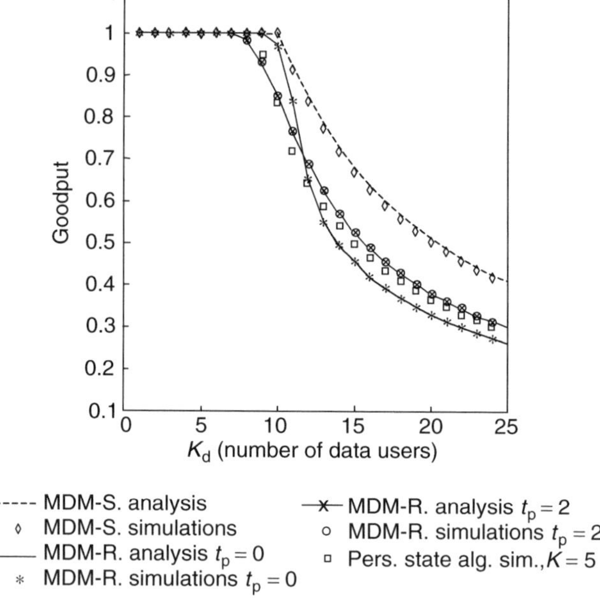

Figure 10.11   Goodput versus number of data users.

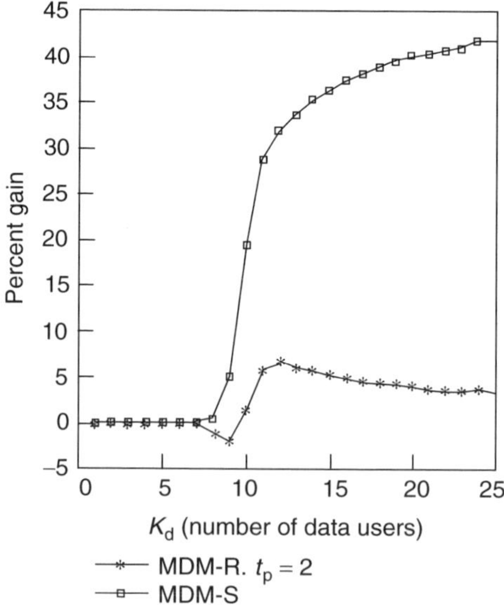

Figure 10.12   Goodput gain percent versus number of users.

10.4 MIXED VOICE/DATA TRANSMISSION USING PRMA PROTOCOL

The presentation in this section is based on Reference [6]. We assume that terminals can send three types of information: 'periodic', 'random' and 'continuous'. Speech packets are always periodic, data packets can be random (isolated packets) or periodic and video packets are continuous as video terminals transmit data in (almost) every slot (assuming a constant bit rate source as a very simple model).

Each downlink (base to mobile station) packet is preceded by feedback based on the result of the most recent uplink transmission. If the base is able to decode the header of one or more arriving packet(s), the feedback identifies the packet sending terminal(s), indicates which of the corresponding packets were received successfully and transmits the permission probabilities for periodic and random information. This information is valid in the corresponding slot in the next frame.

1. *Frames and slots*: The transmission timescale is organized in *frames*, each containing a fixed number of time slots. The frame rate is identical to the arrival rate of voice packets. All transmitters transmit their packets such that they arrive at the BS within the slot boundaries. In contrast to conventional packet reservation multiple access (PRMA), terminals do not classify slots as either 'reserved' or 'available', as the channel access for contending terminals is governed by time-varying permission probabilities.
2. *Reservation mode*: A terminal that generates periodic data switches from contention to reservation mode as soon as a successful packet reception is acknowledged by the BS. It will stay in reservation mode until the last packet of the current spurt is transmitted. The BS counts all the packets sent from periodic terminals in each slot. This can be achieved as long as the headers are detected correctly so that it can compute the permission probability for the same slot in the next frame with the Channel Access Function (CAF), and then transmits this probability in the feedback.
3. *Collisions*: If packets originating from data or video terminals are *corrupted* because of excessive multiple access interference (MAI), they have to be retransmitted. Corrupted voice packets do not have to be retransmitted. They contribute, together with the dropped voice packets, to the total number of lost voice packets.
4. *Contention and packet dropping*: In order to transmit a packet, terminals in contention mode have to perform a Bernoulli experiment with the current permission probability (either for voice p_s or data p_d) as the parameter. They are allowed to transmit a packet if the outcome of the Bernoulli experiment is positive. The terminals attempt to transmit the initial packet of a spurt until the BS acknowledges successful reception of the packet or until the packet is discarded by the terminal because it has been delayed too long. *The maximum packet holding time* of speech packets, D_{max} s, is determined by delay constraints on speech communication.

If a terminal drops the first packet of a spurt, it continues to contend for a reservation to send subsequent packets. It drops additional packets as their holding times exceed D_{max} s. Terminals transmitting *periodic data* packets store packets indefinitely while they contend for reservations ($D_{\max}s = \infty$). *Random* information packets are always sent in contention mode. When a joint CDMA/PRMA (JCP) system becomes congested, the speech packet dropping rate and the data packet delay both increase.

5. *Access for terminals with continuous data*: Terminals with continuous data that do obtain permission to start transmission are allowed to transmit one packet in every slot of every frame; thus they are in a permanent reservation mode. Whether they obtain permission to start transmission depends on the current load of the network and is not to be decided by the MAC layer.

10.4.1 Effects of network congestion

In general, as traffic increases, access to the channel must be restricted in order to avoid excessive packet loss due to MAI, and terminals will encounter delays in gaining access to the channel. Whereas data sources absorb these delays as performance penalties, speech terminals must discard delayed packets since conversations require prompt information delivery. This packet loss occurs at the beginning of talkspurts and is referred to as *front-end clipping*, which impairs the quality of received speech. The amount of front-end clipping is measured by the *packet dropping probability* P_{drop}. Efficient channel access control will have to find a trade-off between P_{drop} and the probability of packet corruption due to MAI, P_{cpted}. As P_{drop} increases, P_{cpted} might increase as well and cause additional speech quality impairment. Assuming that the quality impairments due to P_{drop} and P_{cpted} are perceived in a similar way, then only the sum of these two probabilities, the probability of packet loss P_{loss}, needs to be considered. A key measure of JCP is the number of voice terminals that can share a channel within a given maximum value of P_{loss}.

10.4.2 The channel access function

The permission probabilities for speech, p_{s}, and data, p_{d}, for a given slot in a subsequent frame are set according to the number of periodic users in the current frame. The number of users is related to the permission probability by the CAF shown in Figure 10.13. The purpose of this function is to control the total number of users K in every slot, such that the throughput is maximized without exceeding the p_{loss} limit. The optimal number of simultaneous users K per slot for a system with constant channel load would be

$$K_{\mathrm{opt}} = \max_{K=1,2,\ldots} (K \,|\, 1 - Q_E[K] \le (P_{\mathrm{loss}})\mathrm{req}) \tag{10.42}$$

$Q_E[K]$ is the resulting packet success probability when K simultaneous users are on the channel.

Efficient CAF should enforce a channel load such that most of the slots are loaded with K_{opt} packets. Therefore, the permission probability should be low if a large number of users in reservation mode are already on the channel and zero when K_{opt} in

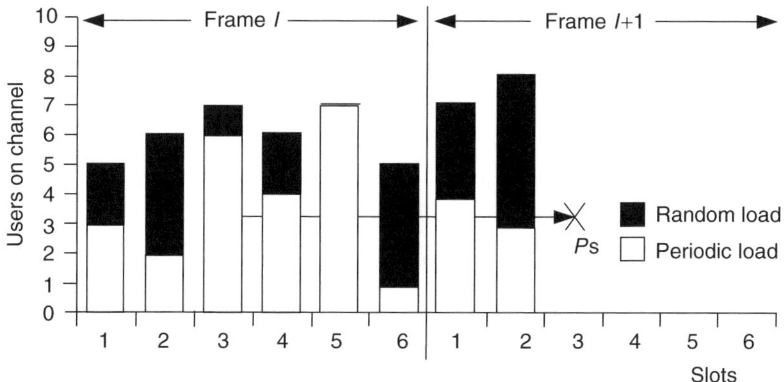

Figure 10.13 The permission probability of slot 3 in frame $I + 1$ is set according to the periodic load in the same slot of the previous frame I.

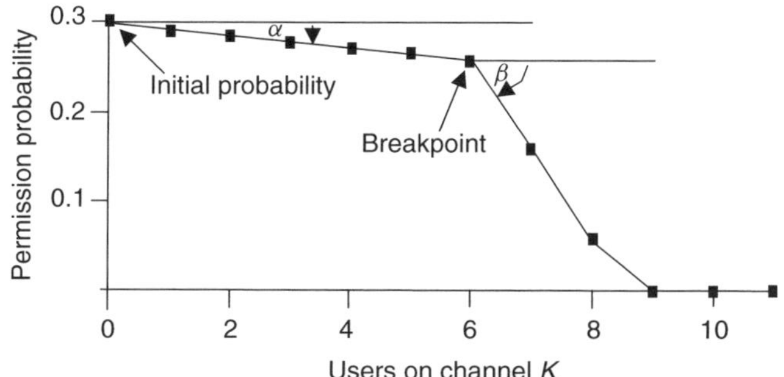

Figure 10.14 An example channel access function [6]. Reproduced from Brand, A. E. and Aghvami, A. H. (1996) Performance of a joint CDMA/PRMA protocol for mixed voice/data transmission for third generation mobile communication. *IEEE J. Select. Areas Commun.*, **14**, 1698–1707, by permission of IEEE, **14**, 1698–1707, by permission of IEEE.

equation (10.42) is exceeded. A heuristic approach function with two linear segments and the following parameters (see also Figure 10.14):

(1) the initial probability p_{si} or p_{di}, (2) the slopes α and β of the two linear segments (probability decrease/additional user) and (3) the position of the breakpoint (in number of users), is used for these purposes. The average signal-to-noise ratio is calculated as [7,8]

$$\overline{SNR} = \sqrt{\frac{P_i}{\left(3N^{-1} \sum_{\substack{k=1 \\ k \neq i}}^{K} P_k + \frac{N_0}{2T}\right)}} \qquad (10.43)$$

for single cell and

$$\overline{SNR} = \sqrt{\frac{3P_0 N}{\underbrace{(K-1)P_0}_{\text{Intracell}} + \underbrace{\sum_{k=1}^{K}\sum_{i=1}^{R} P_{(k,i)_0}}_{\text{Intracell}}}} \qquad (10.44)$$

for a cellular network with intercell interference. In these equations

$$N \rightarrow S_{\text{F}}(\text{spreading factor}) \qquad (10.45)$$

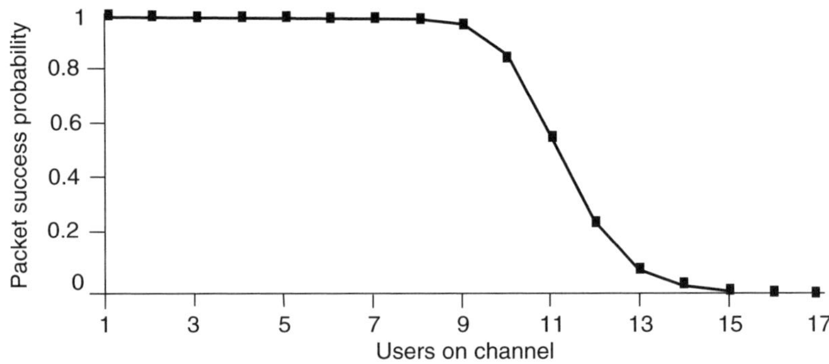

Figure 10.15 Packet success probability for $S_{\text{F}} = 7$, $L = 511$ b and $t = 38$.

Table 10.1 Design parameters of the joint CDMA/PRMA (JCP) protocol.

Definition	Notation	Units	Values
CDMA channel rate	R_{c}	Ships/s^{-1}	3577000
PRMA channel rate after coding	R_{cc}	b/s^{-1}	511000
PRMA channel rate before coding	R_{p}	b/s^{-1}	229000
BCH code	(L, M, t)		511, 229, 38
Source rate (voice terminal)	R_{s}	b/s^{-1}	8000
Source rate (random data terminal)	R_{d}	b/s^{-1}	Variable
Frame duration	T_{f}	s	0.02
Information bits per packet	$R_{\text{s}}T_{\text{f}}$	b	160
Overhead	H	b	69
Slots per frame	N		20
Maximum delay	D_{max}	s	0.02
Speech permission probability	p_{s}		Variable
Data permission probability	p_{d}		Variable
Conversations	M		Variable

For an isolated cell, with equal power reception, a spreading factor $S_F = 7$ and packets of length $L = 511$ b, where a code is employed that can correct up to $t = 38$ errors; $Q_E[K]$ is depicted in Figure 10.15.

For a system specified in Table 10.1 and Figure 10.16(a) [6] and for single cell the results are plotted in Figure 10.16(b). $M_{0.02}$ in Table 10.2 for instance is the maximum number of simultaneous conversations supported with $P_{loss} \leq 0.02$. In Figure 10.16 the following notation is used:

1. *Perfect scheduling*: With the definition of K_{opt} in equation (10.42) and with respect to the shape of $Q_E[K]$, it is apparent that to achieve maximum throughput within a given P_{loss} limit, slots must be loaded with either K_{opt} or $K_{opt} + 1$ packets.

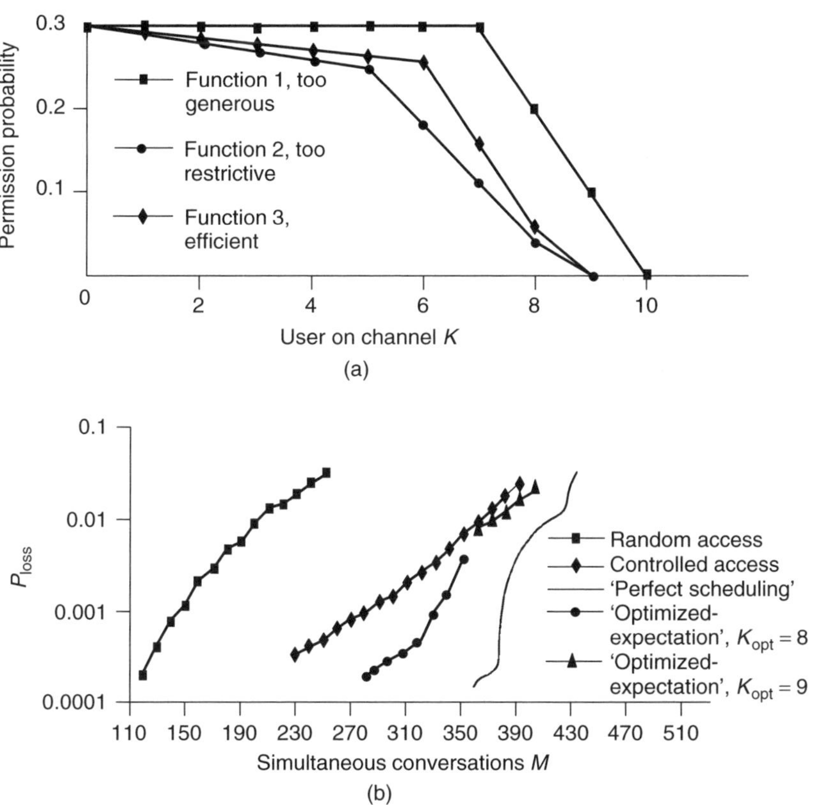

Figure 10.16 (a) Generous, restrictive and efficient channel access functions for voice-only traffic in an isolated cell [6]. Reproduced from Brand, A. E. and Aghvami, A. H. (1996) Performance of a joint CDMA/PRMA protocol for mixed voice/data transmission for third generation mobile communication. *IEEE J. Select. Areas Commun.*, **14**, 1698–1707, by permission of IEEE. (b) Simultaneous conversations M versus average P_{loss} for random access and controlled access in an isolated cell, (voice only).

Table 10.2 Simulation results for voice-only traffic in different environments [6]. Reproduced from Brand, A. E. and Aghvami, A. H. (1996) Performance of a joint CDMA/PRMA protocol for mixed voice/data transmission for third generation mobile communication. *IEEE J. Select. Areas Commun.*, **14**, 1698–1707, by permission of IEEE

Environment	Single cell			Cellular $n = 4$			Cellular $n = 3$		
Protocol	RA	JCP	Gain	RA	JCP	Gain	RA	JCP	Gain
$M_{0.02}$	227	379	67%	180	280	55%	151	220	45%
$M_{0.01}$	201	358	78%	162	266	64%	138	209	51%
$M_{0.001}$	145	277	91%	116	214	84%	100	168	68%

2. *Optimized a posteriori expectation access scheme*: If the number of contending terminals in a particular slot K_{cont} and the number of terminals that will use that slot in reservation mode K_{res} were known, one could choose the permission probability in that slot as

$$p_{\text{s}} = \frac{K_{\text{opt}} - K_{\text{res}}}{K_{\text{cont}}} \tag{10.46}$$

such that $E[K] = K_{\text{opt}}$ in every slot. Although an optimum access scheme would also have to minimize $\text{Var}[K]$, this scheme can be employed as a good benchmark for efficient access control.

Similar results for multiple cell and single cell are compared in Table 10.2. The propagation exponent $n = 4$ and 3.

The results for a mixture of voice/video traffic are shown in Figure 10.17.

For mixed voice, random data traffic results are shown in Figure 10.18 with

$$p_{\text{d}} = c_{\text{p}} \cdot p_{\text{s}} \tag{10.47}$$

Packet success probability in cellular network will depend on propagation exponent n as shown in Figure 10.19. For these reasons, CAF parameter should be modified as shown in Table 10.3 and Figure 10.20. Parameter P_{loss} is now shown in Figure 10.21.

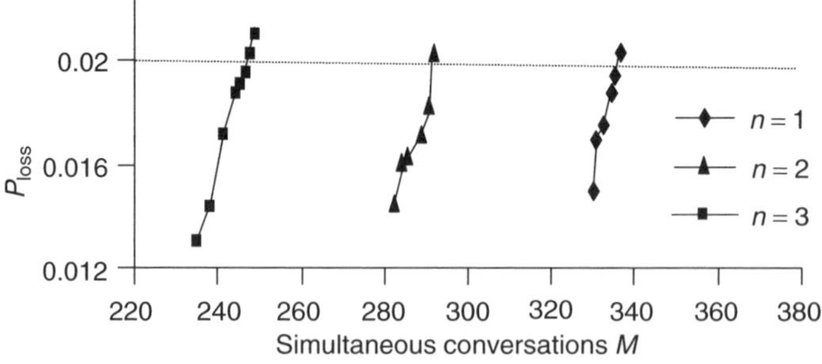

Figure 10.17 Simulation results for mixed voice/video traffic with one, two and three video terminals.

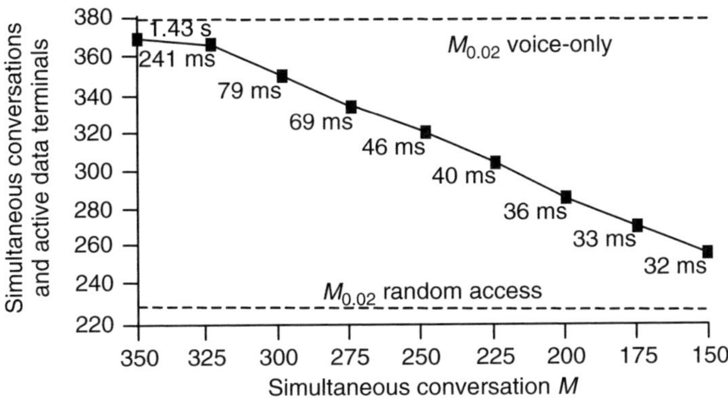

Figure 10.18 Sum of simultaneous conversations and active data terminals versus simultaneous conversations M (with indication of average data packet delays) and $C_p = 0.2$ [6]. Reproduced from Brand, A. E. and Aghvami, A. H. (1996) Performance of a joint CDMA/PRMA protocol for mixed voice/data transmission for third generation mobile communication. *IEEE J. Select. Areas Commun.*, **14**, 1698–1707, by permission of IEEE.

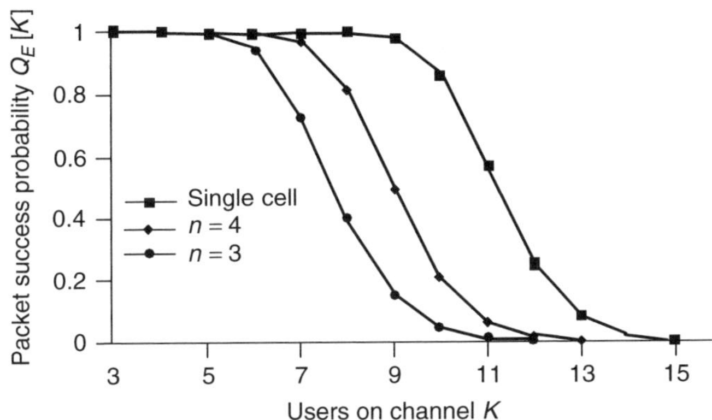

Figure 10.19 Packet success probabilities for a single cell and for a cellular environment, in which $M_{0.02}$ conversations take place in every cell simultaneously.

Table 10.3 Channel access function parameters for different environments

Environment	Single cell	Cellular $n = 4$	Cellular $n = 3$
p_{si}	0.3	0.3	0.3
α	0.007	0.008	0.009
Breakpoint	6	4	3
β	0.1	0.1	0.12

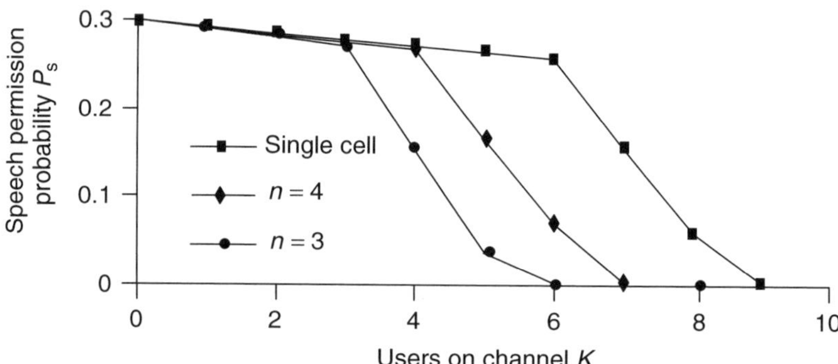

Figure 10.20   Channel access functions employed in a cellular environment.

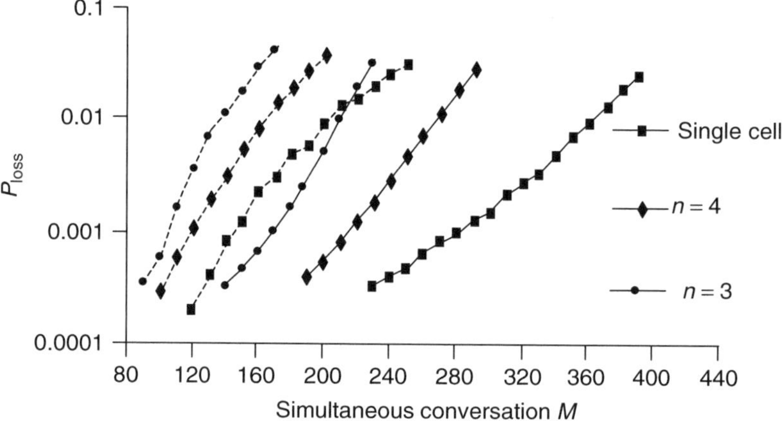

Figure 10.21   Average P_{loss} versus simultaneous conversations M for random access (dashed curves) and controlled access in a cellular environment.

10.5 FUZZY/NEURAL CONGESTION CONTROL

In this section we continue to focus our analysis on the access control in more realistic environment when a cell operates within a cellular network so that the impact of interference from other cells is also present. A frame reservation multiple access (FRMA) is used together with fuzzy logic for interference prediction, access control and performance indication. The general relation between these segments of the system is indicated in Figure 10.22. The analysis in this section is based on Reference [9].

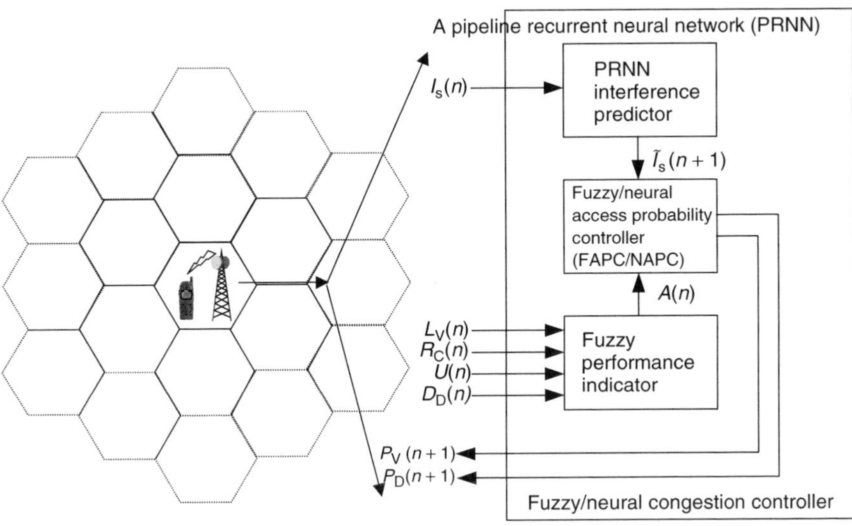

Figure 10.22　A DS-CDMA/FRMA cellular system with the fuzzy/neural congestion controller.

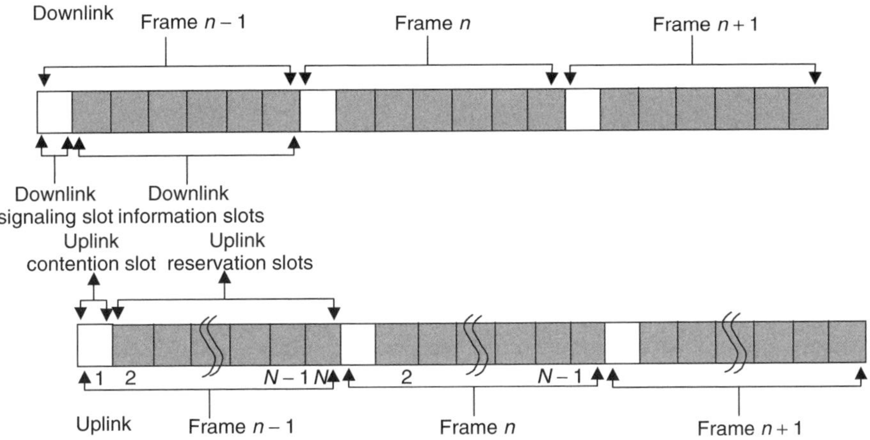

Figure 10.23　Frame structure.

10.5.1 System model

In both uplink and downlink, the DS-CDMA/FRMA protocol has a time-division frame structure, which consists of N slots per frame time T as shown in Figure 10.23. The operation procedures are the same as in the previous section. Each slot has several Code Division Multiple Access (CDMA) code channels for users to transmit their packets.

If a contention user wants to transmit information packets, it first transmits a contention information packet at the contention slot according to its access probability. The voice and data access probabilities for $(n + 1)$th frame are denoted by $P_V(n + 1)$ and $P_D(n + 1)$, respectively.

The radio propagation model here contains two main loss factors: mean path loss and lognormal shadowing, as discussed in Chapter 8. The whole system is assumed to be under perfect power control so that the slow fading can be equalized and thus the received power at the BS has a constant value S. The interference power of user j at any time instant n in BS k is composed of the home cell interference, the first tier adjacent cell interference and the background noise [additive white Gaussian noise (AWGN)], denoted by $I_{H,k}(n)$, $I_{A,k}(n)$ and Ψ, respectively. Home cell interference and the adjacent cell interference are much larger than the background noise, thus we ignore it. The interference power in a basic channel at time instant n, denoted by $I_S(n)$ is the summation of $I_{H,k}(n)$ and $I_{A,k}(n)$. $I_S(n)$ is periodically measured every frame time nT at BS and is chosen as an input variable for the pipeline recurrent neural network (PRNN) interference predictor. Voice source model is characterized as a two-state (talkspurt and silence) Markov chain and will generate one packet in each frame time T. The talkspurt and silence periods are assumed to be exponentially distributed with mean $1/\mu$ and $1/\lambda$, respectively (see Figure 10.3). Data source model is assumed to be a Poisson process with mean arrival rate λ_d. Voice (data) packets will be put into voice (data) queue with capacity $B_V(B_D)$ before being transmitted. If the queue is full or if the packet cannot be successfully received at the base, the packet is considered as dropped.

10.5.2 Fuzzy/neural congestion controller

The building blocks for the fuzzy/neural congestion controller are the PRNN interference predictor, the fuzzy performance indicator and the fuzzy/neural access probability controller as shown in Figure 10.22.

PRNN interference predictor

PRNN is a pipeline structure of recurrent neural network (RNN). It has good prediction capability and fast converges speed, with real-time recurrent learning (RTRL) algorithm [10]. In the PRNN interference predictor, the predicted interference sample at frame $(n + 1)$, $\tilde{I}_S(n + 1)$, can be obtained from p previously measured interference samples $I_S(i)$, $n - p + 1 \leq i \leq n$ and q prediction errors $\tilde{e}(j)$, $n - q + 1 \leq j \leq n$, based on a nonlinear ARMA (NARMA) model of the process.

$$\tilde{I}_S(n + 1) = h[I_S(n), \ldots, I_S(n - p + 1); \tilde{e}(n), \ldots, \tilde{e}(n - q + 1)] \qquad (10.48)$$

where $h(\cdot)$ is an unknown nonlinear function and $\tilde{e}(j) = I_S(j) - \tilde{I}_S(j)$. To approximate the nonlinear function $h(\cdot)$ by RNN with RTRL algorithm, inputs of RNN cannot be error samples [10]. For this reason the above recursive formula is reformulated by using a new function H

$$\tilde{I}_S(n + 1) = H[I_S(n), \ldots, I_S(n - p + 1); \tilde{I}_S(n), \ldots, \tilde{I}_S(n - q + 1)] \qquad (10.49)$$

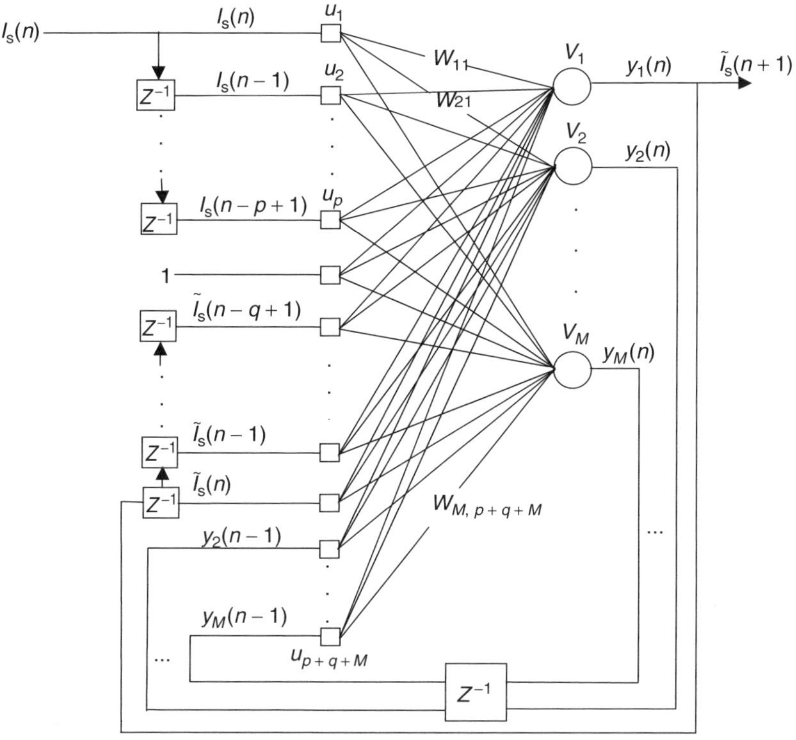

Figure 10.24 The RNN structure.

A fully connected RNN structure has M neurons and $p + q + M$ input nodes as shown in Figure 10.24. The first p input nodes are the external inputs that are the measured interference signals from $I_S(n)$ to $I_S(n-p+1)$. There is a bias input value, which is always 1. The next q input nodes are the predicted signals from $\tilde{I}_S(n)$ to $\tilde{I}_S(n-q+1)$. Finally, $M-1$ feedbacks from neuron outputs, $y_2(n-1) \sim y_M(n-1)$, are also used. In the figure, w_{ji} are weights of the connection from the ith input node to the jth neuron for $1 \le i \le p+q+M$, $1 \le j \le M$.

The jth neuron calculates a weighted sum, denoted by $v_j(n)$ as

$$v_j(n) = \sum_{i=1}^{p+q+M} w_{ji}(n)u_i(n) \tag{10.50}$$

where $u_i(n)$ is the ith input node. Then, it transforms $v_j(n)$ by a sigmoidal activation function $\varphi(\cdot)$ to an output $y_j(n)$

$$y_j(n) = \varphi[v_j(n)] = \frac{1}{1 + \exp[-v_j(n)]} \tag{10.51}$$

After that $\tilde{I}_S(n + 1)$ can be obtained as

$$
\tilde{I}_S(n + 1) = y_1(n) = \varphi \left(\sum_{i=1}^{p} w_{1i}(n) I_S(n + 1 - i) + w_{1,p+1}(n) \right.
$$

$$
\left. + \sum_{i=p+2}^{p+q+1} w_{1i}(n) \tilde{I}_S(n - i + p + 2) + \sum_{i=p+q+2}^{p+q+M} w_{1i}(n) y_{i-p-q}(n - 1) \right)
$$

$$
= \hat{H} \left(I_S(n), \ldots, I_S(n - p + 1), \tilde{I}_S(n), \ldots, \tilde{I}_S(n - q + 1) \right) \qquad (10.52)
$$

where $\hat{H}(\cdot)$ is a nonlinear approximated function of $H(\cdot)$. The incremental change of weight w_{ij} is according to the steepest descent method in the RTRL algorithm (see Chapter 7).

$$
w_{ij}(n + 1) = w_{ij}(n) - \eta \frac{\partial C(n)}{\partial w_{ij}} \qquad (10.53)
$$

where η is the learning rate parameter. $C(n)$ is the cost function defined as

$$
C(n) = \sum_{i=1}^{q} \lambda_n^{i-1} \hat{e}^2(n - i + 1) \qquad (10.54)
$$

where λ_n is the exponential forgetting factor that is bounded in $[0, 1]$.

Fuzzy performance indicator

The performance indicator should include simultaneously the voice packet dropping ratio L_V, the contention corruption ratio R_C, the system utilization U and the data packet D_D. Neither of them can represent the system performance alone without the consideration of others. Fuzzy logic is used to get an overall system performance indication A, on the basis of the four performance measures mentioned above as input linguistic variables. Congestion controller has a concluding performance indication feedback so that it is a closed-loop system and has stable and robust operations. Similarly to discuss on fuzzy logic power control introduced in Chapter 6, we define the term set of L_V as $T(L_V) = \{\text{Low, High}\} = \{\text{Lo, Hi}\}$, R_C as $T(R_C) = \{\text{Little, Big}\} = \{\text{Lt, Bg}\}$, U as $T(U) = \{\text{Small, Large}\} = \{\text{Sm, La}\}$ and D_D as $T(D_D) = \{\text{Short, Long}\} = \{\text{Sh, Lg}\}$. The membership functions (set of values) for $T(L_V), T(R_C), T(U)$ and $T(D_D)$ are defined as $M(L_V) = \{\mu_{\text{Lo}}, \mu_{\text{Hi}}\}$, $M(R_C) = \{\mu_{\text{Lt}}, \mu_{\text{Bg}}\}$, $M(U) = \{\mu_{\text{Sm}}, \mu_{\text{La}}\}$ and $M(D_D) = \{\mu_{\text{Sh}}, \mu_{\text{Lg}}\}$ where

$$
\mu_{\text{Lo}}(L_V) = q(L_V; L_{V,\min}, \text{Lo}_e, 0, \text{Lo}_w)
$$

$$
\mu_{\text{Hi}}(L_V) = q(L_V; \text{Hi}_e, L_{V,\max}, \text{Hi}_w, 0)
$$

$$\mu_{\mathrm{Lt}}(R_{\mathrm{C}}) = q(R_{\mathrm{C}}; R_{\mathrm{C,min}}, \mathrm{Lt}_e, 0, \mathrm{Lt}_w)$$

$$\mu_{\mathrm{Bg}}(R_{\mathrm{C}}) = q(R_{\mathrm{C}}; \mathrm{Bg}_e, R_{\mathrm{C,max}}, \mathrm{Bg}_w, 0)$$

$$\mu_{\mathrm{Sm}}(U) = q(U; U_{\mathrm{min}}, \mathrm{Sm}_e, 0, \mathrm{Sm}_w)$$

$$\mu_{\mathrm{La}}(U) = q(U; \mathrm{La}_e, U_{\mathrm{max}}, \mathrm{La}_w, 0)$$

$$\mu_{\mathrm{Sh}}(D_{\mathrm{D}}) = q(D_{\mathrm{D}}; D_{\mathrm{D,min}}, \mathrm{Sh}_e, 0, \mathrm{Sh}_w)$$

$$\mu_{\mathrm{Lg}}(D_{\mathrm{D}}) = q(D_{\mathrm{D}}; \mathrm{Lg}_e, D_{\mathrm{D,max}}, \mathrm{Lg}_w, 0) \qquad (10.55)$$

and $L_{\mathrm{V,min}}$, $L_{\mathrm{V,max}}$, $R_{\mathrm{C\,min}}$, $R_{\mathrm{C,max}}$, U_{min}, U_{max}, and $D_{\mathrm{D,min}}$, $D_{\mathrm{D,max}}$ are the minimum and maximum possible values for L_{V}, R_{C}, U and D_{D}, respectively. $q(\cdot)$ is trapezoidal function defined as

$$q(x; x_0, x_1, a_0, a_1) = \begin{cases} \dfrac{x - x_0}{a_0} + 1, & \text{for } x_0 - a_0 < x \leq x_0 \\ 1, & \text{for } x_0 < x \leq x_1 \\ \dfrac{x_0 - x}{a_1} + 1, & \text{for } x_1 < x \leq x_1 + a_1 \\ 0, & \text{otherwise} \end{cases} \qquad (10.56)$$

The output linguistic variable is the performance indicator A. The term set of A is defined as $T(A) = \{A_1, A_2, A_3, A_4, A_5, A_6, A_7, A_8\}$, and the membership function of A is denoted by $M(A) = \{\mu_{A_1}, \mu_{A_2}, \mu_{A_3}, \mu_{A_4}, \mu_{A_5}, \mu_{A_6}, \mu_{A_7}, \mu_{A_8}\}$, where

$$\mu_{A_1}(A) = f(A; A_{1,c}, 0, 0)$$

$$\mu_{A_2}(A) = f(A; A_{2,c}, 0, 0)$$

$$\mu_{A_3}(A) = f(A; A_{3,c}, 0, 0)$$

$$\mu_{A_4}(A) = f(A; A_{4,c}, 0, 0)$$

$$\mu_{A_5}(A) = f(A; A_{5,c}, 0, 0)$$

$$\mu_{A_6}(A) = f(A; A_{6,c}, 0, 0)$$

$$\mu_{A_7}(A) = f(A; A_{7,c}, 0, 0)$$

$$\mu_{A_8}(A) = f(A; A_{8,c}, 0, 0) \qquad (10.57)$$

where $f(\cdot)$ is triangular function defined as

$$f(x; x_0, a_0, a_1) = \begin{cases} \dfrac{x - x_0}{a_0} + 1, & \text{for } x_0 - a_0 < x \leq x_0 \\ \dfrac{x_0 - x}{a_1} + 1, & \text{for } x_0 < x \leq x_0 + a_1 \\ 0, & \text{otherwise} \end{cases} \qquad (10.58)$$

Table 10.4 The rule structure for the fuzzy performance indicator

Rule	L_V	R_C	U	D_D	A	
1	Hi	Lt	Sm	Lg	A_1	
2	Hi	Lt	La	Lg	A_1	
3	Hi	Lt	Sm	Sh	A_1	
4	Hi	Lt	La	Sh	A_2	
5	Lo	Lt	Sm	Lg	A_2	
6	Lo	Lt	La	Lg	A_3	
7	Lo	Lt	Sm	Sh	A_3	
8	Lo	Lt	La	Sh	A_4	Target performance
9	Lo	Bg	Sm	Sh	A_5	
10	Lo	Bg	La	Sh	A_6	
11	Lo	Bg	Sm	Lg	A_6	
12	Lo	Bg	La	Lg	A_7	
13	Hi	Bg	Sm	Sh	A_7	
14	Hi	Bg	La	Sh	A_8	
15	Hi	Bg	Sm	Lg	A_8	
16	Hi	Bg	La	Lg	A_8	

The values for $A_{i,c}$ are heuristically set $A_{i,c} = (0.5 + 0.5 \times i)$, $1 \le i \le 8$ to reflect different degrees of the performance indication. $A_{4,c}$ in the middle represents the best performance. Table 10.4 shows the rule structure. These rules are set according to experience and knowledge that the contention corruption ratio R_C and the voice packet dropping ratio L_V have the dominant impact on overall system performance. The max–min interference method is used to calculate the membership value of each term in $T(A)$. Take rules 4 and 5, which have the same term A_2 for example. In the first step, the max–min interference method applies the min operator on membership values of associated term of all the input linguistic variables for each rule. If we denote the weights of rules 4 and 5 by w_4 and w_5, then

$$w_4 = \min[\mu_{Hi}(L_V), \mu_{Lt}(R_C), \mu_{La}(U), \mu_{Sh}(D_D)]$$
$$w_5 = \min[\mu_{Lo}(L_V), \mu_{Lt}(R_C), \mu_{Sm}(U), \mu_{Lg}(D_D)] \qquad (10.59)$$

The max operator on w_4 and w_5 yields the overall membership value of A_2, denoted by

$$w_{A_2} = \max(w_4, w_5) \qquad (10.60)$$

Fuzzy performance indicator uses the center of area defuzzication method to obtain the performance indicator A by combining w_{A_i}, $1 \le i \le 8$

$$A = \frac{\sum_{i=1}^{8} w_{Ai} \times A_{i,c}}{\sum_{i=1}^{8} w_{Ai}} \tag{10.61}$$

Fuzzy access probability controller (FAPC)

FACP takes the predicted interference sample at frame $(n+1)$, $\tilde{I}_S(n+1)$ and the performance indicator at frame n, $A(n)$ as two input linguistic variables. We define term set of $\tilde{I}_S(n+1) \Rightarrow T(\tilde{I}_S) = \{\text{Low, Medium, High}\} = \{\text{Lo, Me, Hi}\}$ and the term set of $A(n) \Rightarrow T(A) = \{\text{Small, Middle, Large}\} = \{\text{Sm, Md, La}\}$. Membership functions for $\tilde{I}_S(n+1)$ and $A(n) \Rightarrow M(\tilde{I}_S) = \{\mu_{\text{Lo}}, \mu_{\text{Me}}, \mu_{\text{Hi}}\}$ and $M(A) = \{\mu_{\text{Sm}}, \mu_{\text{Md}}, \mu_{\text{La}}\}$ where

$$\mu_{\text{Lo}}(\tilde{I}_S) = q(\tilde{I}_S; \tilde{I}_{S,\min}, \text{Lo}_e, 0, \text{Lo}_w), \quad \mu_{\text{Me}}(\tilde{I}_S) = f(\tilde{I}_S; \text{Me}_c, \text{Me}_{w0}, \text{Me}_{w1})$$

$$\mu_{\text{Hi}}(\tilde{I}_S) = q(\tilde{I}_S; \text{Hi}_e, \tilde{I}_{S,\max}, \text{Hi}_w, 0)$$

$$\mu_{\text{Sm}}(A) = q(A; A_{\min}, \text{Sm}_e, 0, \text{Sm}_w), \quad \mu_{\text{Md}}(A) = f(A; \text{Md}_c, \text{Md}_{w0}, \text{Md}_{w1})$$

$$\mu_{\text{La}}(A) = q(A; \text{Lg}_e, A_{\max}, \text{Lg}_w, 0) \tag{10.62}$$

Parameters $\tilde{I}_{S,\min}$, $\tilde{I}_{S,\max}$ and $A_{\min}$, $A_{\max}$ are the minimum and maximum possible values for $\tilde{I}_S$ and A, respectively. The output linguistic variable is here defined as the adjustment amount of $P_V(n)$, denoted by ΔP. The term set for

$$\Delta P \Rightarrow T(\Delta P) = \{\Delta P_1, \Delta P_2, \Delta P_3, \Delta P_4, \Delta P_5, \Delta P_6\}$$

The membership function (the set of values) of

$$\Delta P \Rightarrow M(\Delta P) = \{\mu_{\Delta P_1}, \mu_{\Delta P_2}, \mu_{\Delta P_3}, \mu_{\Delta P_4}, \mu_{\Delta P_5}, \mu_{\Delta P_6}\}$$

Parameters $\mu_{\Delta P_i}(\Delta P), i = 1, \ldots, 6$ are given as

$$\mu_{\Delta P_1}(\Delta P) = f(\Delta P; \Delta P_1, 0, 0), \quad \mu_{\Delta P_2}(\Delta P) = f(\Delta P; \Delta P_2, 0, 0)$$

$$\mu_{\Delta P_3}(\Delta P) = f(\Delta P; \Delta P_3, 0, 0), \quad \mu_{\Delta P_4}(\Delta P) = f(\Delta P; \Delta P_4, 0, 0)$$

$$\mu_{\Delta P_5}(\Delta P) = f(\Delta P; \Delta P_5, 0, 0), \quad \mu_{\Delta P_6}(\Delta P) = f(\Delta P; \Delta P_6, 0, 0) \tag{10.63}$$

where $\Delta P_i, 1 \leq i \leq 6$, is the ith adjustment step. Heuristically we set $-0.125 \leq \Delta P_i \leq 0.125$ and $\Delta P_i = (-0.175 + 0.05 \times i)$ to reflect different degrees of predicted interference and performance indication. As $\tilde{I}_S$ is low and A is in the middle, $\Rightarrow \Delta P = \Delta P_6$, denoting a larger increment for the access probability. If $\tilde{I}_S$ is large and A is large, we select

Table 10.5 The rule structure for FAPC

Rule	I_S	A	ΔP	
1	Hi	La	ΔP_1	
2	Hi	Md	ΔP_1	
3	Hi	Sm	ΔP_2	
4	Me	La	ΔP_2	Target performance
5	Me	Md	ΔP_3	
6	Me	Sm	ΔP_4	
7	Lo	La	ΔP_5	
8	Lo	Md	ΔP_6	
9	Lo	Sm	ΔP_6	

$\Rightarrow \Delta P = \Delta P_1$ denoting a larger decrement for the access probability. These rules are elaborated in Table 10.5.

Max–min interference method is used to calculate the membership value for each term of $T(\Delta P)$ and then apply the center of area for defuzzication. Once ΔP is obtained, $P_V(n+1)$ is given as

$$P_V(n+1) = P_V(n) + \Delta P \tag{10.64}$$

The access probability for data ready contention users $P_D(n+1)$ is obtained by

$$P_D(n+1) = f_d \cdot P_V(n+1) \tag{10.65}$$

where f_d is a real number smaller than 1, denoting voice users have higher access priority than data users (see also previous sections).

Neural-network access probability controller (NAPC)

NAPC adopts radial basis function network (RBFN) shown in Figure 10.25.

The hidden node q in the RBFN performs the normalized Gaussian activation function.

$$z_q \equiv \frac{\exp\lfloor -|\boldsymbol{x} - \boldsymbol{m}_q|^2/2\sigma_q^2 \rfloor}{\sum_{\ell=1}^{k} \exp\left[-|\boldsymbol{x} - \boldsymbol{m}_\ell|^2/2\sigma_\ell^2\right]}, \quad 1 \leq q \leq k \tag{10.66}$$

$\boldsymbol{x}$ the input vector, $\boldsymbol{m}_q(\sigma_q)$ is the mean (variance) of the qth Gaussian function and k is the number of hidden nodes. In this way, hidden node q has its own receptive field center on $\boldsymbol{m}_q$ with size proportional to σ_q, and it will give a maximum response to the input vector closest to $\boldsymbol{m}_q$. For an input vector $\boldsymbol{x} = [\tilde{I}_S(n+1), A(n)]$ lying somewhere in the input space, the receptive fields that are close to it will be properly activated. The

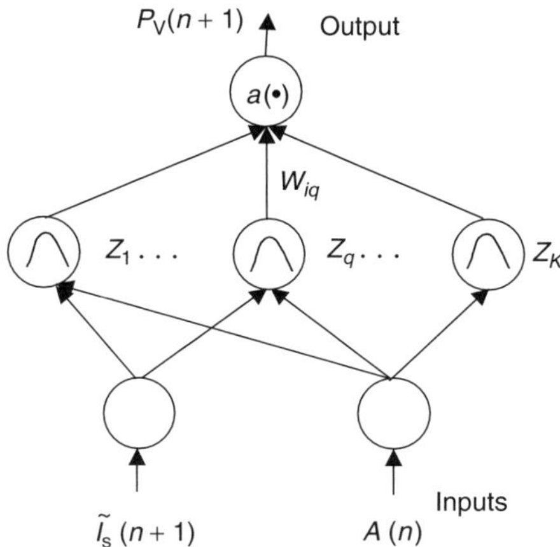

Figure 10.25 The structure of RBFN for NAPC.

output of RBFN, $P_V(n+1)$, is simply the mapping of the weighted sum of the hidden node outputs by

$$P_V(n+1) = a\left(\sum_{q=1}^{k} w_{iq} z_q\right)$$

where $a(\cdot)$ is the output activation function. The function of RBFN is to group the input vectors, which are close to each other, and then teaches every group to which output level it belongs. Parameters σ_ℓ^2 and m_ℓ are trained either by the hybrid learning rule or the error backpropagation rule. The details can be found in a number of textbooks [10].

For illustration purposes the following simulation environment is assumed [9]: $K = 49$, propagation loss exponent, $n = 4$, standard deviation of ξ is 8 dB, $T = 20$ ms, $N = 10$, $1/\mu = 0.44$ s, $1/\lambda = 0.56$ s, $1/\lambda_d = 0.04$, $F = 15$, $T_D = 40$ ms, $B_V = 12$, $B_D = 200$, RBFN output activation function $a(x) = 1/[1 + \exp(-x)]$. The voice source rate is 8 kbps and thus generates 160 information bits per frame time. In addition, 64 header bits and ($L = 511$, $M = 229$, $\zeta = 38$) Bose–Chaudhuri–Hocquenghem (BCH) code is used. The total bandwidth is 3.8325 MHz. We set the packet error probability $P_E{}^*$ and the interference level $I_S{}^*$ to $P_E{}^* = 0.01$ and $I_S{}^* = 16 \times S$. For the system CAF defined in Section 10.4, parameters from Table 10.3 are used with slight modification: $p_{si} = 0.03$, $\alpha = 0.007$, $\beta = 0.08$, breakpoint $= 6$ and f_d is set to be 0.25. The performance indicator includes voice packet dropping ratio L_V, corruption ratio R_C, utilization U and packet delay D_p. These parameters are shown in Figures 10.26 to 10.29.

For the same L_C, fuzzy/neural congestion controller with NAPC provides the higher capacity (see Figure 10.26). In Figure 10.27 one can see that the same method provides by far the best corruption ratio R_C.

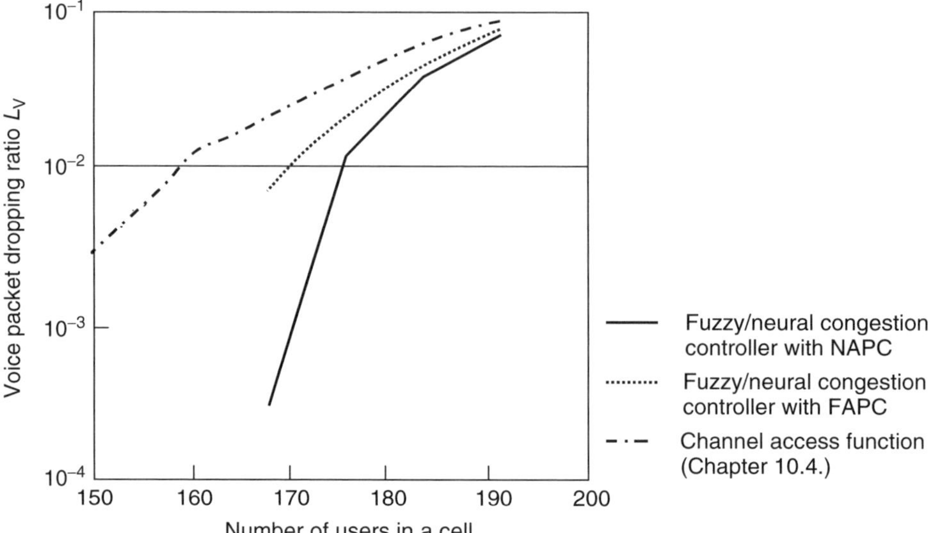

Figure 10.26 The voice packet dropping ratio L_V versus the number of users in a cell.

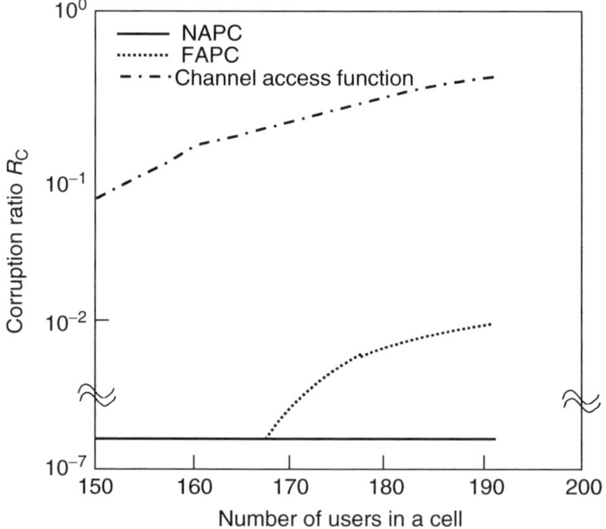

Figure 10.27 The corruption ratio R_C versus the number of users in a cell.

Utilization for NAPC is also the best (see Figure 10.28).

As one could expect, because of the strict control that provides better values for the previous three parameters, the delay in such a system will be increased as shown in Figure 10.29.

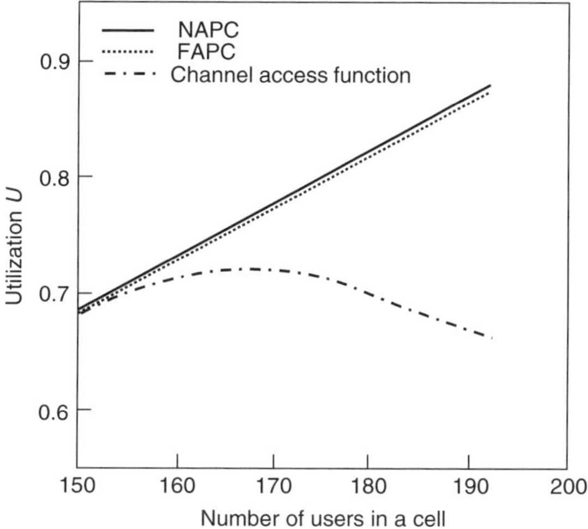

Figure 10.28   The utilization U versus the number of users in a cell.

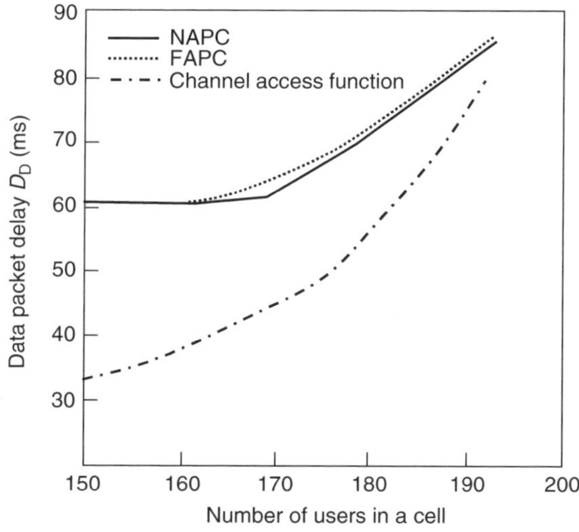

Figure 10.29   The data packet delay D_{D} versus the number of users in a cell.

10.6 ADAPTIVE TRAFFIC ADMISSION BASED ON KALMAN FILTER

A framework for adaptive connection admission in the uplink direction is again based on the estimation of the interference at the BS receivers. This time the estimation algorithm employs a *linear Kalman filter*, which is driven by a measurement of the interference

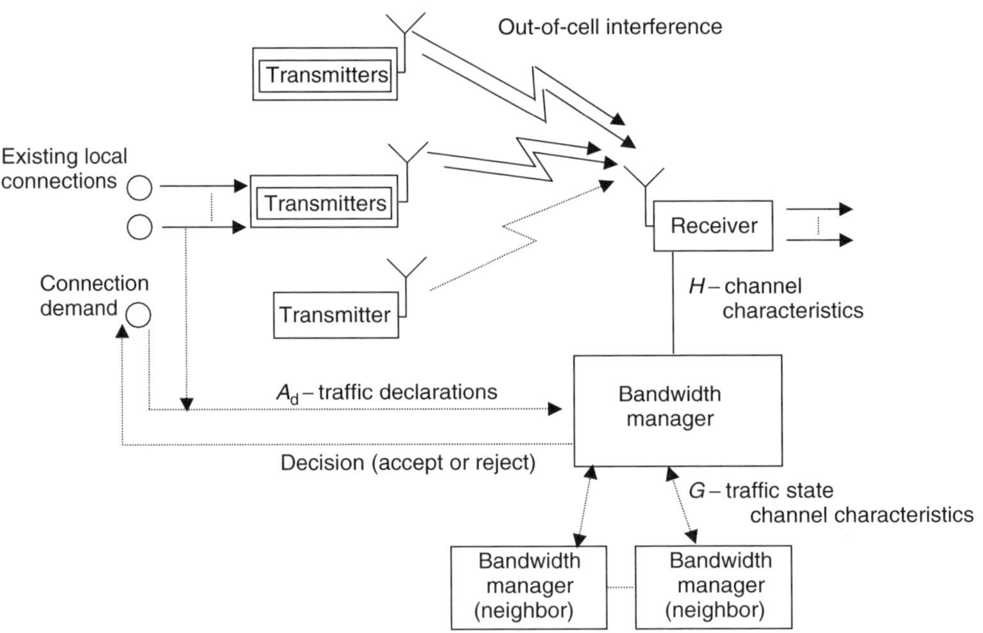

Figure 10.30 Traffic admission environment.

and by predicted traffic parameters of the admitted connections [11]. We discuss several generic variants of the control architecture. They vary from a fixed strategy (FS) with fixed power control (FPC) to an adaptive strategy with full information about network state and adaptive power control. The general block diagram of the control architecture is given in Figure 10.30.

The decision criterion is to accept the call if the outage probability $P\{\mathrm{BER_s} > \varepsilon\} \leq \gamma$.

Three basic categories of information are potentially available to the bandwidth manager. (1) Information about connection traffic parameters. These parameters, a_d, can be defined by the source type, for example, voice connection. Alternatively, it may be that the source itself specifies the traffic characteristics of the connection (e.g. peak rate, burst length, average rate). The traffic parameters resulting from the superposition of all sources served by a BS will be denoted as A_d. In general, the declared parameters are stochastic in nature and should be treated only as estimates since the source itself may not know them exactly in advance. In other words, there is certain declaration error. The range of this error is associated with the source type and can be limited by a policing mechanism at the transmitter. The traffic parameters of the connections within the cell are not sufficient to determine accurately the BER_s since the interference form the out-of-cell connections is an independent random variable. (2) Information obtained by measuring channel characteristics, H, pertinent to BER_s, which are available in the local BS. The most important are the current SIR distribution, $\mathrm{SIR}(t) = S(t)/I(t)$. There is a direct relation between them and $\mathrm{BER}s$ ($\mathrm{SIR}(t) \to \mathrm{BER}_c$ (coded) $\to \mathrm{BER}_s$ (uncoded). From the traffic admission point of view, SIR(t) is the most convenient information to analyze. This follows from the fact that the total interference can be treated as a superposition of the interference caused by each connection signal. In this case, the influence of a new connection on the interference level can be assessed by adding the predicted interference from the new call, $i_p(t)$, to the existing one $I'(t) = I(t) + i_p(t)$. Current value of SIR(t) is not sufficient to make an accurate decision since the decision should provide the required QoS for some period in the future, while SIR(t) may change significantly in this period due to variation in the propagation and traffic parameters. Parameters of the current SIR(t) distribution must be estimated. (3) Information that can be retrieved from neighboring BSs, G. Basically this is the same type of information as in category one and two, except that it may be transformed into more useful form by the bandwidth managers of the neighboring BSs. Two additional factors need to be taken into account when applying the information from the neighboring BSs for traffic admission control. Firstly, there is additional complexity caused by the exchange of a large amount of information between the BSs. Secondly, there is a delay in the information exchange protocol so the information, generally, reflects the state of the network a short time ago. On the basis of the indicated information structure, the bandwidth manager can reject the new call in order to keep QoS at the desired level. It should be mentioned that, in general, the bandwidth manager could use some additional mechanisms to control the QoS level. Most important are mechanisms that can reduce (or increase) the source rate during the connection. In particular, this may be a source with selectable variables or a data source that can vary the transmission rate according to the available bandwidth in the network. In the sequel, we will discuss three different control strategies. Basically for more efficient control we will need more information to make access decisions. The analysis is based on Reference [11].

1. *Fixed strategy (FS)*: The admission decision is based on traffic source declarations only. The bandwidth manager ensures that the declared traffic parameters do not exceed the thresholds $A_T(A_d)$. The thresholds are defined so that the information bit error rate (BER) constraint is met for any source and propagation distribution

$$A_d \leq A_T(A_d) \Rightarrow P\{\text{BER}_s > \varepsilon\} \leq \gamma \tag{10.68}$$

The obvious advantage of the scheme is its simplicity. On the other hand, the algorithm has to be designed for the worst case of source location and propagation distributions, which are not under the BS control so the bandwidth utilization can be far from optimum.

2. *Local adaptive strategy (LS)*: The admission decision is based on both the source traffic declarations and the channel characteristics measured at the local BS H. The threshold A_T adapts to the current propagation conditions at the local BS.

$$A_d \leq A_T(A_d, H) \Rightarrow P\{\text{BER}_s > \varepsilon\} \leq \gamma \tag{10.69}$$

It is clear that the adaptive capabilities of this scheme can provide better bandwidth utilization. At the same time, monitoring of the transmission conditions increases the complexity of the algorithms.

3. *Global adaptive strategy (GS)and soft capacity management*: This is an extension of the LS where the decision is also based on information from the neighboring BSs, G. The idea is to manage the soft capacity too.

$$A_\mathrm{d} \leq A_\mathrm{T}(A_\mathrm{d}, H, G) \Rightarrow P\{\mathrm{BER}_\mathrm{s} > \varepsilon\} \leq \gamma \tag{10.70}$$

The level of traffic that is accepted by the local BSs takes into account the propagation conditions in a larger region. The framework for adaptive traffic admission is presented in Figure 10.31. The objective of the estimation part is to obtain reliable estimates of the state-dependent mean and variance of interference $\hat{I}(y)$, $\hat{V}(y)$.

For as long as the interference level is lower than I_max, the allowable BER will be below the threshold ε.

$$I_\mathrm{s} \leq I_\mathrm{max} \Rightarrow \mathrm{BER}_\mathrm{s} \leq \varepsilon \tag{10.71}$$

The threshold I_max can be updated periodically on the basis of comparison of measured channel BER_c and I_s to compensate for possible model errors. Having determined I_max, the decision criterion based on equation (10.71) is to accept the call if in the new state, for all connections

$$P\{I_\mathrm{s} > I_\mathrm{max}\} \leq \gamma \tag{10.72}$$

If the estimated and the predicted values are exact, the condition (10.72) is fulfilled if

$$\hat{I}(y) + i_\mathrm{p} \leq I_\mathrm{max} - r' \tag{10.73}$$

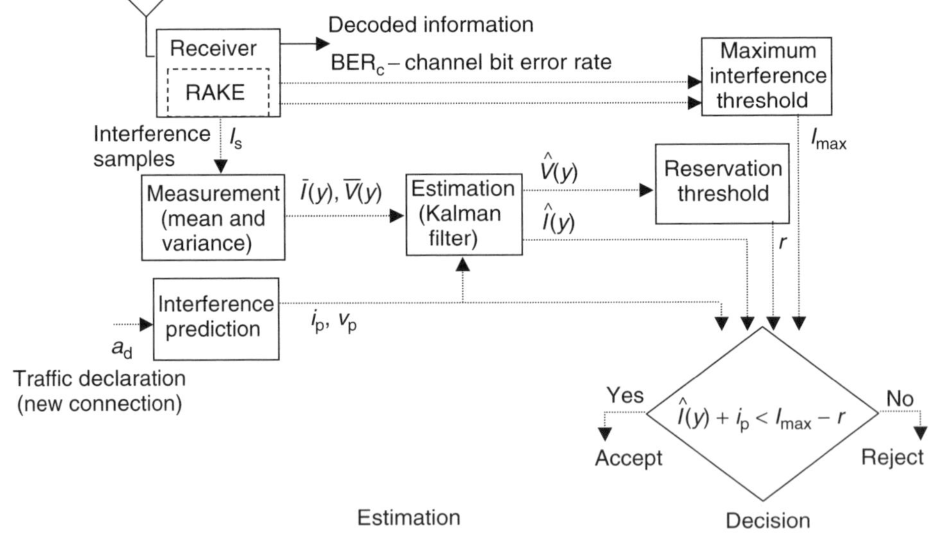

Figure 10.31 Framework for adaptive traffic admission.

where r' is the reservation threshold given by

$$r' = U(\gamma)\sqrt{\hat{V}(y) + v_\mathrm{p}} \tag{10.74}$$

and $U(\gamma)$ is the coefficient derived from the normalized Gaussian distribution.

10.6.1 Receiver structure

Suppose that the correlation receiver with noncoherent square law combining is employed in the uplink direction of the studied CDMA access network. Assume that there are M signaling waveforms used in the transmission of information, namely,

$$s_m(t) = \mathrm{Re}[u_m(t)e^{j2\pi f_c t}], \qquad m = 1, 2, \ldots, M \tag{10.75}$$

where $\{u_m(t)\}$ are equivalent lowpass waveforms. For BPSK/CDMA $u_2 = bc$ where $b = \pm 1$ is binary data and $c = \pm 1$ is the spearing code. The channel through which the signaling waveforms are transmitted is assumed to introduce an attenuation factor α and a phase shift ϕ_m for each signal. Thus the equivalent lowpass received signal, corrupted by additive noise $z(t)$, is represented as

$$r(t) = \alpha e^{-j\phi_m} u_m(t) + z(t) \tag{10.76}$$

In the sequel, we make the following assumptions.
(1) The phases $\{\phi_m\}$ are mutually statistically independent and uniformly distributed; (2) the additive Gaussian noise is white and (3) the M signals are equally likely and have equal energy. The optimum demodulator is the one that computes the M decision variables from

$$U_m = \left| \int_0^T r(t) u_m^*(t)\mathrm{d}t \right| \quad ;m = 1, \ldots, M \tag{10.77}$$

and selects the signal corresponding to the largest decision variable, as shown in Figure 10.32. When broadband signals are used over a frequency-selective slowly fading channel, a RAKE demodulator is used to achieve diversity. The RAKE demodulator combines the outputs of l optimum demodulators with inputs $r[t + (i/W)]$; $i = 0, 1, \ldots, l - 1$, where W is the bandwidth of the input signal (see Chapter 8).

The measured signal energy and interference per symbol are given as

$$E_\mathrm{s} = \frac{1}{n_\mathrm{s}} \sum_{j=1}^{n_\mathrm{s}} \max_k \{d_{j,k}\} - \frac{I_\mathrm{s}}{M - 1} \tag{10.78}$$

$$I_\mathrm{s} = \frac{1}{n_\mathrm{s}} \sum_{j=1}^{n_\mathrm{s}} \sum_{\substack{k=1 \\ k \neq k_\mathrm{max}}}^{M} d_{j,k} \tag{10.79}$$

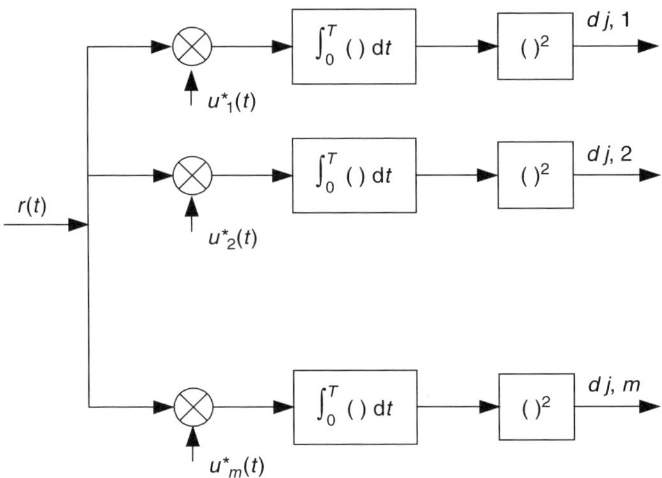

Figure 10.32 Optimum demodulator for noncoherent detection.

where $k_{\max}$ is the index of the largest decision value for the given symbol interval, n_s is the number of symbols in the measurement interval of 1.25 ms (power control command bits are generated every 1.25 ms when fast power control is employed), and $d_{j,k}$ is the kth decision value at the output of the diversity combiner during the jth symbol interval. Assume that E_s and I_s are available since these parameters are used anyway in the power control mechanism.

Two power control schemes are considered:

1. The signal power at the output of the RAKE receiver is kept constant. This scheme is henceforth referred to as FPC.
2. The power control tries to keep the SIR at a certain level, SIRT. If the current SIR is below SIRT, the signal power is increased by 0.5 dB. In the opposite case, the signal power is reduced by 0.5 dB.

10.6.2 Measurement

The estimation process requires measurement of the interference process (FPC) or the interference to signal ratio process (SIR power control). Signal power level is always available. The interference samples I_s can be obtained from the fast power control algorithm. On the basis of these samples, one can evaluate the required values of the interference mean and variance during a particular state following the standard estimation approach:

$$\overline{I} = \frac{\sum I_s}{N} \tag{10.80}$$

$$\overline{V} = \frac{\sum (I_s - \overline{I})^2}{N - 1} \tag{10.81}$$

where N is the number of samples in the state under consideration. The time window for variance measurement should be chosen as

$$T_{\rm v} = \frac{1}{\lambda_{\rm h} + 0.1\lambda_{\rm n}} \tag{10.82}$$

where $\lambda_{\rm h}$ is the connection arrival intensity at the local base and $\lambda_{\rm n}$ is the total connection arrival intensity in the neighboring bases. The idea is to have the window size related to the average interval between two arrivals. The parameter 0.1 is introduced to reduce the weight of arrival in the neighboring stations due to their smaller share in the interference level. The window $T_{\rm v}$ ensures that the accuracy of the measured variance does not degenerate to an unacceptable level.

The *Kalman filter* applied for final estimation of the interference mean and variance requires also information about the measured error variances. The measurement errors for mean and variance of interference are defined as

$$\overline{\delta}^{i} = \overline{I} - I \tag{10.83}$$

$$\overline{\delta}^{v} = \overline{V} - V \tag{10.84}$$

The variance of the error of measured interference variance is given by

$$v_e^{\rm v} = E[\overline{V}^2] - (E[\overline{V}])^2 \tag{10.85}$$

$$E[\overline{V}^2] = E[(I_s - E[I_s])^4] = E[I_s^4] - 4E[I_s^3]E[I_s] + 6E[I_s^2]E^2[I_s] - 3E^4[I_s] \tag{10.86}$$

$$(E[\overline{V}])^2 = (E[I_s^2] - E^2[I_s])^2 = E^2[I_s^2] - 2E[I_s^2]E^2[I_s] + E^4[I_s] \tag{10.87}$$

By substituting equations (10.86 and 10.87) into equation (10.85), we get

$$v_e^{\rm v} = E[I_s^4] - 4E[I_s^3]E[I_s] + 8E[I_s^2]E^2[I_s] - E^2[I_s^2] - 4E^2[I_s] \tag{10.88}$$

For a measurement with N samples, the estimated variance of the interference variance error is given by

$$\overline{v}_e^{\rm v} = \overline{I_s^4} - 4\overline{I_s^3 I_s} + 8\overline{I_s^2} \cdot \overline{I_s^2} - \left(\overline{I_s^2}\right)^2 - 4\overline{I_s^2} \tag{10.89}$$

$$\overline{I_s^j} = \frac{\displaystyle\sum_i I_s^j}{N}, \qquad j = 1, 2, 3, 4 \tag{10.90}$$

In order to avoid a large correlation between the measured interference variance and the estimated error $\overline{v}_e^{\rm v}$, the window for $\overline{v}_e^{\rm v}$ estimation, T_e, is larger than $T_{\rm v}$.

10.6.3 Estimation

Let k be the sequential index of the BS states. The objective of the estimation process is to provide the best estimate of the state average interference, I_k, and interference variance,

V_k. For estimation of these variables, a linear discrete Kalman filter will be used. In general, the Kalman filter provides an optimal least-squares estimate of the system state under the condition that the system is linear and the model and measurement errors are Gaussian random variables. For local strategy (no communication with neighboring BSs), the block diagram of the system model and Kalman filter is shown in Figure 10.33 where X_k denotes either I_k or V_k. As mentioned earlier in Chapter 4, the dynamics of the system is modeled as

$$X_k = F_k X_{k-1} + e_k$$

$$F_k = 1 \pm \frac{x_k}{X_{k-1}} \tag{10.91}$$

where e_k is the model error and x_k denotes either the predicted average interference, i_p, or the predicted interference variance, v_p, of the new $(+)$ or departing $(-)$ connection.

If Y_k is the variance of the measurement error ($\overline{v}_e^i$ or $\overline{v}_e^v$), the Kalman gain is

$$K_k = P_k^e (P_k^e + Y_k)^{-1} \tag{10.92}$$

where

$$P_k^e = F_{k-1}^2 P_{k-1} + Q_k$$

$$P_{k-1} = (1 - K_{k-1}) P_{k-1}^e \tag{10.93}$$

and Q_k is predicted variance of the model error (v_e^i or v_e^v). The system state estimate update is given by

$$\hat{X}_k = \hat{X}_k^e + K_k (Z_k - \hat{X}_k^e) \tag{10.94}$$

where $\hat{X}_k^e = \hat{X}_{k-1} + x_k$ the state estimate extrapolation and Z_k is measurement ($\overline{I}_k$ or $\overline{V}_k$). Note that under the Gaussian assumption of the model and measurement errors, the

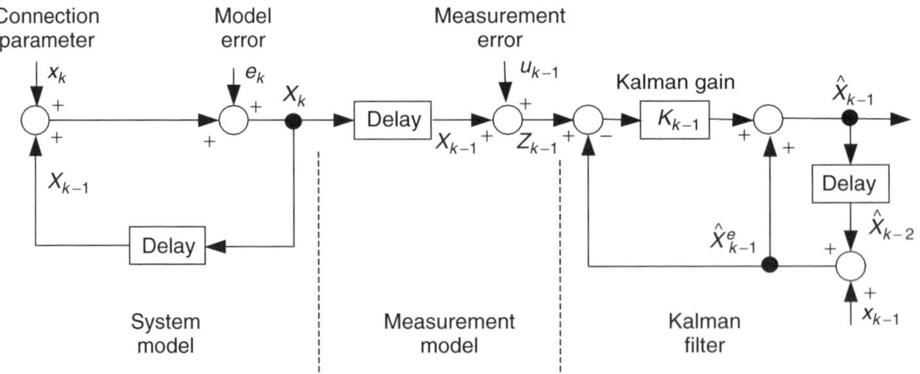

Figure 10.33 System model and Kalman filter.

distribution of the estimation error is also Gaussian with zero mean and variance ($\overline{v}_e^i$ or $\overline{v}_e^v$) defined by error covariance matrix, P_k.

The estimation process for global strategy (communication with neighboring bases) is basically the same with the exception of the transition matrix, which is defined as

$$F_k = 1 \pm \frac{x_k + x_k^n}{X_{k-1}} \tag{10.95}$$

where x_k^n is the predicted change in the average interference, i_p^n, or interference variance, v_p^n, during the state duration in the local BS, caused by state changes in the neighboring BSs. Prediction of x_k^n is based on the state changes in the neighboring BSs during the last state in the local BSs

$$x_k^n = \sum_j \alpha_j^l (\pm x_k^{(j)}) \tag{10.96}$$

where j is the neighboring BS index, $\pm x_k^{(j)}$ denotes the total change in BS j during state k of the local BS, α_j is the coefficient defining the average interference in the local BS caused by a connection in the jth BS, $l = 1$ for the average interference estimation and $l = 2$ for the interference variance estimation. The model error, e_k in equation (10.91), requires some additional discussion. In general, the model error is defined as the difference between the real value in state k and the extrapolation from the previous state

$$e_k = X_k - (X_{k-1} + x_k) \tag{10.97}$$

The evaluation of the model error is based on the assumption that the aggregate interference process can be treated as stationary in two subsequent states. Under this assumption the model error is equal to the prediction error x_k. This error can be caused either by a declaration error in the source parameters or by the nonorthogonality of codes and non-perfect synchronization (two signals with the same power at the BS can cause different levels of interference to the third signal). The variance of these errors can be estimated from statistics. This approach does not take into account possible nonstationarity interference caused by the connection arrivals and departures in the neighboring BSs during the local base state k. In general, the nonstationarity error could be included in the model error. To simplify the analysis, the reservation for this error and the SIR power control error (defined in the next section) are evaluated separately.

10.6.4 Reservation

The main function of the reservation parameter, r in equation (10.93), in the connection acceptance decision is to provide protection against the variability of the estimated interference and possible estimation errors. There are five main factors that influence the value of the reservation parameter.

1. *Estimated interference variance* $\hat{V}$: From the reservation viewpoint, it is important that the interference is assumed to have Gaussian distribution.

2. *Variance of the mean interference estimation error* $\hat{v}_e^i$: According to the Kalman filter model the estimation error has a Gaussian distribution.

3. *Nonstationarity error*: As indicated in the previous section, the estimated interference variance $\hat{V}$ for a given state is used in the admission decision as a prediction for the next state (stationarity assumption). Since on average there are several connection arrivals and departures in the neighboring BS during local base state k, this feature can cause a prediction error, which is henceforth called as nonstationarity error. Note that this effect exists even if the traffic offered to the neighboring stations is stationary.

4. SIR *Power control feedback error*: In the case of SIR power control, the connection acceptance causes certain additional interference variably caused by the power control feedback effect.

5. *Priority for arrivals in the neighboring cells*: As will be illustrated in the next section, the system capacity can be increased when the connection admission in the neighboring cells has priority over the admission of a local connection. This priority can be implemented by increasing the reservation parameter for local connection admission. The reservation level, r_1, for the first two factors can be evaluated from

$$r_1 = U(\gamma)\sqrt{\hat{V} + \hat{v}_e^i} \tag{10.98}$$

where $U(\gamma)$ is the coefficient derived from the normalized Gaussian distribution, which provides that the condition (10.72) is fulfilled if $\hat{I} \le I_{\max} - r_1$. The remaining three factors are taken into account by increasing the reservation level for local connection admission.

$$r = r_1 + r_2 \tag{10.99}$$

Analytical evaluation of the parameter r_2 is complex and requires more information about the traffic process. Here, the optimal values of r_2 are found from numerical experiments. The same approach is used to find the reservation parameter, r_3, for admission decision of connections from neighboring BSs [11].

10.6.5 Connection admission strategies

The three generic strategies are implemented as follows:

1. *Fixed strategy (FS)*: The fixed traffic admission thresholds, $A_T(A_d)$, ensure that

$$A_d \le A_T(A_d) \Rightarrow P\{I_s > I_{\max}\} \le \gamma \tag{10.100}$$

The thresholds are found from numerical experiments under stationary and uniform traffic conditions. This approach gives system capacity larger than the worst-case design (in the case of nonuniform distribution of portables within the cells the system capacity can be smaller).

2. *Local adaptive strategy (LS)*: A new connection is accepted if

$$\hat{I} \le I_{\max} - r_1 - r_2 \tag{10.101}$$

3. *Global adaptive strategy* (*GS*): A new connection is accepted if

$$\hat{I} \leq I_{\max} - r_1 - r_2 \tag{10.102}$$

in the local BS and

$$\hat{I} = \leq I_{\max} - r_1 - r_3 \tag{10.103}$$

in all neighboring BSs. For illustration purposes the following assumptions are used. The network consists of a 7×7 grid of square cells. In order to avoid excessive complexity, the traffic and propagation models take into account only characteristics that are important from the viewpoint of traffic admission. Examples are limited to homogenous connections. Each cell generates a stream of connection demands from portables that obey a Poisson distribution with intensity x_i. The demands are uniformly distributed within the cell. A portable with connection demand chooses the BS providing the best SIR. This BS is referred to as the local BS of the portable. Once the connection is accepted, the terminal generates on–off traffic with activity rate 0.35 and means holding time 3 min (voice or bursty data traffic). The propagation model assumes exponential path loss and lognormal shadow fading. The variance of the lognormal shadowing used in the example is 6 dB. The portable mobility and hand-offs are not modeled. The values of the signals and interference at the BSs are evaluated from the propagation model for average values only. At this stage, the model error, Q_k, is added according to its Gaussian distribution. The variability of the signals, caused by source burstiness, is modeled for the superposition signals at the BSs as a Gaussian noise added to the average value. The performance of the investigated strategies is evaluated on the basis of the traffic carried in the nine central cells. The boundary conditions are provided by the remaining 40 cells that operate under FS and are offered uniform background traffic. The traffic pattern in the nine central cells can be nonuniform and nonstationary. The examples chosen for study are described in Table 10.6 and Figure 10.34 [11].

Table 10.6 Traffic levels [11]. Reproduced from Dziong, Z., Jia, M. and Mermelstein, P. (1996) Adaptive traffic admission for integrated services in CDMA wireless–access networks. *IEEE J. Select. Areas Commun.*, **14**(9), 1737–1747, by permission of IEEE

Examples	A_h	A_l	A_b
SU_2	90	90	90
SU_4	150	150	150
SC_2, NC_2	120	18	90
SC_4, NC_4	180	30	150
SX_2, NX_2	75	0.3	75
SX_4, NX_4	105	15	105

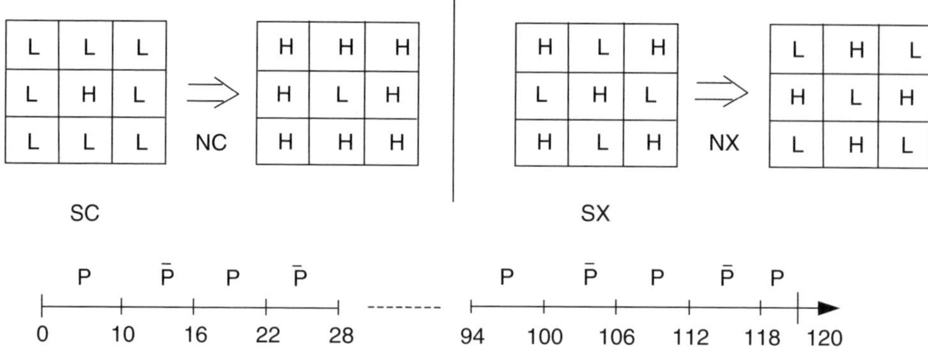

Figure 10.34 Traffic patterns [11]. Reproduced from Dziong, Z., Jia, M. and Mermelstein, P. (1996) Adaptive traffic admission for integrated services in CDMA wireless–access networks. *IEEE J. Select. Areas Commun.*, **14**(9), 1737–1747, by permission of IEEE.

They are denoted by two letters and an index. The first letter indicates whether the traffic is stationary (S) or nonstationary (N). The second letter denotes traffic uniformity: U – uniform, C – nonuniform with one overloaded cell (central) and X – nonuniform with several overloaded cells forming an X pattern. The subscript of each example corresponds to the path loss exponent. The nonuniformity and nonstationarity are realized by two levels of traffic offered to the cells – high (A_h) and low (A_l). These levels together with the background traffic levels A_b are given in Table 10.6.

The traffic patterns for nonuniform cases are presented in Figure 10.34. The same figure illustrates the nonstationary patterns that are achieved by periodical changes (every 6 min) of the original nonuniform pattern, P, to its complement, $\bar{P}$, and vice versa.

Table 10.7 Average number of voice connections per cell with fixed power control [11]. Reproduced from Dziong, Z., Jia, M. and Mermelstein, P. (1996) Adaptive traffic admission for integrated services in CDMA wireless–access networks. *IEEE J. Select. Areas Commun.*, **14**(9), 1737–1747, by permission of IEEE

Fixed strategy		Local strategy		Global strategy		
	A	ΔA	r_2	ΔA	r_2	r_3
SU_2	34.5	+4%	9	+7%	5	1
SU_4	60.3	+10%	9	+12%	5	1
SC_2	11.6	+20%	9	+22%	5	1
SC_4	14.7	+13%	9	+17%	5	1
SX_2	29.9	+3%	9	+8%	5	1
SX_4	50.8	+7%	9	+10%	5	1
NX_2	28.1	+5%	9	+10%	5	1
NX_4	50.8	+6%	9	+8%	5	1

Table 10.8 Mean connections per cell with SIR power control [11]. Reproduced from Dziong, Z., Jia, M. and Mermelstein, P. (1996) Adaptive traffic admission for integrated services in CDMA wireless–access networks. *IEEE J. Select. Areas Commun.*, **14**(9), 1737–1747, by permission of IEEE

Fixed strategy		Local strategy		Global strategy		
	A	ΔA	r_2	ΔA	r_2	r_3
SC_2	11.5	+20%	10	+24%	5	1
SC_4	14.7	+27%	10	+33%	5	1
SX_2	29.3	+15%	10	+31%	5	1
SX_4	51.3	+14%	10	+24%	5	1
NX_2	28.8	+16%	30	+17%	25	21
NX_4	52.2	+10%	30	+17%	25	21

The nonstationary and stationary examples constitute the two extremes of a wide range of possible nonstationary cases with different speed of traffic pattern changes. The QoS in the system is determined by the outage probability defined by equation (10.72). The value of $I_{\max}$ (or equivalently SIR) is a function of the channel and antenna characteristics. Since these parameters are not modeled in this system, $I_{\max} = 30$ is chosen. Because of the nonstationary arrival rates in some of the examples, the outage probability was estimated over 1-min windows. The optimal reservation thresholds, r_2, r_3, (found by exhaustive search) ensure that the outage probability does not exceed the 1% constraint in any of the time intervals. The model error variances are as follows: $v_e^i = 0.25$, $v_e^v = 0.01$. The results presented in this section are based on simulations corresponding to 120 min of system time. Results for the first 8 min of the initial transient period are not included in the statistics. The results are shown in Tables 10.7 and 10.8 [11] for two different power control strategies. Capacity increase up to 30% can be seen.

10.7 SOFT HANDOFF IN CDMA CELLULAR NETWORKS

We will assume that one mobile unit carries one call only. It means that there is no bulk handoff arrival. During the handoff, the mobile combines signals from two BSs soft handoff. Each cell will reserve C_h channels out of a total of C available channels exclusively for handoff calls, a so-called cut-off priority, because a suddenly forced termination during a call session will be more upsetting than a failure to connect. Every handoff requirement is assumed to be perfectly detected and the assignment of the channel is instantaneous if the channel is available. From the intrinsic property of the soft handoff process, a handoff requirement must not be denied immediately when there is no available channel. The handoff requirement shall be put into a queueing list. It is assumed that the allowable maximum queue length is equal to Q.

The call duration time T_c is assumed to be exponentially distributed with mean $\overline{T}_c = \mu_c^{-1}$. A cell is divided into two regions: the normal and the handoff region as shown in Figure 10.35 [12]. A handoff process shall be carried out for any call that enters the handoff region from outside the cell, and the diversity reception from different BSs is possible only in this area. New call and handoff arrivals are assumed to be Poisson distributed with rates Λ_n and Λ_h, respectively. The location of a new generated call is uniformly distributed all over a cell. In the sequel, we use the following notation:

$$\Lambda_{n1} \overset{\Delta}{=} \text{the new call arrival rate in handoff region} = a \cdot \Lambda_n$$

$$\Lambda_{n2} \overset{\Delta}{=} \text{the new call arrival rate in normal region} = (1 - a) \cdot \Lambda_n \qquad (10.104)$$

where

$$a = \frac{\text{area of handoff region}}{\text{area of a cell}} \qquad (10.105)$$

When a new call is generated in the handoff region, it can ask for channel assignment from both the cell of interest and a neighboring cell with which it can communicate. If the cell of interest has a channel available for such a generated call, the new call is considered as a successful new call. If not, the new generated call is said to be blocked from the point of view of the cell of interest. In the latter case, the new call may successfully enter the cellular system by getting a channel assignment from the neighboring cell and becoming a handoff arrival for the cell of interest. If the new call cannot get channel assignment from either of the two cells, the call is said to be blocked by this cellular system, and the mobile unit shall try its attempt later.

The dwell times of a call in two distinct regions are assumed to be exponentially distributed. The mean dwell time in handoff regions is $\overline{T}_{d1} = \mu_d^{-1}$ and that in normal region is $\overline{T}_{d2} = \mu_d^{-1}$. Since the relations between the mean dwell time in the whole cell $\overline{T}_{dc}$ and the mean dwell time in each region shall be a function of the covered area of each region, we can find the related functions. All the mobile units are assumed to be 'random walking' in a cell, which is very suitable for the urban area. According to these

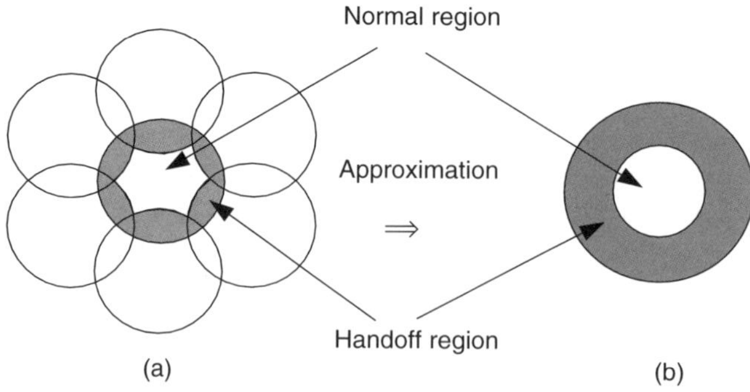

Figure 10.35 Handoff region and normal region of: (a) cellular network and (b) simplified cell.

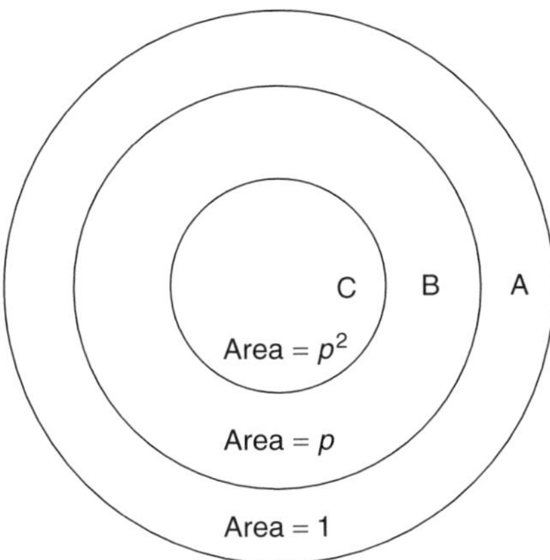

Figure 10.36 Three concentered circles A, B and C [12]. Reproduced from Su, S., Chen, J. and Huang, J. H. (1996) Performance analysis of soft handoff in CDMA cellular networks. *IEEE J. Select. Areas Commun.*, **14**(9), 1762–1769, by permission of IEEE.

related functions, we can get the values of $\overline{T}_{d1}$ and $\overline{T}_{d2}$ for given values of $\overline{T}_{dc}$ and 'a'. To do so, we consider three concentered circles whose areas are 1, p and p^2, respectively, as shown in Figure 10.36 [12] where $0 < p < 1$. Since the circles are in the same regular shape and the mobile units are assumed to be randomly walking in the whole area, we can conclude that the ratio of the average dwell times in two distinct circles shall be a function of the ratio of the area only. This can be represented as

$$T_C = f\left(p^2\right) \cdot T_A = f(p) \cdot T_B = f(p) \cdot f(p) \cdot T_A \qquad (10.106)$$

where T_A, T_B and T_C are average dwell times of three distinct circles A, B and C, respectively and $f(p)$ is the related function. Thus, $f(p^2) = f^2(p)$. Similarly, $f(p^n) = f^n(p)$ for any positive integer n. From this particular formula, we assume $f(p) = x^{\log_{10} p}$, where x is a proper chosen constant.

For the cell of interest as shown in Figure 10.35(b), if the fraction of the occupied area of the handoff region is 'a' then we can conclude the relation between the dwell time of the whole cell $\left(\overline{T}_{dc}\right)$ and the dwell time of the normal region $\left(\overline{T}_{d2}\right)$ is

$$\overline{T}_{d2} = f(1-a) \cdot \overline{T}_{dc} = x^{\log_{10}(1-a)} \cdot \overline{T}_{dc} \qquad (10.107)$$

The computer simulations of $\overline{T}_{d2}/\overline{T}_{dc}$ and $\overline{T}_{d1}/\overline{T}_{dc}$ are shown in Figure 10.37 [12]. From the simulation results $x = 2$ is a proper value in function $f(p)$. Another related function $q(a) = 16^{\log a}$ is a good approximation of the values of $\overline{T}_{d1}/\overline{T}_{dc}$.

10.7.1 System analysis

For the analysis, the birth–death model of the process is used. The state in the process is defined as

$$s = (v_1, v_2, q) \; v_1, v_2 \geq 0 \quad \text{and} \quad 0 \leq q \leq Q \qquad (10.108)$$

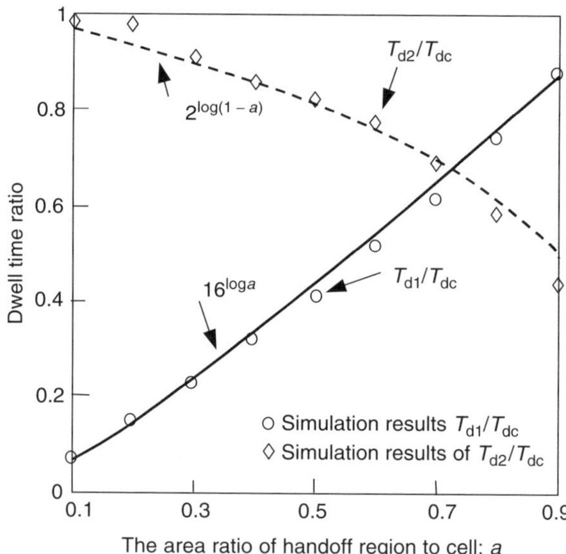

Figure 10.37 Simulation results and approximated relations of the dwell time ratio [12]. Reproduced from Su, S., Chen, J. and Huang, J. H. (1996) Performance analysis of soft handoff in CDMA cellular networks. *IEEE J. Select. Areas Commun.*, **14**(9), 1762–1769, by permission of IEEE.

where v_1, v_2 are the number of active calls in the handoff, normal region and q is the number of mobile units in queue. $N(s)$ is the total number of active channels at state s of a cell, that is, $N(s) = v_1 + v_2$ and $N(s) \leq C$. New calls will be blocked when $N(s) \geq C - C_h$ and the handoff arrival will be put in queue while all the channels are in use, that is, $N(s) \geq C$. Since the queue is not used when $v_1 + v_2 \leq C$ and it is used only when $v_1 + v_2 = C$, the two-dimensional (2D) Markov chain is enough to describe the state transitions. An example with $C = 5$, $C_h = 2$ and $Q = 3$ is shown in Figures 10.38 and 10.39.

There are five conditions that can cause a state transition: New call arrival, call completion, handoff arrival, handoff departure, and region transitions (from the normal region to the handoff region and vice versa). A region transition in which a mobile unit drives into the handoff region from the normal region of the current communicating cell and asks for a channel from the neighboring cell is also a handoff arrival for the neighboring cell. If the mobile unit drives from the handoff region to the normal region in the cell, it is a handoff departure of the neighboring cell. If a mobile unit drives out of the handoff region of a cell, according to its moving direction, it can be a true handoff departure or only region transition of the cell of interest. We assume the rate of the former is $\mu_{do} = (1 - P_b) \cdot \mu_d$ and that of the latter is $\mu_{dk} = P_b \cdot \mu_d$, where the parameter '$P_b$' is the average probability of moving back to the normal region of the cell of interest for a mobile unit in the handoff region. The value of P_b is a function of the value of 'a' and the related function between 'P_b' and 'a' described in the sequel. If a mobile unit is at

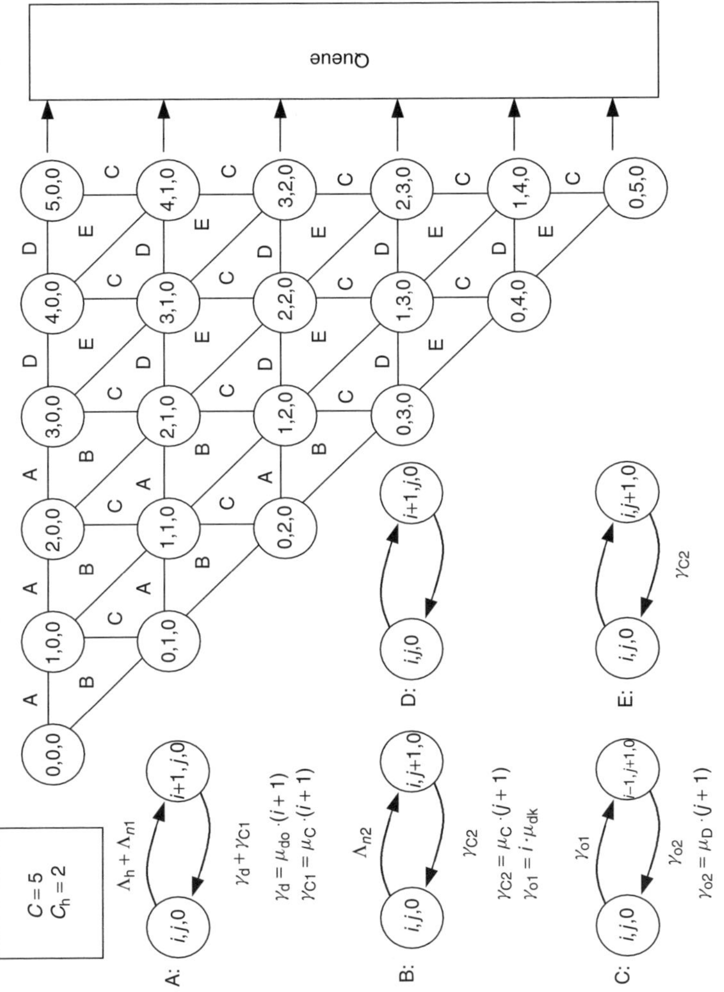

Figure 10.38 The birth–death processes of a cell with $C = 5$ and $C_h = 2$.

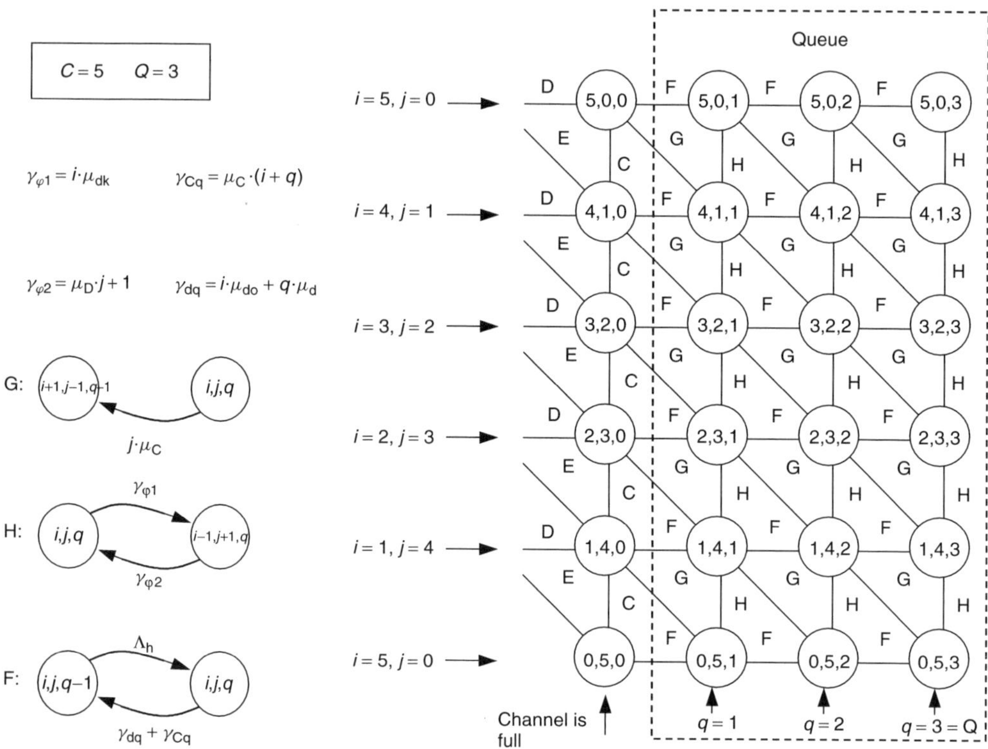

Figure 10.39   The queueing part of the birth–death processes of Figure 10.38 with $Q = 3$.

the point K in the handoff region, as shown in Figure 10.40 [12], the probability that it will drive toward the handoff region is

$$P_b(r_1) = \frac{\theta}{\pi} = \frac{\sin^{-1}(r/r_1)}{\pi} \tag{10.109}$$

The average probability of moving back to the normal region for all the mobile units in the handoff region is

$$P_b = \frac{2}{\pi \cdot (1 - r^2)} \int_r^1 r_1 \cdot \sin^{-1}\left(\frac{r}{r_1}\right) dr_1 \tag{10.110}$$

Figure 10.41 shows the relation between 'a' and the moving back probability 'P_b'.

10.7.2 Flow equilibrium equations and steady-state probabilities

Let $p(i, j, q)$ be steady-state probability of the state $s = (i, j, q)$. There is a flow equilibrium equation for each state, that is, the total rate of flowing into a state will be equal to

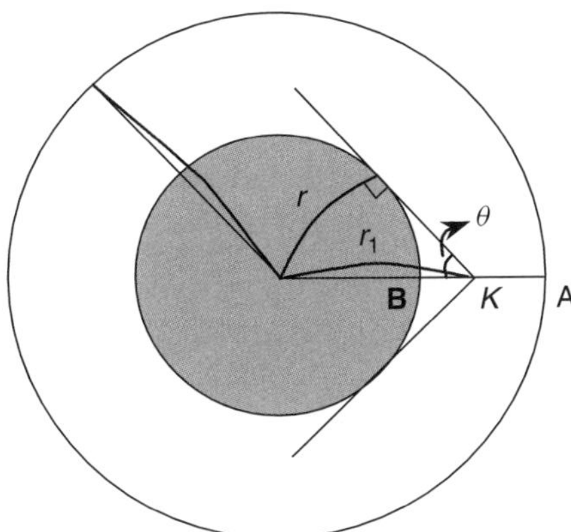

Figure 10.40 The mobile unit is at point K, which has a distance r_1 from the center. The radius of the normal region (inner circle) is r and the radius of the whole cell (outer circle) is one [12]. Reproduced from Su, S., Chen, J. and Huang, J. H. (1996) Performance analysis of soft handoff in CDMA cellular networks. *IEEE J. Select. Areas Commun.*, **14**(9), 1762–1769, by permission of IEEE.

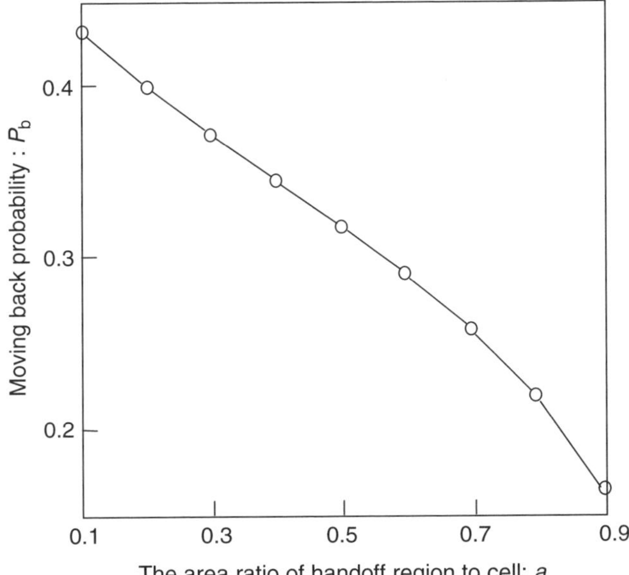

Figure 10.41 The probability P_b versus the ratio a [12]. Reproduced from Su, S., Chen, J. and Huang, J. H. (1996) Performance analysis of soft handoff in CDMA cellular networks. *IEEE J. Select. Areas Commun.*, **14**(9), 1762–1769, by permission of IEEE.

the total rate of flowing out from it. If the total number of states is Ω, there are $\Omega = 1$ linearly independent flow equilibrium equations with parameters like those in Figures 10.38 and 10.39. Besides, the sum of the steady state probabilities is

$$\sum_{i=0}^{C}\sum_{j=0}^{C-i}\sum_{q=0}^{Q} p(i, j, q) = 1. \tag{10.111}$$

Steady-state probabilities are obtained by solving 2D birth–death process from the chosen linearly independent flow equilibrium equations and equation (10.111). The parameters used in the Markov chain, such as $\Lambda_h, \mu_D, \ldots$, and so on are mutually related. Some implicit relations among them are restricted by the system behaviors. A call, which enters the handoff region from the normal region of the current communicating cell, will ask for a channel from the neighboring cell. It is a handoff arrival from the point of view of the neighboring cell. Under the assumption that all the cells have the same probability characteristics in steady state, the parameter shall satisfy the homogenous equation

$$\Lambda_h = \sum_{i=0}^{C}\sum_{j=0}^{C-1} (j \cdot \mu_D) \cdot p(i, j, 0) + \sum_{i=0}^{C}\sum_{q=1}^{Q} (j \cdot \mu_D) \cdot p(i, j, q)|j$$

$$= C - i + \Lambda_{n1} \cdot (1 - P_B) \cdot P_B \tag{10.112}$$

The last term represents the rate of new generated calls that can only get a channel assignment from the neighboring cell and become a handoff arrival for the cell of interest, and P_B is the blocking probability from the cell's point of view, which will be derived in the next section.

10.7.3 Performance measurement

Carried traffic

The carried traffic per cell is defined as the average number of occupied channels in the cell of interest, that is,

$$A_c = \sum_{i=0}^{C}\sum_{j=0}^{C-i} (i + j) \cdot p(i, j, 0) + \sum_{i=0}^{C}\sum_{q=1}^{Q} C \cdot p(i, j, q)|_{j=C-i} \tag{10.113}$$

where Q is the queue length.

Carried handoff traffic

The carried handoff traffic per cell is defined as the average number of occupied channels in the handoff region of the cell of interest, that is,

$$A_{ch} = \sum_{i=0}^{C}\sum_{j=0}^{C-i} i \cdot p(i, j, 0) + \sum_{i=0}^{C}\sum_{q=1}^{Q} i \cdot p(i, j, q)|_{j=C-i} \tag{10.114}$$

Blocking probability

A new call will be blocked by a cell when the total number of used channels $N(s) = (i + j) \geq C - C_h$. The blocking probability from the cell's point of view is given as

$$P_B = \sum_{B_0} p(i, j, q)$$

$$B_0 = \{(i, j, q) | C - C_h \leq (i + j), 0 \leq q \leq Q\} \qquad (10.115)$$

Since a new cell in the handoff region can ask for channel assignment from two cells, its blocking probability from the system's point of view is P_B^2. The average blocking probability from the system's point of view is

$$P_{BS} = \frac{\Lambda_{n1} \cdot P_B^2 + \Lambda_{n2} \cdot P_B}{\Lambda_n} \qquad (10.116)$$

Handoff refused probability

Probability of unsuccessful handoff requirement for three classes of events is defined as

$$P_{HA} = \sum_{i=0}^{C} \sum_{q=0}^{Q} p(i, j, q)|_{j=C-i} \qquad (10.117)$$

P_{HA} is the probability that channels are all occupied when an active mobile unit drives into a cell. In this case, the handoff requirement is unsuccessful at the temporal moment and it may be put into queue for a latter trial or, if the queue is full, it is refused forever. Failure of the handoff requirement does not mean 'immediate communication drop'. If the queue is full and the handoff requirement is refused forever, define P_{HC} for class C such that

$$P_{HC} = \sum_{j=0}^{C-C_h} p(i, j, Q)|_{i=C-j} \qquad (10.118)$$

Define, for class B

$$P_{HB} = \frac{\sum_{j=0}^{C-C_h} \sum_{q=1}^{Q} q \cdot (\mu_c + \mu_d) \cdot p(i, j, q)|_{i=C-j}}{\Lambda_h \cdot (1 - P_{HC})} + P_{HC} \qquad (10.119)$$

The numerator of the first term of P_{HB} means the average leaving rate of the mobile units in queue before being served. Hence, P_{HB} is the total probability that a handoff call does not get any service ultimately from the cell of interest. It is also a sense of refused handoff.

10.7.4 Channel efficiency (trunk resource efficiency)

For assessing the channel efficiency, we define the 'efficiency' of each channel by its 'necessity'. It means that the efficiency of a channel equals one if the corresponding active call holds only one channel and the efficiency reduces to one half if the active call holds two channels. Since we only consider the case that there are at most two different sources in diversity reception in this analysis, each active call in the handoff region will hold at most two channels from two neighboring cells, one from each. We can distinguish the active calls in the handoff region into two classes by the number of its carried channels, as follows:

1. There are three different conditions in which an active call in the handoff region will carry two channels. The following are their generated rates:

$$F_{2n} \stackrel{\Delta}{=} \text{ the generated rate of new calls with two channels}$$

$$= \Lambda_{n1} \cdot (1 - P_{\mathrm{B}})^2 \tag{10.120}$$

$$F_{2h} \stackrel{\Delta}{=} \text{ the arrival rate of handoff calls with two channels}$$

$$= \Lambda_{\mathrm{h}} \cdot (1 - P_{\mathrm{HB}}) \tag{10.121}$$

$$F_{2t} \stackrel{\Delta}{=} \text{ the transition rate of active calls from the normal region}$$
$$\text{that will hold two channels}$$

$$= \Gamma \cdot (1 - P_{\mathrm{HB}}) \tag{10.122}$$

where Γ is the total rate of calls transiting from the normal region to the handoff region

$$\Gamma = \sum_{i=0}^{C} \sum_{j=0}^{C-i} (j \cdot \mu_D) \cdot p(i, j.0) + \sum_{i=0}^{C} \sum_{q=1}^{Q} (j \cdot \mu_D) \cdot p(i, j, q)|_{j=C-i} \tag{10.123}$$

2. Any other active call, which has been acquired by the cell of interest, will hold exactly one channel. The total generated rate of this kind of call in the handoff region is

$$F_1 = \Lambda_{n1} \cdot (1 - P_{\mathrm{B}}) \cdot P_{\mathrm{B}} + \Gamma \cdot P_{\mathrm{HB}} \tag{10.124}$$

The average number of active calls in the handoff region that hold two channels is

$$k_2 = A_{\mathrm{ch}} \cdot \frac{F_{2n} + F_{2h} + F_{2t}}{F_{2n} + F_{2h} + F_{2t} + F_1} \tag{10.125}$$

The total channel efficiency in a cell can be defined as

$$E = 1 - \frac{1}{2} \cdot \left(\frac{k_2}{A_{\mathrm{c}}} \right) \tag{10.126}$$

For illustration purposes we use the following parameters [12]:

$$C = 12, \quad C_{\mathrm{h}} = 2, \quad \mu_{\mathrm{c}} = 0.01, \quad \mu_{\mathrm{dc}} = (T_{\mathrm{dc}})^{-1} = 0.03 \quad \text{and} \quad Q = 4.$$

'a' is fraction of occupied area of the handoff region in a cell.
A diversity reception in the handoff region is assumed.
In numerical examples, two conditions are considered.

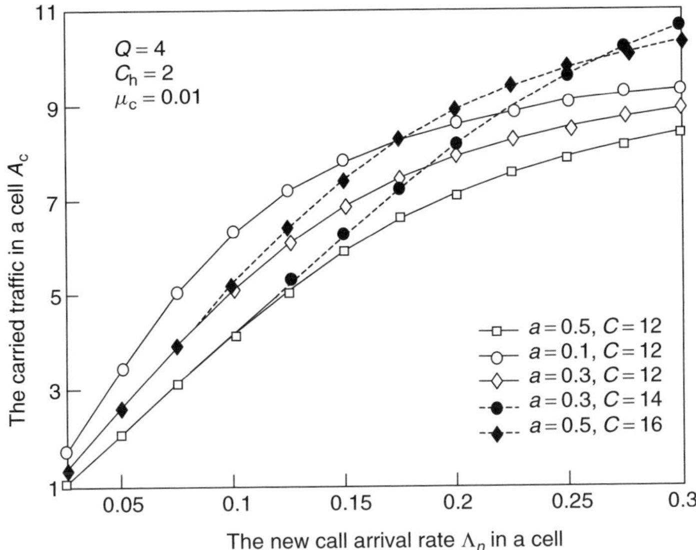

Figure 10.42 The carried traffic in a cell versus the new call arrival rate Λ_n.

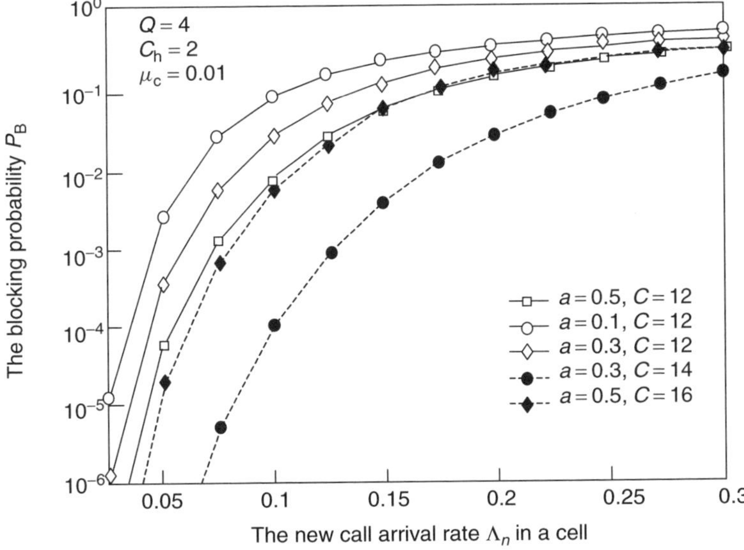

Figure 10.43 The new call blocking probability versus the new call arrival rate Λ_n.

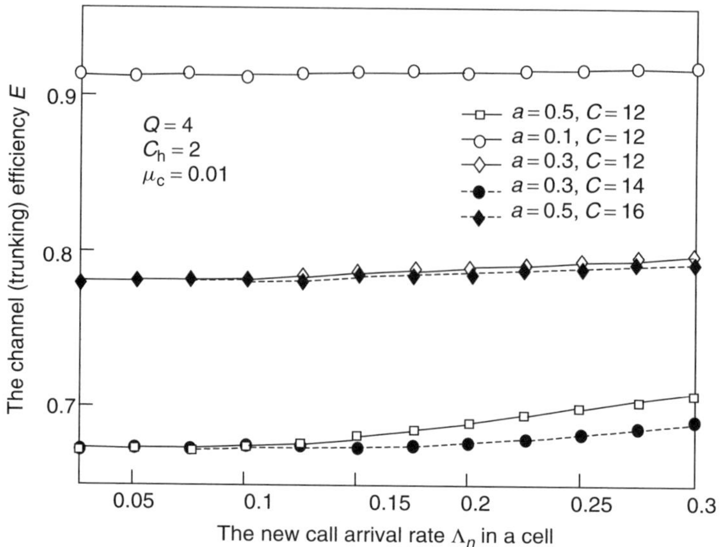

Figure 10.44   The channel efficiency versus the new call arrival rate Λ_n.

Condition 1 Fixed channel capacity $C = 12$ in a cell regardless of the value of 'a'. In this case, the voice/data quality will be better for a larger 'a' because of the higher SIR from the diversity reception in the larger area.

Condition 2 The channel capacity in each cell is increased under the requirement of the same voice/data quality when the value of 'a' becomes larger. Here, we adopt $C = 14, 16$ when $a = 0.3, 0.5$, respectively.

The results are shown in Figures 10.42 to 10.44 [12]. In Figure 10.42 one can see that the carried traffic in a cell will be increased for larger λ_n if C is increased with increasing α.

The blocking probability will be reduced if C is increased with a as shown in Figure 10.43.

As one could expect under the above conditions, if C is increased with larger a, the channel efficiency E would be reduced as shown in Figure 10.44.

10.8 A MEASUREMENT-BASED PRIORITIZATION SCHEME FOR HANDOVERS

In this section, we present a method of improving the QoS in mobile cellular CDMA systems based on prioritization of handover requests. The objective here is to improve the perceived quality of cellular service by minimizing both the probability of forced termination of ongoing calls due to handover failures and the degradation in spectrum utilization. A model is based on a nonpreemptive priority queueing discipline. New calls, which originate within the cell at a Poisson rate, are blocked if all channels are occupied.

Handover requests are queued such that as soon as a channel is available, it is offered to the mobile subscriber with the measurement results closest to the minimum acceptable power level for communication. Handover requests arrive with a Poisson distribution and are queued if no channel is available at the time of arrival. The queue is dynamically reordered as new measurement results are submitted. Service rate is given by channel occupancy time distribution and is assumed to be exponential. The performance criteria of interest are probability of handover failure, probability of call blocking, delay and carried versus offered traffic.

The so-*called 'guard channel' concept*, described in the previous section, offers a means of improving the probability of successful handover by simply reserving a fixed or dynamically adjustable number of channels exclusively for handover requests. Having exclusive handover channels is seen to bear the risk of inefficient spectrum utilization (see results from Figure 10.44). The use of guard channels requires careful determination of the optimum number of these channels, knowledge of the traffic pattern of the area and estimation of the channel occupancy time distributions. Reserving channels for handovers means that fewer channels are being granted to originating calls so that the total carried traffic is reduced. This disadvantage can be overcome to a certain extent by allowing the *queueing of originating call attempts*, which are considerably less sensitive to delays than handovers.

Queueing handover requests, with or without reserving channels for handovers, is another method of reducing the probability of forced termination at the expense of an increased call blocking probability and a decrease in the ratio of carried-to-admitted *traffic*. The reason is that no originating call is granted a channel before the handover requests in the queue are served. Queueing of handover requests is made possible by the existence of the time interval the mobile station (MS) spends in the handover area, where it is physically capable of communicating with both the current and the next BS as shown in Figure 10.45. The fact that a successful handover can take place anywhere during this interval marks a certain amount of tolerance in the delay for the actual channel assignment to the handover request.

10.8.1 Measurement-based prioritization scheme

The handover process is initiated by the issuing of a handover request when the power received by the MS from the BS of a neighboring cell exceeds the power received from the BS of the current cell by a certain amount. This is a fixed value called the handover threshold. For successful handover, a channel must be granted to the handover request before the power received by the MS reaches the receiver threshold, that is, the threshold in the received power, below which acceptable communication with the BS of the current cell is no longer possible. The handover area is the area where the ratio of received power levels from the current and the target BSs is between the handover and the receiver thresholds. If the power level from the current BS falls below the receiver threshold prior to the MS being assigned a channel by the target BS, the call is terminated; therefore, the handover attempt fails.

As previously stated, *queueing of handovers* is possible because the MS spends some period of time in the handover area, during which its communication with the current

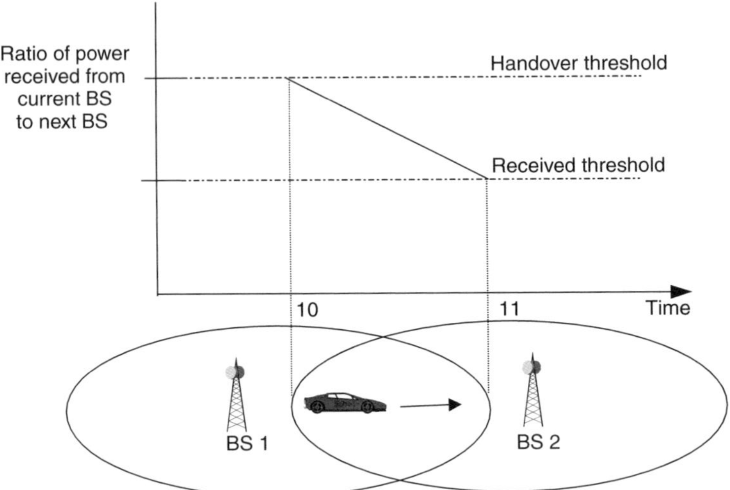

Figure 10.45 Queueing of the handovers during the 'degradation interval' the period in which the MS spends in the handover area where the received power level is between the handover and the receiver thresholds.

BS degrades at a rate depending on its velocity. This degradation is easily monitored by means of radio channel measurements, usually taken by the MS and submitted to the network. The queueing schemes suggested for protection of handovers in Section 10.7 utilize the First-in, First-out (FIFO) queue discipline, which does not consider the rate of degradation in the current radio channel. *Measurement-based prioritized scheme* (MBPS), which is described in further detail below, provides a significantly improved probability of successful handover by taking into account the degradation rate.

In MBPS, if all channels of a cell are occupied, calls originating within that cell are simply blocked and the handover requests to that cell are queued. MBPS is a nonpreemptive dynamic priority discipline. The handover area can be viewed as regions marked by different ranges of values of the power ratio, corresponding to the priority levels such that the highest priority belongs to the MS whose power level is closest to the receiver threshold. On the other end, the MS that has just issued a handover request has the least priority. Obviously, the last comer joins the end of the queue. The power levels are monitored continuously, and the priority of an MS dynamically changes depending purely on the power level it receives while waiting in the queue. A queued MS gains higher priority as its power ratio decreases from the handover threshold to the receiver threshold. The MSs waiting for a channel in the handover queue are sorted continuously according to their priorities. When a channel is released, it is granted to the MS with the highest priority.

MBPS is designed as a handover protection method, which a cellular communication network can utilize along with any channel allocation strategy, with or without the employment of guard channels. The queueing can be performed at the BS or the mobile switching center (MSC) depending on the intelligence distribution between these cellular network components. The basic idea that originating calls are not assigned channels until the queued handover requests are served remains the same. If handover request is queued at the MSC, there is a separate queue maintained for each cell. One should be aware that the employment of guard channels has the effect of reducing the number of handover requests to be queued. However, since we start with a congested cell whose channels are already occupied, the number of guard channels, if any, will not have an impact on the waiting time distributions of the arriving handover requests.

As before *call blocking probability*, P_B, is the probability that a new call cannot enter service because of the unavailability of channels. In MBPS, new calls are served only when a channel is available and no handover request exists in the queue. *Probability of forced termination*, P_F, is the probability that an originated call is eventually not completed because of an unsuccessful handover attempt. P_F, therefore, gives the percentage of handover requests that are not served because the power received by the MS from the current BS approaches the receiver threshold before a channel is granted.

One of the most obvious merits of a cellular network is the total traffic it carries. *Total carried traffic* is the amount of traffic admitted to the cellular network defined by equation (10.113) as opposed to the offered load. In light traffic conditions, the carried traffic can be taken to be equal to the offered traffic. However, in general, the carried traffic is less than the offered load because of blocking of calls and handover failures. The percentage of the offered load that is carried is certainly desired to be as high as possible. This percentage decreases with the increase of offered load, and probabilities of call blocking and handover failure. Another major performance measure is *delay*, as it is in any queueing system. MBPS aims to minimize the time spent by the MS on a poorer communication channel in terms of the power ratio in the handover area before a channel is granted. The aim is to improve perceived quality of cellular service. We call the interval between the submission of the handover request and the actual handover process *the degradation interval*, referring to the fact that communication with the current channel deteriorates during this period. The objective is to minimize the time spent by the MS in higher priority, corresponding to poorer signal reception, by favoring those MSs that receive the lowest power level from their current BS in channel assignment.

For illustration purposes a system with the following parameters is assumed [13]: Channel occupancy times are drawn from an exponential distribution with a mean of 60 s. An occupied channel will be released either because of voluntary termination of a call by the user or because of a handover that is completed with success or failure. Whether the channel was assigned to a call that was initiated within the cell or to a handover from another cell does not have to be taken into consideration. It is also not important whether the channel was released because of voluntary termination of the call by the user or because of a handover to another channel. The memoryless property of the exponential distribution is exploited. The only effect of handovers on the channel occupancy time distribution is that the mean is less than that of call duration distribution, which is generally accepted to be an exponential with a mean of 100 s.

It is accepted that once the handover request is issued the power that the MS receives from the current BS will monotonically degrade as in Figure 10.45. The rate of the degradation, and hence the maximum tolerable degradation interval depends on the velocity of the MS. We assume that the velocities of the MSs have a normal distribution; therefore, the maximum tolerable degradation interval associated with each MS entering the handover area is drawn from a normal distribution with a mean of 10 s and a standard deviation of 2 s. This implies that an MS is physically able to communicate with two BSs during one-tenth to one-fourth of the total channel holding time, which is a reasonable assumption. These values can be altered for different cell sizes and velocity ranges; however, they will not impact the results of comparison of the schemes under consideration.

The radio channel measurements that the prioritization is based on are assumed to be monitored continuously. In the simulation model, each handover request starts a timer. The maximum degradation interval that the handover request can stand is assigned from the normal distribution mentioned above. If the timer of a handover request reaches its maximum tolerance prior to being served, a handover failure has occurred. This introduces a natural boundary on the queue size; therefore, the assumption of an infinite buffer size is not unrealistic. The current power level for each of the queued handover requests is calculated from the timer value divided by the maximum tolerable degradation interval for that MS. This corresponds to a projection onto the power level axis in Figure 10.46. The MS with the lowest power level, that is, closest to the receiver threshold, has the highest priority. Whenever a channel is released, it is given to the MS with the highest priority.

Figure 10.47 presents the probability of forced termination versus offered traffic obtained by simulation [13]. As it was noted earlier, in the nonprioritized call handling scheme, this probability is equal to the probability of call blocking at origination since no distinction between new call attempts and handover requests is being made. The handover prioritization scheme using FIFO queueing (Section 10.7) provides some reduction in the probability of handover failure. As expected, the MBPS (Section 10.8) provides a greater improvement, which becomes more obvious with increasing offered traffic load.

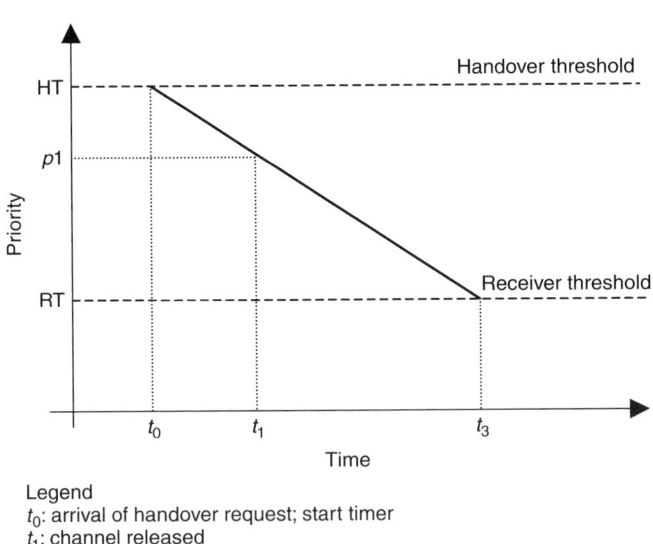

Figure 10.46 Priority profile for the queued handover request.

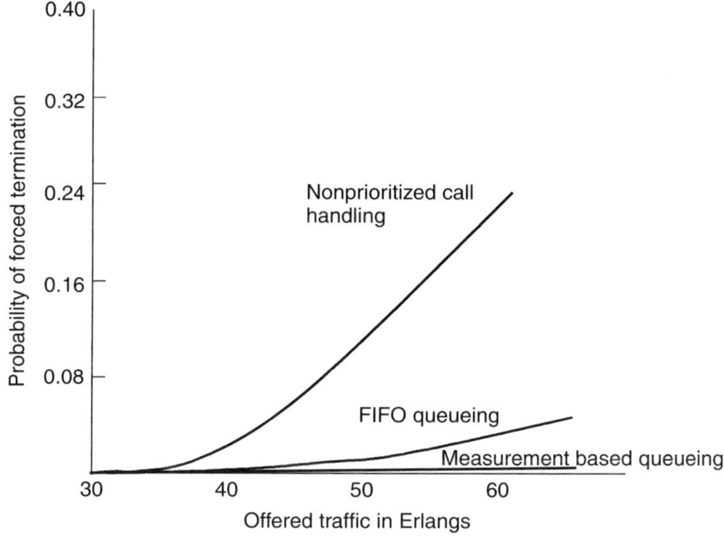

Figure 10.47   Probability of forced termination versus offered load for 50% handover traffic.

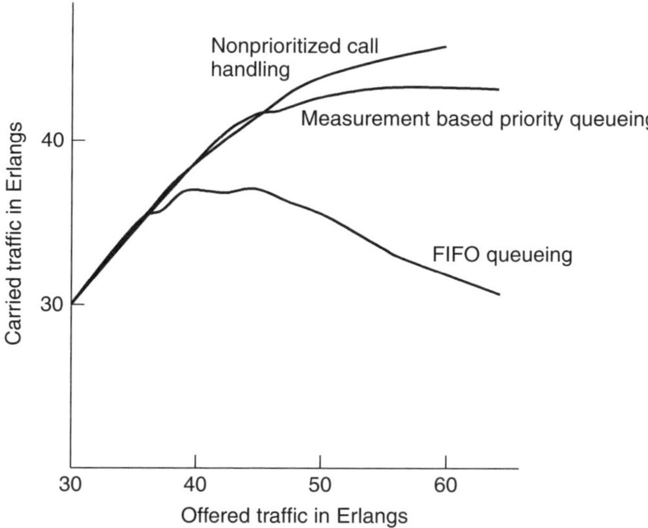

Figure 10.48   Carried traffic versus offered load for 50% handover traffic.

Carried traffic is shown in Figure 10.48 [13]. One can see that MBPS performs better than FIFO system described in Section 10.7.

The same system performs better with respect to delay in handover too, which can be seen from Figure 10.49.

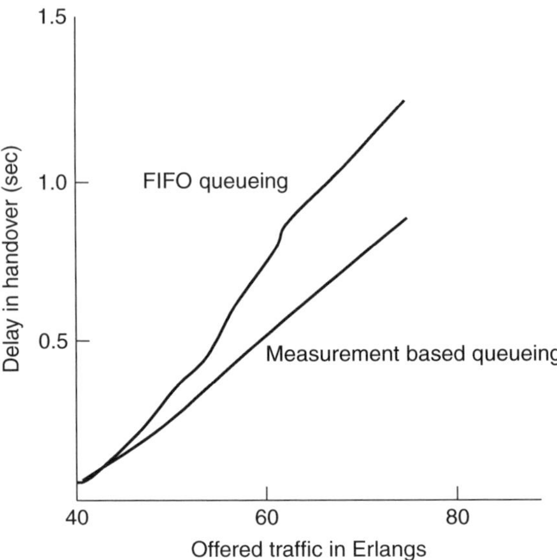

Figure 10.49 Average delay in handover versus offered load for 20% handover traffic.

10.8.2 Analytical modeling

For the nonprioritized call handling scheme, the probability of forced termination equals the probability of call blocking, which is given by the well-known Erlang B formula for the $M/M/c/c$ queue.

$$P_{\mathrm{B}} = P_{\mathrm{F}} = \frac{\frac{(c\rho)^c}{c!}}{\sum_{i=1}^{c} \frac{(c\rho)^i}{i!}}, \qquad \rho = \lambda/c\mu \tag{10.127}$$

In equation (10.127) it is assumed that $1/\mu = 1\,\mathrm{min}$ and, thus, the number of Erlangs of traffic is numerically equal to the sum of the arrival rates of originating calls and handover requests. The number of channels, c, is accepted to be equal to 50. Substituting these values into the above formula and varying the offered traffic, λ/μ, the curve in Figure 10.50 was obtained. This figure shows this curve together with the curve obtained from the simulation of the nonprioritized call handling scheme.

On the basis of Section 10.7, the FIFO queueing scheme can be approximated by an $M/M/c$ queue with nonhomogeneous arrival rates. Until all servers are occupied, the arrival rate is that of the sum of originating calls and handovers, whereas once the number in the system equals the number of servers, only handovers are queued. The blocking probability of originating calls is simply given by the probability of the number in the system being equal to or more than the number of servers, c

$$P_{\mathrm{B}} = \sum_{n=c}^{\infty} p_n \tag{10.128}$$

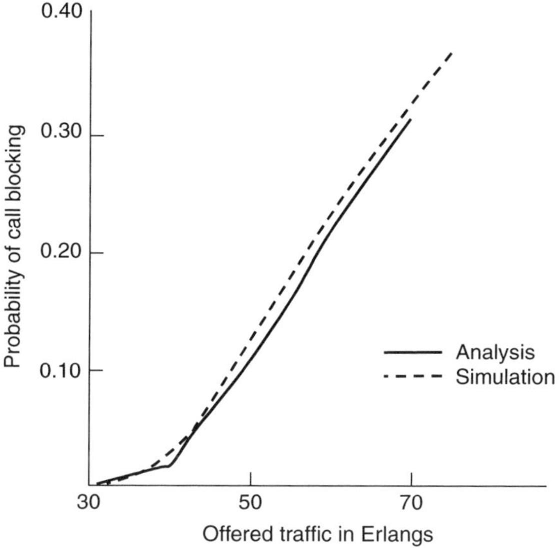

Figure 10.50 Probability of call blocking versus offered traffic for nonprioritized call handling.

and when $n \geq c$

$$p_n = \frac{(\lambda_C + \lambda_H^c)\lambda_H^{n-c}}{c^{n-c}c!\mu^n} p_0 \tag{10.129}$$

where λ_C and λ_H are the arrival rates of new calls and handovers, respectively. Using the above expression for p_n and

$$\sum_0^\infty p_n = 1 \tag{10.130}$$

p_0 is found to be

$$p_0 = \left[\sum_{n=0}^{c-1} \frac{1}{n!} \left(\frac{\lambda_C + \lambda_H}{\mu} \right)^n + \frac{1}{c!} \left(\frac{\lambda_C + \lambda_H}{\mu} \right)^c \left(\frac{c\mu}{c\mu - \lambda_H} \right) \right]^{-1} \tag{10.131}$$

On the basis of the above equations, the probability of call blocking for the FIFO scheme is found to be

$$P_B = \frac{1}{c!} \left(\frac{\lambda_C + \lambda_H}{\mu} \right)^c \frac{c}{c - \lambda_H/\mu} \tag{10.132}$$

Figure 10.51 demonstrates the accuracy of this model by comparing the analytical and simulation results in the case of 20% handover traffic. The average queue size is given by

$$L_q = \sum_{n=c}^\infty (n - c)p_n \tag{10.133}$$

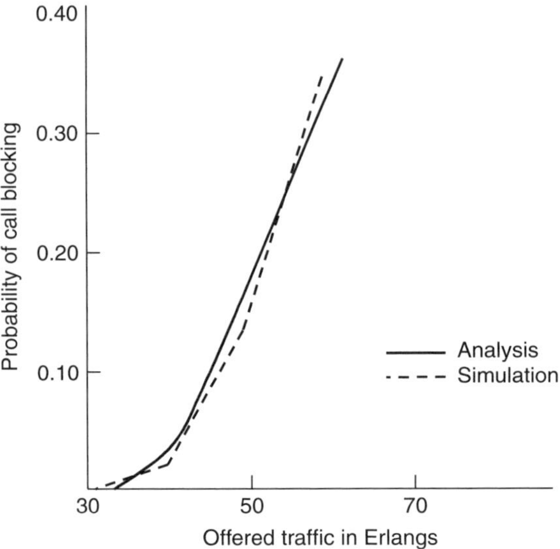

Figure 10.51   Probability of call blocking versus offered traffic for FIFO queueing of handover requests; handover traffic of 20%.

Substituting the expression for p_n,

$$L_q = p_0 \left(\frac{\lambda_C + \lambda_H}{\mu} \right)^c \frac{1}{c!} \frac{\rho}{(1 - \rho)^2}, \quad \rho = \frac{\lambda_H}{c\mu} \tag{10.134}$$

Again, this result is compared to the simulation output in Figure 10.52 [13].

The probability of forced termination, that is, handover failure, is the probability that the waiting time in the queue exceeds the maximum tolerable degradation interval for an MS, which was assumed to be normally distributed with a mean of 10 seconds and standard deviation of 2. In this case we first compute the cumulative queue waiting time distribution $W_q(t) = \text{Probability} \{T_H \leq t)$, where T_H denotes the random amount of time spent in the queue and t stands for any arbitrary point in time. Letting T_d represent the normally distributed degradation interval, the probability of forced termination in a FIFO scheme is given by $P_F = 1 - W_q(T_d)$

$$W_q(t) = \sum_{n=c}^{\infty} [\Pr(n - c + 1) \text{ completions take } \leq t | \text{arrival found } n] p_n + W_q(0) \tag{10.135}$$

where $W_q(0) = \text{Probability} (c - 1 \text{ or less in the system})$, that is,

$$W_q(0) = 1 - \frac{1}{c!} \left(\frac{\lambda_C + \lambda_H}{\mu} \right)^c \left(\frac{c}{c - \lambda_H/\mu} \right) p_0 \tag{10.136}$$

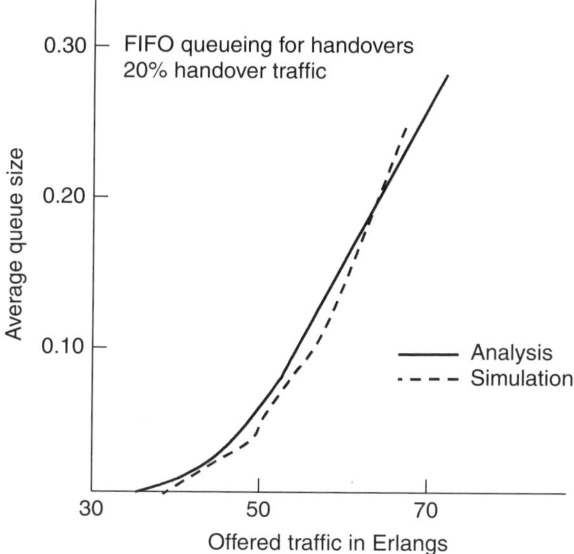

Figure 10.52   Average queue size versus offered traffic for FIFO queueing of handover requests; handover traffic of 20%.

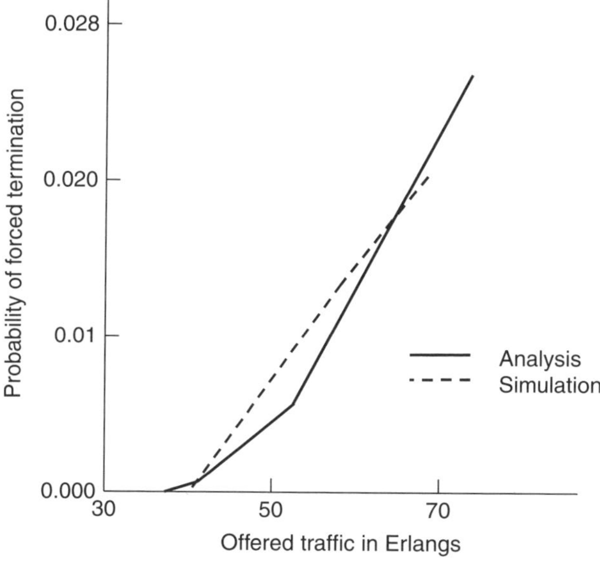

Figure 10.53   Probability of forced termination versus offered traffic for FIFO queueing of handover requests; handover traffic of 20%.

The cumulative queue waiting time distribution is obtained by making the substitution for p_n from equation (10.129) and integrating the service times from 0 to t. The resulting expression is

$$W_{\mathrm{q}}(t) = p_0 \left(\frac{\lambda_{\mathrm{C}} + \lambda_{\mathrm{H}}}{\mu}\right)^c \frac{1}{(c-1)!} \frac{(1 - e^{(\mu c - \lambda_{\mathrm{H}})t})}{(c - \lambda_{\mathrm{H}}/\mu)} + W_{\mathrm{q}}(0) \qquad (10.137)$$

The probability of forced termination is given by

$$P_{\mathrm{F}} = 1 - \int_0^\infty W_{\mathrm{q}}(t) f_{\mathrm{Td}}(t) \, \mathrm{d}t, \ f_{\mathrm{Td}}(t) \text{ is } N(\cdot, \cdot) \qquad (10.138)$$

The integral is numerically computed and the results are graphed in Figure 10.53 together with the simulation results.

A number of related issues to resource management and access control are discussed in References [14–82].

SYMBOLS

$\boldsymbol{P}\{P_i\}$ – transmitted power vector
$\boldsymbol{R}\{R_i\}$ – vector of rates
$\boldsymbol{\Gamma}\{\gamma_i\}$ – vector of required SNRs
$\boldsymbol{p}\{p_i\}$ – power limits
$\boldsymbol{r}\{r_i\}$ – rate limits
$\boldsymbol{h}\{h_i\}$ – channel gains
$\boldsymbol{Q}\{q_i\}$ – received powers
W – bandwidth
N – number of users
η_0 – noise density
v, d (index) – voice, data
$v, V(n)$ – number of active voice users
$d, D(n)$ – number of data users who transmit
$S, S(n)$ – the load
$a_{\mathrm{v}} = W/R_{\mathrm{v}}\gamma_{\mathrm{v}} + 1$
$a_{\mathrm{D}} = W/R_{\mathrm{D}}\gamma_{\mathrm{D}} + 1$
λ – voice model transmission probability from off to on state
μ – voice model transmission probability from on to off state
$P(i/j)$ – Markov model state transition probability
$\Pi_{\mathrm{v}}(k)$ – Markov model steady state probability
P_{out} – outage probability
δ – required threshold on outage probability
π^j – transmission probability
Ω – load threshold
K – transmission probability exponent (j) correction factor
$v(\ )$, ξ – activity indicators for voice, data
$G_{\mathrm{V}}, G_{\mathrm{D}}$ – processing gains

REFERENCES

1. Sampath, A., Kumar, P. S. and Holtzman, J. M. (1995) Power control and resource management for a multimedia CDMA wireless system. *Proc. PIMRC*, Vol. 1, pp. 21–25.
2. Aoki, M. (1971) *Introduction to Optimization Techniques*. The McMillan Book Company, New York.
3. Sampath, A. and Holtzman, J. M. (1997) Access control of data in integrated voice/data CDMA systems: benefits and tradeoffs. *IEEE J. Select. Areas Commun.*, **15**, 1511–1526.
4. Viterbi, A. J. (1994) Capacity of a simple stable protocol for short message service over a CDMA network, in Blahut, R. E. (ed.) *Communications & Cryptography*, Kluwer AP, Boston, pp. 423–429.
5. Comaniciu, C. and Mandayam, N. (2000) Delta modulation based prediction for access control in integrated voice/data CDMA systems. *IEEE J. Select. Areas Commun.*, **18**(1), 112–123.
6. Brand, A. E. and Aghvami, A. H. (1996) Performance of a joint CDMA/PRMA protocol for mixed voice/data transmission for third generation mobile communication. *IEEE J. Select. Areas Commun.*, **14**, 1698–1707.
7. Pursley, M. B. (1977) Performance evaluation for phase-coded spread spectrum multiple-access communication-part I: system analysis. *IEEE Trans. Commun.*, **COM-25**(8), 795–799.
8. Roefs, H. F. A. and Pursley, M. B. (1976) Correlation parameters of random sequences and maximal length sequences for spread-spectrum multiple access communication. *IEEE Canadian Communications Power Conference*, pp. 141–143.
9. Chang, C. J. *et al.* (2000) Fuzzy/neural congestion control for integrated voice and data DS-CDMA/FRMA cellular networks. *IEEE J. Select. Areas Commun.*, **18**(2), 283–294.
10. Lin, C. T. and Lee, C. S. G. (1996) *Neural Fuzzy Systems*. Englewood Cliffs, NJ: Prentice-Hall.
11. Dziong, Z., Jia, M. and Mermelstein, P. (1996) Adaptive traffic admission for integrated services in CDMA wireless–access networks. *IEEE J. Select. Areas Commun.*, **14**(9), 1737–1747.
12. Su, S., Chen, J. and Huang, J. H. (1996) Performance analysis of soft handoff in CDMA cellular networks. *IEEE J. Select. Areas Commun.*, **14**(9), 1762–1769.
13. Tekinay, S. and Jabbari, B. (1992) A measurement-based prioritization scheme for handovers in mobile cellular networks. *IEEE J. Select. Areas Commun.*, **10**(8), 1343–1350.
14. Viterbi, A. J. (1995) *CDMA: Principles of Spread Spectrum Communication*. Reading, MA: Addison-Wesley.
15. Baier, A. *et al.* (1994) Design study for a CDMA-based third-generation mobile radio system. *IEEE J. Select. Areas Commun.*, **12**(4), 733–743.
16. Aghvami, A. H. (1994) Future CDMA cellular mobile systems supporting multi-service operation. *PIMRC '94, WCN*, The Hague, September 1994, pp. 1276–1279.
17. Goodman, D. J. (1991) Trends in cellular and cordless communications. *IEEE Commun. Mag.*, **29**(6), 31–40.
18. Goodman, D. J. (1990) Cellular packet communications. *IEEE Trans. Commun.*, **38**(8), 1272–1280.
19. Goodman, D. J. *et al.* (1989) Packet reservation multiple access for local wireless communications. *IEEE Trans. Commun.*, **37**(8), 885–890.
20. Goodman, D. J. and Wei, S. X. (1991) Efficiency of packet reservation multiple access. *IEEE Trans. Veh. Technol.*, **40**(1), 170–176.
21. Wilson, N. D., Ganesh, R., Joseph, K. and Raychaudhuri, D. (1993) Packet CDMA versus dynamic TDMA for multiple access in an integrated voice/data PCN. *IEEE J. Select. Areas Commun.*, **11**(6), 870–884.
22. Ganesh, R., Joseph, K., Wilson, N. D. and Raychaudhuri, D. (1994) Performance of cellular packet CDMA in an integrated voice/data network. *Int. J. Wireless Inform. Networks*, **1**(3), 199–221.
23. Viterbi, A. J. (1991) Wireless digital communications: a view based on three lessons learned. *IEEE Commun. Mag.*, **29**(9), 33–36.
24. Baier, P. W. (1994) CDMA or TDMA? CDMA for GSM? *PIMRC '94, WCN*, The Hague, September 1994, pp. 1280–1284.

25. Blanz, J., Klein, A., Nasshan, M. and Steil, A. (1994) Performance of a cellular hybrid CIrDMA mobile radio system applying joint detection and coherent receiver antenna diversity. *IEEE J. Select. Areas Commun.*, **12**(4), 568–579.

26. *An Overview of the Application of CDMA to Digital Cellular Systems and Personal Cellular Networks*. San Diego, CA: Qualcomm Incorporated, 1992.

27. Gruber, J. G. and Strawczynski, L. (1985) Subjective effects of variable delay and speech clipping in dynamically managed voice systems. *IEEE Trans. Commun.*, **COM-33**(8), 801–808.

28. Honig, M. L. and Kim, J. B. (1996) Allocation of DS-CDMA parameters to achieve multiple rates and qualities of service. *Proc. GLOBECOM*, Vol. 3, pp. 1974–1978.

29. Sampath, A., Mandayam, N. B. and Holtzman, J. M. (1996) Analysis of an access control mechanism for data traffic in an integrated voice/data wireless CDMA system. *Proc. VTC*, Vol. 3, pp. 1448–1452.

30. Morrow, R. K. and Lehnert, J. S. (1989) Bit-to-bit error dependence in slotted DS/SSMA packet systems with random signature sequences. *IEEE Trans. Commun.*, **37**(10), 1052–1061.

31. Morrow, R. K. (1992) Packet throughput in slotted ALOHA DS/SSMA radio systems with random signature sequences. *IEEE Trans. Commun.*, **40**(7), 1223–1230.

32. Perle, H.-C. and Rechberger, B. (1994) Throughput analysis of direct-sequence CDMA-ALOHA in a near/far environment. *PIMRC '94, WCN*, The Hague, September 1994, pp. 1040–1044.

33. Lin, S. (1970) *An Introduction to Error-Correcting Codes*. Englewood Cliffs, NJ: Prentice-Hall.

34. Wong, W. C. and Goodman, D. J. (1992) A packet reservation multiple access protocol for integrated speech and data transmission. *IEE Proc.-I*, **139**(6), 607–612.

35. *Speech and Voiceband Data Performance Requirements for Future Public Land Mobile Telecommunication Systems (FPLMTS)*, Recommendation ITU-R M.1079, 1994.

36. Gilhousen, K. S., Jacobs, I. M., Padovani, R., Viterbi, A. J., Weaver Jr, L. A. and Wheatley III, C. E. (1991) On the capacity of a cellular CDMA system. *IEEE Trans. Veh. Technol.*, **40**, 303–312.

37. Taaghol, P., Tafazolli, R. and Evans, B. G. (1997) On the reservation multiple access protocols for future mobile communication systems. *Proc. IEEE VTC '97*, Vol. 3, pp. 1523–1527.

38. Narasimhan, P. and Yates, R. D. (1996) A new protocol for the integration of voice and data over PRMA. *IEEE J. Select. Areas Commun.*, **14**, 623–631.

39. Mandayam, N. B. and Holtzman, J. M. (1995) Analysis of a simple protocol for short message service data service in an integrated voice/data CDMA system. *Proc. MILCOM*, San Diego, CA, pp. 1160–1164.

40. Soroushnejad, M. and Geraniotis, E. (1995) Multiple-access strategies for an integrated voice/ data CDMA packet radio network. *IEEE Trans. Commun.*, **43**, 934–945.

41. Haykin, S. and Li, L. (1995) Nonlinear adaptive prediction of nonstationary signals. *IEEE Trans. Signal Process.*, **43**, 526–535.

42. Chang, P. R. and Hu, J. T. (1997) Optimal nonlinear adaptive prediction and modeling of MPEG video in ATM networks using pipeline recurrent neural networks. *IEEE J. Select. Areas Commun.*, **15**, 1087–1100.

43. Williams, R. L. and Zipser, D. (1989) A learning algorithm for continually running fully recurrent neural networks. *Neural Comput.*, **1**, 270–280.

44. Tarraf, A. A., Habib, I. W. and Saadawi, T. N. (1995) Reinforcement learning based neural network congestion controller for ATM networks. *Proc. IEEE MILCOM*, Vol. 2, pp. 668–672.

45. Lin, C. S. and Cheng, Y. H. E. (1994) Radial basis function networks for adaptive critic learning. *Proc. IEEE International Conference on Neural Networks*, Vol. 2, pp. 903–906.

46. Lin, C. T. and Lee, C. S. G. (1994) Reinforcement structure/parameter learning for neural-network-based fuzzy logic control systems. *IEEE Trans. Fuzzy Syst.*, **2**, 46–63.

47. Pursley, M. B. (1997) Performance evaluation for phase-coded spread spectrum multiple-access communication-part I: system analysis. *IEEE Trans. Commun.*, **COM-25**, 795–799.

48. Mermelstein, P., Jalali, A., and Leib, H. (1993) Integrated services in wireless multiple access networks. *Conference Record, ICC '93*, Geneva, Switzerland, pp. 863–867.

49. Jalali, A. and Mermelstein, P. (1994) Effects of diversity, power control, and bandwidth on the capacity of microcellular CDMA wireless systems. *IEEE J. Select. Areas Commun.*, **12**, 952–961.

50. Gilhousen, K. S., Jacobs, L. M., Padovani, R., Viterbi, A. L., Weaver Jr, L. A. and Wheatley III, C. E. (1991) On the capacity of a cellular CDMA system. *IEEE Trans. Veh. Technol.*, **40**(2), 303–312.

51. Yang, W. and Geraniotis, E. (1994) Admission policies for integrated voice and data traffic in CDMA packet radio networks. *IEEE J. Select. Areas Commun.*, **12**(4), 654–664.

52. Zhang, C. G., Hafez, H. M. and Falconer, D. D. (1994) Traffic handling capability of a broadband indoor wireless network using CDMA multiple access. *IEEE J. Select. Areas Commun.*, **12**(4), 645–653.

53. Liu, Z. and Zarki, M. E. (1994) SIR–based call admission control for DSCDMA cellular systems. *IEEE J. Select. Areas Commun.*, **12**(4), 638–644.

54. Lee, W. C. Y. (1991) Overview of cellular CDMA. *IEEE Trans. Veh. Technol.*, **40**(2), 291–302.

55. Ariyavisitakul, S. (1992) SIR-based power control in a CDMA system. *Proc. GLOBECOM*, pp. 868–873.

56. Stuber, G. L., Yiin, L., Long, E. M. and Yang, K. (1991) Outage control in digital cellular systems. *IEEE Trans. Veh. Technol.*, **40**(1), 177–187.

57. Pickholtz, R. L., Milstein, L. B. and Schilling, D. L. (1991) Spread spectrum for mobile communications. *IEEE Trans. Veh. Technol.*, **40**(2), 313–322.

58. Chang, L. F. and Ariyavisitakul, S. (1991) Performance of a CDMA radio communications system with feedback power control and multipath dispersion. *Proc. GLOBECOM*, pp. 1017–1021.

59. Proakis, J. G. (1989) *Digital Communications*. New York: McGraw-Hill.

60. *Mobile Station-Base Station Compatibility Standard for Dual-Mode Wideband Spread Spectrum Cellular System*, EIA/T7A/IS-95 Interim Standard, Telecommunication Industry Association, July 1993.

61. Murase, A., Symington, I. C. and Green, E. (1991) Handover criterion for macro and microcellular systems. *Proc. VTC*, pp. 524–530.

62. Hong, D. and Rappaport, S. S. (1986) Traffic model and performance analysis for cellular mobile radio telephone systems with prioritized and nonprioritized hand-off procedures. *IEEE Trans. Veh. Technol.*, **VT-35**, 77–92.

63. Rappaport, S. S. (1991) The multiple-call hand-off problem in high-capacity cellular communications systems. *IEEE Trans. Veh. Technol.*, **40**, 546–557.

64. Rappaport, S. S. (1991) Modeling the hand-off problem in personal communications networks. *Proc. VTC*, pp. 517–523.

65. Tekinay, S. and Jabbari, B. (1992) A measurement-based prioritization scheme for handover and channel in mobile cellular networks. *IEEE J. Select. Areas Commun.*, **10**(8), 1343–1350.

66. Su, S. L. and Chen, L. S. (1993) Exploration of handoff procedure for mobile cellular communication system. *ISCOM '93*, Taipei, Taiwan, pp. 32–39.

67. Rappaport, S. S. (1993) Blocking, hand-off and traffic performance for cellular communication systems with mixed platforms. *IEE Proc.-1*, **140**(5), 389–401.

68. Purzynski, C. and Rappaport, S. S. (1993) Traffic performance analysis for cellular communication systems with mixed platform types and queued hand-offs. *Proc. VTC*, pp. 172–175.

69. Purzynski, C. and Rappaport, S. S. (1995) Multiple-call hand-off problem with queued hand-offs and mixed platform types. *IEE Proc. -Commun.*, **142**(1), 31–39.

70. Simmonds, C. M. and Beach, M. A. (1993) Network planning aspects of DSCDMA with particular emphasis on soft handoff. *Proc. VTC*, pp. 846–849.

71. Swales, S. C. *et al.* Handoff requirements for a third generation DSCDMA air interface. *IEE Colloquium, Mobility Support Personal Communications*, 1993.

72. Viterbi, A. J., Viterbi, A. M., Gilhousen, K. S. and Zehavi, E. (1994) Soft handoff extends CDMA cell coverage and increases reverse link capacity. *IEEE J. Select. Areas Commun.*, **12**(8), 1281–1288.

73. Tekinay, S. and Jabbari, B. (1991) Handover policies and channel assignment strategies in mobile cellular networks. *IEEE Commun. Mag.*, **29**(11), 42–46.

74. Hong, D. and Rappaport, S. S. (1986) Traffic model and performance analysis for cellular mobile radio telephone systems with prioritized and nonprioritized handoff procedures. *IEEE Trans. Veh. Technol.*, **VT-35**(3), 77–92.

75. Guerin, R. (1986) *Queueing and Traffic in Cellular Radio*. Ph.D. Thesis, Department of Electronics Engineering, California Institute of Technology, Pasadena, CA.

76. Guerin, R. (1987) Channel occupancy time distribution in a cellular radio system. *IEEE Trans. Veh. Technol.*, **35**(3).

77. Viterbi, A. M. and Viterbi, A. J. (1993) Erlang capacity of a power controlled cellular CDMA system. *IEEE J. Select. Areas Commun.*, **11**, 892–900.

78. Viterbi, A. M. and Viterbi, A. J. (1997) Erlang capacity of a power controlled integrated voice and data CDMA system. *Proc. VTC*, Vol. 3, pp. 1557–1561.

79. Ramakrishna, S. and Holtzman, J. (1998) A scheme for throughput maximization in a dual-class CDMA system. *IEEE J. Select. Areas Commun.*, **16**, 830–844.

80. Mandayam, N. B., Holtzman, J. M. and Barberis, S. (1995) Erlang capacity for an integrated voice/data CDMA system with variable bit-rate sources. *Proc. PIMRC*, Vol. 3, pp. 1078–1083.

81. Liu, T. and Silvester, J. A. (1998) Joint admission/congestion control for wireless CDMA systems supporting integrated services. *IEEE J. Select. Areas Commun.*, **16**, 845–857.

82. Lathi, B. P. (1989) *Modern Digital and Analog Communication Systems*. Philadelphia, PA: Holt, Rinehart, and Winston.

11

CDMA packet radio networks

11.1 DUAL-CLASS CDMA SYSTEM

In this chapter we consider some additional details of packet transmission in a Code Division Multiple Access (CDMA) radio network. The approach is very much based on Reference [1]. We start with dual-class traffic and then extend the analysis to multimedia systems. These two classes are characterized by the following set of parameters:

Class 1: The users in this class are delay intolerant. When transmitting information, they require support for a constant bit rate of R_1 bit s^{-1}; they can tolerate a bit error rate (BER) of at most P_{b1}.

Class 2: The users in this class are delay tolerant. When transmitting information, they require support for a bit rate of at least $R_{\min}$ bit s^{-1}; they can tolerate a BER of at most P_{b2}.

When not transmitting information, it is assumed that the users still communicate with the base for synchronization purposes. The bit rate used in this synchronization mode is denoted as R_0 bit s^{-1}, and is referred to as the 'idle rate'. One would expect that $R_0 < R_1$, $R_0 < R_{\min}$; more detailed constraints on R_0 are given later.

In an actual system, Classes 1 and 2 could represent voice and data users, respectively. Minimum rate R_0 would be used in a control channel. According to equation (10.10) a unique solution to a minimum total transmit power problem exists if and only if

$$\sum_{i=1}^{N_1} \frac{1}{\left(\dfrac{W}{R_i^{(1)}} \cdot \dfrac{1}{SIR_i^{(1)}} + 1\right)} + \sum_{j=1}^{N_2} \frac{1}{\left(\dfrac{W}{R_i^{(2)}} \cdot \dfrac{1}{SIR_i^{(2)}} + 1\right)}$$

$$< 1 - \frac{IW}{\min_i\left\{\left[P_i^{\mathrm{peak}} h_i \left(\dfrac{W}{R_i \cdot SIR_i} + 1\right)\right]\right\}_{i=1}^{N_1+N_2}} \tag{11.1}$$

By using the following notation
h_1: (mobile to base) gain of the ith generic user

- $\text{SIR}_i^{(1)} = \gamma_1 = \text{constant} \ \forall \ \text{Class 1 users } i;$
  $\text{SIR } \gamma_1 \Leftrightarrow \text{BER } P_{b1}$
- $\text{SIR}_i^{(2)} = \gamma_2 = \text{constant} \ \forall \ \text{Class 2 users } i;$
  $\text{SIR } \gamma_2 \Leftrightarrow \text{BER } P_{b2}$

minimum total transmit power solution is obtained from equation (11.1) for minimum required rate $R_{\min}.P_i^{\text{peak}} \to \infty \forall i$, constraint (11.1) now becomes

$$\frac{N_1}{\left(\dfrac{W}{R_1 \cdot \gamma_1} + 1\right)} + \frac{N_2^{\max}}{\left(\dfrac{W}{R_{\min} \cdot \gamma_2} + 1\right)} < 1 \qquad (11.2)$$

Given that N_1 Class 1 users are present, one may support at most

$$N_2^{\max} = \left\lfloor 1 + \frac{W}{R_{\min} \cdot \gamma_2} - N_1 \cdot \frac{\left(\dfrac{W}{R_{\min} \cdot \gamma_2} + 1\right)}{\left(\dfrac{W}{R_1 \cdot \gamma_1} + 1\right)} \right\rfloor \quad \text{Class 2 users} \qquad (11.3)$$

Given that N_1 Class 1 and up to $N_2^{\max}$ Class 2 users are present, the transmit powers assigned to the users will then be such that the following hold, (see also Chapter 10, Section 10.1). Each Class 1 user and Class 2 user will achieve an signal-to-interference ratio (SIR) of exactly γ_1 and γ_2, respectively; this follows, as noted before, from the assumption of perfect power control. The BER requirements of users of both classes will therefore also be met with equality. The transmitted powers will be such that their sum is as small as possible; hence, the interference to other cells is minimized. It is assumed that admission control will handle the task of ensuring that the number of users of each class satisfies the constraints in equations (11.2 and 11.3).

11.1.1 Maximization of Class 2 throughput

1. *Mode 1-Unscheduled Class 2 Transmissions*:
All N_2 users are allowed to transmit information, each at rate R_2. The rate R_2 is chosen to be the largest possible so as to satisfy constraint (11.1). Since $N_2 \leq N_2^{\max}$, one has $R_2 \geq R_{\min}$. This transmission mode is very similar to that followed in the present systems. A more efficient (from the point of view of throughput) version of this scheme would allow each Class 2 user to transmit at an appropriate different rate.

2. *Mode 2-Scheduled Class 2 Transmissions*:
The Class 2 transmissions are scheduled in such a way that at any given instant, only $k_2(<N_2)$ of users are transmitting information. The remaining $(N_2 - k_2)$ are in contact

with the base at the idle/synchronization rate R_0 bit s^{-1}. When transmitting information, a Class 2 user is allowed to transmit at a rate R_2^*, which, again, is chosen so as to be the maximum value satisfying constraint (11.1). Thus, assuming a fair division of time, each Class 2 user has a 'duty cycle' of a fraction of time when it is transmitting information, given by $\binom{N_2-1}{k_2-1}/\binom{N_2}{k_2} = (k_2/N_2)$. The remaining fraction of time is spent in maintaining synchronization with the base at a rate R_0. If the transmission rate T of Class 2 for $k_2 = 1$ is T_2, then the throughput gain G measured by the ratio of Mode 2 to Mode 1 throughputs is [1]

$$G = \frac{T_2}{R_2} = \frac{\dfrac{R_0}{R_2}\left(\dfrac{W}{R_2\gamma_2} + 1\right) + N_2\dfrac{W}{R_2\gamma_2}\left(1 - \dfrac{R_0}{R_2}\right)}{N_2\left(\dfrac{W}{R_2\gamma_2} + 1\right) - N_2^2\left(1 - \dfrac{R_0}{R_2}\right)} \tag{11.4}$$

Given R_0, one has the following possibilities:

Case 1 – $R_0 \leq R_{0,\text{upper},1}$: In this case, one has $T \geq R_2$ for any admissible value of k_2, that is, $\forall k_2 \in [1, \ldots, N_2 - 1]$.

Case 2 – $R_{0,\text{upper},1} < R_0 \leq R_{0,\text{upper},2}$: In this case, one has $T \geq R_2$ for the set of k_2 values $k_2 \in \{1, \ldots, k_2^{\max}\}$.
Here, $k_2^{\max} \leq (N_2 - 1)$; also $N_2 \geq 1 \Rightarrow k_2^{\max} \geq 1$.

Case 3 – $R_0 > R_{0,\text{upper},2}$: In this case, one has $T < R_2$ for any admissible k_2. Parameters $R_{0,\text{upper},1}$, $R_{0,\text{upper},2}$ and $k_2^{\max}$ are given as

$$R_{0,\text{upper},1} = \frac{R_2^2}{\left(\dfrac{W}{\gamma_2} + 2R_2\right)}$$

$$R_{0,\text{upper},2} = \frac{N_2 R_2^2}{\left(\dfrac{W}{\gamma_2} + [N_2 + 1]R_2\right)}$$

$$k_2^{\max} = \left\lfloor \frac{N_2 R_2(R_2 - R_0)}{R_0\left(\dfrac{W}{\gamma_2} + R_2\right)} \right\rfloor \tag{11.5}$$

For illustration purposes we use a system [1] with spreading bandwidth $W = 1.23\,\text{MHz}$; Class 1 bit rate $R_1 = 9.6\,\text{kb s}^{-1}$, with a minimum SIR of $\gamma_1 = 7\,\text{dB}$, minimum Class 2 bit rate $R_{\min} = 14.4\,\text{kb s}^{-1}$, with a minimum SIR of $\gamma_2 = 8.5\,\text{dB}$ (7.0795) and idle bit rate $R_0 = 1.2\,\text{kb s}^{-1}$. The Class 2 SIR requirement γ_2 is chosen under the conservative assumption that power control at higher rates might involve higher overheads. The number of Class 1 users N_1 was taken to be the primary variable; on the basis of this, the maximum number $N_2^{\max}$ of Class 2 users permitted was computed according to equation (11.3).

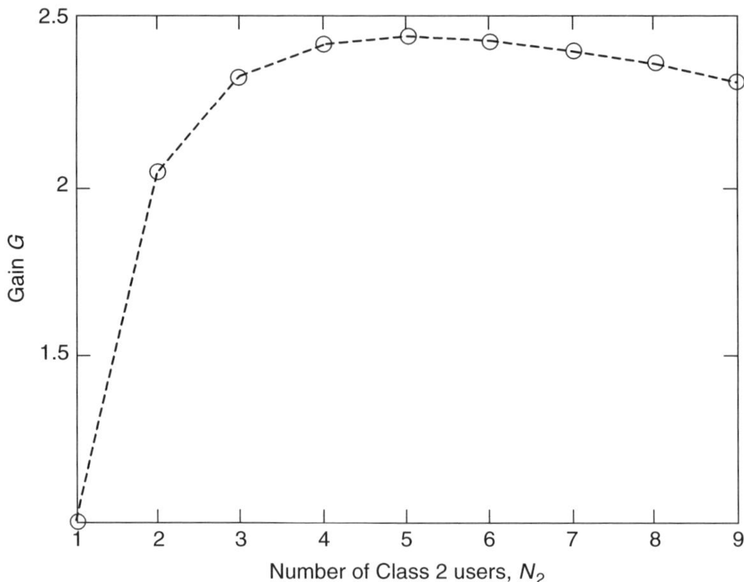

Figure 11.1 Gain $G = T_2/R_2$ versus Class 2 population N_2 with $N_1 = 8$, $N_2^{\max} = 9$.

The number of Class 2 users in the system was then varied from 1 to $N_2^{\max}$, and the corresponding gain G was computed from equation (11.4). The results are plotted in Figure 11.1. One should be aware that for N_1 smaller and smaller, G would be larger and larger. Variation of the synchronization rate limits $R_{0,\text{upper},1}$ and $R_{0,\text{upper},2}$ is shown in Figure 11.2.

In the case of imperfect power control, the initial condition (11.1) should be replaced accordingly (see Section 10.1 of Chapter 10). We use again the spreading bandwidth $W = 1.23\,\text{MHz}$, Class 1 bit rate $R_1 = 9.6\,\text{kbit s}^{-1}$, with the power-controlled target SIR parameter $\gamma_1 = 7\,\text{dB}$, minimum Class 2 bit rate $R_{\min} = 14.4\,\text{kbit s}^{-1}$, with the power-controlled target SIR $\gamma_2 = 8.5\,\text{dB}$ (7.0975) and idle bit rate $R_0 = 1.2\,\text{kbit s}^{-1}$. The results are shown in Figure 11.3.

In the sequel, we consider a situation where an upper limit is placed on the peak interference that a particular cell can create in another. Clearly, such a constraint would translate to peak transmit-power limits on the mobiles in that cell. Also, mobiles located close to the boundary between the cells would have more stringent peak transmit-power limits than those in the interior. Considering the application of the scheduled transmission mode described earlier in such a situation, we note that the presence of constraints on the peak transmit powers translate to constraints on the peak transmission rate, which limits the throughput gains due to scheduling. Thus, in order to better exploit the looser constraints on the Class 2 users in the cell interior, it might be advantageous in such situations to schedule the transmissions of only a certain subset of the Class 2 users in the cell. We use notation $(IW/h_j^{(2)}P_j^{\text{peak}})$ for the ratio of the power received from all

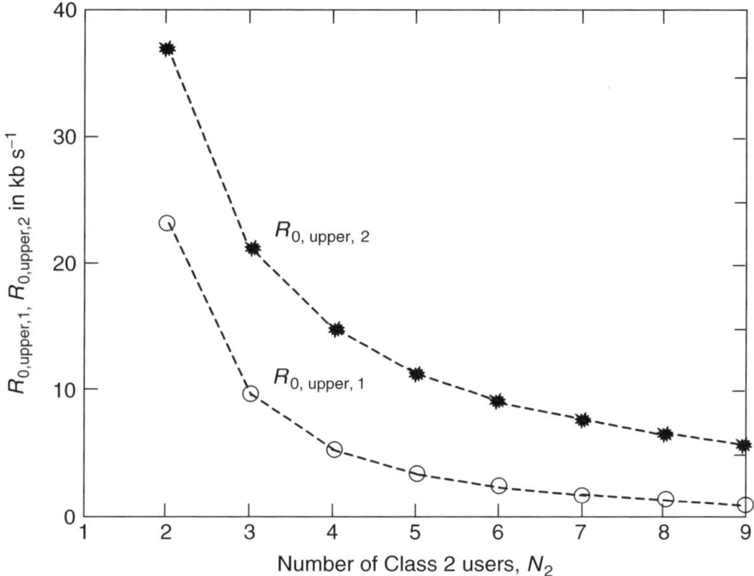

Figure 11.2 The variation of the synchronization rate limits. $R_{0,\text{upper},1}$ and $R_{0,\text{upper},2}$ versus N_2 for the $N_1 = 8$, $N_2^{\max} = 9$.

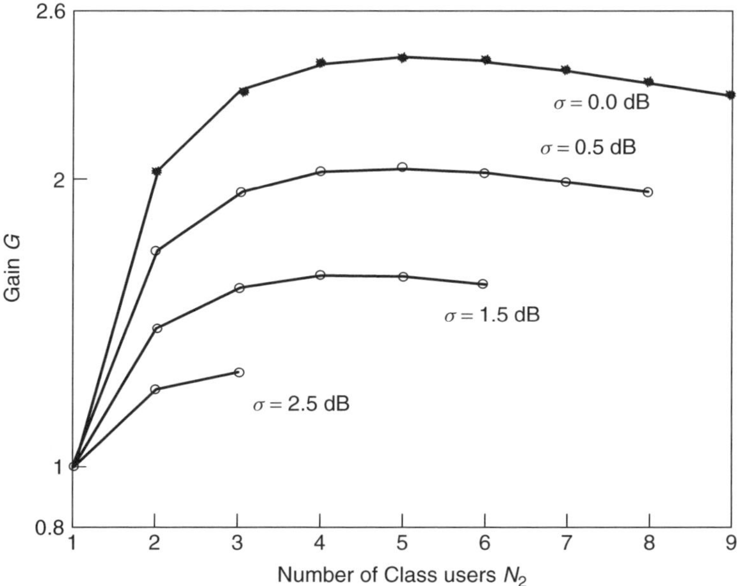

Figure 11.3 Gain $T = T_2/R_2$ versus Class 2 population $N2$ for different values of power control error (τ) and $N_1 = 8$.

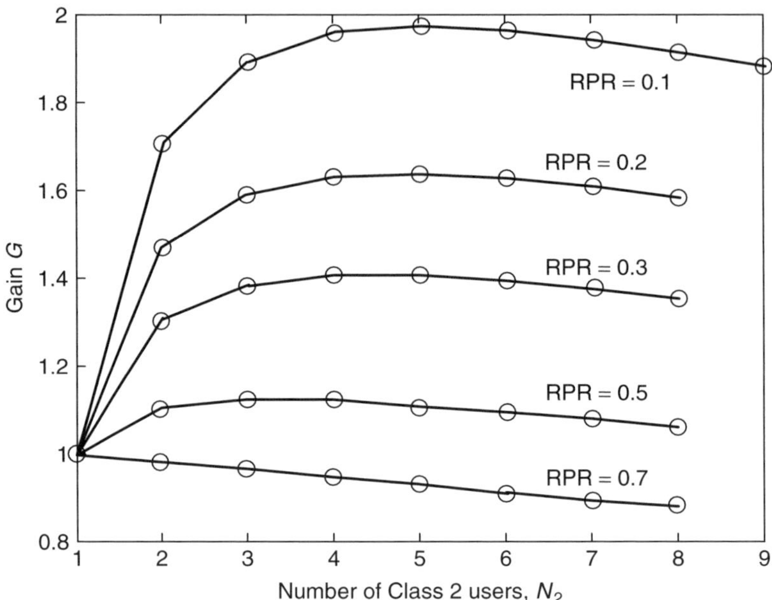

Figure 11.4 The variation of the throughput gain G versus RPR and N_2 for $N_1 = 8$.

sources outside the cell to that of the weakest link user at the base station (BS), assuming that the user is transmitting at its maximum power. It will be referred to as the *received power ratio* (RPR). Figure 11.4 shows gain $G = T_2/R_2$ versus Class 2 population N_2 with $N_1 = 8$ and RPR being a parameter.

11.1.2 Adaptive and reconfigurable transmission

The Class 2 mobiles must be capable of variable rate transmission, dependent on available residual capacity and the base station capable of the corresponding reception. The information transmission rate to be used by the Class 2 users (which was denoted as R_2^*) will be decided by the base station, using information about the population distribution in the system, and in the constrained transmit power case, knowledge of the RPR parameter. In practice, the user-population distribution may be a slowly changing variable. This would imply that Class 2 information rate changes do not have to be effected too often. The Class 2 mobiles transmission rate alternates between the information rate R_2^* and the synchronization rate R_0; hence, there must also be a mechanism to coordinate these rate changes with the base station.

Mechanism to schedule the Class 2 transmission must be available. One option is for Class 2 mobiles to transmit in a 'round-robin' fashion. Another option is that each mobile has a fixed information transmission time of τ units within each cycle time of C units. The actual value of C depends on the amount of buffer space to be provided at the mobile, the average number of Class 2 users expected and so on. Once C is known, an adaptive τ is simply given by $\tau = (C/N_2)$ where, as before N_2 is the number of Class 2 users.

This form of scheduling will also have to be centrally controlled by the base station using information about the user-population distribution. There has to be a mechanism by which the base station informs a particular Class 2 mobile about the cycle time C, its information transmission time τ, as well as its 'place' within the cycle (this is called the slotting mechanism). Although it is desirable to have a fine slotting of the Class 2 users, that is, an arrangement of the slots such that their transmissions do not overlap, it may be difficult to achieve in practice without a significant increase in complexity. In that case, one could resort to coarse slotting in which some (as small as possible) part of the slot assigned to a user overlaps with that assigned to another user. The overlapping portions would then correspond to the case $k_2 = 2$ rather than the desired best case $k_2 = 1$. This would lead to a certain reduction in the throughput gains, but would simplify the implementation of the slotting mechanism.

More sophisticated scheduling schemes, which exploit the traffic characteristics of the Class 2 mobiles, can be designed above the basic scheme. For example, a Class 2 mobile may have no information to transmit in its assigned slot in which case that particular slot could be reassigned to some other user. This would require an adaptive reconfiguration of the upper layers in the network. Such schemes would lead to additional throughput gains. However, the base station would need additional knowledge about the state of the data in the mobiles, and would add some additional scheduling load to the system. Delay constraints need to be incorporated into the scheduling. A Class 2 mobile would need to use significantly higher transmit power in its information transmission slot time τ as compared to the rest of the time in a cycle, in order to maintain the same SIR. During its information transmission slot of time τ, a Class 2 mobile must increase its transmit power to correspond to the rate R_2 and lower it for the rest of the cycle to correspond to the rate R_{02}.

11.2 ACCESS CONTROL FOR WIRELESS MULTICODE CDMA SYSTEMS

We assume a short-term traffic control in the uplink, that is, burst level traffic control. A multicode CDMA system (MC-CDMA) for the integration of multirate, multimedia services is considered [2,3]. In MC-CDMA, a code can be used to transmit information at a basic bit rate. Users (video or data) who need higher transmission rates can use multiple codes in parallel. A two-phase congestion control is used for managing the data traffic and the Packet Error Rate (PER) of Real-Time (RT) traffic. The first congestion control phase makes sure that there is at most a prespecified number of data users who can transmit at a time. For those data users who have been granted the right to transmit (i.e. those who have been assigned CDMA codes), they further follow the second congestion control phase imposed by the BS so as to minimize the impact on the PER of RT traffic.

11.2.1 Call level model

The CDMA system consists of a large number of users who can generate voice calls or data calls. Users with voice calls or data messages will contact the central station (called the base station or BS in the context of cellular networks) by sending reservation requests through the signaling channels. For details on the Universal Mobile Telecommunication System (UMTS) standard, see Chapter 17. On successful reception of a reservation request, the central station will make a decision as to whether the request can be granted the admission control. Depending on whether the system has enough resources in the traffic channel to handle the incoming call, the reservation request will be accepted or rejected, and the user is notified via the downlink (DL) (the forward link). An accepted voice call is assigned a CDMA code immediately, but an accepted data call is first queued at the central station, and will be assigned CDMA codes subject to the congestion control to be elaborated.

The time axis of the CDMA system is fully slotted, with a slot duration equal to the transmission time of a packet that is of the same size for both types of traffic. All users are synchronized at the packet level. The voice generation processes and data calls are Poisson distributed with the average rates of λ_v and λ_d calls per slot, respectively. It is assumed that the duration of a voice call and the data message length are geometrically distributed with an average of $1/\mu$ and L slots, respectively.

11.2.2 Data congestion control scheme

As it was already discussed in Chapter 10, owing to the on–off nature of accepted voice calls, the CDMA channel is not always fully utilized. Therefore, the central station can allow data traffic to dynamically 'steal' the unused channel capacity left by the voice users. It is the congestion control scheme that makes sure that the quality of service (QoS) of the established voice calls is not compromised, that is, the packet error probability remains less than a prespecified value. The congestion control scheme will make sure that among the accepted data message requests queued at the central station, only the first M_d requests are assigned C_d CDMA codes. The data users with assigned CDMA codes start to transmit their outstanding messages with probability p_{start}, which can be determined by the central station according to the state of the system. Upon starting the transmission of its data message, the data user will not stop until the end of the message. Since the data message length is geometrically distributed with an average of L packets, the transmission time of the data message approximately follows a geometric distribution with an average of L/C_d slots.

When a data user advances to the first M_d positions, it is in the standby state, where the user will enter the active state with probability p_{start}. When in the active state, the user will transmit C_d packets in parallel using the C_d spreading codes assigned by the central station. Since the transmission time of a data message is geometrically distributed with mean L/C_d, the transition probability from the active state to the done state is C_d/L. The state diagram of an admitted data user is shown in Figure 11.5. How p_{start} is determined on the basis of some side information will be discussed next.

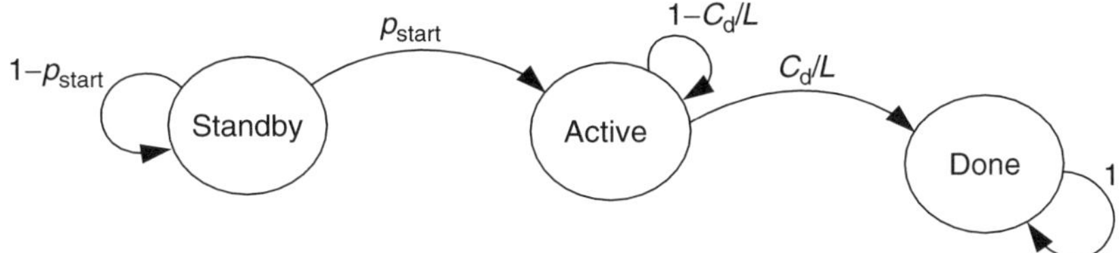

Figure 11.5   State transition diagram of an admitted data user.

11.2.3 Feedback-driven congestion control

In order to fully utilize the unused channel capacity while maintaining the voice PER, p_{start} needs to be dynamically updated (according to the extra information about the system state available to the central station) and broadcast to the data users in the standby state. To set p_{start}, for time-slot $t + 1$, we assume that the central station can obtain some side information about the number of users transmitting in time-slot t. Although there exist many ways to update p_{start} based on the side information $I(N_t^v, N_t^d)$, where N_t^v and N_t^d denote the number of voice and data users that transmits in time-slot t, respectively, we consider a threshold scheme with $I(N_t^v, N_t^d) = (N_t^v, N_t^d)$. Define

$$f(N_t^v, N_t^d) = N_t^v + N_t^d \times C_d \tag{11.6}$$

$f(\cdot)$ gives the number of active codes in the CDMA channel in time slot t, which can be used as a congestion index. Given the side information (N_t^v, N_t^d), the probability p_{start} for time-slot $t + 1$ is

$$p_{start} = \begin{cases} \min\left(1, \dfrac{T - f(N_t^v, N_t^d)}{C_d(\min(M_d, N_t^q) - N_t^d)}\right), \\ \quad \text{if } f(N_t^v, N_t^d) \le T \\ 0, \quad \text{otherwise} \end{cases} \tag{11.7}$$

T is the threshold for congestion control N_t^q is the number of data requests accepted by the system, and $\min(x, y)$ is the minimum of the two arguments. In other words, if the system congestion index does not exceed the threshold T, the data users who are in the standby state should transmit with a probability selected so as to fill available 'slots' on an average. On the other hand, if the system congestion index exceeds T, active data users are allowed to continue their transmission, but other data users standing by should refrain from transmitting. The rationale behind the selection of p_{start} is that $[T - f(N_t^v, N_t^d)]/C_d$ can be thought of as the residual channel capacity, and $[\min(M_d, N_t^q) - N_t^d]$ is the number of data users standing by for transmission. By choosing p_{start} equal to the ratio of the above two factors, the expected number of data users who start to transmit will be equal

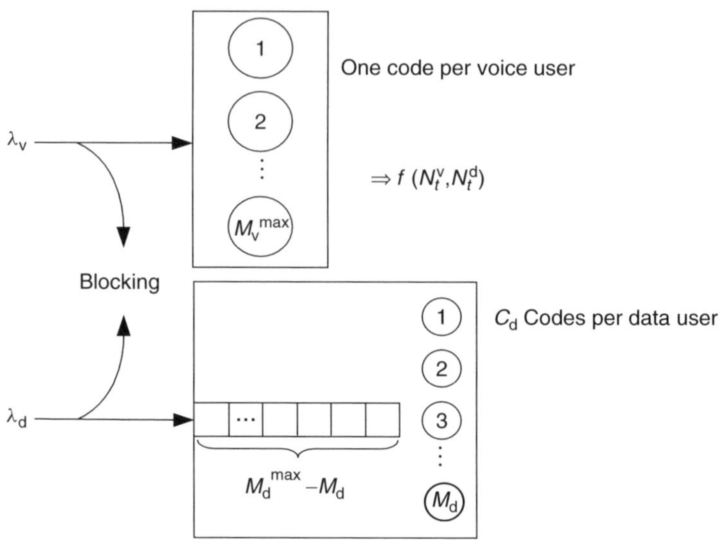

Figure 11.6   System level view of the traffic control mechanism.

Table 11.1 System parameters

	Symbol	Value
Number of voice users admitted to the system	M_v	1–6
Transition probability of a voice source from On-to-Off state	p_{on_off}	1/17
Transition probability of a voice source from On-to-Off state	p_{on_off}	1/22
Slot duration (msec)		20
Data message arrival rate (message/slot)	λ_d	Variable
Average length of data message (packets)	L	10
Capacity of the data message request queue at BS	M_d^{max}	10 or 15
Maximum number of data users in the standby state	M_d	2–8
Number of CDMA codes per data user	C_d	1, 2
Threshold for congestion control	T	4–10
Packet length (bits)		255
Number of correctable bit errors per packet		4
Bit energy to (one-sided) noise spectral density (dB)	E_b/N_0	11
Processing gain	G	64
Rician factor	K_R	$5 - \infty$

to the left over (residual) capacity. The system level view of the traffic control mechanism is shown in Figure 11.6.

For illustration purposes a set of the system parameters from Table 11.1 are used [3].

Data throughput versus data arrival rate is shown in Figure 11.7. Optimal threshold T as a function of M_v is shown in Table 11.2 [3].

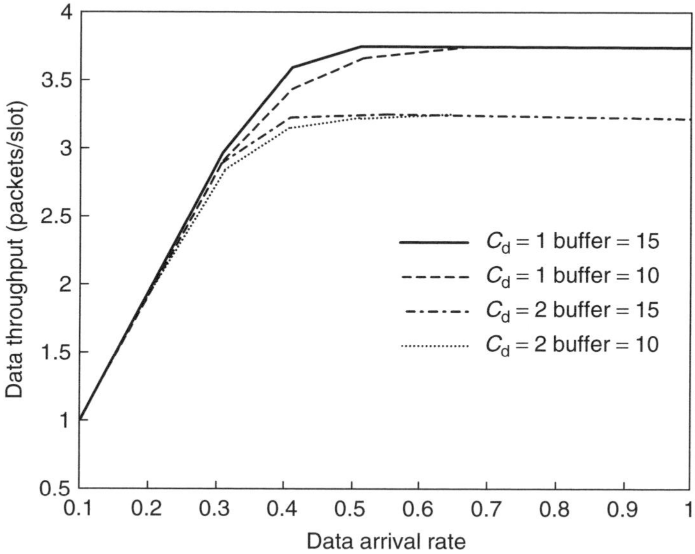

Figure 11.7 Data throughput versus data arrival rate λ_d. This shows that the single code per data user has better performance. A larger buffer in the request queue of the central station can increase the data throughput when the data traffic load is moderate. $M_v = 4$.

Table 11.2 Optimal T as a function of M_v

	$M_v = 1$		$M_v = 4$		$M_v = 6$	
	$C_d = 1$	$C_d = 2$	$C_d = 1$	$C_d = 2$	$C_d = 1$	$C_d = 2$
Threshold T	6.07	5.38	5.46	5.07	5.22	4.86
Maximum data throughput	5.15	4.54	3.79	3.27	2.82	2.45

11.3 RESERVATION-CODE MULTIPLE ACCESS

Most CDMA systems described so far employ the strategy of 'one-code-for-one-terminal' for code assignment. This assignment, though simple, fails to efficiently exploit the limited code resource encountered in practical situations. In this segment, a protocol called *reservation-code multiple access (RCMA)*, which allows all terminals to share a group of spreading codes on a contention basis and facilitates introducing voice/data integrated services into spread-spectrum systems is presented. The RCMA protocol can be applied to short-range radio networks and can be easily, extended to wide area networks if the code-reuse technique is employed. In RCMA, a voice terminal can reserve a spreading code to transmit a multipacket talk spurt while a data terminal has to contend for a code for each packet transmission. As before, the voice terminal will drop a long delayed packet while the data terminal just keeps it in the buffer. Therefore, two performance measures used to assess the proposed protocol are the *voice packet dropping probability* and the *data packet average delay.*

The RCMA is designed as a protocol to control the uplink communication for a star topological network with a small coverage. The control is implemented through three operations. First, the BS synchronizes all the terminals by broadcasting the system timing whereby the terminals can adjust their timing clock. Since the timing error among terminals in a short-range area is small, it is feasible to use the slotted (synchronous) spread-spectrum signaling in the system. Second, the base broadcasts traffic information including the status of transmitted packets and the codes that will be available in the next slot. A short time delay for small coverage makes it possible for the terminal with a transmitted packet to receive feedback information from the base in the same slot as the packet was sent. Finally, by assigning different priority to voice and data, the RCMA system provides a powerful mechanism to integrate the two services.

The speech model introduced in Chapter 10, Section 10.1 is used again with the mean duration of a talkspurt t_1, the mean duration of a gap t_2 and the transition probabilities as shown in Figure 11.8 with

$$\gamma = 1 - \exp(-\tau/t_1); \quad \sigma = 1 - \exp(-\tau/t_2) \tag{11.8}$$

where τ is the width of a slot. The empirical values for t_1 and t_2 are 1 and 1.35 s, respectively (see Chapter 8).

When entering the talking state, the voice signal is encoded into a bit stream by a speech encoder at the bit rate of $R_s = 8\,\text{kb s}^{-1}$. The bit stream is then packetized with a H-bit header added to each packet. The packet header contains address bits, synchronization bits, parity check bits, and control bits. One of the control bits is used to indicate the type of terminal (voice or data). Usually, several packets are continuously generated in the talking state. These packets make up a talkspurt, the length of the talkspurt is a random variable (RV). Let L denote the length of a talkspurt. Then we can write the probability of L as

$$\Pr\{L\} = (1 - \gamma)^{L-1}\gamma \tag{11.9}$$

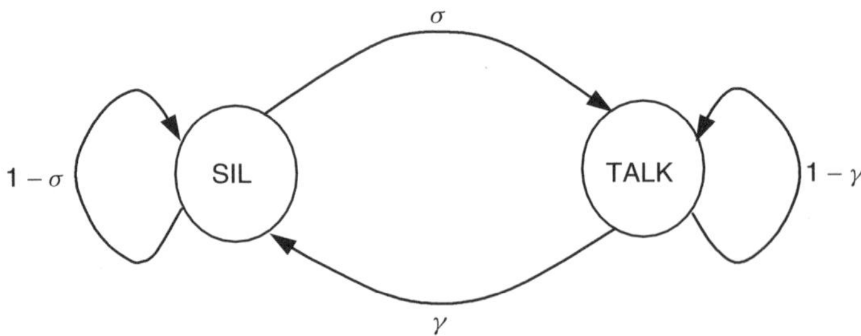

Figure 11.8 Speech on-off model.

A data terminal generally has a bit rate different from a voice terminal and its data stream can be discontinuous. R_d is the average bit rate of the data terminal, assume that $R_d \leq R_s$. The data stream is packetized in the same way as the information stream of a voice terminal. A data packet is independently generated in each slot. σ_d is the probability of generating a packet in a slot. The bit rate of a data terminal is R_s when a data packet is generated and is zero otherwise. Hence, the mean bit rate is

$$R_d = R_s\sigma_d + 0 \cdot (1 - \sigma_d) = \sigma_d R_s \qquad (11.10)$$

In the system, voice requires RT service while data does not. This factor is taken into account in the design of the RCMA protocol. The common point of the two services is that a packet from either voice or data terminal has the same size and is spread by a spreading code, which is chosen by the RCMA protocol. The signal parameters are: R_c, the chip rate of the spreading code, τ, the slot width and G_s, the spread-spectrum processing gain given by

$$G_s = 10\log_{10}[R_c\tau/(R_s\tau + H)] \qquad (11.11)$$

The processing gain increases with the slot width τ, and the maximum processing gain is $10\log_{10}[R_c/R_s]$ when the slot width τ is large so that the effect of the packet header becomes negligible.

11.3.1 Contention and reservation

In an RCMA network, all dispersed terminals in a cell share a group of spreading codes to transmit packetized information over a slotted channel to the BS. Let M_v, M_d be total numbers of voice, data terminals. Among them, b data terminals and c voice terminals are in contention. Each code is identified as 'reserved' or 'available' based on the feedback information (acknowledgment message) from the base at the end of each slot. The feedback information is carried on one or more specific spreading codes on the DL from the base to the terminals (control channel). A 'reserved' code will exclusively serve the terminal that has been granted a reservation in the subsequent slots. An 'available' code can probably be used by any terminal with packets in its buffer but no reserved code in possession.

The RCMA protocol treats voice packets and data packets differently because the two services have different features. Once a voice packet succeeds in contending for an 'available' spreading code, the protocol grants the terminal a reservation for the exclusive use of this code in subsequent slots until all packets in that talkspurt are transmitted. If, a voice packet fails to obtain a spreading code and stays in a 'contending' state for time longer than a preset threshold D_{max} (in second), it has to be dropped.

The threshold is determined by the delay constraints on speech communication and is a design parameter of the RCMA system. The data terminal has to compete for a spreading code to transmit each of its packets. The data packets failing to be transmitted will be held in the buffer without time limitation.

11.3.2 Statistical parameters to describe the RCMA system

As the RCMA traffic becomes congested, voice packets drop frequently and the time delay for data packets increases. The *packet dropping probability* (P_{drop}) for the voice terminal and the *average packet delay* (D_{av}) for the data terminal is measured to evaluate the system performance of the RCMA.

A contending terminal needs 'permission' to use an 'available' spreading code. Let P_v be the probability for a voice terminal to obtain permission and P_d its counterpart for a data terminal. The probabilities P_v and P_d are adjustable parameters in the system design. The remaining parameter that is needed to describe the RCMA system is the total number, N, of spreading codes in a group.

A voice terminal can be in a 'silence,' 'contention,' or 'reservation' state. Given a state of 'reservation,' there are still N possibilities since the terminal can reserve any of the N spreading codes. Accordingly, the total number of states a voice terminal can take is $N + 2$. The state evolution can be described by a Markov model, shown in Figure 11.9, where all the transitions occur at the end of a slot. A terminal moves from the silent state (SIL) to the contending state (CON) when a talkspurt (which consists of multipackets) begins. The probability is σ for the transition from SIL to CON and γ for the reverse transition. Parameters σ and γ take the same values as their counterparts in the speech on-off model shown in Figure 11.8. The base broadcasts feedback messages, keeping all the terminals informed of the reservation status of each code.

A *voice terminal* in the CON state gets a reservation on the ith spreading code and enters the 'Rc_i' state, as a result of the concurrence of the events $\{E_{vk}, k = 0, 1, \ldots, 5\}$ as described below:

E_{v0} terminal does not enter an 'SIL' state at the end of the slot;
E_{v1} code is unreserved;
E_{v2} code happens to be randomly selected by the terminal from a pool of all available codes;

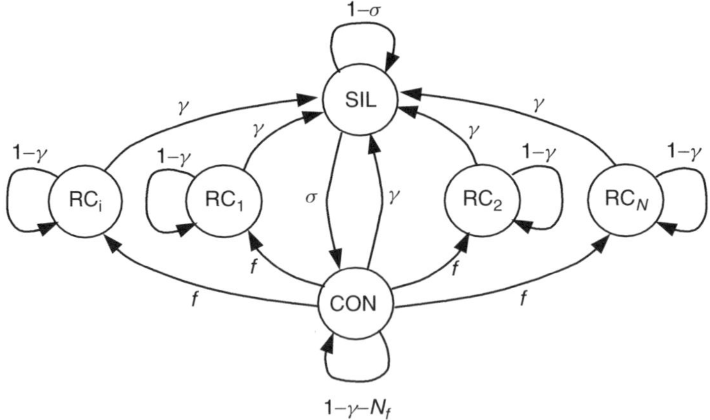

Figure 11.9   Speech subsystem model.

E_{v3} terminal has permission (P_v) to transmit packets;

E_{v4} no other contending voice terminal selects this code or gets permission to transmit a packet even if it selects this code; and

E_{v5} no contending data terminal (with backlogged data packets) selects this code, or obtains permission (P_d) even if selects this code.

A data terminal is not permitted to reserve a spreading code. Hence, it can only stay either in an 'idle' state or in a 'contention' state. The data terminal can store its delayed packets in an infinite buffer. The data terminal remains in the 'contention' state as long as it has at least one packet in its buffer. A data terminal with a nonempty buffer is called a backlogged data terminal. A backlogged data terminal has to contend for a code to transmit each packet; a successful transmission is the result of concurrence of the below five events $\{E_{di}\}$

E_{d0} code is unreserved;

E_{d1} code happens to be selected by the terminal;

E_{d2} terminal has permission (P_d) to transmit;

E_{d3} no contending voice terminal (of c terminals) selects this code, or has permission even if it selects this code; and

E_{d4} no other backlogged data terminal (of $b - 1$ terminals) selects this code, or has permission (P_d) even if it selects this code.

The data subsystem model is shown in Figure 11.10.

For illustration purposes, a microcellular mobile radio network with a hexagonal topology is used [4]. The system employs binary phase-shift keying (PSK) and the total

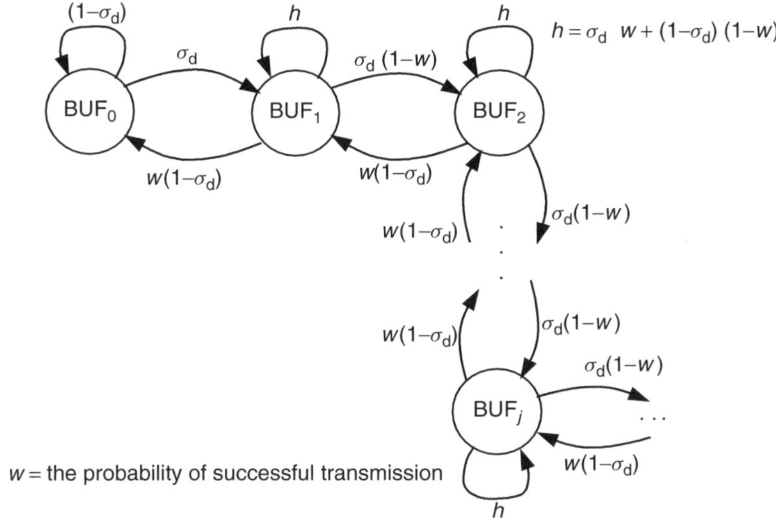

Figure 11.10 Data subsystem model.

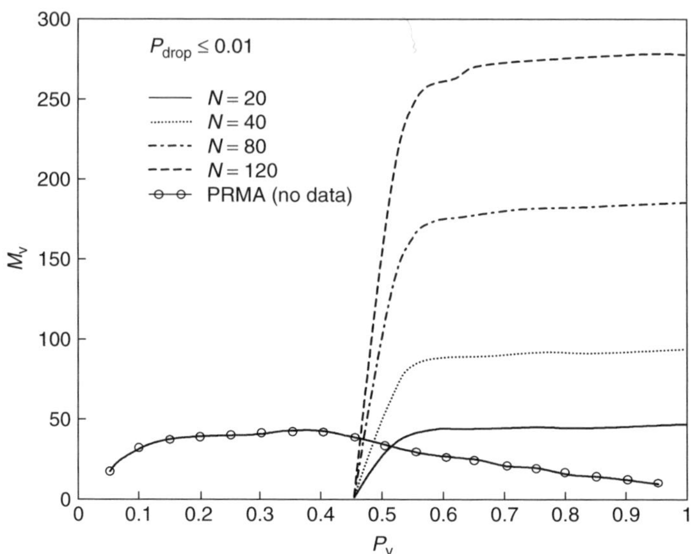

Figure 11.11 Maximum number of voice terminals versus permission probability P_v [4]. Reproduced from Tan, L. and Zhang, Q. T. (1996) A reservation random-access protocol for voice/data integrated spread-spectrum multiple-access systems. *IEEE J. Select. Areas Commun.*, **14**(9), 1717–1727, by permission of IEEE.

bandwidth is 1568 kHz. The seven neighboring cells share the channel bandwidth by using spreading codes with the chip rate of $R_c = 1568\,\mathrm{kb\,s^{-1}}$. The number of the spreading codes is $N = 20, 40, 80, 120$. The channel is slotted with a slot duration $\tau = 20\,\mathrm{ms}$. Voice signals are encoded at the rate of $R_s = 8\,\mathrm{kb\,s^{-1}}$ and segmented into packets of length equal to 224 b plus 64-bit overhead ($H = 64$ b). The bit rate of data source is chosen as $R_d = 1200$ and $2400\,\mathrm{b\,s^{-1}}$. $P_{drop} \leq 0.01$ is assumed and the maximum data packet average delay is $D_{av} \leq 4$ packets. To focus on the performance of the protocol, the channel is assumed to be ideal so that the transmitted packet is free from distortion, noise, and other interferences. A Packet Reservation Multiple Access (PRMA) protocol is chosen for comparative study because of its popularity and good performance. The PRMA protocol is applied to a time division multiple access (TDMA) microcellular system also employing a seven-cell repeat pattern. The seven cells use different carrier frequencies each having a transmission rate of $224\,\mathrm{kb\,s^{-1}}$ so that the total bandwidth is the same as that of the RCMA system, namely, 1568 kHz. The results for an only voice service is shown in Figure 11.11. One can see that M_v increases rapidly in the lower P_v region and becomes basically constant when $P_v > 0.6$. This implies that there is no need to use a random number generator to generate 'permission.' We also observe that the system capacity increases with the available number of spreading codes, N, with M_v being approximately twice as large as N. The PRMA system, though occupying the same system bandwidth, has a much smaller system capacity than the RCMA system. The maximum number of users that PRMA system can support is

42, located in the interval $P_v \in [0.35, 0.40]$. One can see that the RCMA system has significant capacity improvement over the PRMA. When the number N of spreading codes takes the values $= 40$, 80, and 120, the improvement factor is 2.0, 3.9, and 6.0, respectively.

For both data and voice services, M_d as a function of P_d, for three different values of N, is shown in Figure 11.12(a) for $M_v = N$, and Figure 11.12(b) for $M_v = M_d$. The value of P_v was 1 for the RCMA system and 0.35 for the PRMA. The system capacity, again, increases with the number of the available spreading codes. For a given value of N and the data rate, there exists an optimum value for P_d, located in $(0.6, 1]$. The choice of the optimal P_d only results in a slight increase in M_d as compared to that at $P_d = 1$. Thus, for a practical application, we would suggest using $P_d = 1$ instead, so as to simplify the system design since in this case the permission generator can be removed. In Figure 11.12, the system capacity for PRMA has been shown for two different data rates; they are 1200 b s^{-1} and 2400 b s^{-1}, respectively. In any event, the best system capacity that can be supported by the PRMA is less than 24. Clearly, the RCMA system is much

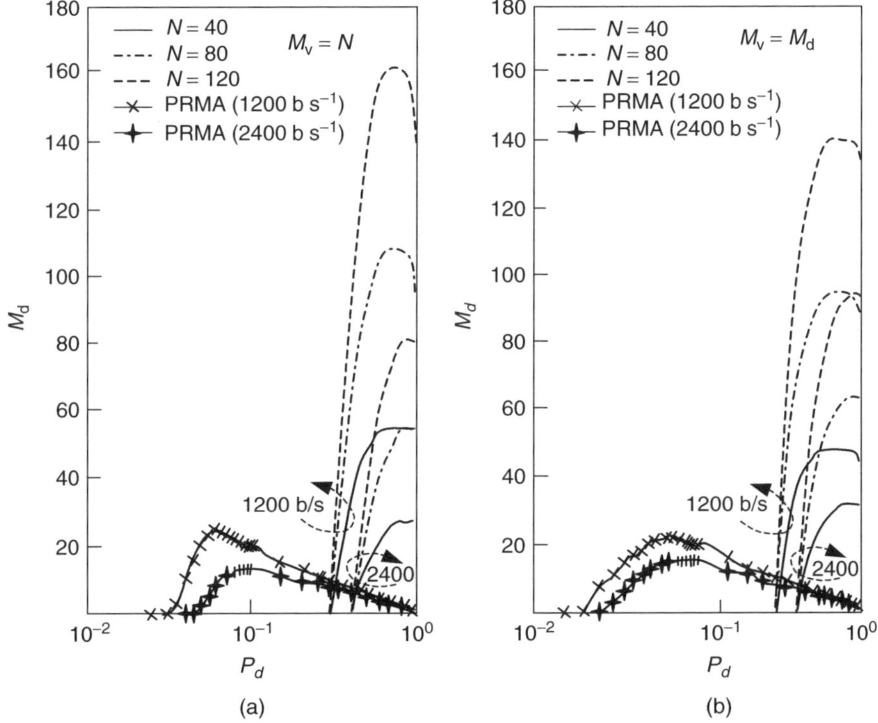

Figure 11.12 Maximum number of data terminals versus permission probability P_d for (a) $M_v = N$ and (b) $N_v = M_d$ [4]. Reproduced from Tan, L. and Zhang, Q. T. (1996) A reservation random-access protocol for voice/data integrated spread-spectrum multiple-access systems. *IEEE J. Select. Areas Commun.*, **14**(9), 1717–1727, by permission of IEEE.

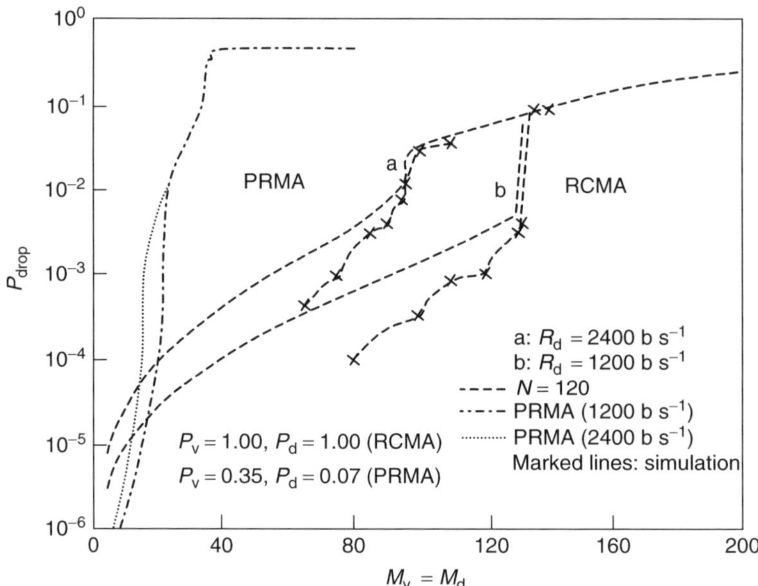

Figure 11.13 Voice packet dropping probability versus the number of terminals assuming that $M_v = M_d$.

superior to its PRMA counterpart in terms of system capacity even when a median size of code set is used.

Figure 11.13 depicts P_{drop} as a function of M_v and M_d. The system parameters are shown in the figure. For the RCMA system, $M_v = M_d$ and $P_v = P_d = 1$ was assumed. For the PRMA system, $P_v = 0.35$ and $P_d = 0.07$ was used. These values were chosen on the basis of previous results, in an attempt to optimize the PRMA system performance for fair comparison. Simulation results for P_{drop} at each point of M_v were obtained after an observation of a long time to ensure good estimation accuracy. Specifically, simulations were performed over 10^6 slots (packets) to obtain P_{drop} when P_{drop} is less than 10^{-3} and over 5×10^5 slots when $P_{drop} \geq 10^{-3}$.

For $N = 120$, one can see significant improvements of RCMA protocol over PRMA. The data *packet average delay* versus M_d (equal to M_v) is plotted in Figure 11.14 where the same conditions for M_v, M_d, P_v, and P_d as for Figure 11.13 were used. For simulation results, each point in Figure 11.14 was obtained over an observation interval of 10^6 slots. Theoretical performance is also included in the figures for comparison.

11.4 MAC PROTOCOL FOR A CELLULAR PACKET CDMA WITH DIFFERENTIATED QoS

In this section, we consider a protocol for handling a variety of multimedia traffic types in an integrated wireless-access network (IWAN). For instance, the protocol is suited for

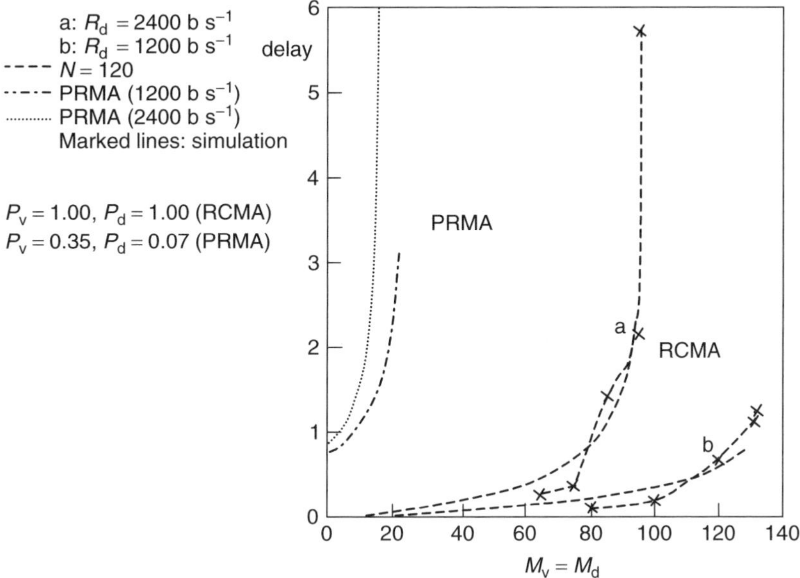

Figure 11.14 Data packet average delay versus the number of terminals assuming that $M_v = M_d$.

carrying multiple traffic types of different priorities, (differentiated QoS). A comparative evaluation of the protocol is done for three different radio frequency bandwidths currently of interest for personal communication services (PCS), that is, 1.25 MHz, 5 MHz, and 10 MHz. The protocol is both robust and flexible for the intended IWAN applications. It offers a significant multiplexing gain as the bandwidth increases. We classify users into Δ types according to their traffic rate. In the case that the traffic of the same rate has different priorities, different traffic types can be created even for the same rate. As the traffic arrives from the source, it is buffered in a finite-length buffer, one for each traffic type as shown in Figure 11.15. The buffer length is B_δ packets for δ-type traffic, where $1 \leq \delta \leq \Delta$. Mobiles receive basic synchronization information from the BS and are able to transmit packets slot-synchronously. If there is any packet in the queue, the user attempts transmission at the beginning of the next slot.

The user can assume three states, 'idle,' 'active,' and 'blocked,' on the basis of the state of the buffer (used or empty) and the success of the previous transmission (see Figure 11.16). If there is no packet queued in the buffer, the user assumes the 'idle' state. An 'active' or 'blocked' user may further assume a 'substate' based on how many packets are queued in the transmission buffer. For instance, being 'active' with two packets in the buffer is a 'substate' that is different from the active substate with three packets in the buffer.

If an 'idle' user's information source generates a packet, the packet is queued in the transmission buffer and the user assumes the 'active' state. An 'active' δ-type user attempts transmission of the head-of-the-queue packet with probability of p_δ, which may differ in value for different δ to assure priority treatment for different queues. A higher p_δ

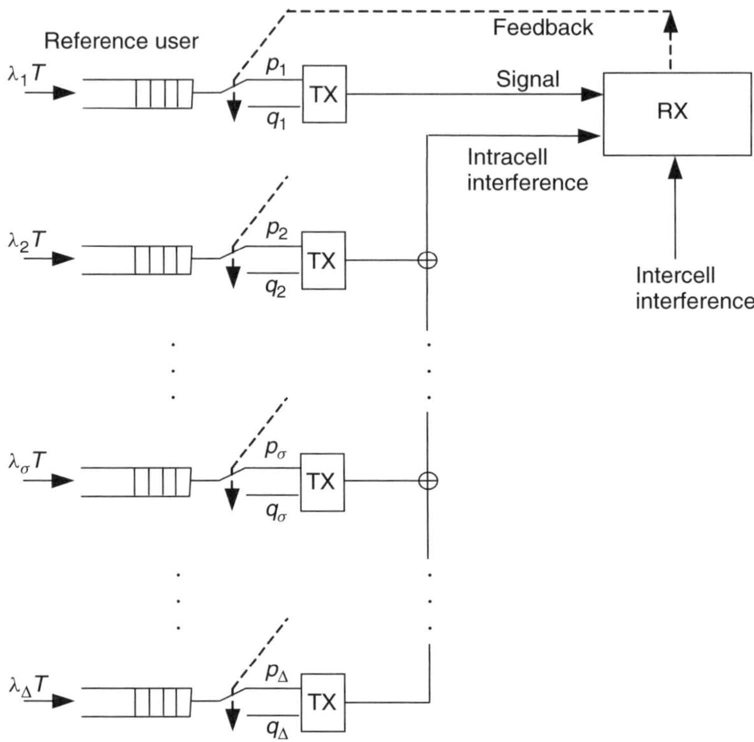

Figure 11.15 System model $\lambda_\delta T$ is the traffic intensity at the δth user.

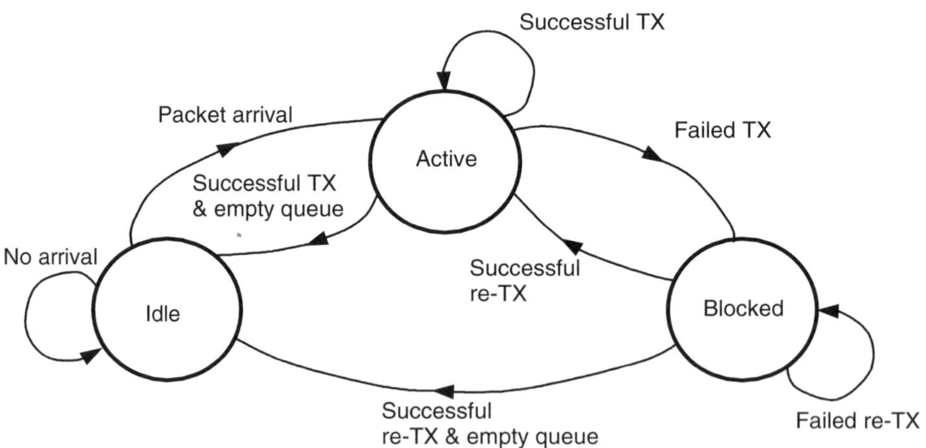

Figure 11.16 User state transition model. (TX = transmission).

corresponds to a higher priority. If the transmission succeeds, the user attempts to transmit the next packet in the queue if it is not empty and the user's state remains unchanged. If the successful transmission emptied the queue and no new packets arrived, the user assumes the 'idle' state. If the transmission failed, the user assumes the 'blocked' state. A δ-type user in the 'blocked' state attempts a retransmission with the probability of q_δ. If the retransmission fails, the user remains in the 'blocked' state. Otherwise, the user assumes the 'idle' or the 'active' state depending on whether the buffer has been emptied or not. For illustration purposes IWAN will carry heterogeneous types of traffic, such as voice, video, data, and interactive data. Therefore in IWAN, the traffic is of mixed type, in general. In order to gain some insight readily, we focus on a two-type case in this section. It may represent an integrated voice and video network. Later on, more general examples will be presented. All users are divided into two major groups ($\Delta = 2$) – high (hi) and low (lo) rate group, each of them may correspond to the video and the voice users, respectively. Given that voice can be transmitted at $8\,\mathrm{kb\,s^{-1}}$ after compression with a voice activity factor of 0.375, it results in an average rate of $3\,\mathrm{kb\,s^{-1}}$. Similarly, given that the low-quality video can be transmitted at an average rate of $64\,\mathrm{kb\,s^{-1}}$, the average rate of high rate traffic, $64\,\mathrm{kb\,s^{-1}}$, is assumed. Both types of traffic are expected to be bursty, that is, it is expected that the packets arrive in batches. The packet arrival is Poisson distributed. Then the probability, r_{lo_i}, that i low rate traffic data packets arrive during one packet transmission is

$$r_{\mathrm{lo}_i} = \frac{(\lambda_{\mathrm{lo}}T)^i}{i!}\mathrm{e}^{-\lambda_{\mathrm{lo}}T} \tag{11.12}$$

where λ_{lo} is the arrival rate of the low rate traffic packets and T is the duration of one packet transmission. Similarly, the probability, r_{hi_i}, that i high rate traffic data packets arrive during one packet transmission is

$$r_{\mathrm{hi}_i} = \frac{(\lambda_{\mathrm{hi}}T)^i}{i!}\mathrm{e}^{-\lambda_{\mathrm{hi}}T} \tag{11.13}$$

where λ_{hi} is the arrival rate of the high rate traffic packets. For illustration, an ATM-size packet [asynchronous transfer mode (ATM) cell] is assumed. The ATM cell consists of 48 bytes of payload and 5 bytes of the cell overhead. This packet size results in the cell rates, $\lambda_{\mathrm{lo}} = 7.81\,\mathrm{cell\,s^{-1}}$ and $\lambda_{\mathrm{hi}} = 166.6\,\mathrm{cell\,s^{-1}}$ for the low and high rate traffic, respectively. A transmission rate of $76.8\,\mathrm{kb\,s^{-1}}$ conveniently accommodates the expected cell rate, resulting in $5.52\,\mathrm{ms}$ of ATM cell transmission duration and leaving about 8% reserved for other purposes such as synchronization and maintenance overheads. The error-control-coding overhead is absorbed in the spreading, which is normal for the Direct Sequence (DS)/CDMA systems. The reserved space for the extra overhead, if unused, is not wasted but is reflected in reduced average transmitted power, thus in reduced mutual interference. The length of buffers is $B_{\mathrm{lo}} = 40$ and $B_{\mathrm{hi}} = 80$ for the low rate and high rate users, respectively. The radio capacity, defined as the maximum number of simultaneously transmitting users per cell for 1% outage probability is given in Table 11.3. The number of users is assumed to be in pairs of high-rate and low-rate users. This assumption allows us to model the case of each user using both types of traffic. Applications such as videophone and teleconferencing tend to warrant this assumption.

Table 11.3 Radio capacity for different bandwidths and resulting coefficient throughput [5]. Reproduced from Pichna, R. and Qiang, W. (1996) A medium-access control protocol for a cellular packet CDMA carrying multirate traffic. *IEEE J. Select. Areas Commun.*, **14**(9), 1728–1736, by permission of IEEE

Bandwidth [MHz]	Radio capacity	Throughput
1.25	1.0096	0.99701
5	4.9435	4.8853
10	11.8	11.7649

Table 11.4 Default cell population and corresponding throughput efficiency and delay for three different bandwidths [5]. Reproduced from Pichna, R. and Qiang, W. (1996) A medium-access control protocol for a cellular packet CDMA carrying multirate traffic. *IEEE J. Select. Areas Commun.*, **14**(9), 1728–1736, by permission of IEEE

Bandwidth [MHz]	Default cell population		Throughput efficiency		Delay [ms]	
	Low rate	High rate	Low rate	High rate	Low rate	High rate
1.25	1	1	0.99999	0.99999	8.57	46.86
5	4	4	0.9999	0.99999	8.54	44.30
10	11	11	0.9999	0.99999	8.56	46.25

The throughput efficiency and expected delay for the specified cell population, the default cell population, are shown in Table 11.4. More details are given in Figures 11.17 to 11.19. The issue of multimedia packet transmission in WCDMA will be addressed in much more detail in Section 11.7.

11.5 CDMA ALOHA NETWORK USING p-PERSISTENT CSMA/CD PROTOCOL

11.5.1 Network description

A network consists of a number of terminals that are equally likely to transmit to one another by using bursty data. In the network, a CDMA ALOHA is used, which allows for an individual 'virtual channel' for each receiving station. With the use of a 'receiver-based code' multiple-access protocol, it is also possible for a station to listen to the channel of the intended receiver before transmission, and also to abort transmission when it detects others transmitting on the same channel. A fully connected, full-duplex, slotted CDMA ALOHA network with channel sensing and collision detection is used. The network is modeled by using a discrete time Markov chain. For a system with a large number of

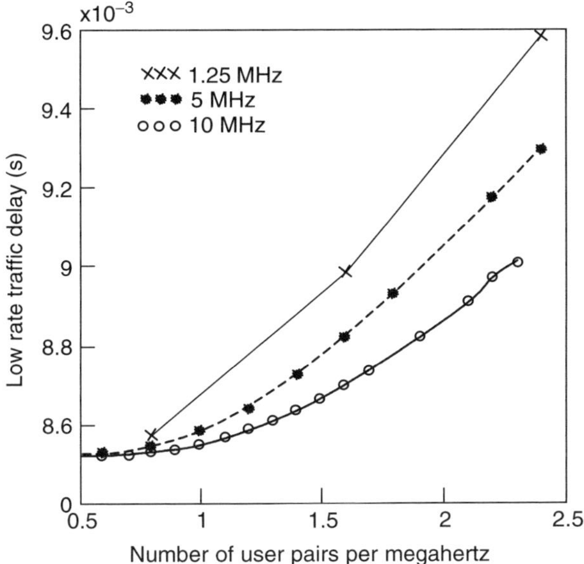

Figure 11.17 Low rate traffic delay versus the number of user pairs (low rate-high rate) per megahertz per cell for the radio outage probability of 1%.

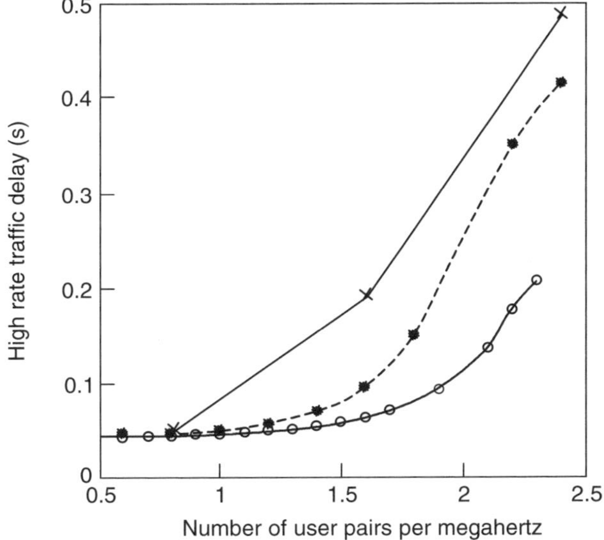

Figure 11.18 High rate traffic delay versus the number of user pairs (low rate–high rate) per megahertz per cell for the radio outage probability of 1%.

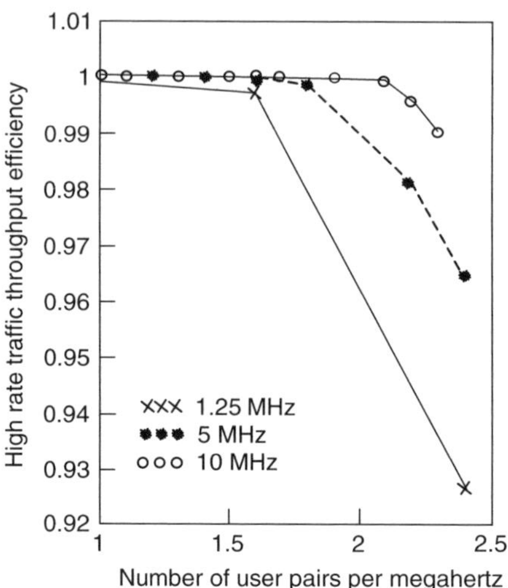

Figure 11.19 High rate traffic throughput efficiency versus the number of users pairs (low rate–high rate) per megahertz per cell for the radio outage probability of 1%.

users, where Markov analysis is impractical, equilibrium point analysis (EPA) is used to predict the stability of the system, and to estimate the throughput as well as the delay performance of the system when it is stable.

A comparison with a simple channel sense multiple access with collision detection (CSMA-CD) network, shows that a substantial improvement in the performance is achieved by this network.

11.5.2 Network mode/p-persistent MAC-protocol

In the network model, the following assumptions are used. Time is divided into minislots and each station is assigned a different spreading sequence or a different chip phase of the same maximal length spreading sequence, if the stations are synchronized, with which it will receive messages. Thus the bandwidth of the system is divided into 'virtual channels,' one for each station. Messages are generated at each of the stations at a similar rate of s messages per minislot independent of other stations and other messages, and each station is equally likely to transmit to any other stations. $s \leq 1$ is assumed, as the stations in the system have only one buffer slot to hold messages. Parameter s can also be interpreted as the message generation probability of each station per minislot. This assumption does not pose a big restriction to the analysis since ALOHA access works well only when $s \ll 1$. The messages are assumed to have a random number of minipackets, each one of a minislot in length, which are geometrically distributed with an average of l.

If a station has no message to transmit, it is said to belong to the idle mode T_0 and has an empty buffer to hold a message. Whenever a message arrives at an idle station during a minislot for station k, the station will listen to the channel of the intended receiver. If the channel is quiet, the station will attempt with probability one to transmit the packet during the next minislot. If no other packets are transmitted to the same receiving station during that same minislot, the station will capture the channel and continue transmitting until the whole message is successfully transmitted. If a collision occurs or if the channel is sensed to be busy, then the station will enter the blocked mode R_k. Blocked stations in R_k will not accept new messages that are generated, and will listen to the channel k during all the following minislots. When the channel is sensed to be free, the blocked station will attempt to transmit with probability p in the following minislot. Since the average length of the messages is l, once a station starts transmitting a message, the probability that the message will be completely sent during each subsequent minislot is $p_t = 1/l$. All timing imperfections, multipath fading, noise, and other sources of interference (such as finite cross-correlation between spread signals meant for different stations) are neglected in this model, and messages sent that do not collide are assumed to be received accurately. The minislots are assumed to be long enough so that acquisition of the spreading sequence can be achieved, collisions of messages can be detected by all stations, and the channels suffering from collisions can be left quiet before the start of the next minislot.

11.5.3 Analysis of the model

In the analysis the following notation is used:

m_0 – The number of idle stations that belong to the idle mode M_0.
m_k – The number of blocked stations belonging to the blocked mode R_k.
$(m_1 t_1 m_2 t_2 \dots m_N t_N)$ – The state of the system during a minislot, the state vector, t_k is 0 or 1, and is represented by a blank or t, depending on whether there is a station that has captured channel k during that minislot.
$\Sigma_{\text{all } i}(m_i + t_i) = N$ – the total number of stations.

All rearrangements of $m_k t_k$ in $(m_1 t_1 n_2 t_2 \dots m_N t_N)$ are substates of the same state and are equally likely since all the stations are statistically identical.

Example, (0 0t 1), (0 1 0t), and (1 0 0t) are all of the same state in a three-user system. They all show that one station is blocked trying to transmit to another while another station has successfully captured a channel. Note that each substate may have several configurations. For example, (0 0t 1) can mean either station 1 is successfully transmitting to station 2, with station 2 blocked trying to transmit to station 3, or station 3 successfully transmitting to station 2 and station 1 blocked trying to transmit to station 3. The details of the Markov model analysis are omitted and the results of the analysis are given in Figures 11.20 and 11.21.

11.5.4 Equilibrium point analysis of multichannel networks

In the analysis presented so far, the amount of computation rises rapidly with the number of users. For the system with just five stations, the number of states is already 71 and the number of substates 1672. A much simpler approach that can be used for large systems is EPA. The idea behind EPA is that in the long run, the system will stay near the states where the number of messages generated will balance the number of messages successfully transmitted. Some results for the system defined above are shown in Figures 11.22 and 11.23 and Tables 11.5 to 11.8.

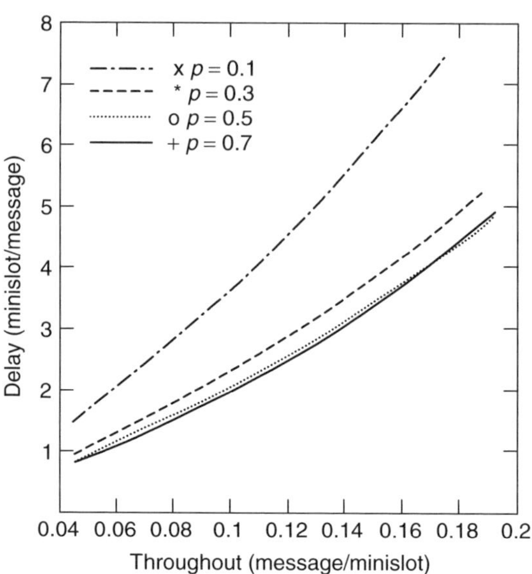

Figure 11.20 Throughput-delay curves for a system with five users using CDMA ALOHA with channel sensing and collision detection. The simulation points are for $s \cdot l = 0.2$, $s \cdot l = 0.5$ and $s \cdot l = 0.8$. The average message length is ten minislots. Simulations are over 200 000 minislots.

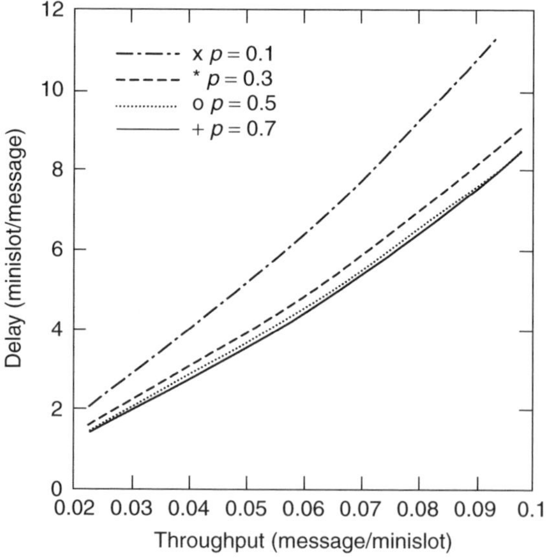

Figure 11.21 Throughput-delay curves for a system similar to that of Figure 11.20, but now with average message lengths of 20 minislots. Simulation points are for $s \cdot l = 0.2$, $s \cdot l = 0.5$ and $s \cdot l = 0.8$. Simulations are done over 200 000 minislots.

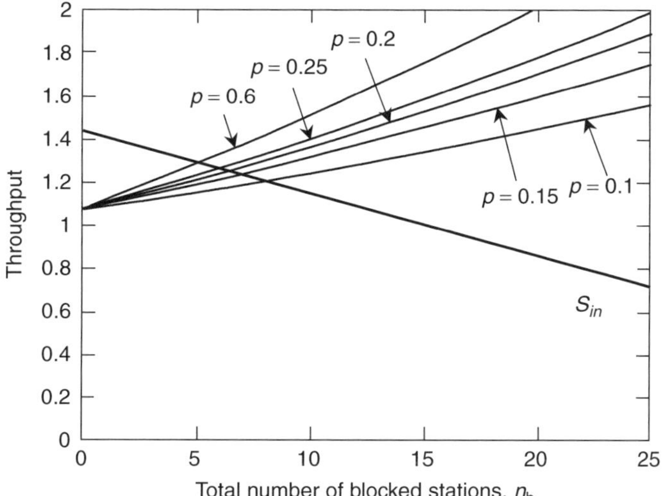

Figure 11.22 Graph of the message input and the average message output curves for $N = 50$, $s = 0.04$ and $l = 10$, for various values of p for a multichannel system [6]. Reproduced from Fook Loong Lo, Tung Sang Ng and Yuk, T. T. (1996) Performance analysis of a fully-connected, full-duplex CDMA ALOHA network with channel sensing and collision detection. *IEEE J. Select. Areas Commun.*, **14**(9), 1708–1716, by permission of IEEE.

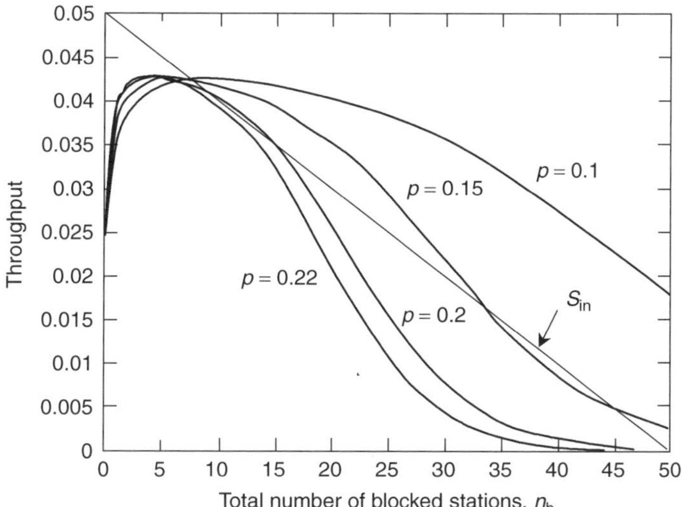

Figure 11.23 Graph of the message input and average message output curves for equation $N = 50$, $s = 0.001$ and $l = 20$ and various values of p, for a single-channel system [6]. Reproduced from Fook Loong Lo, Tung Sang Ng and Yuk, T. T. (1996) Performance analysis of a fully-connected, full-duplex CDMA ALOHA network with channel sensing and collision detection. *IEEE J. Select. Areas Commun.*, **14**(9), 1708–1716, by permission of IEEE.

Table 11.5 Throughputs and delays for a multichannel system with $N = 50$, $s = 0.004$ and $l = 10$ and various values of p. Simulations are done over 100 000 minislots [6]. Reproduced from Fook Loong Lo, Tung Sang Ng and Yuk, T. T. (1996) Performance analysis of a fully-connected, full-duplex CDMA ALOHA network with channel sensing and collision detection. *IEEE J. Select. Areas Commun.*, **14**(9), 1708–1716, by permission of IEEE

s	p	l	S_{cap}-average capture rate	EPA delay	Status	Simulation Throughput	Delay
0.04	0.01	10	1.19	6.78	Stable	1.1783	6.81
0.04	0.15	10	1.23	5.71	Stable	1.2009	6.07
0.04	0.20	10	1.24	5.15	Unstable	1.2170	5.52
0.04	0.25	10	1.25	4.80	Unstable	1.2295	5.25
0.04	0.60	10	1.28	3.97	Unstable	0.0500	965.02

Table 11.6 Throughputs and delays for a single-channel system with $N = 50$, $s = 0.001$ and $l = 20$ and for various values of p. Note that simulations are done over 100 000 minislots, and for unstable systems, the equilibrium values for the first equilibrium point are tabulated [6]. Reproduced from Fook Loong Lo, Tung Sang Ng and Yuk, T. T. (1996) Performance analysis of a fully-connected, full-duplex CDMA ALOHA network with channel sensing and collision detection. *IEEE J. Select. Areas Commun.*, **14**(9), 1708–1716, by permission of IEEE

s	P	l	S_{cap}	EPA Delay	Status	Simulation Throughput	Delay
10^{-3}	0.10	0.20	0.0423	181.1	Stable	0.0429	135.5
10^{-3}	0.15	0.20	0.0424	178.3	Unstable	0.0421	156.1
10^{-3}	0.20	0.20	0.0410	218.3	Unstable	0.00106	3639.0
10^{-3}	0.22	0.20	0.0001	498600	Congested	0.0019	25502

Table 11.7 Throughputs and delays for a multichannel system with $N = 50$, and various values of s, p and l [6]. Reproduced from Fook Loong Lo, Tung Sang Ng and Yuk, T. T. (1996) Performance analysis of a fully-connected, full-duplex CDMA ALOHA network with channel sensing and collision detection. *IEEE J. Select. Areas Commun.*, **14**(9), 1708–1716, by permission of IEEE

s	p	l	S_{cap}	EPA delay	Status	Simulation Throughput	Delay
10^{-3}	0.05	10	0.0495	0.40	Stable	0.050	0.01
10^{-3}	0.05	20	0.0490	1.02	Stable	0.050	0.15
10^{-3}	0.10	10	0.0495	0.20	Stable	0.050	0.00
10^{-3}	0.10	20	0.0490	0.61	Stable	0.050	0.15
2×10^{-3}	0.05	10	0.0979	0.71	Stable	0.100	0.01
2×10^{-3}	0.05	20	0.0958	1.77	Stable	0.100	0.01
2×10^{-3}	0.10	10	0.0979	0.51	Stable	0.100	0.01
2×10^{-3}	0.10	20	0.0959	1.35	Stable	0.098	0.01
10^{-3}	0.20	20	0.0490	0.61	Stable	0.050	0.00
2×10^{-3}	0.20	20	0.0959	1.15	Stable	0.098	0.01

Table 11.8 Throughputs and delays for a single-channel system with $N = 50$ and the same values of s, p and l as in Table 11.7. Note that simulations are done over 100 000 minislots, and for the unstable system, the equilibrium values for the first equilibrium point are used [6]. Reproduced from Fook Loong Lo, Tung Sang Ng and Yuk, T. T. (1996) Performance analysis of a fully-connected, full-duplex CDMA ALOHA network with channel sensing and collision detection. *IEEE J. Select. Areas Commun.*, **14**(9), 1708–1716, by permission of IEEE

s	p	l	S_{cap}	EPA delay	Status	Simulation	
						Throughput	Delay
10^{-3}	0.05	10	0.0487	26.9	Stable	0.0495	17.4
10^{-3}	0.05	20	0.0412	213.6	Stable	0.0418	156.7
10^{-3}	0.10	10	0.0494	13.2	Stable	0.0496	15.8
10^{-3}	0.10	20	0.0423	181.1	Stable	0.0423	148.0
2×10^{-3}	0.05	10	0.0728	186.5	Stable	0.0718	159.1
2×10^{-3}	0.05	20	0.0417	697.8	Stable	0.0420	626.0
2×10^{-3}	0.10	10	0.0720	194.2	Stable	0.0716	167.3
2×10^{-3}	0.10	20	0.0329	1019.2	Stable	0.0335	933.6
10^{-3}	0.20	20	0.0411	215.3	Unstable	0.0057	7712.5
2×10^{-3}	0.20	20	0.0002	249400	Congested	0.0008	62512.9

When the system is lightly loaded, for example, with $s = 0.001, l = 10$, and $p = 0.005$ and 0.10, both the multichannel as well as single-channel CSMA-CD networks can handle almost the maximum message generation rate, $Ns = 0.05$. The delays suffered by the multichannel system, however, are much lower. The difference in performance is even more evident when the system is more heavily loaded. For $s = 0.002, l = 20$, and $p = 0.1$, the single-channel system can handle only about 33% of the maximum possible number of messages generated, with delays suffered by successful messages that go beyond 1000 minislots. The multichannel system, on the other hand, can still handle about 96% of the maximum possible number of messages generated, with successful messages suffering minimal delays. Even when the single-channel system enters the unstable and congested region with $p = 0.20$, the multichannel system can still handle above 95% of the maximum number of messages generated, with negligible message delays.

11.6 IMPLEMENTATION LOSSES IN MAC PROTOCOLS IN WIRELESS CDMA NETWORKS

A number of protocols described in Chapters 10 and 11 depend on the feedback channel state information (FCSI). This section presents an analysis of implementation losses in such systems. Such protocols can be used in different segments (or domains) of the future wireless IP networks including WCDMA cellular networks, wireless LANs and Ad Hoc networks. The motivation behind these protocols is to control the channel access by statically or dynamically changing packet transmission permission probability (TPP), also referred to as persistence probability, depending on FCSI. This is supposed to keep the uplink (UL) load under a certain threshold depending on the QoS requirements. In this section effects of the feedback delay, access delay, and imperfect channel sensing are discussed. These effects are referred to as imperfections. For performance analysis, the system sensitivity functions are introduced. These functions represent the relative performance losses due to the imperfections, which are also referred to as implementation losses. The trade off between throughput, outage probability and packet delay in the presence of imperfections is discussed on the basis of numerical examples for a number of queuing system models.

11.6.1 Network model

Consider the UL of a centralized asynchronous packet radio network, in which N data users (terminals) communicate via central BS using a common DS/CDMA radio channel. The UL and the DL use separate frequency bands. Each user is assigned a unique code sequence for transmitting data packets. The code sequences have N_C chips per bit (processing gain $\equiv N_C$). The packets have equal duration of T_P seconds for their transmission including M bits or MN_C chips. Because of the interference-limited nature of CDMA systems, the radio channel has threshold effects as discussed in Chapter 10. That is, when the number of simultaneous transmissions rises above a certain threshold (effective capacity), collision will occur. The channel threshold can be determined for given QoS requirements. To simplify the system model, the transmit power control is assumed to be perfect. However, the effects of uncertainty in the time-varying radio channel will be investigated in the context of imperfect sensing.

Let K be the channel threshold, which is small compared to the population N, and p_n be the steady-state probability of the system being in state n representing n simultaneous transmissions. Under the perfect power control and homogenous traffic assumptions made above, a packet is considered to be transmitted correctly if, during its transmission period T_P, the system state never exceeds K. The equilibrium system outage probability is defined as the probability that the system state exceeds K. This probability should be kept as low as possible.

11.6.2 Channel access protocol and imperfect sensing

Assume that all data terminals are able to receive the FCSI perfectly in DL. In the FCSI based medium access control (MAC) systems, if a terminal has a packet to send, it uses the latest received FCSI to calculate the TPP probability π_n. It transmits with probability π_n and refrains from transmission with probability $(1 - \pi_n)$. The latest FCSI is supposed to be identical to the current system state n. For the hard-decision system, π_n is defined by

$$\pi_n = \begin{cases} 1, & \text{if } n < K \\ 0, & \text{otherwise} \end{cases} \tag{11.14}$$

For the soft-decision system (one option is discussed in Chapter 10, Section 10.7), π_n has the form shown in Figure 11.24 and can be represented as

$$\pi_n = \begin{cases} 1, & \text{if } 0 \leq n < L \\ \pi^{An-L}, & \text{if } L \leq n < U \\ \pi^{Bn+U}, & \text{otherwise} \end{cases} \tag{11.15}$$

where L, U, A and B are positive parameters, $L \leq K \leq U$ and $0 < \pi < 1$.

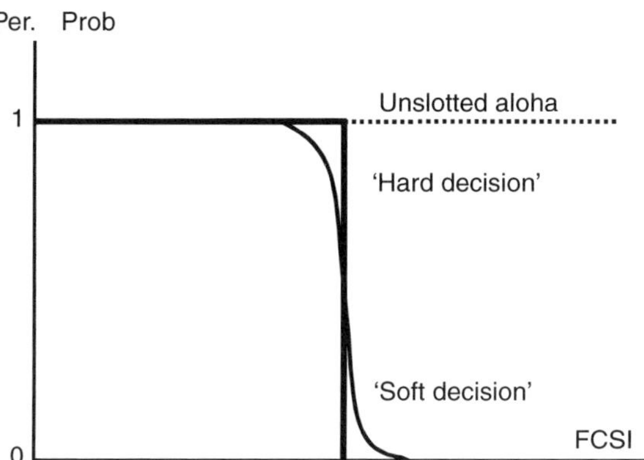

Figure 11.24   The motivation behind the FCSI based MAC protocols.

The alternative is to use the tangent hyperbolic, $tgh(n)$, function as follows:

$$\pi_n = \frac{1}{2}[1 - tgh(an - b)] \tag{11.16}$$

where a and b are also positive parameters acting as A and B listed above. equation (11.16) can be further modified as

$$\pi_n = \frac{1}{2}[1 - tgh_k(\alpha, \beta, n - b)] \quad k = 1, 2 \tag{11.17}$$

where $tgh_k(\alpha, \beta, n), k = 1, 2 \quad \alpha > 0, \beta > 0$ is defined as follows:

$$tgh_1(\alpha, \beta, n) = \frac{\alpha e^n - \beta e^{-n}}{\alpha e^n + \beta e^{-n}}; \quad tgh_2(\alpha, \beta, n) = \frac{\alpha e^{\alpha n} - \beta e^{-\beta n}}{\alpha e^{\alpha n} + \beta e^{-\beta n}} \tag{11.18}$$

Assume that the BS has a bank of matched filters matched to the codes of the users, and that the presence of a particular transmission can be detected upon reception of one bit [7]. For perfect power control and one-class traffic, the number of users characterizes the state of the channel. However, because of numerous certain and uncertain factors affecting the spread-spectrum transmission in unreliable radio channels, the BS eventually makes a wrong decision upon detecting the channel state. In order to be able to quantify the performance losses and also to get a better insight into the system performance in the presence of imperfect sensing of the channel state, we introduce two parameters to specify the imperfect sensing effects. These are: $D \equiv$ Pr { the BS correctly detects the channel state under the threshold K}; $F \equiv$ Pr {the BS falsely detects the channel state over the threshold K}. D and F are mutually dependent and can be also state-dependent determined as in Reference [7], but herein we use them as the parameters taking different constant values to investigate the performance losses. In the presence of imperfect sensing, the channel states may be up to the population N even for the hard-decision system.

11.6.3 Possible extensions of the model

In the case of imperfect power control, a continuum of received signal power level is possible like in Chapter 10 and the overall multiple-access interference level information should be used for access control rather then the number of equivalent users. Thereby, the π_n function can be defined in such way that it adapts to the UL interference distribution, for example. Gaussian, to compensate the error of local-average interference level that dominates the QoS and to benefit from the soft-capacity feature of CDMA systems. Because of limited space these extensions are not considered in this section.

The model can be extended to include multimedia applications. In that case a user may have 2^k higher bit rate with respect to a reference user, resulting in processing gain reduction by the same factor 2^k, which on the other hand requires increase in the power by factor 2^k with respect to the power of reference user. This is equivalent to increasing the number of reference users by the same factor 2^k. In this case the analysis should be modified to allow for bulk arrivals. This issue will be discussed in Section 11.7.

11.6.4 Traffic model and delay effects

The traffic model is based on the following assumptions. For the system offered load, data packets are generated by a finite population of N identical data sources (terminals) according to a state-dependent Poisson process with rate λ_n

$$\lambda_n = \begin{cases} (N - n)\lambda, & \text{if } n \leq N - 1 \\ 0, & \text{otherwise} \end{cases} \tag{11.19}$$

where λ is a packet arrival rate at each terminal, $\lambda / T_P \leq 1$. The terminal has two states: free (idle) and blocked, and no buffer. Upon the arrival of a new packet at a free terminal, the terminal goes to the blocked state immediately and stays in that state until the current packet is successfully transmitted [8]. The blocked terminals attempt to occupy the channel according to the FCSI-based MAC protocol. Erroneous packets defined before must be retransmitted. Those blocked terminals, which cannot transmit their packet within T_P seconds, shall reschedule their packet. The scheduling is randomized according to the Poisson process with the rate also equal to λ so that the overall offered load for the packet-access attempting process is kept the same. The average system offered load defined as the average number of packets generated in a packet interval T_P is

$$G = T_P \sum_{n=0}^{N-1} \lambda_n p_n \tag{11.20}$$

where p_n is the system steady-state probability. The average offered load for the system in the state n is therefore

$$G_n = T_P \sum_{i=n}^{N-1} \lambda_i p_i \tag{11.21}$$

11.6.5 Model validity conditions

It would make sense to use FCSI-based MAC protocols only if the BS was able to provide in time correct FCSI for the terminals to make a decision on their attempt to access the channel. The throughput improvement offered by using FCSI-based MAC protocols will depend on the feedback delay and access delay relatively to the rates at which the channel state varies. If the channel state is changing fast compared to the delays then users can no longer adapt to the channel variations and hence this access control may not result in any significant improvement. To make it work, proper choices and trade- offs of the models and design parameters must be made. These include the service time distribution, the offered load and delay models, the mean values and so on. To be able to trace the changes of the channel states during the service time of a packet, the Erlang type r distribution function with mean $1/\mu \equiv T_P$ is chosen to represent the service. This service process can be seen to be of r identical stages having exponential distribution with mean $1/r\mu$ [9]. During the packet transmission it is assumed that the channel state may vary from stage to stage, but remains unchanged within a stage. Also it is assumed that the BS senses the channel state and broadcasts the updated FCSI once per stage, that is, every T_P/r interval. For the validity of this assumption, r should take such value that the probability of having more than one event (arrival or departure) in T_P/r interval is negligible, at the same time T_P/r must be feasible for the updating process of the FCSI. The reported results in Reference [10] have shown that this assumption is valid. For instance, if $T_P = 20\,\text{ms}$ and $r = 30$, then $T_P/r = 2/3\,\text{ms}$, that is identical to the random-access channel time-slot offset (contention slots) defined in the 3GPP standards for the 3G WCDMA cellular PCS (see Chapter 17). In Reference [10] for a similar feedback control protocol, the assumption that at most one event can occur in a time-slot with zero feedback and access delays was used, which justifies the Markov model for the process. In the sequel, this assumption is referred to as the *validity condition* of the model.

The delays considered in this section are the access delay and the feedback delay. The access delay is defined as a sum of the decision-making processing time at a terminal and the UL propagation delay excluding the random back-off delay. The feedback delay is a sum of the channel-sensing processing time in the BS and the DL propagation delay. To simplify the analysis, we introduce the access timing delay as a sum of the access delay and the feedback delay, which is modeled as an exponentially distributed RV having mean $1/\gamma$ and independent of the service time. This access timing delay is supposed to be not longer than T_P/r interval or $1/r\mu$. Since $1/r\mu$ is considered to be very small, compared to $1/\lambda$ and $1/\mu$, owing to the model validity condition, the effects of the delays can be ignored. In case the access timing delay is not smaller than T_P/r, the effects of delays may cause expansion (increase the number) of the system states. For the analytical method used to derive the performance, the service time distribution as well as TPP of a terminal, at a given system state, should be modified to take into account the delay effects. In Reference [11], a hypothetical packet concept has been developed to take care of access timing delay also resulting in expansion of the channel states. It was assumed that state expansion is a Poisson RV. For this reason in some cases, the average system throughput could even exceed the channel threshold, which cannot happen in practice. Here, because of the validity condition, the number of system states may increase by a factor of $\lfloor r\mu/\gamma \rfloor$ (the largest integer that does not exceed the argument). The TPP for attempting transmission of a packet should now be modified as

$$\overline{\pi}_n = \sum_{i=0}^{m} \pi_{n-i} \mathrm{Pr}\{i \text{ arrivals in } 1/\gamma | \text{ given maximum } m \text{ arrivals in } 1/\gamma\}$$

where $m = \min(n, \lfloor r\mu/\gamma \rfloor)$. That is

$$\overline{\pi}_n = \sum_{i=0}^{m} \pi_{n-i} \frac{(G_a\mu/\gamma)^i}{i!} e^{-G_a\mu/\gamma} \left(\sum_{j=0}^{m} \frac{(G_a\mu/\gamma)^j}{j!} e^{-G_a\mu/\gamma} \right)^{-1} \tag{11.22}$$

where G_a is the average offered load of the system without access control given again by equation (11.20), and π_n is defined as in Figure 11.24.

Meanwhile, each packet generated with rate λ at a terminal will experience the mean delay of $1/\gamma$ before entering service, given no random back-off delay. If we treat the wasted time as a part of the service time, the system can be seen with no delays. The wasted time can be approximated by a minimum of the access timing delay and the time until the next arrival due to random back-off. The average value of this time is equal to $1/(\lambda + \gamma)$. The distribution of the modified service time is the convolution of the service-time distribution and the wasted-time distribution. If the service time were exponentially distributed with mean $1/\mu$, the probability distribution function of the modified service time would be

$$\frac{\mu(\lambda + \gamma)}{\lambda + \gamma - \mu}[e^{-\mu t} - e^{-(\lambda+\gamma)t}] \tag{11.23}$$

Since we assume that $1/\gamma$ is very small compared to $1/\mu$, the above distribution can be approximated as an exponential distribution with mean

$$\frac{1}{\mu'} = \frac{1}{\mu} + \frac{1}{\lambda + \gamma} \tag{11.24}$$

The analysis can be expanded for the hard-decision system in the presence of access timing delay and imperfect sensing. Together with the effect of false detection (probability F) in overload states that is referred to as 'false alarm', the imperfections will cause the expansion of the system states to K'. Recall the validity condition of the traffic model. If only the delays were present (nonzero delays and $D = 1$ and $F = 0$), K' would be $\min(N, K + \lfloor r\mu/\gamma \rfloor)$. If only imperfect sensing was present (zero delays and $D \neq 1$, $F \neq 0$), K' would be up to $\min(N, r)$. Finally, in the presence of all imperfections, K' would be equal to $\min(N, r + \lfloor r\mu/\gamma \rfloor)$. In order to make the analysis more systematic, the system with only delay effects is considered first, then the system with the effects of imperfect sensing and delays. The detailed analysis can be found in Reference [12].

Here we define the sensitivity functions for the system throughput and the average packet delay as the functions of the imperfection parameters, that is, the imperfect sensing probabilities and the normalized access timing delay. The throughput sensitivity S_s is defined as:

$$S_s = \frac{|S - S_0|}{\max(S, S_0)} \tag{11.25}$$

where S is the system throughput in the presence of the corresponding imperfections, S_0 is the throughput of the perfect system. Similarly, the average packet delay sensitivity D_s is defined as:

$$D_s = \frac{|D_P - D_0|}{\max(D_P, D_0)} \tag{11.26}$$

where D_P is the average packet delay in the presence of the corresponding imperfections, and D_0 is the average packet delay without those imperfections.

For illustration purposes the following system parameters are used in Reference [12]: $K = 7$, $N = 30$, $T_P = 20\,\text{ms}$, $r = 30$. This set of the system parameters is chosen in order to be able to present the numerical results compatible with those presented in References [13,14] while the validity condition is still kept. Figure 11.25 presents the performance of the perfect soft-decision system with different persistence probability functions in comparison with the perfect hard decision and the unslotted ALOHA systems. The results show that the soft-decision system outperforms the others. The system throughput improvements for a given delay can be clearly seen from the figure for the

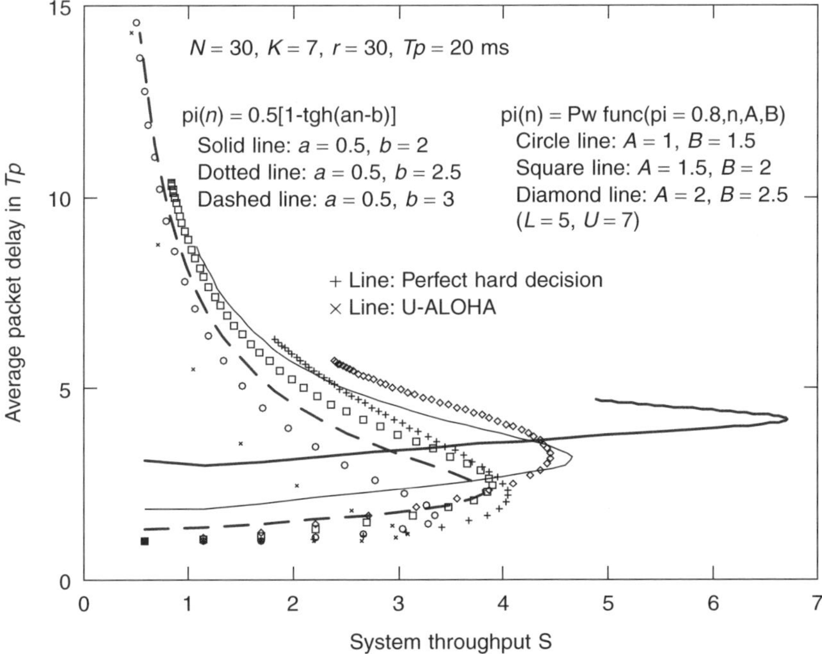

Figure 11.25 The system throughput-delay trade-off for perfect systems.

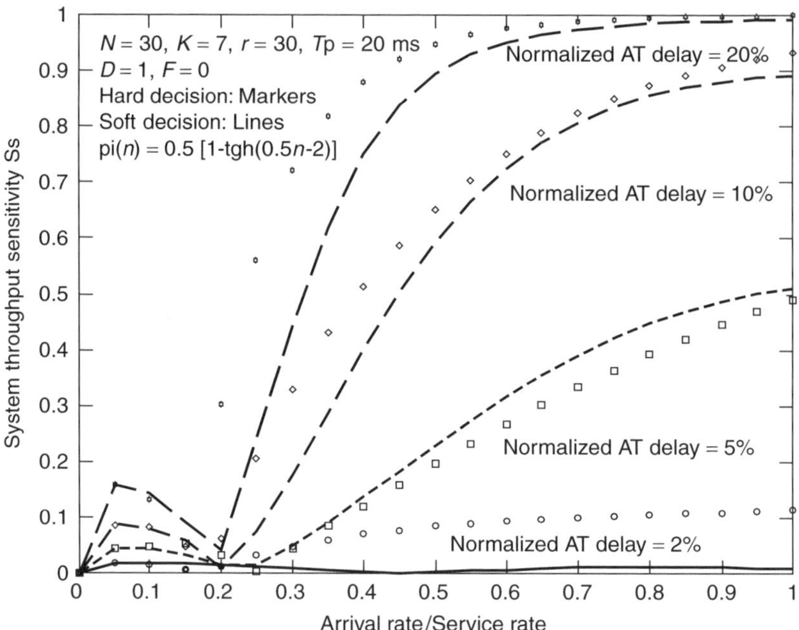

Figure 11.26 The system throughput losses due to the effects of access timing delay.

system having the power persistence function (PwFunction) defined in equation (11.15) with $U = 7$, $L = 5$, $A = 2$, $B = 2.5$, and $\pi = 0.8$. The same conclusion is for the system having the *tgh* persistence function defined in equation (11.16) with $a = 0.5$, $b = 2$, which is the best and will be further discussed in the sequel.

Both protocols show more or less the same throughput sensitivity to the access timing delay (Figure 11.26). In high-loaded systems, more than 90% performance losses can be expected if the normalized access delay is larger than 20%. Figure 11.27 shows that the soft-decision system has more stable delay characteristics. Overall, it has been shown that the systems with the FCSI-based MAC perform much better than the systems without access control even for reasonably high values of imperfect sensing probabilities and delays. A significant improvement in throughput, and more stable performance characteristics under these effects can be obtained by using soft-decision FCSI-based MAC protocol.

11.7 RADIO RESOURCE MANAGEMENT IN WIRELESS IP NETWORKS AND DIFFERENTIATED SERVICES

In this section we summarize the material from this chapter with the focus on possible applications in wireless IP networks.

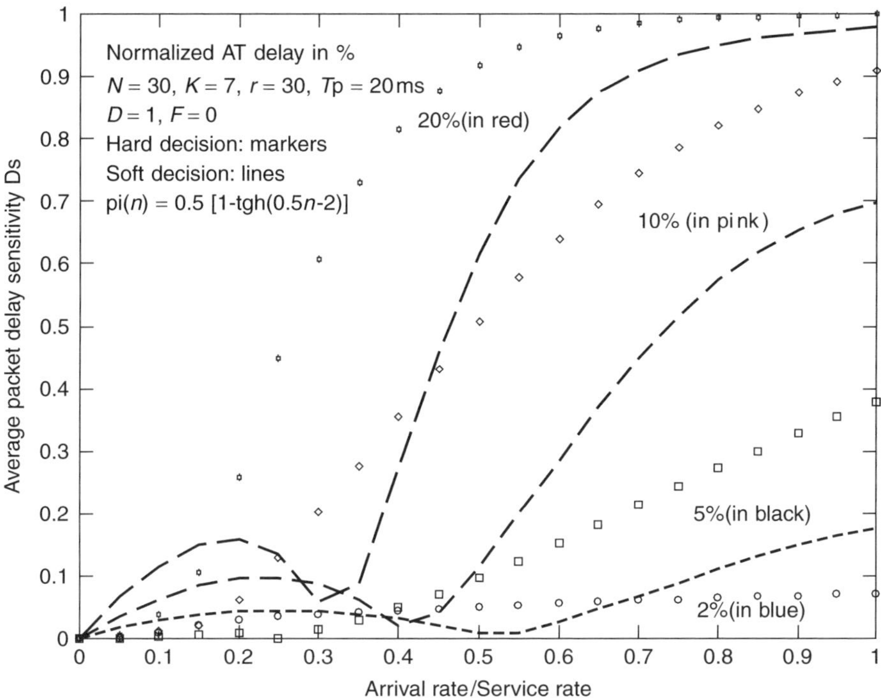

Figure 11.27 The average packet delay losses due to the effects of access timing delay.

The idea behind 'wireless IP' is to use Internet Protocol (IP) technology as common service platform for all mobile services, and also as unified transport platform for a seamless delivery of IP-based services across wired and wireless networks. Recently, a variety of emerging technologies and protocol enhancements has been investigated and designed for providing such IP-based multimedia services to mobile users. These research and standardization efforts are summarized in References [15–17]. Evolution of today's cellular mobile communications and currently being developed third generation systems or in short 3G, are believed to be the most significant segments of future wireless IP networks for reliable mobile access and operations. The radios of those cellular systems are based on WCDMA technology and are likely to remain unchanged or to have minor modifications in the long run because of economic reasons.

11.7.1 Service models

Integration of IP-based multimedia capabilities in the wireless IP leads to unpredictable extensions of mobile applications. The increasing bit rates supported with 3G-radios together with the increasing processing power of mobile terminals gives opportunities for introducing various RT interactive multimedia applications and services to mobile users. In the process of moving toward a seamless convergence of fixed and mobile services or, a coexistence of voice/video and elastic applications like File Transfer Protocol (FTP) or www-browsing over common IP-based platform, it is important to have a robust QoS classification that is capable of providing a fine QoS resolution. In IP-based networks, different applications have different requirements concerning delay-tolerant and error-tolerant characteristics. For examples, voice applications can tolerate low-packet delay and moderate BER while it is vice versa for messaging service. However, it can be expected that QoS parameters of wireless users are more complicated than that of wired users for the same application due to restrictions and limitations of air-interface. The delay that a packet experiences during its delivery in the network is one of the most important QoS parameters for both wired and wireless packet access. The BER parameter is very important in the wireless environment, which is often used as a performance measure of wireless link. Other QoS parameters can be, for example, minimum-guaranteed bit rate, maximum bit rate, maximum packet size and so on.

There are several standardized QoS classifications. Recommendation I.211 [18] of the International Telecommunication Union Telecommunication Sector (ITU-T) has identified two main classes: interactive and distribution services. The interactive services are further classified into conversational (e.g. voice, video, interactive game), messaging (e.g. paging, short message service, e-mail, multimedia message) and retrieval services (e.g. tele-text, Web-browsing, e-shopping). On the other hand, the distribution services are further classified into distribution services without (e.g. broadcasting, multicasting) and with (e.g. uni-casting, video-on-demand) user individual presentation control. The ITU-T classification is application oriented.

Standards of the 3rd Generation Partnership Project (3GPP) and European Telecommunications Standards Institute (ETSI) for UMTS [19,20] define four QoS classes on the basis of network QoS parameters (mainly delay and BER). These are conversational or service Class A, streaming or service Class B, interactive or service Class C, and background or service Class D. The first two classes correspond to RT connection-oriented services with strict-delay constraints, for example, 20 to 50[ms] for Class A and around 50[ms] for Class B. The last two classes are for non-real-time (NRT) connectionless services, where bounded delay constraint is applied for Class C, for example, utmost 300[ms], and low BER is required for Class D, for example, at least 10^{-8}.

Finally, standardization efforts of the Internet Engineering Task Force (IETF) for new IP service models should be mentioned while considering wireless IP networks. IETF introduces two alternative service models: integrated services (IntServ) [21] and differentiated services (DiffServ) [22] to support QoS for multimedia applications. In addition to the best-effort service class of today's Internet, two more classes are introduced in both IntServ and DiffServ models. These are guaranteed class (e.g. RT services) and predictive class (e.g. NRT interactive services) in IntServ and expedited-forwarding class (e.g. strict-delay RT services) and assured-forwarding class (e.g. RT services and NRT interactive services) in DiffServ.

11.7.2 Radio resource management (RRM)

Consider a cellular system or segment of the wireless IP network consisting of N_b cellular access points (CAP) or BS. There are N_c waveforms (or channels) available; N_m possible packet transmission modes, each corresponding to a different bit rate; and N_u active users in the system. Let: $\mathbf{B} = \{1, 2, \ldots, N_b\}$; $\mathbf{C} = \{1, 2, \ldots, N_c\}$; $\mathbf{M} = \{1, 2, \ldots, N_m\}$; and $\mathbf{U} = \{1, 2, \ldots, N_u\}$. To establish a radio link for any communication services in the UL direction, for instance, each active terminal $u_i \in \mathbf{U}$ needs to be assigned to a serving CAP $b \in \mathbf{B}$, suitable waveform or channel(s) $c_i \in \mathbf{C}$, a mode $m_i \in \mathbf{M}$, and transmitter power $P_{tx,i}$. This resource assignment is restricted by limitations of available waveforms or channels in $\mathbf{C}$, transmitter power $P_{tx,i}$ together with tolerable interference level of CAP and allowable modes in $\mathbf{M}$. The main tasks of RRM are to allocate and maintain adequate radio resources for all active radio connections in the system so that the required QoS of mobile users, moving around the service area, can be met once they are admitted into the system. This means that signal to interference + noise ratio (SINR), dominating the quality of radio link in radio transmissions, has to be kept above a target level depending on the transmission mode. Thus, the following inequality must hold for the radio connection between user u_i and serving CAP b using waveform c_i

$$\gamma_i = \frac{P_{tx,i}G_{b,i}}{\displaystyle\sum_{n \neq i} P_{tx,n}\theta_{i,n}G_{b,n} + N_{0b}} \geq \gamma_{i\_\text{target}} \tag{11.27}$$

where γ_i and γ_{i_target} denote actual and target SINR of u_i connection respectively; $G_{b,i}$ denotes the path loss (or link gain) between u_i transmitter and b receiver for all $u_i \in \mathbf{U}$; $\theta_{i,n}$ denotes the cross-correlation between waveform c_i and c_n for all active users; and N_{0b} denotes the thermal noise power received at b.

As the number of users and the demands for services increase, the number of available waveforms or channels becomes insufficient to support QoS requests of all the users complying with equation (11.27). Resource assignment strategies (RAS) provide the means to access the resources of each cell efficiently and assign the available resources to achieve the highest spectrum efficiency. RAS can be classified into static, dynamic and flexible classes. In static RAS class, the resource allocation is often based on *a priori* knowledge of available network resources and resource consumption of users for requested services. This is mainly done during the planning stage, by assigning a set of channels to each cell permanently according to a frequency-reuse pattern, then allowing users to borrow one or several channels from that set for its requested services. The design for static RAS is often simple and distributed, but it has to take into account the 'worst-case' scenarios of network and channel conditions, and also resource consumption in order to guarantee QoS. In the dynamic RAS class, the resource allocation is supposed to make use of available resources fully. There is no need for planning frequency-reuse patterns. The design in most cases relies on complicated online measurements and excessive signaling to keep tracks on changes of network conditions and user-resource consumption for accommodating as many service requests as possible. In practice, adopted RAS combines aspects of both static RAS and dynamic RAS, forming a so-called flexible RAS class. For example, radio resources can be allocated in the static manner up to a certain level, the rest is utilized in the dynamic manner; or a flexible RAS can be based on both static modeling and dynamic measurements of resource and traffic statistics.

Other important RRM functions are call admission control (CAC), packet access control (PAC), power control, and handoff. These mutually support each other and RAS. Technical issues of power control and handoff in wireless IP networks are addressed in Chapters 6 and 10, respectively. The focus of this chapter is on CAC and PAC providing QoS while optimizing GoS (grade of service) and RRU (radio resource utilization).

11.7.3 Overview of call admission control (CAC) policies

In general, CAC policies can be evaluated with respect to the following parameters:

1. *Effectiveness*: based on trade offs between simplicity in implementation and ability in autoconfiguration with various traffic, user and network profiles for maximizing system capacity and revenues. The performance measures are the probabilities of call blocking and handoff failure or some cost functions, for example, GoS defined in detail later.
2. *Reliability*: measured by the probability of loosing communication quality during services for given offered multimedia traffic intensities (MTI).
3. *Robustness*: seen against the need for redesign due to uncertainty and changes of the system parameters.

In the sequel, an elaboration of three different CAC policies, namely static modeling-based complete-sharing CAC (MdCAC), dynamic measurement-based CAC (MsCAC), and flexible system statistics- based soft-decision CAC (SCAC), for UL of CDMA cellular access segments of the wireless IP is presented. QoS differentiation paradigms are introduced and integrated into CAC mechanisms, in which load-based thresholds or fracturing factors are defined for preshaping the traffic (i.e. hard or soft blocking of less important calls).

11.7.4 Principles of MdCAC, MsCAC and SCAC systems

Let us consider UL direction of a cell in a DS/CDMA cellular system consisting of multiple and identical cells. This system is assumed to support M connection-oriented RT service classes and has bandwidth or chip rate W. Let $\mathbf{K} = \{1, 2, \ldots, M\}$ and for all $k \in \mathbf{K}$ let r_k be the bit rate, γ_{target_k} the target SINR of class-k (link-quality requirements), and g_k the processing gain of class-k radio transmissions that depend on the spreading factor, data-source activity factor, diversity-combining gain and so on. Define:

$\mathbf{n} = (n_k, k \in \mathbf{K})$ is the occupancy vector, where n_k is the number of class-k calls in progress;

$\mathbf{w} = (w_k, k \in \mathbf{K})$ is the load vector, where w_k is the average load factor of a class-k call, representing the average resource consumption of radio connections onto system spectrum or radio resources given by $w_k = \gamma_{\text{target}\_k}/g_k$.

C_{a} — the average normalized UL capacity. In the ideal situation, 100% of the available radio spectrum can be utilized resulting in $C_{\text{a}} = 1$. In real situations, C_{a} is strongly dependent on thermal noise and interference from other nearby cells. The quantity of actual capacity is a RV and C_{a} itself can be considered as a bounded stationary RV as well. The 'traditional' average system capacity corresponding to each service class, defined as the largest number of users that the system can handle properly, is given by the ratio C_a/w_k. By modification of results from Chapter 8, the average system capacity C_a is often predicted by using load equation in the boundary condition of tolerable interference level as in References [23,24]:

$$C_a = \frac{1 - \eta}{1 + f} \tag{11.28}$$

where η is the ratio of thermal noise density and UL is the maximum tolerable interference level; f is the coefficient representing the ratio between other- and own-cell interference (for both mean and variance values), which varies with change of propagation parameters, transmitter power control (TPC) inaccuracy, traffic distributions and so on. The fluctuations of f have essential impacts on system capacity resulting in the 'soft-capacity' feature of CDMA systems.

The most straightforward CAC policy is to share 'fixed' C_a units of radio resources in static complete-sharing fashion among active users on the basis of their QoS requests, resulting in MdCAC policy. In the MdCAC system, a call request for class-k service is accepted immediately if utmost $(C_a - w_k)$ units of resources are occupied, otherwise it is rejected. This system may face the following problems. First, the QoS guaranteed scenario results in the 'worst case' of system capacity and RRU. Second, there is a need for redesign of the system-capacity constraint C_a for changes of CDMA radio channel parameters and traffic statistics. Third, advantages of CDMA techniques, such as 'soft-capacity' feature, cannot be exploited.

Results of recent research have emphasized that robust CAC policies may require online measurements (see Chapter 10 and References [25–27] and [28]), resulting in far more complex and high-cost SW/HW implementations [29,30]. The principle of MsCAC is that a new call is accepted immediately into the system if at least the required resource consumption of its service class is available, otherwise it is rejected. The decision is made on the basis of online measurements of related statistics. Latest analysis and simulation results show that when UL traffic is less bursty, the MsCAC has no capacity-gain over the MdCAC described above. It may even suffer degradation due to measurement errors, which agrees with the results presented in References [26,27,30]. Some of these algorithms have been discussed in Chapter 10.

To harmonize the advantages and overcome the problems of static MdCAC and dynamic MsCAC with a flexible system statistics-based SCAC policy, let us first reconsider the resource consumption by using the well-known effective-bandwidth concept. The parameters γ_{target_k}, w_k and f are now treated as stationary RVs and we use the same fonts to denote their mean values. It is assumed that parameters of the requested services and adequate TPC mechanism can provide and ensure *a priori* knowledge and stationary behavior of the resource consumption. Thus, based on the knowledge of local-average SINR (i.e. lognormal RV with standard deviation σ in range of 2[dB] [23,24,31] and cell interference distributions (i.e. normal or Gaussian [23,29,30]), the resource consumption can be modeled as follows. For all $k \in \mathbf{K}$, resource consumption of a class-k connection is a stationary, independent, and bounded normal RV having mean w_k, variance σ_k^2, and falling in between a max- and min-guaranteed resource consumption $w_{u,k}$ and $w_{l,k}$. The system capacity is accordingly modeled with a bounded Gaussian RV having mean

C_a, variance Θ^2, and falling in between C_l and C_u. In theory, C_u can be set up to the isolated-cell capacity of $(1 - \eta)$. In general, the mean, variance and boundary values of link resource consumption and cell capacity can be estimated by using the definition of load vector **w** above and equation (11.28) together with mean, variance and boundaries of SINR and f. Because of the interference-limited nature of CDMA systems the cell shall meet their QoS constraints if

$$c_{\text{other}} + c_{\text{own}} \leq (1 - \eta) \tag{11.29}$$

where c_{other} represents the equivalent UL load factor produced by other-cell interference and c_{own} is the load factor generated by active users in own cell. Define:

$c = \mathbf{n}\mathbf{w}$ the system load state that is identical to c_{own}, where **n** is the occupancy vector and **w** is the load vector defined above.

From equation (11.29) and with Gaussian interference and lognormal local-average SINR, the QoS loss probability (i.e. local outage) conditioned on the system load state c, can be determined by

$$\Pr\{\text{QoS loss}|c\} = Q\left(\frac{1 - \eta - E[Z|c]}{\sqrt{\text{Var}[Z|c]}}\right) \tag{11.30}$$

with:

$$E[Z|c] = c(1 + f)\exp[(\varepsilon\sigma)^2/2]$$

$$\text{Var}[Z|c] = c(1 + f)\exp[2(\varepsilon\sigma)^2]$$

where $Q(x)$ is the standard Gaussian integral function; σ is the standard deviation of lognormal SINR in [dB]; and constant $\varepsilon = \ln(10)/10$. In the long run, with respect to Gaussian-distributed intercell interference we have:

$$\Pr\{c_{\text{other}} \leq 1 - \eta - c\} = 1 - Q\left(\frac{1 - \eta - c - f E[c]}{\sqrt{f \text{Var}[c]}}\right) \tag{11.31}$$

This relation implies that intercell interference has significant impact on the soft capacity and over all communication quality in the long run. Therefore, an optimal way to benefit from this soft-capacity feature is to use flexible CAC mechanisms having decision functions adapted to intercell interference distribution to compensate fluctuations of local-average SINR, hence maintaining the link quality while expanding the admissible region. Initial discussion under the name *Global Access Strategy* is presented in Chapter 10. SCAC policy is a solution that works as follows. Instead of using only the average system capacity C_a as in MdCAC, here we use both the upper and lower limits C_u and C_l, respectively. This will improve the system capacity and the robustness. In the SCAC system, based on equations (11.15 to 11.18), a new call of class-k shall be

1. admitted immediately if the current system load state c is utmost $(C_l - w_k)$,
2. admitted with a probability $\pi_k(c)$, as defined below on the basis of equation (11.31), if c is utmost $(C_u - w_k)$ but larger than $(C_l - w_k)$, and
3. rejected immediately otherwise.

Handoff calls are admitted in a non-preemptive priority discipline, that is, a handoff failure of class-k call occurs only when c is larger than $(C_u - w_k)$. The state-dependent admission probability $\pi_k(c)$ of a new class-k call is based on equations (11.15 to 11.18) and is given by

$$\pi_k(c) = \begin{cases} 1, & \text{if } c + w_k \leq C_l \\ 1 - Q\left(\dfrac{1 - \eta - c - w_k - fE[c]}{\sqrt{f\,\text{Var}\,[c]}}\right), & \text{if } C_l < c + w_k \leq C_u \\ 0, & \text{otherwise} \end{cases} \tag{11.32}$$

Additional options can be used for such decision functions that increase the flexibility of SCAC in tuning the system performance. For instance, to avoid overestimation of C_u that may cause an increase of the system local-outage probability, some adjustments of the mean and variance in $Q(x)$ function, or other functions, such as incomplete-Gamma $\Gamma(v, x)$, can be used to hold back the offered traffic more in high load states. For the latter option, $\pi_k(c)$ for $C_l < c + w_k \leq C_u$ becomes

$$\pi_k(c) = 1 - \Gamma(v, \alpha(1 - \eta - c - w_k)) \tag{11.33}$$

where α and v are implicitly defined as: $fE[c] = v/\alpha$ and $f\text{Var}[c] = v/\alpha^2$. In general, $E[c]$ and $\text{Var}[c]$ can be calculated by using steady-state solutions and inversion techniques. To reduce the computational complexity, the corresponding mean and variance values of MdCAC system can be reused. This is reliable because the MdCAC system can be expected to represent the stationary behavior well.

The SCAC policy has advantages over both MdCAC and MsCAC policies. It is simple, robust and has better RRU. The SCAC can also be combined with measurement-based techniques to reduce the complexity of estimators to compensate the bias of measurements and to enhance performance. Implementation issues will be discussed later on in more detail. Further, SCAC provides better traffic-shaping gain than MdCAC and MsCAC. It gives more chance for calls of high resource-consuming classes to access the system when traffic intensity of lower classes is heavy. This cannot be improved with the other two CAC policies.

11.7.5 QoS differentiation paradigms

Suppose there are J different user classes sorted in the decreasing order of priority, for example, 1 is 'gold', 2 is 'silver' and so on. Let $\mathbf{J} = \{1, 2, \ldots, J\}$. Together with M different service classes sorted in the increasing order of resource consumption, it can be considered as if there are $J \times M$ prioritized traffic classes, which form different group behaviors according to user or service classes. Note that hereafter the term *traffic class* or class-(j, k) is used when QoS differentiation is applied to distinguish from service class-k alone in complete-sharing scenarios. Let us introduce a $J \times M$ prioritized admission probability table given in matrix-form as follows:

$$A_0 = [a_{jk}(c)] \text{ with } j \in \mathbf{J}, k \in \mathbf{K}, 0 \leq a_{jk}(c) \leq 1 \tag{11.34}$$

where $a_{jk}(c)$ is the prioritized admission probability of a new call request from class-j user for class-k service at a given system load state c.

In a hard threshold-blocking case, it is straightforward that $J \times M$ thresholds may be needed, each corresponding to a traffic class. Define a threshold table

$L = [l_{jk}]$ for all $j \in \mathbf{J}$, $k \in \mathbf{K}$, $l_{jk} \geq l_{uv}$ if $j \geq u$ and $k \geq v$.

Thus, $a_{jk}(c)$ in equation (11.34) can be given by

$$a_{jk}(c) = \begin{cases} 1 & \text{if } c + w_k \leq l_{jk} \\ 0 & \text{otherwise} \end{cases} \tag{11.35}$$

The number of necessary thresholds could be reduced significantly if only group behaviors were of interest, for example, J or M thresholds would be needed instead of $J \times M$. For example, upper bound C_u of the system capacity can be used for the 'gold' class access in SCAC system, whereas the 'bronze' class has a blocking threshold of lower bound C_l regardless of requested services.

In the softer blocking case, $a_{jk}(c)$ can take any value in the [0,1] interval depending on the system load state and the priority of requested traffic class. This results in load-based fracturing factors for each traffic class or group behavior. The fractional paradigm on one hand gives operators better flexibility to tune the GoS, and on the other hand allows to unify analysis of a class of guard-resource (or guard-channel) schemes for QoS differentiation and CAC. The hard threshold-blocking case described previously is in fact a special variation of this fractional rule.

Using QoS differentiation, the admission probability of a class-(j, k) call for all $j \in$ $\mathbf{J}$ and $k \in \mathbf{K}$ in SCAC system, conditioned on the system load state c, is given by

$$A_s = [\pi_k(c)a_{jk}(c)] \tag{11.36}$$

where $\pi_k(c)$ is defined in equations (11.32 or 11.33) alternatively; and $a_{jk}(c)$ in equation (11.34). Up to this point, handoff calls have been assumed to have the highest priority regardless of their associated traffic classes. If not so, prioritization of handoff calls can be handled similarly as of new calls. QoS differentiation of the background NRT packet-switched services can be done not only in a blocking fashion but also by granting different throughput-delay connections. A detailed performance analysis of these systems can be found in Reference [32]. For illustration purposes we use a cellular system supporting three RT service classes: Class-1 is 12.2 kbps effective rate (EFR) voice, Class-2 is 64 kbps video and Class-3 is 144 kbps multimedia calls. The parameters are given in Table 11.9. The offered traffic intensities of the three classes are considered in the following proportions: 7:2:1 and 4:3:3 for Class-1, Class-2 and Class-3, respectively. These are called multimedia traffic intensity profiles, MTIP. In other words, let λ be a so-called common divisor of the three-class offered traffic; then 7:2:1 MTIP for instance means that the offered traffic of Class-1 is 7λ [Erlang], Class-2 is 2λ [Erlang] and Class-3 is λ [Erlang]. Thus, the total offered traffic is 10 [Erlang] if $\lambda = 1$.

In scenarios with QoS differentiation, the customers are assumed to be divided into two user classes: business $(j = 1)$ and economy $(j = 2)$ having equal service demands: $\lambda_{l,1k} = \lambda_{l,2k}$ for $k = 1, 2, 3$. The offered traffic patterns given by MTIP above, for example,

Table 11.9 Summary of the system parameters

Name	Definition	Values
W	CDMA chip rate	3.84[Mcps]
f	Coefficient of the average other-to-own cell interference	$40 \pm 15\%$ (microcellular)
η	Constant coefficient of the thermal noise density and maximum tolerable cell interference level	-10[dB]
C_a	Average of system capacity	0.6429
C_l	Lower limit of system capacity	0.5806
C_u	Upper limit of system capacity	0.7200
Θ^2	Variance of system capacity	9e-03
σ	Standard deviation of local-average SINR	1.5[dB]
L	Cell perimeter (microcellular)	200[m]
A	Cell area with hexagon shape	$2.5981e4[m^2]$
$E[v]$	Mean value of users velocity	1[m/s]
r_1	Bit rate for Class-1 service	12.2[kbps] (EFR voice)
$\gamma_{target-1}$	SINR target for Class-1 QoS	6 ± 1[dB] for BER $= 10$e-03
ρ_1	Voice-source activity factor	0.4
w_1	Mean of Class-1 connection effective load factor (CELF)	0.0050
$w_{l,1}$	Lower limit of Class-1 CELF	0.0040
$w_{u,1}$	Upper limit of Class-1 CELF	0.0063
σ_1^2	Variance of Class-1 CELF	2.69e-06
r_2	Bit rate for Class-2	64[kbps] (video)
$\gamma_{target-2}$	SINR target for Class-2 QoS	2.75 ± 0.75[dB], $BER = 10$e-05
ρ_2	Video-source activity factor	1
w_2	Mean of Class-2 CELF	0.0304
$w_{l,2}$	Lower limit of Class-2 CELF	0.0257
$w_{u,2}$	Upper limit of Class-2 CELF	0.0360
σ_2^2	Variance of Class-2 CELF	3e-05
r_3	Bit rate for Class-3	144[kbps] (multimedia)
$\gamma_{target-3}$	SINR target for Class-3 QoS	2 ± 0.5[dB], $BER = 10$e-05
ρ_3	Data-source activity factor	1
w_3	Mean of Class-3 CELF	0.0561
$w_{l,3}$	Lower limit of Class-3 CELF	0.0503
$w_{u,3}$	Upper limit of Class-3 CELF	0.0625
σ_3^2	Variance of Class-3 CELF	7e-05
$1/\mu_1$	Mean call-holding time	120[s]
T_p	TTI of packets	10[ms], 40[ms], 200[ms]
R_p	Bit–rate	32[kbps], 64[kbps], 384[kbps]
γ_p	SINR for the above bit rates	3[dB], 2[dB], 1[dB]

4:3:3, is split in half for each user class resulting in 2:2:1.5:1.5:1.5:1.5 MTIP of 6 traffic classes. The business class is served as long as enough resources are available. In a hard threshold-blocking scenario, the economy users are served only if less than 70% of effective resources are occupied. Parameters $a_{2k}(c)$ in equation (11.35) are used with l_{2k} equal to 0.5 for all k. In a soft-fractional blocking scenario, the economy class can share the resources equally with the business class if less than 65% of effective resources are occupied, that is, c is less than 0.47. Else if c is less than C_l, invoke equation (11.34) with $a_{21}(c) = 0.8, a_{22}(c) = a_{23}(c) = 0.6$. Else if c is less than C_u, $a_{21}(c) = 0.4, a_{22}(c) = a_{23}(c) = 0.3$. Otherwise, $a_{2k}(c) = 0$ for all k.

Figure 11.28 provides a detailed insight into the performance characteristics of the SCAC system in comparison with MdCAC for each and every service class using 7:2:1 MTIP. It shows that the SCAC system suffers from a slight degradation of communication quality or QoS loss, in turn it yields significant improvements in the handoff failure and new-call blocking probability, especially for higher resource-consuming classes. The traffic shaping gain of the SCAC is clearly illustrated. *Figure 11.29* provides an insight into SCAC systems with QoS differentiation in 4:3:3 MTIP. The business class experiences much better GoS than the economy class. The communication quality during the calls is also improved by QoS differentiation paradigms.

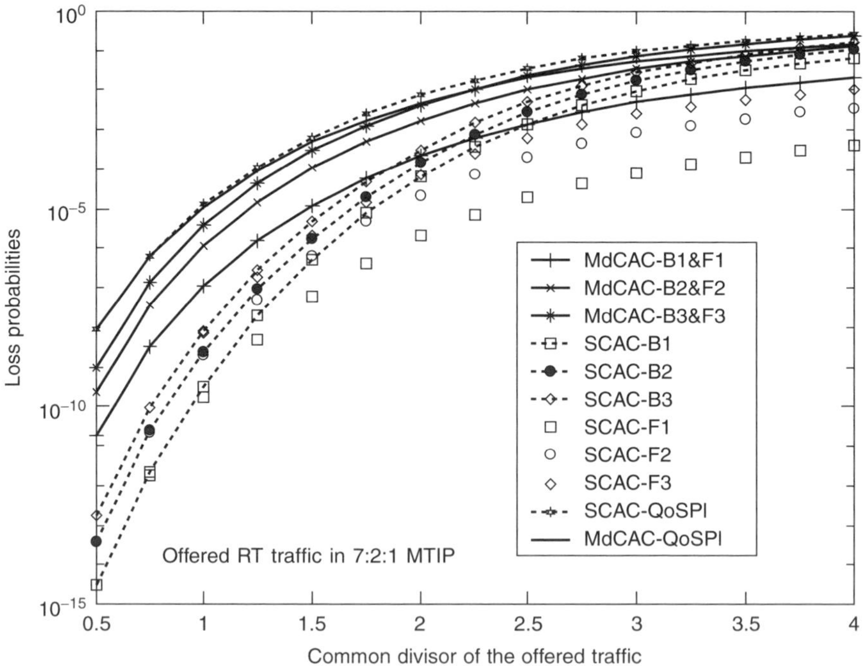

B_k – blocking probability of class-k calls

F_k – the hand of failure probability of class-k calls

Figure 11.28 SCAC versus MdCAC in 7:2:1 MTIP.

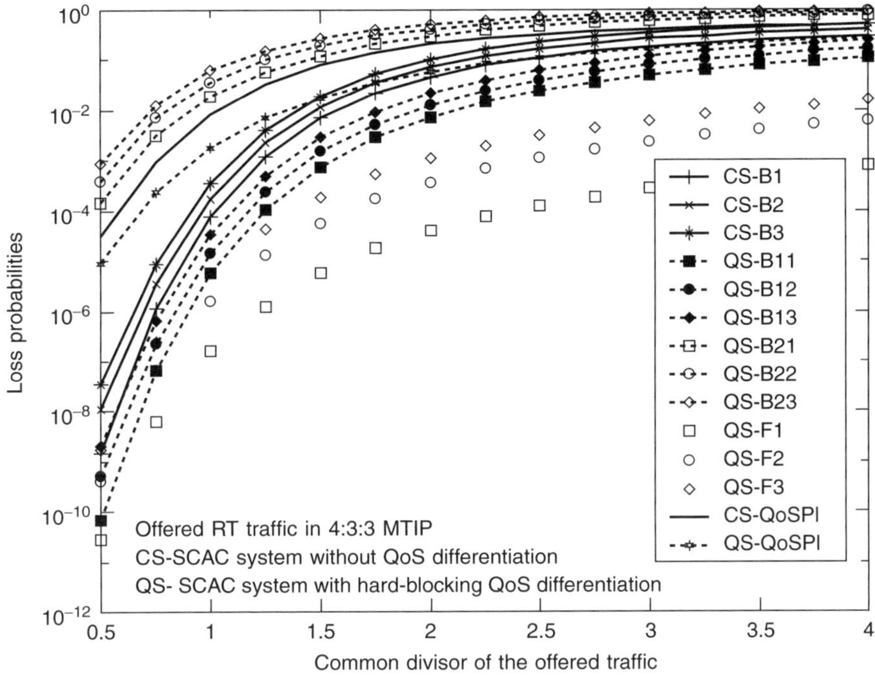

B_k – the new-call blocking probability of class - (j, k)

Figure 11.29 SCAC with QoS differentiation.

The MdCAC and SCAC policies are simple to implement without need for any special SW/HW. However, the modeling parameters are required *a priori* without which the CAC mechanisms cannot operate. Because of the diverse nature of different traffic sources and their often-complex statistics, some of the parameters might be hard to determine efficiently. The soft-decision solutions are believed to offer more flexibility in determining the modeling parameters, thus, good multiplexing gain and robustness can be achieved. On the other hand, implementations of the MsCAC policy require advanced HW and SW to ensure the reliability of measurements. For this reason, it is not a cost-effective solution. Moreover, estimation errors in some circumstances may cause significant degradations of the system performance (see discussion on sensitivity in Section 11.6). But the advantage of MsCAC is that it seems 'insensitive' to the traffic nature and the operation is robust. The network can learn and adapt to the statistics of traffic even when the burstiness of traffic is considered as out of control for the modeling-based systems. To gain the best from all design criteria, a hybrid soft-decision/measurement-based implementation is a reasonable choice. Parameters needed for soft-decision functions, that is, means and variances can rely on auto-regressive measurements. For such solutions, parameters and constraints can simply be thresholds of the UL interference level of cell- and connection-basis, an allowable outage probability, estimates of the current total received interference level with its mean and variance and so on. These are anyhow needed for TPC mechanisms.

11.7.6 Packet access control

Let R_p be the bit rate, L_p the packet length in bits, γ_p the target SINR to meet quality requirements, and T_p the transmit time interval (TTI) of a packet in UL, $T_p = L_p/R_p$. In scenarios without QoS differentiation, TTI is assumed to be a time-slot, $T_p \equiv T_s$.

Let $c(t)$ be the RT system load state at time t; $C(t)$ the system soft capacity at time t; and $z(t)$ the free resources left by the RT traffic in the system at time t. The quasi-stationary behavior of $c(t)$ or $z(t)$ over a sustained period of time is of interest. In the equilibrium condition, we have $z = (C_a - c)$. The process of z and c itself is not Markovian in general, but for larger C_a/w_k it behaves as an approximate Markovian process. For instance, to obtain the equivalent one-dimensional birth–death process for c in complete-sharing systems, we can use the Pascal approximation as in Reference [33]. First, we need to scale the system capacity and resource consumption of each RT service class into integers. To do so we assume $\min\{w_k, k \in \mathbf{K}\} \equiv w_1$ and define the scaled load vector, the scaled system state, and the scaled average system capacity as follows:

$$\mathbf{w}^* = \{w_k^*, k \in \mathbf{K}\} \equiv \{\lfloor w_k/w_1 \rfloor, k \in \mathbf{K}\}; c^* \equiv \lfloor c/w_1 \rfloor; \text{ and } C_a^* \equiv \lfloor C_a/w_1 \rfloor \text{ where } \lfloor x \rfloor$$

is the maximum integer not exceeding the argument. The set of equivalent system states is therefore

$$\Psi^* \equiv \{c^* : c^* = 0, 1, 2, \ldots, C_a^*\}$$

Let the normalized mean service time be $1/\mu \equiv 1$. The equivalent birth–death process at steady state c^* has a death rate of c^* and a birthrate of

$$\lambda(c^*) = v^2/\omega^2 + c^*(1 - v/\omega^2) \text{ with } v = \sum_{k \in \mathbf{K}} w_k^* \alpha_k \quad \text{and} \quad \omega^2 = \sum_{k \in \mathbf{K}} (w_k^*)^2 \alpha_k \quad (11.37)$$

where the normalized $\alpha_k = (\lambda_{l,k} + \lambda_{hl,k})$. The steady-state probability $s(c^*)$ for $c^* \in \Psi^*$ satisfies

$$c^* s(c^*) = \lambda(c^* - 1) \, s(c^* - 1), c^* \geq 1 \text{ and } \sum_{0}^{C_a^*} s(c^*) = 1 \quad (11.38)$$

The quasi-stationary probability of being in load state c^* over a period of time τ can be defined by

$$p(c^*, \tau) = \lim_{t \to \infty} \Pr\{c^*(t + \tau) = c^* | c^*(t) = c^*\} \quad (11.39)$$

This can be approximated by an exponential function [34] as follows:

$$p(c^*, \tau) = e^{-\tau[\lambda(c^*) + c^*]} \quad (11.40)$$

Equation (11.40) for $\tau = T_p$ provides the equilibrium probability that there are maximum $z^* = C_a^* - c^*$ units of resources available for packet transmissions. Let $S(c^*)$ be the state-dependent upper limit number of successful packet transmissions in a given time-slot, thus

$$S(c^*) = p(c^*, T_p) \frac{(C_a^* - c^*)w_1}{R_p \gamma_p / W} \quad (11.41)$$

Equation (11.41) can be used to study trade offs of the design parameters for optimal PAC schemes. For example, let us assume that there are N data users in the system; each generates data packets according to a Poisson process with rate Λ per time-slot, $\Lambda \leq 1$. Thus, there are $N_p = \lfloor N\Lambda \rfloor$ average number of active data sources per time-slot. If each of them attempts to transmit their packets at the beginning of a given time-slot with the probability of $\min(1, S(c^*)/N_p)$, the successful packet transmission probability, that is, Pr{having less than $S(c^*)$ transmissions actually initiated} according to the binomial distribution, is at least 1/2. This is very reasonable compared to ALOHA random access. The average upper limit of UL packet-data throughput denoted by S is given by

$$S = (1 - P_l) \sum_{c^* \in \Psi^*} s(c^*)S(c^*) \tag{11.42}$$

where P_l is the QoS loss probability defined in the previous section. By using Little's formula [35], the average lower limit of packet delay is given by S/N_p time slots.

Figure 11.30 presents the average upper-limit UL data throughput of packet transmissions; also the impacts of RT traffic, bit rates and TTI on the throughput. One can see that throughput characteristics are more sensitive to bit rate than to TTI. Such results can give valuable quantitative input data for the design of effective PAC schemes.

A number of related topics are discussed in References [42–66].

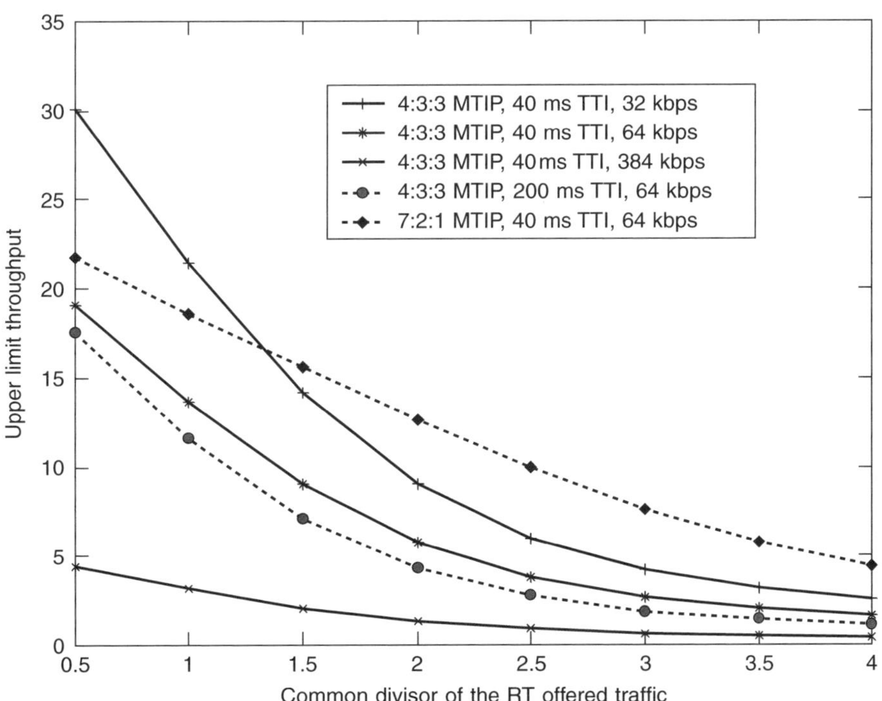

Figure 11.30 Throughput of NRT packet access.

REFERENCES

1. Ramakrishna, S. and Holtzman, J. (1998) A scheme for throughput maximization in dual-class CDMA systems. *IEEE J. Select. Areas Commun.*, **16**(6), 830–845.
2. LI, C. and Gitlin, R. D. *Multi-code CDMA wireless personal communications networks. IEEE ICC '95*, pp. 1060–1064.
3. Liu, T. K. and Silvester, J. (1998) Joint admission/congestion control for wireless CDMA systems supporting integrated services. *IEEE J. Select. Areas Commun.*, **16**(6), 845–858.
4. Tan, L. and Zhang, Q. T. (1996) A reservation random-access protocol for voice/data integrated spread-spectrum multiple-access systems. *IEEE J. Select. Areas Commun.*, **14**(9), 1717–1727.
5. Pichna, R. and Qiang, W. (1996) A medium-access control protocol for a cellular packet CDMA carrying multirate traffic. *IEEE J. Select. Areas Commun.*, **14**(9), 1728–1736.
6. Fook Loong, Lo., Tung Sang Ng. and Yuk, T. T. (1996) Performance analysis of a fully-connected, full-duplex CDMA ALOHA network with channel sensing and collision detection. *IEEE J. Select. Areas Commun.*, **14**(9), 1708–1716.
7. Liu, T.-K. (1996) Performance Modeling and Design Trade-offs in Wireless Networks with Heterogeneous Service Requirements. Ph.D. Dissertation, University of Southern California, CA.
8. Prasad, R. (1996) *CDMA for Wireless Personal Communications*. London: Artech House.
9. Gross, D. and Harris, C. M. (1985) *Fundamentals of Queuing Theory*. New York: John Wiley & Sons.
10. Sampath, A. and Holtzman, J. M. (1997) Access control of data in integrated voice/data CDMA systems: benefits and tradeoffs. *IEEE Select. Areas Commun.*, **15**(8), 1511–1526.
11. Sato, T., Okada, H., Yamazato, T., Katayama, M. and Ogawa, A. (1996) *Performance degradation due to access timing delay on CDMA unslotted ALOHA with channel load sensing. IEEE 5th International Conference on Universal Personal Communications*, Vol. 1, pp. 111–114.
12. Phan, V. V. and Glisic, S. (2001) Estimation of implementation losses in MAC protocols in wireless CDMA networks. *Int. J. Wireless Inform. Networks*, **8**(3), 115–132.
13. Morrow, R. K. and Lehnert, J. S. (1989) Bit-to-bit error dependence in slotted DS/SSMA packet system with random signature sequences. *IEEE Trans. Commun.*, **COM-37**(10), 1052–1061.
14. Holtzman, J. M. (1992) A simple, accurate method to calculate spread-spectrum multiple-access error probabilities. *IEEE Trans. Commun.*, **COM-40**(3), 461–464.
15. Perkins, C. (1998) Mobile networking through mobile IP. *IEEE Int. Comput.*, Tutorial. **2**, 58–69.
16. Macker, J. P., Park, V. D. and Corson, M. S. (2001) Mobile and wireless Internet services: putting the pieces together. *IEEE Commun. Mag.*, **39**(6), 148–155.
17. The Book of Visions 2000–Visions of Wireless World 2000. EC IST-WSI, 2000.
18. ITU-T Recommendation I.211. ISDN Service Capabilities, March 1993.
19. 3G TS 23.107. UMTS: QoS Concept and Architecture, Release 1999.
20. 3G TS 22.105. UMTS: Service Aspects, Services and Service Capabilities, Release 1999.
21. IETF RFC 1633. Integrated Services in the Internet Architecture: An Overview.
22. IETF RFC 2475. An Architecture for Differentiated Services.
23. Viterbi, A. J. (1995) *Principle of Spread Spectrum Communication*. Reading, MA: Addison-Wesley.
24. Holma, H. and Toskala, A. (2000) *WCDMA for UMTS Radio Access for Third Generation Mobile Communications*. New York: John Wiley & Sons.
25. Shin, S. M., Cho, C. H. and Sung, D. K. (1999) Interference-based channel assignment for DS-CDMA cellular systems. *IEEE Trans. Veh. Technol.*, **48**, 233–239.
26. Grossglauser, M. and Tse, D. (1999) A framework for robust measurement-based admission control. *IEEE ACM Trans. Networks*, **7**, 293–309.
27. Breslau, L., Jamin, S. and Shenker, S. (2000) *Comments on the performance of measurement-based admission control algorithms. IEEE InfoCom Conference Proceedings*, pp. 1233–1242.

28. Dimitriou, N., Tafazolli, R. and Sfikas, G. (2000) Quality of service for multimedia CDMA. *IEEE Commun. Mag.*, **38**(7), 88–94.
29. Dziong, Z., Ming, J. and Mermelstein, P. (1996) Adaptive traffic admission for integrated services in CDMA wireless-access networks. *IEEE J. Select. Areas Commun.*, **14**, 1737–1747.
30. Ishikawa, Y. and Umeda, N. (1997) Capacity design and performance of call admission control in cellular CDMA systems. *IEEE J. Select. Areas Commun.*, **15**, 1627–1635.
31. Liu, Z. and Zarki, M. E. (1994) SIR-based call admission control for DS-CDMA cellular systems. *IEEE J. Select. Areas Commun.*, **12**, 638–644.
32. Phan, V. and Glisic, S. (2001) *Radio resource management in CDMA cellular segments of multimedia wireless IP networks. International Symposium on Wireless Personal Multimedia Communications WPMC '01*, Conference Proceedings Invited Paper, September, Alborg, Denmark.
33. Glisic, S. (1996) *Spread Spectrum Systems for Wireless Personal Communications*. London: Artech House.
34. Roberts, J. W. (1981) A service system with heterogeneous user requirements, *Performance of Data Communications Systems and their Applications*, pp. 423–431.
35. Gross, D. and Harris, C. M. (1998) *Fundamentals of Queuing Theory*. New York: John Wiley & Sons.
36. Merakos, L. and Jangi, S. (1992) *Voice packet loses and data integration in reservation random access protocols for wireless access networks. GLOBECOM*, pp. 26–31.
37. Hung, K. W. and Yum, T. S. (1989) *The coded tone sense protocol of multihop spread-spectrum packet radio networks. GLOBECOM*, Dallas, TX, pp. 712–715.
38. Hung, K. W. and Yum, T. S. (1990) *An efficient code assignment algorithm for multihop spread spectrum packet radio networks. GLOBECOM*, pp. 271–274.
39. Sheikh, A., Yao, Y.-D. and Cheng, S. (1994) Throughput enhancement of direct-sequence spread spectrum packet radio networks by adaptive power control. *IEEE Trans. Commun.*, **42**, 884–890.
40. Wu, G., Mukumoto, K. and Fukuda, A. (1994) Performance evaluation of reserved idle signal multiple-access scheme for wireless communication networks. *IEEE Trans. Veh. Technol.*, **43**, 653–658.
41. Jiang, S. and Hsiao, M.-T. T. (1995) Performance evaluation of a receiver based handshake protocol for CDMA networks. *IEEE Trans. Commun.*, **43**, 2127–2138.
42. Goodman, D. J. *et al.* (1989) Packet reservation multiple access for local wireless communications. *IEEE Trans. Commun.*, **37**(8), 885–890.
43. Goodman, D. J. and Wei, S. X. (1989) *Factor affecting the bandwidth efficiency of packet reservation multiple access. 39th IEEE Vehicular Technology Conference*, San Francisco, CA, pp. 292–299.
44. Goodman, D. J. and Wei, S. X. (1991) Efficiency of packet reservation multiple access. *IEEE Trans. Veh. Technol.*, **40**(1), 170–176.
45. Wang, W.-C. (1993) Packet reservation multiple access in a metropolitan microcellular radio environment. *IEEE J. Select. Areas Commun.*, **11**(6), 918–925.
46. Nanda, S., Goodman, D. J. and Timor, U. (1991) Performance of PRMA: a packet voice protocol for cellular system. *IEEE Trans. Veh. Technol.*, **40**(3), 584–598.
47. Mitrou, N. M., Orinos, T. H. D. and Protonotarios, E. M. (1990) A reservation multiple access protocol for microcellular mobile-communication systems. *IEEE Trans. Veh. Technol.*, **39**(4), 340–351.
48. Wong, C. and Goodman, D. J. (1992) A packet reservation multiple access protocol for integrated speech and data transmission. *IEE Proc.*, **139**(6), 607–612.
49. Glisic, S. and Vikstedt, J. (1998) Effect of wireless link characteristics on packet-level QoS in CDMA/CSMA networks. *IEEE Select. Areas Commun.*, **16**(6), 875–889.
50. Glisic, S., Rao, R. and Milstein, L. B. (1990) *The effect of imperfect carrier sensing on nonpersistent carrier sense multiple access. IEEE SUPERCOMM/ICC '90*, Vol. **3**, pp. 1266–1269.
51. Glisic, S. G. (1991) 1-Persistent carrier sense multiple access in radio channels with imperfect carrier sensing. *IEEE Trans. Commun.*, **COM-39**(3), 458–464.

52. Abdelmonem, A. H. and Saadawi, T. N. (1989) Performance analysis of spread spectrum packet radio network with channel load sensing. *IEEE Select. Areas Commun.*, **7**(1), 161–166.
53. Toshimitsu, K., Yamazato, T., Katayama, M. and Ogawa, A. (1994) A novel spread slotted ALOHA system with channel load sensing protocol. *IEEE Select. Areas Commun.*, **12**(4), 665–672.
54. Sato, T., Okada, H., Yamazato, T., Katayama, M. and Ogawa, A. (1996) Throughput analysis of DS/SSMA unslotted ALOHA system with fixed packet length. *IEEE Select. Areas Commun.*, **14**(4), 750–756.
55. Jangi, S. and Merakos, L. (1991) *Performance analysis of reservation random access protocol for cellular packet communications*. GLOBECOM, pp. 895–900.
56. Mitra, D., Reiman, M. I. and Wang, J. (1998) Robust dynamic admission control for unified cell and call QoS in statistical multiplexer. *IEEE J. Select. Areas Commun.*, **16**, 692–707.
57. Evans, J. S. and Everitt, D. (1999) Effective bandwidth-based admission control for multiservice CDMA cellular networks. *IEEE Trans. Veh. Technol.*, **48**, 36–46.
58. Tong, H. and Brown, T. X. (2000) Adaptive call admission control under quality of service constraints: a reinforcement learning solution. *IEEE J. Select. Areas Commun.*, **18**, 209–221.
59. Zander, J. and Kim, S. L. (2001) *Radio Resource Management for Wireless Networks*. London: Artech House.
60. Viterbi, A. J. (1994) Capacity of a simple and stable protocol for short message service over CDMA network, in Blahut, R. (ed.) *Communication and Cryptography*, Kluwer AP, Boston, pp. 423–429.
61. Sampath, A. and Holtzman, J. M. (1997) Access control of data in integrated voice/data CDMA systems: benefits and tradeoffs. *IEEE J. Select. Areas Commun.*, **15**, 1511–1526.
62. Glisic, S. and Phan, V. V. (2000) *Sensitivity function of soft decision carrier sense MAC protocols for wireless CDMA networks with specified QoS*. Invited Paper, *IEEE 11th PIMRC Conference Proceedings*, September.
63. Lin, Y. B., Mohan, S. and Noerpel, A. (1994) Queueing priority channel assignment strategies for PCS handoff and initial access. *IEEE Trans. Veh. Technol.*, **43**(3), 704–712.
64. Geraniotis, E. and Wu, J. (1994) The probability of multiple correct packet receptions in direct-sequence spread-spectrum networks. *IEEE J. Select. Areas Commun.*, **12**, 871–884.
65. Soroushnejad, M. and Geraniotis, E. (1995) Multi-access strategies for an integrated voice/data CDMA packet radio network. *IEEE Trans. Commun.*, **43**, 934–945.
66. Capone, J. M. and Merakos, L. F. (1995) Integrating data traffic into a CDMA cellular voice system. *Wireless Networks*, **1**, 389–401.

12

Adaptive CDMA networks

12.1 BIT RATE/SPACE ADAPTIVE CDMA NETWORK

This section presents a throughput delay performance of a centralized unslotted Direct Sequence/Code Division Multiple Access (DS/CDMA) packet radio network (PRN) using bit rate adaptive location aware channel load sensing protocol (CLSP).

The system model is based on the following assumptions. Let us consider the reverse link of a single-cell unslotted DS/CDMA PRN with infinite population and circle cell coverage centered to a hub station. Users communicate via the hub using different codes for packet transmissions with the same quality of service (QoS) requirements [e.g. the target bit error rate (BER) is 10^{-6}]. The radio packets considered herein are of medium access control (MAC) layer (i.e. MAC frames formed after data segmentations and coding). Packets have the same length of L (bits). The scheduling of packet transmissions, including the retransmissions of unsuccessful packets at mobile terminals, is randomized sufficiently enough so that it is possible to approximate the offered traffic of each user to be the same, and the overall number of packets is generated according to the Poisson process with rate λ. In the sequel, we will use the following notation:

ζ – the path-loss exponent of the radio propagation attenuation in the range of [2, 5]

r – the distance of a mobile terminal from the central hub that is normalized to the cell radius, thus in the range of [0, 1]

R_0 – the primary data rate for given system coverage and efficiency of mobile power consumption;

T_0 – the packet duration (i.e. the time duration needed for transmitting a packet completely) of the primary rate $T_0 = L/R_0$.

The cell area is divided into $M + 1$ rings (M is a natural number representing the spatial resolution) centered to the hub. Let $\mathbf{M} = \{0, 1, \ldots, M\}$; and for all $m \in \mathbf{M}$,

r_m – the normalized radius of the boundary-circle of ring $(m + 1)$ given by $r_m = 2^{-m/\zeta}$, $r_0 = 1$ for the cell-bounding circle and $r_{M+1} = 0$ for the most inner ring;

R_m – the rate of packet transmissions from users in ring $(m + 1)$ given by $R_m = 2^m R_0$, that is, packets from the more inner ring will be transmitted with the higher bit rate;

T_m – the corresponding packet duration $T_m = 2^{-m} T_0$ and also the mean service time of a packet transmission using rate R_m.

For a fixed packet length L, the closer the mobile terminal to the hub, the higher is the bit rate and the shorter is the packet transmission time. In order to ensure the optimal operation of transceivers, the packet duration should be kept not too short, for example, minimum of around 10 ms as the radio frame duration of the 3GPP standards for WCDMA cellular systems. Therefore, a proper trade-off between L, R_0 and M is needed. For example, with $L = 2560$ bits, $R_0 = 32$ kbps and $M = 3$, there are four possible rates for packet transmissions: 32, 64, 128 and 256 kbps with 80, 40, 20 and 10 ms packet duration, respectively. In the absence of shadowing, for the same mobile transmitter power denoted by P from any location in the network, approximated with the spatial resolution described above, the received energy per frame denoted by E is the same:

$$E = \frac{P T_m}{r_m^\zeta} = \frac{P 2^{-m} T_0}{(2^{-m/\zeta})^\zeta} = P T_0 \tag{12.1}$$

This significantly reduces the maximum radiated power into the user direction, reducing the health risk and the interference level produced in the adjacent cell and therefore increasing capacity in the cell. The bit-energy $E_b = E/L$ is also constant. Let

W – the CDMA chip rate, for example, 3.84 Mcps;
g_m – the processing gain of a transmission using rate R_m that is given by $g_m = W/R_m$;
η – the ratio of the thermal noise density and maximum tolerable interference (N_0/I_0);
γ_m – the local average signal to interference plus noise ratio (SINR), also denotes the target SINR for meeting the QoS requirements of transmissions with rate R_m.

The transmitter power control (TPC) is assumed sufficient enough to ensure that the local average SINR can be considered as a lognormal random variable having standard deviation σ in the range of 2 dB. Because the transmitter power of mobile terminals in the rate adaptive system is kept at the norm level (denoted by P above), the dynamic range of TPC can be significantly reduced compared to the fixed rate counterpart for the same coverage resulting in less sensitive operation. Thus, σ of the adaptive system can be expected to be smaller than that in the fixed rate system. Once again, if there were only near–far effects in the radio propagation, due to rate adaptation and perfect TPC, SINR of all transmissions would be the same at the hub. However, the required SINR target of higher bit rate transmissions in DS/CDMA systems tends to be lower for the same BER performance due to less multiple access interference (MAI). For example, the simulation results of Reference [1] show that in the same circumstances the required SINR target for 16-kbps transmissions is almost double that of the 256 kbps transmissions. Thus, less transmitter power is needed for close-in users using higher data rate. The rate/space adaptive transmissions increase the energy efficiency for mobile terminals. It will be shown later that even when the same target SINR was required regardless of the bit rates, the adaptive system still outperforms the fixed counterpart. In the sequel, we will use the following notation:

$\mathbf{n} = \{n_m, m \in \mathbf{M}\}$ is the system state or occupancy vector, where n_m is the number of packet transmissions in progress using rate R_m;

$\mathbf{w} = \{w_m, m \in \mathbf{M}\}$ is the transmission load vector, where $w_m = g_m^{-1}\gamma_m$ represents the average load factor produced by a packet transmission with rate R_m and target SINR γ_m. The higher the bit rate, the more the network resources that will be occupied by the transmission.

$c = \mathbf{nw}$ is the system load state representing MAI in the steady state condition.

It has been shown in Chapter 11 that simultaneous transmissions are considered adequate, that is, meeting the QoS requirements, if MAI satisfies the following condition:

$$\text{MAI} \equiv \sum_{m \in \mathbf{M}} n_m w_m \leq (1 - \eta) \tag{12.2}$$

The task of CLSP is to ensure that the condition (12.2) is always satisfied. Define

$\mathbf{\Omega} = \{\mathbf{n}, \text{condition (12.2) is true}\}$ the set of all possible system states;

$\mathbf{\Psi} = \{c, c = \mathbf{nw} \text{ and } \mathbf{n} \in \mathbf{\Omega}\}$ the set of all possible system load states.

Because of the TPC inaccuracy, the probability that the condition (12.2) is satisfied and the SINR of each packet transmission is kept at the target level, conditioned on the steady system load state c and lognormal SINR can be determined as in Chapter 11, equation (11.30)

$$P_{\text{ok}}(c) = 1 - Q\left(\frac{1 - \eta - E[\text{MAI}|c]}{\sqrt{\text{Var}[\text{MAI}|c]}}\right) \tag{12.3}$$

with

$$E[\text{MAI}|c] = c \exp[(\ln 10/10\sigma)^2/2]$$

$$\text{Var}[\text{MAI}|c] = c \exp[2(\ln 10/10\sigma)^2]$$

where $Q(x)$ is the standard Gaussian integral function, and σ is the standard deviation of lognormal SINR in dB. This is because the system load state c defined above uses the mean (target) values of lognormal SINR for calculating the average load factor of each transmission. The Gaussian integral term $Q(x)$ in equation (12.3) represents the total error probability caused by a sum of lognormal random variable composing the load state.

The above analysis implies that in the equilibrium condition, for a given system load state c, the system will meet its QoS target (e.g. actual bit error probability is less than the target BER of 1e-5) with a probability of $P_{\text{ok}}(c)$. In other words, it will lose its QoS target (actual bit error probability is larger than the target BER of 1e-5) with a probability $1 - P_{\text{ok}}(c)$. As a consequence, each equilibrium system load state c can be modeled with a hidden Markov model (HMM) having two states, namely 'good' and 'bad', which is illustrated in Figure 12.1.

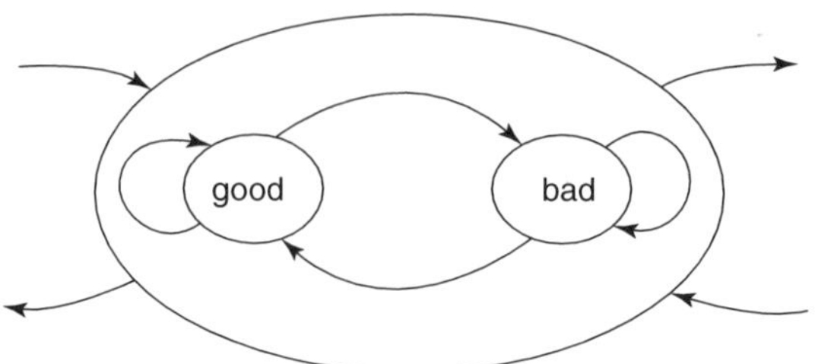

Figure 12.1 Two-state HMM of the system load state.

The stationary probability of HMM state ('good' or 'bad') conditioned on the system load state c is given by

$$\Pr\{\text{'good'}|c\} = P_{\text{ok}}(c) \tag{12.4}$$

$$\Pr\{\text{'bad'}|c\} = 1 - P_{\text{ok}}(c) \tag{12.5}$$

Let us introduce two other parameters for analytical evaluation purposes:

P_{eg} – the equilibrium bit error probability over all 'good' states of the channel, in which the QoS requirements are met. The target BER is supposed to be the worst case of P_{eg}, for example, 1e-5.

P_{eb} – the equilibrium bit error probability over all 'bad' states of the channel, in which the QoS requirements are missed to some extent, for example, $P_{\text{eb}} = 1e - 4$ when the target BER is $1e - 5$. The target BER is therefore the upper bound of P_{eb}.

In the perfect-controlled system, $P_{\text{eg}} = P_{\text{eb}}$ and equal to the target BER. This assumption is widely used in the related publications investigating the system performance on the radio packet level. In this section, the impacts of channel imperfection are evaluated in the context of SINR errors with total standard deviation σ and P_{eb} as a variable parameter representing effects of 'bad' channel condition. Let

$p(c)$ – the steady state probability of being in the system load state $c \in \Psi$;

P_{e} – the equilibrium bit error probability of the system for the actual QoS of packet transmissions. From the above results, we have

$$P_{\text{e}} = \left(\sum_{c \in \Psi} cp(c)\right)^{-1} \left[P_{\text{eg}} \sum_{c \in \Psi} cp(c) \Pr\{\text{'good'}|c\} + P_{\text{eb}} \sum_{c \in \Psi} cp(c) \Pr\{\text{'bad'}|c\} \right] \tag{12.6}$$

It is obvious that in the perfect-controlled system as mentioned above, P_{e} is also equal to the target BER. Let

P_{c} – the equilibrium probability of a correct packet transmission. With employment of forward error correction (FEC) mechanism having the maximum number of correctable bits N_e ($N_e < L$ and dependent on the coding method; $N_e = 0$ when FEC is not used), P_c is generally given by

$$P_c = \sum_{i=0}^{N_e} \binom{L}{i} P_e^i (1 - P_e)^{L-i} \tag{12.7}$$

In the fixed rate system with the same coverage, the primary rate R_0 is used for all packet transmissions. From equation (12.2) under perfect TPC assumption, the channel threshold or system capacity defined as the maximum number of simultaneous packet transmissions can be determined by

$$C_0 = \lfloor (1 - \eta)/w_0 \rfloor \tag{12.8}$$

where $\lfloor x \rfloor$ is the maximum integer number not exceeding the argument.

Thus, with respect to CLSP, the hub senses the channel load (i.e. *MAI*, in general, or the number of ongoing transmissions for the fixed rate system) and broadcasts the control information periodically in a forward control channel. Users having packets to send should listen to the control channel and decide to transmit or refrain from the transmission in a nonpersistent way. The feedback control is assumed to be perfect, that is, zero propagation delay and perfect transceivers in the forward direction. The impacts of system imperfection, such as access delay, feedback delay and imperfect sensing have been investigated in Chapter 11 for the fixed rate systems with dynamic persistent control. Let G – the system offered traffic $G = \lambda T_0$ (the average number of packets per normalized $T_0 \equiv 1$) is kept the same for both adaptive and fixed rate systems for fair comparison purposes. In the adaptive system, $G \equiv \lambda$ is distributed spatially among users that are in different rings.

For $m \in \mathbf{M}$, let

λ_m – be the packet arrival rate from ring $(m + 1)$, which is dependent on λ and the spatial user distribution (SUD) having the probability density function (PDF) $f(r, \theta)$. In general, λ_m is given by

$$\lambda_m = \lambda \int_{r_{m+1}}^{r_m} \int_0^{2\pi} f(r, \theta)\, dr\, d\theta \tag{12.9}$$

For instance, let us assume that the SUD is uniform per unit area in the mobility equilibrium condition. Thus, λ_m can be determined by

$$\lambda_m = \lambda (r_m^2 - r_{m+1}^2) \tag{12.10}$$

For the derivation of the performance characteristics of the rate adaptive CLSP unslotted CDMA PRN, a multirate loss system model of the stochastic knapsack-packing problem [2] can be used. The analysis presented in this section can therefore be used for investigating PRNs supporting multimedia applications and QoS differentiation, where users transmit with different rates depending on the system load state, their potential subscriber class and the required services.

12.1.1 Performance evaluation

Fixed-rate CLSP

The performance characteristics of the unslotted CDMA PRN using fixed-rate CLSP under perfect TPC is given in Chapter 11. Herein, we consider the system with imperfect TPC. Define

n – the number of ongoing packet transmissions in the system or the system state;
p_n – the steady state probability of the system state n;
P_{succ} – the equilibrium probability of successful packet transmissions;
S – the system throughput as the average number of successful packet transmissions per T_0;
D – the average packet delay normalized by T_0;

Using the standard results of the queuing theory for Erlang loss formula [3] with the number of servers set to the channel threshold C_0, the arrival rate of λ and the normalized service rate of $1/T_0 \equiv 1$, we have for the steady state solutions:

$$p_n = \frac{G^n/n!}{\sum\limits_{i=0}^{C_0} G^i/i!} \quad \text{for } 0 \le n \le C_0 \tag{12.11}$$

The equilibrium probability of a successful packet transmission consists of two factors. The first factor is the probability that the given packet is not blocked by the CLSP given by $(1 - B)$, where B is the packet blocking probability:

$$B = p_{C_0} \tag{12.12}$$

The second factor is the equilibrium probability of correct packet transmissions P_{c} given by equation (12.7) with a modification of equation (12.6) as given below:

$$P_{\text{e}} = \left(\sum_{n=0}^{C_0} np_n\right)^{-1}\left[P_{\text{eg}}\sum_{n=0}^{C_0} np_n \Pr\{\text{`good'}|n\} + P_{\text{eb}}\sum_{n=0}^{C_0} np_n \Pr\{\text{`bad'}|n\}\right] \tag{12.13}$$

where similarly to equations (12.4) and (12.5) we have

$$\Pr\{\text{`good'}|n\} = 1 - Q\left(\frac{C_0 - ne^{(\ln 10/10\sigma)^2/2}}{\sqrt{ne^{2(\ln 10/10\sigma)^2}}}\right) \tag{12.14}$$

$$\Pr\{\text{`bad'}|n\} = 1 - \Pr\{\text{`good'}|n\} \tag{12.15}$$

The equilibrium probability of a successful packet transmission, P_{succ}, is now given by

$$P_{\text{succ}} = (1 - B)P_{\text{c}} \tag{12.16}$$

The system throughput is given by

$$S = GP_{\text{succ}} \tag{12.17}$$

The average packet delay is decomposed into two parts: D_b the average waiting time of a packet for accessing the channel including back-off delays and D_r the average resident time of the given packet from the instant of entering to the instant of leaving the system successfully. Formally, the average packet delay (normalized to T_0) is given by

$$D = D_b + D_r \tag{12.18}$$

with

$$D_b = \sum_{i=0}^{\infty} B^i = \frac{B}{1 - B} \tag{12.19}$$

and according to Little's formula [3]

$$D_r = S^{-1} \sum_{n=0}^{C_0} n p_n \tag{12.20}$$

Thus, the performance characteristics can be optimized subject to trade-off of the packet length and the transmission rate. This can be achieved by using adaptive radio techniques for link adaptation.

Rate adaptive CLSP

This system, as mentioned above, can be modeled with a multirate loss network model. It is well known that the steady state solutions of such a system have a product form [2] given by

$$p(\mathbf{n}) = \frac{1}{G_0} \prod_{m \in \mathbf{M}} \frac{\alpha_m^{n_m}}{n_m!} \quad \mathbf{n} \in \Omega \tag{12.21}$$

with

$$G_0 = \sum_{\mathbf{n} \in \Omega} \prod_{m \in \mathbf{M}} \frac{\alpha_m^{n_m}}{n_m!}$$

where $p(\mathbf{n})$ is the steady state probability of having $\mathbf{n}$ transmission combination in the system, $\mathbf{n} \in \Omega$; α_m is the offered traffic intensity from ring $(m + 1)$ using rate R_m. Thus, $\alpha_m = \lambda_m T_m$, where λ_m and T_m are defined above.

For large state sets, that is, large $\mathbf{M}$ and C_0, the cost of computation with the above formulas is prohibitively high. This problem has been considered by many authors, resulting in elegant and efficient recursion techniques for the calculation of the steady system load state and blocking probabilities. The steady state probability $p(c)$ of system load state

$c \in \Psi$ defined above can be obtained by using the stochastic knapsack approximation described in Reference [2]

$$p(c) = \frac{q(c)}{\sum_{c \in \Psi} q(c)} \qquad (12.22)$$

with $q(c)$ given in recursive form as

$$q(c) = \frac{1}{c} \sum_{m \in \mathbf{M}} w_m \alpha_m q(c - w_m) \text{ for } c \in \Psi^+, q(0) = 1 \text{ and } q(-) = 0$$

The equilibrium probability of successful transmissions using rate R_m can be determined similarly to equation (12.16) as

$$P_{\text{succ\_}m} = (1 - B_m)P_c \qquad (12.23)$$

where P_c is given by equation (12.7) with P_e given by equation (12.6) and B_m is the packet blocking probability of transmissions using rate R_m from ring $(m + 1)$

$$B_m = \sum_{c \in \Psi : c > C_0 \, w_0 - w_m} p(c) \qquad (12.24)$$

The system throughput can be given by

$$S = \sum_{m \in \mathbf{M}} \lambda_m P_{\text{succ\_}m} \qquad (12.25)$$

Note that $G = \lambda T_0 \equiv \sum_{m \in \mathbf{M}} \lambda_m$ because of normalized $T_0 \equiv 1$. The average packet delay of this system, similar to equation (12.18), can be obtained by

$$D = \sum_{m \in \mathbf{M}} D_{\text{b\_}m} T_m + D_r \qquad (12.26)$$

with the components

$$D_{\text{b\_}m} = \frac{B_m}{1 - B_m} \qquad (12.27)$$

$$D_r = \left(\sum_{m \in \mathbf{M}} \lambda_m P_{\text{succ\_}m} w_m \right)^{-1} \sum_{c \in \Psi} cp(c) \qquad (12.28)$$

For illustration purposes, the system parameters summarized in Table 12.1 are used [4]. Two simple SUDs are considered: the two-dimensional uniform (per unit area) and the one-dimensional uniform (per unit length) distributions. For the first scenario, the packet arrival rate from ring $(m + 1)$ is given in equation (12.6). For the second scenario, the

Table 12.1 System parameter summary [4]. Reproduced from Phan, V. and Glisic, S. (2002) *Unslotted DS/CDMA Packet Radio Network Using Rate/Space Adaptive CLSP-ICC'02*, New York, May 2002, by permission of IEEE

Name	Definition	Values
W	CDMA chip rate	3.84 Mcps
η	Coefficient of the thermal noise density and max. tolerable interference	-10 dB
ζ	Path-loss exponent	2, 3, 4
L	Packet length	2560, 5120 bits
R_0	Primary rate	32 kbps
γ_0	SINR target of primary rate transmission	3 dB
C_0	Fixed primary rate system capacity	56
$M + 1$	Number of possible rates	4
R_m	Rate of ring 2, 3, 4 for $m = 1, 2, 3$	64, 128, 256 kbps
γ_m	SINR target of R_m rate transmission for $1e - 5$ target BER	3 dB or γ_0 for all rates
σ	Standard deviation of lognormal SINR	1, 2, 3 dB
P_{eg}	Equilibrium bit error probability over 'good' condition	$1e - 5$
P_{eb}	Equilibrium bit error probability over 'bad' condition	$1e - 5, 5e - 4, 1e - 3$

packet arrival rate from ring $(m + 1)$ is given by $\lambda_m = \lambda(r_m - r_{m+1})$ with $r_m = 2^{-m/\zeta}$ and $r_{M+1} = 0$ as defined above. This one-dimensional uniform SUD is often used for modeling the indoor office environment in which users are located along the corridor or the highway. The target SINR is set to 3 dB for all transmissions regardless of the bit rates. This is not taking into account the fact that higher bit rate transmissions need smaller target SINR for the same QoS than the lower bit rate transmissions. The load factor introduced by the transmission is therefore linearly increasing with the bit rate that is compensated by shortening the transmission period with the same factor. Because of this, under perfect-controlled assumption ($P_{eb} = P_{eg}$ set to target BER as explained above), the fixed rate CLSP system could have slightly better multiplexing gain than the adaptive counterpart for the same offered traffic resulting in slightly better throughput as shown in Figure 12.2. In reality, the BER is changing because of the random noise and interference corrupting the packet transmissions. The throughput characteristic of the fixed system worsens much faster than that of the adaptive system because it suffers from higher MAI owing to larger number of simultaneous transmissions and longer transmission period. Further, when the transmission is corrupted, longer transmission period or packet length could cause a drop of the throughput performance and wasting battery energy (Figures 12.2 and 12.6). In any case, the adaptive system has much better packet delay characteristics than the fixed counterpart (Figures 12.2–12.7). The same can be expected for the throughput performance in real channel condition or

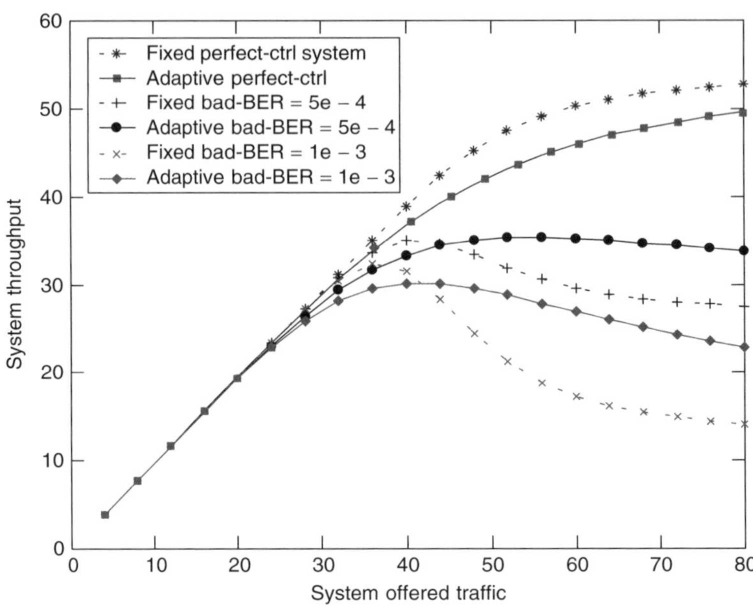

Figure 12.2 Effects of channel imperfection on the throughput performance (two-dimensional uniform SUD, $\zeta = 3$, $\sigma = 2\,\text{dB}$, $L = 2560\,\text{bits}$, $P_{\text{eg}} = 1\text{e} - 5$).

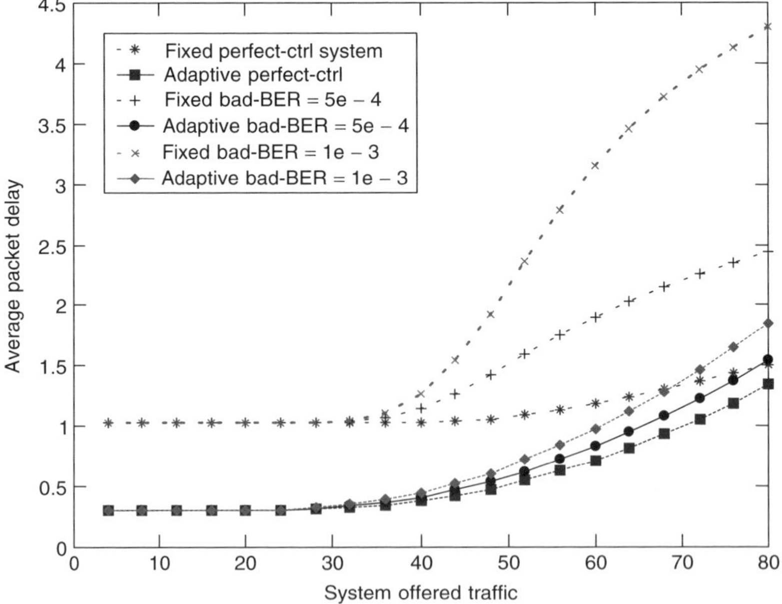

Figure 12.3 Effect of channel imperfection on the packet delay performance (two-dimensional uniform SUD, $\zeta = 3$, $\sigma = 2\,\text{dB}$, $L = 2560\,\text{bits}$, $P_{\text{eg}} = 1\text{e} - 5$).

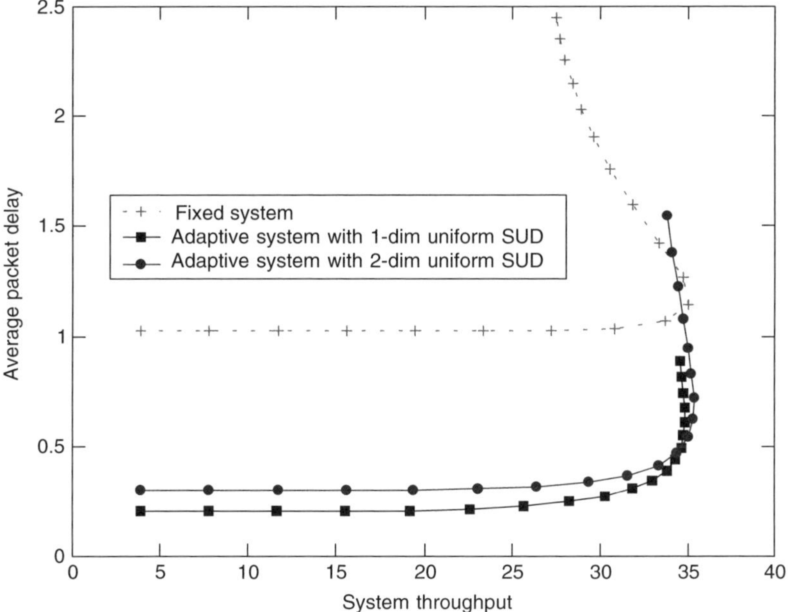

Figure 12.4 Effects of SUD on the performance trade-off ($\zeta = 3$, $\sigma = 2\,\mathrm{dB}$, $L = 2560\,\mathrm{bits}$, $P_{eg} = 1e - 5$, $P_{eb} = 5e - 4$).

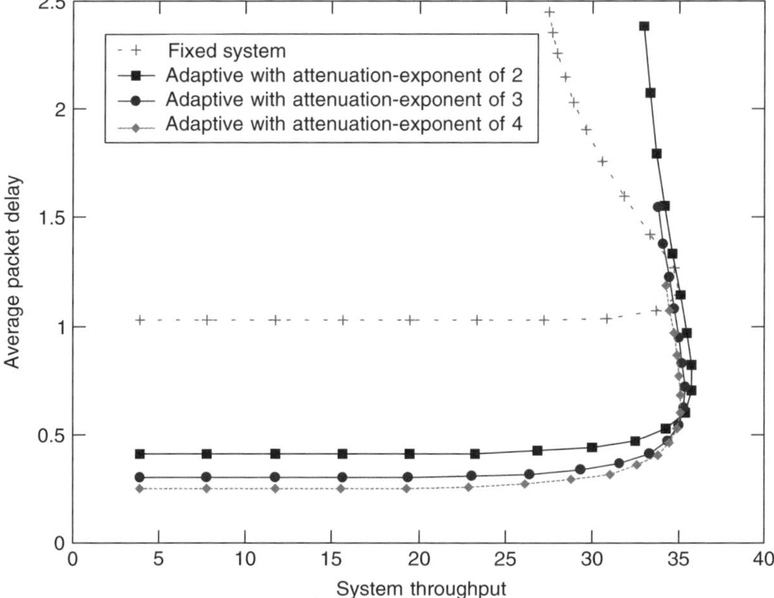

Figure 12.5 Effects of propagation model on the performance trade-off (two-dimensional uniform SUD, $\sigma = 2\,\mathrm{dB}$, $L = 2560\,\mathrm{bits}$, $P_{eg} = 1e - 5$, $P_{eb} = e - 4$).

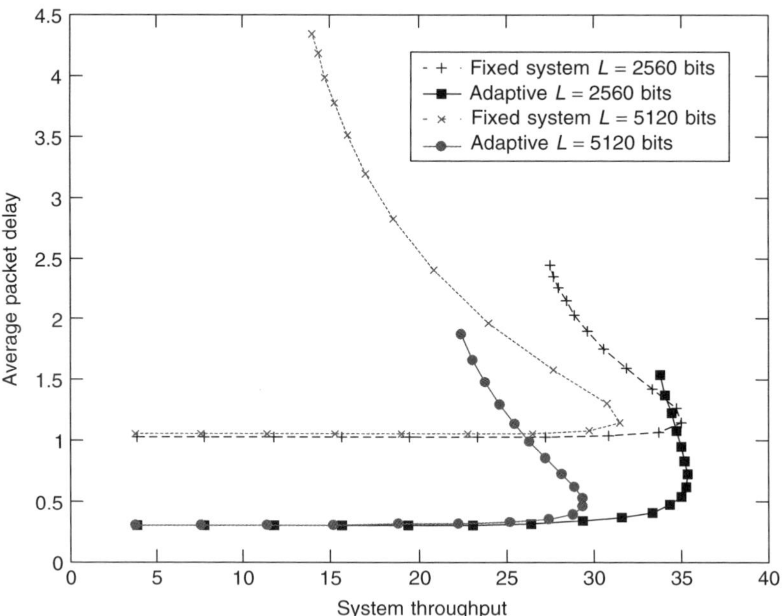

Figure 12.6 Effects of packet length on the performance trade-off (two-dimensional uniform SUD, $\zeta = 3$, $\sigma = 2\,\text{dB}$, $P_{eg} = 1e - 5$, $P_{eb} = 5e - 4$).

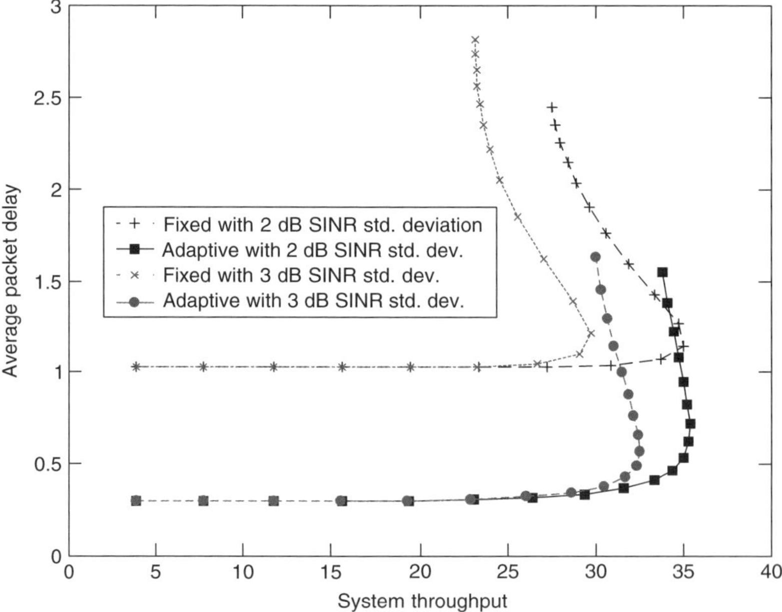

Figure 12.7 Effects of TPC inaccuracy on the performance trade-off (two-dimensional uniform SUD, $\zeta = 3$, $L = 2560\,\text{bits}$, $P_{eg} = 1e - 5$, $P_{eb} = 5e - 4$).

even in ideal channel condition if the advantage of less SINR for higher rate is taken into account. For example, according to Reference [1], target SINR is 1.5 dB for 256 kbps, 2 dB for 128 kbps, 2.5 dB for 64 kbps and 3 dB for 32 kbps. The numerical results for such advantages are not presented in this chapter because of limited space. Overall, the adaptive system outperforms the fixed counterpart. Figures 12.4 to 12.7 show the effects of design and modeling parameters on the performance characteristics. The adaptive system is sensitive to the SUDs (Figure 12.4) and path-loss exponent ζ (Figure 12.5). In the rate/space adaptive systems, spatial positions of the clusters formed by mobile users affect the overall throughput-delay improvement, whereas performance of the fixed rate counterpart is less sensitive to user population profile or does not depend on it given that TPC is perfect. These effects could be desirable if the advantage of less SINR for higher rate was taken into account. That more users are put to more inner rings depending on ζ and SUD boosts up the rate and reduces the time of communications, resulting in better throughput-delay performance. One should keep in mind that larger ζ also causes much larger dynamic range of transmitter power, especially for the fixed system that degrades the TPC performance, significantly resulting in more erroneous packet transmissions thus worse system performance. Figure 12.7 shows the effects of the standard deviation of lognormal SINR (σ) that represents the TPC errors. Although the adaptive system can be expected to have better TPC performance and thus smaller σ, the same value of σ is used for both systems in the numerical examples.

The throughput-delay performance of unslotted DS/CDMA PRNs using rate/space-adaptive CLSP is evaluated against the fixed rate counterpart. The combination of CLSP and adaptive multirate transmissions not only provides a significant performance improvement but also increases the flexibility of access control and reduces the uncertainty of the unslotted DS/CDMA radio channel. Once again, one should be aware that because of equation (12.1) the system is more environment friendly and reduces the level of interference in the surrounding cells too.

12.2 MAC LAYER PACKET LENGTH ADAPTIVE CDMA RADIO NETWORKS

Impacts of packet length on throughput-delay performance of wired/wireless networks have been extensively investigated in the open literature. The packet length optimization problem based on numerous factors is also well elaborated. Let us revisit a standard formula (12.7) for the probability of correct packet transmission determining the throughput characteristic.

It is easy to see from the equation that the smaller L makes the better P_c. In order to have P_c as close to 1 as possible for optimum throughput-delay performance, $P_e L$ needs to be very small compared to max (N_e, 1). In radio transmissions, SINR that is dependent on transmitter power, path loss and MAI, dominates P_e. In the bad channel conditions, P_e can be relatively large and may require impracticably small L in order to meet the performance requirements; otherwise throughput can drop to zero because all packets get corrupted. Meanwhile, mobile terminals are wasting battery energy for having to transmit erroneous packets.

On the other hand, in order to reduce the overhead and improve the goodput, L should be large. This can be seen from the formula for normalized goodput [5],

$$\frac{G}{R_0} = \frac{P_c(L - H)}{L} \tag{12.29}$$

where

G – the goodput, that is, the effective average data rate successfully transmitted excluding protocol overhead;
R_0 – the bit rate for packet transmission;
H – the length of protocol overhead, that is, the total length of packet header and packet tail in (bit).

Equations (12.7) and (12.29) are the basis for the derivation of optimum packet length and packet length adaptation. However, in order to obtain comprehensive and applicable results, further research efforts are required. For the last decade, there have not been many papers actually elaborating the packet length adaptation problem for PRNs. Reference [6] presents a simulation-based study of throughput improvement for a stop-and-wait automatic repeat request (ARQ) protocol using packet length adaptation in mobile packet data transmission. The channel estimation for the adaptation mechanism is based on the number of positive/negative acknowledgements (ACK/NACK). This is a learning-based adaptive process on the data link layer; thus the adaptation can happen even when the radio channel is in good conditions or *vice versa* owing to the bias of the learning toward actual conditions of unreliable radio channel. Reference [5] exploits equations (12.7) and (12.29) as such with no FEC capability to adopt adaptation mechanisms based on estimations of BER or frame error rate (FER). Results are supported by physical measurements with Lucent's WaveLAN radio. No comprehensive channel modeling, derivations and adaptation mechanisms are given despite the fact that equation (12.7) may not be accurate to apply for different fading environments and long-packet applications as targeted with maximum-transmission-unit TCP/IP link in Reference [5]. Technical reasons behind the applicability of equation (12.7) are elaborated in, for example, References [7–11]. The bottom line is that, for robust adaptation, instead of using the uncorrelated formula (12.7), the correlation between channel conditions and packet length in time domain needs to be considered. Moreover, in less correlated fading environments, using suitable FEC channel coding can be a more effective solution. Reference [12] presents a broad adaptive radio framework for energy efficiency of the battery in mobile terminals including packet length adaptation. Similar to Reference [5], Lucent's WaveLAN radio is used to provide results. Although Reference [12] provides valuable insights into adaptive radio problems, no comprehensive mechanisms are given that affect the accuracy and the practicality of the analysis. We should also add here that References [5,6,12] consider the case of noncontention packet access, that is, a single connection-oriented radio link. The packet delay characteristic and the throughput delay trade-off are ignored in References [5,6]. In addition to providing an overview of the existing work in this section, we consider the heavily correlated flat fading, where the error-correcting coding has not yet been effective. Packet length adaptation is used for a multiple access unslotted CLSP/DS-CDMA channel in order to improve the system throughput delay performance and the energy efficiency of mobile terminals.

The adaptation criteria are to eliminate the impacts of fading for an optimal trade-off between throughput, average packet delay and goodput. Two alternative strategies are presented: (A1) keeping the packet length as large as possible to avoid degradation of the goodput while fulfilling the specified QoS requirement, for example, Packet Error Rate (PER); (A2) maximizing the goodput. The correlation between fade duration statistics and packet duration in time domain over a flat Rayleigh-fading channel is studied to ensure the robustness and the practicality of adaptation mechanisms. The chapter also presents comprehensive modeling and analysis tools, taking into account impacts of imperfect power control and user mobility.

12.2.1 Unslotted CLSP/DS-CDMA packet radio access

This section considers the packet radio access in the uplink of a single-cell unslotted DS-CDMA PRN using CLSP with infinite population and circle coverage around a hub station. Mobile users communicate via the hub using different sequences and fixed bit rate R_0 for packet transmissions with the same QoS requirement. Further, the following assumptions are made without loss of generality.

User data are coded and segmented into information blocks. Then a header that contains address, control information and error-correcting control fields is added to each block to form a radio packet, which is sent over the air toward the hub. For a packet length L (bit) including a constant H (bit) of the protocol overhead, define

T – the packet duration, $T = L/R_0$ (ms), also referred to as the packet length in time domain. Thus, T is proportional to L for a given constant bit rate R_0. In practical implementations, for example, according to radios of current 3GPP standards, T should be kept between T_{min} and T_{max} and should take the value of one or multiples 10-ms periods for effective operation of CDMA radios for long-packet duration applications.

In this CLSP system, similar to the model described in Section 12.1, the hub is responsible for sensing the channel load (number of ongoing transmissions) and rejecting further incoming packets when the load is reaching a certain channel threshold by forcing users to refrain from the transmission with feedback control. The hub broadcasts the control information periodically in a forward control channel. Users having packets to send will listen to the control channel and decide to transmit or refrain from the transmission in a nonpersistent fashion. Thus, the 'hidden terminal' problem of distributed carrier sense multiple access (CSMA) systems can be avoided. The feedback control is assumed to be perfect, that is, zero propagation delay and perfect receiving in the forward direction. The impacts of system imperfection, such as access delay, feedback delay and imperfect sensing have been investigated in Chapter 11 for a nonadaptive, perfect power-control system. The channel threshold or system capacity, defined as the maximum number of simultaneous packet transmissions is given by equation (12.8).

The traffic model is based on the assumption that the scheduling of packet transmissions including retransmissions of unsuccessful packets at mobile terminals is randomized sufficiently enough so that the overall number of packets is generated according to the Poisson process with rate λ. Let us define

n – the system state, that is, the number of ongoing packet transmissions in the system. The CLSP is responsible for keeping n under C_0. However, because of imperfect power control, characterized by a lognormal error of average SINR with standard deviation σ (dB), the equilibrium probability that n simultaneous transmissions are not corrupted by the system outage state (i.e. target SINR is kept) can be given by modifying equation (12.3)

$$P_{\text{ok}}(n) = 1 - Q\left(\frac{C_0 - E[\text{MAI}|n]}{\sqrt{\text{Var}[\text{MAI}|n]}}\right) \tag{12.30}$$

where $E[\text{MAI}|n] = n\exp[(\varepsilon\sigma)^2/2]$ and $\text{Var}[\text{MAI}|n] = n\exp[2(\varepsilon\sigma)^2]$ and C_0 is given by equation (12.8), $\varepsilon = \ln(10)/10$, and $Q(x)$ is the standard Gaussian integral.

This equation represents the interference-limited nature of DS-CDMA systems. The smaller L makes the shorter packet transmission duration $T = L/R_0$ and thus the smaller number of simultaneous transmissions n for a given packet arrival rate λ and bit rate R_0. This improves the system outage probability and therefore can be used for adaptation strategy as well. However, in this section CLSP is used to compensate MAI. To simplify the analysis, we assume that all packet transmissions hit by the system outage state are erroneous with Probability 1.

12.2.2 Fading model and impacts on packet transmission

Let us assume that the system operates at 2.4-GHz carrier frequency [industrial scientific and medical (ISM) band] with omnidirectional antenna, 64-kbps packet transmission and 64 spreading factor. The user speed is in the range of 0 to $4\,\text{ms}^{-1}$, which means that it may take at least 32 ms for the user to travel the distance of one wavelength, and the maximum Doppler frequency is up to 32 Hz. This radio channel is modeled as a flat Rayleigh-fading channel, where the fading process is heavily correlated according to the correlation properties presented in References [7,11]. For a certain fade margin, depending on the packet duration T (one or multiple of 10 ms), several fades may occur during the packet transmission period. To determine the probability of correct packet transmission as well as the packet length adaptation criteria, one needs to consider the impacts of fade and interfade duration statistics, and packet lengths. The correlation between them in time domain is illustrated in Figure 12.8. Let us use the following notation:

t_f – the fade duration, that is, the period of time a received signal spends below a threshold voltage R, having PDF $g(t_f)$ and mean t_{f_avrg}

t_{if} – the interfade duration, that is, the period of time between two successive fades, having PDF $h(t_{if})$ and mean t_{if_avrg}

t_{fia} – the fade interarrival time, that is, the time interval between the time instants that two successive fades occur: $t_{fia} = t_f + t_{if}$, having PDF $s(t_{fia})$ and mean t_{fia_avrg}.

For a Rayleigh-fading channel, it has been shown in numerous papers [9,13,14] that t_{f_avrg} and t_{if_avrg} can be approximated as

$$t_{f\_\text{avrg}} = \frac{e^{\rho^2} - 1}{\sqrt{2\pi}\,f_d\rho} \tag{12.31}$$

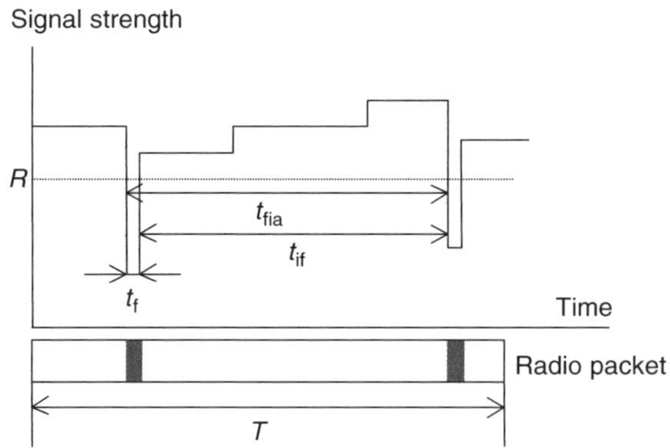

Figure 12.8 Illustration of fading and related time intervals.

$$t_{\text{if\_avrg}} = \frac{1}{\sqrt{2\pi}\, f_d \rho e^{-\rho^2}} \tag{12.32}$$

where f_d is the maximum Doppler frequency; $\rho = R/R_{\text{rms}}$ is the ratio of Rayleigh-fading envelope R and the local root-mean-square (RMS) R_{rms}. The value of ρ is set to $-20\,\text{dB}$ throughout the paper as in Reference [12], that is, the fade margin is equal to 20 dB, and thus $t_{\text{f\_avrg}}/t_{\text{if\_avrg}} = 1\%$. The channel is considered as a slow but deep fading channel.

The closed-form generalized theoretical expressions for fade and interfade duration distributions or PDFs are not available. However, for a deep fading channel, Reference [15] provides an asymptotic formula for the probability that the fade duration t_f lasts for more than t as follows:

$$P_{t_f}\{t_f > t\} = \frac{2}{u} I_1 \left(\frac{2}{\pi u^2} \right) \exp \left(-\frac{2}{\pi u^2} \right) \tag{12.33}$$

where $I_1(z)$ is the modified Bessel function of the first kind and the first order for imaginary argument; and $u = t/t_{\text{f\_avrg}}$ is the normalized fade duration. It has been shown in References [13] and [9] by experiment and simulation that equation (12.33) is valid for the fade margin range of at least 10 dB. The distribution in equation (12.33) has high density around the mean value.

For the interfade duration, the exponential distribution is often assumed, for example, References [9,14] because of the validity of the finite-state Markov channel models [7,8,11]:

$$P_{t_{if}}\{t_{if} \le t\} = 1 - \exp(-t/t_{\text{if\_avrg}}) \tag{12.34}$$

This assumption is not precise in general for correlated fading channels [10,11]. Reference [10] presents an elegant matrix-form for the distribution of level-crossing numbers resulting from a more complicated and sophisticated hidden Markov channel modeling. However, in such a slow fading channel as described above with $t_{\text{f\_avrg}}/t_{\text{if\_avrg}} = 1\%$, the fade duration is very small compared to the interfade duration. Thus, it is reasonable to use equation (12.34) for modeling the interfade duration distribution in our case. The same argument can be found in References [9,15,16]; also supported by simulation in References [9,16].

In general, the probability of correct packet transmission in a fading channel depends on the number of fades occurring during the transmission period, fade duration or the number of bits in each error-block under fade and the error-correcting capability of coding methods. Effects of some specific FEC coding schemes are investigated in References [7,9]. Reference [7] shows that in heavily correlated fading environments the block error rate characteristic is insensitive to error-correcting capability. For the fading model described above, fade duration can be expected in the range of milliseconds and therefore an error-block under fade can contain hundreds of bits depending on the bit rate of packet transmissions. For high bit rate transmissions, a single fade occurring during the transmission period can result in an unsuccessful packet transmission. In accordance with References [14,16], the probability of correct packet transmission, as the function of packet duration T and maximum Doppler frequency f_d, can be given as follows:

$$P_{sf}(T, f_d) = \int_0^\infty \int_T^\infty (t_{if} - T)(t_{if} + t_f)^{-1} h(t_{if}) g(t_f) \, dt_{if} \, dt_f \qquad (12.35)$$

The PDFs $g(t_f)$ and $h(t_{if})$ are obtained from the derivatives of equations (12.33) and (12.34), respectively. By using the argument that fade duration is very small compared to interfade duration, the above integral can be approximated by

$$P_{sf}(T, f_d) = \exp(-T/t_{if\_avrg}) - (T/t_{if\_avrg}) E_1(T/t_{if\_avrg}) \qquad (12.36)$$

$$E_1(y) = \int_y^\infty \exp(-t) t^{-1} \, dt \qquad (12.37)$$

To proceed with our main purpose on investigating benefits of the packet length adaptation, further explanation, derivation and alternative approximation for the probability of correct packet transmission can be found in the appendix. Figure 12.9 shows the impacts of packet duration (i.e. packet length in time domain) and Doppler frequency on $P_{sf}(T, f_d)$ for 2.4-GHz carrier frequency and user speed up to 4 m s^{-1} [17,18].

For a required successful rate of the packet transmission, for example, at least 90%, slow-moving users can transmit with longer packets, while relatively fast-moving users need to keep the packet length at minimum size.

12.2.3 Packet-length adaptation strategy

In this section, practical adaptation strategies and mechanisms are presented for unslotted CLSP/DS-CDMA PRN in the fading environment as described above [17,18]. From equations (12.30) and (12.35), the probability of correct packet transmission, conditioned on system state n, can be given by

$P_c(n) \equiv \Pr\{\text{system not in the outage state } |n\} \Pr\{\text{correct packet transmission under fading conditions}\}$, which is

$$P_c(n) = P_{ok}(n) P_{sf}(T, f_d) \qquad (12.38)$$

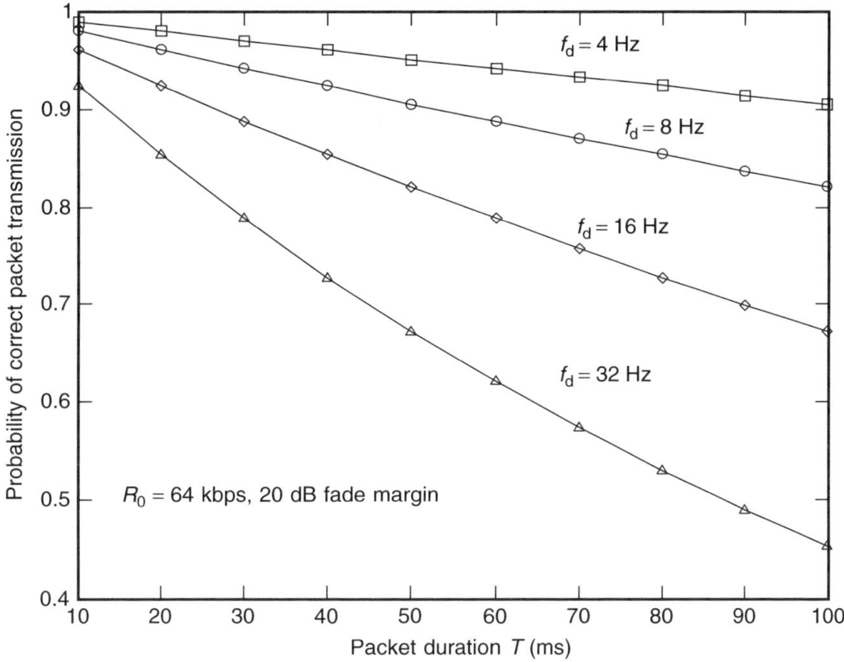

Figure 12.9 Impacts of packet length and Doppler frequency on the probability of correct packet transmission.

The equilibrium probability of correct packet transmission can be obtained by summing up equation (12.38) over all possible system states, that is,

$$P_{\mathrm{c}} = \left(\sum_{n=0}^{C_0} n p_n\right)^{-1} \sum_{n=0}^{C_0} n p_n P_{\mathrm{c}}(n) \tag{12.39}$$

where C_0 is given by equation (12.8) and p_n is the steady state probability of having n simultaneous packet transmissions in the system.

Define

S – the system throughput as the average number of successful packet transmissions per unit time for a given offered traffic

D – the average packet delay

G – the goodput as the effective average data rate put through excluding protocol overhead.

If the criteria for packet-length adaptation were to maximize the throughput S defined above, T or $L = R_0 T$ should be adapted to maximize equation (12.39). Bearing in mind that CLSP has compensating impacts of MAI, a suboptimum solution can be adopted by maximizing $P_{\mathrm{sf}}(T, f_{\mathrm{d}})$ given that T is one or a multiple of 10 ms and limited between

T_{min} and T_{max}. If the protocol overhead was not considered, T_{min} would obviously be the optimum packet size; and so there would be no need for the adaptation. The real packet will have a finite header H; so in this section, the packet-length adaptation criteria are defined on the basis of the dynamic range of Doppler frequency due to the user mobility, subject to an optimal trade-off between throughput, average packet delay and goodput. Two alternative suboptimum strategies are adopted: (A1) keeping the packet length as large as possible for a required percentage of successful packet transmission rate per user transmission to avoid degradation of the goodput; (A2) maximizing the goodput of a single user transmission. The idea behind the second alternative (A2) is used in References [5,8,12,18] for packet-length adaptation and in Reference [14] for packet- length optimization. Except in References [17,18], trade-off between the goodput and the other performance characteristics is not considered, neither are the effects of MAI. Later we will show that (A2) is sensitive to the system load and outage, whereas (A1) has stable characteristics over a much larger range of the offered traffic and overall outperforms both (A2) and fixed systems.

(A1) First alternative: Let us introduce the following system parameter.
β – the required percentage of successful packet transmission rate for a single user transmission, which can be set depending on the nature of the system and the services. For instance, in military or emergency networks, or for providing real-time (RT) and reliable services, β may need to be at least 90%. To satisfy such a requirement with a fixed packet-length system, the packet duration may need to be set to T_{min}, resulting in significant degradation of the goodput. It is desired to adapt the packet duration T to the time-varying channel conditions so that $P_{sf}(T, f_d)$ is kept above β while maximizing the goodput by keeping T as large as possible, given that T needs to be one or a multiple of 10 ms and in the range of $[T_{min}, T_{max}]$. Thus, we can adopt a simple mechanism as follows [17,18]:

$$T = \arg\max\{T : P_{sf}(T, f_d) > \beta, T_{min} \le T = \times 10 \le T_{max}\} \qquad (12.40)$$

where $P_{sf}(T, f_d)$ is given in equation (12.35).

From the results in Figure 12.9 and given $\beta = 90\%$, $T_{min} = 10$ ms and $T_{max} = 80$ ms, equation (12.40) can be simplified as follows. Let us define the set $\mathbf{J} = \{0, 1, 2, 3\}$ and the following notation for $j \in \mathbf{J}$:

f_{d_j} – the boundary value of Doppler frequency range or interval, that is, $f_{d\_0} = 0$ Hz, $f_{d\_1} = 4$ Hz, $f_{d\_2} = 8$ Hz, $f_{d\_3} = 16$ Hz, and $f_{d\_4} = 32$ Hz $\equiv f_{d\_max}$ for a slow-fading channel.
T_j – the adapted packet duration for the interval of $[f_{d\_j}, f_{d\_j+1}]$: $T_j = 2^{-j}T_{max}$, that is, $T_0 = T_{max} = 80$ ms, $T_1 = T_{max}/2 = 40$ ms, $T_2 = T_{max}/4 = 20$ ms, and $T_3 = T_{max}/8 = 10$ ms $\equiv T_{min}$.

Now the packet-length adaptation mechanism can be formulated as follows:

$$\{\text{for all } j \in \mathbf{J}, \text{if } f_{d\_j} < f_d \le f_{d\_j+1} \text{ then } T = T_j \text{ or } L = R_0 T_j\} \qquad (12.41)$$

Our analytical results above are in agreement with simulation results provided by Reference [12] that the packet duration should be smaller than 1/10 of the average interfade duration for 80 to 90% of correct packet transmission rate.

(A2) Second alternative: This alternative is designed to maximize the normalized goodput of a single user transmission, which is defined with the following function of (T, f_d):

$$G_{sf}(T, f_d) = P_{sf}(T, f_d)(R_0 T - H)/R_0 T \qquad (12.42)$$

Figure 12.10 shows the impact of packet duration onto $G_{sf}(T, f_d)$ for different Doppler frequencies. Depending on the maximum Doppler frequency estimate, there is a value of T that optimizes the normalized goodput. The adaptation mechanism can be defined by

$$T = \arg\max\{G_{sf}(T, f_d)$$
$$= P_{sf}(T, f_d)(R_0 T - H)/R_0 T, T_{min} \le T = \times 10[ms] \le T_{max}\} \qquad (12.43)$$

From the results in Figure 12.10 and the notations defined at (A1) above, keeping in mind that $T \in \{10, 20, 40, 80\}$ ms, equation (12.43) can be simplified by: if $f_d < 8$ Hz, $T = T_0 = 80$ ms, else $T = T_1 = 40$ ms. Expression (12.41) can also be used for this case with the following modifications: $\mathbf{J} = \{0, 1\}$; $f_{d\_0} = 0$, $f_{d\_1} = 8$ Hz and $f_{d\_2} = f_{d\_max} = 32$ Hz.

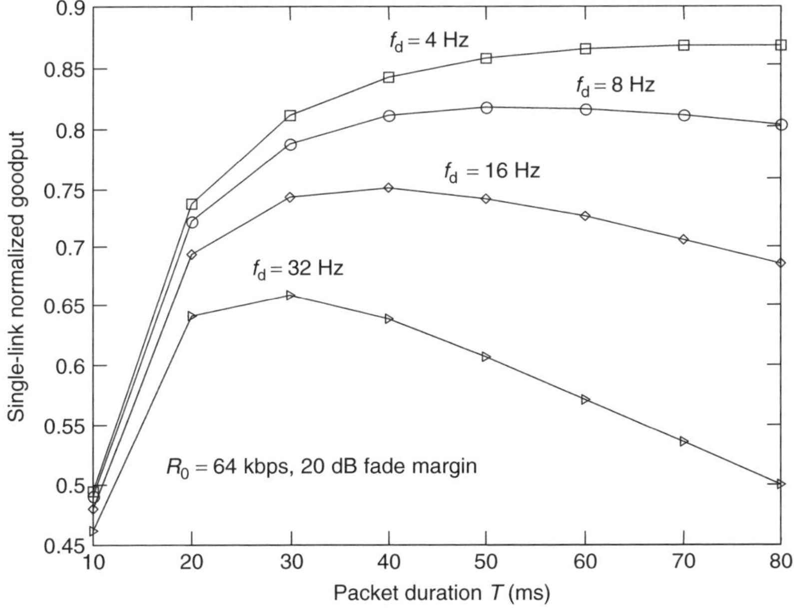

Figure 12.10 Impacts of packet length and Doppler frequency onto normalized goodput.

For the implementation of both alternatives, online estimation of only the maximum Doppler frequency is required at the mobile terminals, which is well known as feasible. Benefits of such simple adaptation mechanisms are evaluated analytically and quantitatively in the following sections through a comparative study of the performance characteristics, such as the system throughput, the average packet delay, the goodput and their trade-offs.

12.2.4 Queuing system model and impacts of Doppler frequency range

From the CLSP packet-access model and the Poisson packet-arrival assumption described in the previous section, a birth–death loss queuing system model can be used for the analysis. The formal notation is Erlang loss $M/D/N/N : D \equiv$ packet duration for deterministic service time, $N \equiv C_0$ for the number of servers and the system capacity. Let

G_0 – the system offered traffic $G_0 = \lambda T_0 \equiv \lambda$, that is, average number of packets per normalized $T_0 \equiv 1$, being kept the same in both adaptive and fixed packet-length systems for comparison purposes.

In adaptive systems, $G_0 \equiv \lambda$ is composed of portions from offered traffic of different packet lengths

λ_j – the arrival rate of T_j-duration packets that is generated by mobile terminals having Doppler frequency estimated in the range of (f_{d_j}, f_{d_j+1}). Given $G_0 \equiv \lambda$ as the system parameter, λ_j is now dependent on λ and the user velocity distribution, or equivalent to that of a so-called Doppler frequency distribution (DfD) determining the probability of the adapting condition $f_{d\_j} < f_d \leq f_{d\_j+1}$. Let us assume that such DfD has PDF $v(f_d)$, where f_d is assumed falling in between f_{d_0} and f_{d_max} for the flat Rayleigh-fading channel. Denote

P_{f_j} – the equilibrium probability that the mobile terminals have Doppler frequency estimated in the range of $[f_{d\_j}, f_{d\_j+1}]$

$$P_{f\_j} = \int_{f_{d\_j}}^{f_{d\_j+1}} v(f_d)\, \mathrm{d}f_d \bigg/ \int_0^{f_{d\_max}} v(f_d)\, \mathrm{d}f_d \tag{12.44}$$

Now λ_j can be given by

$$\lambda_j = \lambda P_{f\_j} \tag{12.45}$$

For instance, if DfD is the uniform distribution in the mobility equilibrium condition, λ_j can be determined by

$$\lambda_j = \lambda(f_{d\_j+1} - f_{d\_j})/f_{d\_max} \tag{12.46}$$

In this section, $v(f_d)$ is modeled with a general Gamma PDF because of its flexibility and richness in modeling [13]

$$v(f_d) = \frac{1}{b^a \Gamma(a)} f_d^{a-1} \mathrm{e}^{-f_d/b} \tag{12.47}$$

where a is the shape-parameter, b is the scale-parameter and $\Gamma(x)$ is the Gamma function. Thus, by adjusting a and b parameters, it is possible to choose a suitable PDF to fit

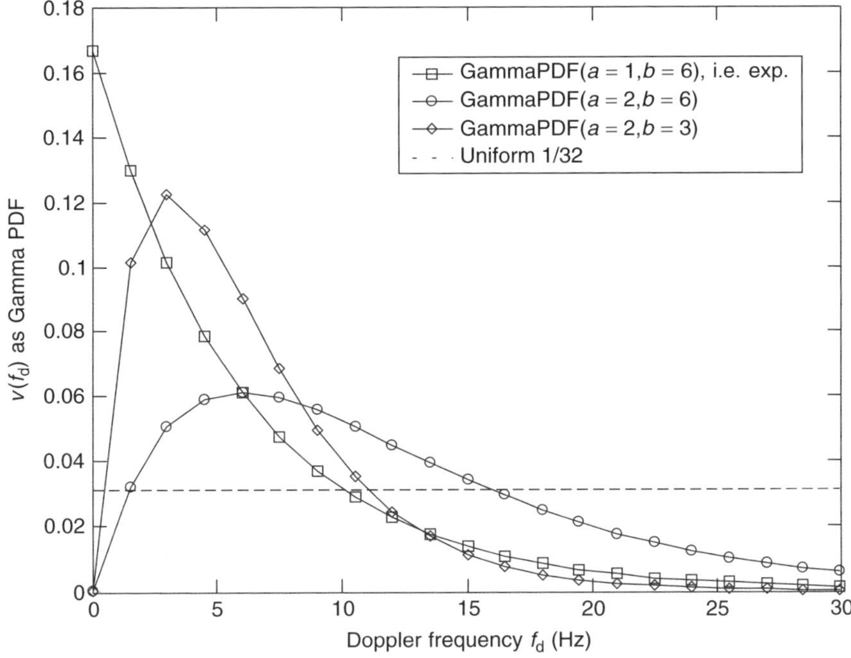

Figure 12.11 Doppler frequency distribution for traffic modeling.

for numerous user-mobility scenarios or expansion of Doppler frequency range. This PDF family also includes the deterministic PDF and the exponential PDF as its special members, where the shape-parameter a is set to infinite or 1, respectively. Figure 12.11 shows some examples of how to choose proper values of a and b. In the case that the majority of the users are moving very slowly, the exponential or the Gamma PDF with $a = 2$, $b = 3$ can be used for modeling, while the uniform or the Gamma PDF with $a = 2$, $b = 6$ can be used for more dynamic mobile systems.

12.2.5 Fixed packet-length system: $T = L/R_0$ constant

The steady state solution for probability p_n, of having n simultaneously transmitting users in the system, can be obtained by using the Erlang loss formula [13]. In the formula, the number of servers should be set to the channel threshold C_0 given in equation (12.8), the packet arrival rate λ, and the normalized packet service time T ($T_0 \equiv 1$). Let $\alpha = \lambda T$, then we have

$$p_n = \frac{\alpha^n}{n!} \left(\sum_{i=0}^{C_0} \frac{\alpha^i}{i!} \right)^{-1} \text{ for } 0 \le n \le C_0 \tag{12.48}$$

Denote

P_{succ}— the equilibrium probability of successful packet transmission that depends on two factors:

a) the probability that the given packet is not blocked by the CLSP given by $(1 - B)$, where B is the packet blocking probability:

$$B = p_{C_0} \tag{12.49}$$

b) the equilibrium probability of correct packet transmissions P_c given by equation (12.39). To calculate it, we first need to modify the factor $P_{sf}(T, f_d)$ in equation (12.38) to include the effects of user mobility or DfD, which is characterized by the PDF $v(f_d)$. Equation (12.38) now becomes

$$P_c(n) = P_{ok}(n) \sum_{f_d} P_{sf}(T, f_d) P(f_d) \tag{12.50}$$

The sum in equation (12.50) is to replace the actual integral of $P_{sf}(T, f_d) v(f_d)$ over f_d, where the estimation of f_d is assumed to take discrete values in $\{0, 1, 2, \ldots, f_{d\_max}\}$, and $P(f_d)$ can be defined similar to equation (12.44) as follows:

$$P(f_d) = \int_{f_d}^{f_d+1} v(f_d) \, df_d \bigg/ \int_0^{f_{d\_max}} v(f_d) \, df_d$$

Thus, the equilibrium probability of successful packet transmission becomes

$$P_{succ} = (1 - B) \left(\sum_{n=0}^{C_0} np(n) \right)^{-1} \sum_{n=0}^{C_0} np(n) P_{ok}(n) \sum_{f_d} P_{sf}(T, f_d) P(f_d) \tag{12.51}$$

The system throughput for the offered traffic G_0 can be given by

$$S = G_0 P_{succ} \tag{12.52}$$

The ratio $S/G_0 \equiv P_{succ}$ is called the normalized throughput. The average packet delay is decomposed into two parts: D_b the average waiting time of a packet for accessing the channel including back-off delays; and D_r the average resident time of the given packet from the instant of entering to the instant of leaving the system successfully. Now the average packet delay, normalized to T_0, is given by

$$D = D_b + D_r \tag{12.53}$$

with

$$D_b = T \sum_{i=0}^{\infty} B^i = \frac{TB}{1 - B} \tag{12.54}$$

and according to Little's formula [3]

$$D_r = S^{-1} \sum_{n=0}^{C_0} np_n \tag{12.55}$$

The system goodput is given by

$$G = S(L - H) \tag{12.56}$$

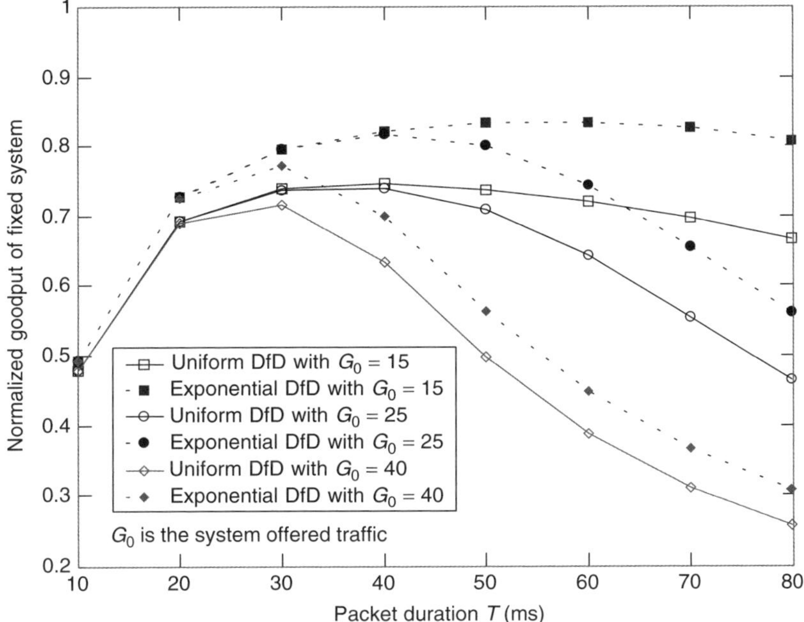

Figure 12.12 Optimal packet length for fixed system.

The normalized goodput can be defined by the ratio $G/(G_0 L) \equiv P_{\text{succ}}(L - H)/L$, which is equivalent to equation (12.29). Figure 12.12 presents the normalized goodput of the fixed system versus packet duration in different mobility scenarios for different system offered traffic G_0. It clearly shows that there is an optimal packet size for the overall system performance that depends on the user mobility scenarios and system load or MAI. Figure 12.12 is different from Figure 12.10 as well as the reported results of link performance in References [5,6,12,14] in the sense that it includes both single link and overall system performance averaged over a given mobility dynamics.

12.2.6 Adaptive packet-length system

$T = T_j$ and $L = T R_0$ if $f_{d-j} < f_d \leq f_{d-j+1}$ for all $j \in \mathbf{J}$

For this system, the steady state solutions can be given by equation (12.48) as well with the following modification of the offered traffic intensity α.

$$\alpha = \sum_{j \in \mathbf{J}} \lambda_j T_j = \lambda \sum_{j \in \mathbf{J}} T_j P_{\text{f}\_j} \tag{12.57}$$

where P_{f_j} is given by equation (12.44). By replacing T with T_j in equation (12.51) and summing up equation (12.51) over $\mathbf{J}$ due to adaptation mechanisms, we can obtain the equilibrium probability of successful packet transmission for the adaptive system

$$P_{\text{succ}} = (1 - B)\left(\sum_{n=0}^{C_0} np(n)\right)^{-1} \sum_{n=0}^{C_0} np(n)P_{\text{ok}}(n) \sum_{j\in\mathbf{J}} \sum_{f_{\text{d}}=f_{\text{d}\_j}}^{f_{\text{d}\_j+1}} P_{\text{sf}}(T_j, f_{\text{d}})P(f_{\text{d}}) \qquad (12.58)$$

Then, the system throughput S can be obtained by using equation (12.52). The normalized average packet delay is given by equation (12.53) with the component D_{r} given by equation (12.55) and with the component D_{b} given by

$$D_{\text{b}} = \sum_{j\in\mathbf{J}} T_j P_{\text{f}\_j} \sum_{i=0}^{\infty} B^i = B/(1 - B) \sum_{j\in\mathbf{J}} T_j P_{\text{f}\_j} \qquad (12.59)$$

Because of the adaptation effects, the system goodput, based on equations (12.42), (12.56) and (12.58), is given by

$$G = G_0(1 - B)\left(\sum_{n=0}^{C_0} np(n)\right)^{-1} \sum_{n=0}^{C_0} np(n)P_{\text{ok}}(n)$$

$$\times \sum_{j\in\mathbf{J}} \sum_{f_{\text{d}}=f_{\text{d}\_j}}^{f_{\text{d}\_j+1}} P_{\text{sf}}(T_j, f_{\text{d}})P(f_{\text{d}})(R_0 T_j - H) \qquad (12.60)$$

The normalized goodput for this adaptive system is defined by the following ratio:

$$G\left(G_0 \sum_{j\in\mathbf{J}} P_{\text{f}\_j} R_0 T_j\right)^{-1} \qquad (12.61)$$

The performance comparison of fixed and adaptive systems will be based on the set of system parameters summarized in Table 12.2 [17,18].

The illustrations include modeling of the user mobility. To investigate the effects of different mobility scenarios, three DfDs are chosen. The exponential distribution with a mean of 6 Hz is used for a low mobile system, the uniform distribution function over 32-Hz range of Doppler frequency for a relatively dynamic mobile system and the Gamma PDF with $a = 2$ and $b = 6$ for an average mobile system (see Figure 12.11). Four possible packet lengths can be used in both fixed and adaptive systems: $T_0 = T_{\max} = 80\,\text{ms}$, $T_1 = T_{\max}/2 = 40\,\text{ms}$, $T_2 = T_{\max}/4 = 20\,\text{ms}$, and $T_3 = T_{\max}/8 = 10\,\text{ms} \equiv T_{\min}$. The equivalent packet lengths in bits with 64-kbps user packet transmission are 5120, 2560, 1280 and 640 bits, respectively. Thus, with a 320-bit packet header and tailer, the load factor of protocol overhead in the worst case can be up to 50%. The adopted alternatives for packet-length adaptation mechanisms, namely (A1) and (A2), are described by equations (12.40–12.41) and (12.42–12.43), respectively.

Figures 12.13 to 12.15 present the effects of packet duration on the performance characteristics of the fixed packet length system: the normalized throughput, average packet delay and goodput versus the offered traffic (i.e. number of packet arrivals per

Table 12.2 System parameter summary

Name	Definition	Values
F_c	Carrier frequency	2.4 GHz
η	Ratio of the thermal noise density and the tolerable interference level	-10 dB
R_0	Bit rate for the packet transmission	64 kbps
G	Processing gain of the R_0 bit rate transmission	64
γ	SINR target of the R_0 bit rate transmission	3 dB
σ	Standard deviation of lognormal average SINR error	2 dB
C_0	Fixed primary rate system capacity	28
T_{min}	Minimum packet duration	10 ms
T_{max}	Maximum packet duration	80 ms
H	Fixed length of the packet header in bits	320
f_{d_max}	Maximum Doppler frequency	32 Hz
ρ	Ratio of R/R_{rms} (the absolute of ρ is the fade margin)	-20 dB
β	Percentage for the desired successful rate of correct packet transmission	90%

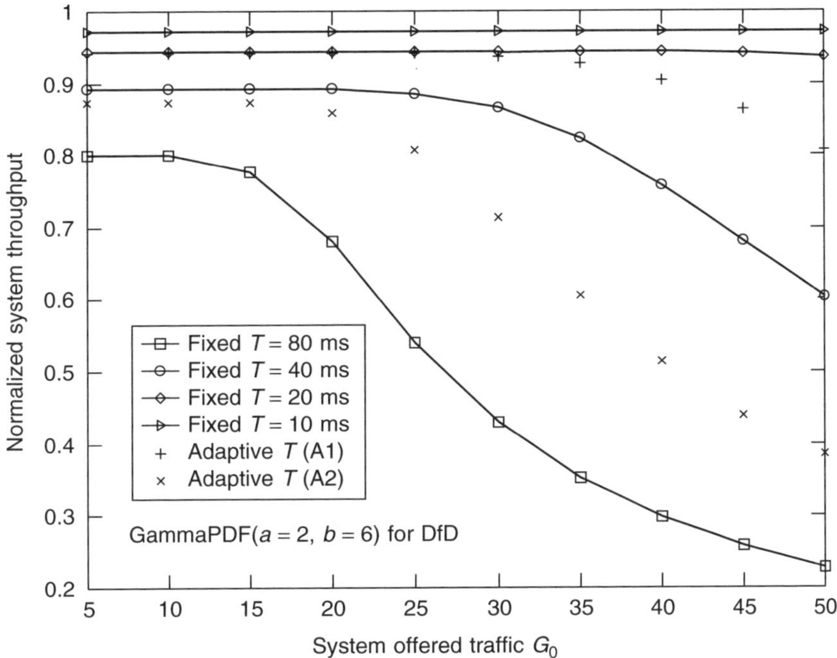

Figure 12.13 Normalized system throughput for comparison.

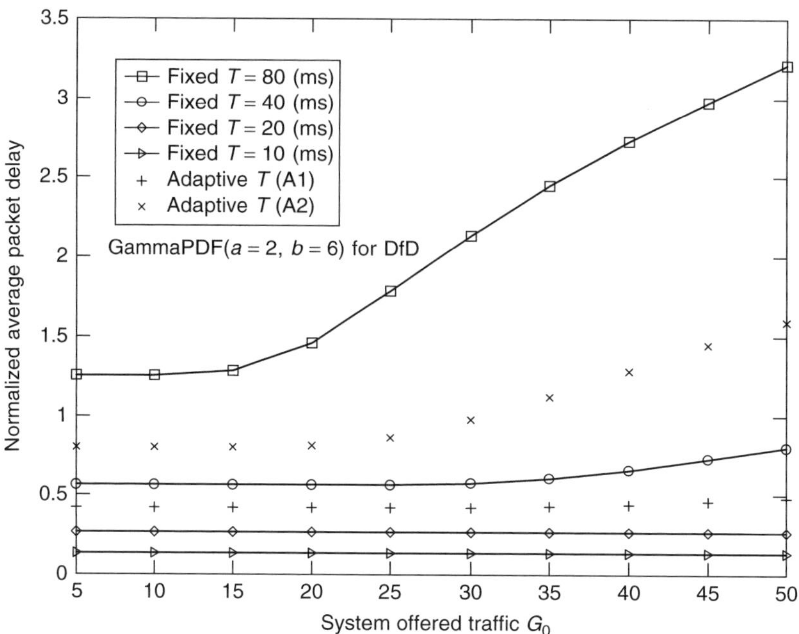

Figure 12.14 Normalized average packet delay.

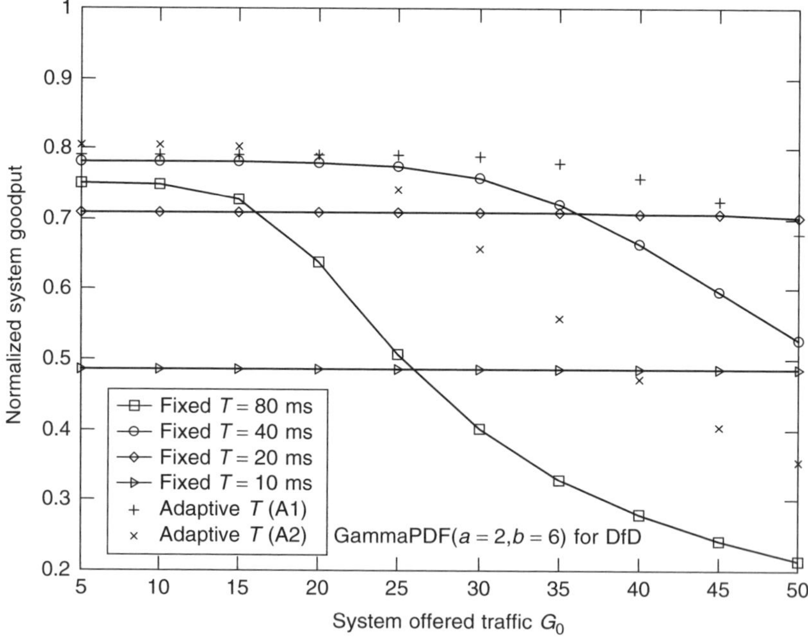

Figure 12.15 Normalized system goodput.

normalized T_0) for the GammaPDF($a = 2$, $b = 6$) DfD scenario. These figures clearly illustrate the dilemma that a small packet-size system can offer much better throughput and delay performance but can suffer from much worse goodput performance as well; and *vice versa* with a large packet-size system. Neither way is sufficient with respect to the efficiency of radio and energy resource utilization in time-varying channels. In a small packet-size system, high activation/deactivation frequency of radio transmissions and heavy protocol overhead can be expected, whereas in a large packet-size system wasting resources for transmitting erroneous packets and retransmissions can be expected. The packet-length adaptive systems, especially the (A1) alternative, can provide a fair gain in the overall system performance: best goodput and a reasonable and stable throughput delay resulting in better resource utilization. The second alternative (A2) provides the best normalized goodput when the offered traffic is less than 20 packets per T_0 (Figure 12.15). However, as the system load becomes heavier, the performance characteristics of the (A2) system worsens rapidly. The reason is that the (A2) alternative, maximizing the goodput of a single user transmission by using large packet sizes, cannot compensate the effects of MAI under imperfect power control in an interference-limited DS-CDMA multiple access channel. This adaptation mechanism is therefore applicable for light loaded situations or systems supporting best-effort IP services that can tolerate moderate delay.

Figures 12.16–12.18 present the effects of different DfD or mobility scenarios on the performance characteristics of both fixed and adaptive packet-length systems. To assure the fairness of performance comparison, for the fixed packet-length system, the packet

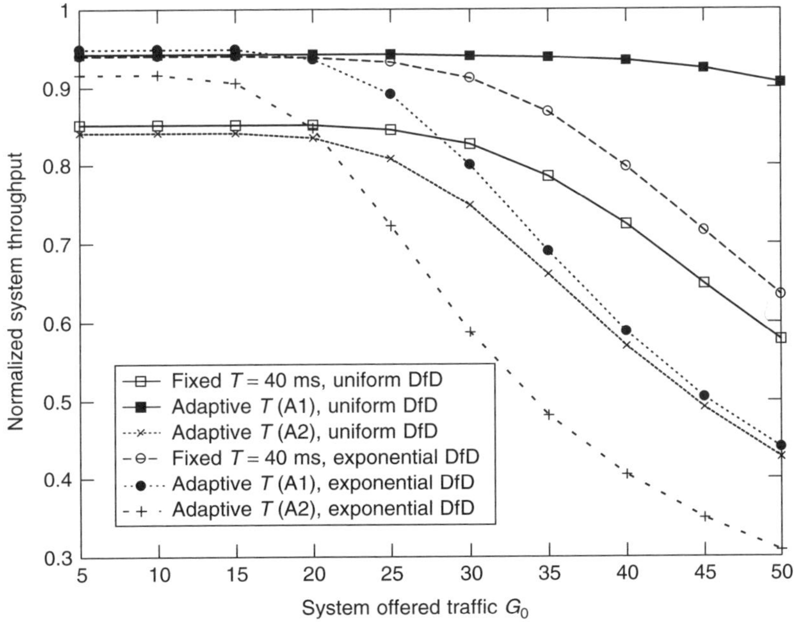

Figure 12.16 Effects of the mobility or DfD to throughput performance in the flat-fading channel.

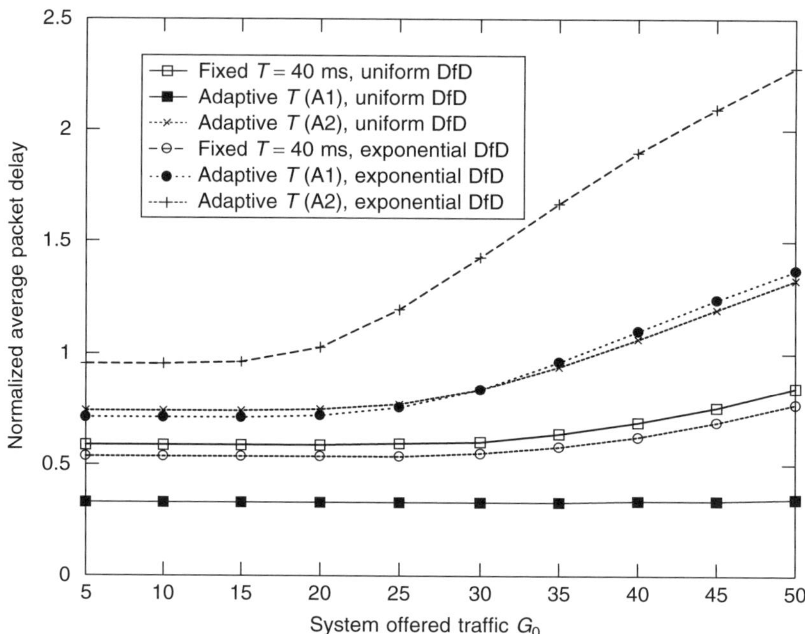

Figure 12.17 Effects of the mobility or DfD to packet-delay performance in the flat-fading channel.

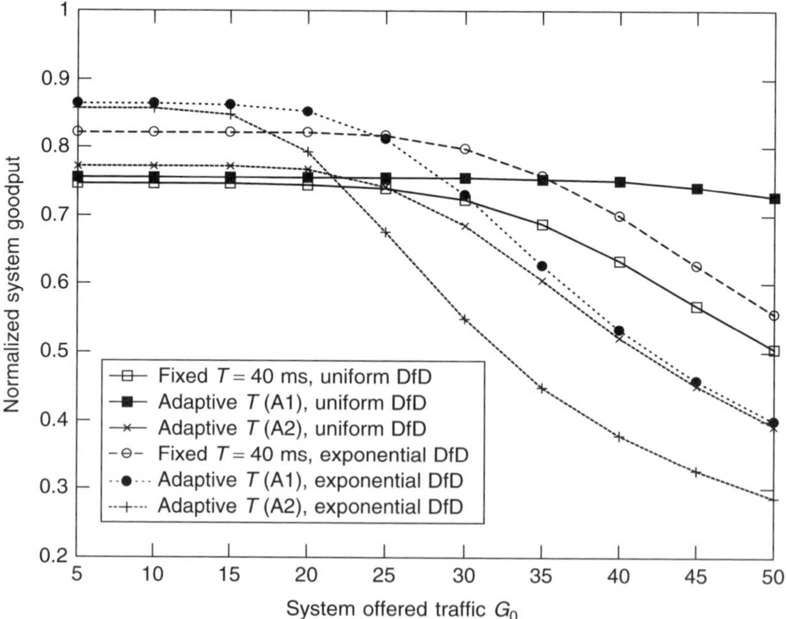

Figure 12.18 Effects of the mobility or DfD to goodput performance in the flat-fading channel.

duration T is set to 40 ms, which provides the best performance of fixed systems as shown in Figures 12.13 to 12.15. Figures 12.16 to 12.18 demonstrate clearly that the adaptations are more effective in relatively dynamic or fast-varying environments. In the uniform DfD scenario, representing a dynamic environment, the adaptive systems outperform the fixed counterpart. In the very slow-varying channel, characterized by the exponential DfD (shown in Figures 12.16 to 12.18 with dotted lines), due to the adaptation mechanisms, the majority of mobile users are transmitting with the maximum-size packets. Thus, in light loaded situations, the best goodput and stable throughput delay can still be expected with adaptive systems as shown in the figures. As the system offered load becomes heavier (i.e. above 20 packets per T_0), there is a drop in performance characteristics of the adaptive systems. This is known as steep brick-wall effects of the fixed system with large packet size [5] (see the square lines in Figures 12.13 to 12.15 for the fixed system with T_0 packet duration). However, the reasons behind these effects are different for the two systems. For the adaptive system, effects of MAI under imperfect power control become more crucial since the majority of packet transmissions spend longer time in the system owing to large packet sizes; and thus larger number of simultaneous transmissions can be expected, resulting in higher outage probability and blocking probability. On the other hand, for the fixed system with large packet size, the effects of fading conditions are as significant as the effects of outage state. Let us notify that as the GammaPDF($a = 2$, $b = 6$) is used for DfD in Figures 12.13 to 12.15 representing a relatively dynamic scenario, no such degradations are experienced with the adaptive system (A1), but only with the T_0 packet duration fixed system.

A number of related topics are discussed in References [19–28].

APPENDIX

Let $g(u)$ be the derivative of the right side of equation (12.33) with respect to u; since $u = t/t_{\text{f\_avrg}}$, we have $g(t_f) = g(u)/t_{\text{f\_avrg}}$; and $h(t_{\text{if}})$ is the derivative of the right-hand side of equation (12.34). Following Siew [14], equation (12.35) can now be represented as

$$P_{\text{sf}}(T, f_d) = \int_0^\infty \int_T^\infty (t_{\text{if}} - T)(t_{\text{if}} + t_f)^{-1} h(t_{\text{if}}) g(t_f) \, dt_{\text{if}} \, dt_f$$

$$= \int_T^\infty h(t_{\text{if}}) dt_{\text{if}} \int_0^\infty g(u) du - \int_0^\infty \int_T^\infty (u t_{\text{f\_avrg}} + T)(t_{\text{if}} + u t_{\text{f\_avrg}})^{-1}$$
$$\times h(t_{\text{if}}) \, dt_{\text{if}} g(u) \, du$$

$$= \exp(-T/t_{\text{if\_avrg}}) - \int_0^\infty \exp(ru)(ru + T/t_{\text{if\_avrg}}) \int_{v + T/t_{\text{if\_avrg}}}^\infty$$
$$\times \exp(-t_{\text{if}})/t_{\text{if}} \, dt_{\text{if}} g(u) \, du$$

where $r = t_{\text{f\_avrg}}/t_{\text{if\_avrg}}$. Since r is close to 0 and u is smaller than 3 with 0.9670 probability due to equation (12.33), ru is close to 0 and therefore

$$P_{\text{sf}}(T, f_d) \cong \exp(-T/t_{\text{if\_avrg}}) - (T/t_{\text{if\_avrg}}) E_1(T/t_{\text{if\_avrg}})$$

where $E_1(y) = \int_y^\infty \exp(-t)/t \, dt$.

Having the same accuracy, we can adopt an alternative and traceable approximation to equation (12.35) as follows. The PDF $s(t_{\text{fia}})$ of the fade interarrival time, $t_{\text{fia}} = t_{\text{f}} + t_{\text{if}}$, can be derived as the convolution of $g(t_{\text{f}})$ and $h(t_{\text{if}})$. Using the argument that the fade duration is very small compared to the interfade duration, we can approximate that $s(t_{\text{fia}})$ as the convolution of $g(t_{\text{f}})$ and $h(t_{\text{if}})$ is also the exponential distribution with mean of $t_{\text{fia\_avrg}} = t_{\text{f\_avrg}} + t_{\text{if\_avrg}}$. Consequently, the fade arrivals can be seen as a Poisson process with the rate of $1/t_{\text{fia\_avrg}}$. Hence, the probability of correct packet transmission, as a function of (T, f_{d}), can be determined by

$$P_{\text{sf}}(T, f_{\text{d}}) \equiv \Pr\{\text{no fade occurs during } T\} = \exp(-T/t_{\text{fia\_avrg}})$$

This simple model does appear to keep the sensitivity of the fading process to Doppler frequency, which is the essential feature, concluded by Reference [7].

REFERENCES

1. Holma, H. and Toskala, A. (2000) *WCDMA for UMTS*. New York: John Wiley & Sons.
2. Ross, K. W. and Tsang, D. H. K. (1989) The stochastic Knapsack problem. *IEEE Trans. Commun.*, **37**(7), 740–747.
3. Gross, D. and Harris, C. M. (1998) *Fundamentals of Queuing Theory*. New York: John Wiley & Sons.
4. Phan, V. and Glisic, S. (2002) *Unslotted DS/CDMA Packet Radio Network Using Rate/Space Adaptive CLSP-ICC '02*, New York, May 2002.
5. Lettieri, P. and Srivastava, M. B. (1998) Adaptive frame length control for improving wireless link range and energy efficiency. *IEEE InfoCom '98*, Vol. 2, pp. 564–571.
6. Hara, S., Ogino, A., Araki, M., Okada, M. and Morinaga, N. (1996) Throughput performance of SAW-ARQ protocol with adaptive packet length in mobile packet data transmission. *IEEE Trans. Veh. Technol.*, **45**(3), 561–569.
7. Zorzi, M., Rao, R. R. and Milstein, L. B. (1996) A Markov model for block errors on fading channels. *IEEE PIMRC '96*, pp. 1074–1078.
8. Wang, H. S. and Chang, P. C. (1996) On verifying the first-order Markovian assumption for a Rayleigh fading channel model. *IEEE Trans. Veh. Technol.*, **45**(2), 353–357.
9. Lai, J. and Mandayam, N. B. (1998) Packet error rate for burst-error-correcting codes in Rayleigh fading channels. *IEEE VTC '98*, Vol. 2, pp. 1568–1572.
10. Turin, W. and Nobelen, R. V. (1998) Hidden Markov modeling of flat fading channels. *IEEE J. Select. Areas Commun.*, **16**(9), 1809–1817.
11. Tan, C. C. and Beaulieu, N. C. (2000) On first-order Markov modeling for the Rayleigh fading channel. *IEEE Trans. Commun.*, **48**(2), 2032–2040.
12. Chien, C., Srivastava, M. B., Jain, R., Lettieri, P., Aggarwal, V. and Sternowski, R. (1999) Adaptive radio for multimedia wireless links. *IEEE J. Select. Areas Commun.*, **17**(5), 793–813.
13. Bodtmann, W. F. and Arnold, H. W. (1982) Fade-duration statistics of Rayleigh distributed waves. *IEEE Trans. Commun.*, **30**(3), 549–553.
14. Siew, C. K. and Goodman, D. J. (1989) Packet data transmission over mobile radio channels. *IEEE Trans. Veh. Technol.*, **38**(2), 95–101.
15. Rice, S. O. (1958) Distribution of the duration of fades in radio transmissions: Gaussian noise model. *Bell Syst. Tech. J.*, **37**, 581–635.
16. Chang, L. F. (1991) Throughput estimation of ARQ protocols for a Rayleigh fading channel using fade- and interfade-duration statistics. *IEEE Trans. Veh. Technol.*, **40**(1), 223–229.

17. Phan, V. (2002) Unslotted DS/CDMA packet radio network using CLSP and packet-length adaptation in Rayleigh fading channel. *IEEE Semiannual Vehicular Technology Conference VTC '02 Spring Conference Proceedings*, May 2002.
18. Phan, V. V. and Glisic, S. (2002) MAC layer packet-length adaptive CLSP/DS-CDMA radio networks: performance in flat Rayleigh fading channel. *IEEE Symposium on Computers and Communications ISCC '02 Conference Proceedings*, Florida, July 2002.
19. Phan, V. and Glisic, S. (2001) Estimation of implementation losses in MAC protocols in wireless CDMA networks. *Int. J. Wireless Inform. Networks*, **8**(3), 115–132.
20. Glisic, S. and Phan, V. V. (2000) Sensitivity function of soft decision carrier sense MAC protocols for wireless CDMA networks with specified QoS. Invited Paper, *IEEE 11th PIMRC Proceedings*, 2000.
21. Yin, M. and Li, V. O. K. (1990) Unslotted CDMA with fixed packet lengths. *IEEE J. Select. Areas Commun.*, **8**(4), 529–541.
22. Toshimitsu, K., Yamazato, T., Katayama, M. and Ogawa, A. (1994) A novel spread slotted ALOHA system with channel load sensing protocol. *IEEE J. Select. Areas Commun.*, **12**(4), 665–672.
23. Sato, T., Okada, H., Yamazato, T., Katayama, M. and Ogawa, A. (1996) Throughput analysis of DS/SSMA unslotted ALOHA system with fixed packet length. *IEEE J. Select. Areas Commun.*, **14**(4), 750–756.
24. Yin, M. and Li, V. O. K. (1990) Unslotted CDMA with fixed packet lengths. *IEEE J. Select. Areas Commun.*, **8**(4), 529–541.
25. Toshimitsu, K., Yamazato, T., Katayama, M. and Ogawa, A. (1994) A novel spread slotted ALOHA system with CLSP. *IEEE J. Select. Areas Commun.*, **12**(4), 665–672.
26. Makrakis, D. and Murthy, K. M. S. (1992) Spread slotted ALOHA techniques for mobile and personal satellite communication systems. *IEEE J. Select. Areas Commun.*, **10**(6), 985–1002.
27. Sato, T., Okada, H., Yamazato, T., Katayama, M. and Ogawa, A. (1996) Throughput analysis of DS/SSMA unslotted ALOHA system with fixed packet length. *IEEE J. Select. Areas Commun.*, **14**(4), 750–756.
28. Glisic, S. and Phan, V. V. (2000) Sensitivity function of soft decision carrier sense MAC protocols for wireless CDMA networks with specified QoS. Invited Paper, *IEEE PIMRC '00*, Vol. 1, pp. 205–211.

13

Multiuser CDMA receivers

In this chapter we present a number of methods for multiple-access interference (MAI) cancelation. MAI is produced by the presence of the other users in the network, which are located in the same bandwidth as our own signal. The common characteristic of all these schemes is some form of joint signal and parameter estimation for all signals present in the same bandwidth. It makes sense to implement this in a Base Station (BS) of a cellular system because all these signals are available there anyway. At the same time this concept will considerably increase the complexity of the receiver. Although very complex, these schemes are being standardized already because they offer significantly better performance. Details can be seen in Chapter 17. Much simpler but less effective solutions feasible for implementations in mobile units are also considered [minimum mean square error (MMSE) type of algorithms].

13.1 OPTIMAL RECEIVER

If user k transmits bit stream b_k, with bit interval T, using spreading sequence s_k, then the low-pass equivalent of the overall signal received in the BS can be represented as [1,2]

$$dr_t = S_t(\mathbf{b}) \, dt + \sigma \, d\omega_t, \quad t \in R \tag{13.1}$$

$$S_t(\mathbf{b}) = \sum_{i=-M}^{M} \sum_{k=1}^{K} b_k(i) s_k(t - iT - \tau_k) \tag{13.2}$$

where K is the number of users, $\boldsymbol{b} = (b_1, b_2, \ldots, b_K)^{\mathrm{T}}$ is the vector of bits of all users and the signal is observed in time interval $[-MT, MT]$. The noise component is represented by the second term of equation (13.1) and τ_k is the delay of signal from user k. On the basis of the likelihood principle described in Chapter 3, the detector selects the vector of bits $\boldsymbol{b}$ that maximizes

$$P[\{r_t, t \in R\}|b] = C \exp[\Omega(\boldsymbol{b})/2\sigma^2] \tag{13.3}$$

where C is a positive scalar independent of b and

$$\Omega(b) = 2 \int_{-\infty}^{\infty} S_t(b)\, \mathrm{d}r_t - \int_{-\infty}^{\infty} S_t^2(b)\, \mathrm{d}t \qquad (13.4)$$

So, the joint maximum likelihood (ML) decision (estimate) for vector b is obtained as

$$\hat{b} = \max_{\text{all } \mathbf{b} \in (+1,\,-1)} \Omega(b) \qquad (13.4\text{a})$$

In other words, vector b that jointly gives the maximum of equation (13.4) is chosen as a joint estimate of bits for all users. The first term in equation (13.4) can be represented as

$$\int_{-\infty}^{\infty} S_t(b)\, \mathrm{d}r_t = \sum_{i=-M}^{M} b^{\mathrm{T}}(i)y(i) \qquad (13.5)$$

where $y(i)$ is a vector with elements $y_k(i)$ representing the output of a matched filter for the ith symbol of the kth user, that is,

$$y_k(i) = \int_{\tau_k+iT}^{\tau_k+iT+T} s_k(t - iT - \tau_k)\, \mathrm{d}r_t \qquad (13.6)$$

The block diagrams of conventional and optimal (ML) detectors are shown in Figures 13.1 and 13.2, respectively.

Without going into the details of evaluating bit error rate (BER) for these detectors, some results are shown in Figures 13.3 to 13.5 [1,2]. To simplify the numerical evaluation, trivial codes shown in Figure 13.3 are used. Such codes also have high (1/3) correlation function so that the effect of optimum detectors are better emphasized. From Figure 13.3 one can see how much optimum (sequence) detector outperforms the conventional detector.

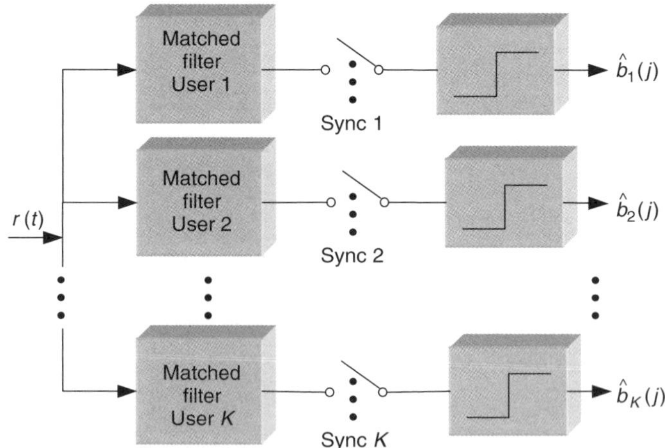

Figure 13.1 Conventional multiuser detector.

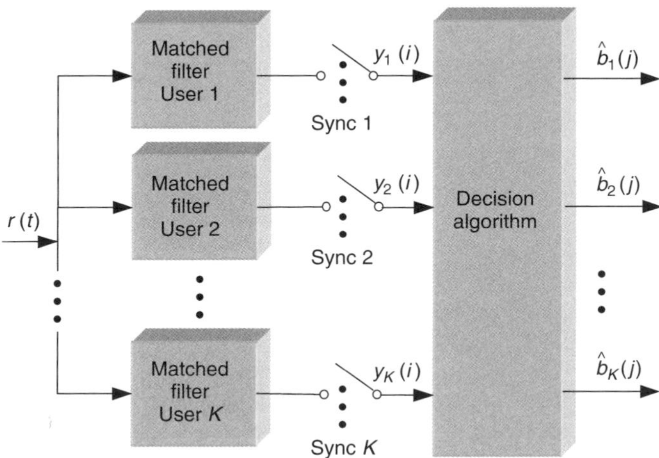

Figure 13.2 Optimum K-user detector for asynchronous multiple-access Gaussian channel.

Figure 13.4 represents the same results for more realistic code, m-sequence of length 31. One can see that the sequence detector performs almost as though only one user is present in the network (single user).

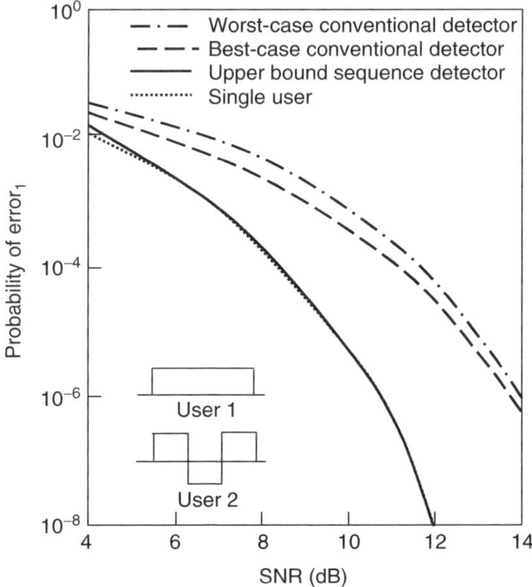

Figure 13.3 Best and worst cases of error probability of User 1 achieved by conventional and optimum detectors.

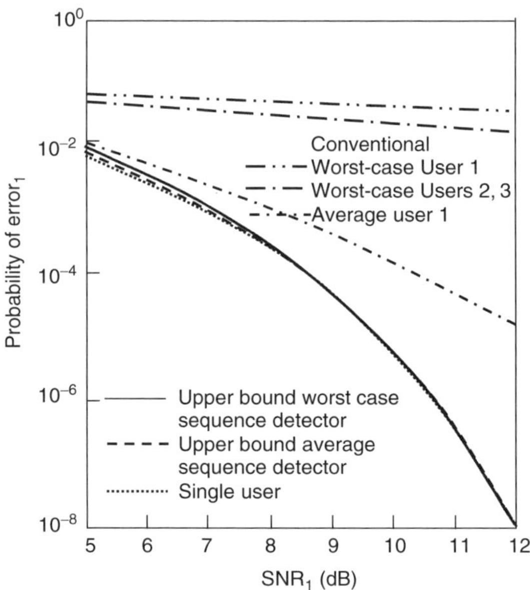

Figure 13.4 Worst-case and average error probabilities achieved by conventional and optimum multiuser detectors with three active users employing m-sequences of length 31.

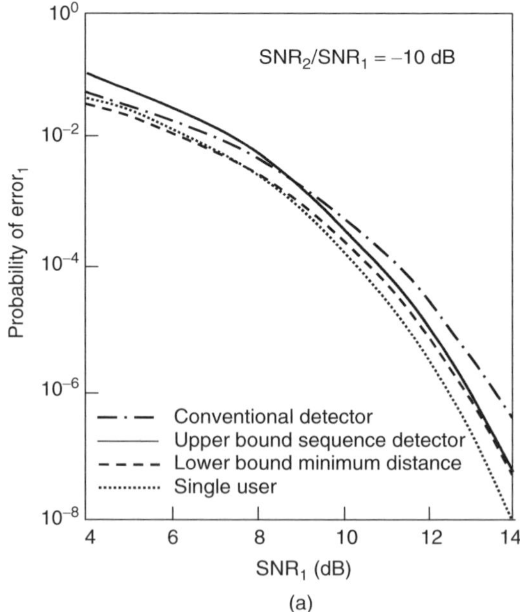

Figure 13.5 Bounds on minimum error probability of User 1. Worst-case delays and two active users: (a) $E_2/E_1 = -10\,\mathrm{dB}$, (b) $-5\,\mathrm{dB}$, (c) $0\,\mathrm{dB}$.

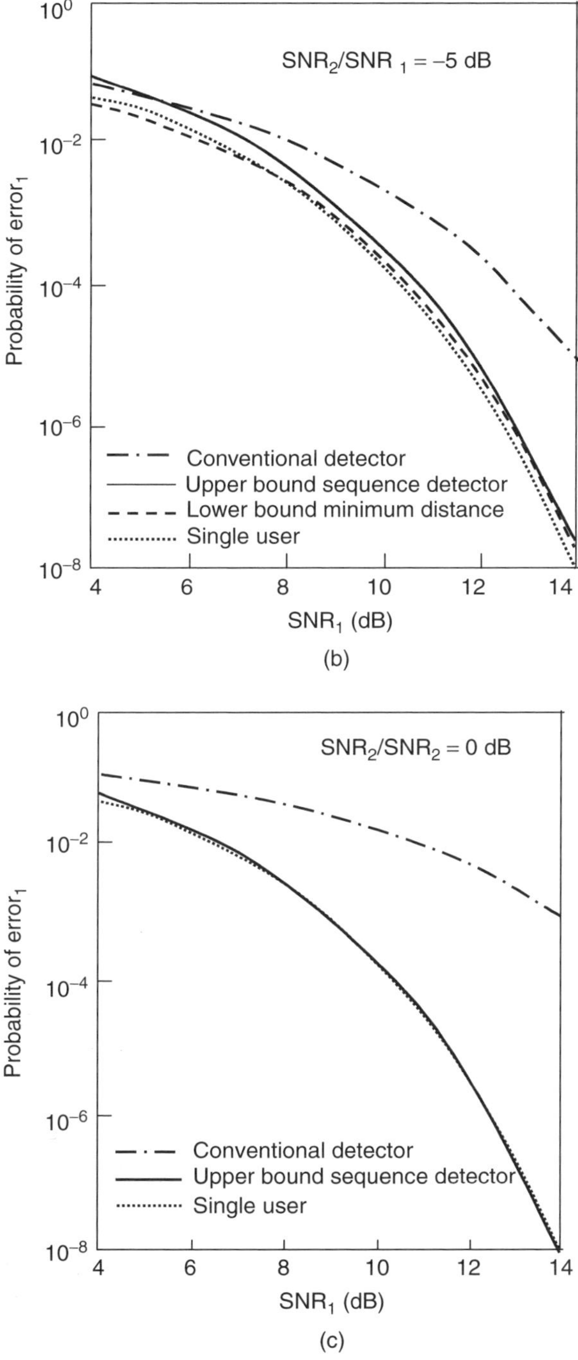

Figure 13.5 (*Continued*).

Figure 13.5(a) to 13.5(c) presents the same results for different near far ratio (NFR) defined as SNR_2/SNR_1. From these figures one can see that the impact of using optimal detector is more evident for larger NFR.

13.2 LINEAR MULTIUSER CDMA DETECTORS

13.2.1 Synchronous CDMA channels

If the signals from different users are received synchronously, equation (13.1) becomes

$$r(t) = \sum_{k=1}^{K} b_k(j) s_k(t - jT) + \sigma n(t)$$

$$t \in [jT, jT + T] \tag{13.7}$$

If we use notation y_k for the output of the matched filter of user k, equation (13.6) becomes

$$y_k = \int_0^T r(t) s_k(t)\, \mathrm{d}t, \quad k = 1, \ldots, K \tag{13.8}$$

and we can write

$$y_1 = \sum_j b_k R_{1j} + n,$$

$$y_2 = \sum_j b_k R_{2j} + n_2$$

$$\vdots$$

$$y_k = \sum_j b_k R_{kj} + n_k \tag{13.9}$$

The vector of these outputs can be presented as

$$y = Rb + n \tag{13.10}$$

where R is the nonnegative definite matrix of cross-correlations between the assigned waveforms:

$$R_{ij} = \int_0^T s_i(t) s_j(t)\, \mathrm{d}t \tag{13.11}$$

Conventional single-user detection can be represented as

$$\hat{b}_k^c = \mathrm{sgn}\, y_k \tag{13.12}$$

The optimum multiuser detector becomes

$$\hat{b} \in \arg\min_{b \in \{-1, 1\}^K} \int_0^T \left[r(t) - \sum_{k=1}^K b_k s_k(t) \right]^2 dt$$

$$= \arg\max_{b \in \{-1, 1\}^K} 2y^T b - b^T R b \tag{13.13}$$

13.2.2 The decorrelating detector

In the absence of noise, the matched filter output vector is $y = Rb$. The detector will perform the following operation $\hat{b} = \text{sgn } R^{-1}y$. Note that the noise components in $R^{-1}y$ are correlated, and therefore sgn $R^{-1}y$ does not result in optimum decisions. It is interesting to point out that this detector does not require knowledge of the energies of any of the active users. To see this, let $\tilde{y}_k = y_k/\sqrt{E_k}$, that is, $\tilde{y}_k$ is the result of correlating the received process with the normalized (unit-energy) signal of the kth user. Then, we have

$$\text{sgn } R^{-1}y = \text{sgn } E^{-1/2}\underline{R}^{-1}E^{-1/2}y$$

$$= \text{sgn } W^{-1/2}\underline{R}^{-1}\tilde{y}$$

$$= \text{sgn } \underline{R}^{-1}\tilde{y} \tag{13.14}$$

where $\underline{R}$ is the cross-correlation matrix of normalized signals and therefore, the same decisions are obtained by multiplying the vector of normalized matched filter outputs by the inverse of the normalized cross-correlation matrix. For an iterative solution of the problem, see Reference [3].

13.2.3 The optimum linear multiuser detector

Linear detector [4] that minimizes the probability of bit error will be referred to as optimum linear multiuser detector. Its operation can be represented as

$$\hat{b} = \text{sgn}(Ty) = \text{sgn}(TRb + Tn) \tag{13.15}$$

We will consider the set $I(R)$ of generalized inverses of the cross-correlation matrix R and analyze the properties of the detector

$$\hat{b} = \text{sgn } R^I y \tag{13.16}$$

in the next chapter. The special case $I(R) = R^{-1}$ is referred to as a decorrelating detector.

13.3 MULTISTAGE DETECTION IN ASYNCHRONOUS CDMA

If the indexing of users is arranged in increasing order of their delays, then the output of the correlator of user k can be represented as

$$z_k^{(i)}(0) = \int_{-\infty}^{\infty} r(t)s_k(t + iT - \tau_k)\,dt$$

$$= \eta_k^{(i)} + \sum_{l=k+1}^{K} R_{kl}(1)b_l^{(i-1)} + \sum_{l=1}^{K} R_{kl}(0)b_l^{(i)} + \sum_{l=1}^{k-1} R_{kl}(-1)b_l^{(i+1)} \quad (13.17)$$

$\eta_k^{(i)}$ is the component of the statistic due to the additive channel noise. In vector notation, letting $z^{(i)}(0) = \lfloor z_1^{(i)}(0), z_2^{(i)}(0), \ldots, z_K^{(k)}(0)\rfloor^{\mathrm{T}}$, we have

$$z^{(i)}(0) = \eta^{(i)} + R(1)b^{(i-1)} + R(0)b^{(i)} + R(-1)b^{(i+1)} \quad (13.18)$$

13.3.1 The multistage detector

The multistage detector [5] recreates the interfering term for each user on the basis of bit estimations in the previous stage (iteration), subtracts the estimated MAI and then makes the new estimate of data that can be represented as

$$\hat{b}_k^{(i)}(m + 1) = \mathrm{sgn}[z_k^{(i)}(m)] \quad (13.19)$$

where

$$z_k^{(i)}(m) = z_k^{(i)}(0) - \sum_{l=k+1}^{K} h_{kl}(1)\hat{b}_l^{(i-1)}(m) - \sum_{l \neq k} h_{kl}(0)\hat{b}_l^{(i)}(m)$$

$$- \sum_{l=1}^{k-1} h_{kl}(-1)\hat{b}_l^{(i+1)}(m) \quad (13.20)$$

The block diagram of multistage multiuser detector (MSMUD) is shown in Figure 13.6.

A detailed implementation of the kth M-stage processor where for each $m = 1, 2, \ldots, M - 1$, $\hat{I}_k^{(i-2m+1)}(m)$ denotes the estimate of the MAI reconstructed in the mth stage on the basis of bit estimates $\hat{b}_j^{(i-2m)}(m - 1), \hat{b}_j^{(i-2m+1)}(m - 1)$ and $\hat{b}_j^{(i-2m+2)}(m - 1)\forall j \neq k$ obtained from the other $K - 1$ processors is shown in Figure 13.7.

An example of probability of error curves is shown in Figure 13.8. All parameters are shown in the figure itself. One can see that even two-stage detector may significantly improve the system performance.

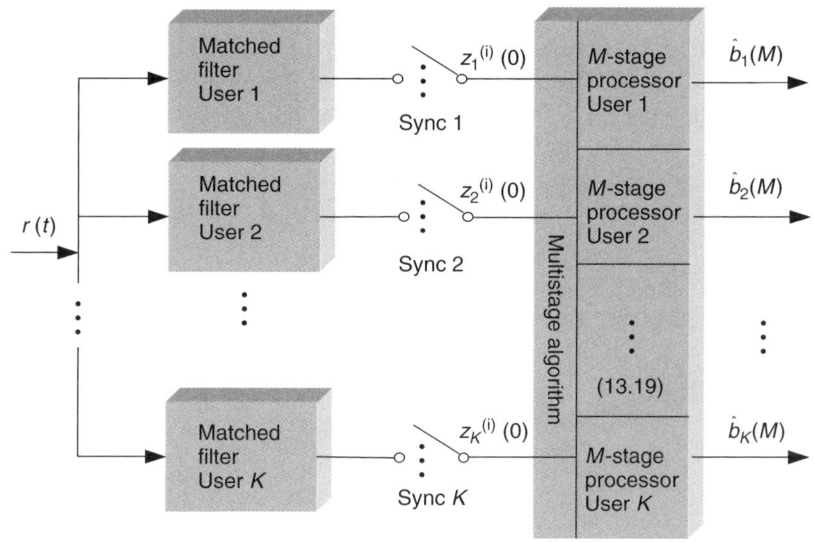

Figure 13.6 The multistage multiuser detector (MSMUD) for the BPSK-CDMA system.

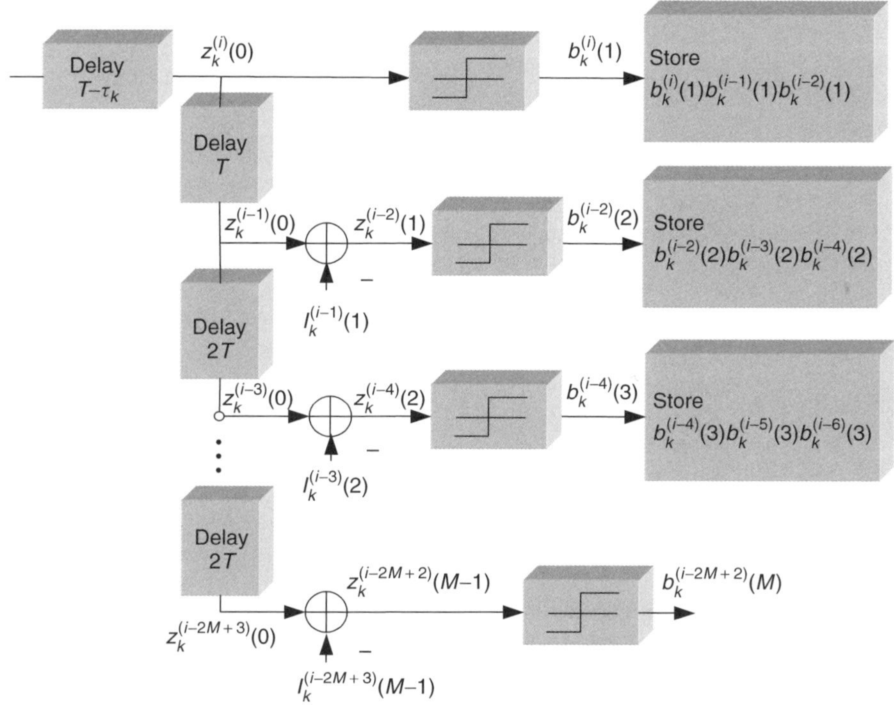

Figure 13.7 A detailed implementation of MSMUD.

In order to further emphasize the role of multiuser detection (MUD) in the presence of near far effect, Figure 13.9 presents BER for the case when the cross-correlation is very high $r_{12} = 1/3$. One can see that when the second user becomes stronger and stronger the improvement compared with a conventional detector is more significant.

This conclusion becomes more and more relevant if either r_{12} is increased, as in Figure 13.10, or SNR is increased, as in Figure 13.11.

Figure 13.12 demonstrates the same results for five users in the network.

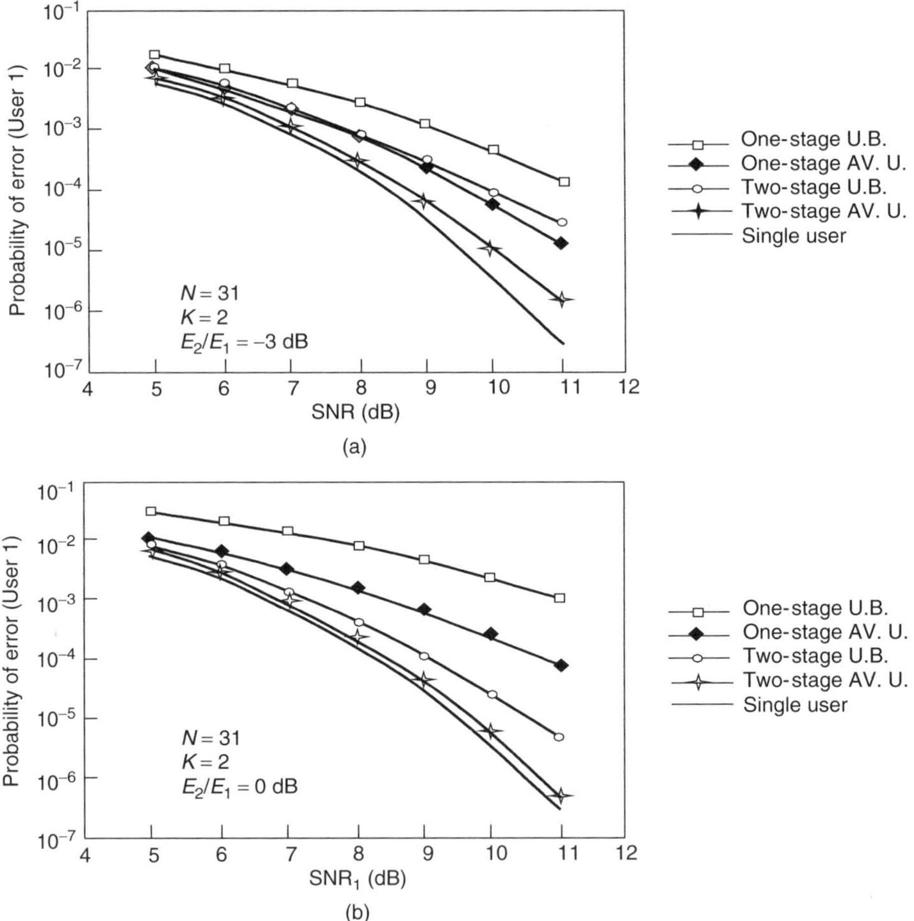

Figure 13.8 A comparison between the worst case and the upper bound of the average error probability of a two-user direct-sequence spread-spectrum system with $N = 31$, for the conventional receiver (CR) and the two-stage receiver and the single-user bit error probability: (a) $E_2/E_1 = -3$ dB, (b) $E_2/E_1 = 0$ dB, (c) $E_2/E_1 = 3$ dB [5]. Reproduced from Varanasi, M. and Aazhang, B. (1990) Multistage detection in asynchronous code division multiple access communications. *IEEE Trans. Commun.*, **38**, 509–519, by permission of IEEE.

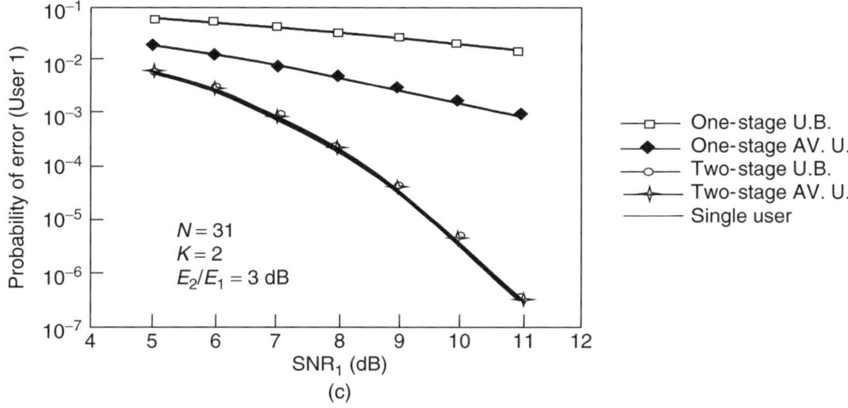

Figure 13.8 (*Continued*).

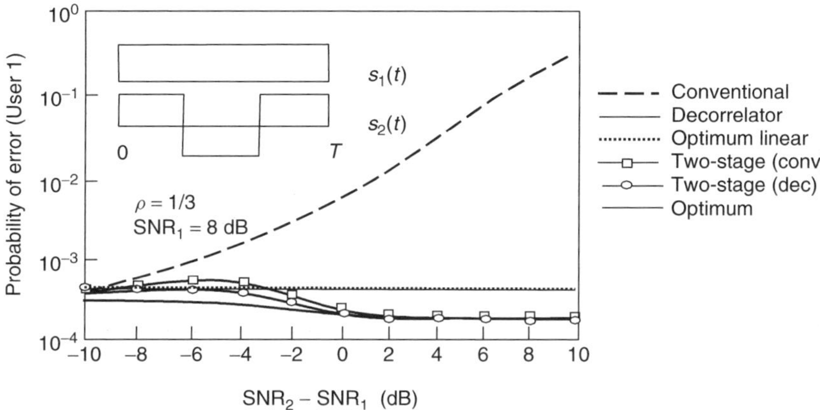

Figure 13.9 Error probability comparison of the linear, two-stage and optimum detectors for a two-user channel with $r_{12} = 1/3$ and SNR of User 1 fixed at 8 dB [6]. Reproduced from Varanasi, M. and Aazhang, B. (1991) Near optimum detection in synchronous code division multiple access systems. *IEEE Trans. Commun.*, **39**, 725–736, by permission of IEEE.

13.4 NONCOHERENT DETECTOR

13.4.1 Conventional noncoherent single-user detector – DPSK

A conventional detector for differential phase keying signals is defined by the following equation

$$\hat{b}_m = \text{sgn}[\text{Re}\{\overline{z_m(-1)}z_m(0)\}]$$

$$z_m(i) = \frac{1}{2}\int_{iT}^{(i+1)T} r(t)\overline{f_m(t-iT)}\,\mathrm{d}t \tag{13.21}$$

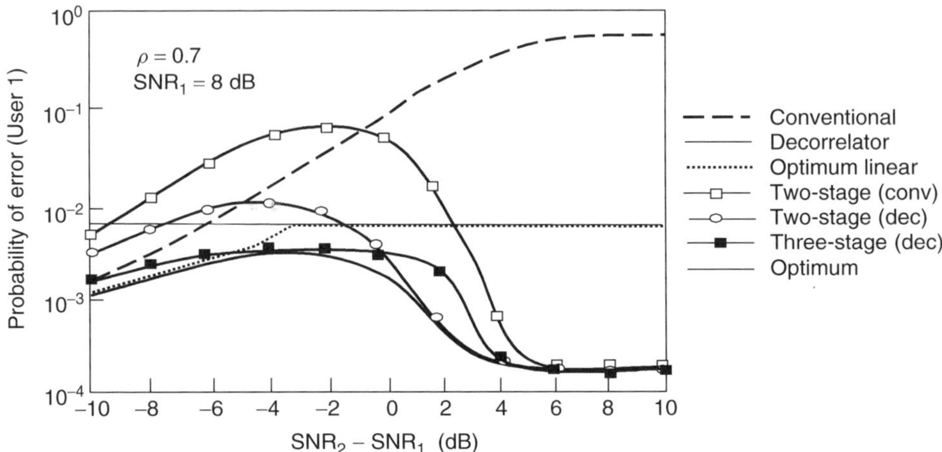

Figure 13.10 Error probability comparison of the linear, three-stage and optimum detectors for a two-user channel with $r_{12} = 0.7$ and SNR of User 1 fixed at 8 dB [6]. Reproduced from Varanasi, M. and Aazhang, B. (1991) Near optimum detection in synchronous code division multiple access systems. *IEEE Trans. Commun.*, **39**, 725–736, by permission of IEEE.

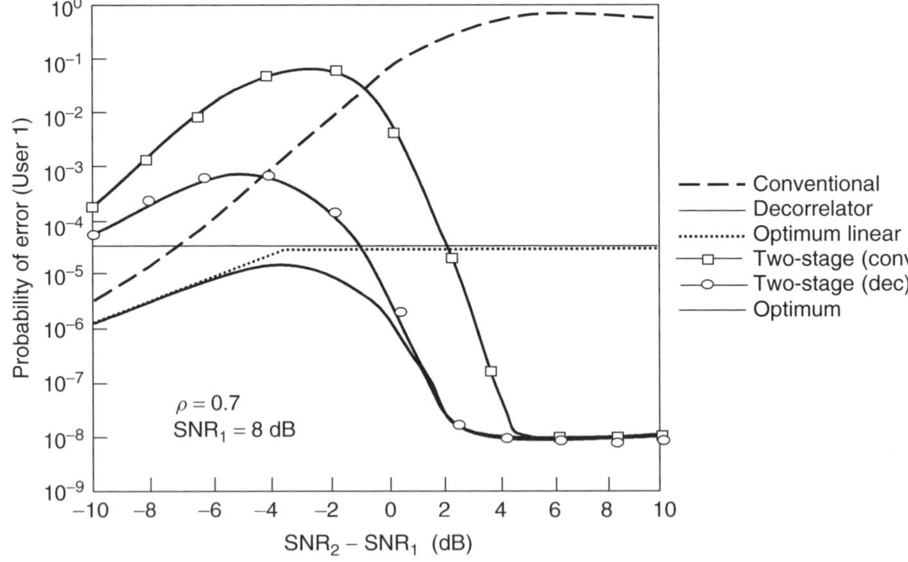

Figure 13.11 Error probability comparison of the linear, two-stage and optimum detectors for a two-user channel with $r_{12} = 0.7$ and SNR of User 1 fixed at 12 dB.

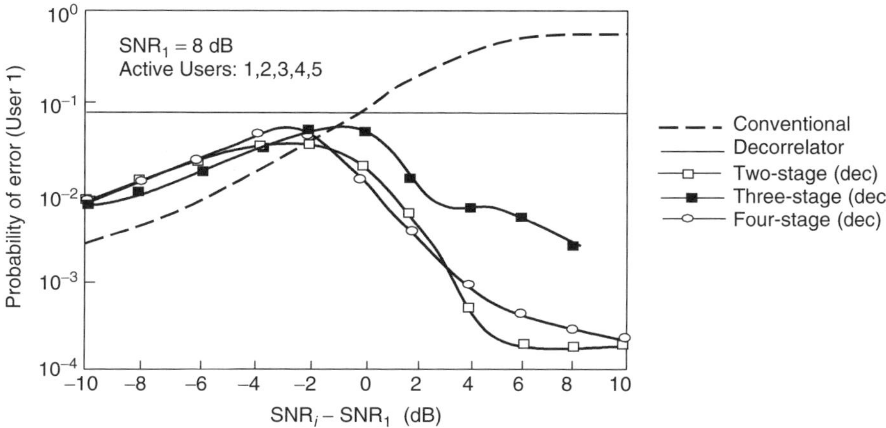

Figure 13.12   Probability of error for five users in the network [6]. Reproduced from Varanasi, M. and Aazhang, B. (1991) Near optimum detection in synchronous code division multiple access systems. *IEEE Trans. Commun.*, **39**, 725–736, by permission of IEEE.

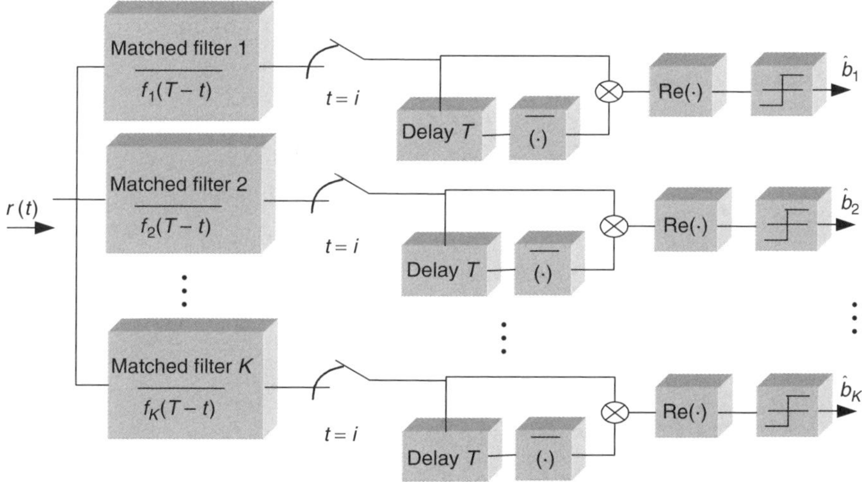

Figure 13.13   Conventional detection: a bank of K single-user DPSK detectors.

where $f_m(t)$ is the signal matched filter function. In the trivial case it is the signal spreading code only. The block diagram is shown in Figure 13.13.

13.4.2 Noncoherent linear multiuser detectors – DPSK

In general, a noncoherent linear multiuser detector for the mth user, denoted by a nonzero transformation $\boldsymbol{h}^{(m)} \in C^K$, is defined by the decision

$$\hat{b}_m = \mathrm{sgn}\left[\mathrm{Re}\left\{\overline{\sum_{k=1}^{K} \overline{h}_k^{(m)} z_k(-1)} \sum_{l=1}^{K} \overline{h}_l^{(m)} z_l(0)\right\}\right] \qquad (13.22)$$

where K is the length of the code.

13.4.3 Decorrelating detectors

A noncoherent decorrelating detector for user m is defined by the decision with the linear transformation $\boldsymbol{h} = \boldsymbol{d}$ where $\boldsymbol{d}$ denotes the complex conjugate of the mth column of a generalized inverse $\boldsymbol{R}^I$ of $\boldsymbol{R}$. If the mth user is linearly independent, it can be shown that $\boldsymbol{R}\overline{\boldsymbol{d}} = \boldsymbol{u}_m$ is the mth unit vector. If all the signature signals are linearly independent, $\boldsymbol{R}^{-1}$ exists and the decorrelating transformation $\boldsymbol{d}$ is uniquely characterized as the complex conjugate of the mth column of the inverse of $\boldsymbol{R}$. The receiver block diagram is shown in Figure 13.14.

For illustration purposes, four users, using Gold sequences from Figure 13.15(a), are considered. Performance results with MU detector are shown in Figure 13.15(b) [7].

13.4.4 Noncoherent detection in asynchronous multiuser channel

The z-transform of equation (13.18) gives

$$\boldsymbol{Z}(z) = \boldsymbol{S}(z) \cdot \hat{\boldsymbol{D}}(z) + \boldsymbol{N}(z) \qquad (13.23)$$

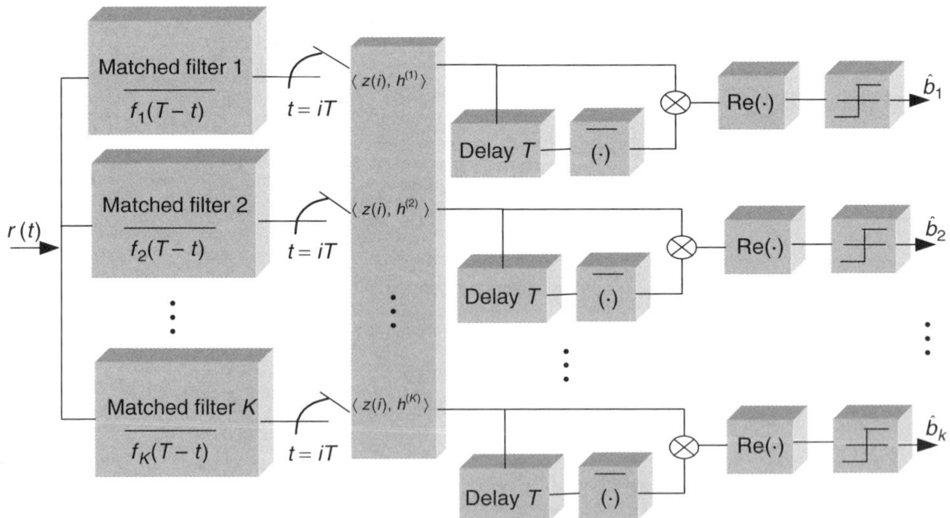

Figure 13.14 Linear multiuser DPSK detector.

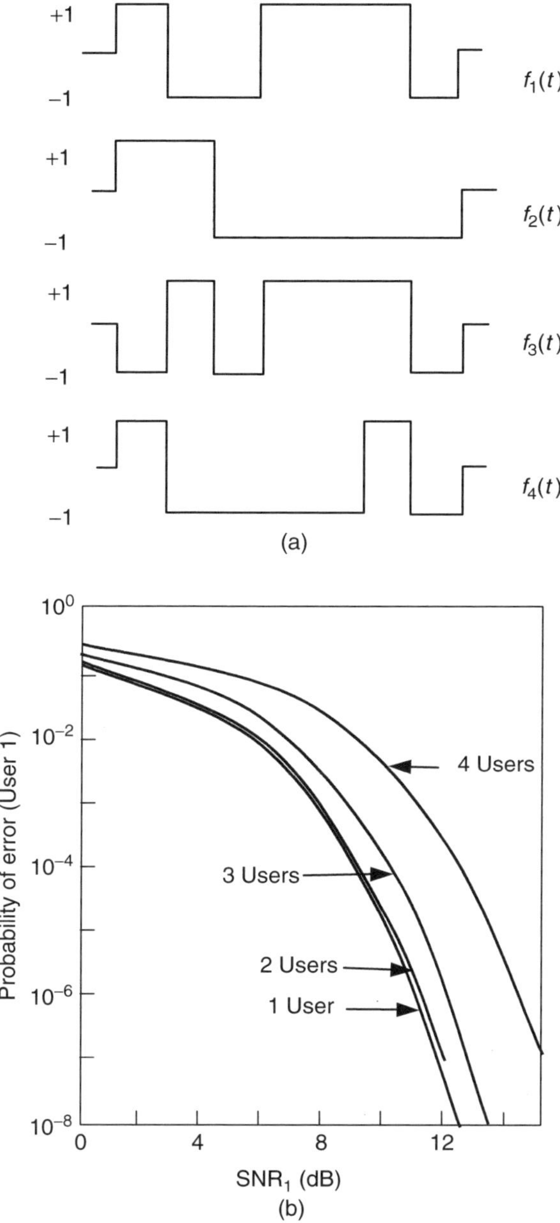

Figure 13.15 (a) Direct-sequence signature signals derived from Gold sequences of length 7 assigned to the four users of a four-user DS-SSMA system. (b) Bit-error rate of first user as a function of the first user's signal-to-noise ratio. These error rates are independent of interfering signal energies and phases.

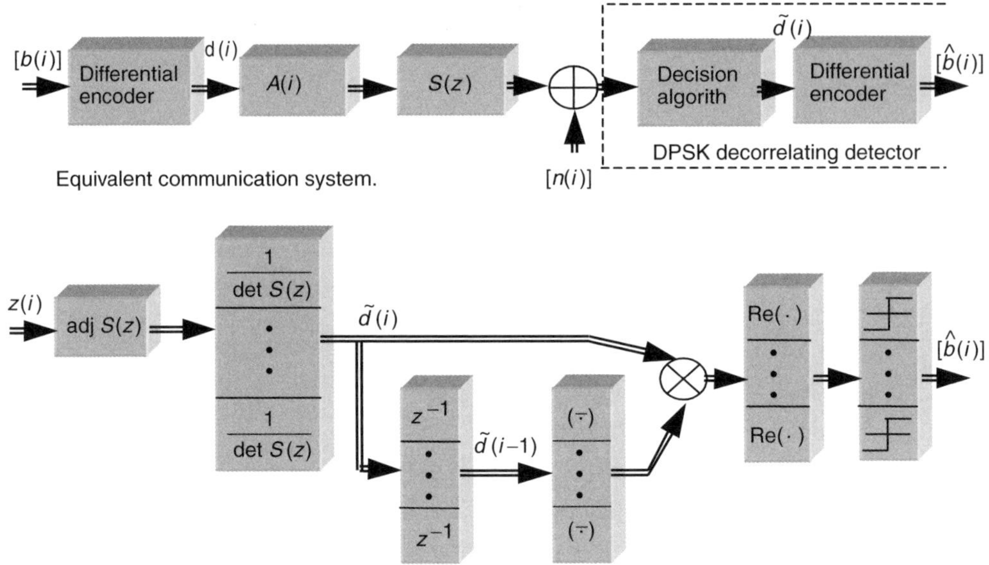

Figure 13.16   Noncoherent decorrelating detector [8].

where

$$S(z) = R(-1)z + R(0) + R(1)z^{-1} \tag{13.24}$$

and $Z(z)$, $\hat{D}(z)$ and $N(z)$ are the vector-valued z-transforms of the matched-filter output sequence, the sequence $\{\hat{d}(l) = A(l)d(l)\}$ and the noise sequence $\{n(l)\}$ at the output of the matched filters. If we define

$$G(z) = [S(z)]^{-1} = \frac{\text{adj } S(z)}{\text{det } S(z)} \tag{13.25}$$

then we have

$$\hat{\mathbf{d}}(z) = G(z)Z(z) \tag{13.26}$$

and

$$\hat{b}(i) = \text{sgn } \text{Re}[\tilde{\mathbf{d}}(i-1) \otimes \tilde{\mathbf{d}}^*(i)] \tag{13.27}$$

The system block diagram is shown in Figure 13.16.

13.5 MULTIUSER DETECTION IN FREQUENCY NONSELECTIVE RAYLEIGH FADING CHANNEL

Topics covered in the previous chapter are now repeated for the fading channel. Previously described algorithms are extended to the fading channel by using as much analogy as

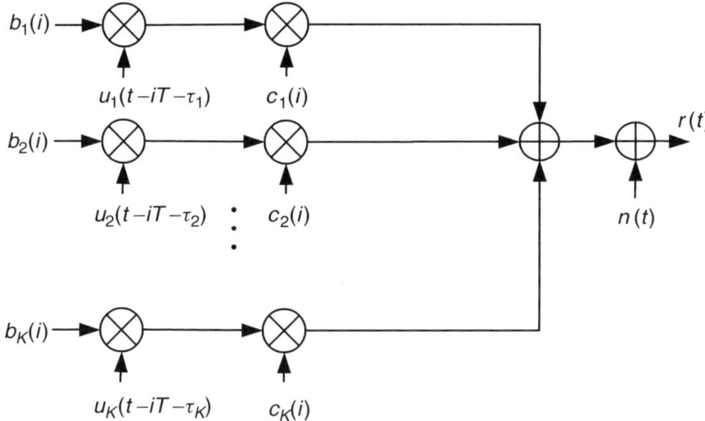

Figure 13.17 Asynchronous CDMA flat Rayleigh fading channel model.

possible in the process of deriving the system transfer functions. In frequency-selective channels, decorrelators are combined with the RAKE type receiver in order to further improve the system performance. A number of simulation results are presented in order to illustrate the effectiveness of these schemes. The concept of this chapter is based on proper understanding of the channel model, which is covered in Chapter 8. The overall system model, including the channel model for frequency-nonselective fading, is shown in Figure 13.17.

Parameters $c_k(i)$ are, for fixed i, independent, zero-mean, complex-valued Gaussian random variables, with variances $\overline{|c_k|^2}$ with independent quadrature components. The time-varying nature of the channel is described via the spaced-time correlation function of the kth channel $\Phi_k(\Delta t)$

$$E\{c_k^*(i)c_k(j)\} = \Phi_k[(j-i)T] \tag{13.28}$$

The received signal at the central receiver can be expressed as

$$r(t) = S(t, \mathbf{b}) + n(t)$$

$$S(t, \mathbf{b}) = \sum_{i=-M}^{M} \sum_{k=1}^{K} b_k(i)c_k(i)u_k(t - iT - \tau_k)$$

$$u_k(t) = \sqrt{E_k}s_k(t)e^{j\phi_k} \tag{13.29}$$

where $u_k(t)$ is referred to as user k signature sequence including signal amplitude (square root of signal energy), code itself and signal phase. By using proper notation, $r(t)$ can be represented as

$$r(t) = \boldsymbol{b}^{\mathrm{T}}\boldsymbol{C}\boldsymbol{u}_t + n(t) \tag{13.30}$$

where

$$b^{\mathrm{T}}[b_1(-M)b_2(-M)\cdots b_K(-M)\cdots b_1(M)b_2(M)\cdots b_K(M)]$$

$$u_t = [u^{\mathrm{T}}(t+MT)\cdots u^{\mathrm{T}}(t-MT)]^{\mathrm{T}}$$

$$u(t) = [u_1(t-\tau_1)\cdots u_K(t-\tau_K)]^{\mathrm{T}}$$

$$C = \mathrm{diag}\ [C(-M)\cdots C(M)]$$

$$C(i) = \mathrm{diag}\ [c_1(i)\cdots c_K(i)] \tag{13.31}$$

13.5.1 Multiuser maximum likelihood sequence detection

By using analogy from the previous section, the likelihood function in this case can be represented as

$$L(b) = 2\,\mathrm{Re}\{b^H y\} - b^H C^H R_u C b \tag{13.32}$$

Upper index ( ) ^{H} denotes conjugate transpose and

$$y = \int_{-\infty}^{+\infty} r(t)C^H u_t^* \,\mathrm{d}t \tag{13.33}$$

represents vector of matched filters outputs. The correlation matrix R_u can be represented as

$$R_u = \int_{-\infty}^{+\infty} u_t^* u_t^{\mathrm{T}} \,\mathrm{d}t = \begin{pmatrix} R_u(0) & R_u(-1) & 0 & \cdots \\ R_u(1) & R_u(0) & R_u(-1) & \cdots \\ & \ddots & \ddots & \\ \cdots & R_u(1) & R_u(0) & R_u(-1) \\ \cdots & 0 & R_u(1) & R_u(0) \end{pmatrix} \tag{13.34}$$

with block elements of dimension $K \times K$

$$\mathbf{R}_u(i-j) = \int_{-\infty}^{+\infty} u^*(t-iT)u^{\mathrm{T}}(t-jT)\,\mathrm{d}t \tag{13.35}$$

and scalar elements

$$[\mathbf{R}_u(i-j)]_{mn} = \int_{-\infty}^{+\infty} u_m^*(t-iT-\tau_m)u_n(t-jT-\tau_n)\,\mathrm{d}t \tag{13.36}$$

13.5.2 Decorrelating detector

If we slightly modify the vector notation, equation (13.30) becomes

$$r(t) = \sum_{i=-M}^{M} s^{\mathrm{T}}(t-iT)E\,\Phi C(i)\mathbf{b}(i) + n(t) \tag{13.37}$$

with normalized signature waveform vector

$$s(t) = [s_1(t - \tau_1)s_2(t - \tau_2) \cdots s_K(t - \tau_K)]^{\mathrm{T}} \tag{13.38}$$

$K \times K$ multichannel matrix

$$C(i) = \mathrm{diag}[c_1(i)c_2(i) \cdots c_K(i)]$$
$$E = \mathrm{diag}(\sqrt{E_1}\sqrt{E_2} \cdots \sqrt{E_K}) \tag{13.39}$$

and matrix of carrier phases

$$\Phi = \mathrm{diag}\ (e^{j\phi_1}e^{j\phi_2} \cdots e^{j\phi_K}) \tag{13.40}$$

$K \times K$ cross-correlation matrices of normalized signature waveforms becomes

$$\mathbf{R}(\ell) = \int_{-\infty}^{+\infty} \mathbf{s}^*(t)\mathbf{s}^{\mathrm{T}}(t + \ell T)\,\mathrm{d}t \tag{13.41}$$

The asynchronous nature of the channel is evident from the matrix elements

$$R_{mn}(\ell) = \int_{\ell T + \tau_m}^{(\ell+1)T + \tau_m} s_m^*(t - \tau_m)s_n(t + \ell T - \tau_n)\,\mathrm{d}t \tag{13.42}$$

Since there is no intersymbol interference (ISI), $\mathbf{R}(\ell) = 0$, $\forall |\ell| > 1$ and $\mathbf{R}(-1) = \mathbf{R}^H(1)$. Because of the ordering of the user, $\mathbf{R}^H(1)$ is an upper triangular matrix with zero elements on the diagonal. The decorrelating detector front end consists of K filters matched to the normalized signature waveforms of the users. The output of this filter bank, sampled at the ℓth bit epoch is

$$\mathbf{y}(\ell) = \int_{-\infty}^{+\infty} r(t)s(t - \ell T)\,\mathrm{d}t \tag{13.43}$$

The vector of sufficient statistics can also be represented as

$$\mathbf{y}(\ell) = \mathbf{R}(-1)E\,\Phi\mathbf{C}(\ell + 1)\mathbf{b}(\ell + 1) + \mathbf{R}(0)E\,\Phi\mathbf{C}(\ell)\mathbf{b}(\ell)$$
$$+ \mathbf{R}(1)E\,\Phi\mathbf{C}(\ell - 1)\mathbf{b}(\ell - 1) + \mathbf{n}_y(\ell) \tag{13.44}$$

The covariance matrix of the matched filter output noise vector sequence, $\{\mathbf{n}_y(\ell)\}$, is given by

$$E\{\mathbf{n}_y^*(i)\mathbf{n}_y^{\mathrm{T}}(j)\} = \sigma^2\mathbf{R}^*(i - j) \tag{13.45}$$

As in equation (13.25), the decorrelator is a K-input K-output linear time-invariant (LTI) filter with transfer function matrix

$$\mathbf{G}(z) = [\mathbf{R}(-1)z + \mathbf{R}(0) + \mathbf{R}(1)z^{-1}]^{-1} \overset{\Delta}{=} \mathbf{S}^{-1}(z) \tag{13.46}$$

The z-transform of the decorrelator output vector is

$$\mathbf{P}(z) = E\,\Phi(\mathbf{Cb})(z) + N_{\mathrm{p}}(z) \tag{13.47}$$

$N_{\mathrm{p}}(z)$ is the z-transform of the output noise vector sequence having power spectral density

$$\sigma^2 \mathbf{S}^{-1}(z) = \sigma^2 \sum_{m=-\infty}^{\infty} \mathbf{D}(m)z^{-m} \tag{13.48}$$

The receiver block diagram is shown in Figure 13.18 for coherent reception and in Figure 13.19 for differential modulation.

Performance results for the two detectors are shown in Figures 13.20 and 13.21. Significant improvement in BER is evident in both figures.

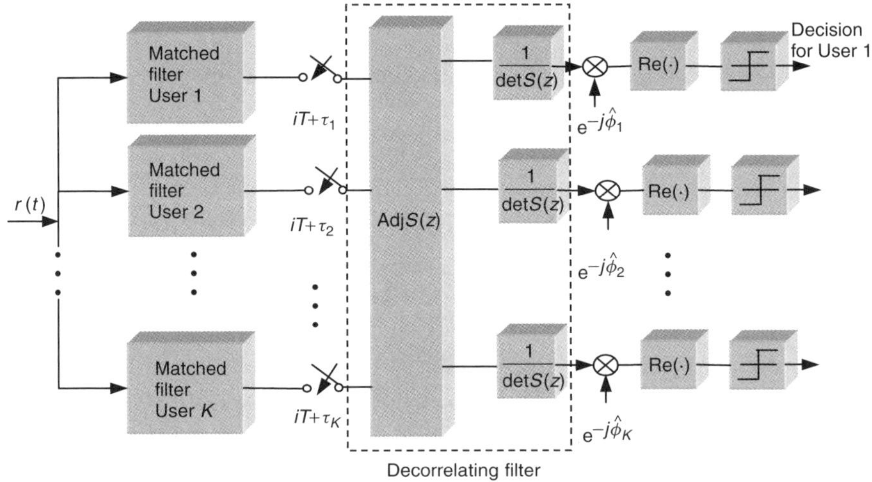

Figure 13.18　Coherent decorrelating multiuser detector.

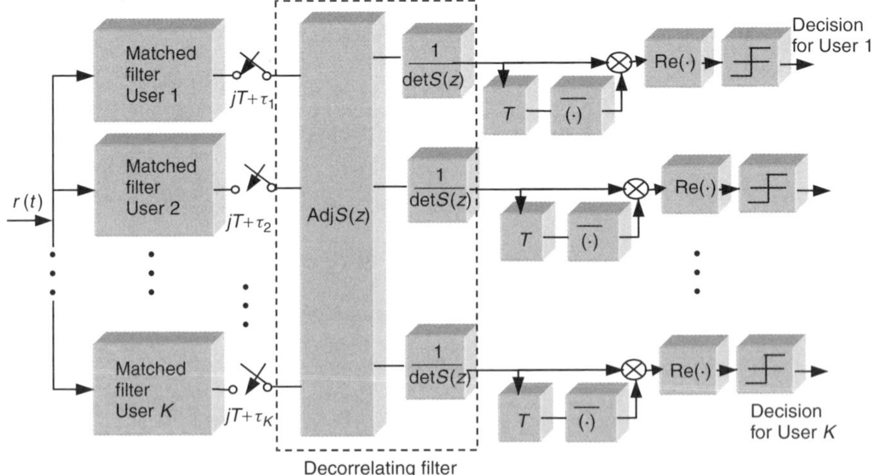

Figure 13.19　Differentially coherent decorrelating multiuser detector.

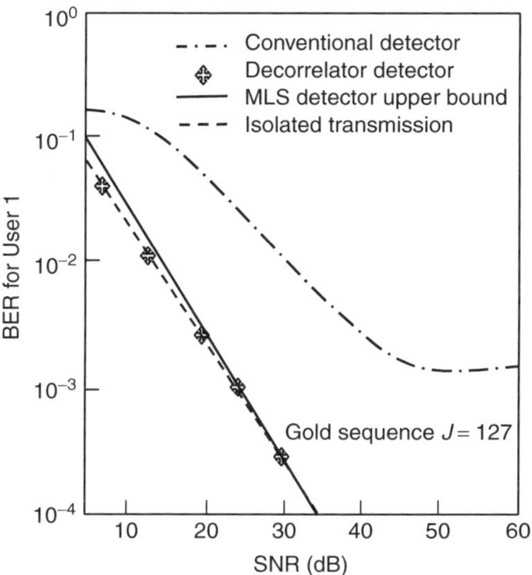

Figure 13.20 Bit error rate of User 1 for the two-user case with Rayleigh-faded paths (same average path strength) and Gold sequences of period $J = 127$ [9]. Reproduced from Zvonar, Z. (1993) *Multiuser Detection for Rayleigh Fading Channel*. Ph.D. Thesis, Department of Electrical and Computer Engineering, Northeastern University, Boston, MA, by permission of IEEE.

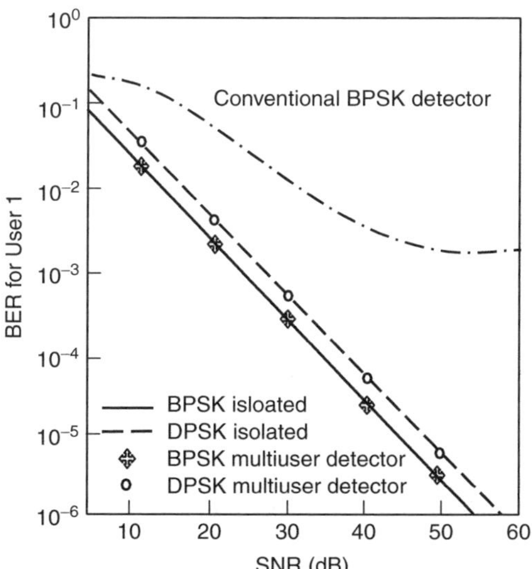

Figure 13.21 Bit error rate of User 1 for two active users with Rayleigh-faded paths (same average path strength) and Gold sequences of period $J = 127$ [9]. Reproduced from Zvonar, Z. (1993) *Multiuser Detection for Rayleigh Fading Channel*. Ph.D. Thesis, Department of Electrical and Computer Engineering, Northeastern University, Boston, MA, by permission of IEEE.

13.6 MULTIUSER DETECTION IN FREQUENCY-SELECTIVE RAYLEIGH FADING CHANNEL

By using an analogy with equation (13.29), the received signal in this case can be represented as

$$r(t) = S(t, \mathbf{b}) + n(t)$$

$$S(t, \mathbf{b}) = \sum_{i=-M}^{M} \sum_{k=1}^{K} b_k(i) h_k(t - iT - \tau_k)$$

$$h_k(t) = c_k(t)^* u_k(t) \tag{13.49}$$

In equation (13.49), $h_k(t)$ is the equivalent received symbol waveform of finite duration $[0, T_k]$ [convolution of equivalent low-pass signature waveform $u_k(t)$ and the channel impulse response $c_k(t)$]. We define the memory of this channel as v, the smallest integer such that $h_k(t) = 0$ for $t > (v + 1)$T, and all $k = 1, \ldots, K$. The impulse response of the kth user channel is given by

$$c_k(t) = \sum_{\ell=0}^{L-1} c_{k,l}(t)\delta(t - \tau_{k,\ell}) \tag{13.50}$$

When the signaling interval T is much smaller than the coherence time of the channel, the channel is characterized as slow fading, implying that the channel characteristics can be measured accurately. Since the channel is assumed to be Rayleigh fading, the coefficients $c_{k,\ell}(t)$ are modeled as independent zero-mean complex-valued Gaussian random processes. In the sequel we use the following notation

$$h_k(t) = \sum_{\ell=0}^{L-1} c_{k,l}(t) u_k(t - \tau_{k,l}) = \mathbf{c}_k^T(t) \mathbf{u}_k(t) \tag{13.51}$$

For the single-user vector of channel coefficients we use

$$\mathbf{c}_k(t) = [c_{k,0}(t) c_{k,1}(t), \ldots, c_{k,L-1}(t)]^T \tag{13.52}$$

and for the signal vector of the delayed signature waveform we use

$$\mathbf{u}_k(t) = [u_k(t - \tau_{k,0}) u_k(t - \tau_{k,1}) \cdots u_k(t - \tau_{k,L-1})]^T \tag{13.53}$$

The equivalent low-pass signature waveform is represented as

$$u_k(t) = \sqrt{E_k} s_k(t) e^{j\phi_k} \tag{13.54}$$

where E_k is the energy, $s_k(t)$ is the real-valued, unit-energy signature waveform with period T and ϕ_k is the carrier phase. In this case the received signal given by equation (13.49) becomes

$$r(t) = S(t, \mathbf{b}) + n(t) = \mathbf{b}^\mathrm{T}\mathbf{h}_t + n(t) \tag{13.55}$$

The equivalent data sequence is as in equation (13.31)

$$\mathbf{b} = [b_1(-M) \cdots b_K(-M) \cdots b_1(M) \cdots b_K(M)]^\mathrm{T} \tag{13.56}$$

The equivalent waveform vector of NK elements is

$$\mathbf{h}_t = [\mathbf{h}^\mathrm{T}(t + MT) \cdots \mathbf{h}^\mathrm{T}(t - MT)]^\mathrm{T} \tag{13.57}$$

with

$$\mathbf{h}(t) = [h_1(t - \tau_1) \cdots h_K(t - \tau_K)]^\mathrm{T} = \mathbf{C}^\mathrm{T}(t)\mathbf{u}(t) \tag{13.58}$$

where

$$\mathbf{C}(t) = \begin{pmatrix} \mathbf{c}_1(t) & 0 & 0 & \cdots \\ & \mathbf{c}_2(t) & 0 & \cdots \\ & & \ddots & \\ \cdots & 0 & 0 & \mathbf{c}_K(t) \end{pmatrix} \tag{13.59}$$

is a $KL \times K$ multichannel matrix. KL is the total number of fading paths for all K users and

$$\mathbf{u}(t) = [\mathbf{u}_1(t - \tau_1) \cdots \mathbf{u}_K(t - \tau_K)]^\mathrm{T} \tag{13.60}$$

is the equivalent signature vector of KL elements.

13.6.1 Multiuser maximum likelihood sequence detection

Log likelihood function in this case becomes

$$L(\mathbf{b}) = 2\,\mathrm{Re}\{\mathbf{b}^H y\} - \mathbf{b}^H H \mathbf{b} \tag{13.61}$$

where superscript 'H' denotes conjugate transpose.

$$y = \int_{-\infty}^{+\infty} r(t)\mathbf{h}_t^* \, \mathrm{d}t \tag{13.62}$$

is the output of the bank of matched filters sampled at the bit epoch of the users. Matrix H is an $N \times N$ block-Toeplitz cross-correlation waveform matrix with $K \times K$ block elements.

$$H(i - j) = \int_{-\infty}^{+\infty} \mathbf{h}^*(t - iT)\mathbf{h}^\mathrm{T}(t - jT) \, \mathrm{d}t \tag{13.63}$$

13.6.2 Viterbi algorithm

Since every waveform $h_k(t)$ is time-limited to $[0, T_k]$, $T_k < (v+1)T$, it follows that $\mathbf{H}(l) = 0$, $\forall |l| > v+1$ and $\mathbf{H}(j) = \mathbf{H}^H(j)$ for $j = 1, \ldots, v+1$.

Because of the ordering of the users, $\mathbf{H}^H(v+1)$ is an upper triangular matrix with zero elements on the diagonal. Provided that knowledge of a channel is available, the MLS detector may be implemented as a dynamic programming algorithm of the Viterbi type. The vector Viterbi algorithm is the modification of the one introduced for M-input M-output linear channels where the dimensionality of the state space is $2^{(v+1)K}$. As in the case of the additive white Gaussian noise (AWGN) channel, a more efficient decomposition of the likelihood function results in an algorithm with a state space of dimension $2^{(v+1)K-1}$.

Frequency-selective fading is described by the wide-sense stationary uncorrelated-scattering model. The bandwidth of each signature waveform is much larger than the coherence bandwidth of the channel, $B_{\mathrm{w}} \gg (\Delta f)_{\mathrm{c}}$. The time-varying frequency-selective channel for each user can be represented as a tapped delay line with tap spacing $1/B_{\mathrm{w}}$, so that equation (13.51) becomes

$$h_k(t) = \sum_{i=0}^{L-1} c_{k,i}(t) u_k\left(t - \frac{i}{B_{\mathrm{w}}}\right) = s_k^T(t) E_k \Phi_k \mathbf{c}_k(t) \tag{13.64}$$

Signature waveform vector may be described as

$$s_k(t) = \left[s_k(t) s_k\left(t - \frac{i}{B_{\mathrm{w}}}\right) \cdots s_k\left(t - \frac{L-i}{B_{\mathrm{w}}}\right) \right]^{\mathrm{T}} \tag{13.65}$$

and

$$E_k = \sqrt{E_k}\mathbf{I}_L$$
$$\Phi_k = \mathrm{e}^{j\phi_k}\mathbf{I}_L \tag{13.66}$$

For a data symbol duration much longer than the multipath delay spread, $T \gg T_{\mathrm{m}}$, any ISI due to channel dispersion can be neglected. On the basis of the above discussion, the channel model is presented in Figure 13.22.

So, the received signal from Figure 13.22 can be represented as

$$r(t) = \sum_{i=-M}^{M} \sum_{k=1}^{K} b_k(i) h_k(t - iT - \tau_k) + n(t) \tag{13.67}$$

In addition, if we use notation

$$b(i) = [b_1(i) b_2(i) \cdots b_K(i)]^{\mathrm{T}}, \quad i = -M, \ldots, M$$
$$s(t) = [s_1^{\mathrm{T}}(t - \tau_1) s_2^{\mathrm{T}}(t - \tau_2) \cdots s_K^{\mathrm{T}}(t - \tau_K)]^{\mathrm{T}}$$
$$E = \mathrm{diag}(E_1, E_2, \ldots, E_K)$$
$$\Phi = \mathrm{diag}(\Phi_1, \Phi_2, \ldots, \Phi_K)$$
$$h^{\mathrm{T}}(t) = [h_1(t - \tau_1) \cdots h_K(t - \tau_K)] = s^{\mathrm{T}}(t) E \Phi C(t) \tag{13.68}$$

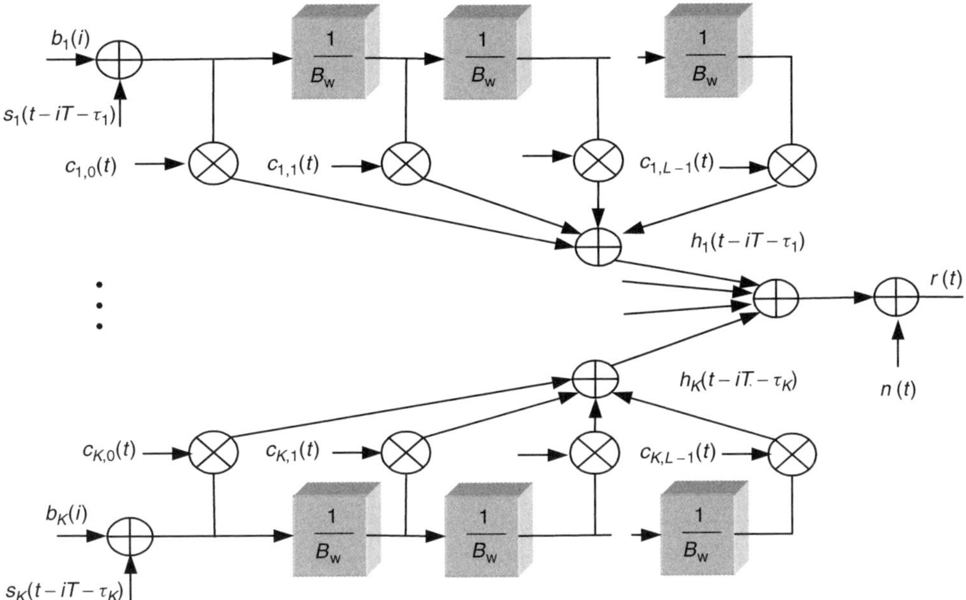

Figure 13.22 A synchronous CDMA frequency-selective Rayleigh fading channel model.

Equation (13.67) becomes

$$r(t) = \sum_{i=-M}^{M} \boldsymbol{h}^{\mathrm{T}}(t - iT)\boldsymbol{b}(i) + n(t) = \sum_{i=-M}^{M} \boldsymbol{s}^{\mathrm{T}}(t)\mathbf{W}\boldsymbol{\Phi}\mathbf{C}(t)\mathbf{b}(i) + n(t) \tag{13.69}$$

We define $KL \times KL$ cross-correlation matrices of normalized signature waveforms,

$$\mathbf{R}(l) = \int_{-\infty}^{+\infty} \boldsymbol{s}(t)\boldsymbol{s}^{\mathrm{T}}(t + lT)\,\mathrm{d}t \tag{13.70}$$

The asynchronous mode is evident from the structure of $L \times L$ cross-correlation matrix between the users m and n,

$$\mathbf{R}_{mn}(l) = \int_{lT+\tau_m}^{(l+1)T+\tau_m} \boldsymbol{s}_m(t - \tau_m)\boldsymbol{s}_n^{\mathrm{T}}(t + lT - \tau_n)\,\mathrm{d}t \tag{13.71}$$

Since there is no ISI, $\mathbf{R}(l) = 0$, $\forall |l| > 1$ and $\mathbf{R}(-1) = \mathbf{R}^H (1)$. Because of the ordering of the users, $\mathbf{R}^H (1)$ is the upper triangular matrix with zero elements on the diagonal.

The front end of the multiuser detector consists of KL filters matched to the normalized properly delayed signature waveforms of the users as shown in Figure 13.23. The output of this filter bank sampled at the bit epochs is given by the vector:

$$\boldsymbol{y}(l) = \int_{-\infty}^{+\infty} r(t)\boldsymbol{s}(t - lT)\,\mathrm{d}t \tag{13.72}$$

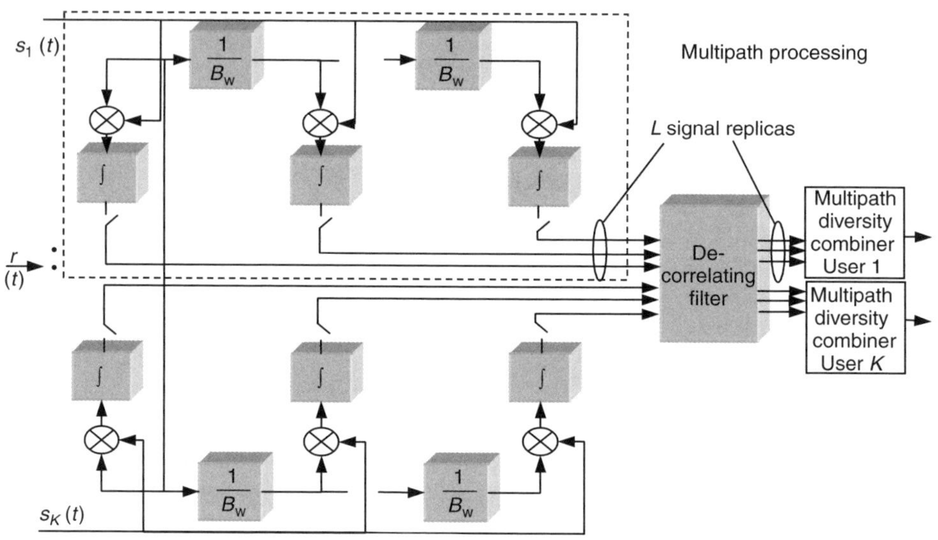

Figure 13.23 Multipath decorrelation.

The vector of sufficient statistics can be also expressed as

$$y(l) = \mathbf{R}(-1)E\,\mathbf{\Phi C}(l+1)\mathbf{b}(l+1) + \mathbf{R}(0)E\,\mathbf{\Phi C}(l)\mathbf{b}(l)$$
$$+ \mathbf{R}(1)E\,\mathbf{\Phi C}(l-1)\mathbf{b}(l-1) + n(l) \tag{13.73}$$

The covariance matrix of the matched filter output noise vector is given by $E[\mathbf{n}^*(i)\mathbf{n}^T(j)] = \sigma^2\mathbf{R}(i-j)$.

Taking the z-transform gives

$$\mathbf{Y}(z) = \mathbf{S}(z)(\mathbf{E\Phi Cb})(z) + \mathbf{N}(z) \tag{13.74}$$

where $(\mathbf{E\Phi Cb})(z)$ is the transform of sequence

$$\left\{\left[\sqrt{E_1}e^{j\phi_1}c_{1,0}(i)b_1(i)\cdots\sqrt{E_1}e^{j\phi_1}c_{1,L-1}(i)b_1(i)\cdots\sqrt{E_K}e^{j\phi_K}c_{K,L-1}(i)b_K(i)\right]^{\mathrm{T}}\right\} \tag{13.75}$$

$\mathbf{S}(z)$ is the equivalent transfer function of the Code Division Multiple Access (CDMA) multipath channel that depends only on the signature waveforms of the users. The multipath decorrelating (MD) filter is a KL-input KL-output LTI filter with transfer function matrix

$$\mathbf{G}(z) \triangleq [\mathbf{S}(z)]^{-1} = \frac{\mathrm{adj}\,\mathbf{S}(z)}{\det\mathbf{S}(z)} = [\mathbf{R}(-1)z + \mathbf{R}(0) + \mathbf{R}(1)z^{-1}]^{-1} \tag{13.76}$$

The necessary and sufficient condition for the existence of a stable but noncausal realization of the decorrelating filter is

$$\det[\mathbf{R}(-1)e^{jw} + \mathbf{R}(0) + \mathbf{R}(1)e^{-jw}]^{-1} \neq 0\,\forall w \in [0, 2\pi] \tag{13.77}$$

The z-transform of the decorrelating detector outputs is

$$\mathbf{P}(z) = (\mathbf{E\Phi Cb})(z) + N_{\mathrm{p}}(z) \tag{13.78}$$

$N_{\mathrm{p}}(z)$ is the z-transform of a stationary, filtered Gaussian noise vector sequence. The z-transform of the noise covariance matrix sequence is equal to

$$\sigma^2 [\mathbf{S}(z)]^{-1} = \sigma^2 \sum_{m=-\infty}^{\infty} \mathbf{D}(m) z^{-m} \tag{13.79}$$

The output of the decorrelating detector containing L signal replicas of user k may be expressed as

$$\boldsymbol{p}_k(l) = \mathbf{c}_k \sqrt{E_k} \mathrm{e}^{j\phi_k} b_k(l) + \mathbf{n}_k(l) \tag{13.80}$$

The noise covariance matrix is given by

$$\sigma^2 [\mathbf{D}(0)]_{kk} = \sigma^2 \frac{1}{2\pi} \int_0^{2\pi} [\mathbf{S}(\mathrm{e}^{-jw})]_{kk}^{-1} \, \mathrm{d}w \tag{13.81}$$

13.6.3 Coherent reception with maximal ratio combining

Since the front end of the coherent multiuser detector contains the decorrelating filter, the noise components in the L branches of the kth user are correlated. The usual approach prior to combining is to introduce the whitening operation in which whitening filter $(\mathbf{T}^H)^{-1}$ is obtained by Cholesky decomposition $[\mathbf{D}(0)]_{kk} = \mathbf{T}^{\mathrm{T}}\mathbf{T}^*$. So, the output of the user of interest is given by

$$\boldsymbol{p}_{kw} = \boldsymbol{f} \sqrt{E_k} \mathrm{e}^{j\phi_k} b_k + \boldsymbol{n}_{kw} = \boldsymbol{p}'_{kw} b_k + \boldsymbol{n}_{kw} \tag{13.82}$$

where

$$\mathbf{f} = (\mathbf{T}^H)^{-1} \mathbf{c}_k \tag{13.83}$$

and $\boldsymbol{n}_{kw}$ is zero mean Gaussian white noise vector with covariance matrix $\sigma^2 \mathbf{I}_L$. The optimal combiner in this situation is the maximal ratio combiner (MRC). The receiver block diagram is shown in Figure 13.24. The output of the MRC can be represented as

$$\hat{b}_k = \mathrm{sgn}\,(\boldsymbol{p}_{kw} \cdot \hat{\boldsymbol{p}}'_{kw}{}^*) \tag{13.83a}$$

13.6.4 Differentially coherent detection with equal gain combining

In this case, differential demodulation is performed in accordance with the following definition

$$d_k = \mathrm{Re} \left\{ \sum_{i=0}^{L-1} \left[\sqrt{E_k} c_{k,i} \mathrm{e}^{j\phi_k} b_{k,i}(l) + n_{k.i}(l) \right] \right.$$

$$\left. \left[\sqrt{E_k} c_{k,i}^* \mathrm{e}^{-j\phi_k} b_{k,i}(l-1) + n_{k.i}^*(l-1) \right] \right\} \tag{13.84}$$

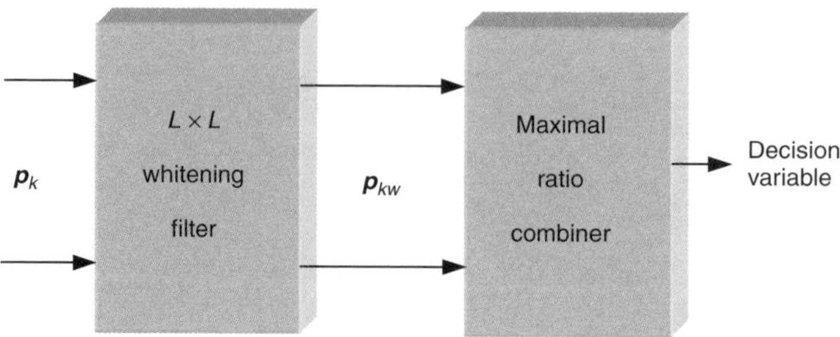

Figure 13.24 Maximal ratio combining after multipath decorrelation for coherent reception.

This can be represented as

$$d_k = \mathbf{v}^H \mathbf{Q} \mathbf{v} \tag{13.85}$$

where

$$v = \begin{pmatrix} \sqrt{w_k}\, c_{k,0} \mathrm{e}^{j\phi_k} b_{k,0}(l) + n_{k.0}(l) \\ \vdots \\ \sqrt{w_k}\, c_{k,L-1} \mathrm{e}^{j\phi_k} b_{k,L-1}(l) + n_{k.L-1}(l) \\ \sqrt{w_k}\, c_{k,0} \mathrm{e}^{j\phi_k} b_{k,0}(l-1) + n_{k.0}(l-1) \\ \vdots \\ \sqrt{w_k}\, c_{k,L-1} \mathrm{e}^{j\phi_k} b_{k,L-1}(l-1) + n_{k.L-1}(l-1) \end{pmatrix} \tag{13.86}$$

and

$$\mathbf{Q} = \begin{pmatrix} 0 & 0.5\mathbf{I}_L \\ 0.5\mathbf{I}_L & 0 \end{pmatrix} \tag{13.87}$$

The receiver block diagram is shown in Figure 13.25.

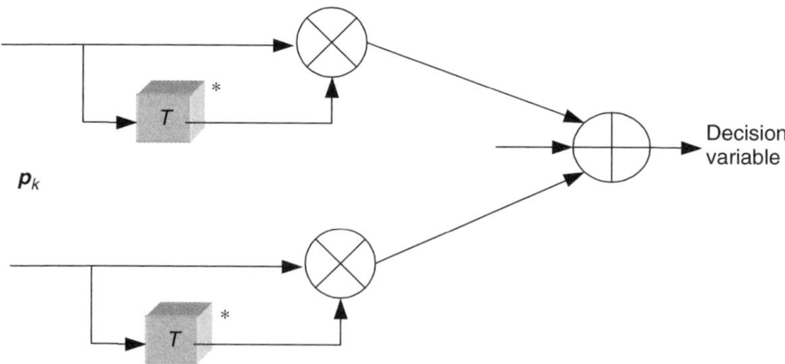

Figure 13.25 Equal gain combining after multipath decorrelation for differentially coherent reception.

For illustration purposes, a CDMA cellular mobile radio system with 1.25-MHz bandwidth and 9600-bps data rate is used. The multipath intensity profile (MIP) of the mobile radio channel is given by

$$r(\tau) = \frac{P}{T_{\mathrm{m}}} e^{-\tau/T_{\mathrm{m}}} \tag{13.88}$$

P is the total average received power and T_{m} is the multipath delay spread. Typical values of the multipath delay spread are $T_{\mathrm{m}} = 0.5\,\mu\mathrm{s}$ for the suburban environment and $T_{\mathrm{m}} = 3\,\mu\mathrm{s}$ for an urban environment. Therefore, we expect the multipath diversity reception with two branches in suburban and four to five branches in an urban setting. For the given parameters, ISI is negligible, and the mobile radio channel can be described as a discrete multipath Rayleigh fading channel with mean-square value of the path coefficients given by

$$\overline{c_{k,l}^2} = \frac{1}{B_{\mathrm{w}}} r\left(\frac{l}{B_{\mathrm{w}}}\right) \quad l = 0, \ldots, L - 1 \tag{13.89}$$

BER for MLS detector and for $L = 1$ and $L = 2$ is shown in Figure 13.26 and for $L = 4$ and $L = 5$ in Figure 13.27. The performance in the case of two users in the network and conventional detector are much worse than in the case in which the MLS multiuser detector is used.

If the number of users is increased from 2 to 5, MLS detector performance will slightly degrade as can be seen in Figure 13.28.

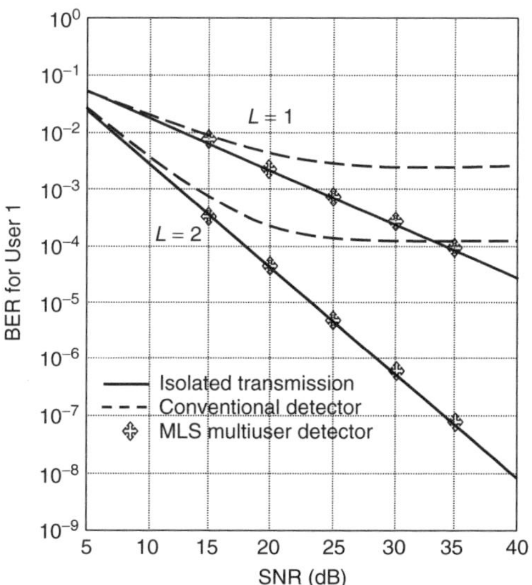

Figure 13.26 Error probability of User 1 when two users are active in a mobile radio channel, spreading ratio $J = 127$, same average path strengths [9]. Reproduced from Zvonar, Z. (1993) *Multiuser Detection for Rayleigh Fading Channel*. Ph.D. Thesis, Department of Electrical and Computer Engineering, Northeastern University, Boston, MA, by permission of IEEE.

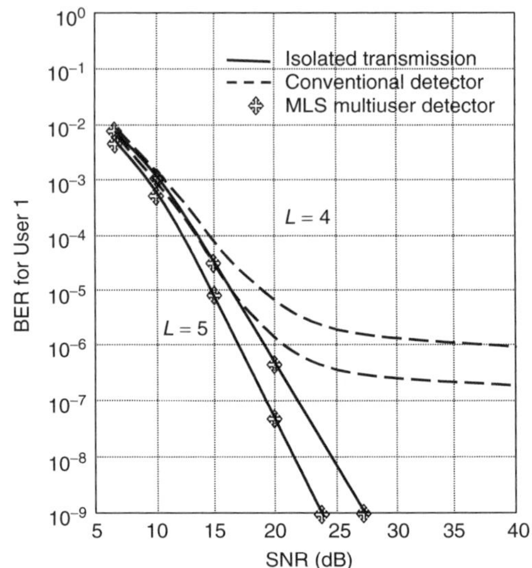

Figure 13.27 Error probability of User 1 when two users are active in an urban mobile radio channel, spreading ratio $J = 127$, same average path strengths [9]. Reproduced from Zvonar, Z. (1993) *Multiuser Detection for Rayleigh Fading Channel*. Ph.D. Thesis, Department of Electrical and Computer Engineering, Northeastern University, Boston, MA, by permission of IEEE.

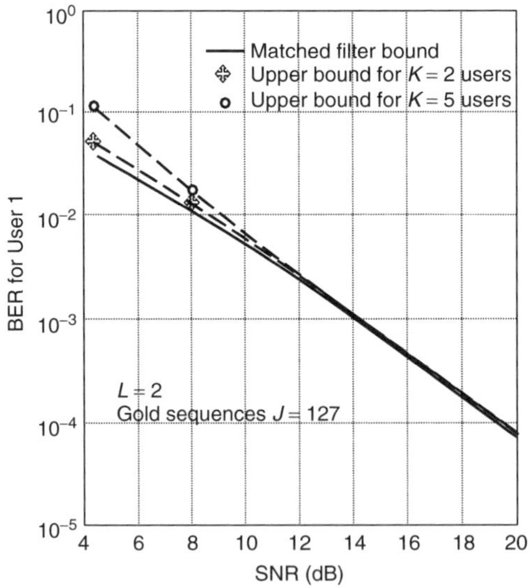

Figure 13.28 Error probability of User 1 for MLS multiuser detector in a suburban mobile radio channel with $L = 2$ resolvable paths [9]. Reproduced from Zvonar, Z. (1993) *Multiuser Detection for Rayleigh Fading Channel*. Ph.D. Thesis, Department of Electrical and Computer Engineering, Northeastern University, Boston, MA, by permission of IEEE.

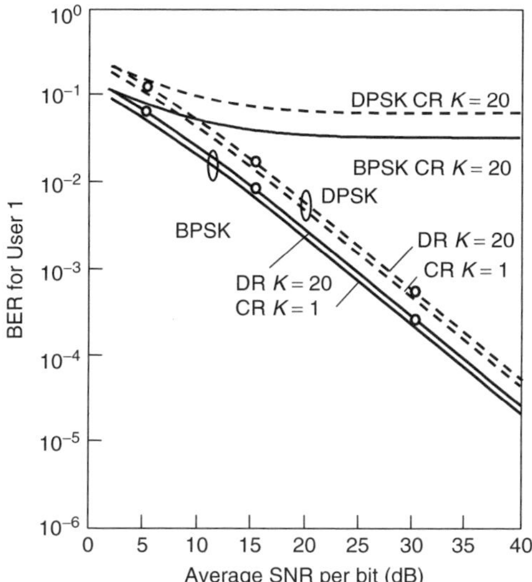

Figure 13.29 Error probability in single-path suburban mobile radio channel with $K = 20$ CDMA users and spreading ratio $J = 127$ [9]. Reproduced from Zvonar, Z. (1993) *Multiuser Detection for Rayleigh Fading Channel*. Ph.D. Thesis, Department of Electrical and Computer Engineering, Northeastern University, Boston, MA, by permission of IEEE.

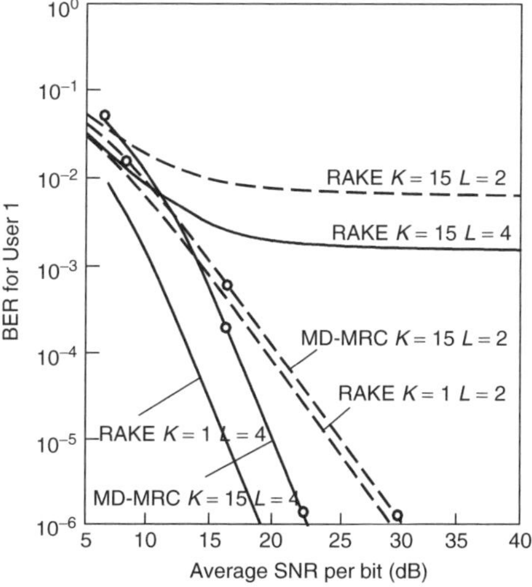

Figure 13.30 Error probability in mobile radio channel with $L = 2$ and $L = 4$ resolvable multipath components, spreading ratio $J = 127$ and $K = 15$ CDMA users [9]. Reproduced from Zvonar, Z. (1993) *Multiuser Detection for Rayleigh Fading Channel*. Ph.D. Thesis, Department of Electrical and Computer Engineering, Northeastern University, Boston, MA, by permission of IEEE.

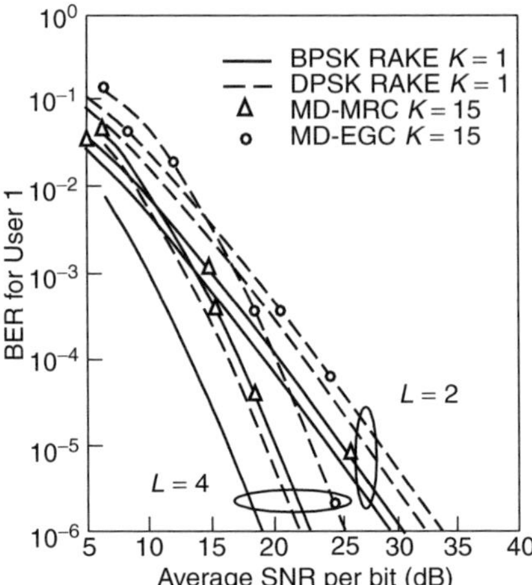

Figure 13.31 Multiuser detectors error probability in mobile radio channel with $L = 2$ and $L = 4$ resolvable multipath components, spreading ratio $J = 127$ and $K = 15$ CDMA users [9]. Reproduced from Zvonar, Z. (1993) *Multiuser Detection for Rayleigh Fading Channel*. Ph.D. Thesis, Department of Electrical and Computer Engineering, Northeastern University, Boston, MA, by permission of IEEE.

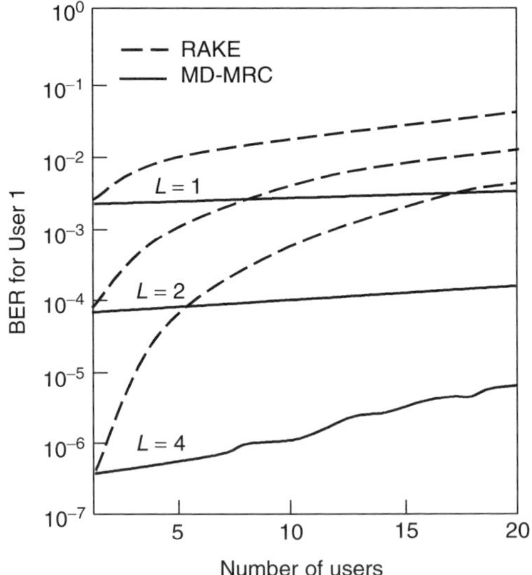

Figure 13.32 Error probability for coherent RAKE and MD-MRC multiuser receiver for different multipath diversity order in mobile radio channel using Gold signature sequences of length $J = 127$, average $i = 20\,$dB [9]. Reproduced from Zvonar, Z. (1993) *Multiuser Detection for Rayleigh Fading Channel*. Ph.D. Thesis, Department of Electrical and Computer Engineering, Northeastern University, Boston, MA, by permission of IEEE.

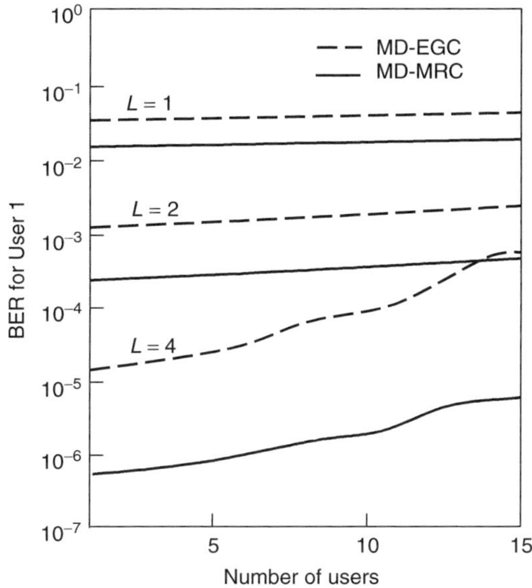

Figure 13.33 Multiuser receivers error probability for different multipath diversity order in mobile radio channel using Gold signature sequences of length $J = 127$, average signal-to-noise ratio (SNR) $= 20\,$dB [9]. Reproduced from Zvonar, Z. (1993) *Multiuser Detection for Rayleigh Fading Channel*. Ph.D. Thesis, Department of Electrical and Computer Engineering, Northeastern University, Boston, MA, by permission of IEEE.

Figure 13.29 demonstrates the same results for a single-path channel with $K = 20$ users. One can see that decorrelating receiver (DR) with $K = 20$ performs almost in the same manner as conventional receiver with $K = 1$.

The same results for $L = 2$ and $L = 4$ are shown in Figure 13.30. Conventional detector is denoted as RAKE and multiuser detector as MD-MRC.

Figure 13.31 also shows the differences between equal gain combiner (EGC) and MRC.

BER versus number of users is shown in Figures 13.32 and 13.33. One can see that for large product LK multiuser detector performance starts to degrade owing to noise enhancement caused by matrix inversion.

A number of additional related issues are discussed in References [10–17].

SYMBOLS

r – received signal
S – signal
b – bits, data
s – sequence
$\Omega(b)$ – likelihood function
$y()$, $z()$ – matched filter (correlator) output

SNR – signal-to-noise ratio
E_b – signal energy per bit
k – user index
K – number of users
y – vector of matched filter outputs
R – matrix of sequence correlation coefficients
$R_{k,l}$ – elements of above matrix
T – linear transformation matrix
$S(z)$ – CDMA channel transfer function (z-transform)
$G(z)$ – channel inverse filter
u – signature sequence
c – channel coefficient
ϕ – signal phase
DR – decorrelating receiver
CR – conventional receiver
MRC – maximum ratio combiner
L – number of paths

REFERENCES

1. Yerdu, S. (1986) Optimum multiuser asymptotic efficiency. *IEEE Trans. Commun.*, **COM-34**, 890–897.
2. Verdu, S. (1986) Minimum probability of error for asynchronous Gaussian multiple-access channels. *IEEE Trans. Inform. Theory*, **IT-32**, 85–96.
3. Tan, P. H. and Rasmussen, L. (2000) Linear interference cancellation in CDMA based on iterative techniques for linear equation systems. *IEEE Trans. Commun.*, **48**(12), 2099–2108.
4. Lupas, R. and Verdu, S. (1989) Linear multiuser detectors for synchronous code division multiple access channels. *IEEE Trans. Inform. Theory*, **35**, 123–136.
5. Varanasi, M. and Aazhang, B. (1990) Multistage detection in asynchronous code division multiple access communications. *IEEE Trans. Commun.*, **38**, 509–519.
6. Varanasi, M. and Aazhang, B. (1991) Near optimum detection in synchronous code division multiple access systems. *IEEE Trans. Cornmun.*, **39**, 725–736.
7. Zvonar, Z. *et al.* (1992) Optimum deletion in a synchronous multiple-access multipath Rayleigh fading charrtels. *Proc. 26th Annual Conference on Information Sciences and Systems*. Princeton, NJ: Princeton University Press.
8. Varanasi, M. (1993) Noncoherent detection in a synchronous multiuser channels. *IEEE Trans. Inform. Theory*, **37**(1), 157–176.
9. Zvonar, Z. (1993) *Multiuser Detection for Rayleigh Fading Channel*. Ph.D. Thesis, Department of Electrical and Computer Engineering, Northeastern University, Boston, MA.
10. Aazhang, B. *et al.* Neural networks for multiuser detection in code division multiple access communications. *IEEE Trans. Commun.*, **COM-40**, 1212–1222.
11. Varanasi, M. and Aazhang, B. (1991) Optimally near-far resistant multiuser detection in differentially coherent synchronous channels. *IEEE Trans. Inform. Theory*, **39**, 1006–1018.
12. Xie, Z. *et al.* (1990) A family of suboptimum detectors for coherent multiuser communications. *IEEE ISAC*, **8**(4), 683–690.
13. Xie, Z. *et al.* (1993) Joint signal detection and parameter estimation in multiuser communications. *IEEE Trans. Commun.*, **41**(7), 1208–1216.

14. Lupas, R. and Verdu, S. (1990) Near-far resistance of multiuser detectors in a synchronous channels. *IEEE Trans. Commun.*, **38**(4), 496–508.
15. Wijayasuriya, S. S. H., Norton, G. H. and McGeehan, J. P. (1992) Sliding window decorrelating algorithm for DS-CDMA receivers. *Electron. Lett.*, **28**, 1596–1598.
16. Wijayasuriya, S. S. H., McGeehan, J. P. and Norton, G. H. (1993) RAKE decorrelating receiver for DS-CDMA mobile radio networks. *Electron. Lett.*, **29**, 395–396.
17. Xie, Z. *et al.* (1990) Multiuser signal detection using sequential decoding. *IEEE Trans. Commun.*, **38**, 578–583.

14

MMSE multiuser detectors

14.1 MINIMUM MEAN-SQUARE ERROR (MMSE) LINEAR MULTIUSER DETECTION

If the amplitude of the user's k signal in equation (13.7) is A_k, then the vector of matched filter outputs $\mathbf{y}$ in equation (13.10) can be represented as

$$\mathbf{y} = \mathbf{RAb} + \mathbf{n} \tag{14.1}$$

where $\mathbf{A}$ is a diagonal matrix with elements A_k

$$\mathbf{A} = \mathrm{diag}\|A_k\| \tag{14.2}$$

If the multiuser detector transfer function is denoted as $\mathbf{M}$, then the minimum mean-square error (MMSE) detector is defined as

$$\min_{M \in R^{K \times K}} E\left\lfloor \|\mathbf{b} - \mathbf{My}\|^2 \right\rfloor \tag{14.3}$$

One can show that the MMSE linear detector outputs the following decisions [1–3]:

$$\hat{b}_k = \mathrm{sgn}\left(\frac{1}{A_k}([\mathbf{R} + \sigma^2 \mathbf{A}^{-2}]^{-1}\mathbf{y})_k\right)$$
$$= \mathrm{sgn}(([\mathbf{R} + \sigma^2 \mathbf{A}^{-2}]^{-1}\mathbf{y})_k) \tag{14.4}$$

The block diagram of a linear MMSE detector is shown in Figure 14.1.

Therefore, the MMSE linear detector replaces the transformation $\mathbf{R}^{-1}$ of the decorrelating detector by

$$[\mathbf{R} + \sigma^2 \mathbf{A}^{-2}]^{-1} \tag{14.5}$$

where

$$\sigma^2 \mathbf{A}^{-2} = \mathrm{diag}\left\{\frac{\sigma^2}{A_1^2}, \ldots, \frac{\sigma^2}{A_K^2}\right\} \tag{14.6}$$

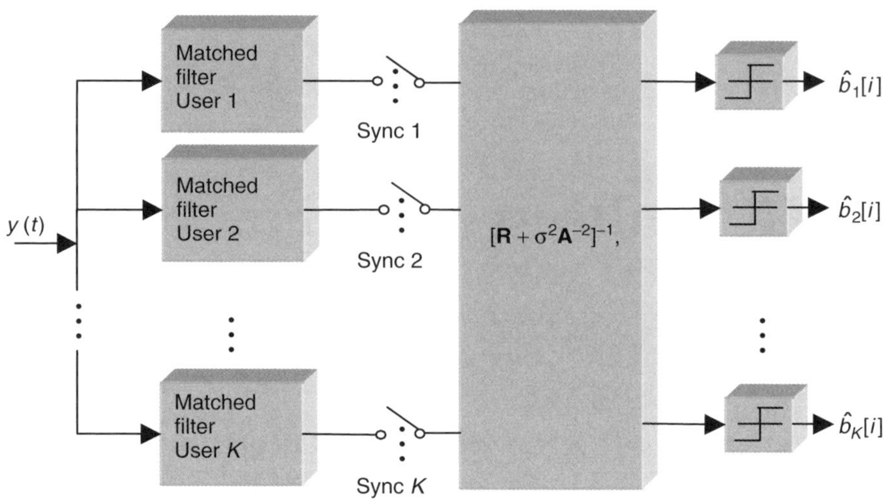

Figure 14.1   MMSE linear detector for a synchronous channel.

As an illustration for the two users case we have

$$[\mathbf{R} + \sigma^2\mathbf{A}^{-2}]^{-1} = \left[\left(1 + \frac{\sigma^2}{A_1^2}\right)\left(1 + \frac{\sigma^2}{A_2^2}\right) - \rho^2\right]^{-1}\begin{bmatrix} 1 + \dfrac{\sigma^2}{A_2^2} & -\rho \\[2ex] -\rho & 1 + \dfrac{\sigma^2}{A_1^2} \end{bmatrix} \tag{14.7}$$

and the detector is shown in Figure 14.2.

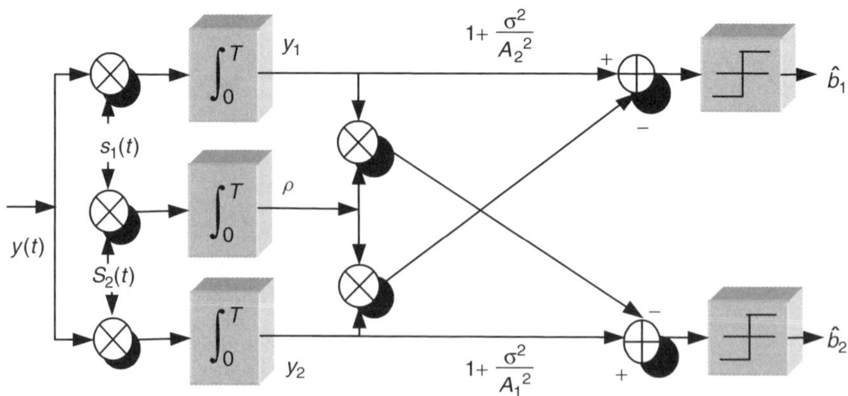

Figure 14.2   MMSE linear receiver for two synchronous users.

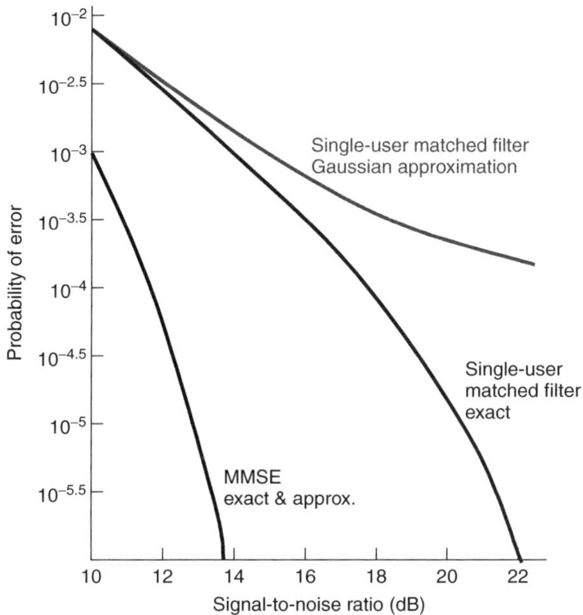

Figure 14.3 Bit-error-rate with eight equal-power users and identical cross-correlations $\rho_{kl} = 0.1$.

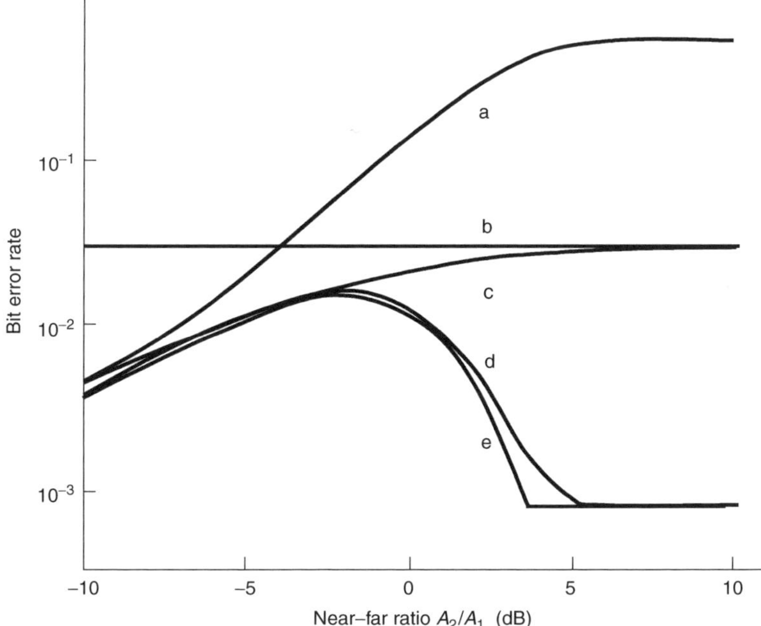

Figure 14.4 Bit-error-rate with two users and cross-correlation $\rho = 0.8$: a – single-user matched filter, b – decorrelator, c – MMSE, d – minimum (upper bound), e – minimum (lower bound).

In the asynchronous case, similar to the solution in Section 13.3 of Chapter 13, the MMSE linear detector is a K-input, K-output, linear, time-invariant filter with transfer function

$$[\mathbf{R}^{\mathrm{T}}[1]z + \mathbf{R}[0] + \sigma^2 \mathbf{A}^{-2} + \mathbf{R}[1]z^{-1}]^{-1} \tag{14.8}$$

Performance results are illustrated in Figures 14.3. and 14.4. As expected, in Figure 14.3, the MMSE detector demonstrates better performance than the conventional detector denoted as a single-user matched filter receiver (MFR).

In Figure 14.4 bit error rate (BER) is presented versus the near–far ratio for different detectors. One can see that MMSE shows better performance than decorrelator. In the figure signal-to-noise ratio (SNR) of the desired user is equal to $10\,\mathrm{dB}$.

14.2 SYSTEM MODEL IN MULTIPATH FADING CHANNEL

In this section the channel impulse response and the received signal will be presented as

$$c_k(t) = \sum_{l=1}^{L_k} c_{k,\ell}^{(n)} \delta(t - \tau_{k,\ell}) \tag{14.9}$$

$$r(t) = \sum_{n=0}^{N_b-1} \sum_{k=1}^{K} \sum_{l=1}^{L} A_k b_k^{(n)} c_{k,l}^{(n)} s_k(t - nT - \tau_{k,l}) + n(t) \tag{14.10}$$

The received signal is time-discretized, by antialias filtering and sampling $r(t)$ at the rate $1/T_\mathrm{s} = S/T_\mathrm{c} = SG/T$, where S is the number of samples per chip and $G = T/T_\mathrm{C}$ is the processing gain. The received discrete-time signal over a data block of N_b symbols is

$$\mathbf{r} = \mathbf{SCAb} + \mathbf{n} \in \mathbf{C}^{SGN_\mathrm{b}} \tag{14.11}$$

where

$$\mathbf{r} = [\mathbf{r}^{\mathrm{T}(0)}, \ldots, \mathbf{r}^{\mathrm{T}(N_\mathrm{b}-1)}]^{\mathrm{T}} \in C^{SGN_\mathrm{b}} \tag{14.12}$$

is the input sample vector with

$$\mathbf{r}^{\mathrm{T}(n)} = \{r[T_\mathrm{s}(nSG + 1)], \ldots, r[T_\mathrm{s}(n + 1)SG]\} \in C^{SG} \tag{14.13}$$

$$\mathbf{S} = [\mathbf{S}^{(0)}, \mathbf{S}^{(1)}, \dots, \mathbf{S}^{(N_b-1)}] \in R^{SGN_b \times KLN_b}$$

$$= \begin{pmatrix} \mathbf{S}^{(0)}(0) & \mathbf{0} & \cdots & & \mathbf{0} \\ \vdots & \mathbf{S}^{(1)}(0) & \ddots & & \vdots \\ \mathbf{S}^{(0)}(D) & \vdots & \ddots & & \mathbf{0} \\ \mathbf{0} & \mathbf{S}^{(1)}(D) & \ddots & \mathbf{S}^{(N_b-1)}(0) \\ \vdots & & \ddots & \ddots & \vdots \\ \mathbf{0} & & \cdots & \mathbf{0} & \mathbf{S}^{(N_b-1)}(D) \end{pmatrix} \tag{14.14}$$

is the sampled spreading sequence matrix, $D = (T + T_m)/T$. In a single-path channel, $D = 1$ due to the asynchrony of users. In multipath channels, $D \geq 2$ due to the multipath spread. The code matrix is defined with several components $(S^{(n)}(0), \dots, S^{(n)}(D))$ for each symbol interval to simplify the presentation of the cross-correlation matrix components. T_m is the maximum delay spread,

$$\mathbf{S}^{(n)} = [\mathbf{s}_{1,1}^{(n)}, \dots, \mathbf{s}_{1,L}^{(n)}, \dots, \mathbf{s}_{K,L}^{(n)}] \in R^{SGN_b \times KL} \tag{14.15}$$

where

$$\mathbf{s}_{k,l}^{(n)} = \begin{cases} \mathbf{0}_{SGN_b \times 1}^{T} & \begin{array}{l} n = 0 \\ \tau_{k,l} = 0 \end{array} \\[2ex] [[s_k[T_s(SG - \tau_{k,l} + 1)]], \dots, s_k(T_sSG)]^T, \mathbf{0}_{(SGN_b - \tau_{k,l}) \times 1}^{T}]^T & \begin{array}{l} n = 0 \\ \tau_{k,l} > 0 \end{array} \\[2ex] [\mathbf{0}_{[(n-1)SG+\tau_{k,l}]\times 1}^{T}, \mathbf{s}_k^{T}, \mathbf{0}_{[SG(N_b-n)-\tau_{k,l}]\times 1}^{T}]^T & \\[1ex] & 0 < n < N_b - 1 \\ (\mathbf{0}_{[SG(N_b-1)+\tau_{k,l}]\times 1}, \{s_k(T_s), \dots, s_k[T_s(SG - \tau_{k,l})]\})^T & n = N_b - 1 \end{cases}$$
$$\tag{14.16}$$

where $\tau_{k,l}$ is the time-discretized delay in sample intervals and

$$\mathbf{s}_k = [s_k(T_s), \dots, s_k(T_sSG)]^T \in R^{SG} \tag{14.17}$$

is the sampled signature sequence of the kth user. By analogy with equation (13.59)

$$\mathbf{C} = \mathrm{diag}\left[\mathbf{C}^{(0)}, \dots, \mathbf{C}^{(N_b-1)}\right] \in C^{KLN_b \times KN_b} \tag{14.18}$$

is the channel coefficient matrix with

$$\mathbf{C}^{(n)} = \mathrm{diag}\left[\mathbf{c}_1^{(n)}, \dots, \mathbf{c}_K^{(n)}\right] \in C^{KL \times K} \tag{14.19}$$

and

$$\mathbf{c}_k^{(n)} = [c_{k,1}^{(n)}, \dots, c_{k,L}^{(n)}]^T \in C^L \tag{14.20}$$

Equation (14.2) now becomes

$$\mathbf{A} = \mathrm{diag}[\mathbf{A}^{(0)}, \ldots, \mathbf{A}^{(N_b-1)}] \in R^{KN_b \times KN_b} \tag{14.21}$$

the matrix of total received average amplitudes with

$$\mathbf{A}^{(n)} = \mathrm{diag}[A_1, \ldots, A_K] \in R^{K \times K} \tag{14.22}$$

Bit vector from equation (13.56) becomes

$$\mathbf{b} = [\mathbf{b}^{T_{(0)}}, \ldots, \mathbf{b}^{T_{(N_b-1)}}]^T \in \aleph^{KN_b} \tag{14.23}$$

with the modulation symbol alphabet $\aleph$ [with binary phase shift keying (BPSK) $\aleph = \{-1, 1\}$] and

$$\mathbf{b}^{(n)} = [b_1^{(n)}, \ldots, b_K^{(n)}] \in \aleph^K \tag{14.24}$$

and $\mathbf{n} \in C^{SGN_b}$ is the channel noise vector. It is assumed that the data bits are independent identically distributed random variables independent from the channel coefficients and the noise process.

The cross-correlation matrix equation (13.70) for the spreading sequences can be formed as

$$\mathbf{R} = \mathbf{S}^T\mathbf{S} \in R^{KLN_b \times KLN_b}$$

$$= \begin{pmatrix}
\mathbf{R}^{(0,0)} & \cdots & \mathbf{R}^{(0,D)} & \mathbf{0}_{KL} \cdots & & \mathbf{0}_{KL} \\
\vdots & \ddots & \ddots & \ddots & & \vdots \\
\mathbf{R}^{(D,0)} & \ddots & \ddots & \ddots & & \mathbf{0}_{KL} \\
\mathbf{0}_{KL} & \ddots & \ddots & \ddots & & \mathbf{R}^{(N_b-D,N_b-1)} \\
\vdots & \ddots & \ddots & \ddots & & \vdots \\
\mathbf{0}_{KL} & \cdots & \mathbf{0}_{KL} & \cdots & & \mathbf{R}^{(N_b-1,N_b-1)}
\end{pmatrix} \tag{14.25}$$

where equation (13.20) now becomes

$$\mathbf{R}^{(n,n-j)} = \sum_{i=0}^{D-j} \mathbf{S}^{T_{(n)}}(i)\mathbf{S}^{(n-j)}(i+j), \; j \in \{0, \ldots, D\} \tag{14.26}$$

and $\mathbf{R}^{(n-j,n)} = \mathbf{R}^{T_{(n,n-j)}}$. The elements of the correlation matrix can be written as

$$\mathbf{R}^{(n,n')} = \begin{pmatrix}
\mathbf{R}_{1,1}^{(n,n')} & \cdots & \mathbf{R}_{1,K}^{(n,n')} \\
\vdots & \ddots & \vdots \\
\mathbf{R}_{K,1}^{(n,n')} & & \mathbf{R}_{K,K}^{(n,n')}
\end{pmatrix} \in R^{KL \times KL} \tag{14.27}$$

and

$$\mathbf{R}_{k,k'}^{(n,n')} = \begin{pmatrix} \mathbf{R}_{k1,k'1}^{(n,n')} & \cdots & \mathbf{R}_{k1,k'L}^{(n,n')} \\ \vdots & \ddots & \vdots \\ \mathbf{R}_{kL,k'1}^{(n,n')} & \cdots & \mathbf{R}_{kL,k'L}^{(n,n')} \end{pmatrix} \in R^{L \times L} \tag{14.28}$$

where equation (13.71) now becomes

$$R_{kl,k'l'}^{(n,n')} = \sum_{j=\tau_{k,l}}^{SG-1+\tau_{k,l}} s_k[T_s(j - \tau_{k,l})]s_{k'}\{T_s[j - \tau_{k'l'} + (n' - n)SG]\} = \mathbf{s}_{k,l}^{T_{(n)}}\mathbf{s}_{k',l'}^{(n')} \tag{14.29}$$

and represents the correlation between users k and k', lth and l'th paths, between their nth and n'th symbol intervals.

14.3 MMSE DETECTOR STRUCTURES

One of the conclusions in Chapter 13 was that noise enhancement in linear Multi-user detection (MUD) causes system performance degradation for large product KL. In this section we consider the possibility of reducing the site of the matrix to be inverted by using multipath combining prior to MUD. The structure is called the postcombining detector and the basic block diagram of the receiver is shown in Figure 14.5 [4].

The starting point in the derivation of the receiver structure is the cost function $E\{|\mathbf{b} - \hat{\mathbf{b}}|^2\}$

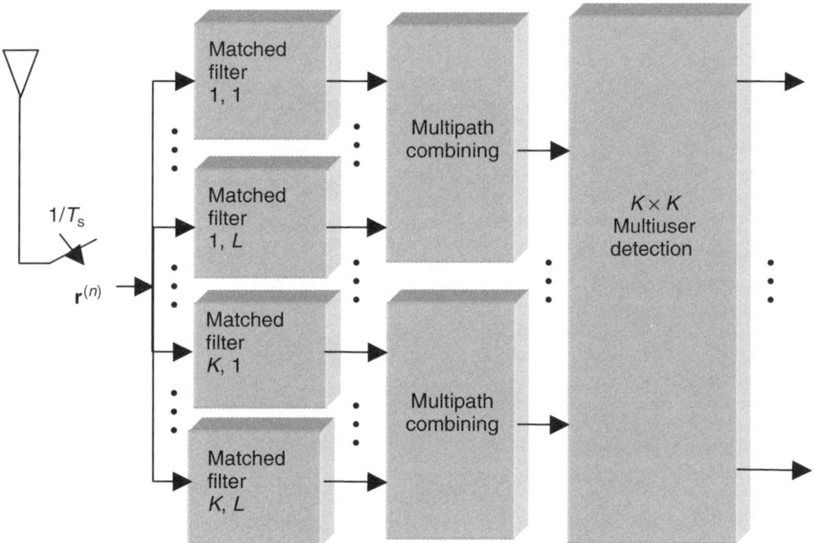

Figure 14.5 Postcombining interference suppression receiver.

where

$$\hat{\mathbf{b}} = \mathbf{L}_{[\text{post}]}^{\mathrm{H}} \mathbf{r} \tag{14.30}$$

The detector linear transform matrix is given as

$$\mathbf{L}_{[\text{post}]} = \mathbf{SCA}(\mathbf{AC}^{\mathrm{H}}\mathbf{RCA} + \sigma^2\mathbf{I})^{-1} \in C^{SGN_b \times KN_b} \tag{14.31}$$

This result is obtained by minimizing the cost function, and derivation details may be found in any standard textbook on signal processing. Here, $\mathbf{R} = \mathbf{S}^{\mathrm{T}}\mathbf{S}$ is the signature sequence cross-correlation matrix defined by equation (14.25). The output of the post-combining LMMSE receiver is

$$\mathbf{y}_{[\text{post}]} = (\mathbf{AC}^{\mathrm{H}}\mathbf{RCA} + \sigma^2\mathbf{I})^{-1}(\mathbf{SCA})^{\mathrm{H}}\mathbf{r} \in C^K \tag{14.32}$$

where $(\mathbf{SCA})^{\mathrm{H}}\mathbf{r}$ is the multipath [maximum ratio (MR)] combined matched filter bank output. For nonfading additive white Gaussian noise (AWGN),

$$\mathbf{L}_{[\text{post}]} = \mathbf{S}(\mathbf{R} + \sigma^2(\mathbf{A}^{\mathrm{H}}\mathbf{A})^{-1})^{-1} \tag{14.33}$$

The postcombining LMMSE receiver in fading channels depends on the channel complex coefficients of all users and paths. If the channel is changing rapidly, the optimal LMMSE receiver changes continuously. The adaptive versions of the LMMSE receivers have increasing convergence problems as the fading rate increases. The dependence on the fading channel state can be removed by applying a precombining interference suppression type of receiver. The receiver block diagram in this case is shown in Figure 14.6 [4].

The transfer function of the detector is obtained by minimizing each element of the cost function

$$E\{|\mathbf{h} - \hat{\mathbf{h}}|^2\} \tag{14.34}$$

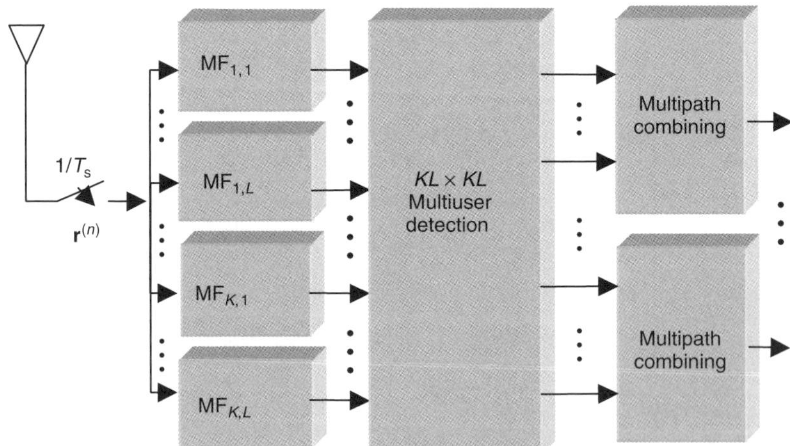

Figure 14.6 Precombining interference suppression receiver.

where

$$\mathbf{h} = \mathbf{CAb} \qquad (14.35)$$

and

$$\hat{\mathbf{h}} = \mathbf{L}_{[\mathrm{pre}]}^{\mathrm{T}}\mathbf{r} \text{ is the estimate} \qquad (14.36)$$

The solution of this minimization is [4]

$$\mathbf{L}_{[\mathrm{pre}]} = \mathbf{S}(\mathbf{R} + \sigma^2 \mathbf{R}_h^{-1})^{-1} \in \mathbf{R}^{SGN_b \times KLN_b} \qquad (14.37)$$

$$\mathbf{R}_h = \mathrm{diag}\left\lfloor A_1^2 \mathbf{R}_{c_1}, \ldots, A_K^2 \mathbf{R}_{c_K} \right\rfloor \in \mathbf{R}^{KLN_b \times KLN_b} \qquad (14.38)$$

$$\mathbf{R}_{c_k} = \mathrm{diag}\left\lfloor E\left\lfloor |c_{k,1}|^2 \right\rfloor, \ldots, E\left\lfloor |c_{k,L}|^2 \right\rfloor \right\rfloor \in \mathbf{R}^{L \times L} \qquad (14.39)$$

$$\mathbf{y}_{[\mathrm{pre}]} = (\mathbf{R} + \sigma^2 \mathbf{R}_h^{-1})^{-1}\mathbf{S}^{\mathrm{T}}\mathbf{r} \in \mathbf{C}^{KL} \qquad (14.40)$$

The two detectors are compared in Figure 14.7. The postcombining scheme performs better.

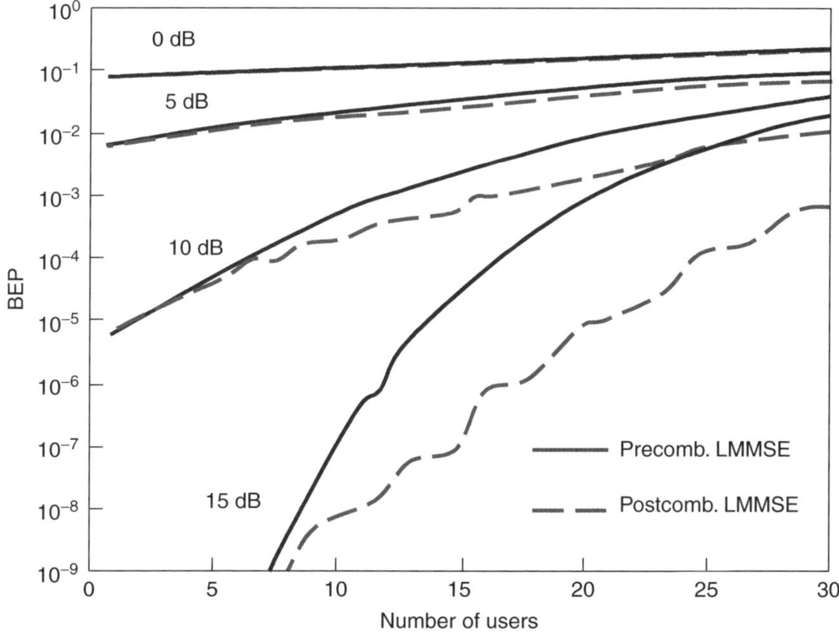

Figure 14.7　Bit error probabilities as a function of the number of users for the postcombining and precombining LMMSE detectors in an asynchronous two-path fixed channel with different SNRs, and bit rate $16\,\mathrm{kb\,s^{-1}}$, Gold code of length 31, $td/T = 4.63 \times 10^{-3}$, maximum delay spread 10 chips [5]. Reproduced from Latva-aho, M. (1998) Advanced Receivers for Wideband CDMA Systems. Ph.D. Thesis, University of Oulu, Oulu, by permission of IEEE.

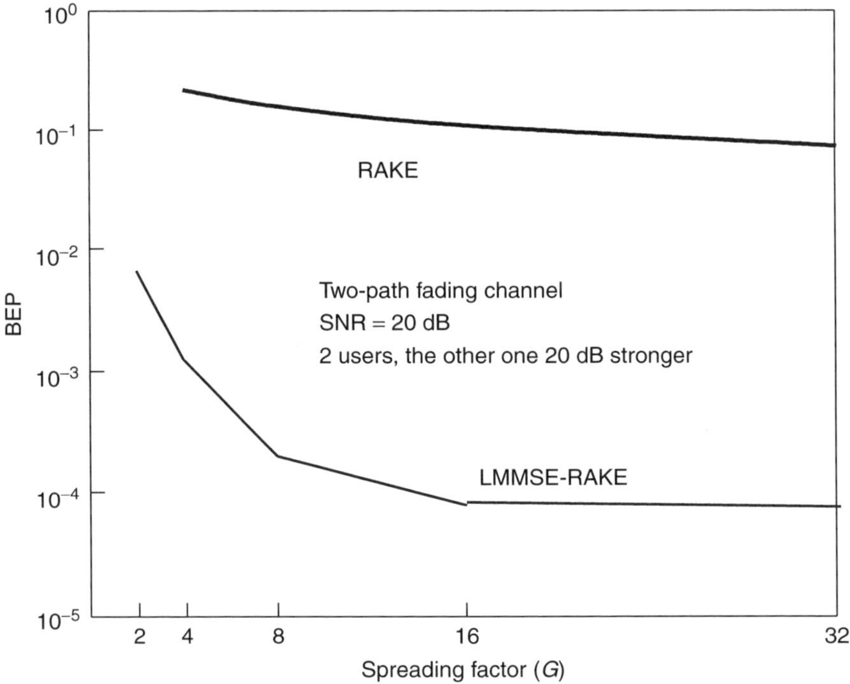

Figure 14.8 Bit error probabilities as a function of the near–far ratio for the conventional RAKE receiver and the precombining LMMSE (LMMSE-RAKE) receiver with a different spreading factor (G) in a two-path Rayleigh fading channel with maximum delay spreads of $2\,\mu$s for $G = 4$, and $7\,\mu$s for other spreading factors. The average signal-to-noise ratio is $20\,$dB, the data modulation is BPSK, the number of users is 2, the other user has 20-dB higher power. Data rates vary from $128\,$kb$\,$s^{-1} to $2.048\,$Mbit$\,$s^{-1}; no channel coding is assumed [5]. Reproduced from Latva-aho, M. (1998) Advanced Receivers for Wideband CDMA Systems. Ph.D. Thesis, University of Oulu, Oulu, by permission of IEEE.

The illustration of LMMSE-RAKE receiver performance in near–far environment is shown in Figure 14.8 [5]. Considerable improvement compared to conventional RAKE is evident.

14.4 SPATIAL PROCESSING

When combined with multiple receiver antennas, the receiver structures may have one of the forms shown in Figure 14.9 [4, 6–8].

The channel impulse response for the kth user's ith sensor can be now written as

$$c_{k,i}(t) = \sum_{l=1}^{L_k} c_{k,l}^{(n)} e^{j2\pi \lambda^{-1} \langle e(\phi_{k,l}), \varepsilon_i \rangle} \delta[t - (\tau_{k,l,i})] \tag{14.41}$$

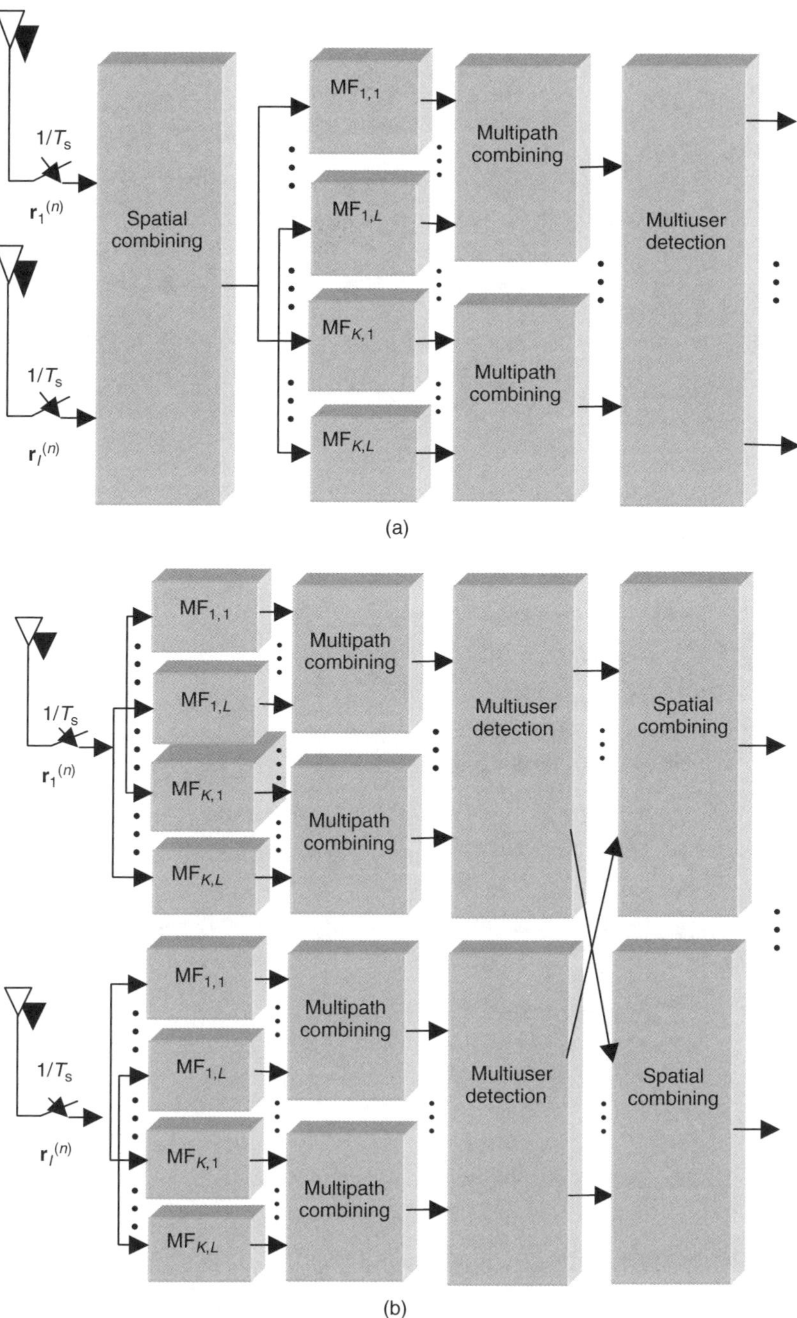

Figure 14.9 (a) The spatial-temporal-multiuser (STM) receiver. (b) TMS receiver. Postcombining interference suppression receivers with spatial signal processing. (c) SMT receiver. (d) MST receiver. Precombining interference suppression receivers with spatial signal processing.

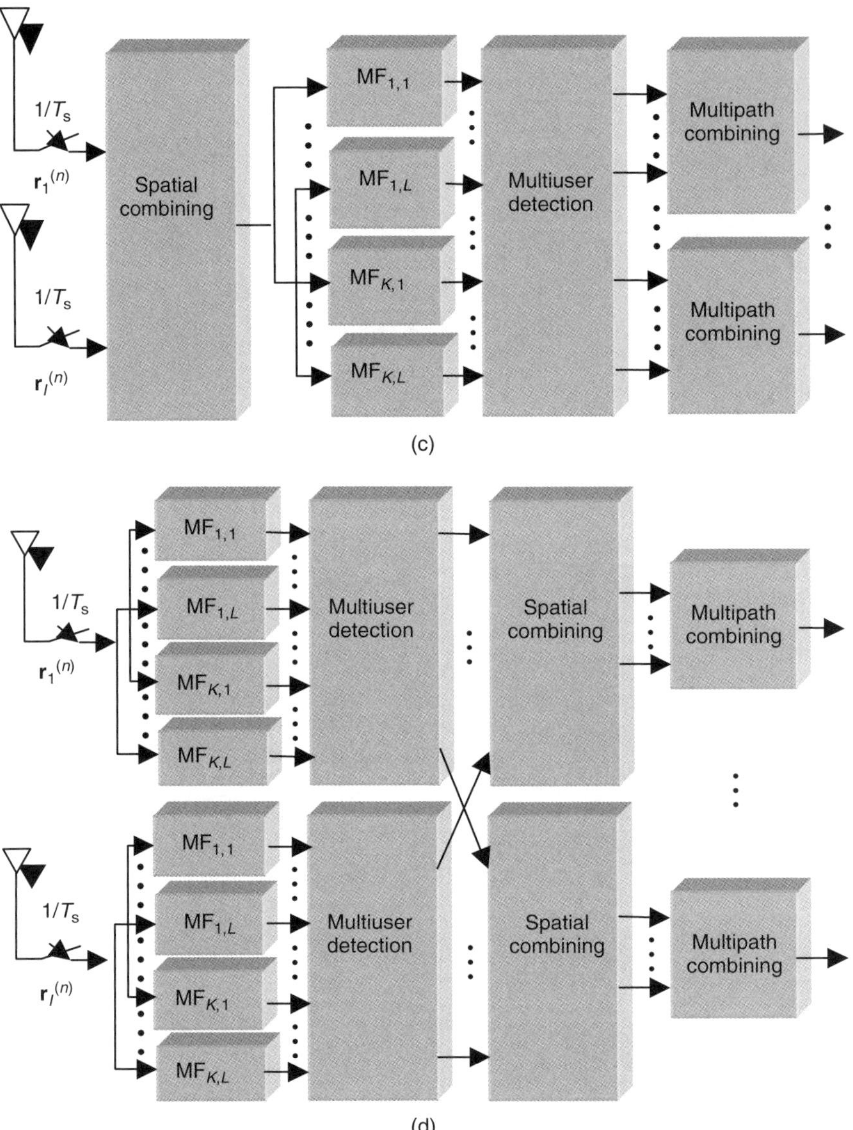

Figure 14.9 *(Continued).*

where L_k is the number of propagation paths (assumed to be the same for all users for simplicity; $L_k = L, \forall k$), $c_{k,l}^{(n)}$ is the complex attenuation factor of the kth user's lth path, $\tau_{k,l,i}$ is the propagation delay for the ith sensor, ε_i is the position vector of the ith sensor with respect to some arbitrarily chosen reference point, λ is the wavelength of the carrier, $e(\phi_{k,l})$ is a unit vector pointing to direction $\phi_{k,l}$ (direction-of-arrival) and $\langle ., . \rangle$ indicates the inner product.

Assuming that the number of propagation paths is the same for all users, the channel impulse response can be written as

$$c_{k,i}(t) = \sum_{l=1}^{L} c_{k,l}^{(n)} e^{j2\pi\lambda^{-1}\langle e(\phi_{k,l}),\varepsilon_i\rangle} \delta(t - \tau_{k,l}) \tag{14.42}$$

The channel matrix for the ith sensor consist of two components

$$\mathbf{C}_i = \mathbf{C}\circ\boldsymbol{\Phi}_i \in C^{KLN_b \times K N_b} \tag{14.43}$$

where $\mathbf{C}$ is the channel matrix defined in equation (14.19). $\circ$ is the Schur product defined as $\mathbf{Z} = \mathbf{X}\circ\mathbf{Y} \in C^{x\times y}$, that is, all components of the matrix $\mathbf{X} \in C^{x\times y}$ are multiplied elementwise by the matrix $\mathbf{Y} \in C^{x\times y}$ and $\boldsymbol{\Phi}_i = \mathrm{diag}(\tilde{\boldsymbol{\phi}}_i) \otimes \mathbf{I}_{N_b}$ with $\tilde{\boldsymbol{\phi}}_i = \mathrm{diag}(\boldsymbol{\phi}_1, \ldots, \boldsymbol{\phi}_K)$, $\boldsymbol{\phi}_k = [\phi_{k,1}, \ldots, \phi_{k,L}]^{\mathrm{T}}$ is the matrix of the direction vectors

$$\boldsymbol{\phi}_i = [e^{j2\pi\lambda^{-1}\langle e(\phi_{1,1}),\varepsilon_i\rangle}, \ldots, e^{j2\pi\lambda^{-1}\langle e(\phi_{K,L}),\varepsilon_i\rangle}]^{\mathrm{T}} \in C^{KL} \tag{14.44}$$

By using the previous notation, one can show that the equivalent detector transform matrixes are given as [4, 6, 7].

$$\mathbf{L}_{[\mathrm{STM}]} = \sum_{i=1}^{I} \mathbf{S}(\mathbf{C}\circ\boldsymbol{\Phi}_i) \cdot \left(\sum_{i=1}^{I} A^{\mathrm{H}}(\boldsymbol{\Phi}_i^{\mathrm{H}}\circ\mathbf{C}^{\mathrm{H}})\mathbf{R}(\mathbf{C}\circ\boldsymbol{\Phi}_i)A + \sigma^2\mathbf{I}\right)^{-1}$$

$$\mathbf{L}_{[\mathrm{SMT}]} = \sum_{i=1}^{I} \mathbf{S}\boldsymbol{\Phi}_i \left(\sum_{i=1}^{I} \boldsymbol{\Phi}_i^{\mathrm{H}}\mathbf{R}\boldsymbol{\Phi}_i + \sigma^2\mathbf{R}_h^{-1}\right)^{-1}$$

$$\mathbf{L}_{[\mathrm{MST}]i} = \mathbf{S}(\mathbf{R} + \sigma^2\mathbf{R}_h^{-1})^{-1}$$

$$\mathbf{L}_{[\mathrm{TMS}]} = \mathbf{SCA}(A\mathbf{C}^{\mathrm{H}}\mathbf{RCA} + \sigma^2\mathbf{I})^{-1}$$

14.5 SINGLE-USER LMMSE RECEIVERS FOR FREQUENCY-SELECTIVE FADING CHANNELS

14.5.1 Adaptive precombining LMMSE receivers

In this case, Mean-Square Error (MSE) criterion $E\{|\mathbf{h} - \hat{\mathbf{h}}|^2\}$ requires that the reference signal $\mathbf{h} = \mathbf{CAb}$ is available in adaptive implementations. For adaptive single-user receivers, the optimization criterion is presented for each path separately, that is,

$$J_{k,l} = E\{|(\mathbf{h})_{k,l} - (\hat{\mathbf{h}})_{k,l}|^2\} \tag{14.45}$$

The receiver block diagram is given in Figure 14.10, [9–17].

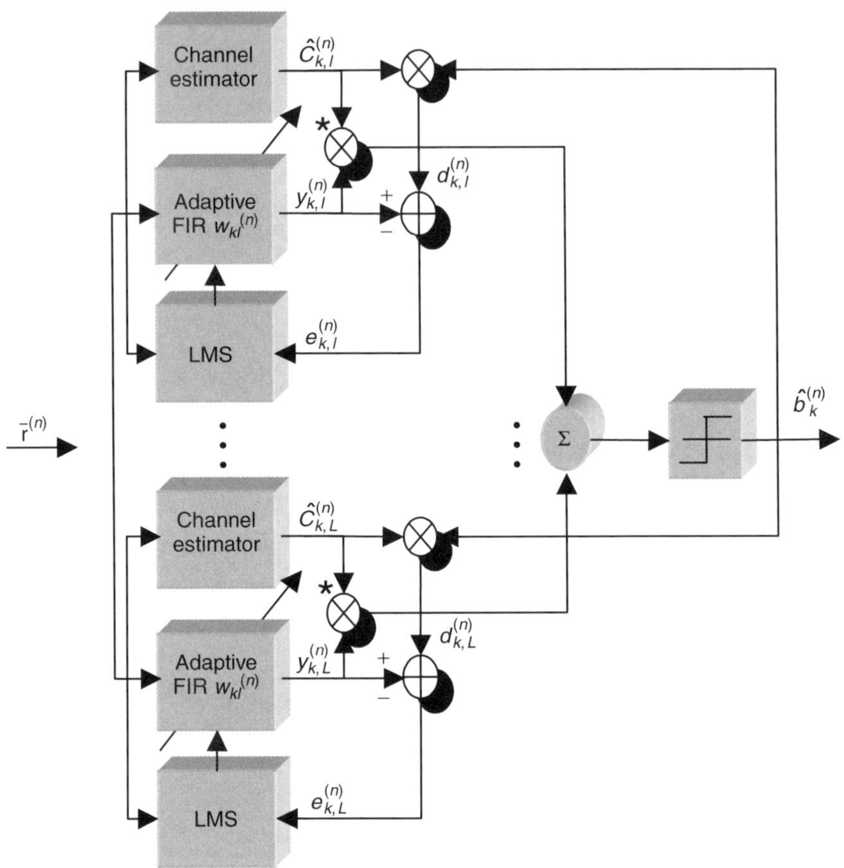

Figure 14.10 General block diagram of the adaptive LMMSE-RAKE receiver.

By using notation

$$\bar{\mathbf{r}}^{(n)} = [\mathbf{r}^{\mathrm{T}(n-D)}, \ldots, \mathbf{r}^{\mathrm{T}(n)}, \ldots, \mathbf{r}^{\mathrm{T}(n+D)}]^{\mathrm{T}} \in \mathbf{C}^{MSG}$$

$$\mathbf{w}_{k,l}^{(n)} = [w_{k,l}^{(n)}(0), \ldots, w_{k,l}^{(n)}(MSG-1)]^{\mathrm{T}} \in \mathbf{C}^{MSG} \tag{14.46}$$

$$y_{k,l}^{(n)} = \mathbf{w}_{k,l}^{H(n)}\bar{\mathbf{r}}^{(n)}$$

the bit estimation is defined as

$$\hat{b}_k^{(n)} = \mathrm{sgn}\left(\sum_{l=1}^{L} \hat{c}_{k,l}^{(n)} y_{k,l}^{(n)}\right) \tag{14.47}$$

The filter coefficients $\mathbf{w}$ are derived using the MSE criterion ($E[|e_{k,l}^{(n)}|^2]$). This leads to the optimal filter coefficients $\mathbf{w}_{[\mathrm{MSE}]k,l} = \mathbf{R}_{\bar{\mathbf{r}}}^{-1}\mathbf{R}_{\bar{\mathbf{r}}d_{k,l}}$ where $\mathbf{R}_{\bar{\mathbf{r}}d_{k,l}}$ is the cross-correlation vector between the input vector $\bar{\mathbf{r}}$ and the desired response $d_{k,l}$ and $\mathbf{R}_{\bar{\mathbf{r}}}$ is the input signal cross-correlation matrix. Adaptive filtering can be implemented by using a number of algorithms.

The steepest descent algorithm

In this case we have

$$\mathbf{w}_{k,l}^{(n+1)} = \mathbf{w}_{k,l}^{(n)} - \mu \nabla_{k,l} \tag{14.48}$$

where ∇ is the gradient of

$$\left(J_{k,l} = E\{|c_{k,l} A_k b_k - \mathbf{w}_{k,l}^{\mathrm{H}} \mathbf{r}|^2\} \right) \tag{14.49}$$

This can be represented as

$$\nabla_{k,l} = \frac{\partial J_{k,l}}{\partial \operatorname{Re}\{\mathbf{w}_{k,l}\}} + j \frac{\partial J_{k,l}}{\partial \operatorname{Im}\{\mathbf{w}_{k,l}\}} = 2 \frac{\partial J_{k,l}}{\partial \mathbf{w}_{k,l}^*} \tag{14.50}$$

If the processing window $M = 1$, we have $\bar{\mathbf{r}}^{(n)} = \mathbf{r}^{(n)} \hat{=} \mathbf{r}$ and equation (14.50) becomes

$$\begin{aligned} \nabla_{k,l} &= -2E \left\lfloor \mathbf{r}(c_{k,l} A_k b_k)^* \right\rfloor + 2E[\mathbf{r}\mathbf{r}^{\mathrm{H}}]\mathbf{w}_{k,l} \\ &= -2\mathbf{R}_{\mathbf{r}d_{k,l}} + 2\mathbf{R}_{\mathbf{r}}\mathbf{w}_{k,l} \end{aligned} \tag{14.51}$$

where $d_{k,l} = c_{k,l} A_k b_k$.

If we assume that $A_k = 1, \forall k$

$$\mathbf{w}_{k,l}^{(n+1)} = \mathbf{w}_{k,l}^{(n)} - 2\mu(\mathbf{R}_{\mathbf{r}d_{k,l}} - \mathbf{R}_{\mathbf{r}}\mathbf{w}_{k,l}^{(n)}) \tag{14.52}$$

As a stochastic approximation, equation (14.51) can be represented as

$$\nabla_{k,l} \approx -2\mathbf{r}(c_{k,l} b_k)^* + 2\mathbf{r}\mathbf{r}^{\mathrm{H}}\mathbf{w}_{k,l}^{(n)} = -2\mathbf{r}(c_{k,l} b_k)^* + 2\mathbf{r}y_{k,l}^*$$

From this equation and assuming that $M > 1$, the least mean square (LMS) algorithm for updating the filter coefficients results in

$$\mathbf{w}_{k,l}^{(n+1)} = \mathbf{w}_{k,l}^{(n)} + 2\mu\bar{\mathbf{r}}^{(n)}(c_{k,l}^{(n)} b_k^{(n)} - y_{k,l}^{(n)})^* \in \mathbb{C}^{MSG} \tag{14.53}$$

We decompose equation (14.53) into adaptive and fixed components as

$$\mathbf{w}_{k,l}^{(n)} = \bar{\mathbf{s}}_{k,l} + \mathbf{x}_{k,l}^{(n)} \in \mathbb{C}^{MSG}$$

where $\mathbf{x}_{k,l}^{(n)}$ is the adaptive filter component and

$$\bar{\mathbf{s}}_{k,l} = [0_{(DSG+\tau_{k,l})\times 1}^{\mathrm{T}}, \mathbf{s}_k^{\mathrm{T}}, 0_{(DSG-\tau_{k,l})\times 1}^{\mathrm{T}}]^{\mathrm{T}}$$

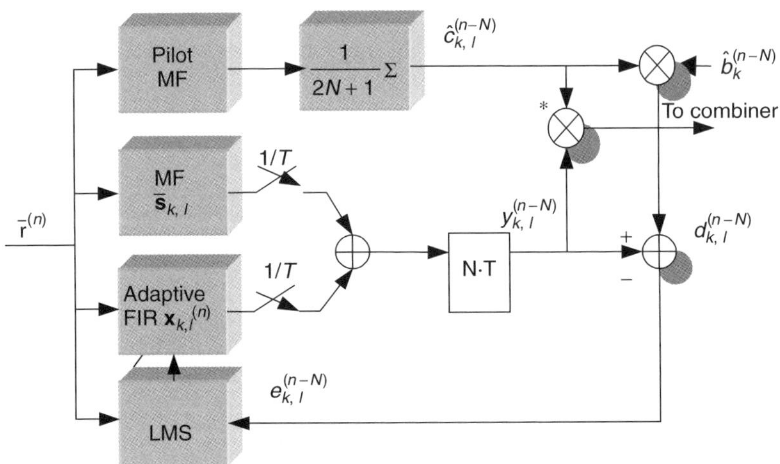

Figure 14.11   Block diagram of one receiver branch in the adaptive LMMSE-RAKE receiver.

is the fixed spreading sequence of the kth user with the delay $\tau_{k,l}$. In this case every branch from Figure 14.10 can be represented as shown in Figure 14.11.

In this case equation (14.53) gives

$$\begin{aligned}
\mathbf{x}_{k,l}^{(n+1)} &= \mathbf{x}_{k,l}^{(n)} - 2\mu_{k,l}^{(n)}(c_{k,l}^{(n)}b_k^{(n)} - y_{k,l}^{(n)})^*\bar{\mathbf{r}}^{(n)} \\
&= \mathbf{x}_{k,l}^{(n)} - 2\mu_{k,l}^{(n)}e_{k,l}^{*(n)}\bar{\mathbf{r}}^{(n)} \\
\mu_{k,l}^{(n)} &= \mu/(\bar{\mathbf{r}}^{\mathrm{H}(n)}\bar{\mathbf{r}}^{(n)}); \quad 0 < \mu < 1 \\
e_{k,l}^{(n)} &= d_{k,l}^{(n)} - y_{k,l}^{(n)}
\end{aligned} \tag{14.54}$$

The reference signal is

$$d_{k,l}^{(n)} = \hat{c}_{k,l}^{(n)}b_k^{(n)} \text{ or } d_{k,l}^{(n)} = \hat{c}_{k,l}^{(n)}\hat{b}_k^{(n)} \tag{14.55}$$

and the channel estimator is using a pilot channel

$$\hat{c}_{k,l}^{(n)} = \frac{1}{2N+1} \sum_{i=-N}^{N} \bar{s}_{p,l}^{\mathrm{T}}\bar{\mathbf{r}}^{(n-i)} \tag{14.56}$$

To illustrate the system operation, the following example is used [5]: Carrier frequency 2.0 GHz, symbol rate 16 kb s^{-1}, 31 chip Gold code and rectangular chip waveform. Synchronous downlink with equal energy two-path ($L = 2$) Rayleigh fading channel with vehicle speeds of 40 km h^{-1} (which results in the maximum normalized Doppler shift of $4.36 \cdot 10^{-3}$) and maximum delay spread of 10 chip intervals. The number of users examined was 1 to 30 including the unmodulated pilot channel. The average energy was the same for the pilot channel and the user data channels. A simple moving average smoother

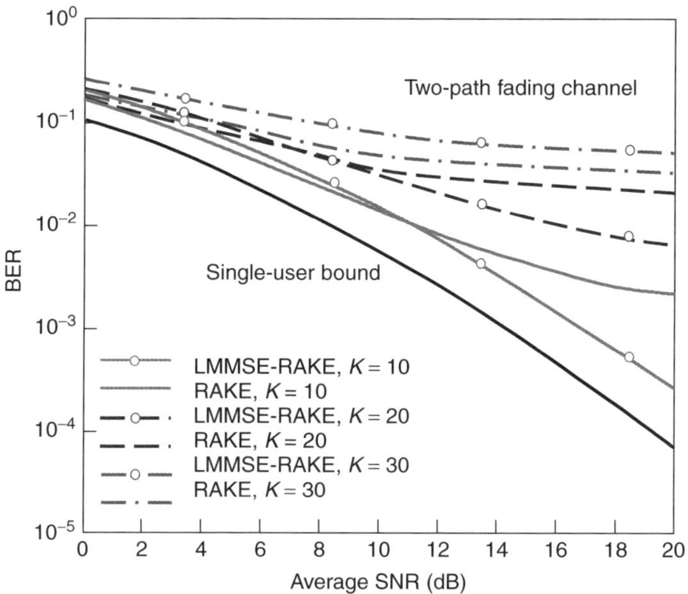

Figure 14.12 Simulated bit error rates as a function of the average SNR for the conventional RAKE and the adaptive LMMSE-RAKE in a two-path fading channel for the vehicle speeds $40\,\mathrm{km\,h^{-1}}$ with different numbers of users [5]. Reproduced from Latva-aho, M. (1998) Advanced Receivers for Wideband CDMA Systems. Ph.D. Thesis, University of Oulu, Oulu, by permission of IEEE.

of length 11 symbols was used in a conventional channel estimator. Perfect channel estimation and ideal truncated precombining LMMSE receivers were used in the analysis to obtain the lower bound for error probability. The receiver-processing window is three symbols ($M = 3$) unless otherwise stated. The adaptive algorithm used in the simulations was normalized LMS with

$$\mu_{k,l}^{(n)} = \frac{1}{100 \cdot (2D+1)SG} (\bar{\mathbf{r}}_{k,l}^{H(n)} \bar{\mathbf{r}}_{k,l}^{(n)})^{-1} \tag{14.57}$$

The simulation results were produced by averaging over the BERs of randomly selected users with different delay spreads.

The simulation results are shown in Figure 14.12. In general, one can notice that the improvement gains are lower than in the case of multiuser detectors.

14.5.2 Blind adaptive receivers

Adaptive LMMSE-RAKE

In this case in equation (14.54) we use estimates of bits $\hat{b}_{k,l}$ instead of $b_{k,l}$ [18–20]

$$\mathbf{x}_{k,l}^{(n+1)} = \mathbf{x}_{k,l}^{(n)} + 2\mu_{k,l}^{(n)} (c_{k,l}^{(n)} \hat{b}_{k,l}^{(n)} - y_{k,l}^{(n)})^* \bar{\mathbf{r}}^{(n)} \tag{14.58}$$

The MSE criterion now gives

$$\mathbf{w}_{[\text{MSE}]k,\ell} = \mathbf{R}_{\bar{r}}^{-1}\mathbf{R}_{\bar{r}d_{k,l}} = \mathbf{R}_{\bar{r}}^{-1}\bar{s}_{k,l}E\left\lfloor|c_{k,l}|^2\right\rfloor \tag{14.59}$$

Similarly, the minimum output energy criteria defined as

$$\text{MOE}(E\left\lfloor|y_{k,l}|^2\right\rfloor) \tag{14.60}$$

gives

$$\mathbf{w}_{[\text{MOE}]k,l} = \mathbf{R}_{\bar{r}}^{-1}\bar{s}_{k,l}/(\bar{s}_{k,l}^{\mathrm{T}}\mathbf{R}_{\bar{r}}^{-1}\bar{s}_{k,l}). \tag{14.61}$$

An implementation example can be seen in Reference [21]. The stochastic approximation of the gradient of equation (14.60) for the MOE criterion gives

$$\nabla_{k,l} = \bar{\mathbf{r}}^{(n)}\bar{\mathbf{r}}^{\mathrm{H}(n)}\mathbf{w}_{k,l} \tag{14.62}$$

If we want to keep the useful signal autocorrelation unchanged, equation (14.61) should be constrained to satisfy $\bar{s}_{k,l}^{\mathrm{T}}\mathbf{x}_{k,l}^{(n)} = 0$. The orthogonality condition is maintained at each step of the algorithm by projecting the gradient onto the linear subspace orthogonal to $\bar{s}_{k,l}^{\mathrm{T}}$. In practice, this is accomplished by subtracting an estimate of the desired signal component from the received signal vector. An implementation can be seen in Reference [22]. So we have

$$\mathbf{x}_{k,l}^{(n+1)} = \mathbf{x}_{k,l}^{(n)} - 2\mu_{k,l}^{(n)}\bar{\mathbf{r}}^{\mathrm{H}(n)}(\bar{s}_{k,l} + \mathbf{x}_{k,l}^{(n)})[\bar{\mathbf{r}}^{(n)} - \mathbf{F}_{k,l}(\mathbf{F}_{k,l}^{\mathrm{T}}\bar{\mathbf{r}}^{(n)})] \tag{14.63}$$

where

$$\mathbf{F}_{k,l} = \begin{pmatrix} \mathbf{0}_{\tau_{k,l}\times1}^{\mathrm{T}}, \mathbf{s}_k^{\mathrm{T}}, \mathbf{0}_{(2DSG-\tau_{k,l})\times1}^{\mathrm{T}} \\ \mathbf{0}_{(SG-\tau_{k,l})\times1}^{\mathrm{T}}, \mathbf{s}_k^{\mathrm{T}}, \mathbf{0}_{((2D-1)SG-\tau_{k,l})\times1}^{\mathrm{T}} \\ \mathbf{0}_{(2DSG+\tau_{k,l})\times1}^{\mathrm{T}}, [s_k(T_\mathrm{s}), \ldots, s_k(T_\mathrm{s}(SG-\tau_{k,l}))] \end{pmatrix}^{\mathrm{T}} \in \mathrm{R}^{MSG\times M} \tag{14.64}$$

is a block diagonal matrix of sampled spreading sequence vectors. Effectively M separate filters are adapted.

Griffiths' algorithm

In this case, instead of assuming that vector $\mathbf{R}_{\bar{r}d_{k,l}}$ is known, the instantaneous estimate for the covariance is used, that is,

$$\mathbf{R}_{\bar{r}} \approx \bar{\mathbf{r}}^{(n)}\bar{\mathbf{r}}^{\mathrm{H}(n)} \tag{14.65}$$

In this case, the cross-correlation is $\mathbf{R}_{\bar{r}d_{k,l}} = E[|c_{k,l}|^2]\bar{s}_{k,l}$, and Griffiths' algorithm results in

$$\mathbf{x}_{k,l}^{(n+1)} = \mathbf{x}_{k,l}^{(n)} + 2\mu_{k,l}^{(n)}(E[|c_{k,l}|^2]\mathbf{F}_{k,l}\mathbf{1}_M - \bar{\mathbf{r}}_{k,l}^{*(n)}(\bar{s}_{k,l} + \mathbf{x}_{k,l}^{(n)})^{\mathrm{H}}\bar{\mathbf{r}}^{(n)}) \tag{14.66}$$

In practice, the energy of multipath components ($E[|c_{k,l}|^2]$) is not known and must be estimated.

Constant modulus algorithm

In this case the optimization criterion is $E[(|y_{k,l}|^2 - \omega)^2]$ where ω is the so-called constant modulus (CM), set according to the received signal power, that is, $\omega = E[|c_{k,l}|^2]$ or $\omega^{(n)} = |c_{k,l}^{(n)}|^2$. By using the CM algorithm, it is possible to avoid the use of the data decisions in the reference signal in the adaptive LMMSE-RAKE receiver by taking the absolute value of the estimated channel coefficients ($|\hat{c}_{k,l}^{(n)}|$) in adapting the receiver. In the precombining LMMSE receiver framework, the cost function for the BPSK data modulation is

$$E[||\hat{\mathbf{h}}|^2 - |\mathbf{h}|^2|^2]$$ (14.67)

The stochastic approximation of the gradient for the CM criterion is

$$\nabla_{k,l}^{(n+1)} = (|y_{k,l}^{(n)}|^2 - |\hat{c}_{k,l}^{(n)}|^2)\bar{\mathbf{r}}^{(n)}\bar{\mathbf{r}}^{H(n)}\mathbf{w}_{k,l}$$ (14.68)

Hence, the constant modulus algorithm can be expressed as

$$\mathbf{x}_{k,l}^{(n+1)} = \mathbf{x}_{k,l}^{(n)} - 2\mu_{k,l}^{(n)}y_{k,l}^{*(n)}(|y_{k,l}^{(n)}|^2 - |\hat{c}_{k,l}^{(n)}|^2)\bar{\mathbf{r}}^{(n)}$$ (14.69)

Constrained LMMSE-RAKE, Griffiths' algorithm and constant modulus algorithm

The adaptive LMMSE-RAKE, the Griffiths' algorithm (GRA) and the constant modulus algorithm contain no constraints. By applying the orthogonality constraint $\bar{\mathbf{s}}_{k,l}^{T}\mathbf{x}_{k,l}^{(n)} = 0$ to each of these algorithms, an additional term $\bar{\mathbf{s}}_{k,l}^{T}\mathbf{x}_{k,l}^{(n)}\bar{\mathbf{s}}_{k,l}$ is subtracted from the new $\mathbf{x}_{k,l}^{(n+1)}$ update at every iteration. The constrained LMMSE-RAKE receiver becomes [23, 24]

$$\mathbf{x}_{k,l}^{(n+1)} = \mathbf{x}_{k,l}^{(n)} + 2\mu_{k,l}^{(n)}(\hat{c}_{k,l}^{(n)}\hat{b}_k^{(n)} - y_{k,l}^{(n)})^*\bar{\mathbf{r}}^{(n)} - \bar{\mathbf{s}}_{k,l}^{T}\mathbf{x}_{k,l}^{(n)}\bar{\mathbf{s}}_{k,l}$$ (14.70)

The GRA and the constant modulus algorithm can also be defined in a similar way.

14.5.3 Blind least squares receivers

All blind adaptive algorithms described in the previous section are based on the gradient of the cost function. In practical adaptive algorithms, the gradient is estimated, that is, the expectation in the optimization criterion is not taken but is replaced in most cases by some stochastic approximation. In fact, the stochastic approximation used in LMS algorithms is accurate only for small step-sizes μ. This results in rather slow convergence, which may be intolerable in practical applications.

Another drawback with the blind adaptive receivers presented above is the delay estimation. Those receiver structures as such support only conventional delay estimation based on matched filtering (MF). The MF-based delay estimation is sufficient for the downlink receivers in systems with an unmodulated pilot channel since the zero-mean multiple-access interference (MAI) can be averaged out if the rate of fading is low enough. If Code Division Multiple Access (CDMA) systems do not have the pilot channel, it would be beneficial to use some near–far resistant delay estimators.

14.5.4 Least square (LS) receiver

One possible solution to both the convergence and the synchronization problems is based on blind linear least square (LS) receivers. Cost function in this case is

$$J_{[\mathrm{LS}]k,l} = \sum_{j=n-N+1}^{n} (c_{k,l}^{(j)} b_k^{(j)} - \mathbf{w}_{k,l}^{\mathrm{H}(n)} \overline{\mathbf{r}}^{(j)})^2 \tag{14.71}$$

N is the observation window in symbol intervals. Filter weights are given as

$$\mathbf{w}_{k,l}^{(n)} = \hat{\mathbf{R}}_{\overline{\mathbf{r}}}^{-1(n)} \overline{\mathbf{s}}_{k,l} \tag{14.72}$$

$\hat{\mathbf{R}}_{\overline{\mathbf{r}}}^{-1(n)}$ denotes the estimated covariance matrix over a finite data block called the *sample-covariance matrix*. This matrix can be expressed as

$$\hat{\mathbf{R}}_{\overline{r}}^{(n)} = \sum_{j=n-N+1}^{n} \overline{\mathbf{r}}^{(j)} \overline{\mathbf{r}}^{\mathrm{H}}(j) \tag{14.73}$$

Analogous to the MOE criterion, the LS criterion can be modified as

$$J_{[LS']k,l} = \sum_{j=n-N+1}^{n} (\mathbf{w}_{k,l}^{\mathrm{H}(n)} \overline{\mathbf{r}}^{(j)})^2, \text{ subject to } \mathbf{w}_{k,l}^{\mathrm{T}} \overline{\mathbf{s}}_{k,l} = 1 \tag{14.74}$$

which results in

$$\mathbf{w}_{k,l}^{(n)} = \frac{\hat{\mathbf{R}}_{\overline{\mathbf{r}}}^{-1(n)} \overline{\mathbf{s}}_{k,l}}{\overline{\mathbf{s}}_{k,l}^{\mathrm{T}} \hat{R}_{\overline{\mathbf{r}}}^{-1(n)} \overline{\mathbf{s}}_{k,l}} \tag{14.75}$$

The adaptation of the blind LS receiver means updating the inverse of the sample-covariance. The blind adaptive LS receiver is significantly more complex than the stochastic gradient-based blind adaptive receivers. Recursive methods, such as the recursive least squares (RLS) algorithm, for updating the inverse and iteratively finding the filter weights are known. Also, the methods based on eigen-decomposition of the covariance matrix have been proposed to avoid explicit matrix inversion.

14.5.5 Method based on the matrix inversion lemma

The general relation

$$(\mathbf{A} + \mathbf{BCD})^{-1} = \mathbf{A}^{-1} - \mathbf{A}^{-1}\mathbf{B}(\mathbf{DA}^{-1}\mathbf{B} + \mathbf{C}^{-1})^{-1}\mathbf{DA}^{-1} \qquad (14.76)$$

becomes

$$\hat{\mathbf{R}}_{\bar{r}}^{-1(n)} = (\hat{\mathbf{R}}_{\bar{r}}^{(n-1)} + \bar{\mathbf{r}}^{(n)}\bar{\mathbf{r}}^{H}(n))^{-1} = \mathbf{R}_{\bar{r}}^{-1(n-1)} - \frac{\mathbf{R}_{\bar{r}}^{-1(n-1)}\bar{\mathbf{r}}^{(n)}\bar{\mathbf{r}}^{H}(n)\mathbf{R}_{\bar{r}}^{-1(n-1)}}{1 + \bar{\mathbf{r}}^{H}(n)\mathbf{R}_{\bar{r}}^{-1(n-1)}\bar{\mathbf{r}}^{(n)}} \qquad (14.77)$$

In time-variant channels, the old values of the inverses must be weighted by the so-called forgetting factor ($0 < \gamma < 1$), which results in

$$\hat{\mathbf{R}}_{\bar{r}}^{-1(n)} = \frac{1}{\gamma}\left(\hat{\mathbf{R}}_{\bar{r}}^{-1(n-1)} - \frac{\hat{\mathbf{R}}_{\bar{r}}^{-1(n-1)}\bar{\mathbf{r}}^{(n)}\bar{\mathbf{r}}^{H}(n)\hat{\mathbf{R}}_{\bar{r}}^{-1(n-1)}}{\gamma + \bar{\mathbf{r}}^{H}(n)\hat{\mathbf{R}}_{\bar{r}}^{-1(n-1)}\bar{\mathbf{r}}^{(n)}}\right) \qquad (14.78)$$

It is sufficient to initialize the algorithm as $\hat{\mathbf{R}}_{\bar{r}}^{-1(0)} = \mathbf{I}$.

For illustration purposes, a number of numerical examples are shown in Figures 14.13 to 4.20 [5] and in Table 14.1. System parameters are shown in the figures.

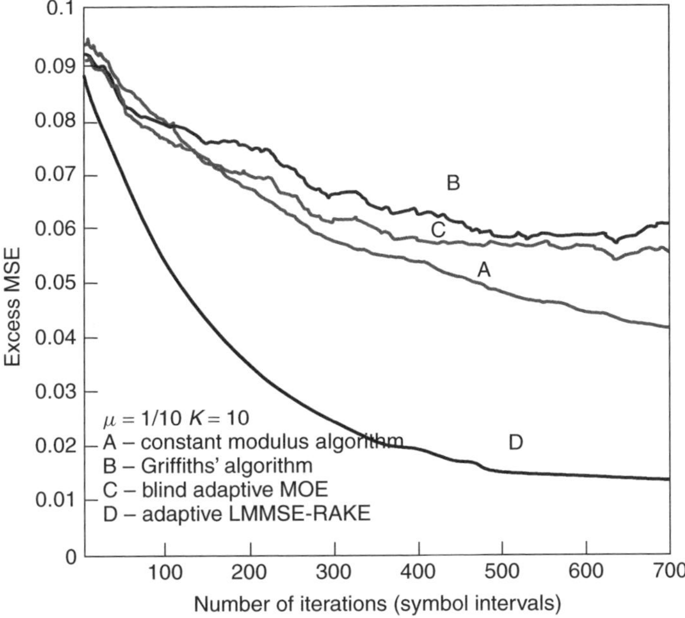

Figure 14.13 Excess mean squared error as a function of the number of iterations for different blind adaptive receivers in a two-path fading channel with vehicle speeds of $40\,\mathrm{km\,h}^{-1}$, the number of active users $K = 10$, $SNR = 20\,\mathrm{dB}$, $\mu = 10^{-1}$ [5]. Reproduced from Latva-aho, M. (1998) Advanced Receivers for Wideband CDMA Systems. Ph.D. Thesis, University of Oulu, Oulu , by permission of IEEE.

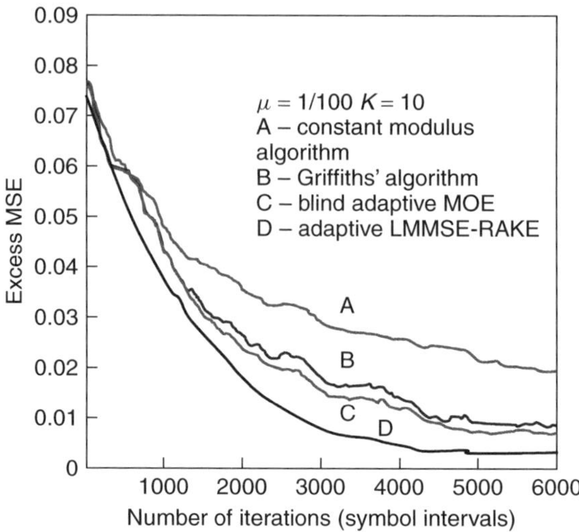

Figure 14.14 Excess mean squared error as a function of the number of iterations for different blind adaptive receivers in a two-path fading channel with vehicle speeds of $40\,\mathrm{km\,h^{-1}}$, the number of active users $K = 10$, $SNR = 20\,\mathrm{dB}$, $\mu = 100^{-1}$ [5]. Reproduced from Latva-aho, M. (1998) Advanced Receivers for Wideband CDMA Systems. Ph.D. Thesis, University of Oulu, Oulu, by permission of IEEE.

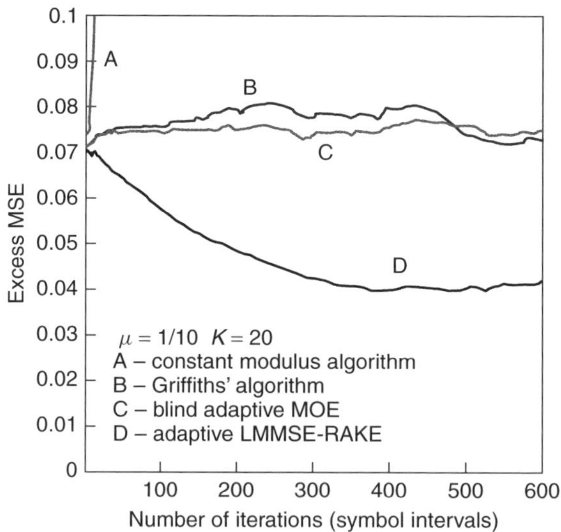

Figure 14.15 Excess mean squared error as a function of the number of iterations for different blind adaptive receivers in a two-path fading channel with vehicle speeds of $40\,\mathrm{km\,h^{-1}}$, the number of active users $K = 20$, $SNR = 20\,\mathrm{dB}$, $\mu = 10^{-1}$ [5]. Reproduced from Latva-aho, M. (1998) Advanced Receivers for Wideband CDMA Systems. Ph.D. Thesis, University of Oulu, Oulu, by permission of IEEE.

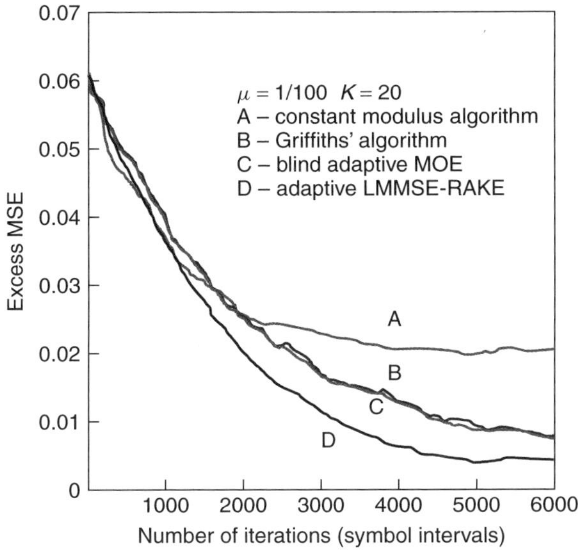

Figure 14.16 Excess mean squared error as a function of the number of iterations for different blind adaptive receivers in a two-path fading channel with vehicle speeds of $40\,\mathrm{km\,h^{-1}}$, the number of active users $K = 20$, $SNR = 20\,\mathrm{dB}$, $\mu = 100^{-1}$ [5]. Reproduced from Latva-aho, M. (1998) Advanced Receivers for Wideband CDMA Systems. Ph.D. Thesis, University of Oulu, Oulu, by permission of IEEE.

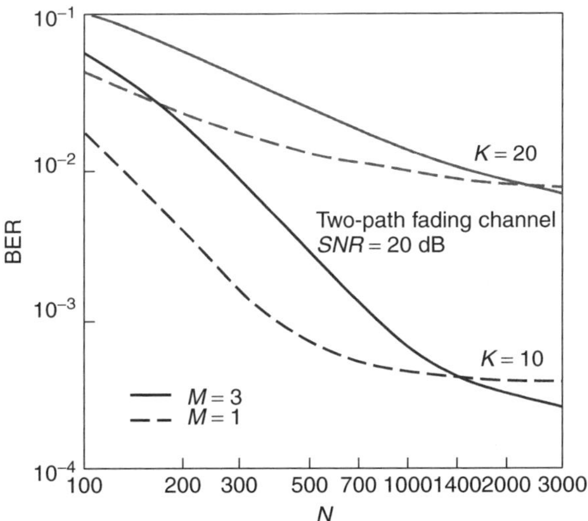

Figure 14.17 BER as a function of the sample-covariance averaging interval for $K = 10$, 20 for receiver spans of one ($M = 1$) and three symbol intervals ($M = 3$) in a two-path fading channel at an SNR of $20\,\mathrm{dB}$ [5]. Reproduced from Latva-aho, M. (1998) Advanced Receivers for Wideband CDMA Systems. Ph.D. Thesis, University of Oulu, Oulu, by permission of IEEE.

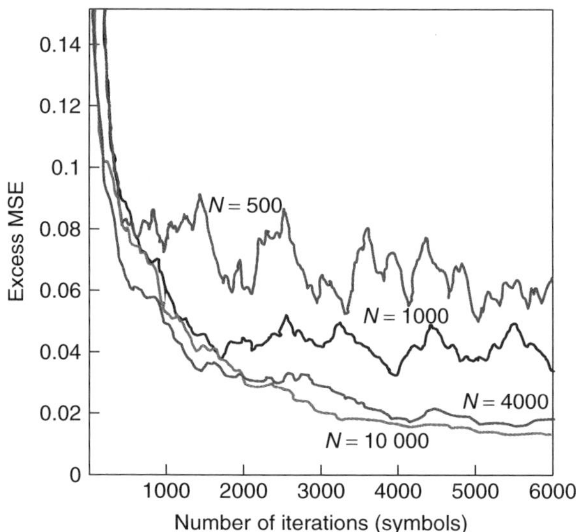

Figure 14.18 Excess mean squared error as a function of the number of iterations for the blind adaptive LS receiver of span three symbol intervals ($M = 3$) with different forgetting factors $(1 - 2/N)$ in a 10-user case at an $SNR = 20$ dB and vehicle speeds of $40\,\text{km}\,\text{h}^{-1}$ [5]. Reproduced from Latva-aho, M. (1998) Advanced Receivers for Wideband CDMA Systems. Ph.D. Thesis, University of Oulu, Oulu, by permission of IEEE.

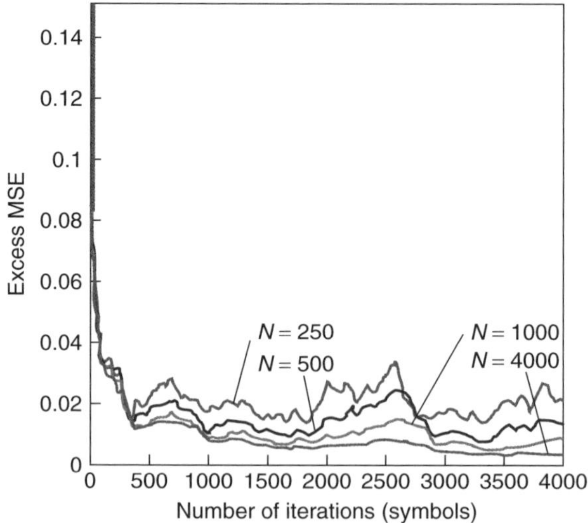

Figure 14.19 Excess mean squared error as a function of the number of iterations for the blind adaptive LS receiver of span one symbol interval ($M = 1$) with different forgetting factors $(1 - 2/N)$ in a 10-user case at an $SNR = 20$ dB and vehicle speeds of $40\,\text{km}\,\text{h}^{-1}$ [5]. Reproduced from Latva-aho, M. (1998) Advanced Receivers for Wideband CDMA Systems. Ph.D. Thesis, University of Oulu, Oulu, by permission of IEEE.

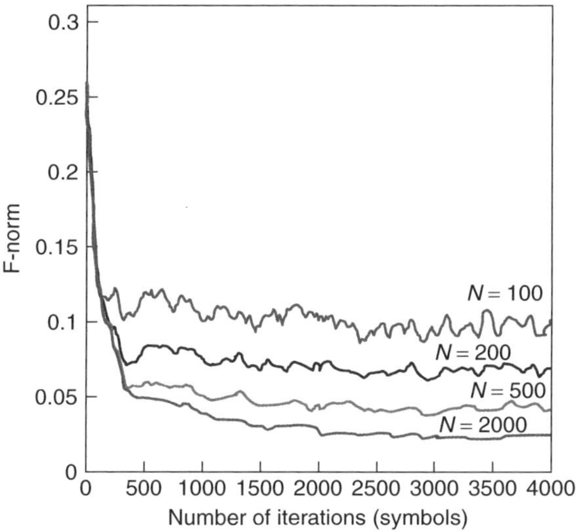

Figure 14.20 Forbenius norm for the iterative inverse updating algorithm in a 10-user case at an SNR of 20 dB and vehicle speeds of 40 km h^{-1} [5]. Reproduced from Latva-aho, M. (1998) Advanced Receivers for Wideband CDMA Systems. Ph.D. Thesis, University of Oulu, Oulu, by permission of IEEE.

Table 14.1 The BERs of different blind adaptive receivers at an SNR of 20 dB in a two-path Rayleigh fading channel at vehicle speeds of 40 km h^{-1}. The acronyms used are adaptive LMMSE-RAKE (LR), adaptive MOE (MOE), Griffiths' algorithm (GRA), constant modulus algorithm with average channel tap powers (CMA2), constrained adaptive LMMSE-RAKE (C-LR), constrained constant modulus algorithm (C-GRA), constrained constant modulus algorithm with average channel tap powers (C-CMA2) and conventional RAKE (RAKE) [5]. Reproduced from Latva-aho, M. (1998) Advanced Receivers for Wideband CDMA Systems. Ph.D. Thesis, University of Oulu, Oulu, by permission of IEEE

Adaptive receiver	$K = 30$		$K = 15$		
	$\mu = 100^{-1}$	$\mu = 10^{-1}$	$\mu = 100^{-1}$	$\mu = 10^{-1}$	$\mu = 2^{-1}$
LR	$4.5 \cdot 10^{-2}$	$3.9 \cdot 10^{-1}$	$6.3 \cdot 10^{-4}$	$7.2 \cdot 10^{-4}$	$3.0 \cdot 10^{-2}$
MOE	$2.8 \cdot 10^{-2}$	$4.2 \cdot 10^{-2}$	$6.0 \cdot 10^{-4}$	$2.1 \cdot 10^{-3}$	$9.1 \cdot 10^{-2}$
GRA	$2.8 \cdot 10^{-2}$	$4.7 \cdot 10^{-2}$	$6.4 \cdot 10^{-4}$	$3.3 \cdot 10^{-3}$	$1.2 \cdot 10^{-1}$
CMA	$3.9 \cdot 10^{-2}$	$4.0 \cdot 10^{-1}$	$1.2 \cdot 10^{-3}$	$2.1 \cdot 10^{-2}$	$5.0 \cdot 10^{-1}$
CMA2	$3.3 \cdot 10^{-2}$	$4.0 \cdot 10^{-1}$	$1.8 \cdot 10^{-3}$	$2.1 \cdot 10^{-2}$	$5.0 \cdot 10^{-1}$
C-LR	$3.2 \cdot 10^{-2}$	$4.2 \cdot 10^{-2}$	$6.3 \cdot 10^{-4}$	$6.4 \cdot 10^{-4}$	$1.9 \cdot 10^{-3}$
C-CMA	$3.3 \cdot 10^{-2}$	$5.0 \cdot 10^{-1}$	$6.1 \cdot 10^{-4}$	$3.8 \cdot 10^{-1}$	$5.0 \cdot 10^{-1}$
C-GRA	$2.8 \cdot 10^{-2}$	$4.2 \cdot 10^{-2}$	$6.1 \cdot 10^{-4}$	$2.3 \cdot 10^{-3}$	$9.7 \cdot 10^{-2}$
C-CMA2	$2.9 \cdot 10^{-2}$	$5.0 \cdot 10^{-1}$	$7.7 \cdot 10^{-4}$	$2.7 \cdot 10^{-1}$	$5.0 \cdot 10^{-1}$
RAKE	$3.1 \cdot 10^{-2}$	$3.1 \cdot 10^{-2}$	$7.1 \cdot 10^{-3}$	$7.1 \cdot 10^{-3}$	$7.1 \cdot 10^{-3}$

In general, one can see that the blind algorithms are inferior compared with the LMMSE-RAKE, using pilot symbols.

A number of related topics are discussed in References [25–44].

SYMBOLS

MMSE – Minimum Mean-Square Error
r – input signal
b – data vector
M – detector matrix
S – signature sequence vector
R – signature sequence cross-correlation matrix
A – diagonal matrix of signal amplitudes
y – vector of matched filter outputs
σ^2 – noise variance
ρ – correlation coefficient
$c_{k,l}$ – channel coefficient for user k and path l
$\tau_{k,l}$ – delay of user k and path l
S – number of samples per chip
N_b – data block of N_b symbols
K – number of users
L – linear detector transformation matrix
post – postcombining
pre – precombining
G – spreading factor
SMT – spatial-multiuser-temporal
MST – multiuser-spatial-temporal
TMS – temporal-multiuser-spatial
h = CAb
$J_{k,l}$ – cost function for user k and path l
MOE – maximum output energy
w – matrix of waiting coefficients
CM – constant module
μ – adaptation step
LS – least square
γ – forgetting factor
MOE – adaptive MOE
GRA – Griffiths' algorithm
CMA2 – constant modulus algorithm with average channel tap powers
C-LR – constrained adaptive LMMSE-RAKE
C-GRA – constrained constant modulus algorithm
C-CMA2 – constrained constant modulus algorithm with average channel tap powers
RAKE – conventional RAKE

REFERENCES

1. Madhow, U. and Honing, M. L. (1994) MMSE interference suppression for direct-sequence spread-spectrum CDMA. *IEEE Trans. Commun.*, **42**(12), 3178–3188.

2. Klein, A., Kaleh, G. K. and Baier, P. W. (1996) Zero forcing and minimum mean-square-error equalization for multiuser detection in code-division multiple access channels. *IEEE Trans. Veh. Technol.*, **45**(2), 276–287.

3. Poor, H. V. and Verdú, S. (1997) Probability of error in MMSE multiuser detection. *IEEE Trans. Inform. Theory*, **43**(3), 858–871.

4. Huang, H. C. (1996) *Combined Multipath Processing, Array Processing, and Multiuser Detection for DS-CDMA Channels*. Ph.D. Thesis, Princeton University, Princeton, NJ.

5. Latva-aho, M. (1998) Advanced Receivers for Wideband CDMA Systems. Ph.D. Thesis, University of Oulu, Oulu.

6. Miller, S. Y. (1989) Detection and Estimation in Multiple-Access Channels. Ph.D. Thesis, Princeton University, Princeton, NJ.

7. Miller, S. Y. and Schwartz, S. C. (1995) Integrated spatial-temporal detectors for asynchronous Gaussian multiple-access channels. *IEEE Trans. Commun.*, **43**, 396–411.

8. Zvonar, Z. (1996) Combined multiuser detection and diversity reception for wireless CDMA systems. *IEEE Trans. Veh. Technol.*, **45**(1), 205–211.

9. Rapajic, P. B. and Vucetic, B. S. (1995) Linear adaptive transmitter-receiver structures for asynchronous SCMA systems. *Eur. Trans. Telecommun.*, **6**(1), 21–27.

10. Miller, S. L. (1995) An adaptive direct-sequence code-division multiple-access receiver for multiuser interference rejection. *IEEE Trans. Commun.*, **43**, 1746–1755.

11. Rapajic, P. B. and Vucetic, B. S. (1994) Adaptive receiver structures for asynchronous CDMA systems. *IEEE J. Select. Areas Commun.*, **12**(4), 685–697.

12. Lee, K. B. (1996) Orthogonalization based adaptive interference suppression for direct-sequence code-division multiple-access systems. *IEEE Trans. Commun.*, **44**(9), 1082–1085.

13. Miller, S. L. (1996) Training analysis of adaptive interference suppression for direct-sequence code-division multiple-access systems. *IEEE Trans. Commun.*, **44**(4), 488–495.

14. Latva-aho, M. and Juntti, M. (1997) Modified adaptive LMMSE receiver for DS-CDMA systems in fading channels. *Proc. IEEE International Symposium on Personal, Indoor, and Mobile Radio Communications (PIMRC)*, Vol. 2, Helsinki, Finland, pp. 554–558.

15. Oppermann, I. and Latva-aho, M. (1997) Adaptive LMMSE receiver for wideband CDMA systems. *Proc. Communication Theory Mini-Conference (CTMC) in conjunction with IEEE Global Telecommunication Conference (GLOBECOM)*, Phoenix, AZ, pp. 133–138.

16. Oppermann, I. and Vucetic, B. S. (1996) Capacity of a coded direct sequence spread spectrum system over fading satellite channels using an adaptive LMS-MMSE receiver. *IEICE Trans. Fundam. Electron. Commun. Comput. Sci.*, **E79-A**(12), 2043–2049.

17. Honig, M. (1998) Adaptive linear interference suppression for packet DS-CDMA. *Eur. Trans. Telecommun.*, **9**(2), 173–181.

18. Hong, M., Madhow, U. and Verdú, S. (1995) Blind adaptive multiuser detection. *IEEE Trans. Inform. Theory*, **41**(3), 944–960.

19. Park, S. C. and Dohery, J. F. (1997) Generalized projection algorithm for blind interference suppression in DS/CDMA communications. *IEEE Trans. Circuits Syst. Part II Analog Digital Signal Process.*, **44**(6), 453–460.

20. Wang, X. and Poor, H. V. (1998) Blind equalization and multiuser detection in dispersive CDMA channels. *IEEE Trans. Commun.*, **46**(1), 91–103.

21. Fanucci, L. *et al.* (2001) VLSI implementation of CDMA blind adaptive interference-mitigating detector. *IEEE JSAC*, **19**(2), 179–190.

22. DeGaudenci, R. *et al.* (1998) Design of a low complexity adaptive interference mitigating detector for DS/SS receiver in CDMA radio networks. *IEEE Trans. Commun.*, **46**(1), 125–134.

23. Schodorf, J. B. and Williams, D. B. (1997) A constrained optimisation approach to multiuser detection. *IEEE Trans. Signal Process.*, **45**(1), 258–262.

24. Iltis, R. A. (1998) Performance of constrained and unconstrained adaptive multiuser detectors for quasi-synchronous CDMA. *IEEE Trans. Commun.*, **46**(1), 135–143.

25. Chu, L. C. and Mitra, U. (1996) Improved MMSE-based multi-user detectors for mismatched delay channels, *Proc. Conference on Information Sciences and Systems (CISS)*. Vol. 1. Princeton, NJ: Princeton University Press, pp. 326–331.

26. Gray, S. D., Preisig, J. C. and Brady, D. (1997) Multiuser detection in a horizontal underwater acoustic channel using array observations. *IEEE Trans. Signal Process.*, **45**(1), 148–160.
27. Brown, T. and Kaveh, M. (1995) A decorrelating detector for use with antenna arrays. *Int. J. Wireless Inform. Networks*, **2**(4), 239–246.
28. Jung, P. and Blanz, J. (1995) Joint detection with coherent receiver antenna diversity in CDMA mobile radio systems. *IEEE Trans. Veh. Technol.*, **44**(1), 76–88.
29. Zecevic, N. and Reed, J. (1997) Blind adaptation algorithms for direct-sequence spread-spectrum CDMA single-user detection. *Proc. IEEE Vehicular Technology Conference (VTC)*, Vol. 3, Phoenix, AZ, pp. 2133–2137.
30. Honig, M., Shensa, M., Miller, S. and Milstein, L. (1997) Performance of adaptive linear interference suppression for DS-CDMA in the presence of flat Rayleigh fading. *Proc. IEEE Vehicular Technology Conference (VTC)*, Vol. 3, Phoenix, AZ, pp. 2191–2195.
31. Berstein, X. and Haimovich, A. M. (1996) Space-time optimum combining for CDMA communications. *Wireless Personal Commun.*, **3**(1–2), 73–89.
32. Ge, H. (1997) Adaptive schemes of implementing the LMMSE multiuser detector for CDMA. *Proc. IEEE International Conference on Communications (ICC)*, Montreal, Canada.
33. Honig, M. (1996) Performance of adaptive interference suppression for DS-CDMA with a time-varying user population. *Proc. IEEE International Symposium on Spread Spectrum Techniques and Applications (ISSSTA)*, Vol. 1, Mainz, Germany, pp. 267–271.
34. Jung, P., Blanz, J., Nasshan, M. and Baier, P. W. (1994) Simulation of the uplink of JD-CDMA mobile radio systems with coherent receiver antenna diversity. *Wireless Personal Commun.*, **1**(2), 61–89.
35. Madhow, U. (1997) Blind adaptive interference suppression for the near-far resistant acquisition and demodulation of direct-sequence CDMA. *IEEE Trans. Signal Process.*, **45**(1), 124–136.
36. Mowbray, R. S., Pringle, R. D. and Grant, P. M. (1992) Increased CDMA system capacity through adaptive cochannel interference regeneration and cancellation. *IEE Proc. I*, **139**, 515–524.
37. Kohno, R., Imai, H., Hatori, M. and Pasupathy, S. (1990) Combination of an adaptive array antenna and a canceller of interference for direct-sequence spread-spectrum multiple-access system. *IEEE J. Select. Areas Commun.*, **8**(4), 675–682.
38. Yoon, Y. C., Kohno, R. and Imai, H. (1993) Cascaded co-channel interference cancellating and diversity combining for spread-spectrum multi-access system over multipath fading channels. *IEICE Trans. Commun.*, **E76-B**(2), 163–168.
39. Saifuddin, A. and Kohno, R. (1996) Performance evaluation of DS/CDMA scheme with diversity coding and MUI cancellation over fading multipath channel. *IEICE Trans. Fundam. Electron. Commun. Comput. Sci.*, **E79-A**(12), 1994–2001.
40. Nelson, L. B. and Poor, H. V. (1996) Iterative multiuser receivers for CDMA channels: an EM-based approach. *IEEE Trans. Commun.*, **44**(12), 1700–1710.
41. Soong, A. C. K. and Krzymien, W. A. (1996) A novel CDMA multiuser interference cancellation receiver with reference symbol aided estimation of channel parameters. *IEEE J. Select. Areas Commun.*, **14**(8), 1536–1547.
42. Miller, S. L. and Barbosa, A. N. (1996) A modified MMSE receiver for detection of DS-CDMA signals in fading channels. *Proc. IEEE Military Communications Conference (MIL-COM)*, Reston, VA, pp. 898–902.
43. Wang, X. and Poor, H. V. (1998) Blind multiuser detection: a subspace approach. *IEEE Trans. Inform. Theory*, **44**(2), 677–690.
44. Juntti, M. and Glisic, S. (1997) Advanced CDMA for wireless communications, in Glisic, S. G. and Leppänen, P. A. (eds) *Wireless Communications: TDMA Versus DCMA*. Kluwer, McLean, pp. 447–490.

15

Wideband CDMA network sensitivity

15.1 THEORY AND PRACTICE OF MULTIUSER DETECTION

Advanced wireless Code Division Multiple Access (CDMA) systems employing multiuser receivers described in Chapters 13 and 14 have gained a lot of attention during the recent years. In the beginning, the research was purely academic. In order to fully exploit CDMA capabilities, the third-generation wideband CDMA networks [1] have been planned to include interference cancellation (IC), as soon as it becomes feasible to implement (for details, see also Chapter 17). A number of overview papers [2–6] discuss basic features of multiuser detection (MUD) and IC techniques. Among linear IC schemes, decorrelating detectors [7–9] belong to a major group of multiuser receivers. As it was discussed in Chapter 13, the basic idea of the decorrelator is to use the inverted cross-correlation matrix to subtract interference caused by other active users, that is, multiple access interference (MAI). With perfect knowledge of code cross-correlations (of all active users), the effect of all MAI can be eliminated at the cost of noise enhancement. One important benefit of the ideal decorrelator is that it does not require knowledge of the users' power levels (or amplitudes) and is thus robust to power fluctuations. On the other hand, the complexity and need for updating the matrix inversion can become a bottleneck at high user populations and fast fading environment. Also, in reality, the cross-correlation matrix must be estimated and that will lead to further imperfections. These imperfections are the focus of this chapter.

Although other options, such as multistage or linear minimum mean square error (LMMSE) detectors discussed in Chapter 14, are also considered for these applications [4], in this chapter we will focus our attention on linear decorrelators and their operation in the presence of imperfections. In practice, performance of these detectors depends critically on the channel parameter estimation and errors that they produce. For these reasons, parameter estimation in multiuser communications has become an important research topic aiming to find feasible solutions for practical MUD applications. The performance of linear decorrelating detectors in the presence of time delay, carrier phase and carrier frequency errors is analyzed in Reference [10]. Methods for low-complexity amplitude and phase estimation with known or unknown delays are analyzed in Reference [11]. Joint signal detection and parameter estimation (amplitudes and phases) is evaluated in Reference [12]. Adaptive symbol and parameter estimation algorithms based on recursive least squares (RLS) and extended Kalman filters (EKF) have been studied in Reference [13]. Sensitivity of multiple-access channels to several mismatches due to imperfect carrier recovery, timing jitter and channel truncation is analyzed in Reference [14]. The impact of timing (delay) errors to the performance of linear MUD receivers is illustrated in References [15,16]. The same authors have also considered the estimation problem in general in Reference [17]. Sensitivity analysis of near–far resistant DS-CDMA receivers to propagation delay estimation errors [18] shows that even quite small errors will destroy the near–far resistance of the decorrelating detector. Joint amplitude and delay estimation is evaluated in Reference [19] by using the EKF and the same authors study quasi-synchronous CDMA systems applying linear decorrelators in Reference [20].

The above references study mainly the link performance with quite limited code lengths and number of users. This is necessary in order to get analytically tractable performance results or to keep complexity of simulations at reasonable levels. For capacity evaluation, however, somewhat releasing assumptions and approximations are usually needed to get results for higher user populations and longer spreading codes. Capacity of the linear decorrelating detector for quasi-synchronous CDMA is evaluated in Reference [21]. Furthermore, comparison against an adaptive receiver employing the minimum mean square error (MMSE) criterion is presented. Outage probability bounds when using a zero forcing detector are presented in Reference [22]. Capacity gains over conventional matched filter–based systems were shown to be significant. A linear interference canceller is applied to microcellular CDMA in Reference [23] in which uplink capacities are estimated for different propagation scenarios.

Besides the multiuser detectors, an advanced CDMA receiver will be using also a RAKE receiver with multipath combining. Linear multipath-decorrelating receivers in frequency-selective fading channels, discussed in Chapter 13, have been compared to the conventional RAKE receiver in Reference [24]. The conclusion was that decorrelators can avoid error floors demonstrated by plain RAKE receivers. These components will be also imperfect. Performance of RAKE combining techniques (selection, equal gain and maximal ratio diversity) in the presence of chip and phase synchronization errors is reported in Reference [25]. As expected, the maximal ratio combiner (MRC) outperforms other combiners and quite drastic capacity losses are seen because of synchronization errors and fading. Diversity methods in Rayleigh fading have also been evaluated and compared in References [26,27].

Capacity evaluation of CDMA networks [28] and comparison of CDMA and time division multiple access (TDMA) systems have been an important and controversial issue. One of the reasons for such a situation is the lack of a systematic, easy to follow mathematical framework for this evaluation. The situation is complicated by the fact that a lot of parameters are involved and some of the system components are rather complex resulting in imperfect operation. The analysis of an advanced CDMA network should in general take all these elements into account including their imperfections and come up with an expression for the system capacity in the form that can be used in practice.

In Reference [29] a systematic analytical framework for the capacity evaluation of wideband CDMA networks with nonlinear IC was presented.

In this chapter we extend the model to networks employing linear decorrelating detectors. This approach should provide a relatively simple way to specify the required quality of MAI canceller (decorrelator) and RAKE receiver in a CDMA system, taking into account all their imperfections.

A flexible, complex signal format is used that enables us to model all currently interesting wideband CDMA standards. For such a signal we first derive a complex decorrelator structure. This is the first time that a complete decorrelator is described for the CDMA signal with two independent data and two independent code streams. This type of the signal is used in all CDMA standards. In the next step all imperfections in the operation of such a system are modeled and analyzed by using the concept of a network sensitivity function. This provides the necessary information to the designer on how much of the theoretically promised ideal performance will be preserved in a real implementation. The theory is general and some examples of practical sets of channel and system parameters are used as illustration.

15.2 SYSTEM MODEL

In order to be able to discuss implementation problems in the existing standards, the complex envelope of the signal transmitted by user $k \in \{1, 2, \ldots, K\}$ in the nth symbol interval $t \in [nT, (n+1)T]$ will be written as

$$s_k = A_k \mathrm{e}^{j} \phi_{k0} S_k^{(n)}(t - \tau_k) \tag{15.1}$$

where A_k is the transmitted signal amplitude of user k, τ_k is the signal delay, ϕ_{k0} is the transmitted signal carrier phase and T is the symbol interval. $S_k^{(n)}(t)$ can be represented by

$$S_k^{(n)}(t) = S_k^{(n)} = S_k = S_{ik} + j S_{qk} = d_{ik} c_{ik} + j d_{qk} c_{qk} \tag{15.2}$$

In this equation, b_{ik} and b_{qk} are two information bits in the I and Q channels, respectively, generated with bit interval T. c_{ik} and c_{qk} are the kth user pseudo-noise codes in the I and Q channels, respectively, generated with chip interval T_c and having a period T/T_c. Equations (15.1) and (15.2) are general and different combinations of the signal parameters cover most of the signal formats of practical interest. For example, in WCDMA/FDD mode the uplink signal format can be expressed as

$$(c_\mathrm{d} d_\mathrm{d} + j c_\mathrm{c} d_\mathrm{c}) c_\mathrm{s} c_\mathrm{sl} \tag{15.3}$$

where c_d and c_c are data and control channel codes, d_d and d_c are data and control channel information bits and c_s and c_sl are scrambling and scrambling long (optional) codes for user k. Equation (15.3) can be written in the form of equation (15.2) with the following mapping:

$$c_{ik} = c_{\mathrm{d}} c_{\mathrm{s}} c_{\mathrm{sl}}$$

$$c_{qk} = c_{\mathrm{c}} c_{\mathrm{s}} c_{\mathrm{sl}}$$

$$d_{ik} = d_{\mathrm{d}}$$

$$d_{qk} = d_{\mathrm{c}} \tag{15.4}$$

Equation (15.3) may be further modified to include complex scrambling codes.

The model of channel impulse response consists of discrete multipath components represented as

$$h_k^{(n)}(t) = \sum_{l=1}^{L} h_{kl}^{(n)} \delta(t - \tau_{kl}^{(n)}) = \sum_{l=1}^{L} H_{kl}^{(n)} e^{j\phi_{kl}} \delta(t - \tau_{kl}^{(n)}) \tag{15.5}$$

$$h_{kl}^{(n)} = H_{kl}^{(n)} e^{j\phi_{kl}} \tag{15.6}$$

where L is the number of multipath components of the channel, $h_{kl}^{(n)}$ and $\tau_{kl}^{(n)} \in [0, T_{\mathrm{m}})$ are the complex coefficient (gain) and delay, respectively, of the lth path of user k at symbol interval with index n and $\delta(t)$ is the Dirac delta function. We assume that T_{m} is the delay spread of the channel. In what follows, indices n will be dropped whenever this does not produce any ambiguity. It is also assumed that $T_{\mathrm{m}} < T$ and an indication for the necessary modifications in the case when $T_{\mathrm{m}} > T$ is provided whenever appropriate. The overall received signal during N_{b} symbol intervals can be represented as

$$r(t) = \mathrm{Re}\left\{ e^{j\omega_0 t} \sum_{n=0}^{N_{\mathrm{b}}-1} \sum_{k} \sum_{l} a_{kl} S_k^{(n)}(t - nT - \tau_k - \tau_{kl}) \right\} + \mathrm{Re}\{z(t) e^{j\omega_0 t}\} \tag{15.7}$$

where $a_{kl} = A_k H_{kl}^{(n)} e^{j\Phi_{kl}} = A_{kl} e^{j\Phi_{kl}}$, $A_k H_{kl}^{(n)} = A_{kl}$, $\Phi_{kl} = \phi_0 + \phi_{k0} - \phi_{kl}$, ϕ_0 is the frequency downconversion phase error, $z(t)$ is a complex zero mean additive white Gaussian noise (AWGN) process with two-sided power spectral density σ^2 and ω_0 is the carrier angular frequency. In what follows, we will drop the noise term for simplicity reason and focus only on proper representation of the MAI. For a correct representation of the overall signal, the noise term will be reintroduced again in equation (15.35). In order to be able to model system imperfections, both I and Q signal components should be represented separately as an explicit function of all parameters that are estimated in the receiver and because these estimates are imperfect they may include some errors. The complex matched filter of user k will create two correlation functions for each path and by omitting the noise terms these signals can be represented as

$$y_{ikl}^{(n)} = \int_{nT+\tau_k+\tau_{kl}}^{(n+1)T+\tau_k+\tau_{kl}} r(t) c_{ik}(t - nT - \tau_k + \tau_{kl}) \cos(\omega_0 t + \tilde{\Phi}_{kl}) \, \mathrm{d}t = \sum_{k'} \sum_{l'} y_{ikl}(k'l') \tag{15.8}$$

where $\tilde{\Phi}_{kl}$ is the estimate of Φ_{kl} and

$$
\begin{aligned}
y_{ikl}(k'l') &= y_{iikl}(k'l') + y_{iqkl}(k'l') \\
&= A_{k'l'}[d_{ik'}\rho_{ik'l',ikl}\cos\varepsilon_{k'l',kl} + d_{qk'}\rho_{qk'l',ikl}\sin\varepsilon_{k'l',kl}]
\end{aligned} \tag{15.9}
$$

and

$$
\begin{aligned}
y_{qkl}^{(n)} &= \int_{nT+\tau_k+\tau_{kl}}^{(n+1)T+\tau_k+\tau_{kl}} r(t)c_{qk}(t - nT - \tau_k + \tau_{kl})\sin(\omega_0 t + \tilde{\Phi}_{kl})\,dt \\
&= \sum_{k'}\sum_{l'} y_{qkl}(k'l')
\end{aligned} \tag{15.10}
$$

$$
\begin{aligned}
y_{qkl}(k'l') &= y_{qqkl}(k'l') + y_{qikl}(k'l') \\
&= A_{k'l'}[d_{qk'}\rho_{qk'l',qkl}\cos\varepsilon_{k'l',kl} - d_{ik'}\rho_{ik'l',qkl}\sin\varepsilon_{k'l',kl}]
\end{aligned} \tag{15.11}
$$

with $\rho_{x,y}$ being the cross-correlation functions between the corresponding code components x and y. A scaling factor $1/2$ is dropped in the above equations for simplicity. Basically by dropping this coefficient for both signal and noise, the signal-to-noise ratio (SNR) that determines the system performance will not change. Each of these components is defined with three indices (i or q, user and path). Parameter $\varepsilon_{a,b} = \Phi_a - \tilde{\Phi}_b$ where a and b are defined with two indices (user and path). Let the L-element vectors $\Im^L(\cdot)$ of matched filter output samples for the nth symbol interval be defined as

$$
\mathbf{y}_{ik}^{(n)} = \Im^L(y_{ikl}^{(n)}) = (y_{ik1}^{(n)}, y_{ik2}^{(n)}, \ldots, y_{ikL}^{(n)})^T \in C^L \tag{15.12}
$$

$$
\mathbf{y}_{qk}^{(n)} = \Im^L(y_{qkl}^{(n)}) = (y_{qk1}^{(n)}, y_{qk2}^{(n)}, \ldots, y_{qkL}^{(n)})^T \in C^L \tag{15.13}
$$

$$
\mathbf{y}_k^{(n)} = \mathbf{y}_{ik}^{(n)} + j\mathbf{y}_{qk}^{(n)} \tag{15.14}
$$

$$
\mathbf{y}^{(n)} = \Im^K(\mathbf{y}_k^{T(n)}) \in C^{KL} \tag{15.15}
$$

$$
\mathbf{y} = \Im^{N_b}(\mathbf{y}^{T(n)}) \in C^{N_b KL} \tag{15.16}
$$

Let in general $\mathbf{R}^{(n)}(i) \in (-1, 1]^{KL \times KL}$ be a cross-correlation matrix with the following partition:

$$
\begin{aligned}
\mathbf{R}^{(n)}(i) &= \begin{pmatrix}
\mathbf{R}_{1,1}^{(n)}(i) & \mathbf{R}_{1,2}^{(n)}(i) & \cdots & \mathbf{R}_{1,K}^{(n)}(i) \\
\mathbf{R}_{2,1}^{(n)}(i) & \mathbf{R}_{2,2}^{(n)}(i) & \cdots & \mathbf{R}_{2,K}^{(n)}(i) \\
\vdots & \vdots & \ddots & \vdots \\
\mathbf{R}_{K,1}^{(n)}(i) & \mathbf{R}_{K,2}^{(n)}(i) & \cdots & \mathbf{R}_{K,K}^{(n)}(i)
\end{pmatrix} \in R^{KL \times KL} \\
&= \Im\Im^K(\mathbf{R}_{k,k'}^{(n)}(i))
\end{aligned} \tag{15.17}
$$

For the final representation of the complex matched filter output signal, given by equations (15.49) and (15.50), we now define four specific matrices of the form given by equation (15.17) with the following notation:

$$\mathbf{R}^{ii(n)}(i) = \Im\Im^{K}(\mathbf{R}_{k,k'}^{ii(n)}(i)) \tag{15.18}$$

$$\mathbf{R}^{qi(n)}(i) = \Im\Im^{K}(\mathbf{R}_{k,k'}^{qi(n)}(i)) \tag{15.19}$$

$$\mathbf{R}^{iq(n)}(i) = \Im\Im^{K}(\mathbf{R}_{k,k'}^{iq(n)}(i)) \tag{15.20}$$

$$\mathbf{R}^{qq(n)}(i) = \Im\Im^{K}(\mathbf{R}_{k,k'}^{qq(n)}(i)) \tag{15.21}$$

where matrices $\mathbf{R}_{k,k'}^{ab(n)}(i) \in R^{L \times L}, \forall k, k' \in \{1, 2, \ldots, K\}$ in equations (15.18) to (15.21) have elements

$$(R_{k,k'}^{ii(n)}(i))_{l,l'} = \cos \varepsilon_{k'l',kl} \times \int_{-\infty}^{\infty} c_{ik}^{(n)}(t - \tau_k - \tau_{kl})c_{ik'}^{(n-i)}(t + iT - \tau_{k'} - \tau_{k'l'})\,dt \tag{15.22}$$

$$(R_{k,k'}^{qi(n)}(i))_{l,l'} = \sin \varepsilon_{k'l',kl} \times \int_{-\infty}^{\infty} c_{qk}^{(n)}(t - \tau_k - \tau_{k,l})c_{ik'}^{(n-i)}(t + iT - \tau_{k'} - \tau_{k'l'})\,dt \tag{15.23}$$

$$(R_{k,k'}^{iq(n)}(i))_{l,l'} = -\sin \varepsilon_{k'l',kl} \times \int_{-\infty}^{\infty} c_{ik}^{(n)}(t - \tau_k - \tau_{k,l})c_{qk'}^{(n-i)}(t + iT - \tau_{k'} - \tau_{k'l'})\,dt \tag{15.24}$$

$$(R_{k,k'}^{qq(n)}(i))_{l,l'} = \cos \varepsilon_{k'l',kl} \times \int_{-\infty}^{\infty} c_{qk}^{(n)}(t - \tau_k - \tau_{k,l})c_{qk'}^{(n-i)}(t + iT - \tau_{k'} - \tau_{k'l'})\,dt$$

$$\forall l, l' \in \{1, 2, \ldots, L\} \tag{15.25}$$

In order to simplify the notation, we present equation (15.22) as $R = \rho \cos \varepsilon$ and its estimation as

$$\hat{R} = \hat{\rho} \cos \hat{\varepsilon} \tag{15.26}$$

In general, the estimated phase difference $\hat{\varepsilon}$ between the two users (e.g. users with index $k = 1$ and $k = 2$) can be represented as

$$\hat{\varepsilon} = \phi_1 - \Delta\phi_1 - \phi_2 - \Delta\phi_2 = \varepsilon + \Delta\varepsilon \tag{15.27}$$

where $\varepsilon = \phi_1 - \phi_2$ and $\Delta\varepsilon = -(\Delta\phi_1 + \Delta\phi_2)$.

The noise samples at the output matched filters for different users are uncorrelated.

So, if $\Delta\phi$ is a process with zero mean and variance σ_ϕ^2, then $\Delta\varepsilon$ is a zero mean process with variance $2\sigma_\phi^2$. The estimated correlation function can be represented as

$$\hat{\rho} = \rho + \Delta\rho \cong \rho + \rho'\varepsilon_\tau = \rho\left(1 + \frac{\rho'\varepsilon_\tau}{\rho}\right) = \rho(1 + s_\rho) \tag{15.28}$$

where ρ' is the slope of the ρ function at the point of zero delay estimation error and

$$\varepsilon_\tau = \Delta\tau_1 - \Delta\tau_2 \tag{15.29}$$

is the difference between the two delay estimation errors. For a given class and code length, ρ' is a parameter [30]. If $\Delta\tau$ is a zero mean variable with variance σ_τ^2, then ε_τ is a zero mean variable with variance $2\sigma_\tau^2$. The second component of equation (15.26) can be represented as

$$\cos\hat{\varepsilon} = \cos(\varepsilon + \Delta\varepsilon) = \cos\varepsilon\cos\Delta\varepsilon - \sin\varepsilon\sin\Delta\varepsilon$$

$$= (1 - \Delta\varepsilon^2/2)\cos\varepsilon - \Delta\varepsilon\sin\varepsilon$$

$$= (1 + s_\varepsilon)\cos\varepsilon \tag{15.30}$$

where

$$s_\varepsilon = -(\Delta\varepsilon^2/2 + \Delta\varepsilon \cdot \tan\varepsilon) \tag{15.31}$$

Now, equation (15.26) becomes

$$\hat{R} = R + \Delta R \tag{15.32}$$

where

$$R = \rho\cos\varepsilon$$

$$\Delta R = \rho\cos\varepsilon(s_\varepsilon + s_\rho + s_\varepsilon s_\rho) \tag{15.33}$$

Whenever $k' \neq k$, the average value of the cross-correlation $\bar{\rho} = 0$ and parameter ΔR can be considered as an additional noise component with a zero mean and variance

$$\sigma_{\Delta R}^2 = \rho^2[(1 + 2\sigma_{\rho'}^2\sigma_\tau^2/\rho^2)(3\sigma_\phi^4 + 2\sigma_\phi^2) + 2\sigma_{\rho'}^2\sigma_\tau^2/\rho^2] \tag{15.34}$$

Similar expressions can be derived for the estimation of equations (15.23) to (15.25).

In the case when multipath delay spread produces severe intersymbol interference (ISI), the overall received signal should be further modified. When the delay spread is limited to less than one symbol interval, then for an asynchronous network the vector equation (15.15) can be expressed as [4]

$$\mathbf{y}^{(n)}(\mathbf{R}, \mathbf{H}, \mathbf{A}, \mathbf{d}) = \mathbf{R}^{(n)}(2)\mathbf{H}^{(n-2)}\mathbf{A}\,\mathbf{d}^{(n-2)} + \mathbf{R}^{(n)}(1)\mathbf{H}^{(n-1)}\mathbf{A}\,\mathbf{d}^{(n-1)}$$

$$+ \mathbf{R}^{(n)}(0)\mathbf{H}^{(n)}\mathbf{A}\,\mathbf{d}^{(n)} + \mathbf{R}^{(n)}(-1)\mathbf{H}^{(n+1)}\mathbf{A}\,\mathbf{d}^{(n+1)}$$

$$+ \mathbf{R}^{(n)}(-2)\mathbf{H}^{(n+2)}\mathbf{A}\,\mathbf{d}^{(n+2)} + \mathbf{n}^{(n)} \tag{15.35}$$

where

$$\mathbf{A} = \text{diag}(A_1, A_2, \ldots, A_K) \in \mathbf{R}^{K \times K} \tag{15.36}$$

is a diagonal matrix of transmitted amplitudes,

$$\mathbf{H}^{(n)} = \mathrm{diag}(\mathbf{H}_1^{(n)}, \mathbf{H}_2^{(n)}, \ldots, \mathbf{H}_K^{(n)}) \in \mathrm{C}^{KL \times K} \tag{15.37}$$

is the matrix of channel coefficient vectors

$$\mathbf{H}_k^{(n)} = (H_{k,1}^{(n)}, H_{k,2}^{(n)}, \ldots, H_{k,L}^{(n)})^{\mathrm{T}} \in \mathrm{C}^L \tag{15.38}$$

$$\mathbf{d}^{(n)} = (d_1^{(n)}, d_2^{(n)}, \ldots, d_K^{(n)})^{\mathrm{T}} \in \Xi^K \tag{15.39}$$

is the vector of the transmitted data and $\mathbf{n}^{(n)} \in \mathrm{C}^{KL}$ is the output vector due to noise. This component is due to processing the second term of equation (15.7) that was dropped for simplicity in derivation of equations (15.8) to (15.34). It is easy to show that $\mathbf{R}^{(n)}(i) = \mathbf{0}_{KL}, \forall |i| > 2$ and $\mathbf{R}^{(n)}(-i) = \mathbf{R}^{T(n+1)}(i)$, where $\mathbf{0}_{KL}$ is an all-zero matrix of size $KL \times KL$. Thus, the concatenation vector of the matched filter outputs (15.16) has the expression

$$y(R, H, A, \boldsymbol{d}) = RHA\boldsymbol{d} + n = RH\boldsymbol{h} + n \tag{15.40}$$

where

$$R = \begin{pmatrix} \mathbf{R}^{(0)}(0) & \mathbf{R}^{T(1)}(1) & \mathbf{R}^{T(2)}(2) & \cdots & \mathbf{0}_{KL} \\ \mathbf{R}^{(1)}(1) & \mathbf{R}^{(1)}(0) & \mathbf{R}^{T(2)}(1) & \cdots & \mathbf{0}_{KL} \\ \mathbf{R}^{(2)}(2) & \mathbf{R}^{(2)}(1) & \mathbf{R}^{(2)}(0) & \cdots & \mathbf{0}_{KL} \\ \vdots & \vdots & \vdots & \ddots & \vdots \\ \mathbf{0}_{KL} & \mathbf{0}_{KL} & \mathbf{0}_{KL} & \cdots & \mathbf{R}^{(N_b-1)}(0) \end{pmatrix} \in \mathrm{R}^{N_b KL \times N_b KL} \tag{15.41}$$

$$H = \mathrm{diag}(\mathbf{H}^{(0)}, \mathbf{H}^{(1)}, \ldots, \mathbf{H}^{(N_b-1)}) \in \mathrm{C}^{N_b KL \times N_b K} \tag{15.42}$$

$$A = \mathrm{diag}(\mathbf{A}, \mathbf{A}, \ldots, \mathbf{A}) \in \mathrm{R}^{N_{st} K \times N_b K} \tag{15.43}$$

$$\mathbf{d} = (\mathbf{d}^{T(0)}, \mathbf{d}^{T(1)}, \ldots, \mathbf{d}^{T(N_b-1)})^{\mathrm{T}} \in \Xi^{N_b K} \tag{15.44}$$

$\mathbf{h} = A\mathbf{d}$ is the data-amplitude product vector and $\boldsymbol{n}$ is the Gaussian noise output vector with zero mean and covariance matrix $\sigma^2 R$. If we define

$$y_{ii} = y(R^{ii}, H, A, \mathbf{d}_i) \tag{15.45}$$

$$y_{qi} = y(R^{qi}, H, A, \mathbf{d}_q) \tag{15.46}$$

$$y_{iq} = y(R^{iq}, H, A, \mathbf{d}_i) \tag{15.47}$$

$$y_{qq} = y(R^{qq}, H, A, \mathbf{d}_q) \tag{15.48}$$

then we have

$$y_i = y_{ii} + y_{qi} \tag{15.49}$$

$$y_q = y_{iq} + y_{qq} \tag{15.50}$$

On the basis of these equations in the sequel, a complex decorrelator receiver structure is derived.

15.2.1 Complex decorrelator

As a starting point, we represent equations (15.49) and (15.50) as

$$y_i = y_{ii} + y_{qi} = \Re^{ii}\mathbf{d}_i + \Re^{qi}\mathbf{d}_q + \mathbf{n}_i \tag{15.51}$$

$$y_q = y_{iq} + y_{qq} = \Re^{iq}\mathbf{d}_i + \Re^{qq}\mathbf{d}_q + \mathbf{n}_q \tag{15.52}$$

where

$$\Re = RHA \tag{15.53}$$

and we use an additional step to produce

$$\Re^{ii-1}y_i = \mathbf{d}_i + \Re^{ii-1}\Re^{qi}\mathbf{d_q} + \Re^{ii-1}\mathbf{n}_i$$

$$\Re^{iq-1}y_q = \mathbf{d}_i + \Re^{iq-1}\Re^{qq}\mathbf{d_q} + \Re^{iq-1}\mathbf{n}_q$$

$$\Re^{qi-1}y_i = \Re^{qi-1}\Re^{ii}\mathbf{d_i} + \mathbf{d_q} + \Re^{qi-1}\mathbf{n}_i$$

$$\Re^{qq-1}y_q = \Re^{qq-1}\Re^{iq}\mathbf{d_i} + \mathbf{d_q} + \Re^{qq-1}\mathbf{n}_q \tag{15.54}$$

From the last set of equations one can show that the data estimates should be obtained as

$$\hat{\mathbf{d}} = \hat{\mathbf{d}}_i + j\hat{\mathbf{d}}_q$$

$$\hat{\mathbf{d}}_\mathbf{q} = \mathrm{sgn}\{\mathbf{D}_{qi}\mathbf{y}_i + \mathbf{D}_{qq}\mathbf{y}_q\}$$

$$\mathbf{D}_{qi} = \{\Re^{qi-1}\Re^{ii} - \Re^{qq-1}\Re^{iq}\}^{-1}\Re^{qi-1}$$

$$\mathbf{D}_{qq} = -\{\Re^{qi-1}\Re^{ii} - \Re^{qq-1}\Re^{iq}\}^{-1}\Re^{qq-1} \tag{15.55}$$

and similarly

$$\hat{\mathbf{d}}_i = \mathrm{sgn}\{\mathbf{D}_{ii}\mathbf{y}_i + \mathbf{D}_{iq}\mathbf{y}_q\}$$

$$\mathbf{D}_{ii} = \{\Re^{ii-1}\Re^{qi} - \Re^{iq-1}\Re^{qq}\}^{-1}\Re^{ii-1}$$

$$\mathbf{D}_{ii} = \{\Re^{ii-1}\Re^{qi} - \Re^{iq-1}\Re^{qq}\}^{-1}\Re^{ii-1} \tag{15.56}$$

Bearing in mind that all current wideband code division multiple access (WCDMA) standards are based on using complex signal formats, future research in the field of multiuser detectors should be focused on the structures defined above.

15.3 CAPACITY LOSSES

The starting point in the evaluation of CDMA system capacity is parameter $Y_{bm} = E_{bm}/N_0$, the received signal energy per symbol per overall noise density in a given reference receiver with index m. For the purpose of this analysis, we can represent this parameter in the general case as

$$Y_{bm} = \frac{E_{bm}}{N_0} = \frac{ST}{I_{oc} + I_{oic} + I_{oin} + \eta_{th}} \tag{15.57}$$

where I_{oc}, I_{oic} and I_{oin} are the power densities of intracell, intercell and overlay type internetwork interference, respectively, and η_{th} is the thermal noise power density. Parameter S is the overall received power of the useful signal and $T = 1/R_b$ is the information bit interval. Contributions of I_{oic} and I_{oin} to N_0 have been discussed in Chapter 8 and in a number of papers, for example, in Reference [28]. In order to minimize repetition in our analysis, we will parameterize this contribution by introducing

$$\eta_0 = I_{oic} + I_{oin} + \eta_{th} \tag{15.58}$$

and concentrate on the analysis of the intracell interference, I_{oc}, in a CDMA network based on advanced receivers using imperfect RAKE, and MAI cancellation based on an imperfect decorrelator. An extension of the analysis to include both intercell and internetwork interference is straightforward. A general block diagram of the receiver is shown in Figure 15.1.

If for user m an L_0-finger RAKE receiver ($L_0 \leq L$) with combiner coefficients w_{mr} ($r = 1, 2, \ldots, L_0$) and an imperfect decorrelator is used, the SNR will become

$$Y_{bm} = \frac{r_m^{(L_0)}}{\varsigma_0 \eta R_b / S} \tag{15.59}$$

where

$$\varsigma_0 = \sum_{r=1}^{L_0} w_{mr}^2 = \mathbf{w}_m \mathbf{w}_m^{\mathrm{T}}; \mathbf{w}_m = (w_{m1}, w_{m2}, \ldots, w_{mL_0}) \tag{15.60}$$

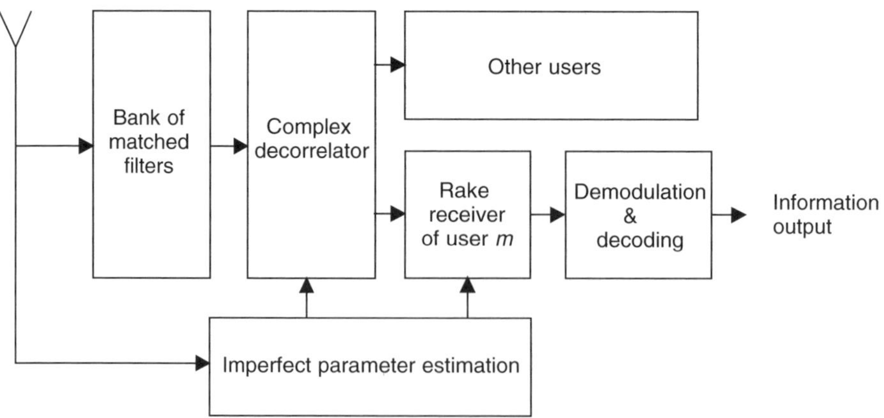

Figure 15.1   General receiver block diagram.

is due to Gaussian noise processing in the RAKE receiver, and the noise density η_0, after decorrelation, becomes η. The relation between these two parameters is elaborated later in equations (15.74) to (15.77). The parameter $r_m^{(L_0)}$ in equation (15.59) is called the *RAKE receiver efficiency* and is given by

$$r_m^{(L_0)} = \left(\sum_{r=1}^{L_0} w_{mr} \cos \varepsilon_{\theta mmr} \sqrt{\alpha_{mr}} \right)^2 = (w_m \cdot \alpha_{mm\sqrt{}})^2 \tag{15.61}$$

with $\alpha_{mm\sqrt{}} = (\cos \varepsilon_{\theta mm1} \sqrt{\alpha_{m1}}, \cos \varepsilon_{\theta mm2} \sqrt{\alpha_{m2}}, \ldots)^{\mathrm{T}}$. Parameter $\varepsilon_{\theta mmr} = \theta_{mmr} - \hat{\theta}_{mmr}$ is the carrier phase synchronization error in receiver m for signal of user m in path r. We will drop index mkl whenever it does not result in any ambiguity. In the sequel we will use the following notation: $\alpha_{kl} = A_{kl}^2/2$, $\hat{A}_{mkl}$ is the estimation of A_{kl} by the receiver m, $\varepsilon_a = \Delta A_{mkl}/A_{kl} = (A_{kl} - \hat{A}_{mkl})/A_{kl}$ is the relative amplitude estimation error and ε_θ is the carrier phase estimation error.

For the equal gain combiner (EGC), the combiner coefficients are given as $w_{mr} = 1$. Having in mind the notation used so far, in the sequel we will drop index m for simplicity. For the maximal ratio combiner (MRC), the combiner coefficients are based on estimates as

$$\hat{w}_r = \frac{\cos \varepsilon_{\theta r}}{\cos \varepsilon_{\theta 1}} \cdot \frac{\hat{A}_r}{\hat{A}_1} \cong \frac{(1 - \varepsilon_{\theta r}^2/2)}{(1 - \varepsilon_{\theta 1}^2/2)} \cdot \frac{A_r(1 - \varepsilon_{ar})}{A_1(1 - \varepsilon_{a1})} \tag{15.62}$$

$$E\{\hat{w}_r\} = w_r(1 - \sigma_{\theta r}^2)(1 + \sigma_{\theta 1}^2)(1 - \varepsilon_{ar})(1 + \varepsilon_{a1}) \tag{15.63}$$

$$E\{\hat{w}_r^2\} = w_r^2(1 - 2\sigma_{\theta r}^2 + 3\sigma_{\theta r}^4)(1 + 2\sigma_{\theta 1}^2 - 3\sigma_{\theta 1}^4)(1 - \varepsilon_{ar})^2(1 + \varepsilon_{a1})^2 \tag{15.64}$$

By using notation $A_r/A_1 = \sqrt{\alpha_r/\alpha_1}$, averaging equation (15.61) gives for EGC

$$E\{r^{(L_0)}\} = E\left[\left(\sum_{r=1}^{L_0} \cos \varepsilon_{\theta r} \sqrt{\alpha_r} \right)^2 \right] \cong E\left[\left(\sum_{r=1}^{L_0} (1 - \varepsilon_{\theta r}^2/2)\sqrt{\alpha_r} \right)^2 \right]$$

$$= \sum_r \sum_{\substack{l \\ l \neq r}} (1 - \sigma_{\theta r}^2)(1 - \sigma_{\theta l}^2)\sqrt{\alpha_r \alpha_l} + \sum_r \alpha_r(1 - 2\sigma_{\theta r}^2 + 3\sigma_{\theta r}^4) \tag{15.65}$$

For MRC the same relation becomes

$$E\{r^{(L_0)}\} = E\left[\left(\sum_{r=1}^{L_0} \hat{w}_r \cos \varepsilon_{\theta r} \sqrt{\alpha_r} \right)^2 \right] \cong E\left[\left(\sum_{r=1}^{L_0} \frac{\alpha_r}{\sqrt{\alpha_1}} \frac{(1 - \varepsilon_{\theta r}^2/2)^2}{(1 - \varepsilon_{\theta 1}^2/2)} \frac{(1 - \varepsilon_{ar})}{(1 - \varepsilon_{a1})} \right)^2 \right]$$

$$= E\left[\sum_{r=1}^{L_0} \frac{\alpha_r^2}{\alpha_1} \frac{(1 - \varepsilon_{\theta r}^2/2)^4}{(1 - \varepsilon_{\theta 1}^2/2)^2} \frac{(1 - \varepsilon_{ar})^2}{(1 - \varepsilon_{a1})^2} \right] + \sum_r \sum_{\substack{l \\ l \neq r}} \frac{\alpha_r \alpha_l}{\alpha_1} (1 - 2\sigma_{\theta r}^2 + 3\sigma_{\theta r}^4)$$

$$\times (1 - 2\sigma_{\theta l}^2 + 3\sigma_{\theta l}^4)(1 + 2\sigma_{\theta 1}^2 - 3\sigma_{\theta 1}^4)(1 - \varepsilon_{ar})(1 - \varepsilon_{al})(1 + \varepsilon_{a1})^2 \tag{15.66}$$

One should note that even when $\hat{w}_1 = 1$, the value of the first term in the above sum is $\cos\varepsilon_{\theta 1}\sqrt{\alpha_1}$, which takes into account the error in the estimation of the phase for the first finger. In order to avoid dealing with the fourth power terms of the type $(1 - \varepsilon_{\theta r}^2/2)^4$ in the evaluation of the first term in equation (15.66), we use limits. For the upper limit we have

$$\varepsilon_{\theta r}^2 \Rightarrow \varepsilon_{\theta 1}^2 \tag{15.67}$$

By using this we have

$$\frac{(1 - \varepsilon_{\theta r}^2/2)^4}{(1 - \varepsilon_{\theta 1}^2/2)^2} \Rightarrow (1 - \varepsilon_{\theta 1}^2/2)^2 \tag{15.68}$$

and the first term becomes [31]

$$\sum_{r=1}^{L_0} \frac{\alpha_r^2}{\alpha_1}(1 - 2\sigma_{\theta 1}^2 + 3\sigma_{\theta 1}^4)\frac{(1 - \varepsilon_{ar})^2}{(1 - \varepsilon_{a1})^2} \tag{15.69}$$

For the lower limit we use

$$\varepsilon_{\theta 1}^2 \Rightarrow \varepsilon_{\theta r}^2 \tag{15.70}$$

and the first term becomes

$$\sum_{r=1}^{L_0} \frac{\alpha_r^2}{\alpha_1}(1 - 2\sigma_{\theta r}^2 + 3\sigma_{\theta r}^4)\frac{(1 - \varepsilon_{ar})^2}{(1 - \varepsilon_{a1})^2} \tag{15.71}$$

For a signal with I and Q components, the parameter $\cos\varepsilon_{\theta r}$ should be replaced by

$$\cos\varepsilon_{\theta r} \Rightarrow \cos\varepsilon_{\theta r} + b\rho\sin\varepsilon_{\theta r} \tag{15.72}$$

where b is the information in the interfering channel (I or Q) and ρ is the cross-correlation between the codes used in the I and Q channels. For small tracking errors, this term can be replaced as

$$\cos\varepsilon_{\theta r} + b\rho\sin\varepsilon_{\theta r} \approx 1 + b\rho\varepsilon - \varepsilon^2/2 \tag{15.73}$$

where the notation is further simplified by dropping the subscript θr. By using equation (15.73) in equations (15.64) to (15.72), similar expressions can be derived for the complex signal format.

We will assume that a linear decorrelator is used for IC in the system. The detector will operate with the inverse of the estimated correlation matrix $\hat{\mathbf{R}}^{-1}$ where the real correlation matrix of the system is $\mathbf{R} = \hat{\mathbf{R}} + \Delta\mathbf{R}$. The elements of $\Delta\mathbf{R}$ have zero mean and variance given by equation (15.34). So, after decorrelation by using $\hat{\mathbf{R}}^{-1}$ the residual noise in the receiver will have variance

$$\text{Var}\left[\hat{\mathbf{R}}^{-1}(\mathbf{n}_r + \mathbf{n})\right] \tag{15.74}$$

where Gaussian noise components of vector $\mathbf{n}$ have variance σ_n^2 and components of residual vector $\mathbf{n}_r$ have variance σ_r^2 that can be approximated as

$$\sigma_r^2 = \sum_{k,l} \alpha_{k,l} \sigma_{\Delta_{k,l}}^2 \tag{15.75}$$

where $\sigma_{\Delta_{k,l}}^2$ is given by equation (15.34) for specific indices k, l. The residual noise is composed of a large number of components with the same distribution that suggests using the central limit theorem to approximate the overall distribution as Gaussian with average variance represented as

$$\overline{\sigma}_r^2 \cong \alpha K L \sigma_\Delta^2 \tag{15.76}$$

where σ_Δ^2 is given by equation (15.34). One should keep in mind that the residual noise $\mathbf{n}_r$ is created in front of the decorrelator. After the decorrelation and the RAKE combiner, the components of the overall noise variance given by equation (15.74) become

$$\sigma^2 \cong \varsigma_0 R_{mm}^+ (\sigma_r^2 + \sigma_n^2) = \sum_r^2 + \sum_n^2 \tag{15.77}$$

where R_{mm}^+ is the mmth component of $\hat{\mathbf{R}}^{-1}$. In this relation, $\sum_r^2$ is the contribution of system imperfections due to the overall noise variance σ^2 and $\sum_n^2$ is the contribution of Gaussian noise after decorrelation. So, the equivalent noise variance is expressed in terms of phase and code delay estimation errors (see equation 15.34). One should notice that the same arguments about using the central limit theorem apply in the case of the noise after decorrelator too because decorrelation is a linear operation. These results should be now used for analysis of the impact of large scale of channel estimators on overall CDMA network sensitivity. A performance measure of any estimator is the parameter estimation error variance that should be directly used in equation (15.77) for equivalent noise variance and equations (15.62) to (15.73) for the RAKE receiver. If joint parameter estimation is used, on the basis of the maximum likelihood (ML) criterion, then the Cramer–Rao bound could be used for these purposes. For Kalman type estimators, the error covariance matrix is available for each iteration of estimation. If each parameter is estimated independently, then for carrier phase and code delay estimation error a simple relation $\sigma_{\theta,\tau}^2 = 1/SNR_{\mathrm{L}}$ can be used where SNR_{L} is the SNR in the tracking loop. For the evaluation of this SNR_{L}, the noise power is in general given as $N = B_{\mathrm{L}} N_0$. In this case, the noise density N_0 is approximated as a ratio of the overall interference plus noise power divided by the signal bandwidth. The loop bandwidth will be proportional to f_{D} where f_{D} is the fading rate (Doppler). If decorrelation is performed prior to parameter estimation, N_0 is obtained from the equivalent noise having the variance defined by equation (15.77). If parameter estimation is used without decorrelation, then the overall noise consists of MAI and Gaussian noise.

For the numerical analysis, further assumptions and specifications are necessary. First of all we need the channel model. The exponential multipath intensity profile (MIP) channel model is a widely used analytical model and is realized as a tapped delay line [32]. It is very flexible in modeling different propagation scenarios. The decay of the profile and the number of taps in the model can vary. Averaged power coefficients in the MIP are

$$\overline{\alpha_l} = \overline{\alpha_0}\, e^{-\lambda l} \quad l, \lambda \geq 0 \tag{15.78}$$

where λ is the decay parameter of the profile. Power coefficients should be normalized as

$$\sum_{l=0}^{L-1} \overline{\alpha_0}\, e^{-\lambda l} = 1 \tag{15.79}$$

For $\lambda = 0$ the profile will be flat. The number of resolvable paths depends on the channel chip rate and this must be taken into account. We will start with the average SNR that can be expressed as

$$\overline{Y_b} = \frac{Sr^{(L_0)}G}{\sigma^2} = \frac{Sr^{(L_0)}(K)G}{\sigma^2(K)} = \overline{Y_b}\,(K) \tag{15.80}$$

where $r^{(L_0)}$ is given in Section 3B, σ^2 is given by equation (15.77) and $G = 1/\rho^2$ is the processing gain. RAKE receiver efficiency and overall noise variance depend on the number of users K. If we accept some quality of transmission, bit error rate (BER) $= 10^{-e}$, that can be achieved with given $SNR = Y_0 = \overline{Y}_b(K = C)$, where C is the system capacity, then in the case of perfect channel estimation we have $C_{\max} = K$, which is the solution to the equation

$$Y_0 = \frac{Sr^{(L_0)}(K)G}{\overset{2}{\underset{n}{\sum}}(K)} \tag{15.81}$$

In the case of imperfect channel estimation we have $C = K$, which is the solution to the equation

$$Y_0 = \frac{Sr^{(L_0)}(K)G}{\overset{2}{\underset{r}{\sum}}(K) + \overset{2}{\underset{n}{\sum}}(K)} \tag{15.82}$$

The system sensitivity function is defined as

$$\mathfrak{R} = \frac{C_{\max} - C}{C_{\max}} \tag{15.83}$$

For illustration purposes, we use system parameters based on UTRA FDD wideband CDMA concept [1] where the chip rate is 4.096 Mchips s^{-1}.

We will see in Chapter 17 that in the latest version of the standard this value is modified to 3.84 Mchips s^{-1}. The chosen coded data rate is 16 kbit s^{-1}, which means that the processing gain is $G = 256$. The required SNR is $Y_0 = 6\,\mathrm{dB}$. Three different exponential channel delay profiles have been used corresponding to $\lambda = 0$, 0.5 and 1. The number of multipaths is assumed to be $L = 4$ for all illustrations. The $K \times K$ cross-correlation matrix $\mathbf{R}$ has been modeled to consist of all ones on the diagonal and constant nonnegative correlation coefficients ρ elsewhere. Variance of the correlation slope represented in equation (15.34) is fixed at $\sigma_{\rho'}^2 = 1$.

Figures 15.2 to 15.4 present capacities as a function of the number of combined RAKE fingers. Figure 15.2 assumes a flat MIP, that is, $\lambda = 0$. Solid lines refer to the maximum capacity C_{max} with no estimation errors. As expected, the capacity will increase when the number of fingers is increased. Higher capacity losses are demonstrated when the correlation between users gets higher. EGC and MRC give identical results because of the equal unit tap weights in the combiner. In Figure 15.3 one can see a difference in the results between EGC and MRC. In this case the channel profile is exponential with $\lambda = 0.5$. Less power is available in the weak multipath components and MRC can cope better with weak taps. The situation is even more critical in Figure 15.4 in which

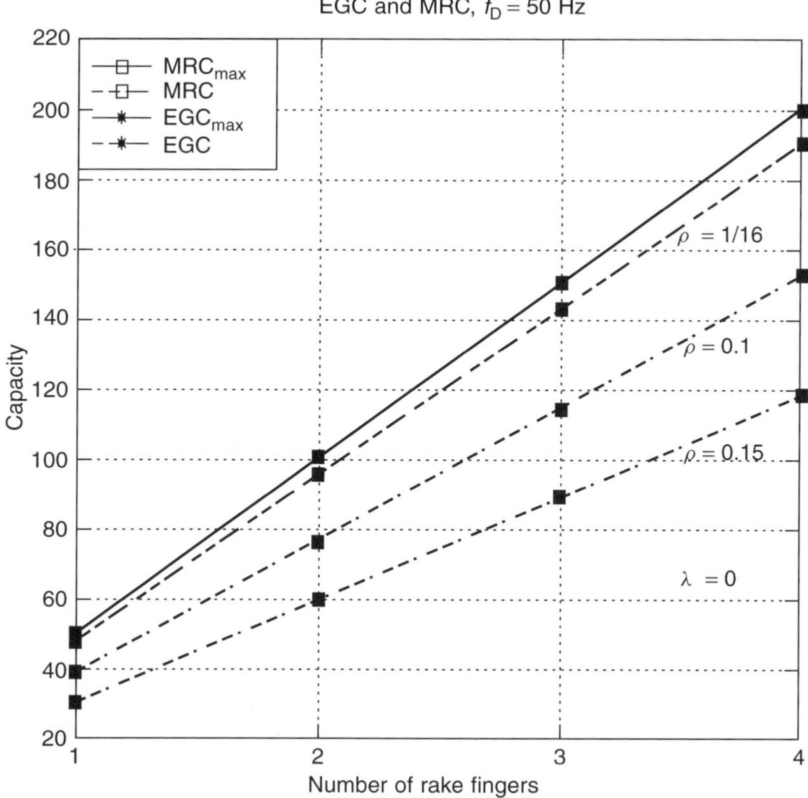

Figure 15.2 Capacity versus the number of RAKE fingers ($\lambda = 0$).

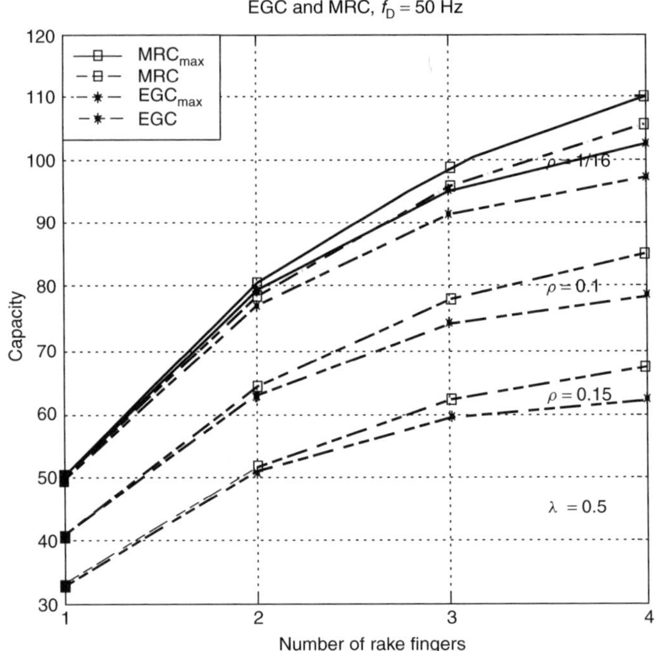

Figure 15.3 Capacity versus the number of RAKE fingers ($\lambda = 0.5$).

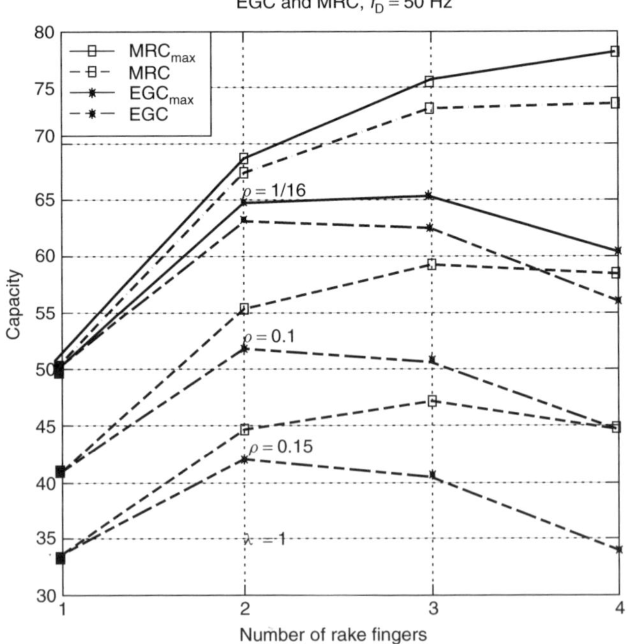

Figure 15.4 Capacity versus the number of RAKE fingers ($\lambda = 1$).

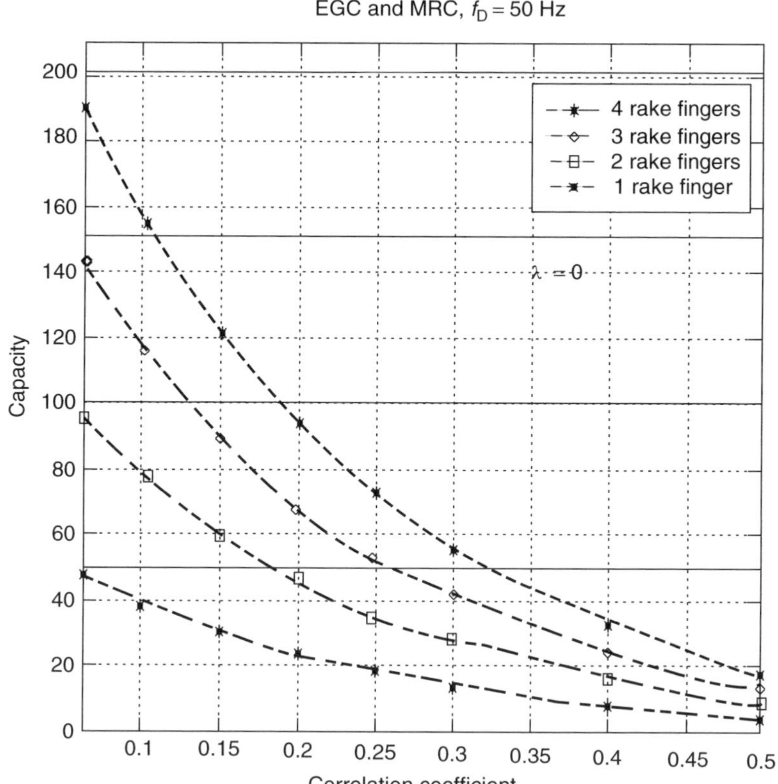

Figure 15.5 Capacity versus the correlation coefficient ($\lambda = 0$).

the capacity starts to decrease with high RAKE finger indices. Capacity loss is higher for EGC.

Capacity versus correlation between active users is plotted in Figure 15.5. Solid horizontal lines represent maximum capacities with $\rho = 1/16$ and no estimation errors. Dashed lines indicate that the capacity with estimation errors ($f_D = 50\,\text{Hz}$) decreases rapidly with increasing correlation. The relative losses in capacity are higher for the RAKE with larger number of fingers.

Capacity versus the correlation coefficients for $L_0 = 4$ and $f_D = 50\,\text{Hz}$ is presented in Figure 15.6 with λ being a parameter. The largest losses in capacity are for $\lambda = 0$.

A better insight into capacity losses can be obtained from Figure 15.7, which presents the network sensitivity function versus the correlation coefficient for $L_0 = 4$. One can see that as much as 90% of the capacity can be lost if correlation coefficients approach value 0.5. In other words the near–far resistance of the decorrelator has been almost completely lost owing to the imperfections in practical implementation.

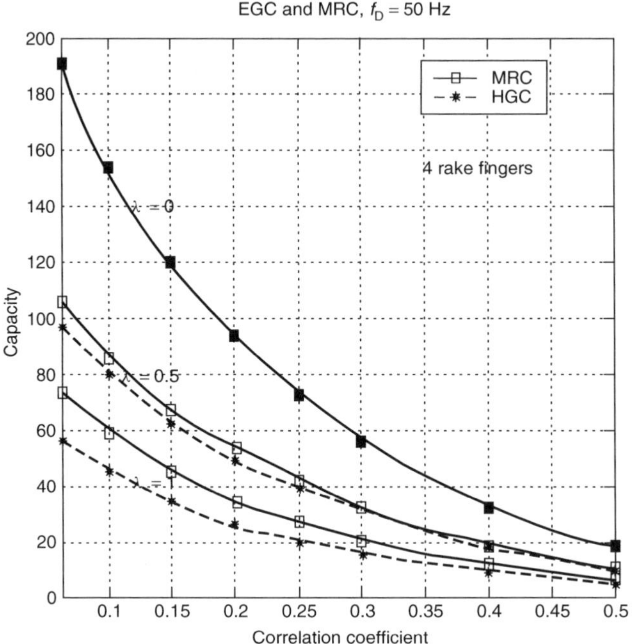

Figure 15.6 Capacity versus the correlation coefficient ($L_0 = 4$).

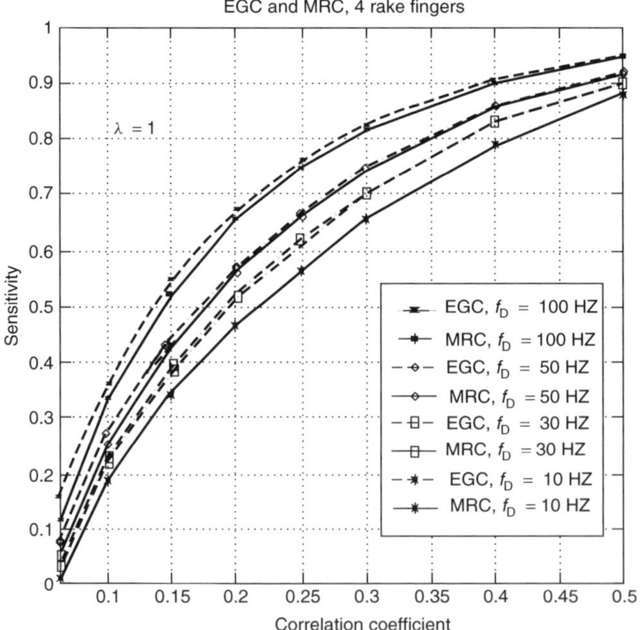

Figure 15.7 Sensitivity versus the correlation coefficient ($L_0 = 4$).

15.4 NEAR FAR SELF-RESISTANT CDMA WIRELESS NETWORK

Two of the most often used versions of spread spectrum systems are frequency hopping (FH) and direct sequence (DS) configurations. Owing to simplicity, DS configuration has been accepted for civil applications in mobile communication systems (e.g. Standards ANSI-95, IS-665, WCDMA UMTS). For large bit rates, DS system will have low processing gain and performance of a RAKE receiver will be considerably degraded. In the presence of near–far effect (imperfect power control or different signal levels due to different data rates), DS systems should use multiuser detectors (optimal or suboptimal structures). Since the optimum DS receiver is difficult to implement and suboptimum schemes are sensitive to implementation imperfections, the communicator may prefer to use a frequency-hopping spread-spectrum (FH-SS) system instead. For these reasons, these systems are used in both military and civil applications. For example, the advanced versions of TDMA (GSM) land mobile communications systems are also using hopping to improve performance in fading channel and to reduce intercell interference.

In order to improve performance in jamming and fading environment, these systems can use diversity. Traditionally, diversity was obtained via multiple hops per information (or coded) symbol. Such a fast hopping makes difficult the synchronization of the carrier phase and, consequently, imposes the use of a noncoherent receiver. Thus, a significant loss in error performance results, owing to both noncoherent demodulation and noncoherent combining of the received diversity replicas. Taking into account these losses and using binary frequency-shift keying (BFSK) modulation, an optimum diversity scheme is analyzed in Reference [33] for the worst-case jammer and with side information on noise and jamming levels. Since optimum diversity has more analytical value than practical existence, the error probability is much higher in practice. In order to recover these performance losses, some authors have studied a solution that makes coherent reception feasible (see References [34–36]). Frequency diversity as used on Rayleigh fading channels, and which differs from the diversity mentioned above, was proposed in Reference [37] to counter band–limited interference. Such a diversity allows one to avoid noncoherent combining loss. In this system, called *frequency-diversity spread spectrum* (FD-SS), the communicator frequency band is partitioned into N disjoint sub-bands on which N replicas of the signal are simultaneously transmitted. However, since FH is considered as mandatory in some applications, some solutions combine both FD-SS and FH-SS systems [33]. The main objective is to guarantee coherent demodulation and to avoid noncoherent combining losses.

If coherency is not feasible, then noncoherent solution is the only option. The effect of partial-band noise jamming on fast frequency-hopped (FFH) BFSK noncoherent receivers with diversity has been examined for channels with no fading [38], as has the effect of partial-band noise jamming on FFH/MFSK (M-ary frequency shift keying) for Ricean fading channels [39]. The performance degradation resulting from both band and independent multitone jamming of FH/MFSK, in which the jamming tones are assumed to correspond to some or all of the possible FH M-ary signaling tones and when thermal and other wideband noise is negligible, is examined in Reference [40]. The effect of tone interference on noncoherent MFSK when AWGN is not neglected is examined for channels with no fading in Reference [41], and the effect of independent multitone jamming on noncoherent FH/BFSK when AWGN is not neglected is examined for channels with no fading in Reference [42].

In this section we present an additional concept of multiple access called τ-CDMA. This system combines the good characteristics of Direct Sequence Spread Spectrum (DSSS) and FH-SS systems. The near far effect is mitigated without the need for complicated multiuser detectors and at the same time simplicity of DSSS system is preserved. There is no need for frequency synthesizer and coherency for coherent RAKE receiver is maintained in a much simpler way than in the FH system. The concept is based on a modification of the DSSS system in which the transmitted waveform includes multiple amplitude and delay replicas of the DSSS signal. The notation *amM-DSDH* will be used for the DS signal that includes m delayed replicas of different amplitudes (a) sent in a limited delay window of M chip intervals. The position of the delay window is hopped (delay hopping) in the range of the code length N. This should provide resistance to near far effect without need for FH. Variable impacts of near far effect, for different positions of the delay window and fading, are simultaneously reduced by interleaving. If the signal energy is split to $m > 1$ separate components making it more vulnerable to noise and fading, the overall flow of useful information will still be increased. The results demonstrate that under the large range of the signal, channel and interference parameters, this system offers better performance without the need for complex multiuser detectors or FH. This makes it applicable in *ad hoc* networks in which a classical base station capable of accepting a lot of processing complexity for multiuser detector is not available.

15.4.1 Signal formats, receiver structures and interference statistics

Coherent CDMA (c-CDMA): For a standard CDMA concept, the simplest form of the overall received signal can be represented as

$$r(t) = \sum_{k'} b_{k'}(t)s_{k'}(t - \tau_{k'}) \cos(\omega t + \theta_{k'}) \tag{15.84}$$

where for the k'th user $b_{k'}$ and $s_{k'}$ are data (bits) and pseudonoise (PN) sequence, respectively. The standard receiver uses coherent despreading and demodulation and will be referred to as coherent CDMA (c-CDMA). Extension to include I and Q signal components is straightforward. If a correlator (composed of a multiply plus integrate) is used for signal despreading, then we will refer to this structure as *correlator receiver* (CR). If a PN matched filter (PNMF) is used at the receiver and if the sequence period $T_s = NT_c$ equals bit period T_b, then at the output of the filter, one correlation pulse generated by the useful signal will appear per bit interval. The correlation pulse will appear each time at the chip interval when the input sequence coincides with the filter coefficients. This will be referred to as *PN matched filter receiver* (PNMFR).

M-ary delay CDMA (Mτ-CDMA): If now instead of sequence $s_{k'}$ a delayed version (cyclic shift) of the same sequence is used, $s_{k'}^{\tau}$, the position of the correlation pulse will depend on the sequence shift $\tau = \mu_{k'}(k) = kT_c$. Equation (15.84) now becomes

$$r(t) = \sum_{k'} b_{k'}(t) s_{k'}(t - \mu_{k'}(k)T_c - \tau_{k'}) \cos(\omega t + \theta_{k'}) \qquad (15.85)$$

If $\mu_{k'}(k) = kT_c$, $k = 0, 1, \ldots, M - 1$ is one out of $M = 2^n$ different adjacent cyclic shifts, then $n = \log_2 M$ additional bits can be transmitted within one symbol interval. The receiver structure is shown in Figures 15.8 to 15.11. In Chapter 8 we have shown that the capacity of a standard coherent CDMA system is roughly

$$K \cong G/y_b \qquad (15.86)$$

where G is the system processing gain and y_b is the SNR needed for a given quality of transmission. In our case, $G = N$. In accordance with the above explanation, capacity of the new CDMA system is additionally increased by a factor

$$K' \cong (G \log_2 M)/y_b' = (N \log_2 M)/y_b' \qquad (15.87)$$

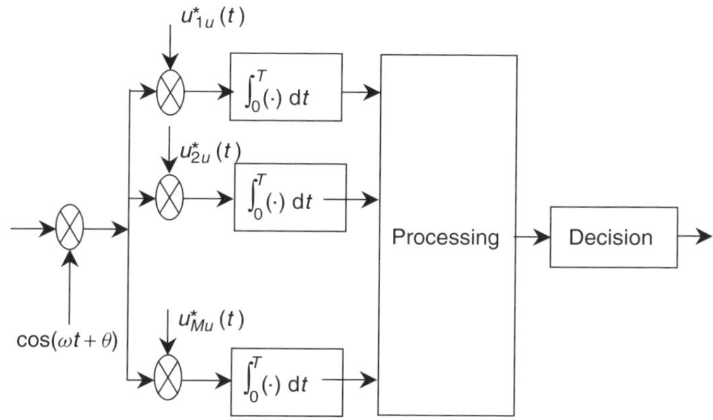

Figure 15.8   Coherent CR detection

For coherent CR detection ($M\tau$-CDMA)

$u_{ku}^* = s(t - kT_c - \tau)$

Processing $=$ calculate$\{U_m\}$ & chose the largest

Coherent CR detection ($mM\tau$-CDMA)

$u_{ku}^* = s(t - kT_c - \tau)$

Processing $=$ calculate $\{U_m\}$ & chose m the largest

Coherent CR detection ($amM\tau$-CDMA)

$u_{ku}^* = s(t - kT_c - \tau)$

Processing $=$ calculate $\{U_m\}$ & chose m the largest in decreasing order

Coherent CR detection ($amM\tau$-CDDHMA)

$u_{ku}^* = s(t - h(i)T_c - kT_c - \tau)$

Processing $=$ calculate $\{U_m\}$ & chose m the largest in decreasing order.

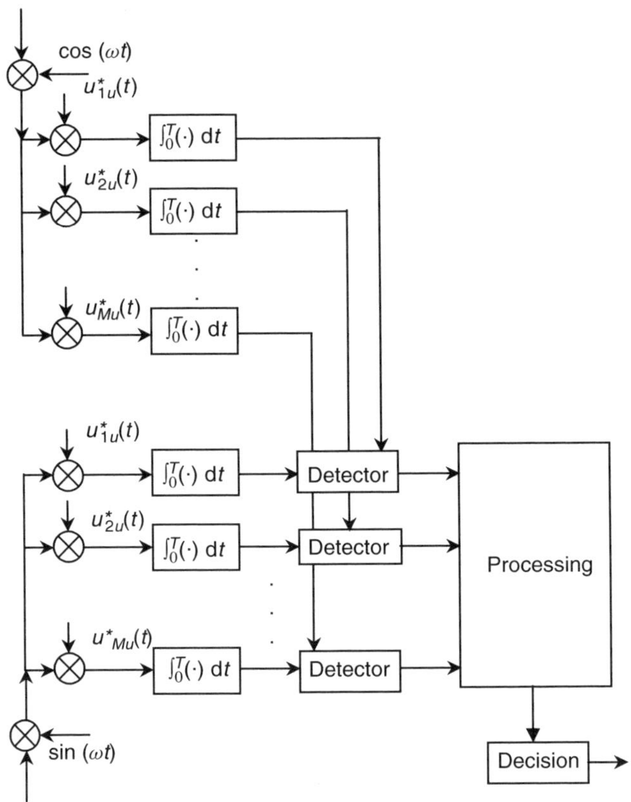

Figure 15.9 Noncoherent CR detection (processing functions defined in Figure 15.8).

where y'_b is the SNR needed for the same BER. Parameter y'_b will depend on the type of demodulator and is the main subject of this chapter.

Differential M-ary delay CDMA (dMτ-CDMA): The previous idea, based on correlation pulse position modulation, can be further modified to include correlation pulse distance modulation that should simplify synchronization.

Multiple M-ary delay CDMA (mMτ-CDMA): Let us suppose that now instead of sending one out of M delayed versions of the signal we send two different delayed replicas simultaneously. If amplitudes are the same, we can form

$$M_2 = \sum_{r=1}^{M-1} r = (M-1)M/2 \tag{15.88}$$

different combinations and send $n_2 = \log(M-1)M/2 = \log M + \log(M-1) - 1$ bits. If M is large, $n_2 \approx 2\log(M-1) \approx 2n$ is almost twice as much as in the case of the simple

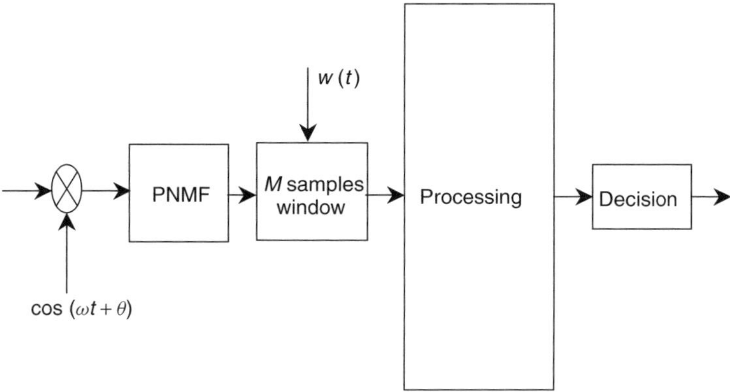

Figure 15.10   Coherent PNMF detection

For Coherent PNMF detection ($M\tau$-CDMA) $w(t)$

Processing = Calculate $\{U_m\}$ & chose the largest

Coherent PNMF detection ($mM\tau$-CDMA) $w(t)$

Processing = Calculate $\{U_m\}$ & chose m the largest

Coherent PNMF detection ($amM\tau$-CDMA) $w(t)$

Processing = Calculate $\{U_m\}$ & chose m the largest in decreasing order

Coherent PNMF detection ($amM\tau$-CDDHMA)

$w(t - h(i)T_\mathrm{c})$

Processing = Calculate $\{U_m\}$ & chose m the largest in decreasing order.

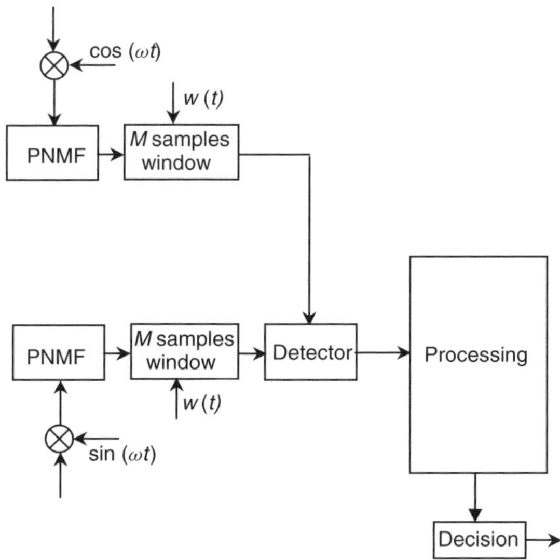

Figure 15.11   Noncoherent PNMF detection (processing functions defined in Figure 15.10).

M-ary modulation. The optimum receiver will now have to find two of the largest samples at the output of the matched filter. If now instead of two, m out of M signal replicas are simultaneously transmitted, we have $mM\tau$-DSSS modulation and corresponding $mM\tau$-CDMA system. The number of transmitted bits per T_s is now further increased. A simple calculus is needed to evaluate capacity improvement in the noise-free channel. Receiver block diagrams are shown in Figures 15.8 to 15.11.

Multiamplitude $mM\tau$-CDMA ($amM\tau$-CDMA):   Let us now go back to the case of $m = 2$. If two delayed signal replicas with different amplitudes A_l ($l = 1$, 2) are used, the number of combinations becomes larger and can be expressed as $M_2(a) = M(M - 1)$. If m replicas each of different amplitude are sent, we have $M_m(a) = M (M - 1)(M - 2)\ldots$ $(M - m + 1)$ and the number of bits transmitted is

$$n_m(a) = \sum_{i=0}^{m-1} \log(M - i) \tag{15.89}$$

If M is large, the previous relation can be approximated as $n_m(a) \approx m \log M = mn$. The name for modulation $amM\tau$-DSSS is self-explanatory. M is the size of the delay window, m the number of simultaneously transmitted replicas and a indicates that all transmitted replicas are of different amplitudes. The receiver block diagram is given in Figure 15.10. Let us assume that we have flat fading. We will characterize the channel coefficient (attenuation) for user k' with $\alpha_{k'}$. In the case of K users, equation (15.85) becomes

$$r(t) = \sum_{l=1}^{m} \sum_{k'=1}^{K} \alpha_{k'} A_{k'l} b_{k'l}(t) s_{k'}(t - \mu_{k'l}(k)T_{\mathrm{c}} - \tau_{k'}) \cos(\omega t + \theta_{k'}) \tag{15.90}$$

In order to better understand the nature of the CDMA interfering signal, let us analyze the output of the correlators from Figure 15.10.

When the input signal of the form given by equation 15.90 is frequency down converted and passed through the matched filters at the reference receiver $k' = r$, the collection (union represented by $\cup$) of the M complex outputs of the filters can be represented as

$$\bigcup_{k\in\{M\}} y_{rk}(t) = \bigcup_{k\in\{M\}} \sum_{l} \sum_{k'} [\alpha_{k'} A_{k'l} b_{k'l}(t) \rho_{k'r}\{t - \mu_{rl}(k)T_{\mathrm{c}}$$

$$+ \mu_{k'l}(k)T_{\mathrm{c}} - \Delta\tau_{k'}\} \exp \mathrm{j}(\theta_{k'} - \hat{\theta}_r)]$$

$$= \bigcup_{kl\in\{m\}} \{y_{rkl}(t) + i_{rk}(t)\} + \bigcup_{k\in\{M-m\}} i_{rk}(t) \tag{15.91}$$

where with notation $\phi_{k'} = \theta_{k'} - \hat{\theta}_r$, $\Delta\tau_{k'} = \tau_{k'} - \hat{\tau}_r$

$$y_{rkl}(t) = \alpha_r A_{rl} b_{rl}(t) \rho_{rr}\{t - \Delta\tau_r\} \exp \mathrm{j}\phi_r$$

$$i_{rk}(t) = \sum_{l} \sum_{\substack{k' \\ lk' \neq lr}} \alpha_{k'} A_{k'l} b_{k'l}(t) \rho_{k'r} \{ t - \mu_{rl}(i) T_{\mathrm{c}} + \mu_{k'l}(i) T_{\mathrm{c}} - \Delta \tau_k' \} \exp \mathrm{j} \phi_k' \qquad (15.92)$$

In this equation, $y_{rkl}(t)$ is the useful signal component carrying the information encoded in $b_{rl}(t)$ and the combination of A_{rl} and $\mu_{rl}(k)$. To simplify notation, this component will be referred to as $y_u(t)$ whenever further specification is not necessary.

Delay hopping amMτ-CDMA (amMτ-CDDHMA):   Parameter $i_{rk}(t)$, in equation (15.92), is the overall interference in the kth branch of the reference receiver generated by all other $K - 1$ users and $(m - 1)$ other signals of the reference user itself. If parameters $\alpha_{k'}$ are very much different (near far effect), $i_{rk}(t)$ [simplified notation $i(t)$] can reach higher values than the signal for some time (delay) slots. If the output of the matched filter for the MFR is sampled every T_{c} seconds ($t_j = jT_{\mathrm{c}}$) and if in F out of N time (delay) slots $i(jT_{\mathrm{c}}) > y_u(jT_{\mathrm{c}})$, we define the near far factor f $= F/N$. For those slots where $i(jT_{\mathrm{c}}) > y_u(jT_{\mathrm{c}})$, the signal cannot be detected. The traditional CDMA system has to use optimum or some form of suboptimum multiuser detector to cope with the near far effect.

In this section we use analogy with the FH system and introduce delay hopping. The basic idea is that if in one symbol interval $i(jT_{\mathrm{c}}) > y_u(jT_{\mathrm{c}})$ at the expected position of the signal correlation peaks, then with hopping the position of the delay window the situation for the next symbol should change. Proper interleaving should correct these occasional symbol errors. For these purposes, we additionally introduce delay in the code by factor $\tau_h = h(t - iT_{\mathrm{s}}) T_{\mathrm{c}} = h(i) T_{\mathrm{c}}$ for $h \in [0, H]$ and $H \in [0, N]$. In this case equations 15.90 and 15.92 become

$$r(t - iT_{\mathrm{s}}) = \sum_{l=1}^{m} \sum_{k'=1}^{K} \alpha_{k'} A_{k'l} b_{k'l}(t) s_{k'}(t - h_{k'}(i) T_{\mathrm{c}} - \mu_{k'l}(k) T_{\mathrm{c}} - \tau_{k'}) \cos(\omega t + \theta_{k'})$$

$$y_{rkl}(t - iT_{\mathrm{s}}) = \alpha_r A_{rl} b_{rl}(t) \rho_{rr} \{ t - \Delta \tau_r \} \exp \mathrm{j} \phi_r$$

$$i_{rk}(t - iT_{\mathrm{s}}) = \sum_{l} \sum_{\substack{k' \\ lk' \neq lr}} [\alpha_{k'} A_{k'l} b_{k'l}(t) \times \rho_{k'r} \{ t - h_r(i) T_{\mathrm{c}}$$

$$- \mu_{rl}(i) T_{\mathrm{c}} + h_{k'}(i) T_{\mathrm{c}} + \mu_{k'l}(i) T_{\mathrm{c}} - \Delta \tau_{k'} \} \exp \mathrm{j} \phi_{k'}] \qquad (15.93)$$

The extension to I and Q signal formats is straightforward. This system will be referred to as *amMτ*-CDDHMA where *amMτ* has the same meaning as before and Code Division Delay Hopping Multiple Access (CDDHMA) stands for code-division delay hopping multiple access. The receiver block diagram is given in Figures 15.8 to 15.11.

15.4.2 CDMA channel models

Gilbert model: As already indicated earlier, for the purposes of this analysis we can use Gilbert model for CDMA interference. In other words, on a given delay slot of T_{c}

seconds wide $i(t)$ is a sum of all phasors in equation (15.92) and can be approximated as an equivalent tone. The amplitude of this tone will depend on the cross-correlations and mutual phases between the interfering users. In real situations, on some delay slots this interference will be low and on others high. This will be approximated by the Gilbert model in which a low-level interfering signal is neglected [which is represented by zero state G(0)] and high interfering signal produces a catastrophic effect and blocks the demodulator G(1). This model is used to characterize the near far effect with the near far factor f defined as the probability that for a given delay τ_k the channel is in the state $G_k(1)$.

Gaussian model: We can approximate $i(t)$ as a Gaussian interference with zero mean and variance

$$\sigma^2 \cong (K - 1)m\alpha^2\overline{A}^2/G \tag{15.94}$$

The model is appropriate when all interfering signals have comparable levels.

CW model: In this model, $i(t)$ is approximated as an equivalent tone (phasor) with random phase $\theta \in [0, 2\pi]$ and power given by equation 15.94. This model is appropriate if there are few predominant interferers.

15.4.3 Performance measure

As a performance measure, we will be discussing symbol error rate or the system efficiency improvement factor defined as [43]

$$E = [(1 - P)n]/[(1 - P_0)n_0] \tag{15.95}$$

In equation (15.95) n is the number of bits per symbol, P is the bit error rate and $(1-P)n$ is the average number of correctly transmitted bits per symbol. The same parameters with index zero refer to the standard modulation ($m = 1$, $M = 1$). By splitting the available signal power on m replicas, the system will be more error prone but the average number of correctly transmitted bits should be still higher. For reasons of simplicity, derivation of symbol error rate for different receiver structures is given in Appendices 1 to 4.

15.4.4 Receiver near far resistance

By definition, near far resistance (*nfr*) is given as

$$nfr = 1 - f \tag{15.96}$$

where f is the probability that for a given delay τ_k the channel is in state $G_k(1)$. For CDDHMA system this parameter can be further improved by modifying the transmit/receive algorithms as defined in the sequel.

Option #1/Interleaving plus block coding: In the standard version of the system, interleaving with depth D and error-correcting coding with block length D and correction capability e are used. The frame error will occur if the number of bad slots ($i > y_u$) out of D is larger than e.

$$f' = \sum_{k=e}^{D} \binom{D}{k} f^k (1-f)^{D-k}; nfr' = 1 - f' \tag{15.97}$$

Option #2/Time diversity: If time diversity is used by repeating the information ($2D + 1$) times and majority logic decision, then

$$f'' = \sum_{k=D+1}^{2D+1} \binom{2D+1}{k} f^k (1-f)^{2D+1-k}; nfr'' = 1 - f'' \tag{15.98}$$

In other words, if the information is repeated $2D + 1$ times, and if the decision is made in favor of the word appearing more than D times, then the final decision will be in error if the word error occurred more than D times. One should be aware that system efficiency defined by equation (15.95) is now reduced by factor $2D + 1$ ($E \Rightarrow E/(2D + 1)$)

Option #3/Combination of coding and time diversity: If options #1 and #2 are combined, which means interleaving plus coding plus time diversity are used, we have

$$f^{1\&2} = f''(f') = \sum_{k=D+1}^{2D+1} \binom{2D+1}{k} f'^k (1-f')^{2D+1-k}; nfr^{1\&2} = 1 - f^{1\&2}$$

Option #4/System with preamble: If a control word of length c together with d data bits in the frame is used in the transmitter, then for $2D + 1$ time diversity we have

$$f''' = f^{2D+1}; nfr''' = 1 - f'''; E \Rightarrow Ed/(c + d)(2D + 1) \tag{15.99}$$

In other words, it is enough to recognize at least one correct preamble, which is an indication that the whole frame is not damaged. Here we make an assumption that MAI has a predominant impact on BER

Option #5 Combination of options #1 and #4: In this case

$$f^{1\&4} = f'''(f') = f'^{2D+1}; \quad nfr^{1\&4} = 1 - f^{1\&4} \tag{15.100}$$

For illustration purposes, efficiency improvements as a function of the number of users for coherent and noncoherent detection, Gaussian interference model and no fading channel are presented in Figure 15.12. The set of curves is obtained for two values of SNR (4 and 16 dB) and a number of combinations for M and m. For high SNR (16 dB), parameter E is larger than one for the large scale of parameters. For smaller K, parameter E is higher if m is larger. If m is lower, the maximum value of E is lower but E remains above one for higher K. Curves for noncoherent detection are presented in the same figure. We can

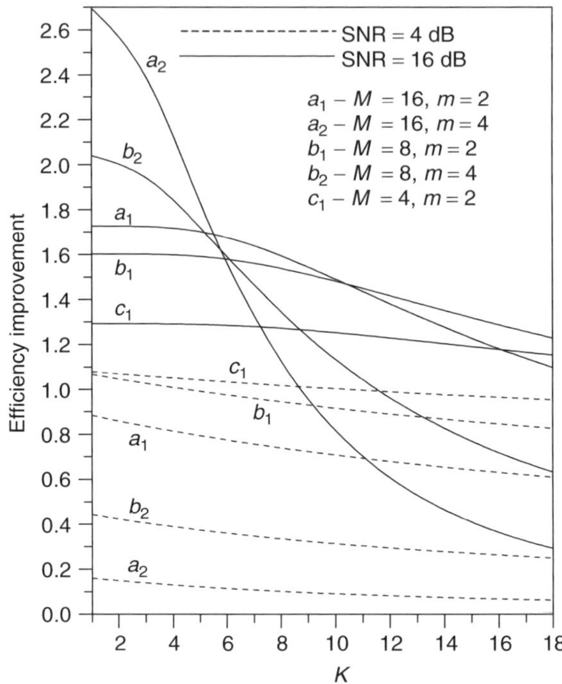

Figure 15.12 Efficiency improvement as a function of the number of users (coherent detection).

see that parameter E for noncoherent detection is lower and lower m is uniformly the better choice. For lower SNR the improvement factor is lower.

Figure 15.13 represents the same results as Figure 15.12 but in the case when the signal is propagating through Rayleigh fading channel. In general, the improvement factor is lower and only for the few users in the channel larger than one.

A three-dimensional figure (Figure 15.14) represents parameter E for $amM\tau$-CDMA system for $m = 2$. Under the assumption that $A_1^2 + A_2^2 = 1$, the maximum value for E is obtained if $A_1^2 \approx 0.75$. This also will depend on K as demonstrated in Figure 15.15.

Efficiency improvement versus K for $amM\tau$-CDMA system is presented in Figure 15.16 for continuous wave (CW) model of MAI. The interpretation of these results is very much similar to those shown in Figures 15.12 and 15.13.

Finally, near far resistance, nfr, is presented in Figure 15.17 with Gilbert model of MAI. This figure is a key result of this analysis. Without going into reasons of what has caused a certain level of near far effect, characterized by the near far factor f, one can see a considerable improvement in the system's near far resistance. One should keep in mind that system efficiency will be also reduced for different solutions as discussed before. If $E < 1$, then $m = 1$ and $M = 1$ should be used ($E = 1$) and still with delay hopping near far self-resistance will be preserved.

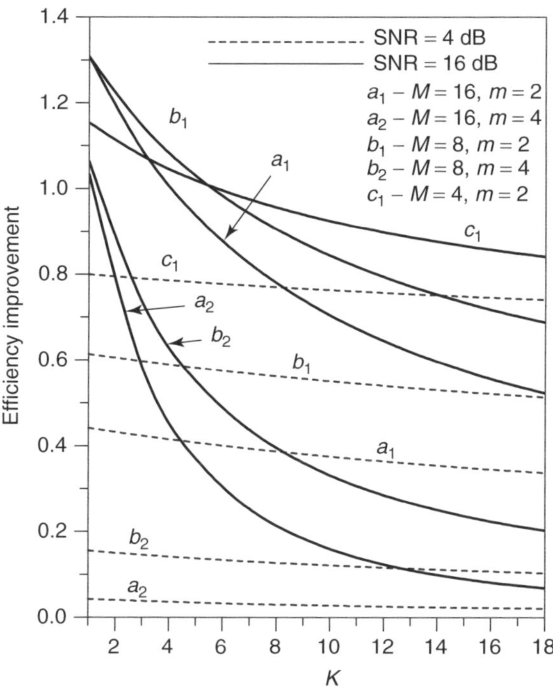

Figure 15.13 Efficiency improvement of $mM\tau$-CDMA system as a function of the number of users (coherent detection, Rayleigh fading of useful signal).

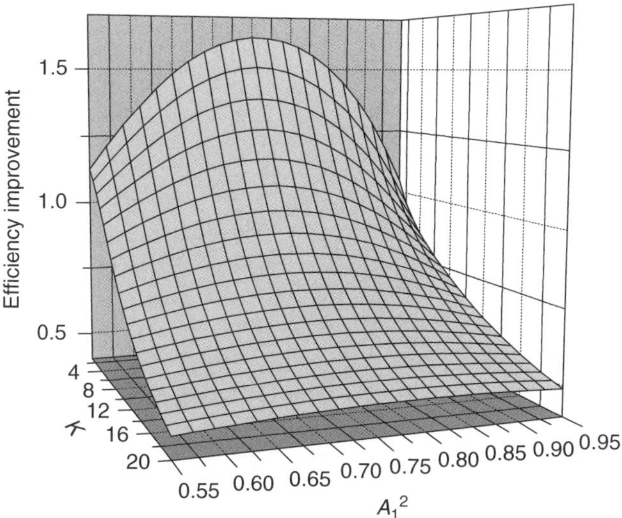

Figure 15.14 Efficiency improvement of $amM\tau$-CDMA system as a function of the number of users and A_1^2 (noncoherent detection, no fading).

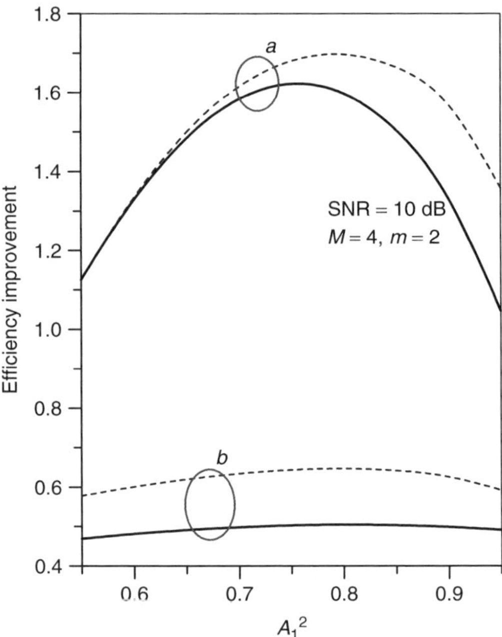

Figure 15.15 Efficiency improvement of $amM\tau$-CDMA system as a function of A_1^2 (no fading).

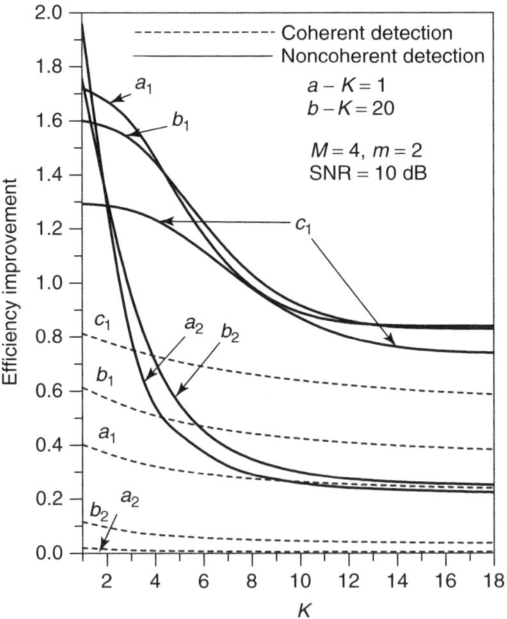

Figure 15.16 Efficiency improvement of $amM\tau$-CDMA system as a function of the number of users (coherent detection, no fading).

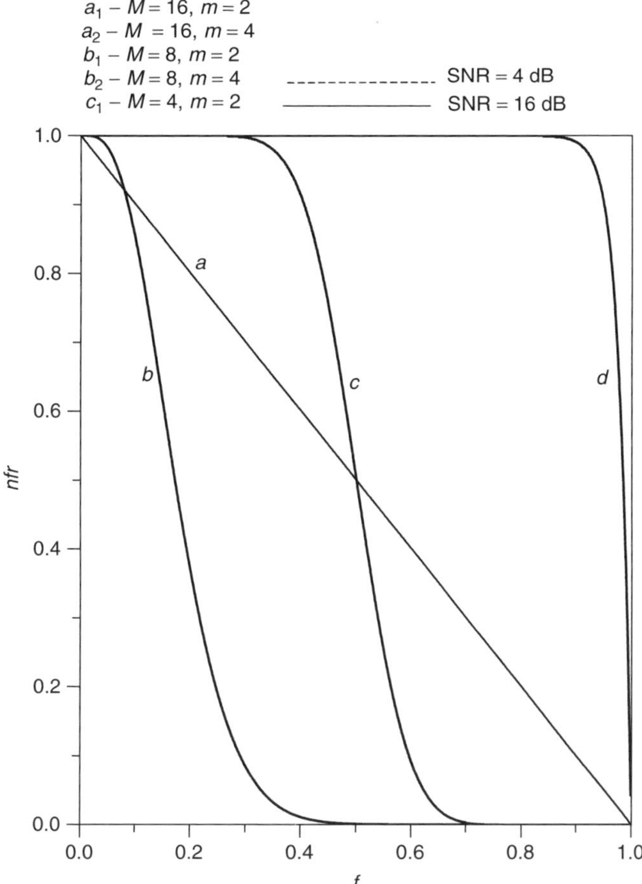

Figure 15.17   Near far resistance as a function of near far factor for the Gilbert model of MAI

$$a - nfr \text{ (equation 15.96)}$$

$$b - nfr' \text{ (equation 15.97)}$$

$$c - nfr'' \text{ (equation 15.98)}$$

$$d - nfr''' \text{ (equation 15.99)}$$

APPENDIX 1 COHERENT DETECTION OF ($mM\,\tau$-CDMA)

A. Gaussian interference model

Signals at the output of integrators are Gaussian-distributed, with the following probability distribution functions (pdfs):

$$p(U_k) = \frac{1}{\sqrt{2\pi(\sigma_n^2 + \sigma_J^2)}} \exp\left(-\frac{(U_k - a_c)^2}{2(\sigma_n^2 + \sigma_J^2)}\right), \quad k = 1, 2, \ldots, m \tag{A1.1}$$

$$p(U_k) = \frac{1}{\sqrt{2\pi(\sigma_n^2 + \sigma_J^2)}} \exp\left(-\frac{U_k^2}{2(\sigma_n^2 + \sigma_J^2)}\right), \quad k = m+1, m+2, \ldots, M \tag{A1.2}$$

where $a_c = \alpha A$ is the received useful signal amplitude, σ_n^2 is the additive Gaussian noise variance and σ_J^2 is the interference power.

The probability of a correct decision is

$$\begin{aligned}
P_c &= P(U_1 > U_{m+1}, U_1 > U_{m+2}, \ldots, U_1 > U_M) \\
&\quad \times P(U_2 > U_{m+1}, U_2 > U_{m+2}, \ldots, U_2 > U_M) \times \cdots \\
&\quad \times P(U_m > U_{m+1}, U_m > U_{m+2}, \ldots, U_m > U_M)
\end{aligned} \tag{A1.3}$$

where

$$\begin{aligned}
&P(U_1 > U_{m+1}, U_1 > U_{m+2}, \ldots, U_1 > U_M) \\
&\quad = \int_{-\infty}^{+\infty} P(U_1 > U_{m+1}, \ldots, U_1 > U_M | U_1) p(U_1)\, \mathrm{d}U_1 \\
&P(U_2 > U_{m+1}, U_2 > U_{m+2}, \ldots, U_2 > U_M) \\
&\quad = \int_{-\infty}^{+\infty} P(U_2 > U_{m+1}, \ldots, U_2 > U_M | U_2) p(U_2)\, \mathrm{d}U_2 \\
&\qquad\qquad\qquad \vdots \quad \vdots \quad \vdots \\
&P(U_m > U_{m+1}, U_m > U_{m+2}, \ldots, U_m > U_M) \\
&\quad = \int_{-\infty}^{+\infty} P(U_m > U_{m+1}, \ldots, U_m > U_M | U_m) p(U_m)\, \mathrm{d}U_m
\end{aligned} \tag{A1.4}$$

Since all m probabilities are equal, we will determine the first one

$$P(U_1 > U_{m+1} | U_1) = \int_{-\infty}^{U_1} p(U_{m+1})\, \mathrm{d}U_{m+1} = \frac{1}{2}\left(1 + \mathrm{erf}\left(\frac{U_1}{\sqrt{2(\sigma_n^2 + \sigma_J^2)}}\right)\right) \tag{A1.5}$$

$$\begin{aligned}
&P(U_1 > U_{m+1}, U_1 > U_{m+2}, \ldots, U_1 > U_M) = \int_{-\infty}^{+\infty} P(U_1 > U_{m+1} | U_1)^{M-m} p(U_1)\, \mathrm{d}U_1 \\
&\quad = \frac{1}{2^{M-m}\sqrt{\pi}} \int_{-\infty}^{+\infty} \mathrm{e}^{-x^2} \left[1 + \mathrm{erf}\left(x + \sqrt{\frac{a_c^2}{2(\sigma_n^2 + \sigma_J^2)}}\right)\right]^{M-m} \mathrm{d}x
\end{aligned} \tag{A1.6}$$

Probability of a correct decision is

$$P_{\mathrm{c}} = \left(\frac{1}{2^{M-m}\sqrt{\pi}} \int_{-\infty}^{+\infty} \mathrm{e}^{-x^2} \left[1 + \mathrm{erf}\left(x + \sqrt{\frac{a_c^2}{2(\sigma_n^2 + \sigma_J^2)}} \right) \right]^{M-m} \mathrm{d}x \right)^m \tag{A1.7}$$

The error probability now becomes

$$P_{\mathrm{s}} = 1 - P_{\mathrm{c}} = F_{\mathrm{coh}}(a_c^2)$$

$$= 1 - \left(\frac{1}{2^{M-m}\sqrt{\pi}} \int_{-\infty}^{+\infty} \mathrm{e}^{-x^2} \left[1 + \mathrm{erf}\left(x + \sqrt{\frac{a_c^2}{2(\sigma_n^2 + \sigma_J^2)}} \right) \right]^{M-m} \mathrm{d}x \right)^m \tag{A1.8}$$

Because of Rayleigh fading, a_c has the following pdf:

$$f(a_c) = \frac{a_c}{\sigma_c^2} \exp\left(-\frac{a_c^2}{2\sigma_c^2} \right) \tag{A1.9}$$

where $2\sigma_c^2$ is the signal power.

After averaging over a_c, the error probability becomes

$$P_{\mathrm{s}} = \int_0^{+\infty} F_{\mathrm{coh}}(x \cdot 2\sigma_c^2) \exp(-x)\, \mathrm{d}x \tag{A1.10}$$

CW interference model

Signals at the output of integrators are Gaussian-distributed, with the following pdfs:

$$p(U_k) = \frac{1}{\sqrt{2\pi\sigma_n^2}} \exp\left(-\frac{(U_k - \sqrt{a_c^2 + a_J^2 + 2a_c a_J \cos\theta})^2}{2\sigma_n^2} \right), \quad k = 1, 2, \ldots, m$$

$$\tag{A1.11}$$

$$p(U_k) = \frac{1}{\sqrt{2\pi\sigma_n^2}} \exp\left(-\frac{(U_k - a_J)^2}{2\sigma_n^2} \right), \quad k = m+1, m+2, \ldots, M \tag{A1.12}$$

where a_J is the interference amplitude, and θ is the interference phase relative to the useful signal, uniformly distributed in 0 to 2π interval. The probability of a correct decision is

$$P_{\mathrm{c}} = P(U_1 > U_{m+1}, U_1 > U_{m+2}, \ldots, U_1 > U_M)$$

$$\times P(U_2 > U_{m+1}, U_2 > U_{m+2}, \ldots, U_2 > U_M) \times \cdots$$

$$\times P(U_m > U_{m+1}, U_m > U_{m+2}, \ldots, U_m > U_M) \tag{A1.13}$$

where

$$P(U_1 > U_{m+1}, U_1 > U_{m+2}, \ldots, U_1 > U_M)$$
$$= \int_{-\infty}^{+\infty} P(U_1 > U_{m+1}, \ldots, U_1 > U_M | U_1) p(U_1) \, dU_1$$

$$P(U_2 > U_{m+1}, U_2 > U_{m+2}, \ldots, U_2 > U_M)$$
$$= \int_{-\infty}^{+\infty} P(U_2 > U_{m+1}, \ldots, U_2 > U_M | U_2) p(U_2) \, dU_2$$

$$\vdots \quad \vdots \quad \vdots$$

$$P(U_m > U_{m+1}, U_m > U_{m+2}, \ldots, U_m > U_M)$$
$$= \int_{-\infty}^{+\infty} P(U_m > U_{m+1}, \ldots, U_m > U_M | U_m) p(U_m) \, dU_m \qquad (A1.14)$$

Since all m probabilities are equal, we will determine the first one

$$P(U_1 > U_{m+1} | U_1) = \int_{-\infty}^{U_1} p(U_{m+1}) dU_{m+1} = \frac{1}{2}\left(1 + \mathrm{erf}\left(\frac{U_1 - a_J}{\sqrt{2\sigma_n^2}}\right)\right) \qquad (A1.15)$$

$$P(U_1 > U_{m+1}, U_1 > U_{m+2}, \ldots, U_1 > U_M | \theta)$$
$$= \int_{-\infty}^{+\infty} P(U_1 > U_{m+1} | U_1)^{M-m} p(U_1) dU_1 = \frac{1}{2^{M-m}\sqrt{\pi}}$$
$$\times \int_{-\infty}^{+\infty} e^{-x^2} \left[1 + \mathrm{erf}\left(x + \sqrt{\frac{a_c^2 + a_J^2 + 2a_c a_J \cos\theta}{2\sigma_n^2}} - \sqrt{\frac{a_J^2}{2\sigma_n^2}}\right)\right]^{M-m} dx$$

$$(A1.16)$$

After averaging over θ, we get

$$P(U_1 > U_{m+1}, \ldots, U_1 > U_M) = \frac{1}{2\pi} \int_0^{2\pi} P(U_1 > U_{m+1}, \ldots, U_1 > U_M | \theta) \, d\theta \quad (A1.17)$$

The error probability is

$$P_s = 1 - P(U_1 > U_{m+1}, \ldots, U_1 > U_M)^m = 1 - F_{\mathrm{cohCW}}(a_c^2) \qquad (A1.18)$$

Because of Rayleigh fading, a_c has the pdf defined by equation (A1.9). After averaging over a_c, the error probability becomes

$$P_s = \int_0^{+\infty} (1 - F_{\mathrm{cohCW}}(x \cdot 2\sigma_c^2)) \exp(-x) \, dx \qquad (A1.19)$$

APPENDIX 2 COHERENT DETECTION OF (*amM* τ-CDMA)

We will consider the case of $m = 2$ and arbitrary M.

Gaussian interference model

Signals at the output of integrators are Gaussian-distributed, with the following pdfs:

$$p(U_k) = \frac{1}{\sqrt{2\pi(\sigma_n^2 + \sigma_J^2)}} \exp\left(-\frac{(U_k - a_{ck})^2}{2(\sigma_n^2 + \sigma_J^2)}\right), \quad k = 1, 2 \tag{A2.1}$$

$$p(U_k) = \frac{1}{\sqrt{2\pi(\sigma_n^2 + \sigma_J^2)}} \exp\left(-\frac{U_k^2}{2(\sigma_n^2 + \sigma_J^2)}\right), \quad k = 3, \ldots, M \tag{A2.2}$$

where a_{ck} is the received useful signal amplitude ($a_{ck} = \alpha A_k$, $A_1^2 + A_2^2 = 1$), σ_n^2 is the additive Gaussian noise variance and σ_J^2 is the interference power. The probability of a correct decision is

$$P_c = P(U_1 > U_2, U_1 > U_3, \ldots, U_1 > U_M) \times P(U_2 > U_3, U_2 > U_4, \ldots, U_2 > U_M) \tag{A2.3}$$

where

$$P(U_1 > U_2, U_1 > U_3, \ldots, U_1 > U_M)$$
$$= \int_{-\infty}^{+\infty} P(U_1 > U_2, \ldots, U_1 > U_M | U_1) p(U_1)\, dU_1$$

$$P(U_2 > U_3, U_2 > U_4, \ldots, U_2 > U_M)$$
$$= \int_{-\infty}^{+\infty} P(U_2 > U_3, \ldots, U_2 > U_M | U_2) p(U_2)\, dU_2 \tag{A2.4}$$

$$P(U_1 > U_2 | U_1)$$
$$= \int_{-\infty}^{U_1} p(U_2) dU_2 = \frac{1}{2}\left(1 + \mathrm{erf}\left(\frac{U_1 - a_{c2}}{\sqrt{2(\sigma_n^2 + \sigma_J^2)}}\right)\right) \tag{A2.5}$$

$$P(U_1 > U_3 | U_1) = \cdots = P(U_1 > U_M | U_1)$$
$$= \int_{-\infty}^{U_1} p(U_3)\, dU_3 = \frac{1}{2}\left(1 + \mathrm{erf}\left(\frac{U_1}{\sqrt{2(\sigma_n^2 + \sigma_J^2)}}\right)\right) \tag{A2.6}$$

$$P(U_2 > U_3 | U_2) = \cdots = P(U_2 > U_M | U_2)$$
$$= \int_{-\infty}^{U_2} p(U_3)\, dU_3 = \frac{1}{2}\left(1 + \mathrm{erf}\left(\frac{U_2}{\sqrt{2(\sigma_n^2 + \sigma_J^2)}}\right)\right) \tag{A2.7}$$

The probability of a correct decision is

$$P_c = \int_{-\infty}^{+\infty} \int_{-\infty}^{U_1} P(U_1 > U_2|U_1) P(U_1 > U_3|U_1)^{M-2}$$

$$\times P(U_2 > U_3|U_2)^{M-2} p(U_1) p(U_2) dU_2 dU_1 \qquad (A2.8)$$

$$P_c = \int_{-\infty}^{+\infty} \int_{-\infty}^{x+\frac{a_{c1}-a_{c2}}{\sqrt{2(\sigma_n^2+\sigma_J^2)}}} \left[\frac{1}{2}\left(1 + \mathrm{erf}\left(x + \frac{a_{c1}-a_{c2}}{\sqrt{2(\sigma_n^2+\sigma_J^2)}}\right)\right)\right]$$

$$\times \left[\frac{1}{2}\left(1 + \mathrm{erf}\left(x + \frac{a_{c1}}{\sqrt{2(\sigma_n^2+\sigma_J^2)}}\right)\right)\right]^{M-2}$$

$$\times \left[\frac{1}{2}\left(1 + \mathrm{erf}\left(x + \frac{a_{c2}}{\sqrt{2(\sigma_n^2+\sigma_J^2)}}\right)\right)\right]^{M-2} e^{-x^2} e^{-y^2} dy\, dx \qquad (A2.9)$$

The error probability now becomes

$$P_s = F_{\mathrm{coha}}(a_{c1}^2, a_{c2}^2) = 1 - P_c \qquad (A2.10)$$

Because of Rayleigh fading, a_{ck} has the following pdf:

$$f(a_{ck}) = \frac{a_{ck}}{\sigma_{ck}^2} \exp\left(-\frac{a_{ck}^2}{2\sigma_{ck}^2}\right), \quad k = 1, 2 \qquad (A2.11)$$

where $2\sigma_{ck}^2$ is the signal power. After averaging over a_{ck}, the error probability becomes

$$P_s = \int_{a_{c1}} \int_{a_{c2}} F_{\mathrm{coha}}(a_{c1}^2, a_{c2}^2) f(a_{c1}) f(a_{c2})\, da_{c1}\, da_{c2} \qquad (A2.12)$$

CW interference model

Signals at the output of integrators are Gaussian-distributed, with the following pdfs:

$$p(U_k) = \frac{1}{\sqrt{2\pi\sigma_n^2}} \exp\left(-\frac{\left(U_k - \sqrt{a_{ck}^2 + a_J^2 + 2a_{ck}a_J\cos\theta}\right)^2}{2\sigma_n^2}\right), \quad k = 1, 2 \quad (A2.13)$$

$$p(U_k) = \frac{1}{\sqrt{2\pi\sigma_n^2}} \exp\left(-\frac{(U_k - a_J)^2}{2\sigma_n^2}\right), \quad k = 3, 4, \ldots, M \qquad (A2.14)$$

where a_J is the interference amplitude, and θ is the interference phase relative to the useful signal, uniformly distributed in 0 to 2π interval. The probability of a correct decision is

$$P_c = P(U_1 > U_2, U_1 > U_3, \ldots, U_1 > U_M) \times P(U_2 > U_3, U_2 > U_4, \ldots, U_2 > U_M)$$

(A2.15)

where

$$P(U_1 > U_2, U_1 > U_3, \ldots, U_1 > U_M)$$
$$= \int_{-\infty}^{+\infty} P(U_1 > U_2, \ldots, U_1 > U_M | U_1) p(U_1) dU_1$$

$$P(U_2 > U_3, U_2 > U_4, \ldots, U_2 > U_M)$$
$$= \int_{-\infty}^{+\infty} P(U_2 > U_3, \ldots, U_2 > U_M | U_2) p(U_2) dU_2$$

(A2.16)

$$P(U_1 > U_2 | U_1, \theta) = \int_{-\infty}^{U_1} p(U_2)\, dU_2$$
$$= \frac{1}{2}\left(1 + \mathrm{erf}\left(\frac{U_1 - \sqrt{a_{c2}^2 + a_J^2 + 2a_{c2}a_J \cos\theta}}{\sqrt{2\sigma_n^2}}\right)\right)$$

(A2.17)

$$P(U_1 > U_3 | U_1, \theta) = \cdots = P(U_1 > U_M | U_1)$$
$$= \int_{-\infty}^{U_1} p(U_3) dU_3 = \frac{1}{2}\left(1 + \mathrm{erf}\left(\frac{U_1}{\sqrt{2\sigma_n^2}}\right)\right)$$

(A2.18)

$$P(U_2 > U_3 | U_2, \theta) = \cdots = P(U_2 > U_M | U_2)$$
$$= \int_{-\infty}^{U_2} p(U_3) dU_3 = \frac{1}{2}\left(1 + \mathrm{erf}\left(\frac{U_2}{\sqrt{2\sigma_n^2}}\right)\right)$$

(A2.19)

Conditional probability of a correct decision is

$$P_c(\theta) = \int_{-\infty}^{+\infty} \int_{-\infty}^{U_1} P(U_1 > U_2 | U_1, \theta) P(U_1 > U_3 | U_1, \theta)^{M-2}$$
$$\times P(U_2 > U_3 | U_2, \theta)^{M-2} p(U_1) p(U_2)\, dU_2\, dU_1$$

After averaging over θ, we get

$$P_c = \frac{1}{2\pi} \int_0^{2\pi} P_c(\theta)\, d\theta$$

(A2.20)

The error probability now becomes

$$P_s = 1 - P_c = 1 - F_{\mathrm{cohCW}a}(a_{c1}^2, a_{c2}^2)$$

(A2.21)

After averaging over a_{ck}, the error probability becomes

$$P_{s} = \int_{a_{c1}} \int_{a_{c2}} (1 - F_{\text{cohCW}a}(a_{c1}^2, a_{c2}^2)) f(a_{c1}) f(a_{c2}) \, \mathrm{d}a_{c1} \, \mathrm{d}a_{c2} \qquad \text{(A2.22)}$$

APPENDIX 3 NONCOHERENT DETECTION OF ($mM\,\tau$-CDMA)

Gaussian interference model

Signals at the output of detectors are Rician- and Rayleigh-distributed, with the following pdfs:

$$p(U_k) = \frac{U_k}{\sigma_n^2 + \sigma_J^2} \exp\left(-\frac{U_k^2 + a_c^2}{2(\sigma_n^2 + \sigma_J^2)}\right) I_0\left(\frac{a_c}{\sigma_n^2 + \sigma_J^2} U_k\right), \quad k = 1, 2, \ldots, m \quad \text{(A3.1)}$$

$$p(U_k) = \frac{U_k}{\sigma_n^2 + \sigma_J^2} \exp\left(-\frac{U_k^2}{2(\sigma_n^2 + \sigma_J^2)}\right), \quad k = m + 1, m + 2, \ldots, M \qquad \text{(A3.2)}$$

Probability of a correct decision is

$$P_{c} = P(U_1 > U_{m+1}, U_1 > U_{m+2}, \ldots, U_1 > U_M)$$
$$\times P(U_2 > U_{m+1}, U_2 > U_{m+2}, \ldots, U_2 > U_M) \times \cdots$$
$$\times P(U_m > U_{m+1}, U_m > U_{m+2}, \ldots, U_m > U_M) \qquad \text{(A3.3)}$$

where

$$P(U_1 > U_{m+1}, U_1 > U_{m+2}, \ldots, U_1 > U_M)$$
$$= \int_0^{+\infty} P(U_1 > U_{m+1}, \ldots, U_1 > U_M | U_1) p(U_1) \, \mathrm{d}U_1$$
$$P(U_2 > U_{m+1}, U_2 > U_{m+2}, \ldots, U_2 > U_M)$$
$$= \int_0^{+\infty} P(U_2 > U_{m+1}, \ldots, U_2 > U_M | U_2) p(U_2) \, \mathrm{d}U_2 \ldots$$
$$P(U_m > U_{m+1}, U_m > U_{m+2}, \ldots, U_m > U_M)$$
$$= \int_0^{+\infty} P(U_m > U_{m+1}, \ldots, U_m > U_M | U_m) p(U_m) \, \mathrm{d}U_m \qquad \text{(A3.4)}$$

Since all m probabilities are equal, we will determine the first one

$$P(U_1 > U_{m+1} | U_1) = \int_0^{U_1} p(U_{m+1}) \mathrm{d}U_{m+1} = 1 - \exp\left(-\frac{U_1^2}{2(\sigma_n^2 + \sigma_J^2)}\right) \qquad \text{(A3.5)}$$

$$P(U_1 > U_{m+1}, U_1 > U_{m+2}, \ldots, U_1 > U_M)$$

$$= \int_0^{+\infty} P(U_1 > U_{m+1}|U_1)^{M-m} p(U_1) \mathrm{d}U_1$$

$$= \sum_{k=0}^{M-m} (-1)^k \binom{M-m}{k} \frac{1}{k+1} \exp\left(-\frac{a_c^2}{2(\sigma_n^2 + \sigma_j^2)} \frac{k}{k+1}\right) \qquad \text{(A3.6)}$$

The probability of a correct decision is

$$P_c = \left(\sum_{k=0}^{M-m} (-1)^k \binom{M-m}{k} \frac{1}{k+1} \exp\left(-\frac{a_c^2}{2(\sigma_n^2 + \sigma_j^2)} \frac{k}{k+1}\right)\right)^m \qquad \text{(A3.7)}$$

The error probability now becomes

$$P_s = 1 - P_c = F_{\text{noncoh}}(a_c^2)$$

$$= 1 - \left(\sum_{k=0}^{M-m} (-1)^k \binom{M-m}{k} \frac{1}{k+1} \exp\left(-\frac{a_c^2}{2(\sigma_n^2 + \sigma_j^2)} \frac{k}{k+1}\right)\right)^m \qquad \text{(A3.8)}$$

In the presence of Rayleigh fading, a_c has pdf given by

$$f(a_c) = \frac{a_c}{\sigma_c^2} \exp\left(-\frac{a_c^2}{2\sigma_c^2}\right), \qquad \text{(A3.9)}$$

and the error probability is

$$P_s = \int_0^{+\infty} F_{\text{noncoh}}(x \cdot 2\sigma_c^2) \exp(-x)\, \mathrm{d}x \qquad \text{(A3.10)}$$

CW interference model

Signals at the output of detectors are Rician-distributed, with the following pdfs:

$$p(U_k) = \frac{U_k}{\sigma_n^2} \exp\left(-\frac{U_k^2 + a_c^2 + a_j^2 + 2a_c a_j \cos\theta}{\sigma_n^2}\right) I_0$$

$$\times \left(\frac{\sqrt{a_c^2 + a_j^2 + 2a_c a_j \cos\theta}}{\sigma_n^2} U_k\right), \qquad k = 1, 2, \ldots, m \qquad \text{(A3.11)}$$

$$p(U_k) = \frac{U_k}{\sigma_n^2} \exp\left(-\frac{U_k^2 + a_j^2}{\sigma_n^2}\right) I_0\left(\frac{a_j}{\sigma_n^2} U_k\right), \qquad k = m+1, m+2, \ldots, M$$

$$\qquad \text{(A3.12)}$$

The probability of a correct decision is

$$P_c = P(U_1 > U_{m+1}, U_1 > U_{m+2}, \ldots, U_1 > U_M)$$
$$\times P(U_2 > U_{m+1}, U_2 > U_{m+2}, \ldots, U_2 > U_M) \times \cdots$$
$$\times P(U_m > U_{m+1}, U_m > U_{m+2}, \ldots, U_m > U_M) \qquad \text{(A3.13)}$$

$$P(U_1 > U_{m+1}, U_1 > U_{m+2}, \ldots, U_1 > U_M)$$
$$= \int_0^{+\infty} P(U_1 > U_{m+1}, \ldots, U_1 > U_M | U_1) p(U_1) \, dU_1$$

$$P(U_2 > U_{m+1}, U_2 > U_{m+2}, \ldots, U_2 > U_M)$$
$$= \int_0^{+\infty} P(U_2 > U_{m+1}, \ldots, U_2 > U_M | U_2) p(U_2) \, dU_2 \ldots$$

$$P(U_m > U_{m+1}, U_m > U_{m+2}, \ldots, U_m > U_M)$$
$$= \int_0^{+\infty} P(U_m > U_{m+1}, \ldots, U_m > U_M | U_m) p(U_m) \, dU_m \qquad \text{(A3.14)}$$

Since all m probabilities are equal, we will determine the first one

$$P(U_1 > U_{m+1} | U_1) = \int_0^{U_1} p(U_{m+1}) dU_{m+1}$$
$$= \int_0^{U_1^2/(2\sigma_n^2)} \exp\left(-x - \frac{a_J^2}{2\sigma_n^2}\right) I_0\left(\frac{a_J}{\sigma}\sqrt{2x}\right) dx = F_1\left(\frac{U_1^2}{2\sigma_n^2}\right) \qquad \text{(A3.15)}$$

$$P(U_1 > U_{m+1}, \ldots, U_1 > U_M | \theta)$$
$$= \int_0^{+\infty} P(U_1 > U_{m+1} | U_1)^{M-m} p(U_1) dU_1 = \int_0^{+\infty} F_1(x)^{M-m}$$
$$\times \exp\left(-x - \frac{a_c^2 + a_J^2 + 2a_c a_J \cos\theta}{2\sigma_n^2}\right) I_0\left(\frac{\sqrt{a_c^2 + a_J^2 + 2a_c a_J \cos\theta}}{\sigma_n}\sqrt{2x}\right) dx$$
$$\text{(A3.16)}$$

After averaging over θ, we get

$$P(U_1 > U_{m+1}, \ldots, U_1 > U_M) = \frac{1}{2\pi} \int_0^{2\pi} P(U_1 > U_{m+1}, \ldots, U_1 > U_M | \theta) d\theta \qquad \text{(A3.17)}$$

The error probability is

$$P_s = 1 - P(U_1 > U_{m+1}, \ldots, U_1 > U_M)^m = 1 - F_{\text{noncohCW}}(a_c^2) \qquad \text{(A3.18)}$$

Because of Rayleigh fading, a_c has the pdf defined by equation. (A3.9). After averaging over a_c, the error probability becomes

$$P_s = \int_0^{+\infty} (1 - F_{\text{noncohCW}}(x \cdot 2\sigma_c^2)) \exp(-x)\, dx \tag{A3.19}$$

APPENDIX 4 NONCOHERENT DETECTION OF ($amM\,\tau$-CDMA)

Gaussian interference model

Signals at the output of detectors are Rician- and Rayleigh-distributed, with the following pdfs:

$$p(U_k) = \frac{U_k}{\sigma_n^2 + \sigma_J^2} \exp\left(-\frac{U_k^2 + a_{ck}^2}{2(\sigma_n^2 + \sigma_J^2)}\right) I_0\left(\frac{a_{ck}}{\sigma_n^2 + \sigma_J^2} U_k\right), \quad k = 1, 2 \tag{A4.1}$$

$$p(U_k) = \frac{U_k}{\sigma_n^2 + \sigma_J^2} \exp\left(-\frac{U_k^2}{2(\sigma_n^2 + \sigma_J^2)}\right), \quad k = 3, 4, \ldots, M \tag{A4.2}$$

The probability of a correct decision is

$$P_c = P(U_1 > U_2, U_1 > U_3, \ldots, U_1 > U_M) \times P(U_2 > U_3, U_2 > U_4, \ldots, U_2 > U_M) \tag{A4.3}$$

where

$$P(U_1 > U_2, U_1 > U_3, \ldots, U_1 > U_M)$$
$$= \int_{-\infty}^{+\infty} P(U_1 > U_2, \ldots, U_1 > U_M | U_1) p(U_1)\, dU_1$$

$$P(U_2 > U_3, U_2 > U_4, \ldots, U_2 > U_M)$$
$$= \int_{-\infty}^{+\infty} P(U_2 > U_3, \ldots, U_2 > U_M | U_2) p(U_2)\, dU_2 \tag{A4.4}$$

$$P(U_1 > U_2 | U_1) = \int_0^{U_1} p(U_2)\, dU_2$$
$$= \int_0^{\frac{U_1^2}{2(\sigma_n^2 + \sigma_J^2)}} \exp\left(-x - \frac{a_{c2}^2}{2(\sigma_n^2 + \sigma_J^2)}\right) I_0\left(\frac{a_{c2}}{\sqrt{\sigma_n^2 + \sigma_J^2}}\sqrt{2x}\right) dx \tag{A4.5}$$

$$P(U_1 > U_3 | U_1) = \cdots = P(U_1 > U_M | U_1)$$
$$= \int_0^{U_1} p(U_3)\, dU_3 = 1 - \exp\left(-\frac{U_1^2}{2(\sigma_n^2 + \sigma_J^2)}\right) \tag{A4.6}$$

$$P(U_2 > U_3 | U_2) = \cdots = P(U_2 > U_M | U_2)$$
$$= \int_0^{U_2} p(U_3)\, dU_3 = 1 - \exp\left(-\frac{U_2^2}{2(\sigma_n^2 + \sigma_J^2)}\right) \tag{A4.7}$$

Conditional probability of a correct decision is

$$P_{\mathrm{c}}(\theta) = \int_{-\infty}^{+\infty} \int_{-\infty}^{U_1} P(U_1 > U_2 | U_1, \theta) P(U_1 > U_3 | U_1, \theta)^{M-2}$$

$$\times P(U_2 > U_3 | U_2, \theta)^{M-2} p(U_1) p(U_2) \, dU_2 \, dU_1 \tag{A4.8}$$

After averaging over θ, we get

$$P_{\mathrm{c}} = \frac{1}{2\pi} \int_0^{2\pi} P_{\mathrm{c}}(\theta) d\theta \tag{A4.9}$$

The error probability now becomes

$$P_{\mathrm{s}} = F_{\mathrm{noncoh}a}(a_{c1}^2, a_{c2}^2) = 1 - P_{\mathrm{c}} \tag{A4.10}$$

After averaging over a_{ck}, the error probability becomes

$$P_{\mathrm{s}} = \int_{a_{c1}} \int_{a_{c2}} F_{\mathrm{noncoh}a}(a_{c1}^2, a_{c2}^2) f(a_{c1}) f(a_{c2}) \, da_{c1} \, da_{c2} \tag{A4.11}$$

CW interference model

Signals at the output of detectors are Rician-distributed, with the following pdfs:

$$p(U_k) = \frac{U_k}{\sigma_n^2} \exp\left(-\frac{U_k^2 + a_{ck}^2 + a_J^2 + 2a_{ck}a_J \cos\theta}{\sigma_n^2} \right)$$

$$\times I_0 \left(\frac{\sqrt{a_{ck}^2 + a_J^2 + 2a_{ck}a_J \cos\theta}}{\sigma_n^2} U_k \right), \quad k = 1, 2 \tag{A4.12}$$

$$p(U_k) = \frac{U_k}{\sigma_n^2} \exp\left(-\frac{U_k^2 + a_J^2}{\sigma_n^2} \right) I_0 \left(\frac{a_J}{\sigma_n^2} U_k \right), \quad k = 3, 4, \ldots, M \tag{A4.13}$$

The probability of a correct decision is

$$P_{\mathrm{c}} = P(U_1 > U_2, U_1 > U_3, \ldots, U_1 > U_M)$$

$$\times P(U_2 > U_3, U_2 > U_4, \ldots, U_2 > U_M) \tag{A4.14}$$

where

$$P(U_1 > U_2, U_1 > U_3, \ldots, U_1 > U_M)$$

$$= \int_{-\infty}^{+\infty} P(U_1 > U_2, \ldots, U_1 > U_M | U_1) p(U_1) \, dU_1$$

$$P(U_2 > U_3, U_2 > U_4, \ldots, U_2 > U_M)$$

$$= \int_{-\infty}^{+\infty} P(U_2 > U_3, \ldots, U_2 > U_M | U_2) p(U_2)\, \mathrm{d}U_2 \tag{A4.15}$$

$$P(U_1 > U_2 | U_1, \theta) = \int_0^{U_1} p(U_2)\, \mathrm{d}U_2 = \int_0^{\frac{U_1^2}{2\sigma_n^2}} \exp\left(-x - \frac{a_{c2}^2 + a_J^2 + 2a_{c2}a_J \cos\theta}{\sigma_n^2}\right)$$

$$\times I_0 \left(\frac{\sqrt{a_{c2}^2 + a_J^2 + 2a_{c2}a_J \cos\theta}}{\sqrt{\sigma_n^2}} \sqrt{2x}\right)\, \mathrm{d}x \tag{A4.16}$$

$$P(U_1 > U_3 | U_1, \theta) = \cdots = P(U_1 > U_M | U_1, \theta) = \int_0^{U_1} p(U_3)\mathrm{d}U_3$$

$$= \int_0^{\frac{U_1^2}{2\sigma_n^2}} \exp\left(-x - \frac{a_J^2}{2\sigma_n^2}\right) I_0\left(\frac{a_J}{\sigma_n}\sqrt{2x}\right)\, \mathrm{d}x \tag{A4.17}$$

$$P(U_2 > U_3 | U_2, \theta) = \cdots = P(U_2 > U_M | U_2, \theta) = \int_0^{U_2} p(U_3)\mathrm{d}U_3$$

$$= \int_0^{\frac{U_2^2}{2\sigma_n^2}} \exp\left(-x - \frac{a_J^2}{2\sigma_n^2}\right) I_0\left(\frac{a_J}{\sigma_n}\sqrt{2x}\right)\, \mathrm{d}x \tag{A4.18}$$

Conditional probability of a correct decision is

$$P_c(\theta) = \int_{-\infty}^{+\infty} \int_{-\infty}^{U_1} P(U_1 > U_2 | U_1, \theta) P(U_1 > U_3 | U_1, \theta)^{M-2}$$

$$\times P(U_2 > U_3 | U_2, \theta)^{M-2} p(U_1) p(U_2)\, \mathrm{d}U_2\, \mathrm{d}U_1 \tag{A4.19}$$

After averaging over θ, we get

$$P_c = \frac{1}{2\pi} \int_0^{2\pi} P_c(\theta)\, \mathrm{d}\theta \tag{A4.20}$$

The error probability now becomes

$$P_s = 1 - P_c = 1 - F_{\mathrm{noncohCW}a}(a_{c1}^2, a_{c2}^2) \tag{A4.21}$$

After averaging over a_{ck}, the error probability becomes

$$P_s = \int_{a_{c1}} \int_{a_{c2}} (1 - F_{\mathrm{noncohCW}a}(a_{c1}^2, a_{c2}^2)) f(a_{c1}) f(a_{c2})\, \mathrm{d}a_{c1}\, \mathrm{d}a_{c2} \tag{A4.22}$$

REFERENCES

1. Special issue on wideband CDMA. *IEEE Commun. Mag.*, **36**(9), 1998, 46–95.
2. Duel-Hallen, A., Holtzman, J. and Zvonar, Z. (1995) Multiuser detection for CDMA systems. *IEEE Personal Commun. Mag.*, **2**, 46–58.
3. Moshavi, S. (1996) Multi-user detection for DS-CDMA communications. *IEEE Commun. Mag.*, **34**, 124–135.
4. Juntti, M. and Glisic, S. (1997) Advanced CDMA for wireless communications, in Glisic, S. G. and Leppänen, P. A. (eds) *Wireless Communications – TDMA Versus CDMA*. Norwell, MA: Kluwer Academic Publishers, pp. 447–490.
5. Woodward, G. and Vucetic, B. S. (1998) Adaptive detection for DS-CDMA. *Proc. IEEE*, **86**(7), 1413–1434.
6. Ojanperä, T. (1997) Overview of multiuser detection/interference cancellation for DS-CDMA. *Proc. IEEE ICPWC '97*, pp. 115–119.
7. Lupas, R. and Verdú, S. (1989) Linear multiuser detectors for synchronous code-division multiple-access channels. *IEEE Trans. Inform. Theory*, **35**(1), 123–136.
8. Lupas, R. and Verdú, S. (1990) Near-far resistance of multiuser detectors in asynchronous channels. *IEEE Trans. Commun.*, **38**(4), 496–508.
9. Verdú, S. (1998) *Multiuser Detection*. New York: Cambridge University Press, Ch. 5, pp. 234–272.
10. Zheng, F.-C. and Barton, S. K. (1995) On the performance of near-far resistant CDMA detectors in the presence of synchronization errors. *IEEE Trans. Commun.*, **43**(12), 3037–3045.
11. Moon, T. K., Xie, Z., Rushforth, C. K. and Short, R. T. (1994) Parameter estimation in a multiuser communication system. *IEEE Trans. Commun.*, **42**(8), 2553–2539.
12. Xie, Z., Rushforth, C. K., Short, R. T. and Moon, T. K. (1993) Joint signal detection and parameter estimation in multiuser communications. *IEEE Trans. Commun.*, **41**(7), 1208–1216.
13. Lim, T. J. and Rasmussen, L. K. (1997) Adaptive symbol and parameter estimation in asynchronous multiuser CDMA detectors. *IEEE Trans. Commun.*, **45**(2), 213–220.
14. Gray, S. D., Kocic, M. and Brady, D. (1995) Multiuser detection in mismatched multiple-access channels. *IEEE Trans. Commun.*, **43**(12), 3080–3089.
15. Parkvall, S., Ström, E. and Ottersten, B. (1996) The impact of timing errors on the performance of linear DS-CDMA receivers. *IEEE J. Select. Areas Commun.*, **14**(8), 1660–1668.
16. Parkvall, S., Ottersten, B. and Ström, E. G. (1995) Sensitivity analysis of linear DS-CDMA detectors to propagation delay estimation errors. *Proc. IEEE GLOBECOM '95*, pp. 1872–1876.
17. Ström, E. G., Parkvall, S., Miller, S. L. and Ottersten, B. E. (1996) Propagation delay estimation in asynchronous direct-sequence code-division multiple access systems. *IEEE Trans. Commun.*, **44**(1), 84–93.
18. Ström, E. G., Parkvall, S., Miller, S. L. and Ottersten, B. E. (1994) Sensitivity analysis of near-far resistant DS-CDMA receivers to propagation delay estimation errors. *Proc. IEEE VTC '94*, pp. 757–761.
19. Iltis, R. A. and Mailaender, L. (1994) An adaptive multiuser detector with joint amplitude and delay estimation. *IEEE J. Select. Areas Commun.*, **12**(5), 774–785.
20. Iltis, R. A. and Mailaender, L. (1996) Multiuser Detection of quasisynchronous CDMA signals using linear decorrelators. *IEEE Trans. Commun.*, **44**(11), 1561–1571.
21. Ghauri, I. and Iltis, R. A. (1997) Capacity of the linear decorrelating detector for QS-CDMA. *IEEE Trans. Commun.*, **45**(9), 1039–1042.
22. Manji, S. and Mandayam, N. B. (1998) Outage probability for a zero forcing multiuser detector with random signature sequences. *Proc. IEEE VTC '98*, pp. 174–178.
23. Kajiwara, A. and Nakagawa, M. Microcellular CDMA system with a linear multiuser interference canceler. *IEEE J. Select. Areas Commun.*, **12**(4), 605–611.
24. Zvonar, Z. and Brady, D. (1996) Linear multipath-decorrelating receivers for CDMA frequency-selective fading channels. *IEEE Trans. Commun.*, **44**(6), 650–653.
25. Sunay, M. O. and McLane, P. J. (1998) Probability of error for diversity combining in DS CDMA systems with synchronization errors. *Eur. Trans. Telecommun.*, **9**(5), 449–463.

26. Eng, T., Kong, N. and Milstein, L. B. (1996) Comparison of diversity combining techniques for Rayleigh-fading channels. *IEEE Trans. Commun.*, **44**(9), 1117–1129.

27. Eng, T., Kong, N. and Milstein, L. B. (1998) Correction to comparison of diversity combining techniques for Rayleigh-fading channels. *IEEE Trans. Commun.*, **46**(9), 1111.

28. Gilhousen, K. S. *et al.* (1991) On the capacity of a cellular CDMA system. *IEEE Trans. Veh. Technol.*, **40**(2), 303–312.

29. Glisic, S. and Pirinen, P. (1999) Wideband CDMA network sensitivity function. *IEEE J. Select. Areas Commun.*, **17**, 1781–1794.

30. Glisic, S. and Vucetic, B. (1997) *CDMA for Wireless Communications*. Boston, MA: Artech House, pp. 1–47.

31. Ziemer, R. E. and Tranter, W. H. (1995) *Principles of Communications, Systems, Modulation, and Noise*. 4th edn. New York: John Wiley & Sons, p. 406.

32. Eng, T. and Milstein, L. B. (1994) Comparison of hybrid FDMA/CDMA systems in frequency selective Rayleigh fading. *IEEE J. Select. Areas Commun.*, **12**(5), 938–951.

33. Lance, E. and Kaleh, G. (1997) A diversity scheme for phase coherent frequency hopping spread spectrum system. *IEEE Trans. Commun.*, **45**(9), 1123–1130.

34. Simon, M. and Polidoros, A. (1981) Coherent detection of frequency-hopped quadrature modulation in the presence of jamming. *IEEE Trans. Commun.*, **COM-29**, 1644–1660.

35. Ghareeb, I. and Yongaçoglu, A. (1996) Performance analysis of frequency-hopped/coherent MPSK in the presence of multitone jamming. *IEEE Trans. Commun.*, **44**, 152–155.

36. Su, C. M. and Milstein, L. B. (1990) Analysis of coherent frequency-hopped spread-spectrum receiver in the presence of jamming. *IEEE Trans. Commun.*, **38**, 715–726.

37. Kaleh, G. K. (1996) Frequency-diversity spread spectrum communication system to counter bandlimited jammers. *IEEE Trans. Commun.*, **44**, 886–893.

38. Lee, J. S., French, R. H. and Miller, L. E. (1984) Probability of error analyses of a BFSK frequency-hopping system with diversity under partialband jamming interference, part I: performance of square-law linear combining soft decision receiver. *IEEE Trans. Commun.*, **COM-32**, 645–653.

39. Robertson, R. C. and Lee, K. Y. (1992) Performance of fast frequency-hopped Rician fading channel with partial-band interference. *IEEE J. Select. Areas Commun.*, **10**, 731–741.

40. Bird, J. S. and Felstead, E. B. (1986) Antijam performance of fast frequency-hopped M-ary NCFSK-AN overview. *IEEE J. Select. Areas Commun.*, **SAC-4**, 216–233.

41. Massaro, M. J. (1975) Error performance of M-ary noncoherent FSK in the presence of CW tone interference. *IEEE Trans. Commun.*, **COM-23**, 1367–1369.

42. Milstein, L. B., Pickholtz, R. L. and Shilling, D. L. (1980) Optimization of the processing gain of an FSK-FH system. *IEEE Trans. Commun.*, **COM-28**, 1062–1079.

43. Glisic, S. G. *et al.* (1987) Efficiency of digital communication systems. *IEEE Trans. Commun.*, **COM-35**(6), 679–684.

44. Juntti, M. (1997) *Multiuser Demodulation for DS-CDMA Systems in Fading Channels*. Acta Universitatis Ouluensis C 106, Doctoral Thesis, Oulu University Press, Finland.

45. Cherubini, G. and Milstein, L. B. (1989) Performance analysis of both hybrid and frequency-hopped phase-coherent spread-spectrum systems, part II: an FH system. *IEEE Trans. Commun.*, **37**, 612–622.

46. Kaleh, G. K. (1997) Performance comparison of frequency diversity and frequency hopping spread spectrum systems. *IEEE Trans. Commun.*, **45**, 910–912.

47. Robertson, R. C. *et al.* (1996) Multiple tone interference of frequency hopped noncoherent MFSK signals transmitted over Rician fading channel. *IEEE Trans. Commun.*, **44**(7), 867–875.

48. Zvonar, Z. and Brady, D. (1994) Multiuser detection in single-path fading channels. *IEEE Trans. Commun.*, **42**(2/3/4), 1729–1739.

49. Mabuchi, T., Kohno, R. and Imai, H. (1993) Multihopping and decoding of error-correcting code for MFSK/FH-SSMA systems. *IEICE Trans. Commun.*, **E76-B**(8), 874–885.

16

Standards

In this chapter we discuss the basic Code Division Multiple Access (CDMA) standards and give a brief history of the standard proposals. We present the main system parameters, which are essential for the understanding of the system concept (common air interface) of each standard. At this stage we use what we have learnt so far in this book to discuss the motivations behind the solutions. We believe that at the end of the book the reader should have the required knowledge to follow the closing discussion on the various choices for the different system parameters, and all the advantages and the drawbacks of these choices.

16.1 IS 95 STANDARD

16.1.1 Reverse link

Available channels in the uplink are shown in Figure 16.1. A block diagram of a reverse channel data path is shown in Figure 16.2. Data and chip rates in different points in the system are indicated on the picture. The voice source is encoded by using variable rate codec with four possible rates of 1.2 to 9.6 kbps. This is the way to exploit voice activity factor.

Prior to modulation, convolutional encoder and block interleaver are used.

Modulation for the reverse CDMA channel is 64-ary orthogonal signaling. One of the possible modulation symbols will be transmitted for each six-code symbol.

$$\text{Modulation symbol number} = c_0 + 2c_1 + 4c_2 + 8c_3 + 16c_4 + 32c_5 \qquad (16.1)$$

c_5 shall represent the last or the most recent and c_0 the first or the oldest binary valued (0 and 1) code symbol of each group of six code symbols that form a modulation symbol. One out of 64 Walsh symbols is transmitted for each different value of equation (16.1). Construction rules for Walsh functions are described in Chapter 2. On the basis of the chip rate 1.22881447 Mchips (indicated in Figure 16.2), the period of time required to transmit a single modulation symbol, referred to as a *Walsh symbol* interval, will be approximately

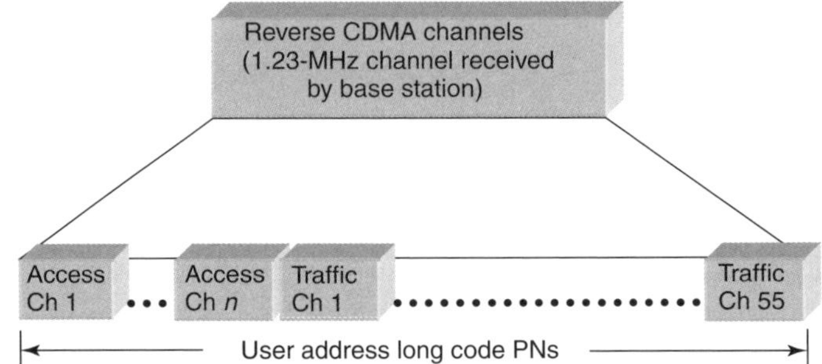

Figure 16.1 Example of logical reverse CDMA channels received at a base station.

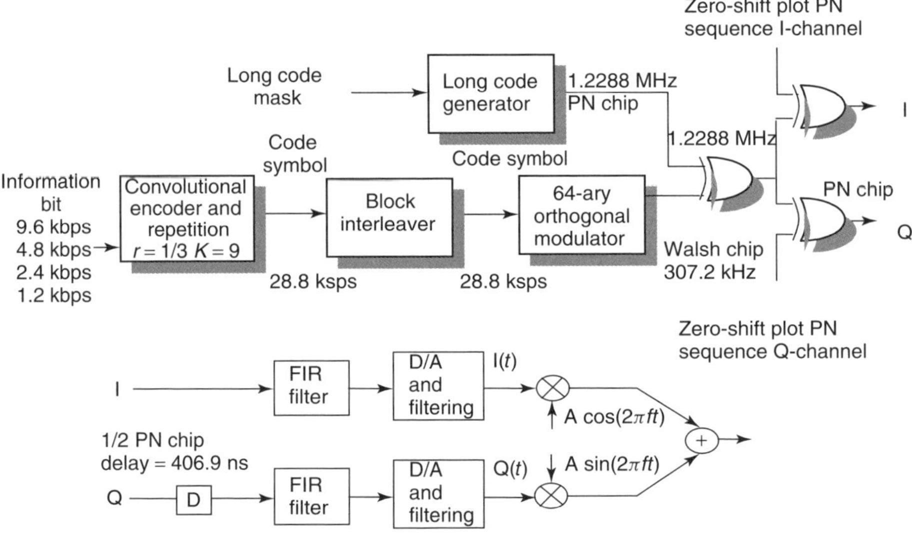

Figure 16.2 Reverse CDMA channel data path example.

equal to $208.333\,\mu s$. The period of time associated with 164th of the modulation symbol is referred to as a *Walsh chip* and will be approximately equal to $3.2552083333\ldots\,\mu s$.

The reverse traffic channel numerology is shown in Table 16.1.

16.1.2 Direct-sequence spreading

The reverse traffic channel and the access channel will be combined with three different pseudonoise (PN) sequences. Data and PN sequence combination involves modulo-2 addition of the encoded, interleaved data stream with two PN code streams, each operating at 1.2288 MHz. The first sequence is referred to as the long code sequence. This sequence shall be a time shift of a sequence of length $2^{42}-1$ chips and shall be generated by a

Table 16.1 Reverse traffic channel numerology

Parameter	Data Rate (bps)				Units
	9600	4800	2400	1200	
PN chip rate	1.2288	1.2288	1.2288	1.2288	Mcps
Code rate	1/3	1/3	1/3	1/3	Bits/code sym
TX duty cycle	100.0	50.0	25.0	12.5	%
Code symbol rate	28 800	28 800	28 800	28 800	Sps
Modulation	6	6	6	6	Code sym/ Walsh sym
Walsh symbol rate	4800	4800	4800	4800	Sps
Walsh chip rate	307.20	307.20	307.20	307.20	kcps
Walsh symbol	208.33	208.33	208.33	208.33	μs
PN chips/code symbol	42.67	42.67	42.67	42.67	PN chip/ code sym
PN chips/Walsh symbol	256	256	256	256	PN chip/ Walsh sym
PN chips/Walsh chip	4	4	4	4	PN chip/ Walsh sym

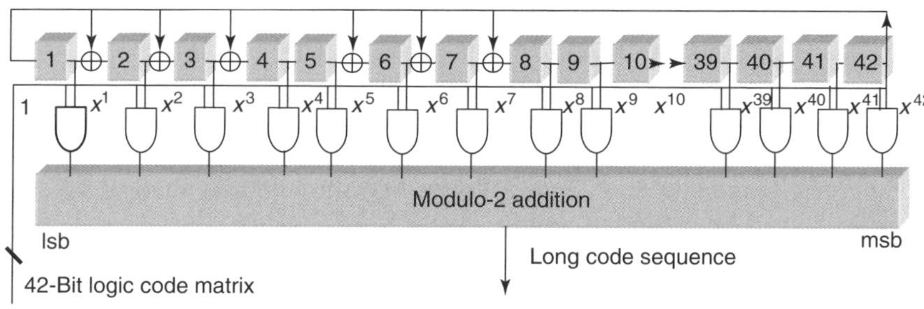

Figure 16.3 Long code generation and masking.

linear generator using the following polynomial:

$$p(x) = x^{42} + x^{35} + x^{33} + x^{31} + x^{27} + x^{26} + x^{25} + x^{22} + x^{21} + x^{19}$$
$$+ x^{18} + x^{17} + x^{16} + x^{10} + x^7 + x^6 + x^5 + x^3 + x^2 + x^1 + 1.$$

The long code will be generated by masking the 42-bit state variables of the generator with a 42-bit mask. The actual PN sequence is generated by the modulo-2 addition of all 42 masked output bits of the sequence generator as shown in Figure 16.3.

16.1.3 Code mask

The structure of the code mask is shown in Figure 16.4. The mask, used for the PN spreading, will vary depending on the channel type on which the mobile station is communicating. Access channel M_{24} through M_{41} shall be set to '1'. M_{19} through M_{23} shall

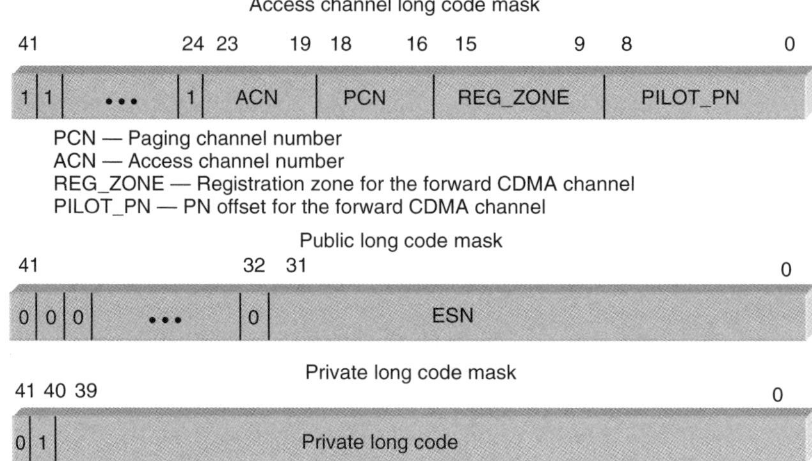

Figure 16.4   Long code mask format.

be set to the access channel number chosen randomly. M_{16} through M_{18} shall be set to the code channel for the associated paging channel (i.e. the range shall be 1 through 7). M_9 through M_{15} shall be set to the REG_ZONE for the current base station (BS). M_0 through M_8 shall be set to the PILOT_PN value for the CDMA channel.

In the reverse traffic channel the mobile station shall use one of two long codes unique to that mobile station: a public long code unique to the mobile station's electronic serial number (ESN) and a private long code unique for each mobile identification number (MIN). The public long code shall be as follows: M_{32} through M_{41} shall be set to '0' and M_0 through M_{31} shall be set to the mobile station's ESN value.

The second and third PN sequences are the I and Q 'short codes'. The reverse access channel and the reverse traffic channel shall be offset quadrature phase shift keying (OQPSK) spread prior to actual transmission. This offset quadrature spreading on the reverse channel shall use the same I and Q PN codes as the forward I and Q PN codes. These codes are of length 2^{15}. The reverse CDMA channel I and Q codes shall be the zero-time offset codes. The generating functions for the I and Q short PN codes shall be as follows:

$$P_I(x) = x^{15} + x^{13} + x^9 + x^8 + x^7 + x^5 + 1$$
$$P_Q(x) = x^{15} + x^{12} + x^{11} + x^{10} + x^6 + x^5 + x^4 + x^3 + 1$$

16.1.4 Data burst randomizer algorithm

The data burst randomizer generates a masking stream of 0 s and 1 s that randomly mask out the redundant data generated by the code repetition. The masking stream pattern is determined by the frame data rate and by the block of 14 bits taken from the long code sequence. These mask bits are synchronized with the data flow and the data is selectively masked by these bits through the operation of the digital filter. The 1.2288-MHz-long code sequence shall be input to a 14-bits shift register, which is shifted at 1.2288 MHz. The contents of this shift register shall be loaded into a 14-bit latch exactly one power control group (1.25 ms) before each reverse traffic channel frame boundary.

$$b_0 b_1 b_2 b_3 b_4 b_5 b_6 b_7 b_8 b_9 b_{10} b_{11} b_{12} b_{13}$$

The binary (0 and 1) contents of this latch shall be denoted as where b_0 shall represent the first bit to enter the shift register and b_{13} shall represent the last (or most recent) bit to enter the sift register. Each 20-ms reverse traffic channel frame shall be divided into 16 equal length (i.e. 1.25 ms) power control groups numbered from 0 to 15. The data burst randomizer algorithm shall be as follows:

Data rate selected: 9600 bps

Frame transmission shall occur on power control groups numbered

$$0, 1, 2, 3, 4, 5, 6, 7, 8, 9, 10, 11, 12, 13, 14, 15$$

Data rate selected: 4800 bps

Frame transmission shall occur on power control groups numbered

$$b_0, 2 + b_1, 4 + b_2, 6 + b_3, 8 + b_4, 10 + b_5, 12 + b_6$$

Data rate selected: 2400 bps

Frame transmission shall occur on power control groups numbered

$$b_0 \text{ if } b_8 = 0, \quad 2 + b_1 \text{ if } b_8 = 1$$
$$4 + b_2 \text{ if } b_9 = 0, \quad 6 + b_3 \text{ if } b_9 = 1$$
$$8 + b_4 \text{ if } b_{10} = 0, \quad 10 + b_5 \text{ if } b_{10} = 1$$
$$12 + b_6 \text{ if } b_{11} = 0, \quad 14 + b_7 \text{ if } b_{11} = 1$$

Data rate selected: 1200 bps

Frame transmission shall occur on power control groups numbered

$$b_0 \text{ if } (b_8 = 0 \text{ and } b_{12} = 0), \quad 2 + b_1 \text{ if } (b_8 = 1 \text{ and } b_{12} = 0)$$
$$4 + b_2 \text{ if } (b_9 = 0 \text{ and } b_{12} = 1), \quad 6 + b_3 \text{ if } (b_9 = 1 \text{ and } b_{12} = 1)$$
$$8 + b_4 \text{ if } (b_{10} = 0 \text{ and } b_{13} = 0), \quad 10 + b_5 \text{ if } (b_{10} = 1 \text{ and } b_{13} = 0)$$
$$12 + b_6 \text{ if } (b_{11} = 0 \text{ and } b_{13} = 1), \quad 14 + b_7 \text{ if } (b_{11} = 1 \text{ and } b_{13} = 1)$$

An example is shown in Figure 16.5.

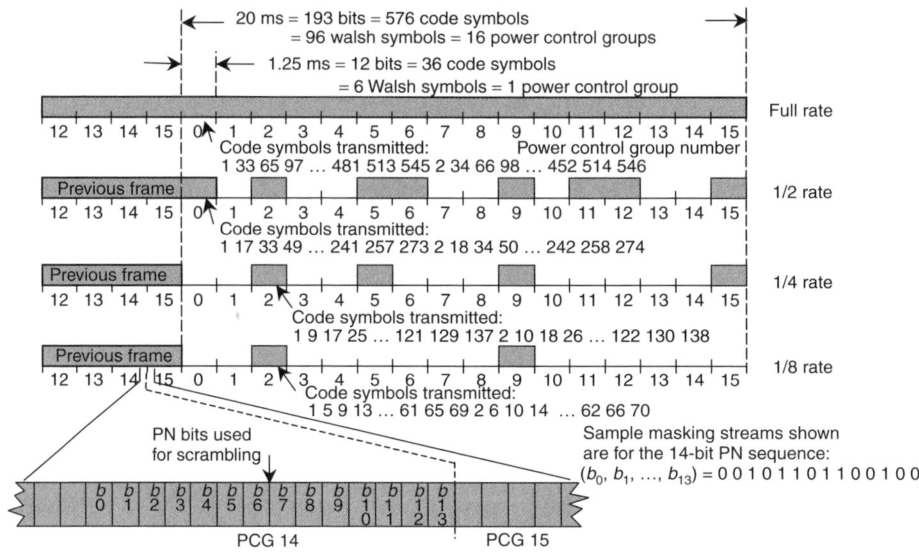

Figure 16.5 Reverse CDMA channel variable data rate transmission example.

16.1.5 Reverse traffic channel frame quality indicator

Each frame of the traffic channel shall include a frame quality indicator. For the default multiplex option's 9600-bps and 4800-bps transmission rates, the frame quality indicator shall be a cyclic redundancy check (CRC). For the 9600-bps and 4800-bps rates, the frame quality indicator (CRC) shall be calculated on all bits within the frame, except the frame quality indicator (CRC) itself and the encoder tail bits. The 9600-bps transmission rate shall use a 12-bit frame quality indicator (CRC), which shall be transmitted within the 192-bit long frame. The generator polynomial for the 9600-bps rate

$$g(x) = x^{12} + x^{11} + x^{10} + x^9 + x^8 + x^4 + x + 1$$

The 4800-bps transmission rate shall use a 8-bit CRC, which shall be transmitted within the 96-bit long frame. The generator polynomial for the 4800-bps rate

$$g(x) = x^8 + x^7 + x^4 + x^3 + x + 1$$

16.1.6 The CRCs procedure

The circuit block diagrams for 9600 and 4800 bps are shown in Figures 16.6 and 16.7, respectively. Initially, all shift register elements shall be set to logical one and the switches shall be set in the up position. The register shall be clocked 172 times (for 192-bit frame) or 80 times (for 96-bit frame) with the traffic or the signaling bits and mode/format indicators as input. The switches shall be set in the down position, and the register shall be clocked an additional 12 times (for 192-bit frame) or 8 times (for 96-bit frame).

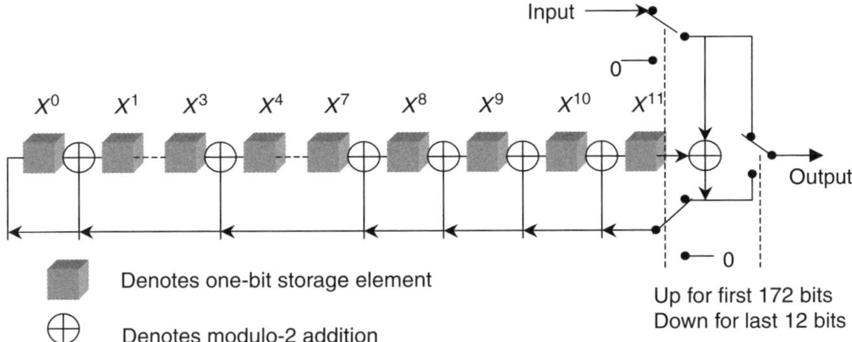

Figure 16.6 Reverse traffic channel frame quality indicator calculation at 9600-bps rate for the default multiplex option(1).

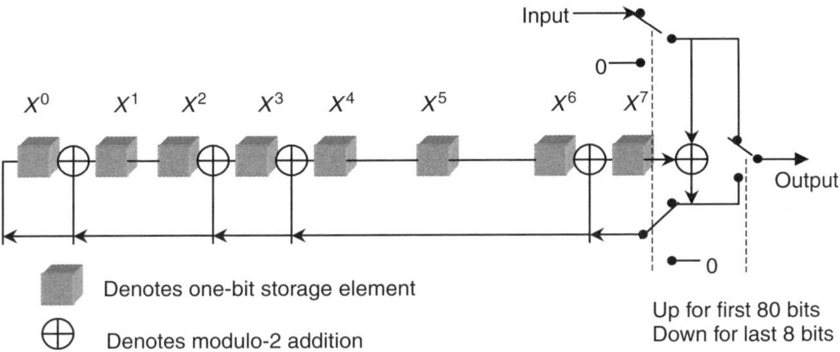

Figure 16.7 Reverse traffic channel frame quality indicator calculation at 4800-bps rate for the default multiplex option(1).

The 12 or 8 additional output bits shall be the check bits. The bits shall be transmitted in the order calculated.

16.1.7 Base station

Transmitter

Each BS within a given system shall use the same CDMA frequency assignments for each of the CDMA channels. The channel structure is shown in Figures 16.8 and 16.9.

Variable data rate transmission

The forward traffic channel shall support variable data rate operation. Four data rates are supported: 9600, 4800, 2400 and 1200 bps. The data rate shall be selectable on a frame-by-frame (i.e. 20-ms) basis without consideration for the rate in the previous or subsequent frames. Although the data rate may vary on a 20-ms basis, the modulation

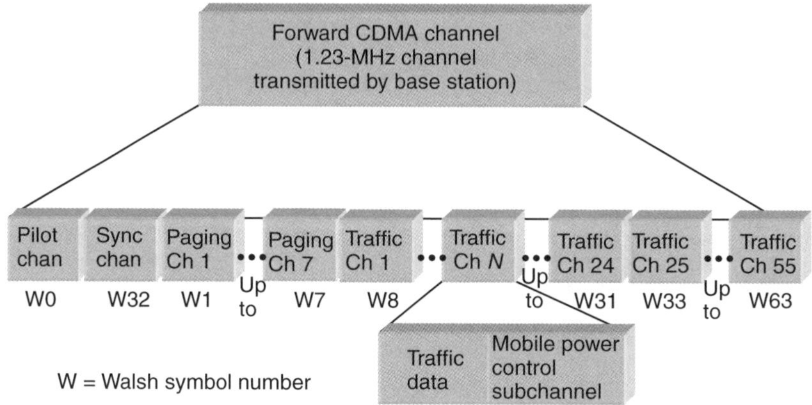

Figure 16.8   Example of a forward CDMA channel transmitted by a base station.

symbol rate is kept constant by code repetition at 19.2 kilo-symbols per second (ksps). The modulation symbols that are transmitted at the lower data rates shall be transmitted using lower energy, as shown in Table 16.2.

Note that all the symbols in the interleaver block are from the same frame. Thus they are all transmitted at the same energy. Power control bits are always transmitted with energy E_b.

Pilot channel

The pilot channel is transmitted at all times by the BS on each active forward CDMA channel. It is an unmodulated spread spectrum signal that is used by a mobile station operating within the geographic coverage area of the base station. It is used by the mobile station to acquire synchronization with the pilot PN sequence, to provide a phase reference and to provide sync channel frame timing.

The acquisition of the pilot channel pilot PN sequence is the first step in the process of the mobile station acquiring the system timing or reacquiring the system timing. Code for the pilot channel shall be a quadrature sequence of length 2^{15} (i.e. 32768 PN chips in length).

Code polynomial

$$P_I(x) = x^{15} + x^{13} + x^9 + x^8 + x^7 + x^5 + 1$$
$$P_Q(x) = x^{15} + x^{12} + x^{11} + x^{10} + x^6 + x^5 + x^4 + x^3 + 1$$

for the in-phase (I) sequence and for the quadrature (Q) phase sequence is used. The length of these sequences is $2^{15}-1$. In order to generate a pilot PN sequence of length 2^{15}, a binary 1 is inserted in the sequence generator output after the contiguous succession of 14 binary 0 outputs (that occurs only once per period of the sequence). The chip rate for the pilot PN sequence shall be 1.2288 Mcps. The pilot PN sequence period is 26.666... ms. Exactly 75 pilot PN sequence repetitions occur every 2 s.

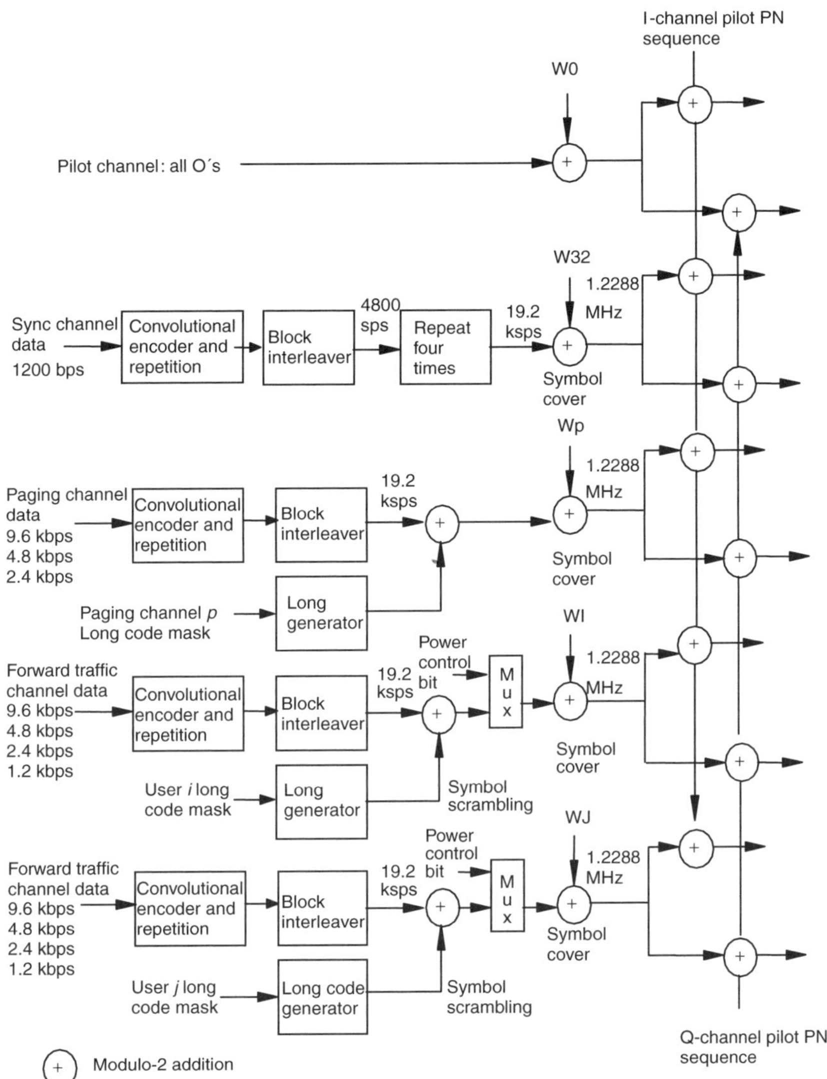

Figure 16.9 Forward CDMA channel structure.

Pilot channel index

Each BS shall use a time offset of the pilot PN sequence to identify its forward CDMA channel. Time offsets may be reused within a CDMA cellular system, so long as the coverage area of the BS emitting a given pilot PN sequence time offset does not overlap the coverage area of another BS using the same pilot PN sequence time offset. Distinct pilot channels shall be designated by an index identifying an offset value from a zero offset pilot PN sequence (in increments of 64 PN chips). The zero offset pilot PN sequence

Table 16.2 Transmitted
symbol energy versus
data rate

Data rate (bps)	Energy per modulation symbol
9600	$E_S = E_b/2$
4800	$E_S = E_b/4$
2400	$E_S = E_b/8$
1200	$E_S = E_b/16$

shall be such that the start of the sequence shall be output at the beginning of every even second in time, referenced to system time. The start of the zero offset pilot PN sequence for either the I or the Q sequence shall be defined as the state of the sequence generator for which the previous 15 outputs were '0'. Five hundred and eleven unique values shall be possible for the pilot PN sequence offset (the offset index of '111 111 111' binary shall bc reserved). The pilot PN sequence offset shall be denoted as a 9-bit binary pilot PN sequence index for a given BS. The timing offset for a given pilot PN sequence shall be equal to the offset index value multiplied by 64 multiplied by the pilot channel chip period (= 813.802 ns). For example, if the pilot PN sequence offset index is 15 (decimal), the pilot PN sequence offset will be $15 \times 64 \times 813.802$ ns = 781.1 µs.

In this case the pilot PN sequence will start 781.1 µs after the start of every even second of the system time. The same pilot PN sequence offset shall be used on all CDMA frequency assignments for a given BS.

The sync channel shall be an encoded, interleaved, modulated direct-sequence spread spectrum signal that is used by mobile stations operating within the geographic coverage area of that BS (a cell or a sector within a cell) to acquire synchronization to the long code sequence and to acquire system timing. Sync channel acquisition is the second step that the mobile station takes in acquiring the system.

Forward traffic channel data scrambler

The forward traffic channel data shall be scrambled by an additional modulo-2 addition operation prior to transmission. This data scrambling shall be performed on the data output from the block interleaver at the 19 200-cps rate. The data scrambling shall be accomplished by performing the modulo-2 addition of the interleaver output symbol with the binary value of the PN chip that is valid at the start of the transmission period for that symbol as shown in Figure 16.10. This sequence generator shall operate at 1.2288-MHz clock rate although only one output of 64 shall be used for data scrambling (i.e. at a 19 200-cps rate). The PN sequence used for data scrambling shall be the decimated version of the sequence used by the mobile station for direct-sequence spreading of the reverse traffic channel (either the public long code or the private long code).

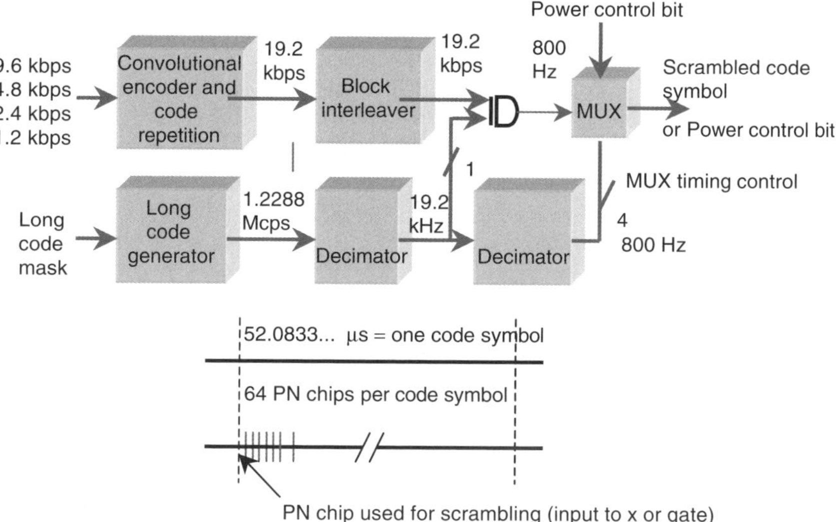

Figure 16.10   Data scrambler timing.

16.2 IS-95B CDMA

IS-95B specifies the high-speed data operation using up to eight parallel codes, resulting in a maximum bit rate of 115.2 kbps. The summary of IS-95A and IS-95B system parameters is given in Table 16.3. The system block diagrams and numerology for rate set 1 and rate set 2 are shown in Figures 16.11 and 16.12, respectively.

16.3 CDMA2000

The goal has been to provide data rates that meet the IMT-2000 performance requirements of at least 144 kbps in a vehicular environment, 384 kbps in a pedestrian environment and 2048 kbps in an indoor office environment. The main focus of standardization has been to provide 144 and 384 kbps with approximately 5-MHz bandwidth. The main parameters of CDMA2000 are listed in Table 16.4.

The two main alternatives for the downlink are multicarrier and direct spread options (see Figure 16.13). Transmission on the multicarrier downlink (nominal 5-MHz band) is achieved by using three consecutive IS 95B carriers where each carrier has a chip rate of 1.2288 Mcps. For the direct spread option, transmission on the downlink is achieved by using a nominal chip rate of 3.6864 Mcps.

16.3.1 Physical channels

Uplink physical channels

In the uplink, there are four different dedicated channels. The fundamental and the supplemental channels carry user data.

Table 16.3 IS-95 interface parameters

Bandwidth	1.25 MHz
Chip rate	1.2288 Mcps
Frequency band uplink	869–894 MHz
	1930–1980 MHz
Frequency band downlink	824–849 MHz
	1850–1910 MHz
Frame length	20 ms
Bit rates	Rate set 1: 9.6 kbps
	Rate set 2: 14.4 kbps
	IS-95B: 115.2 kbps
Speech codec	QCELP 8 kbps
	EVRC 8 kbps
	ACELP 13 kbps
Soft handover	Yes
Power control	Uplink: Open loop + fast closed loop
	Downlink: Slow quality loop
Number of RAKE fingers	4
Spreading codes	Walsh + long M-sequence

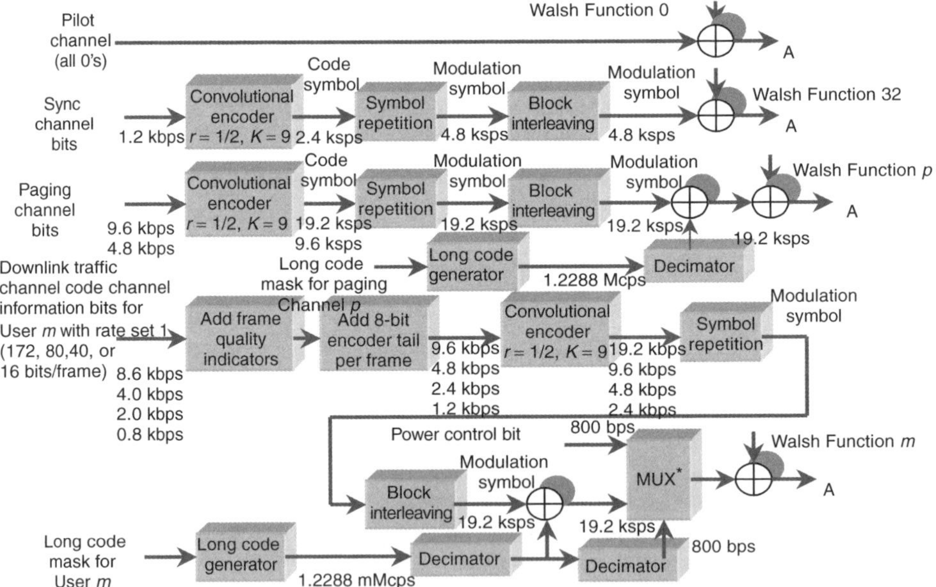

*Power control bits are not multiplexed in for supplemental code channels of the downlink traffic channels.

Figure 16.11 System block diagram for rate set 1.

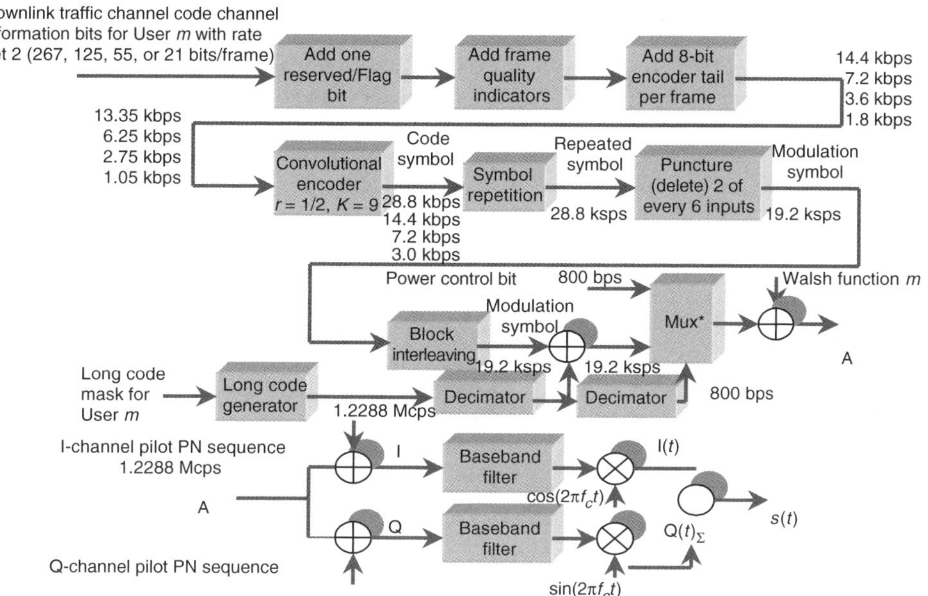

Figure 16.12 System block diagram for rate set 2.

- A dedicated control channel, with a frame length 5 or 20 ms, carries control information such as measurement data, and the pilot channel is used as a reference signal for coherent detection. The pilot channel also carries time-multiplexed power control symbols. Figure 16.14 illustrates the different uplink dedicated channels separated by Walsh codes.

- The reverse access channel (R-ACH) and the reverse common control channel (R-CCCH) are common channels used for communication of layer 3 and medium access control (MAC) layer messages, discussed in Chapters 11 and 12. The R-ACH is used for initial access, while the R-CCCH is used for fast packet access.

- The fundamental channel conveys voice, signaling and low rate data. It will operate at low frame error rate (FER) (around 1%) and it supports basic rates of 9.6 and 14.4 kbps and their corresponding subrates (i.e. rate sets 1 and 2 of IS-95). It always operates in soft handover mode and does not operate in a scheduled manner; thus permitting the mobile station to transmit acknowledgments or short packets without scheduling. This reduces delay and the processing load due to scheduling. Its main difference is using repetition coding rather than gated transmission.

- The supplemental channel provides high data rates. The uplink supports one or two supplemental channels. If only one supplemental channel is transmitted, then the Walsh code (+−) is used on the first supplemental channel. If two supplemental channels are transmitted, then the Walsh code (+ − +−) is used. A repetition scheme is used for variable data rates on the supplemental channel.

Table 16.4 CDMA2000 parameter summary

Channel bandwidth	1.25, 5, 10 and 20 MHz
Downlink RF channel structure	Direct spread or multicarrier
Chip rate	1.2288/3.6864/7.3728/11.0593/14.7456 Mcps for direct spread $n \times 1.2288$ Mcps ($n = 1, 3, 6, 9, 12$) for multicarrier
Roll-off factor	Similar to IS-95
Frame length	20 ms for data and control/5 ms for control information on the fundamental and dedicated control channel
Spreading modulation	Balanced QPSK (downlink)
	Dual-channel QPSK (uplink)
	Complex spreading circuit
Data modulation	QPSK (downlink)
	BPSK (uplink)
Coherent detection	Pilot time-multiplexed with PC and EIB (uplink) Common continuous pilot channel and auxiliary pilot (downlink)
Channel multiplexing in uplink	Control, pilot, fundamental and supplemental code-multiplexed
	I&Q multiplexing for data and control channels
Multirate	Variable spreading and multicode
Spreading factors	4–256
Power control	Open loop and fast closed loop (800 MHz, higher rates under study)
Spreading (downlink)	Variable length Walsh sequences for channel separation, M-sequence 2^{15} (same sequence with time shift utilized in different cells, different sequence in the I&Q channels)
Spreading (uplink)	Variable length orthogonal sequences for channel separation, M-sequence 2^{15} (same for all users, different sequences in the I&Q channels), M-sequence $2^{41} - 1$ for user separation (different time shifts for different users).
Handover	Soft handover
	Interfrequency handover

Downlink physical channels

Downlink has three different dedicated channels and three common control channels. The fundamental and supplemental channels carry user data. The dedicated control channel control messages. The dedicated control channel contains power control bits and rate information. The synchronization channel is used by the mobile stations to acquire initial time synchronization. One or more paging channels are used for paging the mobiles. The pilot channel provides a reference for coherent detection, cell acquisition and handover. In the downlink, CDMA2000 has a common pilot channel, which is used as a reference

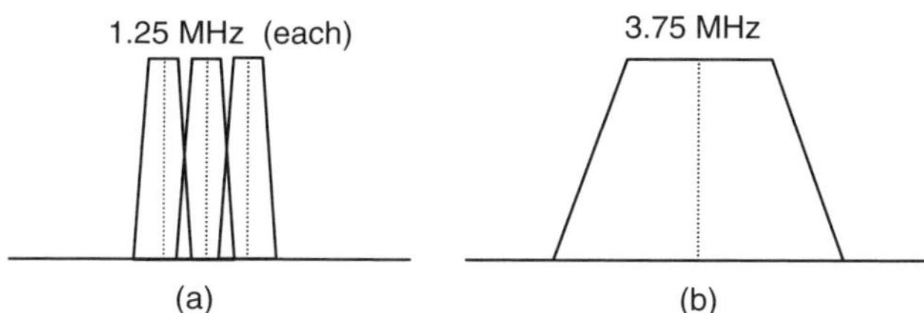

Figure 16.13   Illustration of: (a) multicarrier and (b) direct spread downlink.

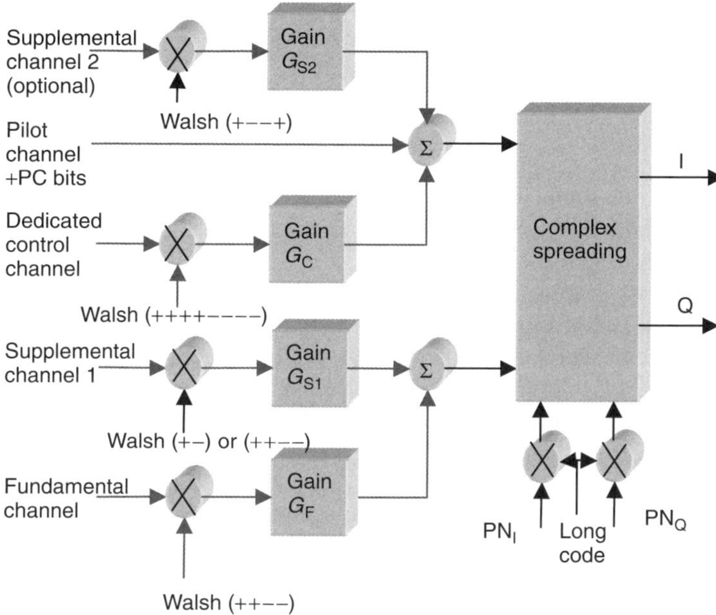

Figure 16.14   The uplink dedicated channel structure.

signal for coherent detection when adaptive antennas are not employed. The pilot channel
is similar to IS-95 (i.e. it is composed of a long PN code and Walsh sequence number 0).
When adaptive antennas are used, auxiliary pilot is used as a reference signal for coherent
detection. Code-multiplexed auxiliary pilots are generated by assigning a different orthogonal code to each auxiliary pilot. This approach reduces the number of orthogonal codes
available for the traffic channels. This limitation is alleviated by expanding the size of
the orthogonal code set used for the auxiliary pilots. Since a pilot signal is not modulated
by data, the pilot orthogonal code length can be extended, thereby yielding an increased
number of available codes, which can be used as additional pilots.

Two alternatives for downlink modulation still exist: direct spread and multicarrier.
The multicarrier transmission principle is illustrated in Figure 16.15.

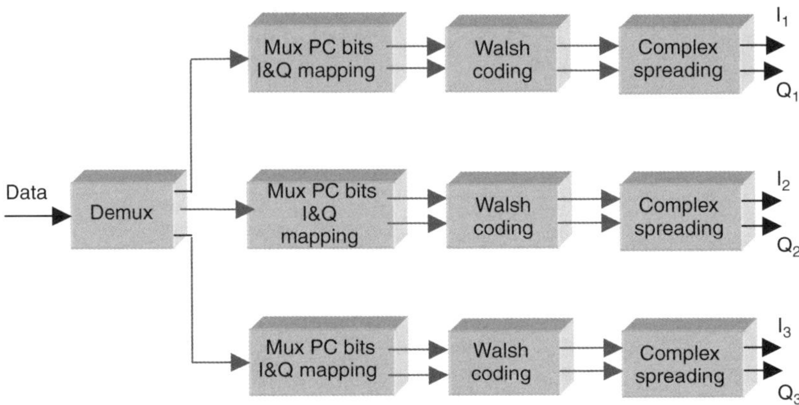

Figure 16.15 Multicarrier downlink.

Spreading On the downlink, the cell separation for CDMA2000 is performed by two M-sequences of length 2^{15}, one for the I channel and one for the Q channel, which are phase-shifted by PN-offset for different cells. During the cell search process, only these sequences need to be searched. Since there is only a limited number of PN sequences, they need to be planned in order to avoid PN confusion. In the uplink, user separation is performed by different phase shifts of M-sequence of length 2^{41}. The channel separation is performed using variable spreading factor Walsh sequences, which are orthogonal to each other. Fundamental and supplemental channels are transmitted with the multicode principle. The variable spreading factor scheme is used for higher data rates in the supplemental channel. Similar to wideband code division multiple access (WCDMA), complex spreading is used. In the uplink, it is used with dual-channel modulation.

Multirate The fundamental and supplemental channels can have different coding and interleaving schemes. In the downlink, high bit rate services with different quality of service (QoS) requirements are code-multiplexed into supplemental channels, as illustrated in Figure 16.16. In the uplink, one or two supplemental channels can be transmitted. The user data frame length of CDMA2000 is 20 ms. For the transmission of control information, 5- and 20-ms frames can be used on the fundamental channel. For the fundamental channel, a convolutional code with constraint length of 9 is used. On supplemental channels, a convolutional code is used up to 14.4 kbps. For higher rates, Turbo codes with constraint length 4 and rate 1/4 are preferred. Rate matching is performed by puncturing, symbol repetition and sequence repetition.

Handover It is expected that soft handover of the fundamental channel will operate similarly to the soft handover in IS-95. In the IS-95, the active set is the set of BSs transmitting to the mobile station. For the supplemental channel, the active set can be a subset of the active set for the fundamental channel. This has two advantages. First, when diversity is not needed to counter fading, it is preferable to transmit from fewer BSs. This increases

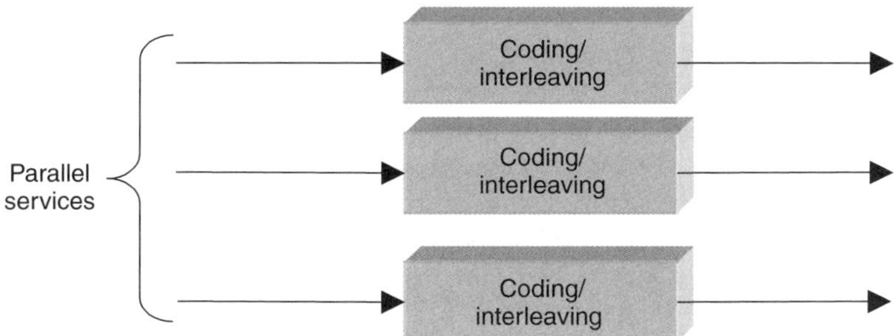

Figure 16.16   Principle of code multiplexing.

the overall downlink capacity. For stationary conditions, an optimal policy is to transmit only from one BS – the BS that would need to radiate the smallest amount of downlink power. For packet operation, the control processes can also be substantially simplified if the supplemental channel is not in soft handover. Maintaining the fundamental channel in soft handover provides the ability to reliably signal the preferred BS to transmit the supplemental channel when channel conditions change.

Packet data CDMA2000 also uses the slotted ALOHA principle for packet data transmission. Instead of fixed transmission power, it increases the transmission power for the random access burst after an unsuccessful access attempt. This is an additional form of adaptive CDMA network described in Chapter 12. When the mobile station has been allocated a traffic channel, it can transmit without scheduling up to a predefined bit rate. If the transmission rate exceeds the defined rate, a new access request has to be made. When the mobile station stops transmitting, it releases the traffic channel but not the dedicated control channel.

After a while, it also releases the dedicated control channel but maintains the link layer and network layer connections in order to shorten the channel setup time when new data need to be transmitted. Short data bursts can be transmitted over a common traffic channel in which a simple automatic repeat request (ARQ) is used to improve the error rate performance.

Transmit diversity The downlink performance can be improved by transmit diversity.

For direct spread CDMA schemes, this can be performed by splitting the data stream and spreading the two streams using orthogonal sequences. For multicarrier CDMA, the different carriers can be mapped into different antennas.

16.4 IS-665 W-CDMA

The origins of IS-665 W-CDMA standard dates back to the broadband CDMA (B-CDMA) proposal by Interdigital. The B-CDMA concept was introduced in 1989. The original bandwidth was as large as 48 MHz and the chip rate 24 Mcps. This system was field-tested with microwave signals in the Personal Communication System (PCS) band. In 1994, the system was tested in the cellular band as American Mobile Phone Systems (AMPS) overlay. Microwave and AMPS carriers were supposed to be filtered out by notch filters. As a result of the US PCS frequency plan, in which 5- and 15-MHz band allocations were adopted, the original B-CDMA proposal was modified to be 5 and 15 MHz and the overlay idea was given up. In 1993, a wideband CDMA scheme was proposed by OKI as a PCS standard. Lockheed Sanders, Interdigital and AT&T cooperated with OKI during the standardization. The proposed W-CDMA system became the TIA/EIA (Electronics Industry Association) IS-665 standard, the T1P1 Trial Use Standard J-STD-015 and the ITU-R Recommendation M.1073. It was also used as a basis for the Core-B proposal in Japan. However, IS-665 W-CDMA has not been commercially successful, and except for trial systems it has not been deployed. Even so, in this section we briefly summarize its basic parameters in order to better understand the evolution path toward wideband CDMA used in Universal Mobile Telecommunication System (UMTS) standard, which is described in the next chapter.

16.4.1 The main parameters of the system

The system has three bandwidths of 5, 10 and 15 MHz to fit into the US PCS spectrum allocations. Basic data rates are up to 16, 32 and 64 kbps. The 64-kbps data rate is transmitted using rate 1/2 convolutional coding, and lower rates use symbol repetition. The Core-B proposal in Japan was an enhanced version of IS-665 W-CDMA offering 128 kbps within a single code and 2 Mbps using multicode transmission. The frame length of the system is 5 ms and interleaving can be over 5, 10 or 20 ms.

IS-665 W-CDMA has open and closed loop power control. The closed loop power control uses variable step sizes of 0.5, 1, 2 and 4 dB, which are adaptively controlled.

IS-665 WCDMA supports the use of interference cancellation. So, even this standard was transparent to the use of the technology described in Chapters 14 and 15. The basic system parameters are summarized in Table 16.5.

The forward link

Pilot channel

- Selection of BS and trigger of handoff.
- Provides a phase reference for coherent demodulation of the other channels.

Table 16.5 IS-655 W-CDMA system parameters

Basic chip rate	4.096, 8.192 and 12.288 Mcps
Base station synchronization	Synchronous
Frame length	5 ms
Multirate/variable rate	Symbol repetition/multicode
Coherent detection UL	Pilot channel
DL	Pilot channel
Power control	2 kbps time-multiplexed

- The pilot code has a period of 81 920 chips (20 ms), which is one part of an M-sequence with a period of $2^{32}-1$ for 5-MHz spread-spectrum modulation.
- Each BS has a unique phase offset of this pilot code.

Sync channel

- Provides frame or slot synchronization, paging channel information and identification of a BS using phase offset of the pilot channel. The slot length is 80 ms.

Paging channel

- Provides broadcast information (phase offset of neighbor BSs and system time).
- Paging, registration, traffic channel assignment, short message service and authentication.
- The slot length is also 80 ms and synchronizes with the slot of the sync channel.

Traffic channel

- Call setup, user traffic, authentication and handoff.

The reverse link

Access channel

- Traffic origination, page response, registration and authentication response.
- The slot width is 80 ms and synchronizes with the paging channel slot.

Traffic channel

- Same as that of the forward link.
- The interleaver span is decided at the initial transmission of traffic channel through the paging channel message.
- The interleaver span is selectable between 5, 10 and 20 ms, decided at the initial transmission of the traffic channel.
- Upon wake-up, the personal system (PS) first finds pilot, sync and paging channels for synchronization.
- Chip and 80-ms slots are synchronized by pilot acquisition and sync channel, respectively.
- Paging channel messages are received by the PS.
- PS messages are transmitted to the BS through the access channel.
- User data is transmitted through traffic channels between the PS and the BS.

System synchronization

The system synchronization consists of chip and slot synchronization.

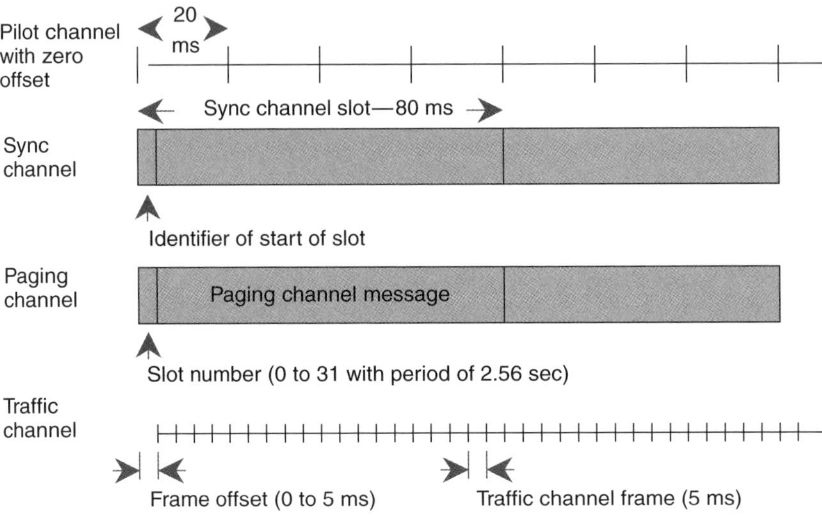

Figure 16.17 Timing relationships among channels.

Figure 16.17. shows the timing relationship among channels for system synchronization.

- All BSs use the same pilot code with a unique phase.
- Code offset between BSs is adjusted by the global positioning system (GPS) timing.
- The default value of code offset is 64 chips in duration, which depends on the cell radius and its layout.
- BS sync and paging channels are synchronized with the pilot channel.
- During a call, each traffic channel has a unique frame offset of from 0 to 5 ms.

Synchronization

While processing the pilot channel, the PS selects a BS with maximum signal power. At the initial search for pilot channel detection, the correlation width is set to 4096 chips. This value is limited by the system search time. After detection of the pilot channel, the RAKE receiver tracks amplitude and phase of the pilot channel for each multipath component, and then demodulates the sync channel coherently. In the estimation of propagation path, a correlation width of 1024 chips is used. During processing of the sync channel, the PS searches the start of slot of sync channel to establish frame synchronization. During processing of paging channel, the PS monitors system information including broadcast information, page and registration. If the PS detects page or origination, the PS sends messages of page response, origination and authentication in the assigned slot on the access channel. The BS assigns a traffic channel and then the PS continuously tracks transmission timing of the frame within one chip using the earliest received multipath component. Selected BSs can search and acquire the reverse pilot channel within a few hundred milliseconds because the reverse traffic channel is synchronized with the forward traffic channel.

Total synchronization time of the BS depends on the search period. If synchronization is missed, the BS initiates a partial search around the system timing reference. A number of Z-search or expanding window search strategies are discussed in Chapter 3. A block diagram of the forward and reverse channel is shown in Figures 16.18 and 16.19, respectively.

Forward error correction and interleaving

The system uses a convolutional code with constraint length of 9 and code rate of 1/2, and a block interleaver. In both forward and reverse links, information and signaling channels are encoded by convolutional codes. Each bit input to the convolutional encoder yields two coded symbols that are separated as in-phase (I) and quadrature-phase (Q) signal. Each forward information signal is interleaved with a 29×10, 13×10 and 6×10 interleaver matrix for input date rates of 64, 32 and 16 kbps, respectively. They are punctured at the rate of 29/32, 26/32 and 24/32, respectively. Each reverse information signal is interleaved with a 32×10, 16×10 and 8×10 interleaver matrix for input data rates of 64, 32 and 16 kbps, respectively.

Spreading modulation/demodulation

The spreading code is a combination code that is generated by PN and orthogonal codes from Walsh/Hadamard sequences to minimize mutual interference between traffic and pilot/sync/paging channels. In the forward link, information, signaling and power control bits are time-division multiplexed (TDM) because the available number of orthogonal codes is limited. After symbol repetition, the information channel is punctured by signaling channel and power control bits, and results in a symbol rate of 64 ksymbols s^{-1}. The forward traffic channel is spread by both PN and orthogonal codes. The PN code is part of an M-sequence with a period of $2^{32}-1$. Each BS is distinguished by the phase offset of the pilot code. All channels of the BS are distinguished by Walsh/Hadamard codes. Voice activity produces a discontinuous transmission of each forward traffic channel. This requires a unique frame offset from 0 to 5 ms in order to randomize the transmission timing and decrease the interference noise of each channel. Figure 16.20 shows the time chart of the discontinuous transmission of the traffic information channel. The randomization is much simpler than that in IS-95 shown in Figure 16.5.

In the reverse link, the input data is spread by the same combination codes described above. The information traffic, pilot and signaling channels are distinguished by Walsh/Hadamard codes. The RAKE receiver configurations are the same in the forward and reverse links. The RAKE receiver detects the pilot channel by a searching technique, and demodulates the access and traffic channels by fingers.

Interference canceller

Required E_b/N_0 for a constant bit error rate (BER) is higher in the reverse than in the forward link. To reduce N_0, an interference cancelation system (ICS) can be used in the

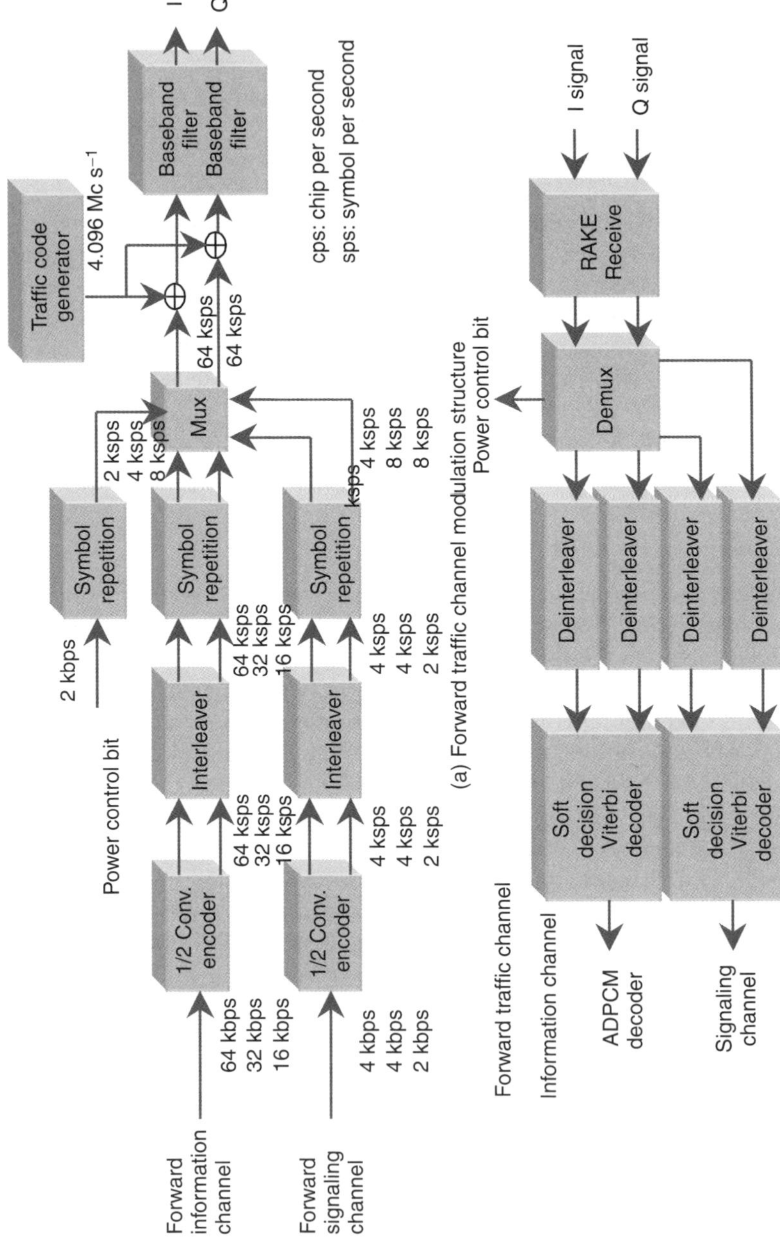

Figure 16.18 Modulation/demodulation diagrams, in forward channel.

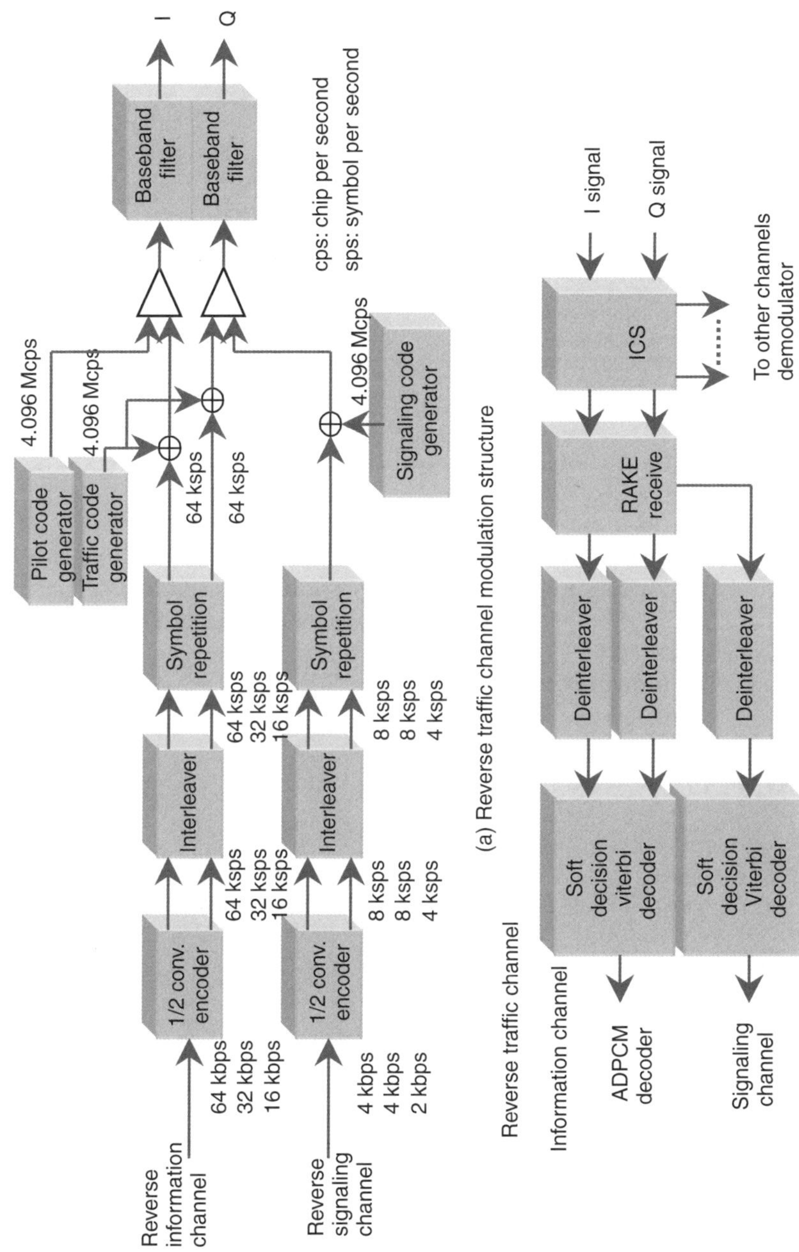

Figure 16.19 Modulation/demodulation diagrams, in reverse channel.

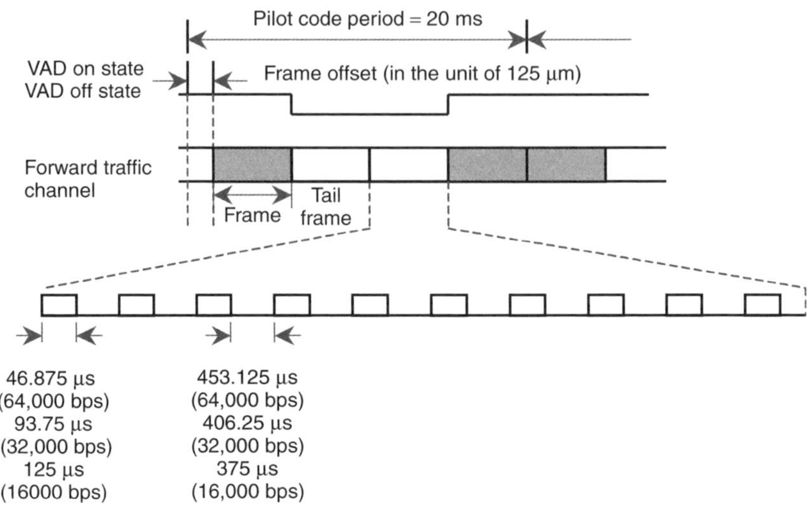

Figure 16.20   Time chart of discontinuous transmission.

reverse link. An ICS scheme has been developed for this purpose. The receive signals of multiple users and corresponding multipath components are estimated by channel signal estimation circuits (CSECs). Computer simulation predicts that the ICS improvement of E_b/N_0 3.2 dB can be achieved.

For further details see References [1–17].

REFERENCES

1. ANSI/TIA/EIA/-95-B, *Mobile Station – Base Station Compatibility Standard for Dual-Mode Wideband Spread Spectrum Cellular Systems*, March, 1999.
2. TIA/EIA/IS-637-A, *Short Message Service for Spread Spectrum Systems*, September, 1999.
3. ANSI/TIA/EIA/IS-683-A, *Over the Air Service Provisioning of Mobile Stations in Spread Spectrum Systems*, June, 1998.
4. TIA/EIA/IS-707-A-1, *Data Services Option Standard for Spread Spectrum Systems – Addendum 1*, December, 1999.
5. TIA/EIA/IS-707-A-2, *Data Services Option Standard for Spread Spectrum Systems – Addendum 2*, June, 2000. Refer also to 3GPP2 C.S0017-0-2.
6. TIA/EIA/IS-801, *Position Determination Service Standard for Dual Mode Spread Spectrum Systems*, November, 1999.
7. ANSI/TIA/EIA-829, *Interoperability Specification for Tandem Free Operation*, June, 2000. Refer also to 3GPP2 A.S0004.
8. TIA/EIA/IS-834, G3G CDMA-DS to ANSI/TIA/EIA-41, March, 2000. Refer also to 3GPP2 C.S0007-0 cdma2000 Wireless IP Network Standard, December, 2000.
9. TIA/EIA/IS-2000-A, *cdma2000 Standards for Spread Spectrum Systems Release A*, March, 2000.
10. TIA/EIA/TSB58-C, *Administration of Parameter Value Assignments for TIA/EIA Spread Spectrum Standards*, May, 2000. Refer also to 3GPP2 CR1001-A.

11. 3GPP TS 25.413 V3.3.0 (Release 1999) Technical Specification Universal Mobile Telecommunications System (UMTS), *UTRAN Iu Interface RANAP Signaling*, September, 2000. Refer also to ETSI TS 125.413 V13.0.
12. 3GPP2 A.S0004, *3GPP2 Tandem Free Operation Specification*, June, 2000. Refer also to ANSI/TIA/EIA-829.
13. 3GPP2 CR1001-A, *Administration of Parameter Value Assignments for cdma2000 Spread Spectrum Standards, Release A*, 14 July, 2000. Refer also to TIA/EIA/TSB58-C.
14. 3GPP2 C.S0002-A, *Physical Layer Standard for cdma2000 Spread Spectrum Systems*, 9 June, 2000. Refer also TIA/EIA/IS-2000(-2)-A.
15. 3GPP2 C.S0007-0, *Direct Spread Specification for Spread Spectrum Systems on ANSI41 (DS-41) Upper Layers Air Interface*, March, 2000. Refer also to TIA/EIA/IS-834.
16. 3GPP2 C.S0017-0-2, *Data Service Options for Spread Spectrum Systems Addendum 2*, 14 January, 2000. Refer also to TIA/EIA/IS-707-A-2.
17. 3GPP2 S.R0005-A, *Wireless Network Reference Model*, July, 1998. Refer also to TIA/EIA/TSB 100.

17

UMTS standard: WCDMA/FDD Layer 1

17.1 TRANSPORT CHANNELS AND PHYSICAL CHANNELS (FDD)

17.1.1 Transport channels

In the terminology used in wireless communications, 'transport channels' are the services offered by Layer 1 to the higher layers. The purpose of this section is to introduce basic terminology and abbreviations used in practice. For more details, see *www.3gpp.org*. We start with the definition of the channels.

- *Dedicated transport channel*
  DCH – Dedicated Channel DCH is a downlink or uplink transport channel that is used to carry user or control information between the network and mobile station.
  DTCH – Dedicated Traffic Channel.
  SDCCH – Stand-Alone Dedicated Control Channel.
  ACCH – Associated Control Channel defined within ITU-R M.1035.
  The DCH is transmitted over the entire cell or over only a part of the cell using lobe-forming antennas.
- *Common transport channels*
  BCCH – Broadcast Control Channel. BCCH is a downlink transport channel that is used to broadcast system and cell-specific information. The BCCH is always transmitted over the entire cell.
  FACH – Forward Access Channel. FACH is a downlink transport channel that is used to carry control information to a mobile station when the system knows the location cell of the mobile station. The FACH may also carry short user packets. The FACH is transmitted over the entire cell or over only a part of the cell using lobe-forming antennas.
  PCH – Paging Channel. PCH is a downlink transport channel that is used to carry control information to a mobile station when the system does not know the location cell of the mobile station. The PCH is always transmitted over the entire cell.

RACH – Random-Access Channel. RACH is an uplink transport channel that is used to carry control information from a mobile station.

The RACH may also carry short user packets. The RACH is always received from the entire cell.

17.1.2 Physical channels

- *The physical resources*
  The basic physical resource is the code/frequency plane.
  On the uplink, different information streams may be transmitted on the I and Q branches. Consequently, a physical channel corresponds to a specific carrier frequency, code, and, on the uplink, relative phase (0 or 90°).
- *Uplink physical channels*
  Dedicated uplink physical channels.
  Uplink DPDCH – The uplink Dedicated Physical Data Channel.
  Uplink DPCCH – The uplink Dedicated Physical Control Channel.
  There may be zero, one, or several uplink DPDCHs on each Layer 1 connection. The Layer 1 control information consists of known pilot bits to support channel estimation for coherent detection, transmit power-control (TPC) commands, an optional transport-format indicator (TFI) and feedback information (FBI) (interference level). The TFI informs the receiver about the instantaneous parameters of the different transport channels multiplexed on the uplink DPDCH. There is one and only one uplink DPCCH on each Layer 1 connection.

Frame structure

Figure 17.1 shows the frame structure of the uplink DPCHs. Each frame of length 10 ms is split into 15 slots, each of length $T_{\mathrm{slot}} = 0.666$ ms, corresponding to one power-control period.

A super frame corresponds to 72 consecutive frames, that is, the super frame length is 720 ms.

The Spreading Factor (SF) may range from 256 down to 4. An uplink DPDCH and uplink DPCCH on the same Layer 1 connection are generally of different rates, that is, they have different SFs. Multicode operation is possible for the uplink DPCHs. When multicode transmission is used, several parallel (6) DPDCHs are transmitted using different channelization codes. There is only one DPCCH per connection.

Common uplink physical channels

PRACH – Physical Random-Access Channel. PRACH is used to carry RACH. It is based on a Slotted ALOHA. The different time offsets called *access slots* are used for transmission of access bursts as shown in Figure 17.2.

Information on the access slots available in the current cell is broadcast on the BCCH. The structure of the Random-Access burst is shown in Figure 17.3.

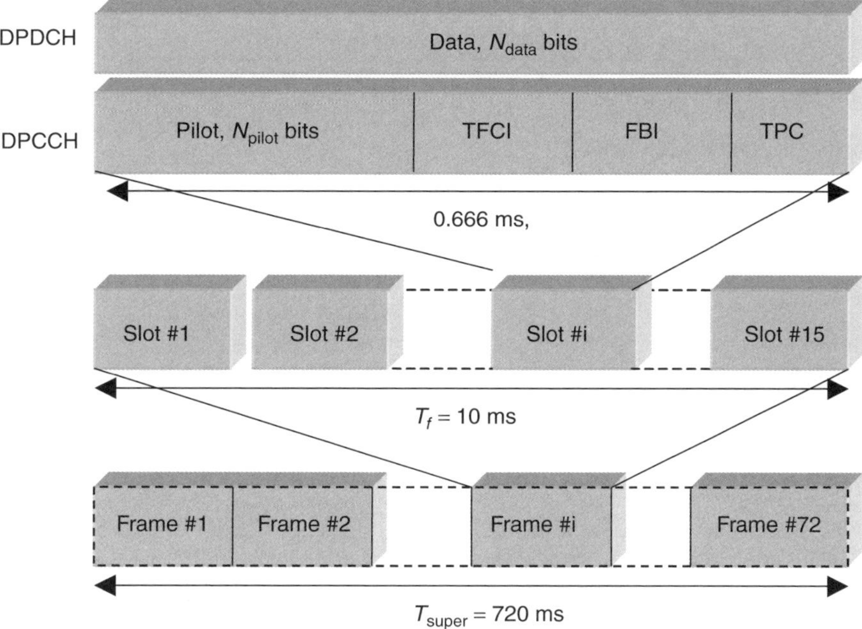

Figure 17.1 Frame structure for uplink DPDCH/DPCCH.

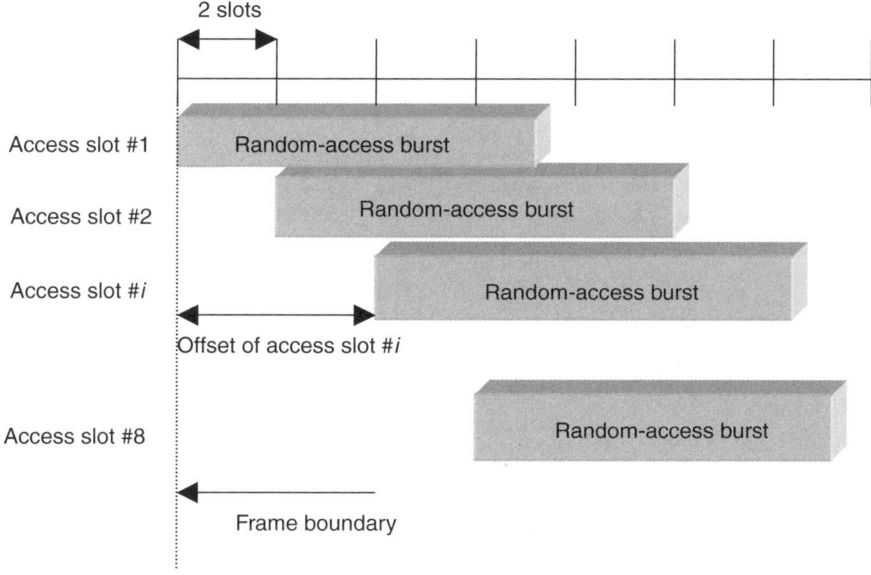

Figure 17.2 Access slots.

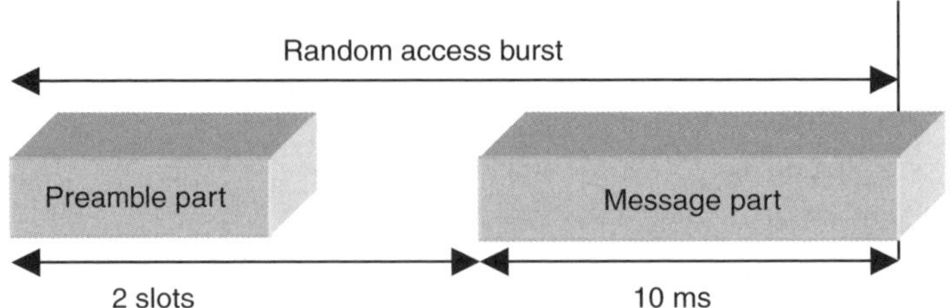

Figure 17.3   Structure of the random-access burst.

- *Preamble part*
  The preamble part of the random-access burst consists of a *signature* of length 16 complex symbols ($\pm 1 \pm j$). Each preamble symbol is spread with a 256-chip real Orthogonal Gold code. There are a total of 16 different signatures, based on the Orthogonal Gold code set of length 16.
- *Message part*
  The message part of the random-access burst has the same structure as the uplink Dedicated physical channel (DPCH). It consists of a data part, corresponding to the uplink DPDCH, and a Layer 1 control part, corresponding to the uplink DPCCH (see Figure 17.4). The data and control parts are transmitted in parallel. The data part carries the random access request or small user packets. The SF of the data part is limited to SF $\in$ {256, 128, 64, 32} corresponding to channel bit rates. The control part carries pilot bits and rate information, using a SF of 256. The rate information indicates which channelization code (or rather the SF of the channelization code) is used on the data part.

Random-access burst

It consists of the following fields (see Figure 17.5):

- Mobile station identification (MS ID) [16 bits]. The MS ID is chosen at random by the mobile station at the time of each Random-Access attempt.

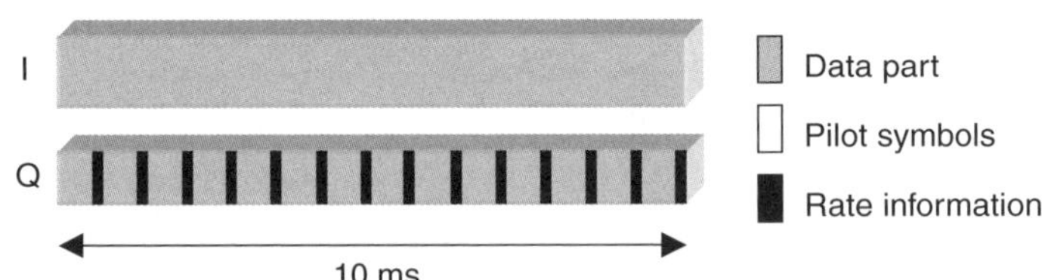

Figure 17.4   The message part of the random-access burst.

Figure 17.5   Structure of random-access burst data part.

- Required Service [3 bits]. This field informs the Base Station (BS) what type of service is required (short packet transmission, DCH set-up and so on).
- An optional user packet.
- A cyclic redundancy check (CRC) to detect errors in the data part of the Random-Access burst [8 bits].

Downlink physical channels

DPCH – Dedicated physical channels. Figure 17.6 shows the frame structure of the downlink DPCH.

In the case of multicode transmission, the slot format is shown in Figure 17.7.

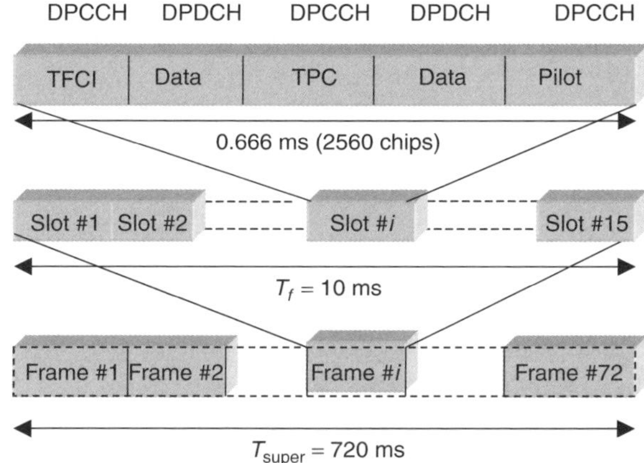

Figure 17.6 Frame structure for downlink DPCH.

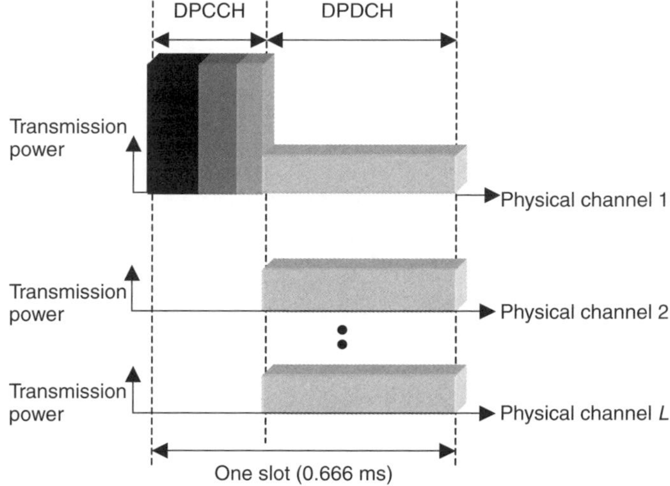

Figure 17.7 Downlink slot format in the case of multicode transmission.

Common physical channels

CCPCH – Primary Common Control Physical Channel. The Primary CCPCH is a fixed rate (30 kbps, SF = 256) downlink physical channel used to carry the BCCH.

Figures 17.8 and 17.9 show the frame structure of the primary and the secondary common control physical channels, respectively. The frame structure differs from the downlink DPCH in that no transmit power control (TPC) command or TFI is transmitted. The only Layer 1 control information is the common pilot bits needed for coherent detection.

The secondary CCPCH is used to carry the FACH and PCH. It is of constant rate. In contrast to the primary CCPCH, the rate may be different for different secondary CCPCH within one cell and between cells, in order to be able to allocate different amounts of FACH and PCH capacity to a cell.

The rate and SF of each secondary CCPCH is broadcast on the BCCH. The set of possible rates is the same as for the downlink DPCH. The FACH and PCH are mapped

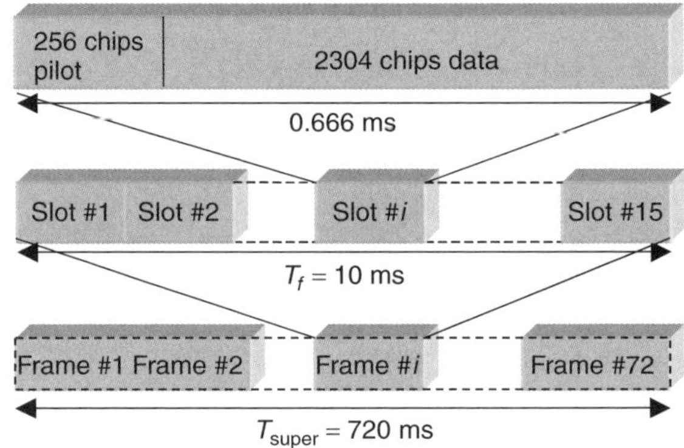

Figure 17.8 Frame structure for primary common control physical channel.

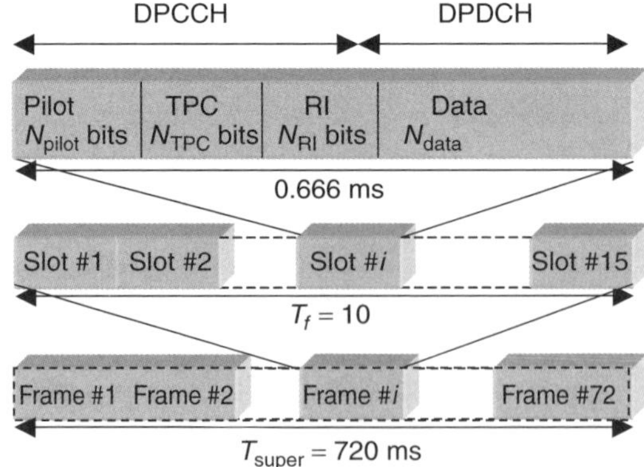

Figure 17.9 Frame structure for secondary common control physical channel.

to separate secondary CCPCHs. The main difference between a CCPCH and a downlink DPCH is that a CCPCH is not power controlled. The main difference between the primary and secondary CCPCH is that the primary CCPCH has a fixed predefined rate while the secondary CCPCH has a constant rate that may be different for different cells, depending on the capacity needed for FACH and PCH. A Primary CCPCH is continuously transmitted over the entire cell while a secondary CCPCH is only transmitted when there is data available and may be transmitted in a narrow lobe in the same way as a DPCH (only valid for a secondary CCPCH carrying the FACH).

17.1.3 Synchronization channel (SCH)

SCH is a downlink signal used for cell search. The SCH consists of two subchannels, the primary and secondary SCH. Details are given in Chapter 3.

17.1.4 Mapping of transport channels to physical channels

Figure 17.10 summarizes the mapping of a subset of transport channels to physical channels.
   For additional details like high-speed downlink shared channel (HSDSCH) and related control channels see www.3qpp.org.

17.1.5 Mapping of PCH to secondary common control physical channel

The mapping method is shown in Figure 17.11. The PCH is divided into several groups in one super frame, and Layer 3 information is transmitted in each group. Each group of PCH shall have information amount worth four slots and consists of a total of six information parts:

– Two Paging Indication (PI) parts – for indicating whether there are MS-terminated calls,
– Four Mobile User Identifier (MUI) parts – for indicating the identity of the paged mobile user.

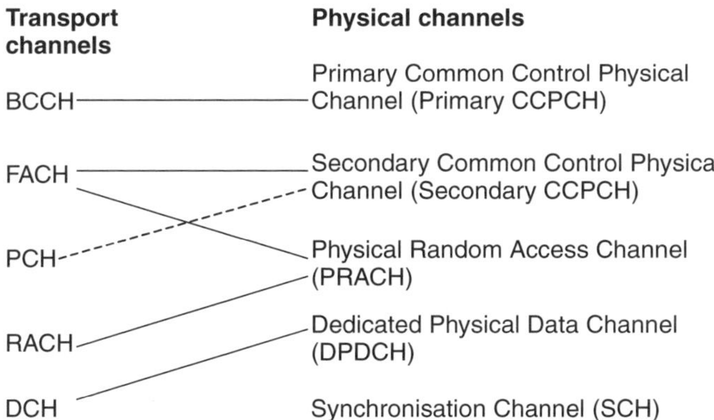

Figure 17.10   Transport channel to physical channel mapping.

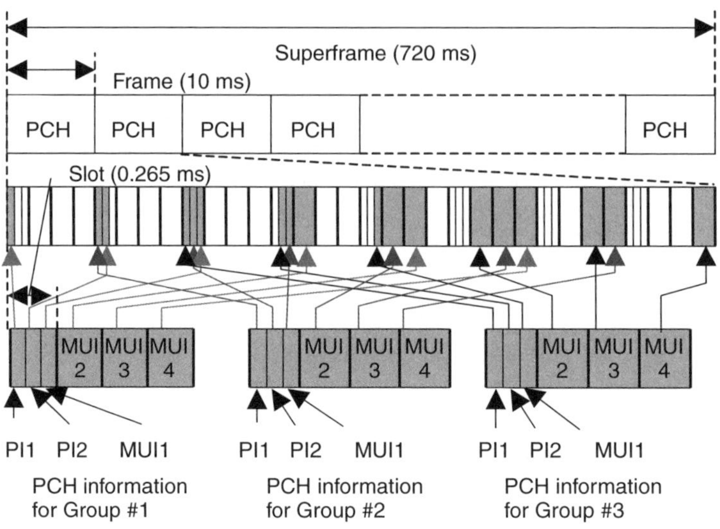

Figure 17.11   PCH mapping method.

In each group, PI parts are transmitted ahead of MUI parts. In all groups, 6 information parts are allocated with a certain pattern in the range of 24 slots. By shifting each pattern by 4 slots, multiple 288 groups of PCH can be allocated on one Secondary Common Physical Channel.

17.2 MULTIPLEXING, CHANNEL CODING AND INTERLEAVING

Multiplexing of transport channels, coding and interleaving for a frequency division duplexing (FDD) system is illustrated in Figure 17.12.

17.2.1 Channel coding

Channel coding is done on a per-transport-channel basis, that is, before transport channel multiplexing. The following options are available (see Figure 17.13):

- Convolutional coding
- Outer Reed–Solomon coding + Outer interleaving + Convolutional coding
- Turbo coding
- Service-specific coding, for example, unequal error protection (UEP) for some types of speech codecs

17.2.2 Convolutional coding

Generator polynomials for the convolutional codes are given below.

Rate	Constraint length	Generator Polynomial 1	Generator Polynomial 2	Generator Polynomial 3
1/3	9	557	663	711
1/2	9	561	753	NA

Rate 1/3 convolutional coding is applied to dedicated transport channels (DCHs) in normal (nonslotted) mode.
Rate 1/2 convolutional coding is applied to DCHs in slotted mode.

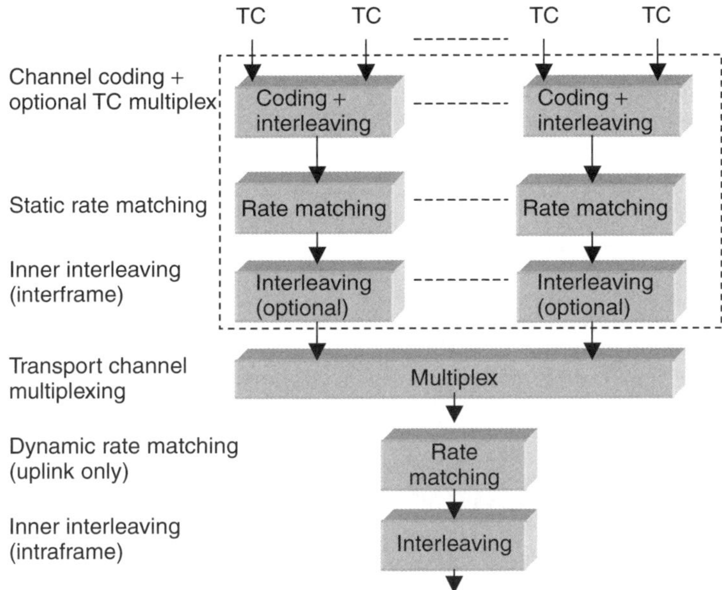

Figure 17.12   Coding and multiplexing of transport channels.

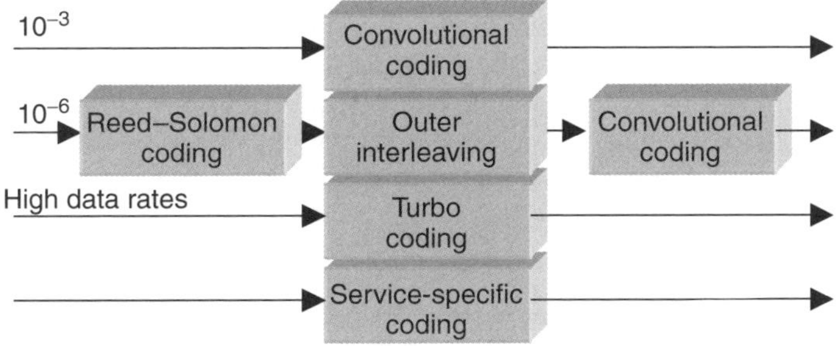

Figure 17.13   Channel coding in UTRA/FDD.

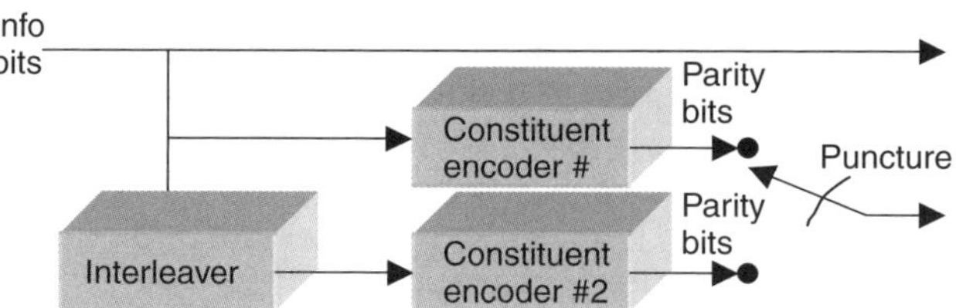

Figure 17.14   Block diagram of a turbo code encoder.

17.2.3 Other types of coding

- *Outer Reed–Solomon coding and outer interleaving*
  The RS-coding is of approximate rate 4/5 using the 256-ary alphabet. The outer inter-leaving is a symbol-based block interleaver with width equal to the block length of the RS code. The interleaver span is variable and can be 10, 20, 40 or 80 ms.
- *Turbo coding*
  The turbo coding is used for high-data rate (above 32 kbps), high-quality services.
  Turbo code of rate 1/3 and 1/2 (for the highest data rates) have replaced the concate-nation of convolutional and RS codes. The block diagram for the basic turbo encoder is shown in Figure 17.14.
- *Service specific coding*
  Additional coding schemes, in addition to the standard coding schemes listed above, can be used. One example is the use of unequal error-protection coding schemes for certain speech codecs.

17.3 SPREADING AND MODULATION

17.3.1 Uplink spreading and modulation

The block diagram of uplink spreading and modulation is shown in Figure 17.15.

For multicode transmission, each additional uplink DPDCH may be transmitted on either the I or the Q branch. For each branch, each additional uplink DPDCH should be assigned its own channelization code. Uplink DPDCHs on different branches may share a common channelization code. The spreading and modulation of the message part of the random-access burst is basically the same as for the uplink DPCHs, in Figure 17.15 in which the uplink DPDCH and uplink DPCCH are replaced by the data part and the control part, respectively. The scrambling code for the message part is chosen on the basis of the base-station-specific preamble code.

17.3.2 Channelization codes

Orthogonal Variable Spreading Factor (OVSF) codes, described in Chapter 2, are used for channelization. All codes within the code tree cannot be used simultaneously by one mobile station.

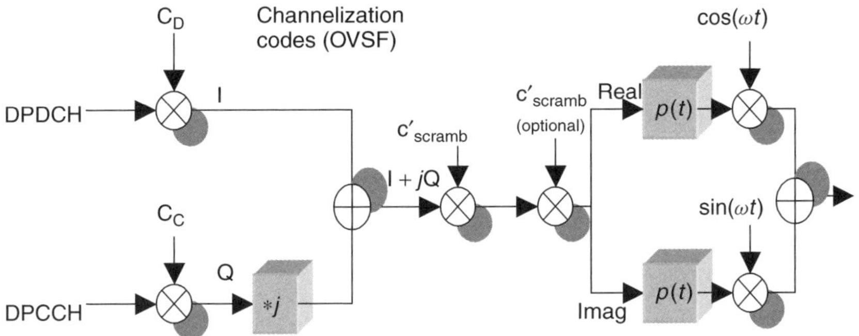

C_D, C_C : channelization codes
c'_{scramb} : primary scrambling code
c''_{scramb} : secondary scrambling code (optional)
$p(t)$: pulse-shaping filter (root raised cosine, roll off 0.22)

Figure 17.15 Spreading/modulation for uplink DPDCH/DPCCH.

A mobile station can use a code if and only if the same mobile station uses no other code on the path from the specific code to the root of the tree or in the subtree below the specific code. This means that the number of available channelization codes is not fixed but depends on the rate and SF of each physical channel. Each connection is allocated at least one uplink channelization code to be used for the uplink DPCCH. In most cases, at least one additional uplink channelization code is allocated for an uplink DPDCH.

Further uplink channelization codes may be allocated if more than one uplink DPDCH is required. All channelization codes used for the DPDCHs must be orthogonal to the code used for the DPCCH.

As different mobile stations use different uplink scrambling codes, the uplink channelization codes may be allocated with no coordination between different connections. The uplink channelization codes are therefore always allocated in a predefined order. The mobile station and network only need to agree on the number and length (SF) of the uplink channelization codes. The exact codes to be used are then implicitly given.

17.3.3 Scrambling codes

Either short or long scrambling codes should be used on the uplink. The short scrambling code is typically used in cells where the BS is equipped with an advanced receiver, such as a multiuser detector or an interference canceler.

With the short scrambling code, the cross-correlation properties between different physical channels and users does not vary in time in the same way as when a long code is used. This means that the cross-correlation matrices used in the advanced receiver do not have to be updated as often as for the long scrambling code case, thereby reducing the complexity of the receiver implementation.

In cells where there is no gain in implementation complexity using the short scrambling code, the long code is used instead because of its better interference averaging properties. For the details of scrambling code construction, see *www.3gpp.org*.

These scrambling codes are designed such that at $N - 1$ out of N consecutive chip times, they produce $+/-90°$ rotations of the In phase $+$ Quadrature (IQ) multiplexed data and control channels. At the remaining 1 out of N chip times, they produce 0, $+/-90°$ or $180°$ rotations.

This limits the transitions of the complex baseband signal that is inputted to the root-raised cosine pulse-shaping filter. This in turn reduces the peak to average ratio of the signal at the filter output, allowing a more efficient power-amplifier implementation.

To guarantee these desirable properties, restrictions on the choice of uplink OVSF codes are also required.

Short scrambling code

For code construction details, see www.3qpp.org. The network decides the uplink short scrambling code. The mobile station is informed about which short scrambling code to use in the downlink Access Grant message, which is the base station response to an uplink Random-Access Request. The short scrambling code may, in rare cases, be changed during the duration of a connection.

Long scrambling code

The long uplink scrambling code is typically used in cells without multiuser detection (MUD) in the BS. The mobile station is informed if a long scrambling code should be used in the Access Grant Message following a Random-Access Request and in the handover message.

Which long scrambling code to use is directly given by the short scrambling code. No explicit allocation of the long scrambling code is thus needed.

Modulation

- *Modulating chip rate*
  The modulating chip rate is 3.84 Mcps. This basic chip rate can be extended to $2 \times$ 3.84 Mcps or $4 \times$ 3.84 Mcps
- *Pulse shaping*
  The pulse-shaping filters are root-raised cosine (RRC) with roll-off $\alpha = 0.22$ in the frequency domain
- *Data modulation*
  For data, quadrature phase shift keying (QPSK) modulation is used. Phase transition restrictions are introduced by the scrambling code design.

17.3.4 Downlink spreading and modulation

Data modulation is QPSK, where each pair of two bits are serial-to-parallel converted and mapped to the I and the Q branches, respectively.

The I and Q branch are then spread to the chip rate with the same channelization code c_{ch} (real spreading) and then scrambled by the same cell-specific scrambling code c_{scramb}

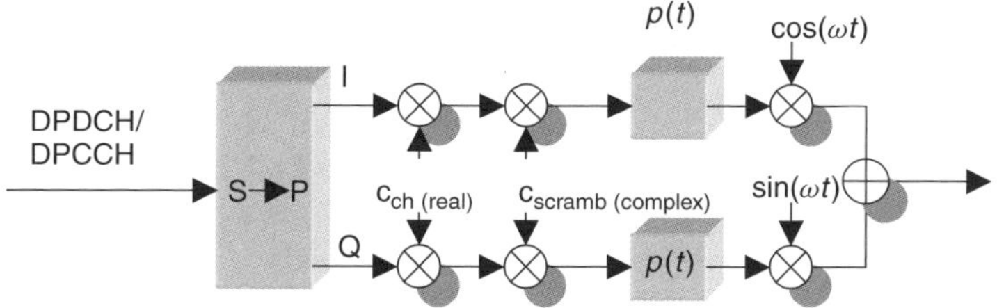

c_{ch} : channelization code,
c_{scramb} : scrambling code
$p(t)$: pulse-shaping filter (root raised cosine, roll-off 0.22)

Figure 17.16 Spreading/modulation for downlink DPCH.

(complex scrambling). The different physical channels use different channelization codes, while the scrambling code is the same for all physical channels in one cell. The system block diagram is shown in Figure 17.16.

The multiplexing of the synchro channel (SCH)

The SCH is only transmitted intermittently (one code word per slot). The SCH is multiplexed *after* the long code scrambling of the DPCH and CCPCH, as shown in Figure 17.17. Consequently, the SCH is *nonorthogonal* to the other downlink physical channels.

For code construction, based on Golay codes, see *www.3gpp.org*.

Modulation

- *Modulating chip rate*
  The modulating chip rate is 3.84 Mcps. This basic chip rate can be extended to $2 \times$ 3.84 Mcps or 4×3.84 Mcps

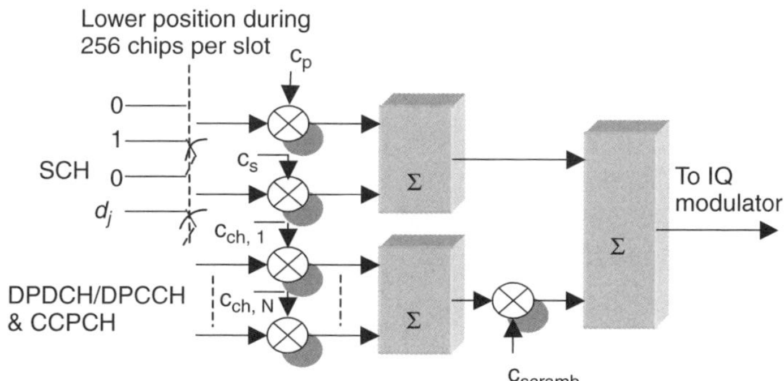

Figure 17.17 Multiplexing of SCH.

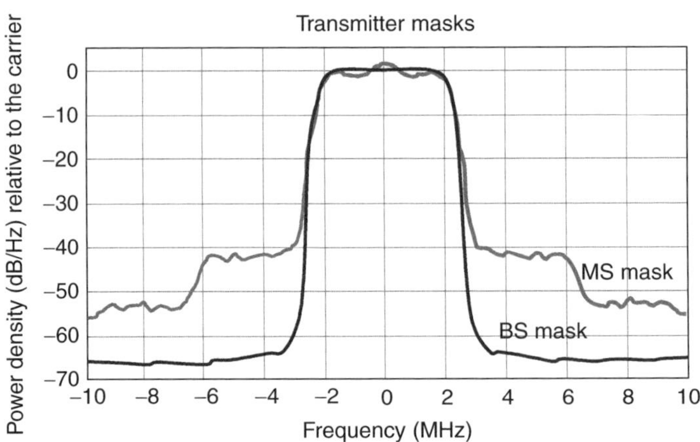

Figure 17.18 Assumed spectrum masks.

- *Pulse shaping*
  The pulse-shaping filters are root-raised cosine (RRC) with roll-off $\alpha = 0.22$ in the frequency domain
- *Modulation*
  For data, QPSK modulation is used

17.3.5 Output RF spectrum emissions

Out-of-band emissions are specified in Figure 17.18.

- *Spurious emissions*
  The limits for spurious emissions at frequencies greater than $\pm 250\%$ of the necessary bandwidth would be based on the applicable tables from the ITU-R recommendation SM.329.

17.4 PHYSICAL LAYER PROCEDURES (FDD)

17.4.1 Uplink power control

- *Closed-loop power control*
  The BS should estimate the received uplink DPCCH power after the RAKE combining of the connection to be power controlled. Simultaneously, the BS should estimate the total uplink received interference in the current frequency band and generate an SIR estimate – SIR_{est}. The BS then generates TPC commands. The algorithms are described in Chapter 6. Upon reception of a TPC command, the mobile station should adjust the transmit power of both the uplink DPCCH and the uplink DPDCH in the given direction with a step of Δ_{TPC} dB. The step size Δ_{TPC} is a parameter that may differ between different cells, in the region of 0.25 to 1.5 dB.

In case of receiver diversity (e.g. space diversity) or softer handover at the BS, the TPC command should be generated after diversity combining.

- *Soft handover*

  In the base stations, a quality measurement is performed on the received signal.

  - In case the quality measurement indicates a value below a given threshold, an increase command is sent to the mobile, otherwise a decrease command is transmitted.
  - All the base stations in the active set send power-control commands to the mobile.

  The mobile compares the commands received from different base stations and increases its power only if all the commands indicate an increased value (this means that all the receivers are below the threshold).

  - In case one command indicates a decreased step (that is, at least one receiver is operating in good condition), the mobile reduces its power.
  - In case more than one decrease command is received by the mobile, the mobile station should adjust the power with the largest step in the 'down' direction ordered by the TPC commands received from each BS in the active set.

  The quality threshold for the BS in the active set should be adjusted by the outer loop power-control (to be implemented in the network node where soft handover combining is performed).

- *Open-loop power control*

  Open-loop power control is used to adjust the transmit power of the physical Random-Access channel. Before the transmission of a Random-Access burst, the mobile station should measure the received power of the downlink Primary CCPCH over a sufficiently long time to remove the effects of the nonreciprocal multipath fading. From the power estimate and knowledge of the Primary CCPCH, transmit power (broadcast on the BCCH), the downlink path loss including shadow fading can be found. From this path-loss estimate and knowledge of the uplink interference level and the required received SIR, the transmit power of the physical Random-Access channel can be determined. The uplink interference level as well as the required received SIR are broadcast on the BCCH.

17.4.2 Downlink power control

- *Closed-loop power control*

  The following steps define the operation of this loop:

  - The downlink closed-loop power control adjusts the base station transmit power in order to keep the received downlink SIR at a given SIR target.
  - The mobile station should estimate the received downlink DPCH power after the RAKE combining of the connection to be power controlled.
  - Simultaneously, the mobile station should estimate the total downlink received interference in the current frequency band.
  - The mobile station then generates TPC commands.
  - Upon reception of a TPC command, the BS should adjust the transmit power in the given direction with a step of Δ_{TPC} dB.

- The step size Δ_{TPC} is a parameter that may differ between different cells, in the range of 0.25 to 1.5 dB.
- In the case of receiver diversity (e.g. space diversity) at the mobile station, the TPC command should be generated after diversity combining.
- *Outer loop* (*SIR target adjustment*)
  - The outer loop adjusts the SIR target used by the closed-loop power control.
  - The SIR target is independently adjusted for each connection on the basis of the estimated quality of the connection.
  - In addition, the power offset between the downlink DPDCH and DPCCH may be adjusted.
  - How the quality estimate is derived and how it affects the SIR target is decided by the Radio Resource Management (RRM), that is, it is not a physical-layer issue.

17.4.3 Random access

The procedure of a Random-Access Request consists of the following:

1. The mobile station acquires synchronization to a base station. Cell acquisition is described in Chapter 3.
2. The mobile station reads the BCCH to get information about
   2.1. the preamble spreading code(s)/message scrambling code(s) used in the cell,
   2.2. the available signatures,
   2.3. the available access slots,
   2.4. the available SFs for the message part,
   2.5. the interference level at the base station,
   2.6. the primary CCPCH transmit power level.
3. The mobile station selects a preamble spreading code/message scrambling code.
4. The mobile station selects a SF for the message part.
5. The mobile station estimates the downlink path loss (by using information about the transmitted and received power level of the primary CCPCH) and determines the required uplink transmit power (by using information about the interference level at the BS).
6. The mobile station randomly selects an access slot and signature from the available access slots and signatures.
7. The mobile station transmits its Random-Access burst.
8. The mobile station waits for an acknowledgment from the base station. If no acknowledgment is received within a predefined time-out period, the mobile station starts again from step 7.

17.4.4 Paging control in idle mode

- *Base station operation*
  Every mobile station belongs to one group. When a paging message should be sent to a mobile, the paging message is transmitted on the PCH in the MUI parts belonging to the terminating mobile's group. The paging message includes the MS ID number

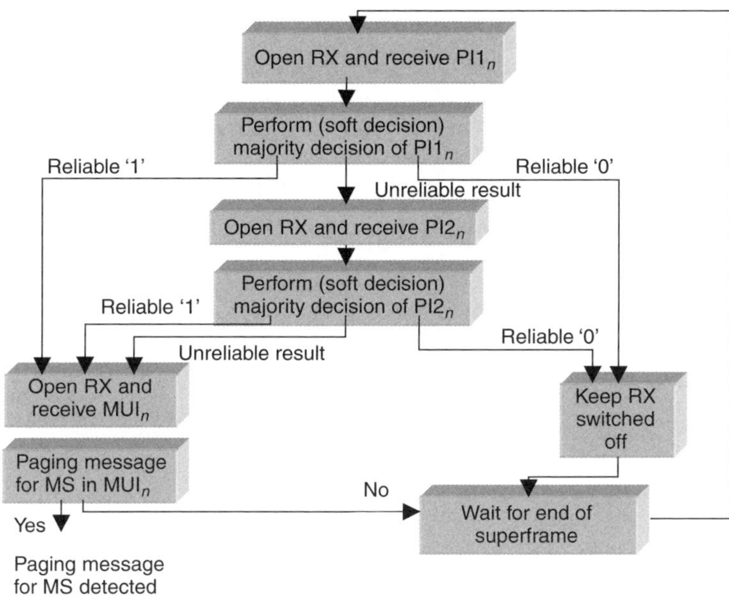

Figure 17.19 Detection of paging messages.

of the mobile station for which the paging message was intended. When an MUI is transmitted, the corresponding PI1 and PI2 fields are also transmitted. The behavior of the base station is described as follows:

- For the PCH of the group, which does not have terminating information, the BS shall transmit the two PI parts (PI1 and PI2) in the PCH as 'all 0'. The MUI part shall not be transmitted.
- For the PCH of the group, which have terminating information, the BS shall transmit the two PI parts (PI1 and PI2) in the PCH as 'all 1'. The MUI part shall be transmitted within the same PCH.

- *Mobile Station Operation*

Detection of paging messages is to open the receiver to detect one or both the paging indicators (PI1 and PI2). If they indicate a paging message for the group the mobile belongs to, the actual paging information part (MUI) is received. When the MUI part is received, the existence of a paging message for the mobile is determined from the information included in the MUI part. The mobile station operation for the detection of paging information in group n is shown in Figure 17.19. PI1$_n$, PI2$_n$, and MUI$_n$ are the PCH components that belong to group n.

For further details see References [1–24].

REFERENCES

1. 3GPP TS 25.211, *Physical Channels and Mapping of Transport Channels onto Physical Channels (FDD)*.

2. 3GPP TS 25.212, *Multiplexing and Channel Coding (FDD)*.
3. 3GPP TS 25.213, *Spreading and Modulation (FDD)*.
4. 3GPP TS 25.214, *Physical Layer Procedures (FDD)*.
5. 3GPP TS 25.215, *Physical Layer – Measurements (FDD)*.
6. 3GPP TS 25.221, *Physical Channels and Mapping of transport Channels onto Physical Channels (TDD)*.
7. 3GPP TS 25.222, *Multiplexing and Channel Coding (TDD)*.
8. 3GPP TS 25.223, *Spreading and Modulation (TDD)*.
9. 3GPP TS 25.224, *Physical Layer Procedures (TDD)*.
10. 3GPP TS 25.301, *Radio Interface Protocol Architecture*.
11. 3GPP TS 25.302, *Services Provided by the Physical Layer*.
12. 3GPP TS 25.303, *UE Functions and Interlayer Procedures in Connected Mode*.
13. 3GPP TS 25.304, *UE Procedures in Idle Mode*.
14. 3GPP TS 25.331, *RRC Protocol Specification*.
15. 3GPP TR 25.922, *Radio Resource Management Strategies*.
16. 3GPP TR 25.923, *Report on Location Services (LCS)*.
17. 3GPP TR 25.401, *UTRAN Overall Description*.
18. 3GPP TS 25.101, *UE Radio Transmission and Reception (FDD)*.
19. 3GPP TS 25.104, *UTRA (BS) FDD; Radio Transmission and Reception*.
20. 3GPP TS 25.133, *Requirements for Support of Radio Resource Management (FDD)*.
21. 3GPP TS 25.225, *Physical Layer – Measurements (TDD)*.
22. 3GPP TS 25.308-520, *High Speed Downlink Packet Access (HSDPA) – Overall Description*.
23. 3GPP TR 25.855, *High Speed Downlink Packet Access – Overall UTRAN Description*.
24. 3GPP TR 25.858, *High Speed Downlink Packet Access – Physical Layer Aspects*.

Index